Lecture Notes in Networks and Systems

Volume 550

The series "Lecture Notes in Networks and Systems" publishes the latest developments in Networks and Systems—quickly, informally and with high quality. Original research reported in proceedings and post-proceedings represents the core of LNNS.

Volumes published in LNNS embrace all aspects and subfields of, as well as new challenges in, Networks and Systems.

The series contains proceedings and edited volumes in systems and networks, spanning the areas of Cyber-Physical Systems, Autonomous Systems, Sensor Networks, Control Systems, Energy Systems, Automotive Systems, Biological Systems, Vehicular Networking and Connected Vehicles, Aerospace Systems, Automation, Manufacturing, Smart Grids, Nonlinear Systems, Power Systems, Robotics, Social Systems, Economic Systems and other. Of particular value to both the contributors and the readership are the short publication timeframe and the world-wide distribution and exposure which enable both a wide and rapid dissemination of research output.

The series covers the theory, applications, and perspectives on the state of the art and future developments relevant to systems and networks, decision making, control, complex processes and related areas, as embedded in the fields of interdisciplinary and applied sciences, engineering, computer science, physics, economics, social, and life sciences, as well as the paradigms and methodologies behind them.

Indexed by SCOPUS, INSPEC, WTI Frankfurt eG, zbMATH, SCImago.

All books published in the series are submitted for consideration in Web of Science.

For proposals from Asia please contact Aninda Bose (aninda.bose@springer.com).

Mostafa Al-Emran · Mohammed A. Al-Sharafi ·
Khaled Shaalan
Editors

International Conference on Information Systems and Intelligent Applications

ICISIA 2022

Springer

Editors
Mostafa Al-Emran
The British University in Dubai
Dubai, United Arab Emirates

Mohammed A. Al-Sharafi
Sunway University
Selangor, Malaysia

Khaled Shaalan
The British University in Dubai
Dubai, United Arab Emirates

ISSN 2367-3370 ISSN 2367-3389 (electronic)
Lecture Notes in Networks and Systems
ISBN 978-3-031-16864-2 ISBN 978-3-031-16865-9 (eBook)
https://doi.org/10.1007/978-3-031-16865-9

This Springer imprint is published by the registered company Springer Nature Switzerland AG
The registered company address is: Gewerbestrasse 11, 6330 Cham, Switzerland

Preface

Information systems (ISs) applications are crucial to every facet of contemporary civilization. These applications have altered the way we engage and interact with one another, get the information we need, work, do business, and manage our social life. The International Conference on Information Systems and Intelligent Applications (ICISIA 2022) is established to be one of the leading international conferences in information systems. The conference brings together information systems academics, scholars, researchers, and practitioners from academia and industry to discuss cutting-edge research in information systems and intelligent applications. ICISIA 2022 aims to discuss fundamental and innovative topics in information systems and their societal impact on individuals and organizations. It mainly focuses on the role of artificial intelligence in organizations, human–computer interaction, IS in education and industry, and IS security, privacy, and trust.

The ICISIA 2022 attracted 109 submissions from 27 different countries worldwide. Out of the 109 submissions, we accepted 60 submissions, which represents an acceptance rate of 55%. Each submission is reviewed by at least two reviewers, who are considered experts in the related submitted paper. The evaluation criteria include several issues, such as correctness, originality, technical strength, significance, presentation quality, interest, and relevance to the conference scope. The conference proceedings is published in *Lecture Notes in Networks and Systems Series* by Springer, which has a high SJR impact.

We acknowledge all those who contributed to the success of ICISIA 2022. We would also like to express our gratitude to the reviewers for their valuable feedback and suggestions. Without them, it was impossible to maintain the high quality and success of ICISIA 2022. As gratitude for their efforts, ICISIA 2022 is partnered with Publons to recognize the reviewers' contribution to peer-review officially. This partnership means that reviewers can opt-in to have their reviews added to their Publons profile.

Dubai, United Arab Emirates
Selangor, Malaysia
Dubai, United Arab Emirates

Mostafa Al-Emran
Mohammed A. Al-Sharafi
Khaled Shaalan

Organization

Conference General Chair

Mostafa Al-Emran, The British University in Dubai, UAE

Honorary Conference Chair

Khaled Shaalan, The British University in Dubai, UAE

Conference Organizing Chair

Mohammed A. Al-Sharafi, Universiti Teknologi Malaysia, Malaysia

Program Committee Chairs

Mohammed A. Al-Sharafi, Universiti Teknologi Malaysia, Malaysia
Mostafa Al-Emran, The British University in Dubai, UAE

Publication Committee Chairs

Mohammed A. Al-Sharafi, Universiti Teknologi Malaysia, Malaysia
Mostafa Al-Emran, The British University in Dubai, UAE

Conference Tracks Chairs

Abdallah Namoun, Islamic University of Medina, Saudi Arabia
Adi A. Alqudah, The British University in Dubai, UAE
Ali Tarhini, Sultan Qaboos University, Oman
Amr Abdullatif Yassin, Ibb University, Yemen
Cham Tat Huei, UCSI University, Malaysia
Heider A. M. Wahsheh, King Faisal University, Saudi Arabia
Ibrahim Arpaci, Bandirma Onyedi Eylul University, Turkey
Kamal Karkonasasi, Universiti Malaysia Kelantan, Malaysia
Khalid Adam, Universiti Malaysia Pahang, Malaysia

Members of Scientific Committee

Abdallah Namoun, Islamic University of Medina, Saudi Arabia
Abdullah Nasser, University of Vaasa, Finland
Abdulmajid Mohammed Aldaba, International Islamic University Malaysia, Malaysia
AbdulRahman Al-Sewari, Universiti Malaysia Pahang, Malaysia
Ahmed M. Mutahar, Management and Science University, Malaysia
Aisyah Ibrahim, Universiti Malaysia Pahang, Malaysia
Akhyari Nasir, University College TATI, Kemaman, Terengganu, Malaysia
Alaa A. D. Taha, University of Mosul, Iraq
Ali Nasser Ali Al-Tahitah, Universiti Sains Islam Malaysia, Malaysia
Ali Qasem Saleh Al-Shetwi, Fahad Bin Sultan University, Saudi Arabia
Ameen A. Ba Homaid, Universiti Malaysia Pahang, Malaysia
Amir A. Abdulmuhsin, University of Mosul, Iraq
Amr Abdullatif Yassin, Ibb University, Yemen
Baraq Ghaleb, Edinburgh Napier University, UK
Basheer Mohammed Al-haimi, Hebei University, Baoding, China
Bokolo Anthony Jnr, Norwegian University of Science and Technology, Norway
Dalal Abdulmohsin Hammood, Middle Technical University, Iraq
Eissa M. Alshari, Ibb University, Yemen
Fadi A. T. Herzallah, Palestine Technical University – Kadoorie, Palestine
Fathey Mohammed, Universiti Utara Malaysia, Malaysia
Garry Wei Han Tan, UCSI University, Malaysia
Gonçalo Marques, Universidade da Beira Interior, Portugal
Hasan Sari, Universiti Tenaga Nasional, Malaysia
Heider A. M. Wahsheh, King Faisal University, Saudi Arabia
Hussam S. Alhadawi, Dijlah University College, Iraq
Hussein Mohammed Esmail Abu Al-Rejal, University Utara Malaysia, Malaysia
Ibrahim Arpaci, Gaziosmanpasa University, Turkey

Joseph Ng, UCSI University, Malaysia
Joshua A. Abolarinwa, Namibia University of Science and Technology, Namibia
Kamal Mohammed Alhendawi, Al-Quds Open University, Faculty of Management, Palestine
Kamal Karkonasasi, Universiti Malaysia Kelantan, Malaysia
Khaled Shaalan, The British University in Dubai, UAE
Marwah Alian, Hashemite University, Jordan
Marwan Saeed Saif Moqbel, Ibb University, Yemen
Mikkay Wong Ei Leen, Sunway University, Malaysia
Mohamed Elwakil, University of Cincinnati, USA
Mohammed A. Al-Sharafi, Universiti Teknologi Malaysia, Malaysia
Mohammed A. Alsaih, University Putra Malaysia, Malaysia
Mohammed Ahmed Talab, Almaarif University College, Iraq
Mohammed Adam Kunna Azrag, Universiti Teknologi MARA (UiTM), Malaysia
Mohammed N. Al-Kabi, Al Buraimi University College, Oman
Mostafa Al-Emran, The British University in Dubai, UAE
Mukhtar A. Kassem, Universiti Teknologi Malaysia, Malaysia
Nejood Hashim Al-Walidi, Sanaa University, Yemen
Noor Akma Abu Bakar, Tunku Abdul Rahman University College (TARC), Malaysia
Noor Al-Qaysi, Universiti Pendidikan Sultan Idris, Malaysia
Noor Suhana Sulaiman, University College TATI, Kemaman, Terengganu, Malaysia
Osama Mohammad Aljarrah, University of Massachusetts Dartmouth, USA
Osamah A. M. Ghaleb, Mustaqbal University, Saudi Arabia
Qasim Al Ajmi, A'Sharqiyah University, Oman
Samer Ali Alshami, Universiti Teknikal Malaysia Melaka, Malaysia
Taha Sadeq, Universiti Tunku Abdul Rahman, Malaysia
Tang Tiong Yew, Sunway University, Malaysia
Vitaliy Mezhuyev, FH JOANNEUM University of Applied Sciences, Austria

Publicity and Public Relations Committee

Hasan Sari, Universiti Tenaga Nasional, Malaysia
Noor Akma Abu Bakar, Universiti Malaysia Pahang, Kuantan, Malaysia

Finance Chair

Taha Sadeq, Universiti Tunku Abdul Rahman, Malaysia

Contents

Why Should I Continue Using It? Factors Influencing Continuance Intention to Use E-wallet: The S-O-R Framework

Aznida Wati Abdul Ghani, Abdul Hafaz Ngah, and Azizul Yadi Yaakop

Abstract Research on e-wallet behaviour has captured the interest of scholars in recent years as a result of the rapid changes in spending patterns. This study aims to investigate e-wallet users' continuance usage intention by incorporating SOR theory. This study illustrates the mediating role of satisfaction and attachment in the relationship between self-congruity and continuance intention to use e-wallets. Through the use of a structured questionnaire, the self-administered data collection reached out to 550 potential respondents across Malaysia. The respondents were chosen using a non-probability purposive sampling technique. In total, 435 replies were evaluated. The analysis was conducted using Smart PLS version 3.3.5. The findings indicate that there is a positive relationship between satisfaction and attachment on one hand and intention to continue use on the other. Additionally, the results proved that attachment and satisfaction sequentially mediated the relationship between self-congruity and continuance usage intention. The conclusions of this study could benefit all stakeholders in Malaysia's Fin-Tech business, particularly those in the e-wallet community.

Keywords E-wallet · Mobile wallet · SOR theory · Self-congruity · Attachment · Satisfaction · Continuance intention to use · Cashless society · Malaysia

1 Introduction

The rise of COVID-19 has coincided with a massive shock to the global economic patterns, gradually altering the ways we live our lives. After nearly two years of living with the virus, we have transformed our old habits of living, playing and shopping in order to comfortably adjust to the new normal. Within a year of the outbreak, the number of electronic wallet transactions had increased by 89% to 468 million [1]. The growing use of e-wallets has exacerbated the divide between traditional and digital retail transaction trends. Apart from health issues, individuals value simple, quick

A. W. A. Ghani (✉) · A. H. Ngah · A. Y. Yaakop
University Malaysia Terengganu, 21300 Kuala Nerus, Terengganu, Malaysia
e-mail: aznidaaghani@gmail.com

M. Al-Emran et al. (eds.), *International Conference on Information Systems and Intelligent Applications*, Lecture Notes in Networks and Systems 550,
https://doi.org/10.1007/978-3-031-16865-9_1

and time-saving transactions. Compared to cash transactions, e-wallet transactions are believed to be safer, more comfortable and handier because users no longer need to carry cash [2]. Everything is stored on their smartphones. E-wallets enable users to store their money through any of the 53 licensed e-wallet providers in Malaysia [3], which can be easily downloaded via the Google Play Store, Huawei App Gallery or Apple App Store.

Apart from the role of e-wallet providers in promoting the numerous benefits of utilising an e-wallet, another critical factor affecting the e-wallet landscape in Malaysia is merchants' adoption of the technology. In 2021, merchant participation in QR payments increased by 57% to 1 million registrations [1], demonstrating their rapid response, whereby they joined the e-wallet ecosystem, further adopting and embracing the trend as more consumers adopted e-wallet usage. While the use of e-wallets is increasing, the technology's future profitability depends on its continued use rather than its initial adoption [4, 5]. Gaining a higher rate of user adoption guarantees a substantial profit for a firm immediately, but maintaining current customers is the measure of a corporation's long-term performance. All this effort and investment in technology would be in vain if e-wallet usage plummeted. The average app user is impatient since so many applications and other forms of media are competing for their attention. According to Ding and Chai [6], three months after installing downloaded apps, only 24% of users continued using them. This reduced to 14% of users after six months and only 4% of users had been retained one year after installation. If app user retention is so low, the mere volume of downloads may become irrelevant.

Retaining consumers is critical for e-wallet providers to prosper and recoup their substantial investment in e-wallet services [5]. Numerous people continue to prefer internet banking to e-wallets [7]. While e-wallets continue to grow in popularity and generate billions of transactions, online banking volumes are also rising; 2021 saw an increase of 36% to 12.1 billion transactions [1]. As previously stated, many users who experimented with e-wallet services then returned to internet banking [8]. Therefore, retaining existing users is crucial [5] to ensuring that government efforts do not fail and that they support the e-wallet providers in avoiding major losses. E-wallet merchants have spent as much as RM 600,000 [9] on each e-wallet app, simply to become involved in these services. All their technological investments would be for nought if e-wallet usage declined. In securing the survival of the e-wallet in Malaysia, it is necessary to identify the variables impacting its continued use. It is appropriate to highlight the factors that contribute to users' continuance intention to use e-wallets regardless the efforts of both the government and the private sector since it is valuable to understand how post-adoption compares to initial adoption [10].

Studies on ensuring continued usage have acquired prominence because previous scholars have shown that interiorised usage cannot be projected using notions such as short-term usage, adoption and acceptance [11]. This had created an urgent need to thoroughly investigate continuous use and the elements impacting it. Thus, this study elucidates the factors driving Malaysians' continuous use of e-wallets. While numerous studies have examined e-wallet user behaviour, the majority of the literature has focused on initial uptake rather than ongoing usage [12]. To the authors' knowledge, little research on the continued use of e-wallets has been conducted. Most previous scholars of continuance usage behaviour adopted the ECM [13], TAM [14], UTAUT [15] or TCT [10] as their underpinning theory.

To the authors' knowledge, there is a scarcity of research using the S-O-R model to assess the drivers that influence the continuity intention to use e-wallets. Ngah et al. [16] argued that the SOR framework provides researchers with greater flexibility to manipulate their research model based on the context of their studies, as long as it reflects the original basic concept of stimulus–organism–response. Earlier research on continuance intention examined S-O-R theory in a variety of contexts, including social media platforms [17] and airline services [18]. The S-O-R model most frequently used to examine the topic of mobile payments centres on usage intention and customer satisfaction [19]. To address this gap, the current study took into account the SOR framework proposed by Mehrabian and Russell [20] to investigate the factors influencing continuance intention to use e-wallets by operationalising self-congruity for the 'stimulus'; attachment and satisfaction for the 'organism'; and continuance intention to use for the 'response'. Additionally, this study contributed to the literature by studying the mediating effects of satisfaction and attachment in the relationship between self-congruity and an e-wallet user's continued intention to use the service.

2 Literature Review

2.1 *The Stimulus-Organism-Response (S-O-R) Framework*

The current study makes use of the Stimulus-Organism-Response approach developed by the environmental psychologists, Mehrabian and Russell [20]. S-O-R theory is fundamentally comparable to the well-known IS processing model, which is composed of the following components: input (stimulus)—process (organism—emotion or cognitive judgement)—output (response). This theory describes how the variables in the model are connected. This idea was first designed to explain how specific environmental inputs boost emotions, resulting in future behaviour. SOR theory has been adapted for use in various research contexts [21, 22]. Additionally, environmental cues impact consumers' experience-based judgements to produce a unique response in a specific setting, and this theory has been frequently applied in research. The theory's adaptability enables researchers to construct and expound new models of consumer behaviour that are based on the SOR approach (Fig. 1).

The S-O-R paradigm has already been used to explain consumer loyalty [23], purchase intention [24] and engagement [21], among others. Meanwhile, research on volunteerism [25], technology adoption [26] and other contexts has proven the predictive abilities of SOR theory. Nonetheless, the objectives of several studies

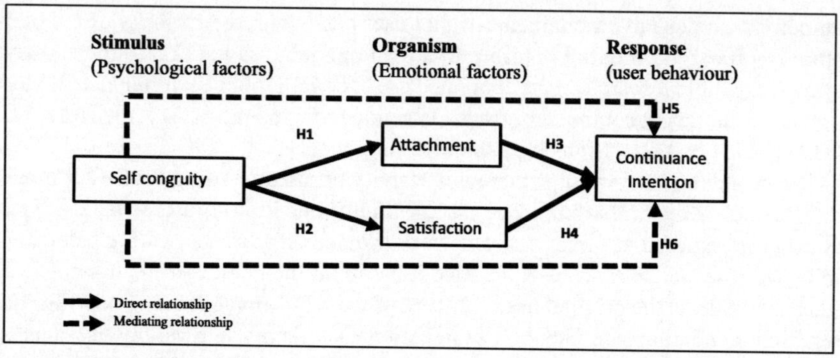

Fig. 1 Theoretical framework

have been to evaluate the continued use of a product based on SOR theory, as well as highlight continuance usage intention based on SOR theory [17]. In these cases, scholars modified the environmental stimuli and emotional evaluations to the study context. The psychological concept (self-congruity) of e-wallet services was used to represent "environmental stimuli" in this study; these are critical aspects since they reflect the user's early adoption phase of e-wallet usage. The term "organism" refers to an organism's internal state, which is influenced by environmental stimuli [20]. Baghozzi [27] defined the term "organism" as an individual's internal states of emotion, perception, and affection that influence future behaviour. Similarly, the study characterised satisfaction and attachment as internal states impacted by an individual's self-congruity. Additionally, the SOR model postulated that the organism had an indirect effect on the relationship between stimulus and response [28]. The character of an organism affects a response. For e-wallet users, satisfaction and attachment have a considerable influence on e-wallet users' decisions to continue using the e-wallet or switch to alternative mobile payment methods that offer comparable services; thus, the likelihood of switching to other ways is high. Moreover, this study used S-O-R as a crucial explanatory framework for analysing the aspect of human behaviour to anticipate cognitive judgement and future, or post-adoption, behaviour. Due to the study's sequential effect on the user's psychological acceptance stage, as well as on their emotions, feelings and behaviours, the SOR approach was applicable. Using the SOR model as a guide, the current study proposes a theoretical basis on which to explain the influence of self-congruity stimuli on attachment and satisfaction, which subsequently affects continuing intention.

Self-congruity. Self-congruity is defined as the alignment between the product image and the customer's self-image [29], that is, the congruence between the consumer's self-concept and the perceived image of the product or service [34, 35]. Self-congruity occurred in the current study when users pictured how the product image reflected their own image. Specifically, e-wallet customers seeking a meaningful and personal connection with their e-wallet app were found to align the e-wallet's image with their own. The stronger the self-congruence between a user's actual self-image and his/her actual activity, the more likely the user is to be driven to engage in future action [30]. Additionally, congruity with a product contributes to the development of pleasant sensations among users, both pre- and post-purchase [29, 31]. Kim and Thapa [30] observed that self-congruity improved satisfaction. When users believe products or services are consistent with their self-image, they demonstrate increased product engagement, brand loyalty and brand relationship quality [32–34]. Japutra et al. [35] established that self-congruity has a direct influence on brand attachment. As a result, the following hypotheses were advanced:

H1. Self-congruity has a positive impact on attachment.

H2. Self-congruity has a positive impact on satisfaction.

Attachment. Brand attachment has been lauded as a critical concept in marketing literature due to its ability to accurately reflect a consumer's emotional bond to a product over time [36]. Just as humans have a natural need to build attachments with other people [37], they also develop attachments to services for various reasons. This bond has an effect on behaviour, which in turn increases brand loyalty and consumer lifetime value [38]. The current research defines attachment as a unique and strong bond of self-connection between users and e-wallets. This unique bond is formed through a psychological and technological connection (as a result of positive experiences with e-wallet services), coupled with their personal attitudes towards e-wallet usage behaviour. Attachment fosters a strong rapport between both players (users and e-wallets apps), to the point at which both parties become emotionally invested and willing to invest additional resources - such as energy and time - to preserve the connection [39]. Previous research studied the influence of attachment on consumer behaviour, more precisely, on the proclivity to maintain usage [40]. Consumers who have a strong brand attachment are unwilling to swap brands and display a greater propensity to persist with their initial choice. Additionally, Cao et al. [39] revealed that of all the predictors, emotional attachment had the strongest relationship with continuing intention. As a result, the following hypothesis was advanced:

H5. Attachment has a positive impact on continuance usage intention.

Satisfaction. Satisfaction is a critical sign of success in the e-commerce ecosystem [41]. User satisfaction reflects users' confidence in the service's ability to elicit happy emotions [42]. Users' satisfaction with e-wallet usage is a result of their interactions with the services, and this influences their future behaviours. Satisfied users are more likely to shop again and refer the business to others [43], and they are also expected to continuously sustain the use of technology [44], whereas dissatisfied users will abandon the retailer with or without complaints. Taking into account the perspectives and concepts of these preceding researchers, "satisfaction" can be described in the context of the present study as the degree to which e-wallet services meet the users' anticipated outcomes following their use of e-wallets, which in turn encourages the continued use of e-wallets. As Bhattacherjee [4] noted, user satisfaction is a significant element in determining continuation or behavioural intention, a finding that Chuah et al. [45] and Veeramootoo et al. [46] corroborated. Additionally, Hepola et al. [47] discovered that satisfaction is a superior predictor of future service use intention. Thus, the authors hypothesised the following:

H6. Satisfaction has a positive impact on continuance usage intention.

Mediating Effect (Attachment and Satisfaction). Despite the possibility of interaction between variables based on the SOR model, insufficient research has been undertaken on the influence of mediating factors (attachment, satisfaction) on the relationships between self-congruity and continued intention to use e-wallets. Mediating factors in a causal chain are variables that relate antecedent variables to outcomes [48, 49]. The exploration of theoretical mediating factors is common in business research and social sciences [49], since most business researchers integrate mediation and/or moderation into their research frameworks [25, 48]. In the present study, mediation can be recognised in the idea that self-congruity and the intention to continue using e-wallets are influenced by attachment and satisfaction. As such, mediation is considered the underlying mechanism and process connecting the antecedents and consequences [48]. Based on the prior discussion, the current study demonstrates that the existing literature presents consistent associations between self-congruity and attachment [35], attachment and continuance intention [39], self-congruity and satisfaction [30], as well as satisfaction and continuance intention [50]. There is a persistent positive association between self-congruity and intention to continue use [51]. Thus, it is hypothesised that attachment and satisfaction mediate the relationship between self-congruity and intention to continue use. In light of this, the following hypotheses were proposed:

H5. The relationship between self-congruity and continuance intention is mediated by satisfaction.

H6. The relationship between self- congruity and continuance intention is mediated by attachment.

3 Methodology

The present study used a quantitative technique to test the hypotheses and determine whether they fit within the research framework. Due to the individual being used as the unit of analysis and the lack of a complete sample frame, a self-administered survey questionnaire was employed in conjunction with a purposive, non-probability sampling approach. The data was acquired from Malaysians who were experienced e-wallet customers. All possible respondents over the age of 15 were given the survey questionnaires, together with a cover letter. This was undertaken in retail premises, as well as among public and social groups. Only those who volunteered to participate in the survey received the survey questions. Of the 600 surveys issued, exactly 550 responses were received. However, 115 of the returned questionnaires were discarded due to their poor data quality, such as respondents answering with a straight line or providing partial responses. Thus, 435 responses were usable, amounting to a response rate of 79%. In the remaining questionnaire sets, there was no missing data. A priori power analysis was done using G*Power 3.1 before the commencement of data collection to estimate the minimum sample size necessary to establish the appropriate statistical power required to explain the model's interactions [52]. The results showed that a sample size of 67 participants with two predictors would be necessary to attain 80% power at a medium effect size (0.15) and a 0.05 confidence level. A total of 435 responses obtained were deemed sufficient for testing the study model.

Since this study used the SmartPLS (partial least squares) 3.3.5 programme [53], all the measuring items for each construct were adapted from prior researchers. Even though several of the items had previously been used in other research studies, the authors justified their inclusion in the context of the present study without changing its original purpose. The self-congruity (SC) measuring items were adapted from Sharma et al. [51], those for satisfaction (SAT) and continuance intention (CI) from Rahi et al., [54] and the items for attachment (ATC) from Pedeliento et al. [55].

4 Data Analysis

In terms of age, 54.7% of the 435 people who responded were in the 15 to 25 age group. Females comprised 60.7% of the total number, while 73.% had incomes below RM 3,170 per month. According to respondents' e-wallet profiles, 56.8% were interested in adopting e-wallets as a result of the government's e-wallet incentive programmes, 85.5% had used e-wallets for less than two years and 77.2% had acquired no more than two e-wallets.

The study's predictive nature necessitated the use of SmartPLS software [53, 56]. The current research evaluated multivariate skewness and kurtosis in accordance with the published recommendations [57, 58]. Mardia's multivariate kurtosis was $\beta = 47.572701$, $p < 0.01$, while Mardia's multivariate skewness was $\beta = 5.313324$, $p < 0.01$. These values suggested that the data was somewhat non-normal. As a consequence, since the data did not match the criterion for normality, it was suitable to perform the analysis using SmartPLS [53].

The common method variance issue may arise when only one source is used to obtain the data, with the conclusion potentially being affected [59]. Consequently, the authors addressed this problem using both forms of analysis, procedural and statistical. The authors assessed the study's constructs using a different anchor scale in the procedural approach [60, 61]. The intention to keep utilising the product was gauged using a seven-point Likert scale. The remainder of the constructs were rated on five-point Likert scales. According to the recommendations by Kock [62] and Ngah [59], full collinearity should be evaluated to limit the likelihood of common method bias. Through this method, each variable was regressed against a common variable. When employing a single source of data, a VIF score of less than five indicates that bias is not a serious problem in the research [63]. The VIF values were less than five (attachment = 3.448, continuance intention = 2.073, satisfaction = 3.268, self-congruity = 2.109), which suggested that this study was not significantly affected by CMV.

4.1 Measurement Model

The authors posited argued for a two-step approach to SEM analysis that incorporates both a measurement and a structural model. The measurement model's convergent and discriminant validity had to be determined before the study could proceed to the subsequent phase of using the structural model to test the hypothesis. As stated by Hair et al. [64], it is possible to achieve convergent validity when the loading and average variance explained (AVE) values exceed 0.5; moreover, the composite reliability must be over 0.7. Each of these sets of values was utilised in the evaluation of construct validity illustrated in Table 1. As indicated in Table 1, all the results were larger than the minimum values stated in the literature, indicating that the study's convergent validity had been demonstrated.

It is necessary to determine discriminant validity once convergent validity has been confirmed. According to Franke and Sarstedt [65], discriminant validity is shown when the heterotrait-monotrait ratio (HTMT) is less than 0.90. Table 2 shows that each HTMT value was below the most conservative value stipulated, demonstrating that the discriminant validity of the study had not been compromised [65].

Table 1 Convergent validity

Construct	Item	Loading	CR	AVE
Attachment	ATC1	0.937	0.964	0.870
	ATC2	0.933		
	ATC3	0.927		
Continuance intention	CI1	0.932		
	CI2	0.969	0.978	0.937
	CI3	0.968		
Satisfaction	SAT1	0.967		
	SAT2	0.937	0.963	0.897
	SAT3	0.959		
Self-congruity	SC1	0.944		
	SC2	0.921	0.952	0.868
	SC3	0.939		
	SC4	0.935		

Table 2 Discriminant validity: HTMT ratio

	Attachment	Continuance Intention	Satisfaction	Self Congruity
Attachment				
Continuance Intention	0.688			
Satisfaction	0.861	0.670		
Self Congruity	0.706	0.672	0.697	

4.2 Assessment of Structural Model

The structural model analysis step is where hypotheses are tested within the study framework. The proposed hypotheses in the research model were tested using a bootstrapping approach with 500 samples [66]. Experts have recommended that four aspects of the research hypotheses should be evaluated: i) the hypotheses should be reflected in the study's direction; ii) t-values should be ≥ 1.645; iii) p-values should be ≤ 0.05; and iv) the confidence interval of the study should contain no zero values between the lower (LL) and upper levels (LL) [67, 68].

The direct effect results suggest that attachment ($b = 0.414$, $p < 0.001$) and satisfaction ($b = 0.302$, $p < 0.001$) were both positively related to continuance intention. Hence, H3 and H4 were supported. Self-congruity was positively related to both attachment ($b = 0.662$, $p < 0.001$) and satisfaction ($b = 0.651$, $p < 0.001$), proving that H1 and H2 were also supported. In terms of the effect size values, the definition given by Cohen [69] was that small (S) effect sizes were 0.02, medium (M) effect sizes were 0.15 and large (L) effect sizes were 0.35. Table 3 indicates that the effect sizes were either small or large for each supported hypothesis.

The mediation analysis revealed that two hypotheses were supported: H5 and H6. These hypotheses were bootstrapped using the methodology described by Preacher

Table 3 Hypothesis testing

Relationship	Beta	SE	T Values	P Values	LL	UL	VIF	F2	Decision
H3: ATC → CI	0.414	0.064	6.438	0.001	0.31	0.524	2.978	0.108 (S)	Supported
H4: SAT → CI	0.302	0.068	4.466	0.001	0.18	0.4	2.978	0.057 (S)	Supported
H1: SC → ATC	0.662	0.038	17.604	0.001	0.596	0.725	1.000	0.782 (L)	Supported
H2: SC → SAT	0.651	0.038	17.003	0.001	0.586	0.707	1.000	0.735 (L)	Supported
H6: SC → SAT → CI	0.196	0.048	4.119	0.001	0.101	0.285			Supported
H5: SC → ATC → CI	0.274	0.051	5.432	0.001	0.174	0.371			Supported

and Hayes [70]. Table 3 indicates the significant relationship that self-congruity had with continuance intention when either attachment (SC → ATTC → CI; $b = 0.274$, $p < 0.001$) or satisfaction were the mediators (SC → SAT → C1: $b = 0.196$, $p < 0.001$). It can be reasonably concluded that the analysis featured the effects of mediation since it was indicated by the confidence intervals that no zero values straddled the lower level and the upper levels.

5 Discussion

Overall, the contribution the current research makes to the literature is to enhance the knowledge of the different features that impact the continued intention to utilise e-wallets. This area is especially apposite, so it should be researched further by academics in the context of service settings. This study examined how both satisfaction and attachment were related to users' decisions to keep utilising e-wallets, with this association revealed to be positively affected. These outcomes aligned with those of Raman and Aashish [71], who reported that a contented user tends to exhibit favour towards a service and an intention to keep utilising it. A delighted user would return to e-wallet-based applications on a continuous basis. As a result, this pleasant sensation will inspire their future continued adoption behaviour to be more favourable. This finding is consistent with Khayer and Bao [72]. E-wallet companies must have an innovative and compelling strategy for attracting customers to their apps. Promoting additional privileges and advantages will attach multiple pull factors that help to retain current users and entice new ones to use the services.

Psychological elements (self-experience and self-congruity) as stimuli agents in the SOR framework introduced another dimension to the discussion on e-wallet behaviour research. Zou et al. [73] stated that attachment and self-congruity are positively affected by one's own experience. If a customer is an existing e-wallet user whose self-experience while using the service has been positive, they are likely to develop associations with the service based on emotion. As users of the e-wallet community, prior experience with e-wallets enables users to justify their self-image in terms of the prevalent picture of e-wallet users.

Researchers have indicated that self-experience and continued use are directly related [74]. Nevertheless, as the results of this study confirm, two pairs of successive mediators affect the way self-experience and continuation intention are related: self-congruity and attachment (H8), in addition to self-congruity and satisfaction (H9). These results show the crucial roles played by self-congruity, attachment and satisfaction when assessing how continued e-wallet use is related to self-experience and intention.

6 Theoretical and Practical Contributions

This research contributes essential aspects of the theory concerning the intention to keep utilising e-wallets, while the findings could be generalised to different forms of post-adoption actions in numerous situations involving innovative information technology. The study has the potential to contribute to the body of knowledge in relevant disciplines involving technology adoption, both theoretically and practically. The authors have identified a lack of literature that uses the SOR model to explore continuing intention to use e-wallets. Interestingly, the current data supports each of the proposed hypotheses. However, future studies of moderating factors may offer more fascinating outcomes that add a new dimension to the topic. An important way this study contributes is by validating the ways self-congruity and attachment act as mediators when examining how self-experience and continuance intention are associated.

This study has practical ramifications that can help enhance future e-wallet usage. Favourable attitudes to the use of e-wallets should be developed, which could be achieved by encouraging those who provide e-wallet services to focus on progressing the practical advantages of the e-wallets so they meet the user's expectations and raise their satisfaction. Service providers must invest more money in research and development to ensure that e-wallet systems provide all of the functionality and features requested by users. It would be useful to utilise social media platforms to undertake instructive and engaging campaigns to promote e-wallet services. As Malaysia moves closer to becoming a cashless society, the government must continue to foster the growth of the FinTech industry by streamlining licensing procedures and providing tax benefits. To ensure a significant impact on post-adoption behaviour, it is also recommended to explore cost-free marketing plans that promote both electronic and positive word-of-mouth (EWOM and WOM) among current e-wallet users [75, 76]. This strategy would be effective for increasing peer awareness of the distinct features and benefits of e-wallet systems in comparison to other forms of digital payment. This would have a substantial effect on users' behaviours, potentially assisting in the retention of current users and attracting new ones.

7 Conclusion, Limitations and Future Research

While contributing to important elements of the theory and exploring the outcomes for management, the current research involved various limitations. To begin with, because this study focused on end-users, the external validity of the findings was compromised. Future studies should examine the e-wallet community as a whole, including e-wallet providers and e-wallet intermediates (merchants), to gain a better knowledge of the circumstances. Second, research on post-adoption behaviour involving e-wallets needs to be expanded across regions, age groups, locations and socioeconomic positions to provide relevant data and aid in the development of successful marketing designs to meet specific demands. Such details could be used to instil in users stronger emotional attachments to e-wallet use, which should hasten the transfer to cashless transactions. Third, as the current study used a cross-sectional approach, mono-method bias may be an issue; hence, future research should employ a qualitative or longitudinal strategy to elicit more detailed information about e-wallet users' persistence behaviour. Fourth, while this study considered the key characteristics of continuing e-wallet usage, the inclusion of service security and social dimensions as potential predictors of ongoing usage behaviour could produce exciting results. Last, the model outlined and explained in this article could be expanded in future research. One way to achieve this would be to include and discuss the self-congruity dimension (actual and ideal self-congruity).

References

1. Malaysia Ministry Of Finance (2021) Malaysian Banking and Finance Summit 2021. https://www.mof.gov.my/portal/en/news/speech/malaysian-banking-and-finance-summit-2021. Accessed 6 Feb 2022
2. Hanafi WNW, Toolib SN (2020) Influences of perceived usefulness, perceived ease of use, and perceived security on intention to use digital payment : a comparative study among Malaysian younger and older adults. Int J Bus Manage 3(1):15–24
3. Bank Negara Malaysia (2021) List of non-bank e-money issuers. https://www.bnm.gov.my/non-bank-e-money-issuers. Accessed 1 Jul 2021
4. Bhattacherjee A (2001) Understanding information systems continuance: An expectation-confirmation model. MIS Q Manag Inf Syst 25(3):351–370. https://doi.org/10.2307/3250921
5. Foroughi B, Iranmanesh M, Hyun SS (2019) Understanding the determinants of mobile banking continuance usage intention. J Enterp Inf Manag 32(6):1015–1033. https://doi.org/10.1108/JEIM-10-2018-0237
6. Ding Y, Chai KH (2015) Emotions and continued usage of mobile applications. Ind Manag Data Syst 115(5):833–852. https://doi.org/10.1108/IMDS-11-2014-0338
7. J.P. Morgan Insight (2020) E-commerce payments trends: Malaysia e-commerce insights. JPMorgan Chase. https://www.jpmorgan.com/europe/merchant-services/insights/reports/malaysia. Accessed 3 Jul 2021
8. Yang HL, Lin RX (2017) Determinants of the intention to continue use of SoLoMo services: consumption values and the moderating effects of overloads. Comput Human Behav 73:583–595. https://doi.org/10.1016/j.chb.2017.04.018
9. Emizen Tech (2020) How to develop an e-wallet mobile? Cost & key features wallet. Emizen Tech. https://www.emizentech.com/blog/e-wallet-mobile-app-development.html. Accessed 20 Jul 2021
10. Daragmeh A, Sági J, Zéman Z (2021) Continuous intention to use e-wallet in the context of the COVID-19 pandemic: integrating the Health Belief Model (HBM) and Technology Continuous Theory (TCT). J Open Innov Technol Mark Complex 7(2):132. https://doi.org/10.3390/joitmc7020132
11. Al-Sharafi MA, Arshah RA, Abu-Shanab EA (2017) Factors affecting the continuous use of cloud computing services from expert's perspective. ENCON 2017-2017 IEEE Region 10 Conference, vol 2017-Decem, p 986–991. https://doi.org/10.1109/TENCON.2017.8228001
12. Abdul-Halim NA, Vafaei-Zadeh A, Hanifah H, Teoh AP, Nawaser K (2021) Understanding the determinants of e-wallet continuance usage intention in Malaysia. Qual Quant. https://doi.org/10.1007/s11135-021-01276-7Understanding

13. Kumar A, Adlakaha A, Mukherjee K (2018) The effect of perceived security and grievance redressal on continuance intention to use M-wallets in a developing country. Int J Bank Mark 36(7):1170–1189. https://doi.org/10.1108/IJBM-04-2017-0077
14. Garrouch K (2021) Does the reputation of the provider matter? A model explaining the continuance intention of mobile wallet applications. J Decis Syst 30:150–171. https://doi.org/10.1080/12460125.2020.1870261
15. Phuong NND, Luan LT, Van Dong V, Khanh NLN (2020) Examining customers' continuance intentions towards e-wallet usage: The emergence of mobile payment acceptance in Vietnam. J Asian Financ Econ Bus 7(9):505–516. https://doi.org/10.13106/JAFEB.2020.VOL7.NO9.505
16. Ngah AH et al (2022) The sequential mediation model of students' willingness to continue online learning during the COVID-19 pandemic. Res Pract Technol Enhanc Learn 17:1–17. https://doi.org/10.1186/s41039-022-00188-w
17. Gogan ICW, Zhang Z, Matemba ED (2018) Impacts of gratifications on consumers' emotions and continuance use intention: an empirical study of Weibo in China. Sustain 10(9):3162. https://doi.org/10.3390/su10093162
18. Perumal S, Ali J, Shaarih H (2021) Exploring nexus among sensory marketing and repurchase intention: application of S-O-R Model. Manag Sci Lett 11:1527–1536. https://doi.org/10.5267/j.msl.2020.12.020
19. Chen SC, Chung KC, Tsai MY (2019) How to achieve sustainable development of mobile payment through customer satisfaction: the SOR model. Sustain 11(22):1–16. https://doi.org/10.3390/su11226314
20. Mehrabian A, Russell JA (1974) An approach to environmental psychology. The MIT Press
21. Cho WC, Lee KY, Yang SB (2019) What makes you feel attached to smartwatches? The stimulus–organism–response (S-O-R) perspectives. Inf Technol People 32(2):319–343. https://doi.org/10.1108/ITP-05-2017-0152
22. Zhu L, Li H, Wang FK, He W, Tian Z (2020) How online reviews affect purchase intention: a new model based on the stimulus-organism-response (S-O-R) framework. Aslib J Inf Manag 72(4):463–488. https://doi.org/10.1108/AJIM-11-2019-0308
23. Herrando C, Jiménez-Martínez J, Martín-De Hoyos MJ (2019) 'Social Commerce Users' optimal experience: stimuli, response and culture. J Electron Commer Res 20(4):199
24. Fu S, Yan Q, Feng GC (2018) Who will attract you? Similarity effect among users on online purchase intention of movie tickets in the social shopping context. Int J Inf Manage 40:88–102. https://doi.org/10.1016/j.ijinfomgt.2018.01.013
25. Ngah AH, Rahimi AHM, Gabarre S, Saifulizam NIFC, Aziz NA, Han H (2021) Voluntourism sustainability: a case of Malaysian east coast island destinations. Asia Pacific J Tour Res 26(12):1364–1385. https://doi.org/10.1080/10941665.2021.1983622
26. Luqman A, Masood A, Weng Q, Ali A, Rasheed MI (2020) Linking excessive SNS use, technological friction, strain, and discontinuance: the moderating role of guilt. Inf Syst Manag 37(2):94–112. https://doi.org/10.1080/10580530.2020.1732527
27. Baghozzi R (1986) Principles of marketing management. Science Research Associates, Chicago, IL, USA
28. Tuan Mansor TM, Mohamad Ariff A, Hashim HA, Ngah AH (2020) External whistleblowing intentions of auditors: a perspective based on stimulus–organism–response theory. Corp Gov: Int J Bus Soc 22(4):871–897. https://doi.org/10.1108/cg-03-2021-0116
29. Sirgy MJ (2018) Self-congruity theory in consumer behavior: a little history. J Glob Sch Mark Sci 28(2):197–207. https://doi.org/10.1080/21639159.2018.1436981
30. Kim M, Thapa B (2018) The influence of self-congruity, perceived value, and satisfaction on destination loyalty: a case study of the Korean DMZ. J Herit Tour 13(3):224–236. https://doi.org/10.1080/1743873X.2017.1295973
31. Widjiono LM, Japarianto E (2014) Analisa Pengaruh self image congruity, retail service quality, Dan customer perceived service quality Terhadap repurchase intention Dengan customer satisfaction Sebagai Variabel intervening Di Broadway Barbershop Surabaya. J Manaj Pemasar 9(1):35–42. https://doi.org/10.9744/pemasaran.9.1.35-42

32. Gwinner KP, Eaton J (1999) Building brand image through event sponsorship: the role of image transfer. J Advert 28(4):47–57. https://doi.org/10.1080/00913367.1999.10673595
33. Peters S, Leshner G (2013) Get in the game: the effects of game-product congruity and product placement proximity on game players processing of brands embedded in advergames. J Advert 42(2–3):113–130. https://doi.org/10.1080/00913367.2013.774584
34. Phua J, Kim J (2018) Starring in your own Snapchat advertisement: influence of self-brand congruity, self-referencing and perceived humor on brand attitude and purchase intention of advertised brands. Tele Inf 35(5):1524–1533. https://doi.org/10.1016/j.tele.2018.03.020
35. Japutra A, Ekinci Y, Simkin L (2019) Self-congruence, brand attachment and compulsive buying. J Bus Res 99:456–463. https://doi.org/10.1016/j.jbusres.2017.08.024
36. Japutra A, Ekinci Y, Simkin L (2014) Exploring brand attachment, its determinants and outcomes. J Strateg Mark 22(7):616–630. https://doi.org/10.1080/0965254X.2014.914062
37. Bowlby J (1970) Attachment and loss. Br J Sociol 21(1):111–112
38. Park CW, MacInnis DJ, Priester J, Eisingerich AB, Iacobucci D (2010) Brand attachment and brand attitude strength: conceptual and empirical differentiation of two critical brand equity drivers. J Mark 74(6):1–17. https://doi.org/10.1509/jmkg.74.6.1
39. Cao YY, Qin XH, Li JJ, Long QQ, Hu B (2020) 'Exploring seniors' continuance intention to use mobile social network sites in China: a cognitive-affective-conative model. Univers Access Inf Soc 21:71–92. https://doi.org/10.1007/s10209-020-00762-3
40. Dwivedi A, Johnson LW, Wilkie DC, De Araujo-Gil L (2019) Consumer emotional brand attachment with social media brands and social media brand equity. Eur J Mark 53(6):1176–1204. https://doi.org/10.1108/EJM-09-2016-0511
41. Shin JI, Chung KH, Oh JS, Lee CW (2013) The effect of site quality on repurchase intention in Internet shopping through mediating variables: the case of university students in South Korea. Int J Inf Manage 33(3):453–463. https://doi.org/10.1016/j.ijinfomgt.2013.02.003
42. Udo GJ, Bagchi KK, Kirs PJ (2010) An assessment of customers' e-service quality perception, satisfaction and intention. Int J Inf Manage 30(6):481–492. https://doi.org/10.1016/j.ijinfomgt.2010.03.005
43. Trivedi SK, Yadav M (2017) Predicting online repurchase intentions with e-Satisfaction as mediator: a study on Gen Y. J Inf Knowl Manag Syst 40(3):427–447. https://doi.org/10.1108/VJIKMS-10-2017-0066
44. Al-Sharafi MA, Al-Qaysi N, Iahad NA, Al-Emran M (2021) Evaluating the sustainable use of mobile payment contactless technologies within and beyond the COVID-19 pandemic using a hybrid SEM-ANN approach. Int J Bank Mark 40(5):1071–1095. https://doi.org/10.1108/IJBM-07-2021-0291
45. Chuah SHW, Rauschnabel PA, Marimuthu M, Thurasamy R, Nguyen B (2017) Why do satisfied customers defect? A closer look at the simultaneous effects of switching barriers and inducements on customer loyalty. J Serv Theory Pract 27(3):616–641. https://doi.org/10.1108/JSTP-05-2016-0107
46. Veeramootoo N, Nunkoo R, Dwivedi YK (2018) What determines success of an e-government service? Validation of an integrative model of e-filing continuance usage. Gov Inf Q 35(2):161–174. https://doi.org/10.1016/j.giq.2018.03.004
47. Hepola J, Leppäniemi M, Karjaluoto H (2020) Is it all about consumer engagement? Explaining continuance intention for utilitarian and hedonic service consumption. J Retail Consum Serv 57:102232. https://doi.org/10.1016/j.jretconser.2020.102232
48. Aguinis H, Edwards JR, Bradley KJ (2017) Improving our understanding of moderation and mediation in strategic management research. Organ Res Methods 20(4):665–685. https://doi.org/10.1177/1094428115627498
49. MacKinnon DP (2015) Mediating variable, 2nd edn., vol. 15. Elsevier
50. C. Peng, Z. OuYang, and Y. Liu, 'Understanding bike sharing use over time by employing extended technology continuance theory', *Transp. Res. Part A Policy Pract.*, vol. 124, pp. 433–443, 2019. https://doi.org/10.1016/j.tra.2019.04.013.
51. Sharma TG, Hamari J, Kesharwani A, Tak P (2020) Understanding continuance intention to play online games: roles of self-expressiveness, self-congruity, self-efficacy, and perceived risk. Behav Inf Technol 41(2):348–364. https://doi.org/10.1080/0144929X.2020.1811770

52. Hair JF, Hult GTM, Ringle CM, Sarstedt M (2017) A primer on partial least squares structural equation modeling (PLS-SEM), Second Edition. SAGE, Los Angeles
53. Ringle CM, Wende S, Becker JM (2015) In press—journal of rheology. SmartPLS 3:1–16
54. Rahi S, Khan MM, Alghizzawi M (2021) Extension of Technology Continuance Theory (TCT) with Task Technology Fit (TTF) in the context of Internet banking user continuance intention. Int J Qual Reliab Manage 38(4):986–1004. https://doi.org/10.1108/IJQRM-03-2020-0074
55. Pedeliento G, Andreini D, Bergamaschi M, Salo J (2016) Brand and product attachment in an industrial context: the effects on brand loyalty. Ind Mark Manag 53:194–206. https://doi.org/10.1016/j.indmarman.2015.06.007
56. Hair JF, Risher JJ, Sarstedt M, Ringle CM (2019) When to use and how to report the results of PLS-SEM. Eur Bus Rev 31(1):2–24. https://doi.org/10.1108/EBR-11-2018-0203
57. Halimi MF, Gabarre S, Rahi S, Al-Gasawneh JA, Ngah AH (2021) 'Modelling Muslims' revisit intention of non-halal certified restaurants in Malaysia, J Islam Mark. https://doi.org/10.1108/JIMA-01-2021-0014
58. Ngah AH, Kamalrulzaman NI, Ibrahim F, Osman NAA, Ariffin NA (2021) The effect of soft skills, ethics, and value on the willingness of employers to continue recruiting UMT graduates. Manag Sci Lett 11:1689–1698. https://doi.org/10.5267/j.msl.2020.12.002
59. Ngah AH, Kim HD, Hanafiah RM, Salleh NHM, Jeevan J, Asri NM (2019) Willingness to pay for Halal transportation cost: the stimulus-organism-response model. Int J e-Navig Marit Econ 12:11–21
60. Podsakoff PM, MacKenzie SB, Podsakoff NP (2012) Sources of method bias in social science research and recommendations on how to control it. Annu Rev Psychol 63:539–569. https://doi.org/10.1146/annurev-psych-120710-100452
61. Ngah AH, Gabarre S, Han H, Rahi S, Al-Gasawneh JA, Park SH (2021) Intention to purchase halal cosmetics: Do males and females differ? A multigroup analysis. Cosmetics 8(1):1–14. https://doi.org/10.3390/cosmetics8010019
62. Kock N (2015) Common method bias in PLS-SEM. Int J e-Collab 11(4):1–10. https://doi.org/10.4018/ijec.2015100101
63. Kock N (2015) Common method bias in PLS-SEM: a full collinearity assessment approach. Int J e-Collab 11(4):1–10. https://doi.org/10.4018/ijec.2015100101
64. Hair JF, Babin BJ, Krey N (2017) Covariance-based structural equation modeling in the journal of advertising: review and recommendations. J Advert 46(1):163–177. https://doi.org/10.1080/00913367.2017.1281777
65. Franke G, Sarstedt M (2019) Heuristics versus statistics in discriminant validity testing: a comparison of four procedures. Internet Res 29(3):430–447. https://doi.org/10.1108/IntR-12-2017-0515
66. Hair J, Hollingsworth CL, Randolph AB, Chong AYL (2017) An updated and expanded assessment of PLS-SEM in information systems research. Ind Manag Data Syst 117(3):442–458. https://doi.org/10.1108/IMDS-04-2016-0130
67. Ngah AH, Thurasamy R, Mohd Salleh NH, Jeevan J, Md Hanafiah R, Eneizan B (2021) Halal transportation adoption among food manufacturers in Malaysia: the moderated model of technology, organization and environment (TOE) framework. J Islam Mark. https://doi.org/10.1108/jima-03-2020-0079
68. Tuan Mansor TM, Ariff AM, Hashim HA, Ngah AH (2021) 'Whistleblowing intentions among external auditors: an application of the moderated multicomponent model of the theory of planned behavior. Meditari Account Res 30(5):1309–1333. https://doi.org/10.1108/MEDAR-07-2020-0948
69. Cohen J (1992) A power primer, July
70. Preacher KJ, Hayes AF (2008) Asymptotic and resampling strategies for assessing and comparing indirect effects in multiple mediator models. Behav Res Methods 40(3):879–891. https://doi.org/10.3758/BRM.40.3.879
71. Raman P, Aashish K (2021) To continue or not to continue: a structural analysis of antecedents of mobile payment systems in India. Int. J. Bank Mark 39(2):242–271. https://doi.org/10.1108/IJBM-04-2020-0167

72. Khayer A, Bao Y (2019) The continuance usage intention of Alipay: Integrating context-awareness and technology continuance theory (TCT). Bottom Line 32(3):211–229. https://doi.org/10.1108/BL-07-2019-0097
73. Zou Y, Meng F, Li Q (2021) Chinese diaspora tourists' emotional experiences and ancestral hometown attachment. Tour Manag Perspect 37:100768. https://doi.org/10.1016/j.tmp.2020.100768
74. Mombeuil C, Uhde H (2021) Relative convenience, relative advantage, perceived security, perceived privacy, and continuous use intention of China's WeChat Pay: a mixed-method two-phase design study. J Retail Consum Serv 59:102384. https://doi.org/10.1016/j.jretconser.2020.102384
75. Matute J, Polo-Redondo Y, Utrillas A (2016) The influence of EWOM characteristics on online repurchase intention. Online Inf Rev 40(7):1090–1110. https://doi.org/10.1108/oir-11-2015-0373
76. Bulut ZA, Karabulut AN (2018) Examining the role of two aspects of eWOM in online repurchase intention: an integrated trust–loyalty perspective. J Consum Behav 17(4):407–417. https://doi.org/10.1002/cb.1721

The Impact of Artificial Intelligence and Supply Chain Resilience on the Companies Supply Chains Performance: The Moderating Role of Supply Chain Dynamism

Ahmed Ali Atieh Ali, Zulkifli B. Mohamed Udin, and Hussein Mohammed Esmail Abualrejal

Abstract In light of the information revolution, this study aims to clarify the impact of artificial intelligence and supply chain resilience on the supply chain performance of engineering, electrical, and information technology companies registered with the Jordan Chamber of Industry. This study expands knowledge by exploring the relationships between artificial intelligence and the moderating supply chain dynamism. This study looks at artificial intelligence as an important resource, in addition to resilience supply chains, an important resource in raising the supply chain performance for companies. The questionnaire was conducted via e-mail and the study sample included (208) companies registered with the Jordanian Chamber of Industry and Commerce. The data was analyzed using the smart (Pls) software and its direct link with artificial intelligence and supply chain resilience. In addition, the analysis shows that there is a direct relationship between the mediating variables supply chain dynamism and supply chain resilience and supply chain performance. These results provide an insight into the relationship between artificial intelligence and supply chains, and the Moderating variable on the performance of a company's supply chains, which may be an entry point for companies to enhance their performance due to the importance of this sector to the Jordanian economy.

Keywords Artificial intelligence · Supply chain Resilience · Supply chain performance · Supply chain dynamism · Engineering electrical · Information technology · Jordan

A. A. A. Ali (✉)
School of Technology and Logistics Management, Universiti Utara Malaysia (UUM), Sintok, Kedah 06010, Malaysia
e-mail: ahmadaliatiehali@gmail.com

Z. B. M. Udin · H. M. E. Abualrejal
School of Technology Management and Logistic, College of Business, Universiti Utara Malaysia, Sintok, Kedah 06010, Malaysia
e-mail: zulkifli@uum.edu.my

H. M. E. Abualrejal
e-mail: abualrejal@uum.edu.my

M. Al-Emran et al. (eds.), *International Conference on Information Systems and Intelligent Applications*, Lecture Notes in Networks and Systems 550,
https://doi.org/10.1007/978-3-031-16865-9_2

1 Introduction

Since the inception of the concept of artificial intelligence, it has become a top priority for businesses, and this priority has been largely driven by the obtainability of big data and the emergence of advanced infrastructure technology, where technology is required to improve business in all sectors, including the supply chain [1]. As more data connected to (big data) analytics became available, predictive analytics was used to investigate the causes of supply chain interruptions, resulting in improved supply chain performance. According to a recent survey published by the Gartner Foundation, the number of firms using artificial intelligence has increased by 270 per cent in only four years [2]. While there are numerous doubts regarding artificial intelligence's potential economic worth, firms that have started to implement it has seen significant improvements in their performance. However, owing to rising demand and interruptions in global supply networks, the contemporary supply chain is more complicated than previous supply systems, necessitating the digitization of the supply chain to tackle such problems [3].

Traditional businesses seek ways to enhance their supply networks' performance and the coordination of supply chain participants [4]. Supply chain resilience may help reduce the risks of supply chain disruptions. defined supply chain resilience as a feature of a supply network that allows it to return to its original form after a brief time of disturbance. Whereas, with the outbreak of the Corona pandemic, Supply chain resilience has risen to the top of the discussion, since supply chain resilience is concerned with the supply chain's capacity to respond to current actions [5].

As a result, the creation, coordination, and management of information are essential aspects for supply chain recovery, according to the Resilient Supply Chains reports, as organizations must adopt a dynamic and innovative approach to managing their chain, and risks and threats must be viewed as opportunities to develop in the face of highly disruptive networks and threats [6]. which, in turn, will have an effect on the SC performance [7].

Today's supply chains are getting more dynamic in today's world settings and with technology change[8], need continuous information as external and internal threats to carry on to stifle their performance [9]. Furthermore, [10] have identified dynamism of the environment as an important component toward consider when addressing performance-related concerns. As a result, knowing the connections between AI, supply chain resilience, and supply chain performance is crucial, and the links are predicted to provide useful information about how performance capabilities should be developed.

Based on the above-mentioned research gaps, the following research questions emerge:

RQ1. Is there a relationship between Artificial intelligence (AI) and Supply chain resilience on supply chain performance?

RQ2. Does supply chain dynamism influence the relation between Artificial intelligence AI and Supply chain resilience in Supply Chain Performance?

In addressing the research questions, Based on the findings, we develop a research framework. "(OIPT)", where the study population is from the engineering, electrical, and information technology industries registered with the Jordanian Chamber of Industry, which numbered 453 and the study sample reached 208, where this sector The volume of exports is 465.8 million Jordanian dinars annually, and it employs 31,725, which is the total number of employees working in the Jordan Chamber of Industry [11]. It is expected and its growth in this sector in the coming years, which shows great importance to this sector and to see it, and to help in ways to develop this sector in particular in Jordan and the world in general, and from here stems The importance of the study that deals with "the impact of artificial intelligence and supply chain resilience on supply chain performance: Moderating Dynamic Supply Chain".

2 Literature Review

2.1 Artificial Intelligence

During the last two decades, many organizations have attempted to digitize their operations, and the industry has just recently become a business bumblebee [12]. For a long time, artificial intelligence has been regarded as one of the most important technologies for facilitating machine-to-machine communication [13]. Because the supply chain encompasses a range of complicated jobs, artificial intelligence might help to streamline operations by resolving issues faster and more accurately while also processing large amounts of data [14]. Although (AI) isn't new concept, it has only recently been recognized for its potential in a variety of applications, including supply chain management [15]. To predict issues, artificial intelligence can provide smart and quick decision-making in the supply chain. As a result, through on-time and undamaged delivery, a proactive AI system contributes to improved service quality and customer satisfaction [16]. Artificial intelligence (AI) automates compliance, cost-cutting and enhancing the efficiency of a supply chain network [17]. In today's developing business environment, artificial intelligence greatly influences the predictive skills necessary for demand forecasting. Conversations with AI-powered bots may be personalized, making client contact more efficient. These bots, which are backed up by echo users and customer service representatives, can help track the status of an item's delivery [18].

2.2 Supply Chain Resilience

Defining supply chain resilience is the ability of a supply chain to withstand unexpected, disruptive events and swiftly recover to its prior level of performance or to a new level necessary to sustain expected operational market, and financial performance to build a strong supply chain, businesses must identify and assess the nodes for hazards, occurrence frequency, severity, and how these hazards might be detected [19]. Businesses use a range of strategies to keep their supply networks healthy. During the early phases of the Corona pandemic, certain supply chains identified inventory and capacity stores as a source of resilience. Others depended on underused production capacity for goods, while others relied on underutilized production capacity for other commodities [10]. Certain supply chains have profited from resilience as a result of multisourcing plans as compared to a single source of supplies [20]. The corona-virus pandemic has also emphasized the need for near-shoring in order to reduce geographic dependence on global networks [21]. Local supply networks provide for improved inventory management and quicker delivery of products to clients [22]. The more local the network, the more likely manufacturing technologies will be effectively harmonized, allowing for a more seamless flow of commodities across the network [23].

2.3 *Supply Chain Performance*

Overall, supply chain performance (SCP) is defined as the advantages gained from supply chain operations' efficiency and adaptability in an ever-changing environment [24]. It measures how well a company's supply chain meets the expectations of its customers in terms of product obtainability, while also keeping "costs" to a minimum. [25]. SCP and its precursors have been extensively studied in the past. In order to succeed in business and the marketplace, organizations must have strong supply chain resilience. According to [26], SCP includes resource efficiency, output effectiveness, and adaptability performance at the organizational level (agility). Customer's value, such as quality, pricing, and delivery time, can be created more efficiently, effectively, and quickly; supply chain performance (SCP) can continue to create value in a chaotic and uncertain environment [24].

2.4 *The Moderation of Supply Chain Dynamism*

Supply chains are becoming more dynamic, [7] Supply chain dynamism is defined as the use of the transformative pace of change in goods and supply chain processes in business conditions and technology. SC working in a dynamism environment face many "internal and external" challenges that reduce their effectiveness, which requires a continuous flow of information [27].

Based on [28], three indicators can be used to assess dynamic supply chains: the income generated from goods and services, the speed of process innovation, and the level of product innovation. According to [29], organizations must have a complete comprehension of the breadth of supply chain dynamism in order to create more resilient methods and improve supply chain performance. OIPT advocates for how supply chain dynamics affect supply chain practice and information sharing. Supply chain dynamism boosts the efficiency of its numerous components, including[30]. Another study found that supply network dynamism positively influences Disruption of the supply chain as well as SC resilience [27]. SC resilience, which has been demonstrated to precede supply chain dynamism, has an impact on a firm financial performance. The association between the integration and performance of the supply chain was demonstrated to be mitigated by supply chain dynamics [7].

3 Conceptual Model

Performance in information processing is based on the needs and abilities of the organization, according to OIPT, which is a group of people, The relationship between information processing skills and the consequences associated with them may be altered by supply chain unpredictability [31]. So, supply chains need to be able to communicate with stakeholders in a proactive way to improve visibility and traceability in the supply chain. The data analytics ability is thought of as a way to process information based on "OIPT" from the literature, with the effect on supply chain performance being looked at [9]. According to Galbraith 1974, in theory, organizations might choose to use "mechanistic" organizational resources instead of information, this could help them lessen their reliance on information or improve their ability to process information.

Importantly, according to OIPT, businesses must handle information with rising uncertainty in order to maintain a given degree of performance. A highly necessary organizational competency is the ability to process information in the face of risk, volatility, and dynamism [32]. In this research, artificial intelligence is defined as an information-processing tool that should be built from the ground up to eliminate functioning challenges and uncertainty.

In addition, the "OIPT" says that businesses should build "ability buffers" and be able to process data to deal with supply chain interruptions [32]. OIPT's assumptions are supported by a number of different theories. According to Wamba, AI is a resource that can be used to support higher-order capabilities like SC resilience and SC performance [33]. OIPT theory suggests that organizations should align their information processing capabilities with customer demand. This is what they should do [34]. In this view, supply chain resilience may be linked to better supply chain performance if the amount of information that can be changed matches the amount of supply chain disruptions. We want to fill up the gaps and limits of these ideas by offering a complete theoretical basis based on the OIPT. AI, Supply Chain Resilience and Supply Chain Performance will benefit from this background. Resilience and its impact on supply chain performance are the focus of this research. Supply Chain Dynamism and artificial intelligence (AI) will be used to demonstrate this (Fig. 1).

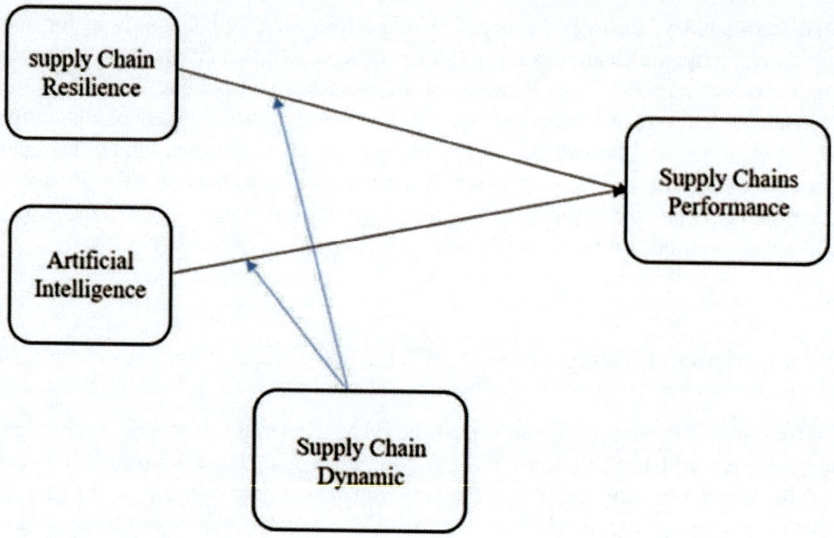

Fig. 1 Framework of study

4 Research Method

The data for this study were collected using a questionnaire, which is a quantitative research method. The research was conducted in previous studies to determine the factors of the current study, and these factors were determined by artificial intelligence and Resilience supply chains, and they were referred to in our current study as independent variables, the mediator variable dynamic supply chains, and the dependent variable, supply chains performance. Questionnaire questions were built on previous studies where artificial intelligence [35]. SC resilience [8, 36] and mediator supply chains dynamics [35, 37]. Supply chain performance [32] and the study population consists of 453 facilities specialized in engineering, electrical and information technology industries based on Jordan Chamber of Industry [11]. The study sample, based on [38], consisted of 208 managers and executives of these companies in Jordan.

5 Data Analysis

We utilized the SmartPLS 3.3.2 version of partial least squares (PLS) modelling. In order to evaluate the study's premise, the researchers used a two-stage technique. The measurement model, which includes convergent and discriminant validity, is the first step. It will move on to testing hypotheses and making a structural model after the validity of their claims has been proven.

Table 1 Summary of the "factor loadings"

ITEMS	Factor loadings	Cronbach's Alpha	Composite reliability	Average variance extracted (AVE)
SCR1	0.761	0.899	0.926	0.716
SCR2	0.771			
SCR3	0.825			
SCR4	0.915			
SCR5	0.942			
AI1	0.771	0.847	0.867	0.567
AI2	0.666			
AI3	0.809			
AI4	0.66			
SCD1	0.842	0.878	0.914	0.726
SCD2	0.832			
SCD3	0.879			
SCD4	0.867			
SCP1	0.891	0.883	0.918	0.738
SCP2	0.849			
SCP3	0.833			
SCP4	0.863			

For starters, convergent validity examines whether an item really measures the latent variable it promises to [39].

The assessment of the measuring model entails the analysis of the link between each construct and its items. The reflective measurement model investigation includes the assessment of "indicator loading," indicator reliability, internal consistent reliability, "convergence validity", and discriminant validity. When it comes to indicator loading, the conventional rule of thumb is 0.708 or greater [40]. According to hair [41], in social science research, it is common to identify weaker item loading and delete items with low loading. Furthermore, it is permissible to consider eliminating items with an outer loading of between "0.4 and 0.7" if doing so improves the value of composite reliability and the average variance extracted (AVE) [41]. Table 1 shows a summary of the "factor loadings".

5.1 *Structural Model*

The structural model is tested After developing the measurement model for reliability and validity. Analyzing structural models entails evaluating how effectively the theory or ideas are empirically supported by the facts and, as a result, deciding whether the hypothesis is empirically proven (Table 2).

Table 2 Fronell-Larcker

ITEMS	Artificial intelligence	Supply chain resilience	Supply chains performance
Artificial intelligence	**0.756**		
Supply chain resilience	0.331	**0.846**	
Supply chains performance	0.346	0.77	**0.859**

5.2 *Demographic Information of Respondents (Table 3).*

Table 3 Demographic information of respondents

1. Characteristic	2. Frequency	3. Percentage
4. Gender	5.	6.
7. Male	8. 176	9. 84.6
10. Female	11. 32	12. 15.4
13. Age	14.	15.
16. less than 27	17. 10	18. 4.8
19. 27-less than 35	20. 28	21. 13.5
22. 35-less than 45	23. 93	24. 44.7
25. 45 and above	26. 77	27. 37.0
28. Education	29.	30.
31. Diploma	32. 5	33. 2.4
34. Undergraduate degree	35. 141	36. 67.8
37. Postgraduate degree (Master/PhD)	38. 62	39. 29.8
40. Experience	41.	42.
43.. less than 10	44. 28	45. 13.5
46. 10-less than 15	47. 36	48. 17.3
49. 15-less than 20	50. 70	51. 33.7
52. 20-less than 25	53. 58	54. 27.9
55. 25 and above	56. 16	57. 7.7
58. Specialization	59.	60.
61. Engineering	62. 152	63. 73.1
64. Business Administration	65. 46	66. 22.1
67. Other	68. 10	69. 4.8

5.3 Hypotheses Testing

The PLS Algorithm function was used to examine the path coefficient in the structural model. For regression analysis, the SmartPLS 3.0 model's path coefficient is equivalent to the usual beta weight. From –1 to + 1, the estimated path coefficients vary from a strong positive association to one that's strongly negative, while a path coefficient near to zero implies that there's no relationship at all. It is shown in Table 4 that the path coefficient, standard error, T-Value, P-Value and significance level of the analysis were all tested for statistical significance (Fig. 2).

For the purpose of determining the accuracy of predictions, the findings of R2 are shown in Table 5. The correlation coefficient (R2) for Supply Chain Performance is 0.267. These findings confirm that explanatory factors account for more than 26% of variances.

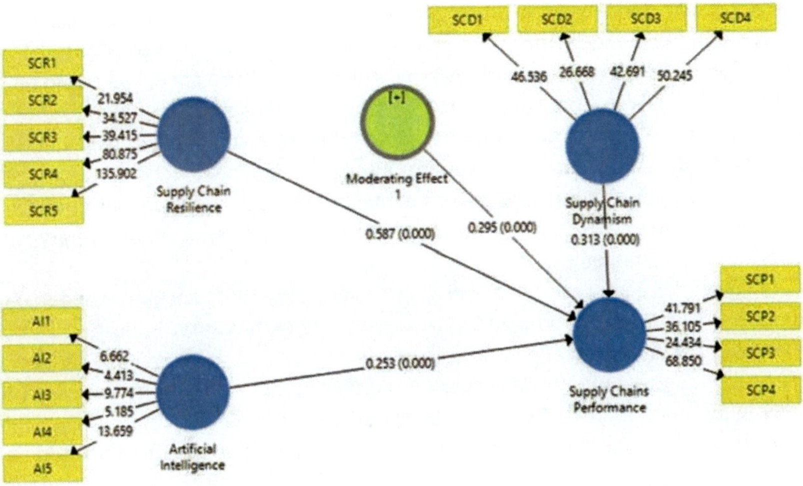

Fig. 2 The results of the structural model

Table 4 PLS-SEM path coefficients results

Hypo	Relationships	Std. beta	Std. error	T-value	P-values	Decision
H1	Artificial Intelligence → Supply chains performance	0.253	0.036	6.988	0.000	Supported
H2	Moderating effect → Supply chains performance	0.295	0.048	6.172	0.000	Supported
H3	Supply chain dynamic → Supply chains performance	0.313	0.056	5.576	0.000	Supported
H4	Supply chain resilience → Supply chains performance	0.587	0.065	9.03	0.000	Supported

Table 5 R^2 adjusted

Variable	R^2	R^2 Adjusted
Supply chains performance	0.267	0.256

6 Discussion and Conclusion

This study answers many calls made by many studies to examine the relationship between artificial intelligence and supply chain performance, and it is clear from previous studies that the Moderator supply chain dynamism has an important role in promoting supply chains resilience, which is reflected in the supply chains performance, and the study identified the relevant characteristics Among the variables using the theory (OPIT), and according to this study, companies that deal with information and in an increasing degree of uncertainty must maintain a certain degree of performance. On the supply chain performance, this illustrates the importance of supply chains resilience in those companies, where Resilience incorporate work functions reduces the degree of uncertainty and enhances performance, and based on the results of the analysis, there is no direct relationship between the Moderator Supply Chain dynamism and artificial Intelligence, and these results agreed with the study [2] and a study [35]. Hence, companies may need to consider a correct supervisory approach to ensure that the applications of smart The study recommends conducting more research to understand the impact of artificial intelligence on the supply chain performance in companies.

However, there are several caveats that must be taken into account when making inferences from this study's results. When answering certain survey questions, survey takers may not know exactly what information is needed. It does open the door to further investigation into the link between AI and supply chain performance, particularly in terms of the moderator effects that affect the dynamic nature of the supply chain. It will be important to know to see whether the answers of this study can be applied to other nations, given this research only attentive on Jordanian engineering, power, and information technology enterprises. It is possible to undertake comparable research in other sectors such as relief organizations in order to boost the generalizability of the present study because of the limited sample size.

References

1. Belhadi A, Mani V, Kamble SS, Khan SAR, Verma S (2021) Artificial intelligence-driven innovation for enhancing supply chain resilience and performance under the effect of supply chain dynamism: an empirical investigation. Ann Oper Res 2021:1–26
2. Gartner (2022) No title. https://www.gartner.com/en/documents/3897266/2019-cio-survey-cios-have-awoken-to-the-importance-of-ai
3. Dubey R, Gunasekaran A, Childe SJ et al (2020) Big data analytics and artificial intelligence pathway to operational performance under the effects of entrepreneurial orientation and environmental dynamism: a study of manufacturing organisations. Int J Prod Econ 226:107599

4. Choi T, Wallace SW, Wang Y (2018) Big data analytics in operations management. Prod Oper Manag 27(10):1868–1883
5. Grover P, Kar AK, Dwivedi YK (2020) Understanding artificial intelligence adoption in operations management: insights from the review of academic literature and social media discussions. Ann Oper Res 308:177–213
6. Ivanov, D, Dolgui A, Das A, Sokolov B (2019) Handbook of ripple effects in the supply chain, vol 276, Springer
7. Dubey R et al (2020) Big data analytics and artificial intelligence pathway to operational performance under the effects of entrepreneurial orientation and environmental dynamism: a study of manufacturing organisations. Int J Prod Econ 226:107599
8. Belhadi A, Zkik K, Cherrafi A, Yusof SM, El fezazi S (2019) Understanding Big Data analytics for manufacturing processes: insights from literature review and multiple case studies. Comput Ind Eng 137:106099
9. Belhadi A, Kamble S, Fosso Wamba S, Queiroz MM (2022) Building supply-chain resilience: an artificial intelligence-based technique and decision-making framework. Int J Prod Res 60(14):4487–4507
10. Queiroz MM, Ivanov D, Dolgui A, Fosso Wamba S (2020) Impacts of epidemic outbreaks on supply chains: mapping a research agenda amid the COVID-19 pandemic through a structured literature review. Ann Oper Res 2022:1–38
11. Jordan Chamber of Industry (2022) No title. https://www.jci.org.jo/Chamber/Sector/80066/-The_engineering_electrical_andinfor-mationtechnology_industries?l=en
12. Wollschlaeger M, Sauter T, Jasperneite J (2017) The future of industrial communication: automation networks in the era of the internet of things and industry 4.0. IEEE Ind Electron Mag 11(1):17–27
13. Guzman AL, Lewis SC (2020) Artificial intelligence and communication: a human–machine communication research agenda. New Media Soc 22(1):70–86
14. Schniederjans DG, Curado C, Khalajhedayati M (2020) Supply chain digitisation trends: an integration of knowledge management. Int J Prod Econ 220:107439
15. Huin S-F, Luong LHS, Abhary K (2003) Knowledge-based tool for planning of enterprise resources in ASEAN SMEs. Robot Comput Integr Manuf 19(5):409–414
16. Toorajipour R, Sohrabpour V, Nazarpour A, Oghazi P, Fischl M (2021) Artificial intelligence in supply chain management: a systematic literature review. J Bus Res 122:502–517
17. Treleaven P, Batrinca B (2017) Algorithmic regulation: automating financial compliance monitoring and regulation using AI and blockchain. J Financ Transform 45:14–21
18. Huang M-H, Rust RT (2021) A strategic framework for artificial intelligence in marketing. J Acad Mark Sci 49(1):30–50
19. Dubey R, Altay N, Gunasekaran A, Blome C, Papadopoulos T, Childe SJ (2018) Supply chain agility, adaptability and alignment: empirical evidence from the Indian auto components industry. Int J Oper Prod Manag 38(1):129–148
20. Jeble S, Kumari S, Venkatesh VG, Singh M (2020) Influence of big data and predictive analytics and social capital on performance of humanitarian supply chain: Developing framework and future research directions. Benchmarking 27(2):606–633
21. Kano L, Oh CH (2020) Global value chains in the post-COVID world: governance for reliability. J Manag Stud 57(8):1773–1777
22. Sundarakani B, Pereira V, Ishizaka A (2021) Robust facility location decisions for resilient sustainable supply chain performance in the face of disruptions. Int J Logis Manag 32(2):357–385
23. Adobor H (2020) Supply chain resilience: an adaptive cycle approach. Int J Logis Manag 31(3):443–463
24. Chowdhury MMH, Quaddus M, Agarwal R (2019) Supply chain resilience for performance: role of relational practices and network complexities. Supply Chain Manag 24(5):659–676
25. Tarafdar M, Qrunfleh S (2017) Agile supply chain strategy and supply chain performance: complementary roles of supply chain practices and information systems capability for agility. Int J Prod Res 55(4):925–938

26. Khan A, Bakkappa B, Metri BA, Sahay BS (2009) Impact of agile supply chains' delivery practices on firms' performance: cluster analysis and validation. Supply Chain Manag Int J, 14(1):41–48
27. Yu W, Jacobs MA, Chavez R, Yang J (2019) Dynamism, disruption orientation, and resilience in the supply chain and the impacts on financial performance: a dynamic capabilities perspective. Int J Prod Econ 218:352–362
28. Zhou H, Benton WC (2007) Supply chain practice and information sharing. J Oper Manag 25(6):1348–1365
29. Lee HY, Seo YJ, Dinwoodie J (2016) Supply chain integration and logistics performance: the role of supply chain dynamism. Int J Logis Manag 27:668–685
30. Cegarra-Navarro JG, Soto-Acosta P, Wensley AKP (2016) Structured knowledge processes and firm performance: the role of organizational agility. J Bus Res 69(5):1544–1549
31. Wong CWY, Lirn T-C, Yang C-C, Shang K-C (2020) Supply chain and external conditions under which supply chain resilience pays: an organizational information processing theorization. Int J Prod Econ 226:107610
32. Srinivasan R, Swink M (2018) An investigation of visibility and flexibility as complements to supply chain analytics: an organizational information processing theory perspective. Prod Oper Manag 27(10):1849–1867
33. Kavota JK, Kamdjoug JRK, Wamba SF (2020) Social media and disaster management: case of the north and south Kivu regions in the Democratic Republic of the Congo. Int J Inf Manage 52:102068
34. Tushman ML, Nadler DA (1978) Information processing as an integrating concept in organizational design. Acad Manag Rev 3(3):613–624
35. Dubey R et al (2020) Big data analytics and artificial intelligence pathway to operational performance under the effects of entrepreneurial orientation and environmental dynamism: a study of manufacturing organisations. Int J Prod Econ 226:107599
36. Altay N, Gunasekaran A, Dubey R, Childe SJ (2018) Agility and resilience as antecedents of supply chain performance under moderating effects of organizational culture within the humanitarian setting: a dynamic capability view. Prod Plan Control 29(14):1158–1174
37. Belhadi A, Zkik K, Cherrafi A, Sha'ri MY (2019) Understanding big data analytics for manufacturing processes: insights from literature review and multiple case studies. Comput Ind Eng 137:106099
38. Krejcie RV, Morgan DW (1970) Determining sample size for research activities. Educ Psychol Meas 30(3):607–610
39. Hair JF, Hult TM, Ringle CM, Sarstedt M (2017) A primer on partial least squares structural equation modeling (PLS-SEM). Sage, Thousand Oaks, p 165
40. Fornell C, Larcker DF (1981) Evaluating structural equation models with unobservable variables and measurement error. J Mark Res 18(1):39–50
41. Hair JF, Hult GTM, Ringle CM, Sarstedt M (2014) A primer on partial least squares structural equation modeling (PLS-SEM). Eur J Tour Res 6(2):211–213

Acceptance of Mobile Banking in the Era of COVID-19

Bilal Eneizan, Tareq Obaid, Mohanad S. S. Abumandil, Ahmed Y. Mahmoud, Samy S. Abu-Naser, Kashif Arif, and Ahmed F. S. Abulehia

Abstract The prevalence of the COVID-19 pandemic and the impact of lockdown initiatives to curb the spread of the disease have had a significant effect on daily human activities and the global economy in general, and the operations of the banking sector in particular. Few studies have been carried out on the factors that affect the acceptance of mobile banking especially during and after the COVID-19 pandemic. Thus, the aim of this current research is to identify the drivers of mobile banking usage intention among banking customers in Palestine during the current pandemic. For this purpose, a total of 290 people were surveyed using an electronic questionnaire. The study's conceptual model was analyzed using structural equation modeling. The findings showed that Attitude significantly affects intention, the intention was revealed to significantly affect adoption, PBC significantly affects intention, PEOU does not affect attitude, PR was also found to have no significant effect on intention, PU significantly affects attitude as well as intention, PEOU significantly affects PU,

B. Eneizan (✉)
Business School, Jadara University, Irbid, Jordan
e-mail: Bilalmomane@gmail.com

T. Obaid · A. Y. Mahmoud · S. S. Abu-Naser
Faculty of Engineering and IT, Alazhar University, Gaza, Palestine
e-mail: tareq.obaid@alazhar.edu.ps

A. Y. Mahmoud
e-mail: ahmed@alazhar.edu.ps

S. S. Abu-Naser
e-mail: abunaser@alazhar.edu.ps

M. S. S. Abumandil
Faculty of Hospitality, Tourism and Wellness, Universiti Malaysia Kelantan, Kelantan, Malaysia
e-mail: mohanad.ssa@umk.edu.my

K. Arif
Faculty of Management Science, Shaheed Zulfiqar Ali Bhutto Institute of Science and Technology, Karachi, Pakistan

A. F. S. Abulehia
School of Accountancy, Universiti Utara Malaysia, 06010 Sintok, Kedah, Malaysia

M. Al-Emran et al. (eds.), *International Conference on Information Systems and Intelligent Applications*, Lecture Notes in Networks and Systems 550,
https://doi.org/10.1007/978-3-031-16865-9_3

SN significantly affects intention, and finally, trust was revealed to significantly affect intention.

Keywords COVID-19 pandemic · Mobile banking · Consumer behavior

1 Introduction

People's daily activities have been substantially transformed with the advent of mobile devices, most notably in relation to doing financial transactions. In recent years, there is an upward trend in mobile banking usage in various industries. WorldPay reported that mobile banking makes up 22% of the 2019 global points of sale spending, which is expected to rise further to 29.6% in 2023 (Worldpay, Global Payments Reports). Studies such as that of [1] had examined the intention of users to adopt mobile banking in various contexts. Yet, there remains a lack of variation in the determinants and theoretical proof from various standpoints particularly in the context of a pandemic [2].

COVID-19 first emerged in December 2019 when 66,243,918 cases were confirmed worldwide along with 1,528,984 deaths [55]. Owing to the disease's high transmission rate, the WHO had declared it a global pandemic and issued recommendations for social distancing measures to reduce human contact [3, 4].

The prevalence of the pandemic and the impact of the lockdown initiatives to curb the spread of the disease have had a significant effect on daily human activities and the global economy in general, and the operations of the banking sector in particular. In response, governments worldwide had offered public guarantees on bank loans and ratified moratoriums to facilitate clients that are short of liquidity. In terms of the banking sector, usage of digital channels and digital payment methods began to intensify thus changing consumer behaviour along with the easing of a number of regulatory and supervisory condition [5]. But such shifts also come with various challenges including impacts on operational resilience and a higher rate of non-performing loans.

The shift in consumer behaviour and how businesses are run due to the pandemic has become a major concern for companies and financial industries worldwide. Not only do they have to take short-term actions to accommodate the changes, they also have to formulate medium- and long-term strategies to ensure future sustainability.

The Palestinian retail banking sector also feels the pinch of the shift in consumer behaviour due to the pandemic. The nation's retail banking consumers had no other choice but to shift to using digital channels. The findings of this study justify the usage of the Technology Acceptance Model (TAM). This study assumes that banking customers in Palestine will likely adopt mobile banking services due to the various uncertainties and trepidations caused by the current global pandemic.

Few studies have been carried out on the acceptance of mobile banking in developing countries especially in Palestine during and after the COVID 19 pandemic.

Therefore, this study attempts to reinvestigate the factors that affect the acceptance of mobile banking by the customers during and after COVID 19 pandemic.

2 Background and Hypothesis Development

The lockdown and social distancing orders that had been established to curb the coronavirus from spreading had directly resulted in an increase in online activities. In the context of Palestine, [6] highlighted the major shift to online shopping while predicting that pre-pandemic norms are unlikely to make a comeback. In this light, current business banking models will be greatly affected particularly in terms of the distribution channels [7].

To predict the impact of the pandemic on the behaviour of banking consumers in terms of their acceptance and utilization of mobile banking services, this study begins by discussing the evolution of technology. Numerous companies in various fields have invested in technology as a means for gaining competitive advantage [8]. Today, most organizations and individuals prefer to use digital technology [9]. There is increased business competition in today's globalized era with more confidential data being transmitted via online channels [4, 10]. The greater the usage of digital banking technology, the more digital channels are being used for disseminating product and service information as well as for socializing [11].

Consumer behaviour with regards to technology acceptance and usage can be explicated via the Theory of Reasoned Action (TRA), the Theory of Planned Behaviour (TPB) and the Technology Acceptance Model (TAM). Fishbein and (Fishbein) introduced the TRA which asserts that intention significantly determines behaviour [12]. Attitude as mediated by subjective norms will also affect behaviour. Apart from that, the model also highlights the significance of cognition, affect and conation in influencing behavior [12].

The TAM is an extension of the TRA, developed to model information technology adoption by users [13]. The model incorporates the constructs of perceived usefulness and perceived ease of use for predicting usage behavioral intention. According to Davis [13], perceived usefulness entails the degree of one's belief that the usage of a certain system will boost his/her performance. Meanwhile, perceived ease of use entails the degree of one's belief that the usage of a given system will only need minimal effort. The user will develop the intention to use the system repeatedly once its usefulness and ease of use are established [14]. If not, they will opt for another system that fulfills those necessities. A system is considered useful when it does not require too much time to understand and navigate. Hence, perceived ease of use positively affects perceived usefulness as proven by a number studies such as that of [15].

Meanwhile, the TPB predicts and explains human behavior in relation to the usage of information technology [16]. This theory asserts that actual behavior is determined by intention, of which is affected by the factors of attitude, subjective norms, and perceived behavioral control. Behavioral intention measures the extent of an individual's disposition to conduct a given activity. The person's attitude (A) describes his/her assessment of the said action. In addition, attitude directly affects the depth of the action and its perceived consequence. Subjective norm (SN) refers to the social or organizational pressure put upon the person to conduct the said action. According to Shaikh and Karjaluoto [17], SN significantly affects mobile banking usage hence making it a major determinant of mobile banking adoption [18].

Trust (TR) has been indicated as a factor in driving user intention to use a certain technology [18, 19]. This has been linked to the substantial unpredictability of e-banking services and the high-risk nature of financial services [20].

When a person feels worried, concerned, uncomfortable, uncertain and cognitively conflicted about using e-payment channels, they will hence refrain from doing so. According to Aldammagh et al. [6], e-payments in financial business such as the E-PaySIMTM E-payment are under the governance of Bank Negara. Trust and perceived risk significantly affect consumers' evaluation of their relationship with a certain bank [21]. In light of this, the current study incorporates both the aforesaid constructs into its research model.

The extended TAM is a combination of the original TAM and the TPB i.e., with the integration of the constructs of trust and perceived risk. Quan et al., [22] found that a combined TAM and TPB model is suitable for evaluating the intention and adoption of mobile service. Both the TAM and TPB are commonly employed for exploring IT and e-service usage [23]. However, both have not been able to provide consistent evidence in predicting behaviours [23]. Hence, many studies are now merging the two models in investigating IT and e-service adoption. The integrated model has been proven to have greater exploratory capability [24]. As this current research concentrates on mobile banking adoption, the incorporation of the TAM and the TPB must be comprehensive enough to be able to predict the users' behavioral intention to adopt mobile banking.

2.1 Perceived Usefulness

For this current study, perceived usefulness entails the extent of the customer's belief that mobile banking usage will boost the performance of his/her banking activity [24]. Perceived usefulness has been indicated to positively affect the intention and attitude to adopt mobile banking [25]. Other studies had documented the significant positive effect of perceived usefulness on the attitude and intention to use mobile banking [26]. Apart from that, perceived usefulness has also been shown to mediate the relationship between attitude and intention. Thus, the higher the perception of the usefulness of mobile banking, the more likely for a customer to use the system. In light of the discussions above, this current study develops the hypothesis below:

H1: Perceived usefulness significantly affects mobile banking intention in the backdrop of the COVID-19 pandemic.

H2: Perceived usefulness significantly affects mobile banking attitude in the backdrop of the COVID-19 pandemic.

2.2 *Perceived Ease of Use*

Perceived ease of use entails the extent of the customer's belief that mobile banking usage would be free from effort [6]. This construct has been identified as a major determinant driving the intention to adopt the new technology [27]. Several studies had indicated this construct's effect on mobile banking intention, Kaur and Malik [28] asserted that this construct can boost the intention to conduct various mobile banking transactions. Mobile banking users prefer mobile banking menus that are simple, memorable, and functional to their needs [25]. In light of all the above, this study develops the hypothesis below:

H3: Perceived ease of use significantly affects mobile banking attitude in the backdrop of the COVID-19 pandemic.

Perceived ease of use positively affects perceived usefulness as proven by a number studies such as that of [29].

H4: Perceived ease of use significantly affects usefulness of mobile banking.

2.3 *Attitude and Behavioral Intention*

The construct of attitude has been indicated to pose a direct and significant effect on the behavioral intention to adopt an e-business service [8]. Cudjoe et al. [30] revealed that attitude poses a positive effect on the behavioral intention of customers to shop online. Meanwhile, Aboelmaged and Gebba [31] explored the effect of attitude on wireless technology adoption. Shaikh and Karjaluoto [17] proved the significant correlation between attitude and mobile banking usage intention. In light of all the above, this study developed the hypothesis below:

H5: Attitude positively affects mobile banking intention during the COVID-19 pandemic.

2.4 *Subjective Norms and Behavioral Intention*

This construct refers to the normative social belief which drives a person to conduct a given behavior. Such social pressure comes from people whom the person deems as important (Fishbein). According to Van et al. [32], social pressure significantly drives internet usage. [33] highlighted subjective norms as a factor influencing individual attitude. Marinkovic and Kalinic [34] also revealed that this construct drives internet banking usage intention. In light of all the above, this current study develops the hypothesis below:

H6: Subjective norm significantly affects mobile banking intention in the backdrop of the COVID-19 pandemic.

2.5 *Perceived Behavioral Control and Behavioral Intention*

This construct refers to people's degree of perception regarding their capability of performing a certain behavior. People would be more willing to perform a certain behavior of which they can control, and vice versa. Hence, a person who perceives him/herself to be adequately competent in performing a given behavior would have a higher intention of actually doing so.

This construct has been indicated as a major determinant of technology usage intention [34, 35]. An individual with a high perception of his/her capability to use an e-business system will also show a higher inclination to actually use the system. According to Luo et al. [36], one's perceived behavioral control drives one's intention to engage in online activities. In light of all the above, this current study develops the hypothesis below:

H7: Perceived behavior control significantly drives mobile banking intention in the backdrop of the COVID-19 pandemic.

2.6 *Trust and Behavioral Intention*

This construct refers to the user's confidence in a given mobile banking system's capability to deliver its promised services [37]. Trust significantly elevates the customer's confidence that his/her needs will be fulfilled [38, 39]. Trust also minimizes the risk of bank-customer conflict [40]. Trust and risk come hand-in-hand when making decisions. A poorly developed technology results in higher risks and lower customer satisfaction, and eventually lower intention to use the said technology. Based on all the above, this current study develops the hypothesis below:

H8: Trust significantly affects mobile banking intention in the backdrop of the COVID-19 pandemic.

2.7 *Perceived Risk of COVID-19 and Behavioral Intention*

According to Bauer [41], this construct affects the behavioral intention of customers to make a purchase. The construct has also been indicated to significantly affect technology usage intention [42]. With regards to digital payments, privacy and security risks have been identified by many studies as the main risk affecting mobile payment usage intention [43]. So far, scarce research had been conducted on the effect of 'disease risk' on digital payment usage intention. The COVID-19 pandemic had hindered the usage of physical money among consumers in Malaysia and Indonesia [44], but boosted the usage of e-wallets instead. The usage of mobile payment channels has also been identified as a preventive measure against the spread of the disease [45] as recommended by the WHO which encourages people to make contactless

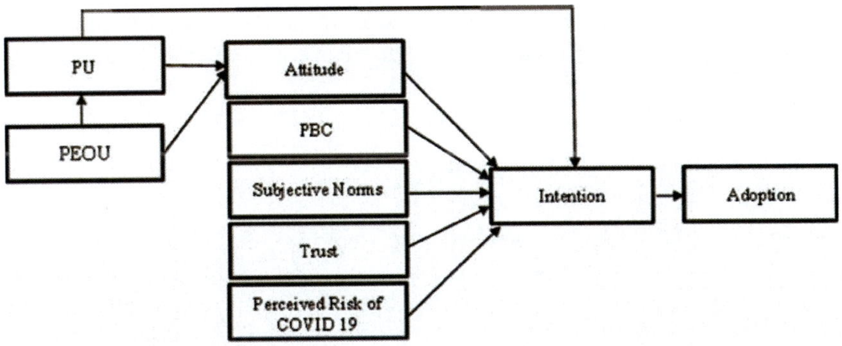

Fig. 1 The conceptual framework

payments in view of the current pandemic [46]. In light of all the above, this current study develops the hypothesis below:

H9: Perceived risk of covid 19 significantly affects mobile banking intention in the backdrop of the COVID-19 pandemic.

2.8 *Intention and Actual Adoption*

The TPB predicts and explains human behavior in relation to the usage of information technology [16]. This theory asserts that actual behavior is determined by intention, of which is affected by the factors of attitude, subjective norms, and perceived behavioral control. In light of all the above, this current study develops the hypothesis below:

H10: Behavioral intention significantly affects mobile banking actual adoption in the backdrop of the COVID-19 pandemic (Fig. 1).

3 Methodology

This study is quantitative in nature. The needed data was collected using close-ended questionnaires distributed to 290 respondents, with questions regarding the previously discussed constructs. Relevant statistical tools were used to determine the questionnaire's reliability and validity. The measurement of the items in the questionnaire was done using a 5-point Likert scale whereby 1 = strongly disagree and 5 = strongly agree.

Random sampling was applied in this study. 310 questionnaire was distributed among banks customers in Palestine. 290 questionnaire was received and valid. The unit of analysis was the banks customer who familiar with mobile banking.

Data analysis was performed using PLS-SEM, specifically Smart PLS 3 [47]. This is a variance-based structural equation modeling technique which aids the analysis of complex models with multiple relationships. Its aim is to predict and test the developed hypotheses, and eventually provide empirical proof of the findings.

4 The Instrument of the Study

The items of COVID 19 risk were adopted from [48], the items of attitude and intention were adopted from [49], the items of ease of use, usefulness, and trust were adopted from [50], the items of perceived behavioral control were adopted from [51], and the items of subjective Norms were adopted from [52], the items of actual adoption were adopted from [52].

5 Results and Analysis

The gathered data was analysed utilizing the partial least squares (PLS) modeling i.e. Smart PLS version 3.3.2. There were two stages involved in the hypothesis testing, beginning with the assessment of the measurement model and subsequently the structural model. The measurement model assessment entails determining the con-vergent and discriminant validity. Convergent validity confirms whether an item is measuring the latent variable that it is supposed to measure [53]. This involves the measurements of: i) the loadings which must be higher than 0.7, ii) the average variance extracted (AVE) which must be higher than 0.5, and iii) the compo-site reliability (CR) which must be higher than 0.7. As shown in Table 1 below, all the

Table 1 Convergent validity

Construct	Item	Loading	CR	AVE
Adoption	**AD1**	0.923	0.944	0.850
	AD2	0.924		
	AD3	0.920		
Attitude	**AT1**	0.957	0.970	0.914
	AT2	0.955		
	AT3	0.957		
Intention	**IN1**	0.897	0.915	0.783
	IN2	0.882		
	IN3	0.875		
Perceived behavioral control	**PBC1**	0.885	0.903	0.699
	PBC2	0.822		

(continued)

Table 1 (continued)

Construct	Item	Loading	CR	AVE
	PBC3	0.826		
	PBC4	0.809		
Perceived ease of use	**PEOU1**	0.940	0.935	0.827
	PEOU2	0.918		
	PEOU3	0.869		
Perceived risk	**PR1**	0.903	0.942	0.844
	PR2	0.939		
	PR3	0.914		
Perceived usefulness	**PU1**	0.849	0.916	0.784
	PU2	0.859		
	PU3	0.945		
Subjective norms	**SN1**	0.819	0.891	0.672
	SN2	0.823		
	SN3	0.840		
	SN4	0.795		
Trust	**TRS1**	0.829	0.890	0.669
	TRS2	0.816		
	TRS3	0.794		
	TRS4	0.831		

loadings, AVE and CR values are greater than the set thresholds suggested by [53]. Hence, convergent validity is confirmed for this research.

As proposed by [54], the AVE square root for each one of the constructs must be higher than their correlation coefficient to confirm discriminant validity. As shown in Table 2 below, this condition has been fulfilled.

Table 2 Discriminant validity

	ATT	Adoption	Intention	PBC	PEOU	PR	PU	SN	Trust
ATT	0.956								
Adoption	0.311	0.922							
Intention	0.443	0.306	0.885						
PBC	0.280	0.255	0.455	0.836					
PEOU	0.151	0.111	0.364	0.064	0.909				
PR	0.192	0.197	0.271	0.075	0.157	0.919			
PU	0.384	0.323	0.809	0.398	0.293	0.234	0.886		
SN	0.268	0.162	0.415	0.434	0.099	0.021	0.354	0.820	
Trust	0.287	0.204	0.511	0.320	0.164	0.226	0.404	0.330	0.818

Table 3 Relationships results

	Original sample (O)	Sample mean (M)	Standard deviation (STDEV)	T Statistics (|O/STDEV|)	P values	Result
ATT → Intention	0.099	0.098	0.035	2.846	**0.005**	Accepted
Intention → Adoption	0.306	0.311	0.055	5.588	**0.000**	Accepted
PBC → Intention	0.089	0.093	0.036	2.455	**0.014**	Accepted
PEOU → ATT	0.043	0.042	0.055	0.772	**0.440**	Rejected
PR → Intention	0.060	0.060	0.033	1.836	**0.067**	Rejected
PU → ATT	0.371	0.374	0.051	7.221	**0.000**	Accepted
PU → Intention	0.631	0.629	0.032	19.537	**0.000**	Accepted
PEOU → PU	0.293	0.298	0.055	5.354	**0.000**	Accepted
SN → Intention	0.072	0.072	0.032	2.269	**0.024**	Accepted
Trust → Intention	0.161	0.162	0.034	4.676	**0.000**	Accepted

6 Structural Model

Bootstrapping was also carried out to determine the p-values. Table 3 below shows the findings for the hypotheses testing i.e. attitude significantly affects intention as $p < 0.005$; intention significantly affects adoption as $p < 0.000$; PBC significantly affects intention as $p < 0.014$; PEOU significantly affects attitude as $p < 0.772$; PR has no significant effect on intention as $p < 0.067$; PU significantly affects attitude as $p < 0.000$; PU has a significant effect on intention as $p < 0.000$; PEOU significantly affects PU as $p < 0.000$; SN significantly affects intention as $p < 0.024$, and trust significantly affects intention as $p < 0.000$.

7 Discussion, Conclusion, Recommendations, and Implications

This current study primarily aims to predict mobile banking usage intention using the integrated model of TAM and TPB, which incorporates the constructs of trust and perceived risk of COVID 19. It was found that attitude significantly affects intention ($p < 0.005$), which is in line with the result of [24]. Thus, the first hypothesis is accepted i.e. attitude positively affects intention. Next, intention was revealed to significantly affect adoption ($p < 0.000$). PBC significantly affects intention ($p < 0.014$), which is agreed by [41] in regards to online adoption. Meanwhile, PEOU does not affect attitude ($p < 0.772$).PR was also found to have no significant effect on intention ($p < 0.067$). PU significantly affects attitude ($p < 0.000$) as well as intention ($p < 0.000$), which is in line with the finding of [56]. Likewise, PEOU significantly affects PU ($p < 0.000$), which is corroborated by the finding of [32], in the context of online technology usage. Next, SN significantly affects intention ($p < 0.024$), which is in line with the results of [17, 29] in the context of mobile banking usage intention. Finally, trust was revealed to significantly affect intention ($p < 0.000$), which is in agreement with the assertion of [15, 20] who proved the significant roles of trust and perceived credibility in driving mobile banking usage among bank customers in Iran.

The current study intends to determine the drivers of mobile banking usage intention among bank customers in Palestine. Towards that end, the study applies the TPB (ATT, PBC, SNs) and extends it by incorporating the construct of Trust (PT). As identified in past literature [40], PT significantly affects intention. In light of this finding, banks are recommended to focus on ways to improve customer trust such as by establishing a trademark that signifies trustworthiness, improving the proof of security for the mobile application, and boosting service quality. Such measures may lead to the attainment of new customers and retention of current ones. In addition, the study also demonstrated that SN negatively affects the intention of customers to adopt mobile banking, as Palestinians tend to stay true to their culture. Future studies may want to examine this construct and identify the factors influencing it. As the current findings are only applicable for the Palestinian banking sector, future studies should explore trust in affecting intention in the context of other sectors and nations to enable the generalizability of the findings.

The current study contributes to the literature on mobile banking technology acceptance by investigating the effect of usefulness, ease of use, attitude, Perceived Behavioral Control, Subjective Norms, perceived risk of covid 19, trust on the intention to use the mobile banking in Palestine. On the other hand, the current study combined TAM and TPB model for evaluating the intention and adoption of mobile service [22].

The practical implications of the current study, the bank mangers in Palestine may use the results of the current study to enhance the acceptance of mobile banking by the customers. On the other hand, the benefit of using the mobile banking technology is reducing the cost on the bank by encourage the customers dealing with the technology rather than the human. Finally, bank mangers seek to enhance the customers satisfaction, therefore, providing a new technology to the customers will help the mangers to enhance the satisfaction of customers towards the banks.

References

1. Di Pietro L, Guglielmetti Mugion R, Mattia G, Renzi MF, Toni M (2015) The integrated model on mobile payment acceptance (IMMPA): an empirical application to public transport. Transp Res Part C Emerg Technol 56:463–479. https://doi.org/10.1016/j.trc.2015.05.001
2. Dahlberg T, Guo J, Ondrus J (2015) A critical review of mobile payment research. Electron Comm Res Appl 14(5):265–284. https://doi.org/10.1016/j.elerap.2015.07.006
3. Tang B, Bragazzi NL, Li Q, Tang S, Xiao Y, Wu J (2020) An updated estimation of the risk of transmission of the novel coronavirus (2019-nCov). Infect Dis Model 5:248–255. https://doi.org/10.1016/j.idm.2020.02.001

4. Obaid T et al (2022) Factors contributing to an effective e-government adoption in Palestine. Lecture Notes Data Eng Commun Technol 127:663–676. https://doi.org/10.1007/978-3-030-98741-1_55
5. Ali AAA, Abualrejal HME, Mohamed Udin ZB, Shtawi HO, Alqudah AZ (2021) The role of supply chain integration on project management success in Jordanian engineering companies. In: International conference on emerging technologies and intelligent systems, pp 646–657
6. Aldammagh Z, Abdaljawad R, Obaid T (2020) Factor driving e-learning adoption in Palestine: an integration of technology acceptance model and is success model. Financ Internet Q 17(1):41–49. https://doi.org/10.2478/fiqf-2021-0005
7. Baicu CG, Gârdan IP, Gârdan DA, Epuran G (2020) The impact of COVID-19 on consumer behavior in retail banking. Evidence from Romania. Manag Mark 15(s1):534–556. https://doi.org/10.2478/mmcks-2020-0031
8. Aldammagh Z, Abdeljawad R, Obaid T (2021) Predicting mobile banking adoption: an integration of TAM and TPB with trust and perceived risk. Financ Internet Q 17(3):35–46. https://doi.org/10.2478/fiqf-2021-0017
9. Sun T, Zhang WW, Dinca MS, Raza M (2021) Determining the impact of Covid-19 on the business norms and performance of SMEs in China. Econ Res Istraz 35:2234–2253. https://doi.org/10.1080/1331677X.2021.1937261
10. Yang S, Lu Y, Chau PYK (2013) Why do consumers adopt online channel? An empirical investigation of two channel extension mechanisms. Decis Support Syst 54(2):858–869. https://doi.org/10.1016/j.dss.2012.09.011
11. Dincă VM, Dima AM, Rozsa Z (2019) Determinants of cloud computing adoption by Romanian SMES in the digital economy. J Bus Econ Manag 20(4):798–820. https://doi.org/10.3846/jbem.2019.9856
12. Wu LY, Chen KY, Chen PY, Cheng SL (2014) Perceived value, transaction cost, and repurchase-intention in online shopping: a relational exchange perspective. J Bus Res 67(1):2768–2776. https://doi.org/10.1016/j.jbusres.2012.09.007
13. Davis FD (1989) Perceived usefulness, perceived ease of use, and user acceptance of information technology. MIS Q Manag Inf Syst 13(3):319–339. https://doi.org/10.2307/249008
14. Eneizan BM, Wahab KA, Zainon MS, Obaid TF (2016) Prior research on green marketing and green marketing strategy: critical analysis. Singapor J Bus Econ Manag Stud 51(3965):1–19. https://doi.org/10.12816/0033265
15. Jouda H, Abu Jarad A, Obaid T, Abu Mdallalah S, Awaja A (2020) Mobile banking adoption: decomposed theory of planned behavior with perceived trust. In: The 1st International Conference on Information Technology & Business ICITB2020. https://doi.org/10.2139/ssrn.3660403
16. Ajzen I (2002) Perceived behavioral control, self-efficacy, locus of control, and the theory of planned behavior. J Appl Soc Psychol 32(4):665–683. https://doi.org/10.1111/j.1559-1816.2002.tb00236.x
17. Shaikh AA, Karjaluoto H (2015) Mobile banking adoption: a literature review. Telemat Inf 32(1):129–142. https://doi.org/10.1016/j.tele.2014.05.003
18. Alrifai K, Kalloub M, Musabeh A (2020) Financial development, economic growth and welfare: evidence from emerging countries. Pressacademia 9(2):118–131. https://doi.org/10.17261/pressacademia.2020.1218
19. Alalwan AA, Dwivedi YK, Rana NPP, Williams MD (2016) Consumer adoption of mobile banking in Jordan: examining the role of usefulness, ease of use, perceived risk and self-efficacy. J Enterp Inf Manag 29(1):118–139. https://doi.org/10.1108/JEIM-04-2015-0035
20. Hanafizadeh P, Behboudi M, Abedini Koshksaray A, JalilvandShirkhani Tabar M (2014) Mobile-banking adoption by Iranian bank clients. Telemat Inf 31(1):62–78. https://doi.org/10.1016/j.tele.2012.11.001
21. Al-alak BA (2014) Impact of marketing activities on relationship quality in the Malaysian banking sector. J Retail Consum Serv 21(3):347–356. https://doi.org/10.1016/j.jretconser.2013.07.001

22. Quan S, Hao C, Jianxin Y (2010) Factors influencing the adoption of mobile service in China: an integration of TAM. J Comput 5(5):799–806. https://doi.org/10.4304/jcp.5.5.799-806
23. Chen SC (2012) To use or not to use: understanding the factors affecting continuance intention of mobile banking. Int J Mob Commun 10(5):490–507. https://doi.org/10.1504/IJMC.2012.048883
24. Glavee-Geo R, Shaikh AA, Karjaluoto H (2017) Mobile banking services adoption in Pakistan: Are there gender differences? Int J Bank Mark 35(7):1088–1112. https://doi.org/10.1108/IJBM-09-2015-0142
25. Sudarsono H, Nugrohowati RNI, Tumewang YK (2020) The effect of Covid-19 pandemic on the adoption of internet banking in Indonesia: Islamic Bank and conventional bank. J. Asian Financ Econ Bus 7(11):789–800. https://doi.org/10.13106/jafeb.2020.vol7.no11.789
26. Vuković M, Pivac S, Kundid D (2019) Technology acceptance model for the internet banking acceptance in split. Bus Syst Res 10(2):124–140. https://doi.org/10.2478/bsrj-2019-022
27. Al-Emran M, Granić A, Al-Sharafi MA, Ameen N, Sarrab M (2020) Examining the roles of students' beliefs and security concerns for using smartwatches in higher education. J Enterp Inf Manag 34(4):1229–1251. https://doi.org/10.1108/JEIM-02-2020-0052
28. Kaur A, Malik G (2019) Examining factors influencing Indian customers' intentions and adoption of internet banking: extending TAM with electronic service quality. Innov Mark 15(2):42–57. https://doi.org/10.21511/im.15(2).2019.04
29. Koenig-Lewis N, Palmer A, Moll A (2010) Predicting young consumers' take up of mobile banking services. Int J Bank Mark 28(5):410–432. https://doi.org/10.1108/02652321011064917
30. Cudjoe AG, Anim PA, Tetteh Nyanyofio JGN (2015) Determinants of mobile banking adoption in the Ghanaian banking industry: a case of access bank Ghana limited. J Comput Commun 3(2):1–19. https://doi.org/10.4236/jcc.2015.32001
31. Aboelmaged M, Gebba TR (2013) Mobile banking adoption: an examination of technology acceptance model and theory of planned behavior. Int J Bus Res Dev 2(1):35–50. https://doi.org/10.24102/ijbrd.v2i1.263
32. Van HN, Pham L, Williamson S, Chan CY, Thang TD, Nam VX (2021) Explaining intention to use mobile banking: integrating perceived risk and trust into the technology acceptance model. Int J Appl Decis Sci 14(1):55–80. https://doi.org/10.1504/IJADS.2021.112933
33. Ali MA (2021) Impact of Islamic financial literacy, subjective norms, risk perception and perceived behavioral control on adoption of Islamic Banking in Pakistan. Rev Gestão Inovação e Tecnol 11(3):220–233. https://doi.org/10.47059/revistageintec.v11i3.1929
34. Marinkovic V, Kalinic Z (2017) Antecedents of customer satisfaction in mobile commerce: exploring the moderating effect of customization. Online Inf Rev 41(2):138–154. https://doi.org/10.1108/OIR-11-2015-0364
35. Baptista G, Oliveira T (2015) Understanding mobile banking: the unified theory of acceptance and use of technology combined with cultural moderators. Comput Human Behav 50:418–430. https://doi.org/10.1016/j.chb.2015.04.024
36. Luo X, Li H, Zhang J, Shim JP (2010) Examining multi-dimensional trust and multi-faceted risk in initial acceptance of emerging technologies: an empirical study of mobile banking services. Decis Support Syst 49(2):222–234. https://doi.org/10.1016/j.dss.2010.02.008
37. Bashir I, Madhavaiah C (2015) Consumer attitude and behavioural intention towards Internet banking adoption in India. J Indian Bus Res 7(1):67–102. https://doi.org/10.1108/JIBR-02-2014-0013
38. Al-Emran M, Al-Maroof R, Al-Sharafi MA, Arpaci I (2020) What impacts learning with wearables? An integrated theoretical model. Interact Learn Environ, pp 1–21
39. Al-Sharafi MA, Al-Qaysi N, Iahad NA, Al-Emran M (2022) Evaluating the sustainable use of mobile payment contactless technologies within and beyond the COVID-19 pandemic using a hybrid SEM-ANN approach. Int J Bank Mark 40(5):1071–1095. https://doi.org/10.1108/IJBM-07-2021-0291
40. Gefen D, Benbasat I, Pavlou PA (2008) A research agenda for trust in online environments. J Manag Inf Syst 24(4):275–286. https://doi.org/10.2753/MIS0742-1222240411

41. Bauer RA (1960) Consumer behavior as risk taking. In Cox DF (ed) Risk taking and information handling in consumer behavior, p 389–398
42. Hu Z, Ding S, Li S, Chen L, Yang S (2019) Adoption intention of fintech services for bank users: an empirical examination with an extended technology acceptance model. Symmetry (Basel) 11(3):340. https://doi.org/10.3390/sym11030340
43. Singh N, Sinha N, Liébana-Cabanillas FJ (2020) Determining factors in the adoption and recommendation of mobile wallet services in India: analysis of the effect of innovativeness, stress to use and social influence. Int J Inf Manag 50:191–205. https://doi.org/10.1016/j.ijinfomgt.2019.05.022
44. Aji HM, Berakon I, Md Husin M (2020) COVID-19 and e-wallet usage intention: a multigroup analysis between Indonesia and Malaysia. Cogent Bus Manag 7(1):180–4181. https://doi.org/10.1080/23311975.2020.1804181
45. Sreelakshmi CC, Prathap SK (2020) Continuance adoption of mobile-based payments in Covid-19 context: an integrated framework of health belief model and expectation confirmation model. Int J Pervasive Comput Commun 16(4):351–369. https://doi.org/10.1108/IJPCC-06-2020-0069
46. Welsch G, Dürr A, Thiesse F (2020) A consolidated business model canvas of blockchain-based fintech startups: evidence from initial coin offerings. In Proceedings of the 15th International Conference on Business Information Systems 2020 "Developments, Opportunities and Challenges of Digitization", WIRTSCHAFTSINFORMATIK 2020, pp 189–194. https://doi.org/10.30844/wi_2020_b5
47. Ringle CM, Wende S, Becker JM (2015) SmartPLS 3. SmartPLS GmbH. J Serv Sci Manag 10(3):32–49. https://scholar.google.com/scholar?hl=en&as_sdt=0%2C5&q=%28Ringle%2C+Wende%2C+and+Becker+2015%29&btnG=#d=gs_cit&t=1652799706568&u=%2Fscholar%3Fq%3Dinfo%3AfRlanSDTLFUJ%3Ascholar.google.com%2F%26output%3Dcite%26scirp%3D0%26hl%3Den
48. Yıldırım M, Güler A (2020) Factor analysis of the COVID-19 perceived risk scale: a preliminary study. Death Stud 46(5):1065–1072. https://doi.org/10.1080/07481187.2020.1784311
49. Venkatesh V, Morris MG, Davis GB, Davis FD (2003) User acceptance of information technology: toward a unified view. MIS Q Manag Inf Syst 27(3):425–478. https://doi.org/10.2307/30036540
50. Eappen N (2019) Mobile wallet adoption in India: impact of trust and information sharing. South Asian J Manag 26(1):32
51. Alam SS, Omar NA, Mohd Ariffin AA, Nikh Ashim NMH (2018) Integrating TPB, TAM and DOI theories: an empirical evidence for the adoption of mobile banking among customers in Klang valley, Malaysia. Int J Bus Manag Sci 8(2):385–403
52. Altin Gumussoy C, Kaya A, Ozlu E (2018) Determinants of mobile banking use: an extended TAM with perceived risk, mobility access, compatibility, perceived self-efficacy and subjective norms. In Industrial Engineering in the Industry 4.0 Era, Springer, p 225–238
53. Thiele KO, Sarstedt M, Ringle CM (2016) Mirror, mirror on the wall: a comparative evaluation of six structural equation modeling methods. Dev Mark Sci Proc Acad Mark Sci 45(5):991–992. https://doi.org/10.1007/978-3-319-26647-3_212
54. Fornell C, Larcker DF (1981) Evaluating structural equation models with unobservable variables and measurement error. J Market Res 18(1):39. https://doi.org/10.2307/3151312
55. World Health Organization (2020) Laboratory testing for coronavirus disease 2019 (COVID-19) in suspected human cases: interim guidance, 2 March 2020 (No. WHO/COVID-9/laboratory/2020.4). World Health Organization
56. Hanudin A, Baba R, Muhammad MZ (2007) An analysis of mobile banking acceptance by Malaysian customers. Sunway Acad J 4:1–12

Comparing Accuracy Between SVM, Random Forest, K-NN Text Classifier Algorithms for Detecting Syntactic Ambiguity in Software Requirements

Khin Hayman Oo

Abstract Software requirements are ambiguous due to the ambiguity of natural language in general. The ambiguity of the requirements leads to software development errors. As a result, a multitude of approaches and techniques for detecting ambiguity in software requirements have emerged. This study used three supervised ML algorithms that used Bag-of-Words features to detect grammatical ambiguity in software requirements: support vector machine (SVM), random forest (RF), and k-nearest neighbours (KNN). RF had the highest accuracy of 86.66%, followed by SVM (80%) and KNN (70%).

Keywords SVM · Random forest · KNN · Bags-of-words

1 Introduction

Requirements engineering is the process of gathering, analysing, specifying and validating user requirements. Software requirements are gathered with both functional and non-functional requirements constructed into a document called "software requirements specification" (SRS) [11, 17]. The majority of SRSs are written in natural language, which is inherently ambiguous [8, 12, 22]. There are four main types of ambiguities commonly found in SRS: lexical, syntactic, semantic and pragmatic [3, 8, 27]. Lexical ambiguity occurs when a word can be translated in multiple ways. Syntactic ambiguity occurs when a sentence may be translated in multiple ways due to ambiguous sentence structure and grammar. Semantic ambiguity occurs when there is more than one interpretation of a given context of a sentence. Pragmatic ambiguity occurs when a sentence has a lot of meanings in the context where it is stated [12]. This study focuses on detecting syntactic ambiguities in SRS because most ambiguities in requirements are found to be syntactic in nature due to multiple

K. H. Oo (✉)
Department of Computer Science, International Islamic University Malaysia, Jalan Gombak, 53100 Kuala Lumpur, Malaysia
e-mail: khinhaymanoo@gmail.com

M. Al-Emran et al. (eds.), *International Conference on Information Systems and Intelligent Applications*, Lecture Notes in Networks and Systems 550,
https://doi.org/10.1007/978-3-031-16865-9_4

ambiguous words in requirements documents [12], while other types of ambiguities focus on the meaning of the sentence.

There are three approaches for detecting ambiguities in SRS: manual, semi-automatic using natural language processing (NLP) techniques and semi-automatic using machine learning (ML) algorithms [18]. In previous study, we compared the accuracy of ML approaches to detecting syntactic ambiguity in SRS, such as naive Bayes (NB), support vector machines (SVM), random forest (RF) and k-nearest neighbours (KNN) algorithms, with manual approaches [19]. The results show that the NB text classifier detects syntactic ambiguities in SRS more accurately (77%) than manual approaches (50%) and in less time. Hence, this research has further investigated to detect syntactic ambiguity in software requirements using other machine learning techniques like SVM, Random Forest and KNN and comparing the accuracy result between them. The details explanation of SVM, Random Forest and KNN are described in below related work section.

2 Related Work

Text classification is the process of categorising data based on training data. There are three approaches to text classification: supervised, unsupervised and semi-supervised. In the supervised technique, all training data is manually labelled based on expert knowledge and the text classifier algorithm predicts the output based on the training data. In the unsupervised technique, all training data is unlabelled and the text classifier algorithm predicts output based on the inherent structure of the training data. Semi-supervised techniques combine supervised and unsupervised techniques, which means that some data are labelled while others are not [23, 26]. The three supervised ML algorithms investigated in this study are SVM, RF and KNN.

2.1 *Support Vector Machine Text Classifier Algorithm*

The SVM model employs POS tagging to label every element in the corpus as a word. SVM then assigns weight to the corpus. The weight is compared to the threshold value. An error is detected if the weight is either too large or too difficult to identify using SVM [16]. As a supervised learning method, SVM needs training and testing data sets. The training data of features vector $X_i \in R^L$ with label $y_i \in \{+1, -1\}$ are mapped into a high-dimensional space by a nonlinear function $\phi(x)$, but linear separated them. Equation 1 is formulated to solve quadratic programming problems and supports the separation of the optimal hyperplane.

$$\frac{minimize1}{\alpha_1 \ldots\ldots \alpha_l 2}\sum_{i,j=1}^{l} \alpha_i.\alpha_j.y_i.y_j K(X_i, X_j) - \sum_{i=1}^{l} \alpha_i, \quad (1)$$

Equation 1 can be represented a $0 \le \alpha_i \le C$ 888 $(1 \le i \le l), \sum_{i=1}^{l} \alpha_i.y_i = 0$, where $K(X_i, X_j)$ is the inner product of the nonlinear function or kernel function $K(X_i, X_j) = \phi(X_i).\phi(X_j)$ and C is the constant that controls the training errors and becomes the upper bound of α_i. For a testing example, x is used and y is the label decided by summing the inner product of the testing example. The training example is weighted by α_i as follows:

$$y = sgn\left(\sum_{i=1}^{l} \alpha_i \cdot y_i \cdot K(x_i,)\right), \quad (2)$$

where b is a threshold value. SVM assigns a weight α_i to each training example. Errors are detected when the weights for the testing example are too large or too difficult for the SVM classifier to identify.

An SVM text classification model was developed to classify Indonesian textual information on the web for tropical diseases such as dengue fever, malaria and bird flu, which have become epidemics in Indonesia [29]. The researchers used the proposed model for web mining since it is the best hyperplane in input space called the "structural minimization principle" from statistical learning theory. Users can avoid the dimensionality problem by downloading spatiotemporal information about tropical diseases extracted by the SVM from the internet. The classifier was created in the Indonesia language and the words were tokenized to segment them. The sentences were lemmatized or stemmed to convert the tokens to a standard form. The redundant words in each category, such as names, places and foreign words, were then removed. The words were labelled with numbers or indexing, such that words with the same root were labelled with the same number. As a result of indexing, the number of distinct words was reduced from over 11,000 to 3713. The accuracy of the SVM classifier was compared to that of the NB, KNN and C4.5 Decision Tree classifiers. It was found that SVM achieved the highest accuracy of 92.5%, followed by NB (90%), C4.5 Decision Tree (77.5%) and KNN (49.17%) [29].

The SVM text classification model was used to classify input text data into one of two groups [25]. There are two stages in text classification: training and classification. The training process is conducted using a supervised learning method, with the first set of training data (text) and the second set of data labels. When the training process is completed, users can test any new input text to classify text data based on the training data to complete the classification process. In the future, the researchers plan to use the SVM text classification model to classify more than two classes [25].

SVM, neural networks and KNN were used to predict students at risk based on a survey of 800 students. The accuracy, sensitivity and specificity of the models were assessed. SVM achieved the highest accuracy of 86.7%, while neural networks outperformed SVM and KNN in terms of sensitivity. All three models performed well in terms of specificity. This study can help educators identify at-risk students early on [28].

2.2 Random Forest Text Classifier Algorithm

The hyperparameters of RF are the same as those of a decision tree or a bagging classifier. RF comprises multiple individual trees. Each tree votes on an overall classification for the given set of data, and the classification with the most votes is chosen by RF. Each decision is built from a random subset of the training dataset [13]. The decision tree algorithm also advanced the search from a general to a specific search for a feature by adding the most useful features to a tree structure. In the learned decision tree process, each feature is selected during the search process, which is signified by a node, and each node represents a selective point between numbers of unlike possible values for a feature. This process is repeated until a decision tree can account for all training examples [21]. RF selects observations and features at random to build several decision trees before averaging the results [6]. The RF algorithm text classifier can be represented as follows [14]:

$$l(y) = argmax_C\left(\sum_{n=1}^{N} I_{h_n(y)=c}\right) \quad (3)$$

where I is the indicator function and h_n is the n-th tree of the RF. RF has an internal mechanism for estimating generalisation errors, also known as out-of-bags (OOB) errors. In the bootstrap sample, two-thirds of the original data cases are used to build each tree, while one-third of the OOB data instances are classified from the constructed tree and tested for performance. The OOB error estimate is the averaged prediction error for each training case y using only trees that do not include y in their bootstrap sample. The training data sets are then placed in each tree for computing the proximity matrix between training data based on the pairs of cases that end up in the same terminal node of a tree for the RF construction.

A feature weighting method was developed using an RF classifier called the weighting tree RF (WTRF) to reduce error in large data sets and compare it to Breiman's RF (BRF) and tree RF (TRF) [30]. The classification performance of WTRF was compared to that of other ML algorithms, including SVM, NB, decision trees and KNN. Based on the findings, WTRF reduces more errors than TRF and BRF, and it outperforms other ML algorithms in text classification [30].

The RF algorithm was used for text classification, with six authentic text data sets chosen for their diversity in terms of the number of features, data volume and number of classes [32]. Their dimensions range from 2000 to 8460, the number of documents ranges from 918 to 18,772, and the minority category rate ranges from 0.32 to 6.43%. Fbis, Re0, Re1, Oh5, and Wapdatasets were also used as classification data sets, and [9] pre-processed them successfully. The documents were classified using the RF algorithm with the new feature weighting method and the tree selection method to categorise text documents. The researchers were able to reduce generalisation errors and improve classification performance by employing the RF algorithm. This study shows that RF algorithm could classify large datasets in various applications [32].

Supervised ML text classification algorithms were used to detect ambiguities in SRS written in the Malay language. The algorithm contains two main steps: document collection and text processing. In document collection, four SRSs written in the Malay language were collected and processed, including data labelling (ambiguous or non-ambiguous) and data cleansing (removing tables and irrelevant parts of documents). Text processing involves expected output activity that includes classification features and labels using a dictionary. Text classification consists of two parts: univariate analysis and classification algorithm selection. The predictive power of predictors (feature words from the data sets) was measured and used to assess classification performance in univariate analysis using WEKA. The classification algorithms tested are OneR, NB, logistic regression, KNN, decision table, decision stump, J84, RF and random tree. RF achieved the highest accuracy of 89.67% in identifying ambiguous requirements in the Malay language, followed by random tree (80.89%), J48 (82.67%), logistic regression (80.94%), NB (80.22%), decision table (78.06%), oneR (78.06%), decision stump (77.17%) and KNN (71.89%) [20].

The mining of student information system records was used to investigate academic performance prediction at a private university in the United Arab Emirates [24]. The RF algorithm was used to predict academic performance since it is the most appropriate data mining technique. The student data sets were divided into four major categories: demographics, past performance, course and instructor information, and general student information. This research can assist higher education institutions in identifying weaknesses and factors influencing students' performance, which can also serve as a warning sign for students' failures and poor academic performance [24].

2.3 K-Nearest Neighbours Text Classifier Algorithm

The KNN algorithm is a non-parametric method for classification and regression. The input to both the classification and regression processes is the k-closest training example in the feature space. For KNN classification, the object is classified based on the highest vote of its neighbours, as well as the object assigned to the class most common among its k neighbours, such that k is a positive but small integer ($k = 1$), then the object was assigned to the class of the single nearest neighbour [2]. The output of KNN regression is the object's property value, which is calculated based on the average of the values of its k-nearest neighbours. The KNN text classifier's algorithm is as follows [2]:

$$Sim_{Text}(D_i, D_j) = \frac{\sum_{k=1}^{m}(W_{ik} \times W_{jk})}{\sqrt{\sum_{k=1}^{m}(W_{ik})^2 \times \sum_{k=1}^{m}(W_{jk})^2}}, \quad (4)$$

where D_i is the test document, D_j is the training document, W_{ij} is the weight of the k-the element of the term vector D_i, W_{jk} is the weight of the k-th element of the term vector D_j and m is the number of distinct terms in the documents that represent the category.

A novel approach to automatically detect nocuous ambiguities in requirements specifications was developed using the KNN algorithm [5]. The emphasis was on coordinating ambiguities, which occur when a sentence contains more than one "and" or "or" word. The data for this study came from ambiguous phrases in a corpus of requirements specifications, as well as a collection of associated human judgments about how they should be interpreted. In terms of detecting coordinating ambiguities, the proposed algorithm was found to be 75% accurate.

The KNN text classifier algorithm was used to implement an automatic approach to identifying nocuous ambiguity in natural language software requirements [31]. The nocuous ambiguity occurs when the same text is interpreted differently by different readers, particularly in the case of pronouns. String-matching, grammatical, syntactic and semantic feature sets were used to construct the training data, such as "*Y if both noun phrases (NPs) contain the same string after the removal of non-informative words, else N*", " *Y if both NPs contain the same headword, else N*", "*Y if one NP is the PP attachment of the other NP, else N*", and so on. The completed training data set was built on the studied human judgments and heuristics that model those judgments. WEKA was used to put the algorithm through its paces. With a precision of 82.4% and a recall of 74.2%, KNN outperforms other ML algorithms for identifying nocuous ambiguity, followed by NB (73.6%), decision tree (70.39%), LogitBoost (72.09%), and SVM (70.16%) [31].

The KNN text classifier was used to improve automatic text categorization for Arabic text [2]. The unstemmed and stemmed data from the TREC-2002 dataset were used to remove prefixes and suffixes. Punctuation marks, diacritics, non-letters and stop words were removed from the Arabic text during text preprocessing, as were words with fewer than three letters. In the first experiment with trigram, the KNN classifier achieved the highest accuracy with I_{new} (80%), followed by Dice and Jaccard (77.91% each) and cosine (77.91%); while with Bag-of-Words (BoW), cosine achieved the highest accuracy (86.09%), followed by I_{new} (84.43%), Jaccard (83.75%) and Dice (83.5%). In the second experiment with trigrams, cosine achieved the highest accuracy (87.55%), followed by Jaccard (4.28%), Dice (84.02%) and I_{new},(81.73%); while with BoW, cosine achieved the highest accuracy (88.2%), followed by Dice (87%), Jaccard (86.34%) and I_{new}, (86.02%). In the third experiment with trigrams, I_{new}, achieved the highest accuracy (91%), followed by cosine (88.5%), Jaccard and Dice (89% each); while with BoW, I_{new} achieved the highest accuracy (92.6%), followed by cosine (88.8%), Jaccard (88.6%) and Dice (88.3%). It was discovered that I_{new} outperforms cosine, Dice and Jaccard, with BoW results outperforming trigram results [2].

3 Methodology

This section outlines how SVM, RF and KNN were used to accomplish the research goal. These ML algorithms are well-known for their high performance and have been widely used for text classification (see Fig. 1).

3.1 Data Collection

Forty requirements sentences were selected from fifteen existing SRSs based on five rules serve as training data for setting the training and testing processes of using the ML algorithm in detecting syntactic ambiguities in SRS. Natural Language SRS [3] and Ambiguity Handbook [7] were used to develop the five rules. As shown in Table 1, these rules are fundamental to English grammar and aids requirements engineers, practitioners and researchers avoid ambiguity when constructing requirements statements. Sentence structure and grammar are critical in requirements documents because the majority of ambiguities in requirements documents are syntactic in nature as a result of multiple ambiguous words [12].

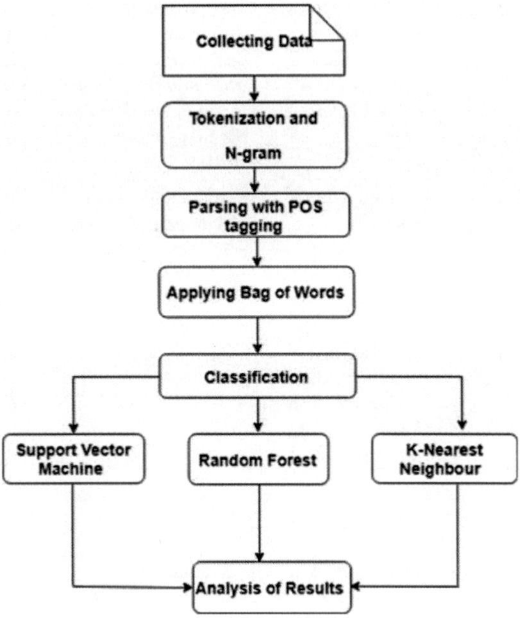

Fig. 1 The process of the supervised ML algorithm

Table 1 Syntactic ambiguity rules

Rule	Rule name	Description	Example	Remarks
Rule 1	**Coordinating Conjunction Ambiguity**	More than one conjunction "and" or "or" is used in a sentence	**Sentence 1:** *"The user shall be able to register for an account by entering their full name* **and** *email address* **or** *password."*	In Sentence 1, both "**and**" and "**or**" are used, resulting in Rule 1 ambiguity
Rule 2	**Multiple Sentences Ambiguity**	Occur when more than one subject or main verb is used in a sentence	**Sentence 2:** *"The store manager* **selects** *a product item and* **changes** *its sales price."*	In Sentence 2, more than one main verbs "**selects**" and "**changes**" are used, resulting in Rule 2 ambiguity
Rule 3	**Referential Ambiguity (pronoun in a single sentence)**	Occurs when "it" and "they" can refer to more than one element	**Sentence 3:** *"The user may access the system whenever* **they** *desire."*	In Sentence 3, the pronoun "**they**" is used, resulting in Rule 3 ambiguity. (Other examples of pronouns: **it, they. them, their, those**)
Rule 4	**Quantification Ambiguity**	Scope ambiguity occurs when two quantifiers can express the same scope over each other in various ways	**Sentence 4: "All** *data in the system database will be backed up every 24 h and the backup copies are stored in a secure location which is not in the same building as the system."*	In Sentence 4, "**all**" is used, resulting in Rule 4 ambiguity. The words **all, each** and **every** can be substituted for one another
Rule 5	**Subjective Sentences Ambiguity**	Occurs when a sentence expresses a personal opinion or feeling. In other words, a sentence that contains either "adjective" and "adverb" or "as adjective as"	**Sentence 5:** *"Users should be able to display who is* **currently** *on leave using date."* **Sentence 6:** *"Software tasks should be* **as synchronous as** *possible."*	In Sentence 5, the adverb "**currently**" is used, resulting in Rule 5 ambiguity In Sentence 6, The phrase "**as synchronous as**", which is in the form of "as adjective as," resulting in Rule 5 ambiguity

3.2 Tokenization and n-gram Representation

Tokenization is a method of segmenting each word in a text-based data set. An n-gram is a frame of length n that moves over a consecutive sequence of tokens in a text. The n-gram of size one is called a "unigram", the n-gram of size two is called

Table 2 n-gram representation

<table>
<tr><th>n-gram</th><th>Example</th><th>Remarks</th></tr>
<tr><td>Unigram</td><td>The |cashier | and| the |customer| try |again</td><td>Rule 1 and Rule 2 cannot be applied
Rule 3, Rule 4 and Rule 5 can be applied</td></tr>
<tr><td>Bigram</td><td>The cashier | and the| and again | the customer| customer try| try again| again and</td><td>Rule 1 and Rule 2 cannot be applied
Rule 3, Rule 4 and Rule 5 can be applied</td></tr>
<tr><td>Trigram</td><td>The cashier and | cashier and the | and the customer| the customer try| customer try again | try again and |again and again</td><td>Rule 1 and Rule 2, cannot be applied
Rule 3, Rule 4 and Rule 5 can be applied</td></tr>
<tr><td>Quadrigram</td><td>The cashier and the| cashier and the customer | and the customer try | the customer try again |customer try again and |try again and again</td><td>Rule 1 and Rule 2 cannot be applied
Rule 3, Rule 4 and Rule 5 can be applied</td></tr>
<tr><td>5-g</td><td>The cashier and the customer| cashier and the customer try| and the customer tries again | the customer tries again and| customer tries again and again</td><td>Rule 1 and Rule 2 cannot be applied
Rule 3, Rule 4 and Rule 5 can be applied</td></tr>
<tr><td>6-g</td><td>The cashier and the customer try| cashier and the customer try again| and the customer try again and |the customer try again and again</td><td>Rule 1, Rule 2, Rule 3, Rule 4 and Rule 5 can all be applied</td></tr>
</table>

a "bigram," the n-gram of size three is called a "trigram," the n-gram of size four is called a "quadrigram" and the n-gram of size five or higher is called an "n-gram" [33]. Table 2 lists the description of n-gram. Many studies on text processing have used bigram up to 5-g [4, 10, 15]. In our study, we used 6-g because the sentences were long enough to capture the rules. See Table 2 for a remark on applying rules in n-grams.

3.3 *Part-Of-Speech Tagging*

After implementing 6-g on the sentences, each word is tagged using part-of-speech (POS) tagging. POS tagging has been used in many natural language processing tasks because it displays word categories such as noun (NN), verb (VB) and adjective (JJ) [1, 8, 17] (see Fig. 2).

The ***DT*** cashier ***NN*** and ***CC*** the ***DT*** customer ***NN*** try ***VB*** again ***RB*** and ***CC*** again ***RB***

Fig. 2 POS tagging in a sentence

Table 3 6-g based POS patterns for ambiguity rules

1	Rule 1	**Coordinating Conjunction Ambiguity**	**[*] CC [*] CC [*] [*]**
2	Rule 2	**Multiple Verbs Ambiguity**	**[*] VB CC VB [*] [*]**
			[*] VBG CC VBG [*] [*]
			[*] VBN CC VBN [*] [*]
			[*] VBD CC VBD [*] [*]
			[*] VBZ CC VBZ [*] [*]
3	Rule 3	**Referential Ambiguity**	**[*] PRP [*] [*] [*] [*]**
			[*] PRP$ [*] [*] [*] [*]
4	Rule 4	**Quantification Ambiguity**	**[*] DT [*] [*] [*] [*]**
5	Rule 5	**Subjective Sentence Ambiguity**	**[*] JJ [*] [*] [*] [*] [*] RB JJ IN [*] [*]**
			[*] IN JJ IN [*] [*]
			The square bracket with an asterisk (*) represents optional **Any** or **None**

The square bracket with an asterisk (*) represents optional Any or None

The words are replaced with the related POS tag since we only need the sentence structure and grammar to detect syntactic ambiguity. As a result, we obtained 6-g-based POS patterns for ambiguity rules, as shown in Table 3.

This is how the POS patterns and ambiguity rules are used to label the 6-g (see Fig. 3).

If multiple rules apply to a sentence, all but one are replaced with "XX" so that only one rule is applied to the training sentence. For example, if two rules coordinate both conjunction ("CoorConj") and multiple verbs ("MultiVerbs") in the same sentence, one rule is replaced with "XX" (see Fig. 4). The use of "XX" is intended to avoid changing the categories of each word rather than removing multiple rules from each sentence.

```
train = [("PRP$ CC PRP$ NN NN CC ","CoorConj"),
("VB CC VB NN NNS VBP", "MultiVerbs"),
("DT NNP IN PRP VBZ VBN", "Pronouns"),
("DT CD NNS CC DT NN", "Quantifier"),
("RB JJ IN JJ JJ", "Subjective"),
("DT NNP CC DT NN VB", "Other")]
```

Fig. 3 Labelled data sets

Before

("CC DT NN VB RB CC", "Rule 1, Rule 4, Rule 5")

After

("CC XX NN VB XX CC", "Rule 1")
("XX DT NN VB XX XX", "Rule 4")
("XX XX NN VB RB XX", "Rule 5")

Fig. 4 Simplifying multiple rules by replacing the rules with "XX"

3.4 Bag-Of-Words Representation

BoW is a simple and flexible model for extracting features from the text for modelling in ML. For example, BoW assumes $w = (w_1, w_2, \ldots, w_k, \ldots, w_v)$ where v the number of unique words in the document collection is. In BoW, requirements document $d_i = (w_{i1}, w_{i2}, \ldots, w_{ik}, \ldots, w_{in})$ where w_{ik} is the frequency of k-th word in the requirements documents d_i [22]. After parsing the requirements collection to extract unique words, BoW removes stop words and words that appear only once. For example, term frequency (TF) counts the frequency of terms, which is often divided by the document length to normalise because a term appears much more frequently in log documents than in shorter documents. Inverse document frequency (IDF) is a measure of how common a word appears across documents. TF-IDF is used to measure the number of times a word appears in a document as its value in each unique word, which corresponds to a feature with $TF(w_i, d_i)$. Each word is weighted by TF-IDF, which is used for exchanging information based on a predefined set of features. IDF improves performance by refining the document representation because it can be calculated from w_i, which is calculated from the document frequency $DF(w_i)$, where w_i is the number of documents in which words occurs [22]. For example:

a. Collect the number of documents, $d_i = (w_{i1}, w_{i2}, \ldots, w_{ik}, \ldots, w_{in})$ where w_{ik} is the frequency of the k-th word in the requirements documents d_i.

```
corpus = ["XX CC XX NN NN CC",
    "VBZ XX NN NN CC VBZ",
    "XX NNP IN PRP XX XX",
    "MD XX XX TO XX DT",
    "NNS MD XX RB JJ IN",
    "NNP NNP XX XX NN NN"]
```

b. Call function of BoW using CountVectorizer().

```
vectorizer = CountVectorizer()
```

c. Transform corpus to X.
d. Extract or print unique words.

X = ['CC', 'DT', 'IN', 'JJ', 'MD', 'NN', 'NNP', 'NNS', 'PRP', 'RB', 'TO', 'VBZ', 'XX']

e. Transform X into array form to weight each word by using *TF-IDF*

["XX CC XX NN NN CC"], [2 0 0 0 0 2 0 0 0 0 0 0 2]

["VBZ XX NN NN CC VBZ"], [1 0 0 0 0 2 0 0 0 0 0 2 1]

["XX NNP IN P RP XX XX"], [0 0 1 0 0 0 1 0 1 0 0 0 3]

["MD XX XX TO XX DT"], [0 1 0 0 1 0 0 0 0 0 1 0 3]

["NNS MD XX RB JJ IN"], [0 0 1 1 1 0 0 1 0 1 0 0 1]

["NNP NNP XX XX NN NN"], [0 0 0 0 0 2 2 0 0 0 0 0 2]

3.5 Classification of Rules Using SVM, RF and KNN

The training data was completed after applying "XX" replacement for multiple rules. Following that, the ML algorithms, SVM, RF and KNN, predict based on the BoW features. The BoW training data was used to generate the testing data (requirements statements). This study used 30 requirements statements from original requirements statements derived from the combinations of 15 existing SRS for testing data. The Python GUI programme was used, and users can enter new requirements statements and click the check button to view the results. Six outcomes were obtained from our experiments: Rule 1 (CoorConj), Rule 2 (MultiVerbs), Rule 3 (Pronoun), Rule 4 (Quantifier), Rule 5 (Subjective), and "Other" (see Fig. 5). The system removes words that are frequently used in requirements statements during testing with new requirements statements as shown in Table 4. These words are not ambiguous, but in the training data, they represent the same POS taggers.

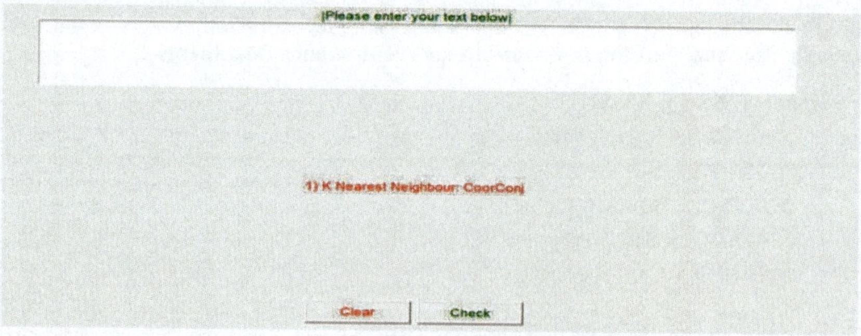

Fig. 5 Sample output

Table 4 Eliminated words

No	Eliminated words	POS tagger	Rules applied
1	But	CC (coordinating conjunction)	Rule 1 (coordinating conjunction)
2	Allow, provide, have, be, do	VB (verbs, base form)	Rule 2 (multiple verbs)
	Has, is, does	VBZ (third person singular present)	
	Had, was, were, did	VBD (verbs past tense)	
	Am, are	VBP (non-person singular present 3rd)	
3	A, an, the, both	DT (determiner)	Rule 4 (quantifier)
4	Not	RB (adverb)	Rule 5 (subjective)

4 Results and Discussion

Ambiguity in requirements is common and costly. It is critical for requirements engineers to resolve ambiguity in requirements statements before delivering the software project to stakeholders, or to construct requirements statements that are free of ambiguity. In this study, we used SVM, RF and KNN to detect syntactic ambiguity in requirements statements. We ran two experiments with all three ML algorithms, one before and one after eliminating Rule 2since Rule 2 stipulates that there should be a conjunction (CC) between verbs (eg: ***VB CC VB, VBZ CC VBZ, VBD CC VBD, VBN CC VBN***). We evaluated the accuracy of SVM, RF and KNN when Rule 2 was used and when it was not used. By including Rule 2 in the first experiment, syntactic ambiguity in requirements statements was detected. As shown in Table 5, RF achieved the highest accuracy (86.66%) in identifying syntactic ambiguity in requirements statements, followed by SVM (80%) and KNN (50%). In the second experiment without Rule 2, RF achieved the highest accuracy (83.33%) in identifying syntactic ambiguity in requirements statements, followed by SVM (70%) and KNN (56.66%). In either experiment, with or without Rule 2, there were no significant differences in identifying syntactic ambiguity. It can be concluded that SVM, RF and KNN classified Rule 2 correctly.

Table 5 Classification results

Algorithm	Accuracy before removing Rule 2 (%)	Accuracy after removing Rule 2 (%)
SVM	80	70
RF	86.66	83.33
KNN	50	56.66

Table 6 Analysis results

No	New sentence	Actual rules	SVM	RF	KNN
1	The DT System NNP ***caches VBZ each DT*** sale NN and ***CC writes VBZ them PRP*** into IN the DT Inventory NNP as ***RB*** soon RB as IN ***it PRP*** is VBZ available JJ again RB	Rule 2 (VBZ CC BZ) Rule 3 (PRP) Rule 4 (DT) Rule 5 (RB)	Rule 5	Rule 3	Rule 3
2	The DT Store NNP Manager NNP ***chooses VBZ*** the DT product NN items NNS to DT order NN ***and CC enters VBZ*** the DT ***corresponding JJ*** amount NN	Rule 2 (VBZ CC BZ) Rule 5 (JJ)	Rule 1 (X)	Other (X)	Rule 1 (X)
3	A DT report NN including VBG ***all DT available JJ*** stock NN items NNS in IN the DT store NN is VBZ displayed VBN	Rule 4 (DT) Rule 5 (JJ)	Rule 4	Rule 4	Rule 4

5 Conclusion and Future Work

In this study, SVM, RF and KNN were used to analyse the results of our experiments using various rules. The algorithms can correctly classify the rule in Sentences 1 and 3, but not in Sentence 2 (see Table 6). In the future, we will investigate other ML algorithms and use a semi-automated approach based on NLP techniques to detect syntactic ambiguity in SRS.

References

1. Al-Emran M, Zaza S, Shaalan K (2015) Parsing modern standard Arabic using treebank resources. In: International Conference on Information and Communication Technology Research (ICTRC), IEEE Abu Dhabi, United Arab Emirates, pp 80–83
2. Alhutaish R, Omar N (2015) Arabic text classification using k-nearest neighbour algorithm. Int Arab J Inf Technol 12(2):190–195
3. Berry DM, Kamsties E, Krieger MM (2003) From contract drafting to software specification: linguistic sources of ambiguity. Los Angeles, CA, USA
4. Cavnar WB, Trenkle JM (1994) N-gram-based text categorization. In: Proceedings of SDAIR-94, 3rd Annual Symposium on Document Analysis and Information Retrieval, CiteSeer, pp 161–175
5. Chantree F, Nuseibeh B, De Roeck A, Willis A (2006) Identifying nocuous ambiguities in natural language requirements. In: 14th IEEE International Requirements Engineering Conference (RE'06), IEEE, Minneapolis/St. Paul, MN, USA

6. Introduction to Random Forest Algorithm. https://towardsdatascience.com/introduction-to-random-forest-algorithm-fed4b8c8e848.
7. Fabbrini F, Fusani M, Gnesi S, Lami G (2001) The linguistic approach to the natural language requirements quality: benefit of the use of an automatic tool. In: Proceedings 26th Annual NASA Goddard Software Engineering Workshop, IEEE Greenbelt, MD, USA, pp. 95–105
8. Gleich B, Creighton O, Kof L (2010) Ambiguity detection: towards a tool explaining ambiguity sources. Part of the Lecture Notes in Computer Science book series (LNPSE), vol. 6182, Springer, pp 218–232
9. Han EHS, Karypis G (2002) Centroid-based document classification: analysis and experimental results. In: European conference on principles of data mining and knowledge discovery, Springer, Berlin, Heidelberg, pp 424–431
10. Houvardas J, Stamatatos E (2006) N-gram feature selection for authorship identification. In: International Conference on Artificial Intelligence: Methodology, Systems, and Applications, Springer, Berlin, Heidelberg, pp 77–86
11. Hussain I, Ormandjieva O, Kosseim L (2007) Automatic quality assessment of SRS text by means of a decision-tree-based text classifier. In: Seventh International Conference on Quality Software, IEEE, Portland, OR, USA, pp 209–218
12. Kamsties E, Berry DM, Paech B (2001) Detecting ambiguities in requirements documents using inspections. In: Proceedings of the First Workshop on Inspection in Software Engineering, pp 68–80
13. Klassen M, Paturi N (2010) Web document classification by keywords using RFs. In: International Conference on Networked Digital Technologies, Springer, pp 256–261
14. Liparas D, HaCohen-Kerner Y, Moumtzidou A, Vrochidis S, Kompatsiaris I (2014) News articles classification using RFs and weighted multimodal features. In: Information Retrieval Facility Conference, Springer, Heidelberg, pp 63–75
15. Mansur M, Uz-Zaman N, Khan M (2006) Analysis of n-gram based text categorization for Bangla in a newspaper corpus. Doctoral dissertation, BRAC University
16. Nakagawa T, Matsumoto Y (2002) Detecting errors in corpora using support vector machines. In: Proceedings of the 19th International Conference on Computational Linguistics, ACM Digital Library, pp 709–715
17. Nigam A, Arya N, Nigam B, Jain D (2012) Tool for automatic discovery of ambiguity in requirements. Int J Comput Sci Iss 9(5):350–356
18. Oo KH, Nordin A, Ismail AR, Sulaiman S (2018) An analysis of ambiguity detection techniques for software requirements specification. Int J Eng Technol 7:501–505
19. Oo KH, Nordin A, Ismail AR, Sulaiman S (2018) An approach to detect syntactic ambiguity using Naïve Bayes (NB) Text Classifier for Software Requirements. In: Proceedings of the 11th Edition of Postgraduate Research Workshop (PRW) at SOFTEC Asia Conferences
20. Osman MH, Zaharin MF (2018) Ambiguous software requirement specification detection. In: Proceedings of the 5th International Workshop on Requirements Engineering and Testing – RET '18, pp 33–40
21. Pedersen T (2001) A decision tree of bigrams is an accurate predictor of word sense. In: Proceedings of the Second meeting of the North American Chapter of the Association for Computational Linguistics on Language technologies, pp 1–8
22. Polpinij J, Ghose A (2008) An automatic elaborate requirement specification by using hierarchical text classification. In: International Conference on Computer Science and Software Engineering, pp 706–709
23. Rajeswari RP, Juliet K, Aradhana D (2017) Text classification for student data set using naive Bayes classifier and KNN classifier. Int J Comput Trends Technol 43(1):8–12
24. Saa AA, Al-Emran M, Shaalan K (2019) Mining student information system records to predict students' academic performance. In: The International Conference on Advanced Machine Learning Technologies and Applications (AMLTA2019), Springer, pp 229–239
25. Sarkar A, Chatterjee S, Das W, Datta D (2015) Text classification using support vector machine. Int J Eng Sci Invent 4:33–37

26. Sharma R, Bhatia J, Biswas KK (2014) Machine learning for constituency test of coordinating conjunctions in requirements specifications. In: Proceedings of the 3rd International Workshop on Realizing Artificial Intelligence Synergies in Software Engineering, pp 25–31
27. Singh S, Saikia P, Chandra L (2015) Ambiguity in requirement engineering documents: importance, approaches to measure and detect, challenges and future scope. Int J Adv Res in Comput Sci Softw Eng 5(10):791–798
28. Wahdan A, Hantoobi S, Al-Emran M, Shaalan K (2021) A review of learning analytics studies. Recent advances in technology acceptance models and theories, Springer
29. Wulandini F, Nugroho AS (2009) Text classification using support vector machine for web mining based spatio temporal analysis of the spread of tropical diseases. In: International Conference on Rural Information and Communication Technology, pp 189–192
30. Xu B, Guo X, Ye Y, Cheng J (2012) An improved RF classifier for text categorization. J Comput 7(12):2913–2920
31. Yang H, De Roeck A, Gervasi V, Willis A, Nuseibeh B (2011) Analyzing anaphoric ambiguity in natural language requirements. Requirements Eng 16(3):163
32. Zakariah M (2014) Classification of large datasets using random forest algorithm in various applications: survey. Certif Int J Eng Innov Technol 9001(3):2277–3754
33. Maroulis G (2014) Comparison between maximum entropy and naïve bayes classifiers: case study; appliance of ML algorithms to an Odesk's corporation dataset. Thesis submitted in partial fulfilment of the requirements of Edinburgh Napier University, pp 35–36

Environmental Concern in TPB Model for Sustainable IT Adoption

Nishant Kumar, Ranjana Dinkar Raut, Kamal Upreti, Mohammad Shabbir Alam, Mohammed Shafiuddin, and Manvendra Verma

Abstract Rapid advancement in technology and continuous environmental degradation has attracted the attention of practitioners toward sustainable solutions. This study aims to investigate educated millennial beliefs and behavior toward sustainable IT practices. The Theory of Planned Behavior (TPB) model deployed in the study was extended through perceived environmental responsibility. A survey was conducted to examine the sustainable IT adoption behavior of millennial in the National Capital Region, Delhi India. Variance based partial least square structure equation modeling was employed to evaluate the hypothesized model. Findings of the study confirm environmental concern (ER) a precursor for attitude (ATT), perceived behavioral control (PBC), and subjective norm (SN). Further, there is a significant positive influence of ATT, PBC, and SN on the adoption intention of sustainable IT practices, followed by the effect of adoption intention on actual adoption behavior. Study

N. Kumar
School of Business and Management, CHRIST University, Bangalore, India
e-mail: nishantkumar00@gmail.com

R. D. Raut
Department of Applied Electronics, Sant Gadge Baba University, Amravati, Maharashtra, India
e-mail: ranjanaraut@sgbau.ac.in

K. Upreti (✉)
Department of Computer Science and Engineering, Dr. Akhilesh Das Gupta Institute of Technology and Management, Delhi, India
e-mail: kamal.upreti@adgitmdelhi.ac.in

M. S. Alam
Razak Faculty of Technology and Informatics, Universiti Teknologi Malaysia, Kuala Lumpur, Malaysia

M. Shafiuddin
Oman College of Management and Technology, Halban, Oman
e-mail: mohammed.shafiuddin@omancollege.edu.om

M. Verma
Delhi Technological University, Shahbad Daulatpur, Bawana Road, Delhi, India
e-mail: mv075415@gmail.com

M. Al-Emran et al. (eds.), *International Conference on Information Systems and Intelligent Applications*, Lecture Notes in Networks and Systems 550,
https://doi.org/10.1007/978-3-031-16865-9_5

disseminates valuable insights to policymakers and marketers to formulate strategies and policies to attain sustainability through sustainable IT practices.

Keywords Sustainable information technology · Theory of planned behavior · Environmental concern · Millennial · Structure equation modeling

1 Introduction

Increasing pollution, climate change, and concern over energy consumption are the major drivers for sustainable development. With the advent of technology and rapid economic development, people are getting worried about its impact on the environment [1]. Information technology has been observed as a technique to channelize the energy of a system and enhance operational efficiency [2]. Undoubtedly in the recent past, there has been huge investment in technological advancement front by institutions and simultaneously it has increased its impact on the environment. High volume IT computing devices have increased energy consumption and become a liability for institutions. Sustainability in information technology (IT) is the most debated word to overcome environmental deterioration. Sustainable information technology (IT) mainly focuses on the usage, design, and disposing of IT computing devices in such a way that it creates a minimum hindrance to the environment [3]. The level of awareness and adoption of Sustainable information technologies is in the nascent stage as rightly pointed that there exist four gaps namely; "knowledge gaps, practice gaps, opportunity gaps, and knowledge- doing gaps" [4].

Sustainable IT practices rely on enhancing the effectiveness and efficiency of IT computing devices. Most of the research studies on the topic of sustainable IT are based on developed countries and very limited research studies are on developing nations. India is one of the fastest-growing economies with an attractive business opportunity for the global market is facing a concern regarding the disposal of electronic waste. Individual actions toward such practices are imperative to observe to evaluate the success of sustainable IT practices. The paucity of research studies focusing on individual adoption intention and actual adoption behavior provides a strong foundation for this research. To bridge the literature gap, this study attempts to investigate educated millennial belief and behavior towards sustainable IT practices through the Theory of Planned Behavior (TPB).

This chapter is organized into seven sections. Section one covers the introduction and objectives. The second section reviews the theoretical background of the study followed by the conceptual framework developed in section three. Survey instrument and study methodology are described in the fourth section. Data analysis and empirical findings of the study are covered under section five. A detailed discussion of the study finding and its implication are included in the sixth part of the chapter and the study is concluded under section seven.

2 Theoretical Background

Practitioners have started focusing on sustainable IT solutions and identified ecological efficiency and ecological effectiveness as the drivers for the adoption of Sustainable IT practices. Ecological efficiency relies on operational cost reduction and ecological effectiveness focuses on the establishment of sustainable value and belief system [5]. Adoption of Sustainable IT practices not only reduces operational cost but also helps to build a better corporate image, employee motivation, and employee retention [6, 7]. It is also evident to state that the effectiveness of the Sustainable IT solution depends on individual value, belief, and intention toward its actual adoption [8]. Hence, it becomes indispensable for organizations to comprehend individual intention towards sustainable IT practices to attain ecologically sustainable strategic disclosure [9]. Theory of Reasoned Action (TRA) propounded by Fishbein & Ajzen [10] was found to be a fundamental framework to understand human behavior based on their intention to accept or reject a given action. However, the intention was defined by two decision variables: attitude and subjective norm. Attitude is defined as a list of believes that may transform intention to carry out the act. Subjective norm is in line with the normative belief that the social circle determines individual intention to perform the act. Behavioral intention is an individual's willingness to perform the act [11]. Theory of Planned Behavior (TPB) is an extension to TRA and included the third construct perceived behavior control, which states an individual's comfort or discomfort while performing the act. TPB further explains that the strength of individual intention to perform an act determines actual behavior to perform the act. The strength of intention to perform an act is directly proportional to the attitude towards the behavior, subjective norm, and perceived behavioral control. TPB model has further extensively also studied behavioral intention, pro-environmental behavior and has been proved as an effective model [12, 13].

3 Conceptual Framework

Model adapted from literature is an extension of TPB. The conceptual framework represented in Figure 1 mainly includes six constructs; Environmental concern (ER), Attitude (ATT), Subjective norms (SN), Perceived behavioral control (PBC), Adoption intention (AI), Actual adoption behavior (AAB). The emotional involvement of an individual in environmental problems explains Environmental concern [14]. Individuals with a high, level of environmental concern is likely to have a positive attitude towards the sustainable environmental act [15]. Furthermore, literature provides sufficient empirical evidence for the influence of perceived environment on attitude, subjective norm, and perceived behavioral control [16].

H1: ER has a significant positive influence on individual ATT towards sustainable IT practices.

H2: ER has a significant positive influence on individual SN towards sustainable IT practices.

H3: ER has a significant positive influence on individual PBC towards sustainable IT practices.

H8: ER has a significant positive influence on individual AI towards sustainable IT practices.

Attitude is described as a positive or negative opinion towards conducting a particular behavior and found to be a significant predictor to get engaged in environment-friendly act [17]. Regarding sustainable IT practices, individuals with a higher level of awareness and favorable attitude are more intended towards the adoption of such technologies [18]. Subjective norm signifies the influence of family, friends, and colleagues on individual intention to perform specific behavior [19]. An individual with more social pressure is more likely to get liked to sustainable IT practices. An individual does buy sustainable products if he has the wish, time, and convenience of purchasing it. A self-motivated consumer will buy sustainable products if has enough resources for buying them [20]. Perceived behavioral control is considerably and positively linked with purchase intention and this has been proved in past studies [21]. The behavioral intention has a significant direct influence on actual behavioral intention [22].

H4: ATT has a significant positive influence on AI towards sustainable IT practices.

H5: SN has a significant positive influence on AI towards sustainable IT practices.

H6: PBC has a significant positive influence on AI towards sustainable IT practices.

H7: AI has a significant positive influence on AAB of sustainable IT practices.

Based on the above discussion, Fig. 1 shows the relationship between constructs and the proposed hypothesis in the study.

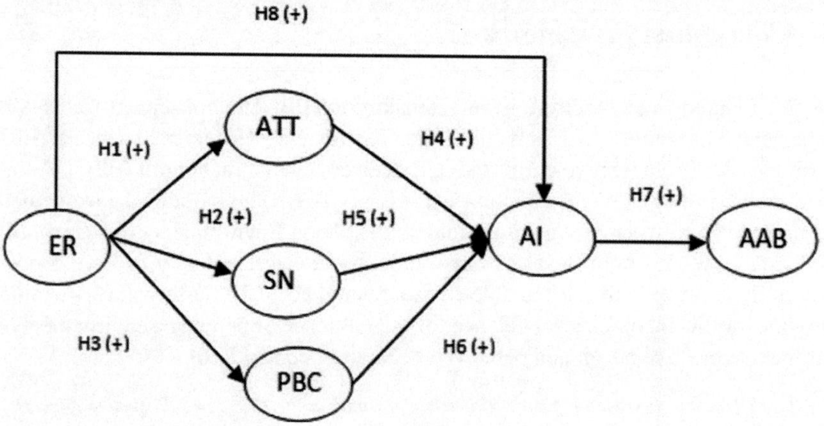

Fig. 1 Conceptual framework

4 Methodology

The scale validated from the literature has been adopted for the study. The latent construct perceived environmental concern (ER) has been used in addition to the existing TPB model constructs; attitude (ATT), subjective norm (SN), perceived behavioral control (PBC), purchase intention (PI), and actual adoption behavior (AAB) towards sustainable IT practice to extend the existing TPB model. Scale items for attitude were measured on a semantic differential scale (extremely bad-1 and extremely good-7) and other scale items were anchored on a seven-point scale (1- strongly agree to 7-strongly disagree) to record the responses. Descriptions of latent constructs are detailed in Table 1.

A questionnaire was designed as a survey instrument to target respondents within the age group of 18-34 years. The educated millennial (18-34 years) has the ability to simplify and better comprehend eco-friendly products [27]. The questionnaire administered in the Delhi-National Capital Region (India) consisted of 24 statements. Respondents were assured that their response is going to be used for academic research. Convenience sampling was used to select the sample from the population. The research instrument was circulated through various social media platforms to reach respondents. A total of 400 questionnaires were circulated and 205 complete

Table 1 Description of latent construct

Sr. No	Construct	Scale	Refere nces
1	ER: Environmental concern	ER1: Express your level of willingness to save the environment	[23]
2		ER2: Environment protection is our responsibility	
3		ER3: I am concerned with environmental degradation and impact on health	
4		ER4: Sustainable IT solution should be opted	
5	ATT: Attitude	ATT1:Performing sustainable IT practices are extremely bad (1)/extremely good (7)	[17]
6		ATT2:Performing sustainable IT practices are extremely undesirable (1)/extremely desirable (7)	
7		ATT3:Performing sustainable IT practices are extremely unenjoyable (1)/extremely enjoyable (7)	
8		ATT4:Performing sustainable IT practices are extremely unfavorable (1)/extremely favorable (7)	

(continued)

Table 1 (continued)

Sr. No	Construct	Scale	Refere nces
9	SN: Subjective Norm	SN1: People close to me think that I should use sustainable IT practices	[20, 24]
10		SN2: People opinion I value most suggest that I should use sustainable IT practices	
11		SN3: Favorable nature of my friend & colleagues influences me to use sustainable IT practices	
12		SN4: Many people like me using sustainable IT practices	
13	PBC: Perceived behavioral control	PBC1: I am willing to use sustainable IT practices	[16, 25]
14		PBC2: If it is up to me, then I would use sustainable IT practices	
15		PBC3: It seems that using sustainable IT practices are not in my control	
16		PBC4: I would use sustainable IT practices	
17	AI: Adoption intention	AI1: I will intend to use sustainable IT practices	[8, 20]
18		AI2: I will put required efforts to use sustainable IT practices	
19		AI3: I would plan to use sustainable IT practices	
20		AI4: I believe in using sustainable IT practices for environmental benefits	
21	AAB: Actual adoption behavior	AAB1: I often use sustainable IT practices	[8, 26]
22		AAB2: I select IT solutions based on their manufacturing nature based on the environment	
23		AAB3: I prefer sustainable IT solutions over non environmentally friendly IT products	
24		AAB4: I use to have sustainable IT solutions irrespective of their cost	

in all aspects yielding a response 51% rate, was retained for analysis. Out of the total respondents, 84 (41%) were male and 121(59%) were female. The majority of respondents were graduates (49%) followed by postgraduates (27%) and others. Data descriptive also confirms that more than 80% of the respondents included in the study are involved in one or other form with sustainable IT practices. This indicates that respondents pay attention to sustainable IT practices.

5 Analysis

Data analysis was performed using SPSS 21.0 and Smart-PLS 2.0. Descriptive and demographic of respondents were ascertained through SPSS 21.0. Smart-PLS 2.0 was employed for two-step model testing i.e., measurement model assessment, and structural model assessment.

5.1 Measurement Model Assessment

Composite reliability (CR) was determined to evaluate the internal consistency of items among the construct. The identified CR value for AAB, AI, PBC, SN, ATT, and ER represented in Table 2 were well above the defined threshold of 0.7 [28]. As represented in Figure 2 all item loadings were more than 0.7 and establish indicator reliability [29]. The AVE value for all latent variable AAB, AI, PBC, SN, ATT, and ER were found to be more than 0.5 [30] and $\sqrt{}$AVE values of each construct represented in the diagonal of Table 2 exceed the off-diagonal values in the correlation matrix provides evidence for convergent and discriminant validity in data [31].

Table 2 Reliability and validity result

	AAB	AI	PBC	SN	ATT	ER
AAB	**0.912**					
AI	0.511	**0.817**				
PBC	0.499	0.510	**0.770**			
SN	0.374	0.479	0.482	**0.790**		
ATT	0.532	0.617	0.493	0.492	**0.838**	
ER	0.473	0.386	0.467	0.342	0.578	**0.725**
CR	0.937	0.858	0.835	0.809	0.904	0.845
AVE	0.833	0.669	0.628	0.586	0.704	0.578

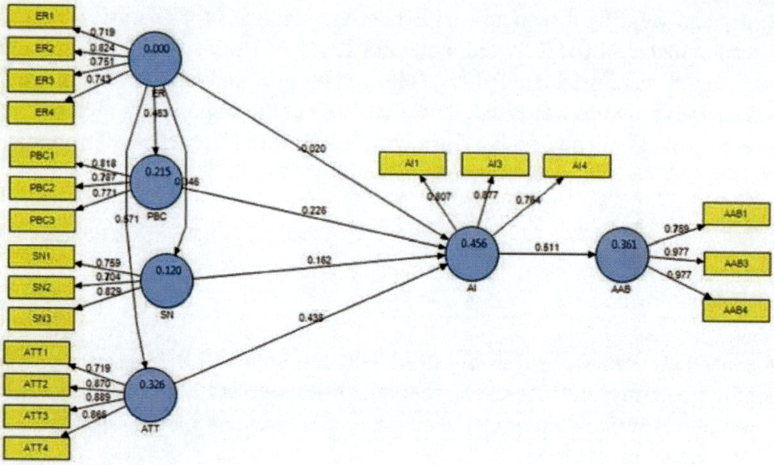

Fig. 2 Result structural model

5.2 *Structural Model Assessment*

Proposed hypothesized relationships were tested and a result for the same is represented in Fig. 2. Non-parametric bootstrapping was employed to examine the roposed relationship among constructs. The finding of the study does not provide much empirical support for the effect of individual environmental cerncern on the adoption intention of sustainable IT practices (H7: $\beta = -0.020$, $t = 0.979$). The causal effect of other constructs was supported. PBC has strong positive significant influence on ATT (H1: $\beta = 0.571$, $t = 18.736$) followed with PBC (H3: $\beta = 0.463$, $t = 12.903$) and SN (H2: $\beta = 0.346$, $t = 8.469$). The decision variable ATT (H4: $\beta = 0.438$, $t = 11.825$) was found to be the dominating factor towards AI of sustainable IT practices followed by PBC (H6: $\beta = 0.226$, $t = 7.602$) and SN (H5: $\beta = 0.162$, $t = 5.589$). The result of the study also confirms a significant positive direct influence of AI (H8: $\beta = 0.511$, $t = 19.984$) towards AAB of sustainable IT practices. The predictive ability of the model was measured through R^2 value against the suggested range of 0.19, 0.33, and 0.67 indicating weak, moderate, and substantial effect, respectively [32]. The extended TPB model exerts a significant moderating influence on adoption intention and actual adoption behavior.

6 Discussion and Implication

The theory of planned behavior (TPB) has been widely used in studies with sustainable adoption intention. The attitude-behavior gap was studied in developing countries to understand sustainable consumption behavior [33]. The moderating role of gender in the TPB framework was also studied to examine student adoption intention for biodegradable drinking straw [34]. TPB was extended with sustainable government support, environmental concern, and sustainable engagement to study sustainable building practices, and further benefits were also highlighted for integrating IOT with sustainable building [35, 36].

Literature mainly addresses the issue of sustainable IT adoption from an organizational perspective but limited research focuses on its adoption from an individual behavioral perspective [37]. The study intended to scrutinize the application of the extended TPB model for anticipating the sustainable IT adoption behavior of millennials. Study outcome indicates that attitude, perceived behavioral control, subjective norm are the significant predictors of sustainable IT adoption intension followed with actual adoption behavior, whereas the additional construct added to model perceived environmental concern also significantly predicts attitude, subjective norm, and perceived behavioral control of youth towards sustainable IT practices. Environmental concern has shown a relation to environmental education in different countries and cultural scenarios [38, 39, 42]. The strongest significant influence of environmental concern on attitude explains that millennial are well aware of their responsibility towards the environment that helps in developing a favorable attitude towards adoption intention of sustainable IT solutions. The responsibility of millennial towards the environment also influences perceived behavioral control and subjective norms. Further, it is also noteworthy that being responsible towards the environment is not sufficient enough for developing favorable adoption intention for sustainable IT applications as the relationship between environmental concern and adoption intention was found to be insignificant. Environmental concern act as a precursor for attitude, perceived behavioral control, and subjective norm of the millennial generation. The individual attitude was found to be the strongest predictor for sustainable IT adoption intention. Sustainable IT attitude refers to "sentiments, values, and norms with climate change, eco-sustainability, and IT's role" and level of awareness for the effect of information technology on surroundings [12, 18, 41]. Attitude or belief mainly represents that individuals concern about the environment which directs to pro-environmental behavior. Marketers should take this opportunity for extensive product promotion and deploy marketing strategies to build a strong product image for endpoint users. The significant positive influence of Perceived behavioral control on sustainable IT adoption intention indicates that youth with resources, time, and willingness to adopt sustainable computing depends on the availability of sustainable solutions for its purchase. Marketers should bridge the gap between consumers and sustainable solutions through better visibility, end-user benefits, and affordable offerings. Subjective norms were reported as a week antecedent to adoption intention as compared to attitude and perceived behavioral control. The significant positive influence of subjective norm indicates that youth give due importance to the opinion of friends and colleagues. Sustainable IT adoption also has a significant positive influence on actual adoption behavior. The findings of the study are in line with the outcomes of [8, 16, 40]. Businesses involved in sustainable IT solutions should start a sustainable campaign and educate individuals that how the adoption of sustainable IT solutions can curb pollution. Companies should also focus on the fact of how the actual adoption of sustainable IT practices can ease their day to day life activities. It is also evident from the findings that millennials are inclined toward a sustainable environment and believe in a clean healthy life. The proposed extended framework disseminates valuable insights to policymakers and marketers to formulate sustainable strategies and policies for future market opportunities.

7 Conclusion

This study incorporated the TPB model with perceived environmental concern to examine millennial sustainable IT adoption intention. The study outcome reveals that the respondents are well aware of their responsibility towards the environment and feel it is an important attribute to develop a favorable attitude, subjective norms, and behavioral control towards sustainable IT adoption. This research makes a novel contribution by extending the TPB model to identify the attitude-behavior gap while opting for sustainable IT practices. Furthermore, in the future, the research can be extended by examining the relationship between personality traits, social, economic, cultural factors, and sustainable technology adoption.

References

1. Ling-Yee L (1997) Effect of collectivist orientation and ecological attitude on actual environmental commitment: the moderating role of consumer demographics and product involvement. J Int Consum Mark 9(4):31–53
2. Capra E, Francalanci C, Slaughter SA (2012) Is software "green"? Application development environments and energy efficiency in open source applications. Inf Softw Technol 54(1):60–71
3. Melville NP (2010) Information systems innovation for environmental sustainability. MIS Q 34(1):1–21
4. Jenkin TA, McShane L, Webster J (2011) Green information technologies and systems: employees' perceptions of organizational practices. Bus Soc 50(2):266–314
5. Molla A, Abareshi A (2012) Organizational green motivations for information technology: empirical study. J Comput Inf Syst 52(3):92–102
6. del Río GP (2005) Analysing the factors influencing clean technology adoption: a study of the Spanish pulp and paper industry. Bus Strateg Environ 14(1):20–37
7. Fernández E, Junquera B, Ordiz M (2003) Organizational culture and human resources in the environmental issue: a review of the literature. Int J Hum Resour Manag 14(4):634–656
8. Chow WS, Chen Y (2009) Intended belief and actual behavior in green computing in Hong Kong. J Comput Inf Syst 50(2):136–141
9. Molla A, Abareshi A (2011) Green IT adoption: a motivational perspective. PACIS 2011:137
10. Fishbein M, Ajzen I (1977) Belief, attitude, intention, and behavior: an introduction to theory and research. Philos Rheto 10:1–12
11. Ajzen I (1991) The theory of planned behavior. Organ Behav Hum Decis Process 50(2):179–211
12. Zahedi S, Batista-Foguet JM, van Wunnik L (2019) Exploring the public's willingness to reduce air pollution and greenhouse gas emissions from private road transport in Catalonia. Sci Total Environ 1(646):850–861
13. Kaiser FG (2006) A moral extension of the theory of planned behavior: norms and anticipated feelings of regret in conservationism. Person Individ Differ 41(1):71–81

14. Lai OK (2000) Greening of Hong Kong? Forms of manifestation of environmental movements. Dyn Soc Move Hong Kong 2000:259–295
15. Attaran S, Celik BG (2015) Students' environmental responsibility and their willingness to pay for green buildings. Int J Sustain Higher Educ 16:327–340
16. Shukla S (2019) A study on millennial purchase intention of green products in India: applying extended theory of planned behavior model. J Asia-Pacific Bus 20(4):322–350
17. Kim Y, Han H (2010) Intention to pay conventional-hotel prices at a green hotel—a modification of the theory of planned behavior. J Sustain Tour 18(8):997–1014
18. Molla A, Abareshi A, Cooper V (2014) Green IT beliefs and pro-environmental IT practices among IT professionals. Inf Technol People 27:129–154
19. Biswas A, Roy M (2015) Green products: an exploratory study on the consumer behaviour in emerging economies of the East. J Clean Prod 15(87):463–468
20. Paul J, Modi A, Patel J (2016) Predicting green product consumption using theory of planned behavior and reasoned action. J Retail Consum Serv 1(29):123–134
21. Maichum K, Parichatnon S, Peng KC (2016) Application of the extended theory of planned behavior model to investigate purchase intention of green products among Thai consumers. Sustain 8(10):1077
22. Gholami R, Sulaiman AB, Ramayah T, Molla A (2013) Senior managers' perception on green information systems (IS) adoption and environmental performance: results from a field survey. Inf Manag 50(7):431–438
23. Kim Y, Choi SM (2005) Antecedents of green purchase behavior: an examination of collectivism, environmental concern, and PCE. ACR North Am Adv 32:592–599
24. Arvola A, Vassallo M, Dean M, Lampila P, Saba A, Lähteenmäki L, Shepherd R (2008) Predicting intentions to purchase organic food: the role of affective and moral attitudes in the theory of planned behaviour. Appetite 50(2–3):443–454
25. Armitage CJ, Conner M (1999) The theory of planned behaviour: assessment of predictive validity and perceived control. Br J Soc Psychol 38(1):35–54
26. Jaiswal D, Kant R (2018) Green purchasing behaviour: a conceptual framework and empirical investigation of Indian consumers. J Retail Consum Serv 1(41):60–69
27. Chan RY (2001) Determinants of Chinese consumers' green purchase behavior. Psychol Mark 18(4):389–413
28. Nunnally JC (1994) Psychometric theory 3E. Tata McGraw-hill education
29. Hair JF, Black WC, Babin BJ, Anderson RE, Tatham RL (1998) Multivariate data analysis. Prentice hall, Upper Saddle River, NJ
30. Fornell C, Larcker DF (1981) Evaluating structural equation models with unobservable variables and measurement error. J Mark Res 18(1):39–50
31. Chiu CM, Wang ET (2008) Understanding web-based learning continuance intention: the role of subjective task value. Inf Manag 45(3):194–201
32. Chin WW (1998) The partial least squares approach to structural equation modeling. Mod Methods Bus Res 295(2):295–336
33. Emekci S (2019) Green consumption behaviours of consumers within the scope of TPB. J Cons Mark 36:417
34. Hassan NN, Kadir JM, Abd Aziz NN (2020) Examining a TPB model towards intention to use biodegradable drinking straw using PLS-SEM. Environ Behav Proc J 5(15):1–6
35. Saleh RM, Al-Swidi A (2019) The adoption of green building practices in construction projects in Qatar: a preliminary study. Manag Environ Qual Int J 30:1238–1255
36. Jain R, Goel V, Rekhi JK, Alzubi JA (2020) IoT-based green building: towards an energy-efficient future. In: Green building management and smart automation 2020, IGI Global, pp 184–207
37. Ojo AO, Raman M, Downe AG (2019) Toward green computing practices: a Malaysian study of green belief and attitude among information technology professionals. J Clean Prod 1(224):246–255
38. Hanson-Rasmussen NJ, Lauver KJ (2018) Environmental responsibility: millennial values and cultural dimensions. J Global Resp 9:6–20

39. Kumar N, Upreti K, Upreti S, Shabbir Alam M, Agrawal M (2021) Blockchain integrated flexible vaccine supply chain architecture: excavate the determinants of adoption. Hum Behav Emerg Technol 3(5):1106–1117
40. Kumar N, Upreti K, Mohan D (2022) Blockchain adoption for provenance and traceability in the retail food supply chain: a consumer perspective. Int J E-Bus Res 18(2):1–7
41. Al Shamsi JH, Al-Emran M, Shaalan K (2022) Understanding key drivers affecting students' use of artificial intelligence-based voice assistants. Educ Inf Technol 1:1–21
42. Arpaci I, Al-Emran M, Al-Sharafi MA, Shaalan K (2021) A novel approach for predicting the adoption of smartwatches using machine learning algorithms. In: Recent advances in intelligent systems and smart applications 2021. Springer, Cham, pp 185–195

The Role of Artificial Intelligence in Project Performance in Construction Companies in Palestine

Koutibah Alrifai, Tareq Obaid, Ahmed Ali Atieh Ali, Ahmed F. S. Abulehia, Hussein Mohammed Esmail Abualrejal, and Mohammed Bassam Abdul Raheem Nassoura

Abstract This study aims to see the role of artificial intelligence in the project performance of construction companies in Palestine. The study community consists of construction engineering offices in Palestine, and the sample of the study reached 183 engineering offices and hence the importance of the research. This study relied on the knowledge management theory (KMT) model. From this point of view, this study came. It is worth mentioning that the results of the study indicated that there is a relationship between artificial intelligence and the performance of construction companies in Palestine. The researcher recommends conducting further research related to artificial intelligence taking into account the current variables of the study.

Keywords Artificial intelligence · Project performance · Construction companies · Palestine

K. Alrifai (✉)
Graduate School of Social Sciences, Yeditepe University, Istanbul, Turkey
e-mail: qs-93@hotmail.com

T. Obaid
Faculty of Engineering and IT, Alazhar University, Gaza, Palestine
e-mail: tareq.obaid@alazhar.edu.ps

A. A. A. Ali
School of Technology and Logistics Management, Universiti Utara Malaysia UUM, Sintok, Kedah
06010, Malaysia

A. F. S. Abulehia
School of Accountancy, Universiti Utara Malaysia UUM, Sintok, Kedah 06010, Malaysia

H. M. E. Abualrejal
School of Technology Management and Logistic, College of Business, Universiti Utara Malaysia UUM, Sintok, Kedah 06010, Malaysia
e-mail: abualrejal@uum.edu.my

M. B. A. R. Nassoura
School of Business, Universiti Utara Malaysia UUM, Sintok, Kedah 06010, Malaysia

M. Al-Emran et al. (eds.), *International Conference on Information Systems and Intelligent Applications*, Lecture Notes in Networks and Systems 550,
https://doi.org/10.1007/978-3-031-16865-9_6

1 Introduction

The use of information technology, robots, and other emerging technologies to design and build has long been a dream of architects, engineers, and researchers. Their intellectual understandings of what might be done and their visionary perspectives of the future of the building Outstripped the practical, technological, economic, cultural, and/or organizational restrictions that had to be overcome in order for them to be realized.

For a project to be a success, each stakeholder's expectations must be met, regardless of who the owner, planner, engineer, contractor, or operator is [1]. In the construction industry, project success is measured by factors such as budget, timeliness, quality, and safety [2]. Construction Project Success Survey was developed by [3] to identify critical success factors before the start of a project and to evaluate the degree of success achieved at its conclusion. Cost, scheduling, performance, and safety are just some of the objective and subjective factors that go into the evaluation process.

Different factors have an effect on project outcomes at different points in time. Several factors must be considered when predicting project outcomes at different periods in the project's timeline [3, 4]. Project managers will need to use a dynamic prediction approach to keep track of project progress. However, several time-dependent elements impact the project's results at each time point. Furthermore, such factors are unknown owing to the nature of the building sector [5]. It's never simple to anticipate the project's conclusion dynamically under such complicated and variable situations. Human specialists may assess the success of a project based on their expertise; however, the value of these assessments is limited by their subjective cognitions and/or limited knowledge. Building computer systems that solve problems intelligently by replicating the human brain is what AI is all about. Artificial intelligence (AI) technology offers strategies for computer programs to do many jobs that people are presently superior at [6]. As a result, AI paradigms are suitable for resolving project management issues [7].

Several obstacles have slowed the construction industry's growth and resulted in very low productivity levels when compared to other industries, such as manufacturing [8]. However, the construction industry is among the least digital in the world, with most stakeholders admitting a long-standing resistance toward transformation [9]. Project management becomes increasingly difficult and time-consuming due to the lack of digitalization and the industry's largely manual nature [10]. The absence of appropriate digital expertise and technology adoption in the construction industry has been linked to some issues, such as project delays, poor quality performance, inaccurate decision-making, and poor productivity, health, and safety performance [11]. It has become evident in recent years that the construction industry has to embrace digitalization and rapidly enhance its technological capabilities in light of the challenges of current labor shortages caused by the COVID-19 epidemic [12]. Digital technology has made substantial contributions to improving company operations, service procedures and industrial efficiency in recent years, compared to traditional methods. AI techniques have improved automation and delivered superior competitive advantages, compared with traditional methods [13]. An example of a real-world application of artificial intelligence (AI) is the employment of artificial intelligence subfields such as machine learning, natural language processing, robotics, computer vision and optimisation. Automation, data-driven technologies, and advanced artificial intelligence (AI) processes in manufacturing are all part of Sector 4.0, a term used to describe the fourth industrial revolution [12].

As a consequence, understanding the interconnections and performance is critical and the links are expected to give important information on how to increase performance capabilities.

The following research question stems from the aforementioned research gaps:

RQ1. Is there a relationship between AI and performance?

In addressing the research questions, we develop a research framework based on the Knowledge management theory (KMT), where the study population is from the construction engineering companies and offices in Palestine registered with the Palestinian Engineers Syndicate, which numbered 384 offices the study sample reached 183.

2 Literature Review

2.1 Artificial Intelligence

John McCarthy introduced artificial intelligence as a topic [14] and it was publicly established at Dartmouth Conference in 1956. [15]. AI is a branch of computer science that aims to create artificial intelligence, but it has lately received a lot of attention [16]. Artificial intelligence (AI) is the copying of social intelligence processes (learning, reasoning, and self-correction) by computers. The processing powers, potential, and genetic algorithms of AI are the most well-known aspects of the technology [17] Speech recognition algorithms, expert systems, and machine vision are some of the most common AI applications [18].

Philosophy, literature, imagination, computer science, electronics, and engineering innovations all contributed to the notion of constructing robots with human-like intellect. Alan Turing's intelligence test was a watershed moment in AI because it goes beyond previous theological and mathematical assumptions concerning the potential of sentient computers. Sixty years later, intelligent computers are exceeding humans in a variety of categories, including learning, thanks to significant developments in other technologies like large data and computing power [19]. "AI is the study of how to make robots perform things that humans do better at the present" accurately defines the notion of AI. The goal of AI is to have robots function at the same level as humans, It refers to creating machines that can handle a variety of complicated issues in several disciplines, operate themselves independently, and have their own thoughts, anxieties, emotions, strengths, weaknesses, and dispositions, according to

[20]. This is still an important AI objective, but achieving it has proven tough and elusive. Artificial intelligence (AI) is focused on creating robots that can outperform humans in a variety of fields [21].

2.2 Project Performance

Although project management has a long history, it is just now beginning to emerge as a distinct subject with its own theoretical base [22]. It lacks a consistent metric for project success and failure, in particular.

It's very uncommon for construction projects to suffer from delays and cost overruns, which have captivated the curiosity of both construction professionals and academics. For example, [23]. cited financial and payment concerns, poor contract management, changes in site circumstances, and material shortages as four of the primary reasons for schedule delays and cost overruns in their study [24]. Obiad found that delays due to design modifications, low labor productivity, inadequate planning, and resource shortages are the most common causes of time overruns, while material price increases, erroneous material assessment, and project complexity are the most common causes of cost overruns. According to [25], the chief reasons for schedule delays and cost overruns include payment challenges, poor contractor management, material procurement issues, limited technical expertise, and an increase in material prices. [26], on the other hand, have looked at the most common causes of quality defects, including human error and poor workmanship. These studies have a pessimistic view of project outcomes. Time, cost, and quality have been identified as three of the most critical characteristics for evaluating the success of building projects, thanks in part to the work of researchers like [27–29].

To ensure successful project completion, project managers and governing bodies are typically presented with prescriptive lists of critical success aspects, failure factors, or risk factors. This line of research is important because it identifies important prerequisites and motivators for a project's success, but it does not provide a precise definition of that achievement (although the factors identified may indirectly point to relevant criteria). In the second stream, the emphasis is on identifying extra contingency elements that may impact project performance or need special management attention in order to avoid negative repercussions. Some academics refer to these traits as "dimensions" of project success. A few examples include the scope and kind of the project, where it is in the life cycle [30], the degree of complexity in the project management process [31], and the focus on strategic vs operational goals [32]. In this research, new project elements are identified. Depending on the project environment and the variables that are controlled, may have a major impact on the project's overall success. Project success measures, however, are not expressly stated in this stream. As part of the evaluation process, a third stream is typically used to determine whether or not a project is a success or failure. The third stream is concerned with the criteria for determining whether or not a project is successful. The first two streams are intertwined. An understanding of how a project's success or failure is defined is essential so that project effort may be allocated where it will have the most impact on meeting performance goals [33].

2.3 Conceptual Model

Knowledge Management Theory (KMT) is a prominent method for improving an organization's competitive position [34]. Organizational learning may assist in the development of new knowledge or the provision of insights that can influence behaviour. There are two types of knowledge: explicit and tacit. Explicit knowledge is information that has been recorded in some form (files, databases, manuals, etc.), while tacit knowledge is only found among personnel and is a valuable resource in any organization. An organization's performance may be greatly improved by accessing, sharing, and implementing both explicit and tacit information [35]. Knowledge is a significant aspect of manufacturing in the twenty-first century. In order to increase profit margins, each company must learn and adapt to an unpredictable market. As a result, every organization's knowledge-based strategy is a critical pillar. It might be used to conceptualize new ideas for managing knowledge based on sound argumentation and help construct knowledge management approaches that can predict the outcome of a process [36]. Relationship marketing is now part of KMT [37]. Although knowledge management and relationships are two independent ideas, research on both demonstrates that both necessitate communication.

However, Relationship management relies heavily on customer data, and knowledge management is a promising interface for future research, although knowledge management and relationships are two distinct ideas, both need communication, as shown by the literature on both [37]. Because of the economic strength of its transactions, B2B marketing has lately gotten a lot of attention [38]. In the recent decade, theoretical advances have been made, and future studies should concentrate on innovation, customer connections, data analytics, and using technology to improve revenue as well as the manufacturing environment. This research intends to respond to previous researchers' requests and construct a theoretical model based on the prior conversations.

In this context, the model for this research takes into account, AI technologies, and organizational performance. KMT may be used to deal with knowledge dispersion and growth, as well as to examine the knowledge characteristics of a relationship in order to better manage it [39]. Knowledge is a combination of contextual data, expert expertise, and value that may lead to innovation [40]. Organizational performance and creativity may both benefit from knowledge management. Knowledge management enablers are variables that may help you improve your knowledge management duties. Organizational structure, culture, and technology are the main forces behind this. It's worth noting how important information technology has been in removing communication obstacles. The purpose of information technology is to facilitate reciprocal learning, knowledge sharing, and communication amongst individuals [41]. To increase organizational performance, the knowledge

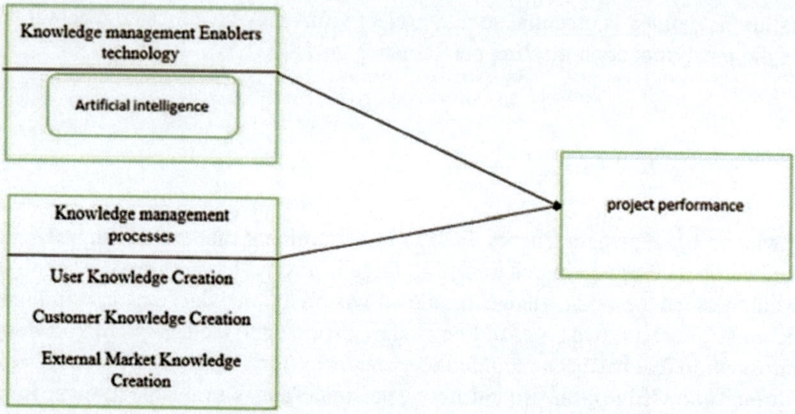

Fig. 1 Conceptual model of the research

management process includes activities such as gathering, designing, managing, and communicating knowledge (Fig. 1).

3 Research Method

The data for this study were collected by using a questionnaire, which is quantitative research conducted in previous studies to determine the factors in the present research. The current study population is construction engineering in Palestine registered with the Palestinian Engineers Syndicate. The factors were adapted and modified to suit this study. The Likert scale ranged from 1 (strongly disagree) to 5 (strongly agree). The study sample consisted of 183 Managers/CEO/CFO of these companies in Palestine.

4 Data Analysis

The author's utilized SMART PLS for data analysis. In order to evaluate the study's premise, the researchers used a two-stage technique. The measurement model, which includes convergent and discriminant validity, is the first step. The investigation will go on to hypothesis testing after the validity has been confirmed. Assessment of the measurement model's convergent and discriminant validity is part of the process. Whether an item measures the latent variable or not, the design for assessment is confirmed by its convergent validity [42]. If the loadings are above 0.5 and the average variance extracted (AVE) is above 0.5, as well as the composite reliability (CR) which is above 0.7, then this is acceptable. There are no CR or AVE values in

Table 1 that fall below the predefined criteria provided by Table 1. For this study, convergent validity has been established.

Table 1 Convergent validity

Constructs	Items	Factor loadings	Cronbach's alpha	CR	(AVE)
Artificial intelligence	AI-1	0.614	**0.903**	**0.919**	**0.511**
	AI-2	0.743			
	AI-3	0.676			
	AI-4	0.743			
	AI-5	0.718			
	AI-6	0.752			
	AI-7	0.687			
	AI-8	0.739			
	AI-9	0.813			
	AI-10	0.657			
	AI-11	0.697			
User knowledge creation	USK-1	0.720	**0.814**	**0.865**	**0.517**
	USK-2	0.675			
	USK-3	0.683			
	USK-4	0.767			
	USK-5	0.741			
	USK-6	0.727			
Customer knowledge creation	CKC-1	0.614	**0.818**	**0.860**	**0.519**
	CKC-2	0.630			
	CKC-3	0.628			
	CKC-4	0.674			
	CKC-5	0.632			
	CKC-6	0.649			
	CKC-7	0.675			
	CKC-8	0.768			
External market knowledge creation	EMK-1	0.798	**0.808**	**0.873**	**0.633**
	EMK-2	0.759			
	EMK-3	0.834			
	EMK-4	0.789			
Project performance	PP-1	0.701	**0.849**	**0.882**	**0.623**
	PP-2	0.654			
	PP-3	0.619			
	PP-4	0.550			

(continued)

Table 1 (continued)

Constructs	Items	Factor loadings	Cronbach's alpha	CR	(AVE)
	PP-5	0.739			
	PP-6	0.759			
	PP-7	0.678			
	PP-8	0.680			
	PP-9	0.674			

CR, Composite reliability; AVE, Average variance extracted

Table 2 HTMT

	Artificial intelligence	Customer knowledge creation	External market knowledge creation	Project performance	User knowledge creation
Artificial intelligence	**0.715**				
Customer knowledge creation	0.479	**0.660**			
External market knowledge creation	0.400	0.561	**0.795**		
Project performance	0.549	0.543	0.613	**0.675**	
User knowledge creation	0.252	0.496	0.492	0.595	**0.719**

Constructs' correlation coefficients must be greater than their correlation square root in order to demonstrate discriminant validity, according to [43]. Table 2 above shows that this criterion has been met.

5 Hypotheses Testing

The PLS Algorithm function was used to investigate the route coefficient in the structural model. In regression analysis, the path coefficient of the SmartPLS 3.0 model is equivalent to the conventional beta weight. The estimated path coefficients vary from –1 to +1, and a path coefficient close to zero suggests that the two variables have no relationship at all. The study's path coefficient, standard error, t-statistic, and significance level were all checked for statistical significance, as shown in Table 3. Path coefficient of the research hypotheses.

A tenfold method was used by [44] to assess the predictive relevance of PLS prediction. The predictive relevance of PLS-LM is verified if there is a little difference between the items; on the other hand, if there is a significant difference, it is not. Although the predictive value is limited if the majority of differences are small, the reverse is true if the most of differences are high (Fig. 2).

Table 3 Path coefficient and p-value

Hypo	Relationships	Std. beta	Std. error	T-value	P-values	Decision
H1	Artificial intelligence → Project performance	0.321	0.054	5.993	0.000	Supported
H2	Customer knowledge Creation → Project performance	0.060	0.060	0.991	0.322	Not supported
H3	External market knowledge creation → Project performance	0.281	0.061	4.638	0.000	Supported
H4	User knowledge creation → Project performance	0.346	0.057	6.067	0.000	Supported

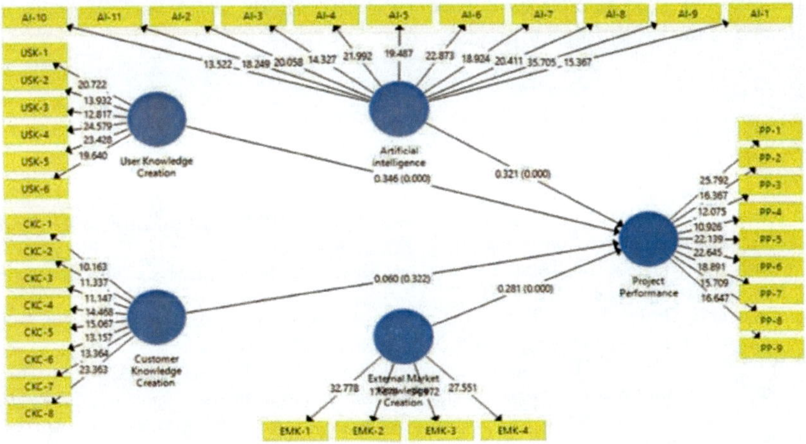

Fig. 2 Hypothesis testing

6 Discussion and Conclusion

We began by interviewing small company owners and managers in Gaza, a territory known as the Palestinian Territories. Therefore, the results may differ from country to country. A study like a multigroup analysis is needed to compare the data across states and even countries in order to make comparisons (groups). Second, Organizational competitiveness may be boosted using the KMT (Knowledge Management Theory) [34]. Organizational knowledge may aid in the creation of new information or the dissemination of visions that have the potential to alter behaviour [11]. Models were utilized to create the research's theoretical model. As a consequence, qualitative in-depth research methodologies may be employed in future studies to add more components. To round things off, the theoretical model built in this study looked at the influence of independent characteristics on big data adoption and how that adoption affected company performance. It appears that Customer Knowledge Creation has a major impact on the Project Performance, and these results are consistent with the study [40]. Personalization resources can help organizations leverage, acquire, and use AI technologies effectively by creating an atmosphere that encourages employees to do so. Organizational willingness to accept AI input and approval stages must be built and actively engaged in by managers, who are expected to do so. According to the findings of the research, Identifying the most critical factors that contribute to a product or service's success is the first step. Managers should then be given the chance to influence what steps should be taken. Manage each influence component and how to change conventional decision-making environments. One of the most essential aspects of the workshop is to help the project's management team members develop a collaborative and disciplined decision-making culture, which will inevitably be reflected in the project's performance.

References

1. Sanvido V, Grobler F, Parfitt K, Guvenis M, Coyle M (1992) Critical success factors for construction projects. J Constr Eng Manag 118(1):94–111
2. Hughes SW, Tippett DD, Thomas WK (2004) Measuring project success in the construction industry. Eng Manag J 16(3):31–37
3. Blundell R, Griffith R, Van Reenen J (1999) Market share, market value and innovation in a panel of British manufacturing firms. Rev Econ Stud 66(3):529–554
4. Ali AA, Abualrejal HM, Mohamed Udin ZB, Shtawi HO, Alqudah AZ (2021) The role of supply chain integration on project management success in Jordanian engineering companies. In: International Conference on Emerging Technologies and Intelligent Systems. Springer, Cham, pp 646–657
5. Barraza GA, Back WE, Mata F (2000) Probabilistic monitoring of project performance using SS-curves. J Constr Eng Manag 126(2):142–148
6. Haykin S (1999) Neural networks: a guided tour. Soft Comput Intell Syst theory Appl 71:71–80
7. Ko DG (2002) Determinants of knowledge transfer in enterprise resource planning implementation. University of Pittsburgh
8. Lenfle S, Loch C (2017) Has megaproject management lost its way. In: The Oxford handbook of megaproject management, pp 21–38

9. Dacre N, Senyo PK, Reynolds D (2019) Is an engineering project management degree worth it? Developing agile digital skills for future practice. Eng Educ Res Netw 2019:1–8
10. Pospieszny P, Czarnacka-Chrobot B, Kobylinski A (2018) An effective approach for software project effort and duration estimation with machine learning algorithms. J Syst Softw 137:184–196
11. Ku ECS, Hsu SF, Wu WC (2020) Connecting supplier–supplier relationships to achieve supply chain performance of restaurant companies. J Hosp Tour Insights 3(3):311–328
12. Khalfan M, Ismail M (2020) Engineering projects and crisis management: a descriptive study on the impact of COVID-19 on engineering projects in Bahrain. In: 2020 Second International Sustainability and Resilience Conference: Technology and Innovation in Building Designs (51154), pp 1–5
13. Donkers T, Loepp B, Ziegler J (2017) Sequential user-based recurrent neural network recommendations. In: Proceedings of the eleventh ACM conference on recommender systems, pp 152–160
14. McCarthy PJ, Hovey RJ, Ueno K, Martell AE (1955) Inner complex chelates. I. Analogs of bisacetylacetoneethylenediimine and its metal chelates1, 2. J Am Chem Soc 77(22):5820–5824
15. Moor J (2006) The Dartmouth College artificial intelligence conference: the next fifty years. AI Mag 27(4):87
16. Ivanov D, Dolgui A, Das A, Sokolov B (2019) Handbook of ripple effects in the supply chain, vol. 276, Springer International
17. Baryannis G, Validi S, Dani S, Antoniou G (2019) Supply chain risk management and artificial intelligence: state of the art and future research directions. Int J Prod Res 57(7):2179–2202
18. Obaid T, Eneizan B, Naser SS, et al (2022) Factors contributing to an effective e-government adoption in Palestine. In: International Conference of Reliable Information and Communication Technology. Springer, Cham, pp 663–676
19. Abualrejal HM, Alqudah AZ, Ali AA, Saoula O, AlOrmuza TK (2021) University Parcel centre services quality and users' satisfaction in higher education institutions: a case of Universiti Utara Malaysia. In: InInternational Conference on Emerging Technologies and Intelligent Systems. Springer, Cham, pp 885-895
20. Pennachin C, Goertzel B (2007) Contemporary approaches to artificial general intelligence. Artif Gen Intell 2007:1–30
21. Emmert-Streib F, Yang Z, Feng H, Tripathi S, Dehmer M (2020) An introductory review of deep learning for prediction models with big data. Front Artif Intell 3:4
22. Söderlund J (2004) Building theories of project management: past research, questions for the future. Int J Proj Manag 22(3):183–191
23. Mansfield NR, Ugwu OO, Doran T (1994) Causes of delay and cost overruns in Nigerian construction projects. Int J Proj Manag 12(4):254–260
24. Obaid T (2015) The impact of green recruitment, green training and green learning on the firm performance: conceptual paper. Int J Appl Res 1(12):951–953
25. Frimpong Y, Oluwoye J, Crawford L (2003) Causes of delay and cost overruns in construction of groundwater projects in a developing countries; Ghana as a case study. Int J Proj Manag 21(5):321–326
26. Love PED, Li H (2000) Quantifying the causes and costs of rework in construction. Constr Manag Econ 18(4):479–490
27. De Wit A (1988) Measurement of project success. Int J Proj Manag 6(3):164–170
28. Munns AK, Bjeirmi BF (1996) The role of project management in achieving project success. Int J Proj Manag 14(2):81–87
29. Chua DKH, Kog Y-C, Loh PK (1999) Critical success factors for different project objectives. J Constr Eng Manag 125(3):142–150
30. Pinto JK, Mantel SJ (1990) The causes of project failure. IEEE Trans Eng Manag 37(4):269–276
31. Shenhar AJ, Wideman RM (1996) Improving PM: linking success criteria to project type. Proc Proj Manag 96:71–76
32. Shenhar AJ, Poli M, Lechler T (2000) A new framework for strategic project management. Management of Technology VIII, University of Miami, Miami, FL

33. Lo S, Li X, Henzl MT, Beamer LJ, Hannink M (2006) Structure of the Keap1: Nrf2 interface provides mechanistic insight into Nrf2 signaling. EMBO J 25(15):3605–3617
34. Dwivedi YK, Venkitachalam K, Sharif AM, Al-Karaghouli W, Weerakkody V (2011) Research trends in knowledge management: analyzing the past and predicting the future. Inf Syst Manag 28(1):43–56
35. DeTienne KB, Jackson LA (2001) Knowledge management: understanding theory and developing strategy. Compet Rev Ann Int Bus J 11:1–11
36. Baskerville R, Dulipovici A (2006) The theoretical foundations of knowledge management. Knowl Manag Res Pract 4(2):83–105
37. Rowley J (2004) Just another channel? Marketing communications in e-business. Mark Intell Plan 22:24–41
38. Cortez RM, Johnston WJ (2017) The future of B2B marketing theory: a historical and prospective analysis. Ind Mark Manag 66:90–102
39. Powell JH, Swart J (2010) Mapping the values in B2B relationships: a systemic, knowledge-based perspective. Ind Mark Manag 39(3):437–449
40. Abubakar AM, Elrehail H, Alatailat MA, Elçi A (2019) Knowledge management, decision-making style and organizational performance. J Innov Knowl 4(2):104–114
41. Dubey R et al (2020) Big data analytics and artificial intelligence pathway to operational performance under the effects of entrepreneurial orientation and environmental dynamism: a study of manufacturing organisations. Int J Prod Econ 226:107599
42. Hair Jr JF, Hult GT, Ringle CM, Sarstedt M (2017) A primer on partial least squares structural equation modeling (PLS-SEM). Sage, Thousand Oaks, p 165
43. Fornell C, Larcker DF (1981) Evaluating structural equation models with unobservable variables and measurement error. J Mark Res 18(1):39–50
44. Shmueli G et al (2019) Predictive model assessment in PLS-SEM: guidelines for using PLSpredict. Eur J Mark 53:2322–2347

Say Aye to AI: Customer Acceptance and Intention to Use Service Robots in the Hospitality Industry

Zufara Arneeda Zulfakar, Fitriya Abdul Rahim, David Ng Ching Yat, Lam Hon Mun, and Tat-Huei Cham

Abstract As industrial revolution 4.0 is introduced, many turn into the use of technology and artificial intelligence (AI) in creating a new competitive advantage to business as well as create an automation that ease the operations and increases profitability. With the important contributions of the tourism and hospitality industry, the advantages of using AI can be benefited. Consequently, it is vital for business to understand customers perception towards the use of AI and services robots. Thus, this study aims to investigate the relationship of eight items under three elements of the service robot acceptance model with the acceptance and intention to use service robots. The results of the study show that the perceived usefulness, trust and rapport are significantly related to acceptance of service robots, which in turn, positively related to intention to use them. Implications of the research findings are discussed to support the notion should the industry Say Aye to AI.

Keywords Artificial intelligence · Service robots · Customer acceptance · Technology adoption · Tourism · Marketing

1 Introduction

The industry revolution 4.0 (IR 4.0) has triggered various technology advancement especially in line with the growth of artificial intelligence (AI). This includes the

Z. A. Zulfakar (✉) · F. A. Rahim · D. N. C. Yat · L. H. Mun
Faculty of Accountancy and Management, Universiti Tunku Abdul Rahman, Kajang, Selangor, Malaysia
e-mail: zufara@utar.edu.my

F. A. Rahim
e-mail: fitriya@utar.edu.my

D. N. C. Yat
e-mail: ngcy@utar.edu.my

T.-H. Cham
UCSI Graduate Business School, UCSI University, Kuala Lumpur, Malaysia

M. Al-Emran et al. (eds.), *International Conference on Information Systems and Intelligent Applications*, Lecture Notes in Networks and Systems 550,
https://doi.org/10.1007/978-3-031-16865-9_7

introduction of big data, automation, robots and both artificial and virtual reality [1–3]. This growth can be seen to exists significantly in the tourism industry. With the industry being a trillion-dollar industry [4], it has become vital for industry players to venture into new developments to boost tourists interest as well as enhancing the services provided to customers [1].

AI were used as several methods of assistance through mobile and online applications, self-service technologies and kiosks and more advanced ones such as robots [1]. Few hotels around the world have been using robots as part of their frontline workers such as Henn-na Hotel in Japan, Yotel worldwide hotels which includes few locations in USA and Singapore and Motel One in Germany [5]. These robots mainly perform functions using internet of things (IoT) technology to automatically perform functions of receptionists and room services [3]. The use of AI in the hospitality industry has proven to significantly providing positive results through its functions in improving services for companies and better satisfying customers [1].

Despite starting prior to the COVID-19 pandemic, AI has been essential in aiding in several operations as well as flattening the curve of the spread. AI has played an important role especially as seen in the healthcare sector through big data providing information to detect emerging risks as well as the use of several technological devices and robots for enrichment of social distancing measures [6]. The latter was one of the key actions taken to reduce the number of COVID-19 cases [7].

As a result of the COVID-19 pandemic, tourism industry has been realized as one of the most affected and is in dire need of revival [8, 9]. The quick spread of the virus induced fear amongst tourists due to social interactions during their travels. Thus, AI could provide benefits in relation to the social distancing practices as well as in performing dirty and dangerous functions such as cleaning and sanitizing [10, 11] as well as reducing the spread of viruses as "robots do not sneeze" [12].

Accordingly, this paper focuses on the use of service robots in the frontline services in which it refers to the use of technology and AI in the interaction, communication, and delivery of services to the visitors of hotels [13]. There have been various research studying the impact of AI in economies in Asia especially in China and Singapore [7]. Previous study has gone into understanding the perception and acceptance of various AI technology in hospitality industry [14–17]. Nonetheless, this area is still relatively new especially in the tourism industry in Malaysia with the first robot hotel initiated by EcoWorld's Eco Nest serviced apartments in Johor [18].

With this, the aim of this study is to explore the acceptance and intention to use service robots by understanding the perception on functional, social-emotional and relational elements. Due to it being a new technology, many may not have had the experiences of using such technology leading to several fears in terms of security risk as well as other technophobia [1, 19]. It is imperative for industry players to understand the level of acceptance of such technology before deciding to implement it [20]. Hence, this study aims to understand if the adoption of service robots is feasible in the hospitality industry in Malaysia and hopes to provide insights to hotels in Malaysia to enhance the services provided.

2 Literature Review

Technology-based study have been applying the Technology Acceptance Model (TAM) which focuses on perceived usefulness and perceived ease-of-use [21, 22]. Subsequently, studies have created an extension through the service robot acceptance model (sRAM) to include other elements that would describe service robots' functions better through social-emotional and relational elements [13, 23]. This extension is to include the elements of social cues and behaviour expected from customers upon using service robots [24].

Under the functional elements, the use of service robots may provide customers with perceived ease-of-use (PEU) and perceive usefulness (PU) defined respectively as how customers perceived the freedom and effortlessness in the usage of service robots and the level of benefits perceived to be obtained in using the technology in enhancing the job done [13]. Subjective subject norm (SSN) is defined as how one behaves based on how important people in their lives influences their decisions [25].

As part of the social-emotional elements, perceived humanness (PH) refers to the concept of anthropomorphism in which objects, in this case the service robots have human-like behaviors through in their movements and functions [26]. On another note, if robots are not able to be identical visually to a human, perceived social interactivity (PSI) are expected from robots in providing appropriate emotions and actions like those of social and human norms [13]. As service robots are meant to replace the humans, perceived social presence (PSP) are the sensation in which the robots give customers the feeling of having and getting the company of a human assistant [27].

Relational elements involve trust (T) and rapport (R) of customers towards the service robots. Trust is defined on a literal sense as a set of beliefs and the dependency of one to rely on another in risky situations [28]. Trust on technology may be towards the technology itself or the providers [28]. As a consequent of trust, customers will start to develop a relationship or bond with the technology forming a rapport [29].

The three elements above are tested against the acceptance (A) and intention to use (ITU) service robots. Acceptance is defined as customers willingness to use or purchase a particular service or product [30] while ITU relates to the process in which the decision is made to use a specific product or service offered, in this case, service robots [31–33].

2.1 Functional Elements and Acceptance of Service Robots

As mentioned earlier, AI is implemented by businesses to ease their operations. For that to be achievable, customers must be able to have a perception towards the usefulness, ease-of-use and social symbolic benefits that are obtain from the functions of the technology for them to be able to accept it [34]. It has been proven that if the functions of the technology able to provide significant aids and benefits to its users, thus the likeliness to adopt it would increase [13]. Hence, this study hypothesized that:

H1: There is a significant relationship between functional elements—(a) perceived ease-of-use, (b) perceived usefulness and (c) subjective social norms with acceptance of service robots.

2.2 *Social-Emotional Elements and Acceptance of Service Robots*

With advancement in technology, robots are almost indistinguishable from humans with previous studies noting that some customers could not be sure if they are communicating with a human or a chatbot [13]. Customers mostly feel confidence in using technology and service robots when social-emotional elements that are obtained upon communicating with humans are received when dealing with technology [26].

The more human-like the robot is, the more significant the tolerance towards the robots. Therefore, this study proposed that

H2: There is a significant relationship between social-emotional elements—(a) perceived humanness, (b) perceived social interactivity and (c) perceived social presence with acceptance of service robots.

2.3 *Relational Elements and Acceptance of Service Robots*

Relational elements were previously found as key reason for acceptance [28]. In any forms of acceptance, a general sense of trustworthiness and connection is needed in ensuring that the relationship between technology and the users are positive [13]. These elements were recognized to have direct influence towards the willingness to accept technology [35]. Once a user has a sense of trust and bond with those technology that they are using, they can reduce fear and technophobia as the foreign feeling is avoided [19]. With that, this study hypothesized that

H3: There is a significant relationship between relational elements—(a) trust and (b) rapport with acceptance of service robots.

2.4 *Acceptance and Intention to Use Service Robots*

Previous studies have found relationships between the three elements above with the acceptance of technology amongst various users and customers [15, 23, 36]. Such elements influence emotions and perception of customer towards AI which increases the willingness to undertake the functions of the technology [16]. Due to the acceptance, the intention to use is highly likely as indicated by research in

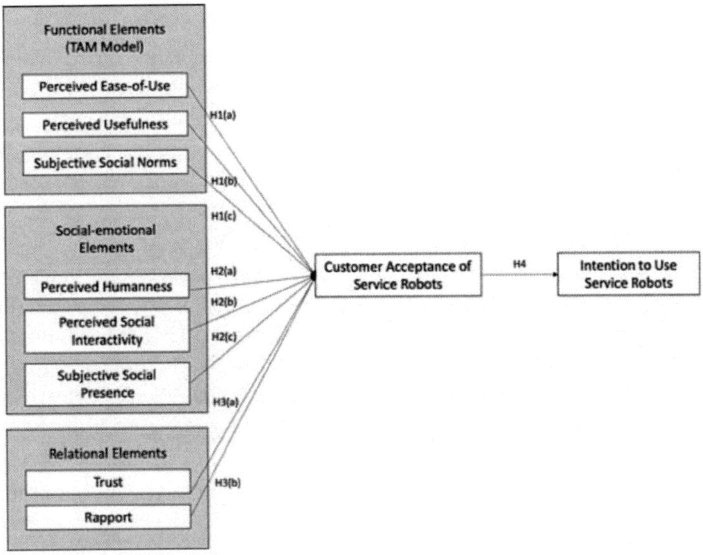

Fig. 1 Research model

the past [27, 37]. It can be summarized that with willingness and acceptance, users would have the intent to use such technology which would be a positive indicator for implementation in the industry. Hence, this study believes that

H4: There is a significant relationship between acceptance and intention to use service robots.

Figure 1 represents the research framework of this study.

3 Research Method

This study targeted to obtain responses from Malaysians above 18 years old. As this is a perception study, the respondents are not necessarily those who have experienced service robots but are those who have knowledge of what the technology is.

Questionnaires were distributed via Microsoft Forms through various social media platforms. The self-administered questionnaires contained several sections starting with Section A covering filtering questions to ensure only those above 18 years of age are able to answer the questionnaire.

It is followed by Section B which measured the independent variables beginning with functional elements covering perceived ease-of-use, perceived usefulness, and subjective social norms; social-emotional elements covering perceived humanness, perceived social interactivity and perceived social presences; and lastly relational elements covering trust and rapport. All sixteen measurement items were adapted from [15].

Section C measured the mediating variable—acceptance of service robots with six measurement items adapted from [15, 16]. Lastly Section D covers the dependent variable—intention to use service robots measuring three items adapted from [36]. Items in Section B, D and D were measured using a 5-point Likert Scale.

Following data cleaning and editing, 243 responses were useable and analyzed using Partial Least Squares Structural Equation Modelling (PLS-SEM) through the Smart PLS 3.3.7 software. PLS-SEM is commonly used in determining causal-predictive relationship between both independent and dependent variables [38]. The analysis started with evaluating the model through reliability and validity of the construct and related indicators in which outer loadings of above an ideal amount of 0.78 indicates that the variables explain more than 50% of the indicator variances, signifying reliability [38]. As per Table 1, all the reflective indicators signify an acceptable item reliability. Furthermore, the composite reliability (CR) values of all indicators are all above 0.60 which is considered as acceptable in exploratory research [38]. All the average variance extracted (AVE) of the indicators are above 0.60 denoting that the construct explains at least 50% of the variances of the items.

As for the discriminant validity, this study analyzed the heterotrait-monotrait (HTMT) ratio to identify the mean value of items correlations across constructs [38]. Based on Table 2, the discriminant validity is present for all items except for between rapport and perceived social interactivity. Additionally, the inner variance inflation factor (VIF) of relationship of variables resulted with PEU (1.639); PU (1.692); SSN (1.594); PH (1.646); PSI (2.288); PSP (1.804); T (1.744) and R (2.141) in which all values are below 3, which does not indicate any possible or probable collinearity issues [38]. Furthermore, the adjusted R^2 value of 0.455 for the relationship of independent variables and acceptance and 0.583 for the relationship of acceptance and intention to use signaling that the model shows a substantial fit to explain the variables [38] (Table 3).

Table 1 Result of convergent validity and internal consistency assessment of variables

	Items	Loadings	AVE	CR
PEU	2	0.846–0.900	0.763	0.865
PU	2	0.894–0.928	0.830	0.907
SSN	2	0.911–0.923	0.840	0.913
PH	2	0.867–0.925	0.804	0.891
PSI	2	0.834–0.879	0.734	0.846
PSP	2	0.537–0.966	0.611	0.744
T	2	0.903–0.907	0.818	0.900
R	2	0.845–0.917	0.778	0.875
A	6	0.741–0.868	0.648	0.917
ITU	3	0.861–0.871	0.752	0.901

Table 2 Result of discriminant validity (HTMT ratio)

	A	ITU	PEU	PU	SSN	PH	PSI	PSP	T
A									
ITU	0.884								
PEU	0.620	0.582							
PU	0.634	0.605	0.610						
SSN	0.390	0.325	0.251	0.550					
PH	0.240	0.237	0.333	0.343	0.424				
PSI	0.666	0.500	0.745	0.596	0.566	0.737			
PSP	0.344	0.250	0.273	0.431	0.648	0.826	0.809		
T	0.687	0.645	0.686	0.647	0.401	0.359	0.769	0.353	
R	0.559	0.420	0.521	0.529	0.685	0.691	0.921	0.868	0.528

Table 3 Result of hypotheses testing

	Standardized estimate (β)	t-value	p-value	Hypothesis
H1(a): PEU → A	0.136	1.890	**0.059**	**Rejected**
H1(b): PU → A	0.238	4.205	**0.000**	**Supported**
H1(c): SSN → A	0.011	0.182	**0.856**	**Rejected**
H2(a): PH → A	-0.128	1.782	**0.075**	**Rejected**
H2(b): PSI → A	0.140	1.820	**0.069**	**Rejected**
H2(c): PSP → A	0.007	0.106	**0.916**	**Rejected**
H3(a): T → A	0.269	3.898	**0.000**	**Supported**
H3(b): R → A	0.162	2.196	**0.029**	**Supported**
H4: A → ITU	0.765	22.660	**0.000**	**Supported**

4 Discussion and Conclusion

The results of this study showed that as part of the functional elements only perceived usefulness has a significant relationship with acceptance [15]. It is obvious that willingness to accept is dependent on whether users feel like the technology would provide benefits and are useful to them. On another hand, perceived ease-of-use and subjective social norms are not necessarily pertinent to influence the perception of users [16]. Users are mainly concerned that in terms of functionality, importantly, such technology must be useful and beneficial to them as users for them to be able to accept the use of the service robots. However, it is not necessarily be one that is easy to use as well as boosting their social status as it will not impact their decisions on acceptance due to the lack of relationship between these variables.

It is seen that the social-emotional elements are not an essential element in determining the acceptance level of service robots as are all rejected. As stated earlier, AI technology are created and implemented to provide sufficient functional benefits and in most cases are created with the look of machines rather than humans [39]. This shows that customers do not necessarily concern on the humanness and societal being of the robots in influencing their willingness to accept the technology. This is due to the understanding that service robots are machine and not humans, and despite of the fact, customers may still have tendency to accept the technology.

The relational elements, both trust and rapport have a significant relationship in influencing acceptance of service robots. As mentioned earlier, these elements are fundamental in ensuring that customers feel comfortable with the technology [15]. These elements are closely related to technophobia [19]. Customers need to know that the technology in which the service robots operate are reliable and would not expose them to any risks that induces fear and reduced acceptance of the technology.

Finally, customers acceptance does have a significant relationship with intention to use service robots. It is apparent that with the acceptance, users will have the motivation to use the services if they are made available to them. Thus, industry players should utilize the availability of the technology such as service robots in handling the frontline functions in their hotels. Customers seemed to have an intention to adopt the technology if they can accept it from the usefulness that it provides and the connection that they are able to form with the service robots. Thus, from a practical standpoint, it is important for providers of service robots in the industry to ensure that the technology would be useful for them by providing enough function to help users to deal with as many things as possible.

Additionally, on top of the implications towards players of the industry, this paper contributes to the gap to the literature on acceptance of AI in Malaysia especially in the hospitality and tourism industry. Future research may obtain information on the actual usage of the technology focusing on samples that have had experience with such technology. Comparisons may also be made between those with and without experience in using service robots. As such technology is new in Malaysia, this paper does come with limitations, however, it does provide an important insight for industry players, adding literature to the gap of studies in Malaysia and may trigger more research of this area for hospitality industry in Malaysia.

References

1. Bulchand-Gidumal J (2020) Impact of artificial intelligence in travel, tourism, and hospitality. In: Handbook of e-Tourism. Springer International, Cham, pp 1–20
2. Li JJ, Bonn MA, Ye BH (2019) Hotel employee's artificial intelligence and robotics awareness and its impact on turnover intention: The moderating roles of perceived organizational support and competitive psychological climate. Tour Manag 73:172–181
3. Samala N, Katkam BS, Bellamkonda RS, Rodriguez RV (2020) Impact of AI and robotics in the tourism sector: a critical insight. J Tour Futures 8:73–87
4. World Travel & Tourism Council, Economic Impact Reports, https://wttc.org/Research/Economic-Impact. Accessed 2 May 2022
5. Travel Channel, 10 hotels that have robot employees, https://www.travelchannel.com/interests/gear-and-gadgets/photos/10-hotels-that-are-using-robots. Accessed 2 May 2022

6. Mahomed S (2020) COVID-19: The role of artificial intelligence in empowering the healthcare sector and enhancing social distancing measures during a pandemic. S Afr Med J 110(7):610–613
7. Haseeb M, Mihardjo LW, Gill AR, Jermsittiparsert K (2019) Economic impact of artificial intelligence: New look for the macroeconomic assessment in Asia-pacific region. Int J Comput Intell Syst 12(2):1295
8. Helble M, Fink A (2020) Reviving tourism amid the COVID-19 pandemic. ADB Briefs 1(150):1–13
9. Sharma GD, Thomas A, Paul J (2021) Reviving tourism industry post-COVID-19: a resilience-based framework. Tour Manag Persp 37:100786
10. Mukherjee S, Baral MM, Venkataiah C, Pal SK, Nagariya R (2021) Service robots are an option for contactless services due to the COVID-19 pandemic in the hotels. Decision 48(4):445–460
11. Seyitoğlu F, Ivanov S (2021) Service robots as a tool for physical distancing in tourism. Curr Issue Tour 24(12):1631–1634
12. The Star: 'Robots don't sneeze': Hotels, hospitals, offices turning to delivery bots during coronavirus pandemic. https://www.thestar.com.my/tech/tech-news/2020/10/12/robots-dont-sneeze-hotels-hospitals-offices-turning-to-delivery-bots-during-coronavirus-pandemic. Accessed 18 Febr 2022
13. Wirtz J, Patterson PG, Kunz WH, Gruber T, Lu VN, Paluch S, Martins A (2018) Brave new world: service robots in the frontline. J Serv Manag 29(5):907–931
14. Belanche D, Casaló LV, Flavián C (2021) Frontline robots in tourism and hospitality: service enhancement or cost reduction? Electr Mark 31(3):477–492
15. Fernandes T, Oliveira E (2021) Understanding consumers' acceptance of automated technologies in service encounters: drivers of digital voice assistants adoption. J Bus Res 122:180–191
16. Gursoy D, Chi OH, Lu L, Nunkoo R (2019) Consumers acceptance of artificially intelligent (AI) device use in service delivery. Int J Inf Manage 49:157–169
17. Park S (2020) Multifaceted trust in tourism service robots. Ann Tour Res 81:1–33
18. Malay Mail: EcoWorld to unveil Malaysia's first robot hotel. https://www.malaymail.com/news/money/2019/02/17/ecoworld-to-unveil-malaysias-first-robot-hotel/1723905. Accessed 2 Sept 2022
19. Oh C, Lee T, Kim Y, Park SH, Kwon S, Suh B (2017) Us vs. them: understanding artificial intelligence technophobia over the Google DeepMind Challenge Match. In: Conference on Human Factors in Computing Systems—Proceedings, pp 2523–2534
20. Cham TH, Low SC, Lim CS, Aye AK, Ling RL (2018) Bin: preliminary study on consumer attitude towards fintech products and services in Malaysia. Int J Eng Technol 7(2):166–169
21. Godoe P, Johansen TS (2012) Understanding adoption of new technologies: technology readiness and technology acceptance as an integrated concept. J Eur Psychol Stud 3:38
22. Al-Emran M, Granić A (2021) Is it still valid or outdated? A bibliometric analysis of the technology acceptance model and its applications from 2010 to 2020. In: Recent advances in technology acceptance models and theories. Springer, Cham
23. Stock RM, Merkle M (2017) A service robot acceptance model: user acceptance of humanoid robots during service encounters. In: 2017 IEEE International Conference on Pervasive Computing and Communications Workshops, PerCom Workshops, pp 339–344
24. Stock RM, Merkle M (2018) Can humanoid service robots perform better than service employees? A comparison of innovative behavior cues. In: Proceedings of the Annual Hawaii International Conference on System Sciences, 2018-January(February), pp 1056–1065
25. Venkatesh V, Davis FD (2000) Theoretical extension of the technology acceptance model: four longitudinal field studies. Manag Sci 46(2):186–204
26. Tinwell A, Grimshaw M, Williams A (2011) The uncanny wall. Int J Arts Technol 4(3):326–341
27. Heerink M, Krose B, Evers V, Wielinga B (2008) The influence of social presence on acceptance of a companion robot by older people. J Phys Agents 2(2):33–40
28. Siau K, Wang W (2018) Building trust in artificial intelligence, machine learning, and robotics. Cutter Bus Techno J 31(2):47–53

29. Gremler DD, Gwinner KP (2000) Customer-employee rapport in service relationships. J Serv Res 3(1):82–104
30. Schmidt DM, Brüderle P, Mörtl M (2016) Focusing aspects of customer acceptance for planning product-service systems—A case study from construction machines industry. Procedia CIRP 50:372–377
31. Abdul Rahim F, Goh PJ, Cheah LF (2019) Malaysian coffee culture: attributes considered to purchase coffee beverages. J Mark Adv Pract 1(1):50–62
32. Cham TH, Ng CKY, Lim YM, Cheng BL (2018) Factors influencing clothing interest and purchase intention: a study of Generation Y consumers in Malaysia*. Int Rev Retail Distrib Cons Res 28(2):174–189
33. Abdul Rahim F, Zulfakar ZA, Rusli KA (2021) Halalan Toyyiban: the mediating effect of attitude on Muslim's purchase intention towards imported Halal food in Malaysia. J Mark Adv Pract 3(2):60–75
34. McLean G, Osei-Frimpong K (2019) Hey Alexa … examine the variables influencing the use of artificial intelligent in-home voice assistants. Comput Hum Behav 99:28–37
35. Heerink M, Krose B, Evers V, Wielinga B (2010) Assessing acceptance of assistive social agent technology by older adults: The Almere model. Int J Soc Robot 2(4):361–375
36. de Kervenoael R, Hasan R, Schwob A, Goh E (2020) Leveraging human-robot interaction in hospitality services: incorporating the role of perceived value, empathy, and information sharing into visitors' intentions to use social robots. Tour Manag 78:104042
37. Lu L, Cai R, Gursoy D (2019) Developing and validating a service robot integration willingness scale. Int J Hosp Manag 80:36–51
38. Hair JF, Risher JJ, Sarstedt M, Ringle CM (2019) When to use and how to report the results of PLS-SEM. Eur Bus Rev 31(1):2–24
39. Huang MH, Rust RT (2018) Artificial intelligence in service. J Serv Res 21(2):155–172

Ontology Integration by Semantic Mapping for Solving the Heterogeneity Problem

Moseed Mohammed, Awanis Romli, and Rozlina Mohamed

Abstract In recent years, ontology integration has received an increased focus in ontology engineering. Ontology integration is a complex process that has some difficulties such as semantic heterogeneity. The goal of this research is to use semantic mapping to reduce integration complexity and solve semantic heterogeneity. What is ontology engineering? What difficulties haven't been solved until now by ontology integration? What is the effective role of semantic mapping in semantic heterogeneity? This research seeks to address these questions. The expected contribution of this research is to build a comprehensive view of ontology integration and support interoperability. The significance of using semantic mapping to improve interoperability on ontology integration is confirmed by researchers.

Keywords Ontology engineering · Ontology integration · Semantic mapping · Interoperability

1 Introduction

Ontology is a formal specification of conceptualizations and formal explanation of knowledge [1]. Ontology is created in a branch of artificial intelligence for knowledge-based systems and established to retrieve information problems [2]. Ontology is generally used in several areas such as semantic web [3], engineering systems [4], software engineering [5], healthcare information [6], IoT technology [8], library system [9], knowledge organisation [10], decision-making method [11], and manufacturing systems [12], as ontology decreases the difficulty of information and increases its association [13] as well as eases information sharing. Ontology is used to solve the interoperability problems of multiple domains [14] and create a knowledge-based system [15]. The significance of using semantic mapping to improve interoperability in different areas is confirmed by researchers [16–18].

M. Mohammed (✉) · A. Romli · R. Mohamed
Faculty of Computing, Universiti Malaysia Pahang, Pahang, Malaysia
e-mail: qutamee@yahoo.com

M. Al-Emran et al. (eds.), *International Conference on Information Systems and Intelligent Applications*, Lecture Notes in Networks and Systems 550,
https://doi.org/10.1007/978-3-031-16865-9_8

Ontology integration is a procedure to integrate two or more ontologies to build a new integrated ontology [17]. Most present ontology integration methods are restricted for matching between two ontologies [18], and only a few methods manage more than two ontologies simultaneously [19]. There are two basic stages for ontology integration which are the matching stage and merging stage. Ontology integration has been studied over the past two decades, but it remains a stimulating job, where the applications of ontology integration have been greatly benefited from in the biomedical area [20] and the Internet of Things [21]. This paper is focused on heterogeneity problems in ontology integration. There are two types of heterogeneity in ontology integration, which are schema heterogeneity [22] and semantic heterogeneity [23]; however, the researchers have not focused on semantic heterogeneity [24]. Ontology matching is a real method to address the problem of ontology heterogeneity [25]. Ontology matching is the greatest solution to the heterogeneity problem because it detects matches between semantically related entities in ontologies [20]. Most existential matching solutions depend on schema-level much more than data-level [26]. The goal of this research is to use semantic mapping to reduce integration complexity and solve the heterogeneity. Semantic mapping between concepts is very significant for integration [27], but it is the largest share of unresolved problems and not used much due to their need for a complex process [20]. Syntactic measures are the most similarity used because it is easy for implementation [20]; structural measures are also used while semantic measures are not used much due to their want for difficult operations [28]. This paper is organised as follows. Section 2 defines the study methodology. Section 3 describes the ontology engineering background, explains ontology and the ontology development process. Section 4 presents the concepts used in the integration of ontologies, which are the matching and merging of ontology. Section 5 describes the different existing tools of ontology integration. Section 6 draws the conclusion of this paper.

2 Study Methodology

The guideline that was used to perform the review in this paper was to search for proceedings from conferences and journal papers in Google Scholar, Scopus, and Web of Science. The articles focused on the background of ontology engineering, ontology integration, and semantic mapping. The selected articles were deemed eligible based on their appropriate studies to provide answers to the research questions presented in this research, which are: What is ontology engineering? What difficulties have not been solved until now by ontology integration? What is the effective role of semantic mapping in semantic heterogeneity?

3 Ontology Engineering

Ontology is a set of axioms that explains and describes domain entities [26]. Ontology is a 5-tuple O = (C, P, I, Λ, Γ) [20], where C is a set of classes, P is a set of properties, I is a set of individuals, Λ is a set of axioms, and Γ is a set of annotations. Table 1 describes in detail the components of ontology. Ontology engineering is a branch of knowledge engineering that studies ontology building methods and methodologies [29]. Ontology engineering studies the ontology development process [30], ontology life cycle, ontology construction methods [31], ontology integration [27], and languages that support them. Ontology integration is a significant subject of interest in ontology engineering, as referred to in the next section. Ontology language is a formal language for coding ontology and the user is able to inscribe strong formal representations of domains. There are several languages for ontology, such as Resource Description Framework (RDF) [32], RDF Schema (RDFS) [33], and Ontology Web Language OWL [34].

Table 1 describes the components of ontology which is a set of objects that has static and dynamic parts. The static part of ontology concerns the structure that is modelled within a particular field such as classes and properties, and the dynamic part revolves around reasoning, inferences, and deriving new facts from already known facts such as axioms and rules.

Table 1 The ontology components

Item	Description
Classes	Set of objects that are grouped according to common features
Properties	Set of features or characteristics of the object
Individuals	Set of instances of classes in the real world which are also called terms
Relations	Set of relationships that provides logical connections between individuals or classes that describe the relation between them
Axioms	Set of axioms used for checking the consistency of ontology or inferencing new information based on rules in a logical form
Annotations	Set of annotations that provides metadata for information to be understood
Function	Set of structures molded by definite relationships that may replace individual terms with extra complex terms
Restrictions	Set of official declarations that describe what must be true for some declarations to be measured true
Rules	Set of sentences (if–then statements) which defines inferences that are extracted by confirmation

4 Ontology Integration

Ontology integration is a critical task in ontology engineering. Ontology integration is the procedure to merge two or more ontologies with the goal of building a new integrated ontology [27]. There are many terms regarding ontology integration such as matching, merging, mapping, and relationship that are unclear and at times unused. So, Table 2 provides a description for each term. Ontology integration includes three different cases [27]: (1) Develop a new ontology by reusing ontologies; (2) Create a new unified ontology by integrating different ontologies; and (3) Integrate various ontologies into a single application to describe or apply a knowledge-based system.

Ontology integration approaches contain two basic stages [11]: First, a matching stage that resolves differences by recognising semantic similarity between the different elements. Second, the merging stage that achieves the outcome of the matching stage by merging or linking matching elements to create a new united vision. Ontology matching approaches are simple matching [35] and complex matching [36]. Ontology merging approaches are simple merge [26], full merge [18], and symmetric merge [37]. Ontology integration has been widely and effectively applied in biomedical [23] and the Internet of Things, while there is a great lack in manufacturing [18].

4.1 *Ontology Matching*

Ontology matching is the method of identifying the semantic correspondences of entities in different ontologies. Similarity measure is critical for matching ontology methods [24]. There are three categories of similarity measures as shown in Table 3, which are syntactic measure, structure measure, and linguistic measure. These will be presented in detail in the next section.

Table 2 Ontology integration terms

Terms	Description
Matching	Determining the semantic matches of entities in different ontologies, which is an active way to address the problem of ontological heterogeneity
Merging	Building complete ontology by integrating knowledge from other ontologies
Mapping	Mapping an equivalence correspondence which named mapping rules when they are read as ontological declarations or axioms
Relation	Giving a correspondence for integral relation such as the equivalence, subsumption, and disjointness

Table 3 Describes similarity measures categories

Author	Measure 1	Measure 2	Measure 3
[36]	Terminological mapping	Structural mapping	Semantic mapping
[20]	Syntactic measure	Taxonomy measure	Linguistic measure
[38]	Statistics techniques	Logic techniques	Linguistics techniques
[39]	Terminological techniques	Structural techniques	Semantic techniques
[40]	Syntactic similarity	Structural similarity	Linguistic measure
[41]	Syntactic techniques	Lexical techniques	Semantic techniques
[42]	Syntactic measure	Structural measure	Linguistic Semantic

Table 4 Ontology integrating tools

Tools	Description
GTM	Graph Theory Model is a division of separate mathematics which are education graph models and their characteristics. Graphs are mathematical network like models collected of two sets, V (set of apices/nodes) and E (set of edges/arcs)
CBM	Context-Based Measure is to match big rule ontologies, where the measurement of lexical similarity in ontology matching is performed using WordNet
ANN	Artificial neural networks are computational systems stimulated by the human brain. It has proven its suitability for ontology matching
Protégé	Protégé is a tool used for matching ontologies to get similar classes, objects, and instances

4.1.1 Syntactic-Based Measures

There are two syntactic measures that are mostly used which are String Metric for Ontology Alignment (SMOA) [43] and Levenshtein [20]. Assumed two strings × 1 and × 2, the SMOA similarity is defined as follows:

SMOA(×1, × 2) = comm(×1, × 2) – diff(×1, × 2) + winklerImpr(×1, × 2)(1).

where comm(×1, × 2) stands for the common length of × 1 and × 2, while diff(× 1, ×2) for the different lengths and winklerImpr(×1, × 2) is the improved approach proposed in [43].

4.1.2 Linguistic-Based Measures

Linguistic similarity between two strings is determined by considering semantic relationships (such as synonyms and hypernym) that typically require the use of thesaurus and dictionaries. WordNet is widely used as an electronic vocabulary database that collects all meanings of different words [24]. For example, two words d1 and d2, Linguistic Similarity (d1, d2) equals:

1 if Words D1 and D2 Are Synonyms in Wordnet.
0.5, if word d1 is the hypernym of word d2 or the opposite is true in Wordnet.
0, otherwise.

4.1.3 Structure-Based Measures

Structure-based measures are to make full use of the ontology hierarchy relation to determine the similarity between two entities by considering the similarity of their neighbours (parents, children, and siblings) [44] or have similar instances [42]. For example, if entities e1 in Q1 and e2 in Q2 are properly matched, then the neighbours of e1 are probable match neighbours of e2. When the correspondences linking the neighbours of e1 and e2 have a self-assurance rate, the correspondence (e1 $\equiv$ e2) may be correct. Semantic mapping between concepts is very significant for integration [27]. Syntactic measures are the most similarity used because it is easy for implementation. Structural measures are also used while semantic measures are not used due to their want for complex operations.

4.2 *Semantic Mapping*

Semantic mapping of a particular correspondence can be a relationship [26], like equivalence relationship ($\equiv$), subsumption relationship ($\sqsupseteq$ or $\sqsubseteq$), disjointness relationship ($\perp$), and overlap relationship ($\between$). Relationships are identified by the next signs: " $=$ " (is equivalent to), " $>$ " (includes or is more general than), " $<$ " (is included by or is more specific than), and "%" (disjointness with).

4.2.1 Equivalence Relationship

The equivalence relationship among two classes C and D indicates that all cases of C are also cases of D, which means that together, the classes have a similar set of entities. The equality relationship that holds between two properties P1 and P2 means that an individual x is linked to an individual or literal data together by P1 and P2. Equivalence relationship between two entities z and w means that entity z is same/equivalent/duplicate to entity w.

4.2.2 Subsumption Relationship

An implicit relationship between classes C and D means that the set of cases of C is a subgroup/super group of the set of cases of D. Subsumption relationship land among two properties P1 and P2 means that if an entity z is linked by P1 to an entity or a data accurate w, then z is linked by P2 to w.

4.2.3 Disjointness Relationship

A disjointness relationship between two classes C and D means that cases of C are absolutely not cases of D. A dissociation relationship between two properties P1 and P2 means that no entity z is linked to a single individual or literal data by P1 and P2.

Equivalence and disjointness are the simplest types of relations, then comes the subsumption relations [45]. Equivalence and subsumption are the simplest relationships, followed by disjointness relationship [46]. Integration approaches must deal with a variety of semantic relationships.

4.3 Ontology Merging

The merging phase is the process of merging the nominated input ontologies into an integrated ontology. The goal of merging is to build a more comprehensive ontology on a topic, and to gather knowledge in a coherent way from other ontologies on the same topic [27]. There are three kinds of ontology merging which are simple merge that is bridge ontology, full merge that is semantically equal, and symmetric merge that is really ontology enhancement. Ontology merging facilitates creating an ontology, support assistance, and growth semantic interoperability. The main violations in ontology merging are [46] incoherence, inconsistency, and redundancy (structural and relational). Ontology incoherence means that there are unsatisfying classes and properties in merging ontology, which reduces its performance and makes it unclear and unusable. An inconsistency in integrated ontology occurs as a result of unintended repercussions of logical inferences that are still hard to discover, understand, clarify, and fix in advance. Structural redundancy or semantic redundancy happens in class hierarchy, where more than one path exists from the root to the leaf. Relational redundancy occurs due to the complete merge of entities or by the adding of equality relationships that connect diverse entities in merging ontology.

5 Ontology Integrating Tools

Several tools have been developed to integrate ontology, particularly for the matching process, such as Graph Theory Model (GTM) [47], Context-Based Measure (CBM) [48], Artificial Neural Networks (ANN) [28], and Protégé [49], as shown in Table 4.

6 Conclusion

This paper aims to review ontology integration and some related features that belong to the field of ontology matching. The paper reviewed literature on ideas, methods, several subjects, and future work in the ontology integration field. Most present ontology integration methods are restricted for matching between two ontologies, as only a few methods can manage more than two ontologies simultaneously. The greatest research work in the field of ontology matching remains concentrated on identifying simple equality correspondences among ontological entities which are the easy cases of ontological matching. Limited systems attempt to discover additional difficult correspondences or account for unequal relationships, like subsumption and disjointness. This study is expected to contribute to building a comprehensive view of ontology integration and interoperability support in many areas.

Acknowledgements The research reported in this study is conducted by the researchers at University Malaysia Pahang (UMP), it is funded by FRGS/1/2018/TK10/UMP/02/3 grant. The researchers would like to thank Ministry of Higher Education and UMP for supporting this research.

References

1. Ren G, Ding R, Li H (2019) Building an ontological knowledgebase for bridge maintenance. Adv Eng Softw 130:24–40
2. Huang X, Zanni-Merk C, Crémilleux B (2019) Enhancing Deep Learning with Semantics: an application to manufacturing time series analysis. Proc Comput Sci 159(2018):437–446
3. Zhang J, Li H, Zhao Y, Ren G (2018) An ontology-based approach supporting holistic structural design with the consideration of safety, environmental impact and cost. Adv Eng Softw 115:26–39
4. Shang Z, Wang M, Su D (2018) Ontology based social life cycle assessment for product development. Adv Mech Eng 10(11):1–17
5. Karray MH, Ameri F, Hodkiewicz M, Louge T (2019) ROMAIN: towards a BFO compliant reference ontology for industrial maintenance. Appl Ontol 14(2):155–177
6. Otte JN, Kiritsi D, Ali MM, Yang R, Zhang B, Rudnicki R, Rai R, Smith B (2019) An ontological approach to representing the product life cycle. Appl Ontol 14(2):179–197
7. Slimani T (2014) A study on ontologies and their classification. Recent Adv Electr Eng Educ Technol 2014:86–92
8. Mohammed M, Romli A, Mohamed R (2021) Existing semantic ontology and its challenges for enhancing interoperability in IoT environment. In: 2021 International Conference on Software Engineering & Computer Systems and 4th International Conference on Computational Science and Information Management (ICSECS-ICOCSIM). IEEE, pp. 22–26
9. Hobbs J, Fenn T (2019) The design of socially sustainable ontologies. Philos Technol 32(4):745–767
10. Mohd M, Bilo M, Louge T, Rai R, Hedi M (2020) Computers in industry ontology-based approach to extract product's design features from online customers' reviews. Comput Ind 116:103175
11. Cheng H, Zeng P, Xue L, Shi Z, Wang P, Yu H (2016) Manufacturing ontology development based on Industry 4.0 demonstration production line. In: 2016 Third International Conference on Trustworthy Systems and their Applications (TSA), IEEE. pp 42–47
12. He Y, Hao C, Wang Y, Li Y, Wang Y, Huang L (2020) An ontology-based method of knowledge modelling for remanufacturing process planning. J Clean Prod 258:120952
13. Ostad-Ahmad-Ghorabi H, Rahmani T, Gerhard D (2013) An ontological approach for the integration of life cycle assessment into product data management systems. In: CIRP Design 2012. Springer, London, pp 249–256

14. AN MM, Romli A, Mohamed R (2021) Eco-ontology for supporting interoperability in product life cycle within product sustainability eco-ontology for supporting interoperability in product life cycle within product sustainability. In: IOP conference in series of materials science engineering
15. Mohammed M, Romli A, Mohamed R (2021) Using ontology to enhance decision-making for product sustainability in smart manufacturing. In: 2021 international conference on intelligent technology, system and service for internet of everything (ITSS-IoE). IEEE, pp 1–4
16. Okikiola FM, Ikotun AM, Adelokun AP, Ishola PE (2020) A systematic review of health care ontology. Asian J Res Comput Sci 5(1):15–28
17. Salman R (2020) Literature review to compare efficiency of various machine learning algorithms in predicting chronic kidney disease (CKD), pp 1–4
18. Ocker F, Vogel-Heuser B, Paredis CJJ (2022) A framework for merging ontologies in the context of smart factories. Comput Ind 135:103571
19. Babalou B, König-Ries S (2020) Towards building knowledge by merging multiple ontologies with co merger. arXiv Prepr 2020
20. Xue X, Yang C, Jiang C, Tsai P, Mao G, Zhu H (2021) Optimizing ontology alignment through linkage learning on entity correspondences. Complexity 1:2021
21. de Roode M, Fernández-Izquierdo A, Daniele L, Poveda-Villalón M, García-Castro R (2020) SAREF4INMA: a SAREF extension for the industry and manufacturing domain. Semantic Web 11(6):911–926
22. Li L (2018) China's manufacturing locus in 2025: With a comparison of "Made-in-China 2025" and "Industry 4.0". Technol Forecast Social Change 135:66–74
23. Xingsi X (2019) An automatic biomedical ontology meta-matching technique. J Netw Intell 4(3):109–113
24. Zhu H, Xue X, Jiang C, Ren H (2021) Multiobjective sensor ontology matching technique with user preference metrics. Wireless Commun Mobile Comput 2021:5594553
25. Xue X, Wang H, Zhang J, Huang Y, Li M, Zhu H (2021) Matching transportation ontologies with Word2Vec and alignment extraction algorithm. J Adv Transp 2021:4439861
26. Osman I, Yahia SB, Diallo G (2021) Ontology integration: approaches and challenging issues. Inf Fusion 71:38–63
27. Salamon JS, Reginato CC, Barcellos MP (2018) Ontology integration approaches: a systematic mapping. In: ONTOBRAS 2018, 161–172
28. Salamon JS, Reginato CC, Barcellos MP (2018) Ontology integration approaches: a systematic mapping. In: ONTOBRAS, pp 161–172
29. Mohammed M, Romli A, Mohamed R (2021) Eco-design based on ontology: Historical evolution and research trends. In: AIP Conference Proceedings, vol. 2339, AIP Publishing LLC
30. Fernández-Izquierdo A, García-Castro R (2022) Ontology verification testing using lexico-syntactic patterns. Inf Sci 582:89–113
31. Tartir S, Arpinar IB, Sheth AP (2010) Ontological evaluation and validation. In: Theory and applications of ontology: Computer applications. Springer, Dordrecht, pp 115–130
32. Berners-Lee T, Chen Y, Chilton L, et al (2006) Tabulator: exploring and analyzing linked data on the semantic web. In: Proceedings of the 3rd international semantic web user interaction workshop, vol. 2006, p 159
33. Fonseca FT, Egenhofer MJ, Davis CA, Borges KAV (2000) Ontologies and knowledge sharing in urban GIS. Comput Environ Urban Syst 24(3):251–272
34. Lemaignan S, Siadat A, Dantan JY, Semenenko A (2006) MASON: a proposal for an ontology of manufacturing domain. In: IEEE Workshop on Distributed Intelligent Systems: Collective Intelligence and Its Applications (DIS'06). IEEE, pp 195–200
35. Aldana-montes JMJF (2011) Evaluation of two heuristic approaches to solve the ontology meta-matching problem. Knowl Inf Syst 26(2):225–247
36. Hooi YK, Hassan MF, Shariff AM (2014) A survey on ontology mapping techniques. Adv Comput Sci Appl 2014:829–836

37. Raunich S, Rahm E (2012) Towards a benchmark for ontology merging. In: OTM Confederated International Conferences "On the Move to Meaningful Internet Systems". Springer, Berlin, Heidelberg, pp 124–133
38. Konys A (2018) Knowledge systematization for ontology learning methods. Proc Comput Sci 126:2194–2207
39. Gracia J, Kernerman I, Bosque-Gil J (2017) Toward linked data-native dictionaries. In: Electronic Lexicography in the 21st Century: Lexicography from Scratch. Proceedings of the eLex 2017 conference, pp 19–21
40. Lv Y, Xie C (2010) A framework for ontology integration and evaluation. In: 2010 third international conference on intelligent networks and intelligent systems. IEEE, pp 521–524
41. Châabane S, Jaziri W, Gargouri F (2009) A proposal for a geographic ontology merging methodology. In: 2009 International Conference on the Current Trends in Information Technology (CTIT). IEEE, pp 1–6
42. Pileggi SF, Crain H, Yahia SB (2020) An ontological approach to knowledge building by data integration. In: International Conference on Computational Science. Springer, Cham, pp 479–493
43. Stoilos G, Stamou G, Kollias S (2005) A string metric for ontology alignment. In: International semantic web conference. Springer, Berlin, Heidelberg, pp 624–637
44. Ju SP, Esquivel HE, Rebollar AM, Su MC (2011) CreaDO—A methodology to create domain ontologies using parameter-based ontology merging techniques. In: 2011 10th Mexican International Conference on Artificial Intelligence. IEEE, pp 23–28
45. Cheatham M, Pesquita C (2017) Semantic data integration. In: Handbook of big data technologies. Springer, Cham, pp 263–305
46. Solimando A, Guerrini G, Jiménez-ruiz E (2017) Minimizing conservativity violations in ontology alignments: algorithms and evaluation. Knowl Inf Syst 51(3):775–819
47. Petrov P, Krachunov M, Todorovska E, Vassilev D (2012) An intelligent system approach for integrating anatomical ontologies: an intelligent system approach for integrating anatomical. Biotechnol Equip 26(4):3173–3181
48. Ndip-agbor E, Cao J, Ehmann K (2018) Towards smart manufacturing process selection in cyber-physical systems. Manuf Lett 17:1–5
49. Kumar J, Reddy S (2013) Implementation of ontology matching using Protégé. Int J Comput Appl Technol Res 2(6):723–725

Sentiment Analysis Online Tools: An Evaluation Study

Heider A. M. Wahsheh and Abdulaziz Saad Albarrak

Abstract A sentiment analysis tool interprets text chats and assesses each opinion's style, purpose, and feeling. The tool can better understand the context of users' discussions, allowing the client service team to classify client feedback accurately. This is especially valuable for companies that actively address clients' inquiries and complaints on social media, live chat, and email. Despite its vitality for business, there is still a challenge to decide the sentiment behind the content, especially for the Arabic language. Although most are not available for public usage, many sentiment analysis models and tools are developed in the literature. However, there is a lack of research identifying these tools' practicality for the Arabic language. This paper investigates two pure online Arabic sentiment analysis tools by employing a sizeable Arabic dataset in the experiments. Prediction quality measurements were utilized to assess these tools. The yielded results recommended Sentest SA as a promised tool for detecting sentiment analysis polarity for the preprocessed Arabic social network contents.

Keywords Sentiment analysis · Polarity · Prediction quality measurements · Experimental evaluation

1 Introduction

Social networks investigation has appeared as one of the most general research ideas, mainly due to the extensive daily social media posts. Powerful subproblems of social networks study contain sentiment analysis (SA) and intent detection on social network content [1]. Social networks are websites that provide billions of web users to share a common interest [2]. Social networks allow users to share files, photos, and

H. A. M. Wahsheh (✉) · A. S. Albarrak
Department of Information Systems, College of Computer Science and Information Technology, King Faisal University, P.O. Box 400, Al-Ahsa 31982, Saudi Arabia
e-mail: hwahsheh@kfu.edu.sa

A. S. Albarrak
e-mail: barrakas@kfu.edu.sa

M. Al-Emran et al. (eds.), *International Conference on Information Systems and Intelligent Applications*, Lecture Notes in Networks and Systems 550,
https://doi.org/10.1007/978-3-031-16865-9_9

videos, create posts, send messages, and conduct instant messaging conversations [3]. In recent years, social networking occupied a significant position in the virtual world. The increased number of social network users led to dynamically changing decision-making factors [2]. The diversity of social networks includes different platforms such as Snapchat, Instagram photo-sharing, Twitter, and Facebook. Various services and tools attracted many Internet users to use it to become the largest social media platform worldwide. Today, many worldwide who share and use the Internet for the first time have become a social networking platform, especially Twitter of the first experiences in using the Internet in general [4]. No one can deny the impact of social networking powers and growth on all countries, especially during the COVID-19 pandemic. It affects society, economy, politics, education, and other pillars of nation-building [5, 6]. Approximately 90% of Arab youths utilize social networks, compared to the international population usage of under 60%, according to [7].

Moreover, recent studies of the Arab world highlighted that 75% of social network users' consumption on Facebook, Instagram, Twitter, and TikTok had increased due to social distancing during and behind the COVID-19 pandemic [4]. Many organizations focus on collecting and extracting users' opinions for different fields, especially marketing and advertising, to understand the impact of these ideas on economists and public relations [8]. Sentiment analysis belongs to the data mining area that aims to understand, analyze, and extract the users' needs from their social comments or tags [9]. These days, sentiment analysis is considered the primary source of accurate information from many people without asking them to fill out direct surveys [10].

Sentiment analysis is one of the most fulfilled tasks in natural language processing (NLP), The significant difference between Arabic and English NLP is the preprocessing phase [11–13]. There are multiple developed sentiment analysis prototypes and models in the literature, but most are not available for use [14–17]. Despite the Arabic language being one of the world's most spoken languages, it receives little attention regarding online sentiment analysis tools and APIs [18, 19]. Two previous studies [18, 19] presented comparisons of online tools that support the Arabic language. This study investigates two pure online Arabic sentiment analysis tools [20, 21]. A sizeable Arabic dataset was applied in the experiments. Prediction quality measurements evaluated the results to find the best recommended online Arabic sentiment analysis tool among several data collections.

The remainder of this study is organized as follows. Section two presents the research methodology. Section three explores the experiments and evaluation performance. Section four illustrates the discussion. Section five concludes the paper and suggests future work.

2 Research Methodology

The primary purpose of this paper is to assess two pure Arabic sentiment analysis online tools among several datasets. The framework includes the following steps:

- Collect a dataset of Arabic social networks (i.e., Facebook and Twitter) textual content that uses Modern Standard Arabic (MSA) and slang in the Arabic language.
- Perform preprocessing steps and construct Arabic polarity (positive, negative, and neutral) lexicons and employ them to perform the class labeling of the collected dataset automatically.
- Conduct experiments to test two SA tools: Sentest [20] and Mazajak [21].
- Evaluate and compare Sentest [20] and Mazajak [21] results using prediction quality measurements.
- Discuss the yielded results and highlight the recommendation for Arabic sentiment analysis.

2.1 Arabic Social Networks Dataset Description

Some of the earlier studies in the literature collected data and labeled the polarity as negative, positive, or neutral manually [22, 23]. With the increase in the volume of comments and posts on social networks in various fields, and to evaluate the polarity performance of pure online Arabic sentiment analysis tools, it has become necessary to collect data automatically and determine its polarity. A crawler is developed to automatically build an Arabic social networks dataset of 21,000 Arabic comments. This crawler targets Twitter and Facebook users' tweets, posts, and comments depending on specific keywords related to the COVID-19 pandemic [5, 6]. The collected dataset contains modern textual standard Arabic (MSA) reviews, Arabic dialects (i.e., Jordan and Gulf countries), and emoticons. The total number of positive, negative, and neutral reviews was distributed equally to have a balanced dataset, with 7000 reviews for each polarity. The significant difference between Arabic and English NLP is the preprocessing phase [11–13]. For our collected dataset, we perform several preprocessing steps as follows [14]:

- Remove non-Arabic text, symbols, and punctuations.
- Normalize similar characters (i.e. (Alif, "أ، إ، ى، آ") to (Bare Alif, "ا"), (Taa', haa', "ة، ه") to (Haa', "ه"), (Yaa', "ى، ي ء") to (Yaa', "ي").
- Remove Kashida (extended letter): refers to ("Tatweel" تطويل," or "lengthened") which is a style of explanation in the Arabic language and some other scripts. The Unicode standard sets code point U + 0640, and it expands the length of particular words by using the elongation (ـ) in a font. For example, the term (Nice "جميـــل") is converted to the same term (Nice "جميل"), same meaning but without the lengthened.
- Remove Arabic stop words.
- Tokenize Arabic text.

We employed the polarity lexicons, including text and emoticons collected in [3], as 1000 positive, 1000 negative, and 350 neutral words/phrases. We set an algorithm to automatically label the collected dataset according to the polarity lexicons, mainly based on Term Frequency (*TF*). Figure 1 explores the adopted polarity algorithm.

In this paper, we considered every emoticon as a single feature; our lexicons convert the polarity for the words or phrases if the negation keywords [14] such as: (no, "لا ("and (not, "لم ") appeared in the text before them.

```
Input:
TR: Textual Review
PL: Set of Positive lexicon with emoticons.
NL: Set of Negative lexicon with emoticons.
NUL: Set of Neutral lexicon with emoticons.
Output:
PO: Polarity Outcome.
Initialization:
P_TF= 0, where P_TF is the TF for the positive review.
N_TF= 0, where Neg_TF is the TF for the negative review.
Neut_TF_W = 0, where Neut_TF is the TF for neutral review.
Begin
    Read TR
    For each TR:
    Remove punctuations from Arabic characters.
    Remove stop words.
    Normalize similar characters.
    Remove anextended letter
    Divide TR into w word tokens.
    For each w, Search for similar w in PL, NL.
    If w in PL, then
            P_TF = P_TF + 1
            PO= Positive
        Else If w in NL then
            N_TF = N_TF + 1
            PO = Negative
              Else
            Neut_TF = Neut_TF +1
            PO= Neutral
              End If
    End If
    End For
    If (P_TF> N_TF) then
        PO=Positive
    Else If (N_TF> P_TF) then
        PO = Negative
    Else
        PO = Neutral
    End If
    Write PO to the final result file.
    End For
    End
```

Fig. 1 Textual reviews polarity algorithm

Fig. 2 The Sentest main interface

2.2 *Arabic Sentiment Analysis Online-Tools*

This subsection presents the two pure Arabic Sentiment Analysis Online tools: Sentest [20] and Mazajak [21]. These two tools are dedicated only to the Arabic language, not like the previous studies in the literature [18, 19]. Sentest is a part of the Arabic Tools collection specializing in analyzing sentiments in Arabic texts. It categorizes results into three groups: positive, negative, or neutral, depending on the analysis of the entered text. It gives a percentage certainty of the decision of each sentence [20]. Figure 2 presents the simple main interface of Sentest, with an example of positive polarity (احسنتم شكرا لكم) means (Well done, thank you), which yielded 100 percent.

Mazajak is a free online Arabic sentiment analyzer based on a deep learning model which conducts accurate outcomes among several Arabic dialect datasets [21]. The Mazajak tool indicates one of three sentiment classification classes (positive, negative, neutral). Figure 3 presents the simple main interface of Mazajak, with an example of neutral polarity (لقاحات كورونا متعددة.) means (There are Multiple corona vaccines.).

Both Sentest and Mazajak have friendly and straightforward interfaces. Still, Mazajak offers several features, such as testing the sentiment analysis for each sentence or file of several sentences or submitting a Twitter account and getting an analysis of the user account. Mazajak allows user feedback after deciding on the polarity [21].

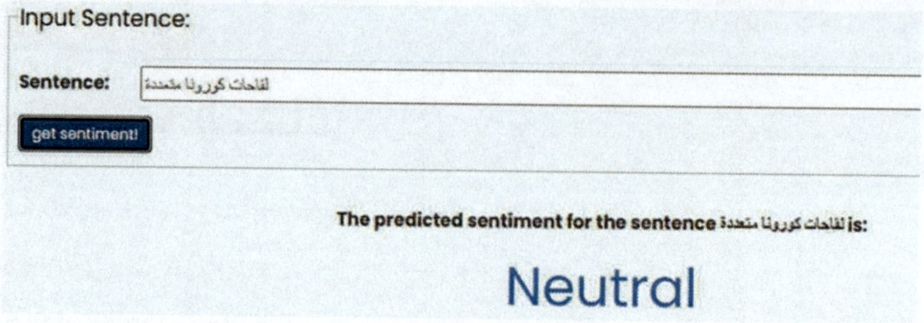

Fig. 3 Mazajak main interface

3 Experiments and Evaluation Performance

To evaluate the SA online tools' performance, we used the following measurement: Accuracy, True Positive (TP), True Negative (TN), False Positive (FP), False Negative (FN), Precision, Recall, and F-Measure (F-M) as shown in formulas (1)–(4) [1].

$$Accuracy_i = \frac{\text{Correctly Predicted}(TP + TN)}{\text{Total no of observations } (TP + FP + TN + FN)} \tag{1}$$

$$Recall_i = \frac{\text{Correctly predicted positive obsevations } (TP)}{\text{Actual observations } (TP + FN)} \tag{2}$$

$$Precision_i = \frac{\text{Correctly predicted positive values } (TP)}{\text{Total no of predictive positive observations } (TP + FP)} \tag{3}$$

$$F - measure = \frac{2(Recall * \Pr ecision)}{(Recall + \Pr ecision)} \tag{4}$$

The overall results showed that Sentest accuracy is better than Mazajak by more than 7%, yielding 84.76% and 77.34%, respectively, as shown in Tables 1 and 2.

In the detailed results, we can find that because of configuring the used dataset by removing normalization and Kashida, Sentest recognized all polarity classes with high accuracy results. Sentest incorrectly identified any keywords change as neutral

Table 1 Detailed results for Sentest SA Tool

Class	TP	FP	Precision	Recall	F-M
Positive	0.858	0.138	0.756	0.858	0.804
Negative	0.732	0.075	0.830	0.732	0.778
Neutral	0.953	0.016	0.968	0.953	0.961
Weighted AVG	0.848	0.076	0.852	0.848	0.848

Table 2 Detailed results for Mazajak SA Tool

Class	TP	FP	Precision	Recall	F-M
Positive	0.978	0.311	0.611	0.978	0.752
Negative	0.432	0.011	0.950	0.432	0.594
Neutral	0.911	0.017	0.963	0.911	0.937
Weighted AVG	0.774	0.133	0.842	0.774	0.761

without removing normalization or Kashida of the dataset. Mazajak tool was capable of classifying positive successfully with an accuracy of 97.8%, which is better than Sentest (85.8%). Mazajak tool detected neutral polarity with close accuracy results of Sentest, as 91.1% and 95.3%, respectively.

In contrast, Mazajak could not obtain high accuracy results for detecting negative class and yielded only 43.2%, as shown in Table 2. This might be due to the Mazajak dealing with the negation keywords and failing to consider them to convert positive words to negative meaning if they are used within content. We examine the highest overall TP values, precision, recall, and F-measure when comparing the tools.

On the other hand, the FP rate should be minimized. According to this, Tables 1 and 2 present that Sentest obtained better outcomes for all classes. The weighted average results recorded 0.848 for TP for both recall and 0.858 for precision. F-measure yielded 0.848 and less than 0.076 for the FP.

4 Discussion

The vogue of free online SA online tools and the minor studies prove that the reality of these tools rises to the present work. A larger Arabic dataset conducted the evaluation comparisons of two free Arabic SA online tools. We notice that Sentest did not perform normalization or Kashida, which are considered one of the main steps in the preprocessing phase for Arabic sentiment analysis research. We have already configured the used dataset with normalization and Kashida preprocessing before conducting the experiments. Otherwise, the Sentest would not have achieved good results for positive and negative classes since it does not preprocess the content and considers it neutral even if the content is positive or negative. The main important feature of Sentest is that it considers the Arabic negation words to represent all the words that negation features. Arabic negation keywords such as: (no, لا, and, not, لم) convert the sentiment polarity state to an opposite form.

The Sentest is missing features that make it more attractive to other researchers, such as allowing reading from a file. Mazajak appears more professional in design and accepts tasks from files or Twitter accounts. It adopts deep learning models and does not need to pre-configure data about the normalization process. The negative side of Mazajak did not convert the meaning when Arabic negation keywords appeared in the content. The most serious issue in social networks is that some content includes spam (irrelevant) information [24]. A large percentage of news in Arabic provides utterly false statements over social networks. There are several studies conducted and dedicated to the content of Arabic spam [25–32]. In these studies, the researchers underline the Arabic spam techniques such as keyword stuffing and attractive words. They mainly used spam links, content features, and behavior by machine and deep learning models to filter and detect these reviews. Further Sentiment analysis online tools should consider adopting the promising models and topic-reviews similarity approaches for spam detection methods. Exceptionally, spam content could be harmful not with false information but by propagating malicious content over social networks [33].

5 Conclusion and Future Works

Sentiment analysis is the main issue of text classification, and many algorithms attempt to categorize and identify the opinions into three main polarity types: positive, negative, and neutral. Using an Arabic social network data collection consisting of 21,000 tweets/comments, the study examines two online Arabic sentiment analysis tools. Prediction quality measurements were employed to evaluate these tools, and the obtained outcomes recommended the Sentest tool as a promised tool to be used if the text is preprocessed. Future work could expand the effort by utilizing additional commercial online tools among several datasets. Moreover, we aim to extend the study by using statistical parsing [34] and functional lexical grammar methods [35]. As well as discussing multiple social media topics such as news and sports will add valuable contributions [36].

Acknowledgements The authors acknowledge King Faisal University for the financial support.

References

1. Aakanksha S, Sinha GR, Bhatia S (2021) New opportunities for sentiment analysis and information processing. IGI Global
2. Al-Kabi M, Alsmadi I, Khasawneh RT, Wahsheh H (2018) Evaluating social context in Arabic opinion mining. Int Arab J Inf Technol 15:974–982
3. Al-Kabi MN, Wahsheh HA, Alsmadi IM (2016) Polarity classification of Arabic sentiments. Int J Inf Technol Web Eng 11:32–49
4. Radcliffe D, Abuhmaid H (2021) How the Middle East used social media in 2020. Available at SSRN 3826011
5. Mansoor M, Gurumurthy K, Prasad V, et al (2020) Global sentiment analysis of COVID-19 tweets over time. arXiv preprint arXiv:2010.14234
6. Alamoodi A, Zaidan BB, Zaidan AA, et al (2021) Sentiment analysis and its applications in fighting COVID-19 and infectious diseases: a systematic review. Expert Syst Appl 167:114155
7. Doaa Soliman (2021) Social media: a decade of leading change in the Arab world. https://p.dw.com/p/3ukhk. Accessed 1 Apr 2022

8. Valle-Cruz D, Fernandez-Cortez V, López-Chau A, Sandoval-Almazán R (2022) Does twitter affect stock market decisions? Financial sentiment analysis during pandemics: a comparative study of the H1N1 and the covid-19 periods. Cogn Comput 14:372–387
9. Lee SW, Jiang G, Kong HY, Liu C (2021) A difference of multimedia consumer's rating and review through sentiment analysis. Multimedia Tools Appl 80:34625–34642
10. Ligthart A, Catal C, Tekinerdogan B (2021) Systematic reviews in sentiment analysis: a tertiary study. Artif Intell Rev 54:4997–5053
11. Dolianiti FS, Iakovakis D, Dias SB (2019) Sentiment analysis on educational datasets: a comparative evaluation of commercial tools. Educ J Univ Patras UNESCO Chair
12. Khasawneh RT, Wahsheh HA, Alsmadi IM, AI-Kabi MN (2015) Arabic sentiment polarity identification using a hybrid approach. In: 2015 6th International Conference on Information and Communication Systems (ICICS). IEEE, pp 148–153
13. Al-Kabi M, Al-Qudah NM, Alsmadi I, Dabour M, Wahsheh H (2013) Arabic/English sentiment analysis: an empirical study. In: The Fourth International Conference on Information and Communication Systems (ICICS 2013), pp 23–25
14. Al-Kabi MN, Gigieh AH, Alsmadi IM, Wahsheh HA, Haidar MM (2014) Opinion mining and analysis for Arabic language. Int J Adv Comput Sci Appl 5:181–195
15. Khasawneh RT, Wahsheh HA, Al-Kabi MN, Alsmadi IM (2013) Sentiment analysis of Arabic social media content: a comparative study. In: 8th International Conference for Internet Technology and Secured Transactions (ICITST-2013). IEEE, pp 101–106
16. Al-Kabi M, Gigieh A, Alsmadi I, Wahsheh H, Haidar M (2013) An opinion analysis tool for colloquial and standard Arabic. In: The Fourth International Conference on Information and Communication Systems (ICICS 2013), pp 23–25
17. Al-Ayyoub M, Khamaiseh AA, Jararweh Y, Al-Kabi MN (2019) A comprehensive survey of Arabic sentiment analysis. Inf Process Manag 56:320–342
18. Rabab'Ah, AM, Al-Ayyoub M, Jararweh Y, Al-Kabi MN (2016) Evaluating sentistrength for Arabic sentiment analysis. In: 2016 7th International Conference on Computer Science and Information Technology (CSIT). IEEE, pp 1–6
19. Khafajeh H (2020) Opinion mining: How efficient are Online classification tools? Int J Emerg Trends Eng Res 8:557–567
20. Ali Salhi (2022) Arabic tools. https://www.arabitools.com/sentest.html. Accessed 1 Apr 2022
21. Farha IA, Magdy W (2019) Mazajak: an online Arabic sentiment analyser. In: Proceedings of the Fourth Arabic Natural Language Processing Workshop, pp 192–198
22. Al-Kabi M, Al-Ayyoub M, Alsmadi I, Wahsheh H (2016) A prototype for a standard Arabic sentiment analysis corpus. Int Arab J Inf Technol 13:163–170
23. Abdulla NA, Ahmed NA, Shehab MA, Al-Ayyoub M (2013) Arabic sentiment analysis: Lexicon-based and corpus-based. In: 2013 IEEE Jordan conference on applied electrical engineering and computing technologies (AEECT). IEEE, pp. 1–6
24. Najadat H, Alzubaidi MA, Qarqaz I (2021) Detecting Arabic spam reviews in social networks based on classification algorithms. Trans Asian Low-Res Lang Inf Process 21:1–13
25. Al-Kabi MN, Alsmadi IM, Wahsheh HA (2015) Evaluation of spam impact on Arabic websites popularity. J King Saud Univ Comput Inf Sci 27:222–229
26. Al-Kabi MN, Wahsheh HA, Alsmadi IM (2014) OLAWSDS: an online Arabic web spam detection system. Int J Adv Comput Sci Appl 5:105–110
27. Wahsheh HA, Al-Kabi MN, Alsmadi IM (2013) SPAR: a system to detect spam in Arabic opinions. In: 2013 IEEE Jordan Conference on Applied Electrical Engineering and Computing Technologies (AEECT). IEEE, pp 1–6
28. Alsmadi I, Al-Kabi MN, Wahsheh H, Bassam B (2013) Video spam and public opinion in current middle eastern conflicts. Int J Soc Netw Mining 1:318–333
29. Saeed RM, Rady S, Gharib TF (2022) An ensemble approach for spam detection in Arabic opinion texts. J King Saud Univ Comput Inf Sci 34:1407–1416
30. Abu-Salih B, Qudah DA, Al-Hassan M et al (2022) An intelligent system for multi-topic social spam detection in microblogging. arXiv preprint arXiv:2201.05203

31. Wahsheh HA, Al-Kabi MN, Alsmadi IM (2013) A link and content hybrid approach for Arabic web spam detection. Int J Intell Syst Appl 5(1):30–43
32. Sahoo SR, Gupta BB, Peraković D, Peñalvo FJG, Cvitić I (2022) Spammer detection approaches in online social network (OSNs): a survey. In: Sustainable Management of Manufacturing Systems in Industry 4.0. Springer, Cham, pp 159–180
33. Alsmadi M, Alsmadi I, Wahsheh HA (2022) URL links malicious classification towards autonomous threat detection systems. International Conference on Emerging Technologies and Intelligent Systems. Springer, pp 497–506
34. Al-Emran M, Zaza S, Shaalan K (2015) Parsing modern standard Arabic using Treebank resources. In: 2015 International Conference on Information and Communication Technology Research (ICTRC). IEEE, pp 80–83
35. Salloum SA, Al-Emran M, Shaalan K (2016) A survey of lexical functional grammar in the Arabic context. Int J Comput Netw Technol 4(3)
36. Salloum SA, Al-Emran M, Shaalan K (2017) Mining text in news channels: a case study from Facebook. Int J Inf Technol Lang Stud 1(1):1–9

Building Machine Learning Bot with ML-Agents in Tank Battle

Van Duc Dung and Phan Duy Hung

Abstract In recent years, Deep Reinforcement Learning has made great progress in video games, including Atari, ViZDoom, StarCraft, Dota2, and so on. Those successes coupled with the release of the ML-Agents Toolkit, an open-source that helps users to create simulated environments, shows that Deep Reinforcement Learning can now be easily apply to video games. Therefore, stimulating the creativity of developers and researchers. This research aspires to develop a new video game and turn it into a simulation environment for training intelligent agents. Experienced it with tuning the hyperparameters to make the agent getting the best performance for a final commercial video game product.

Keywords Reinforcement learning · Proximal policy optimization · ML-agents · Tank-game

1 Introduction

Reinforcement learning (RL), one of a training method of machine learning that is inspired by the way in which humans and animals learn and adapt to the environment. The basic working principle of this method is based on the reward and agent received through the results of a sequence of actions. That is to say, the agent learns by trial and error, and the reward guidance behavior obtained through interaction with the environment aims to make the Agent get the maximum reward [1]. In some aspects, it is comparable to supervised learning in that developers must offer algorithms well defined goals as well as set rewards and punishments. Therefore, explicit programming is a more mandatory requirement. In the process of training, the algorithm will be provided with very little information. So RL usually has a longer time to reach the

V. D. Dung (✉) · P. D. Hung
Computer Science Department, FPT University, Hanoi, Vietnam
e-mail: dungvdhe141196@fpt.edu.vn

P. D. Hung
e-mail: hungpd2@fe.edu.vn

M. Al-Emran et al. (eds.), *International Conference on Information Systems and Intelligent Applications*, Lecture Notes in Networks and Systems 550,
https://doi.org/10.1007/978-3-031-16865-9_10

optimal solution than other methods. In this way, RL improves the strategy mainly through its experience in exploring the environment and making mistakes [2].

In recent years, we have seen many breakthroughs in artificial intelligence. Almost 25 years ago, an AI had defeated the strongest chess player for the first time in history, surprising the whole world [3]. Twenty years later, in 2016, AphalGO, a computer once again beat humans at Go. A board game whose total number of moves could be more than the number of atoms in the universe, a thing that was once thought to be impossible [4]. Not stopping there, two years later, OpenAIFive was developed to play a game even more hardened: Dota2. A real-time strategy game with a complexity of several tens to several hundred times Go and chess [5]. OpenAI has opened a new era for the artificial intelligence industry with many possibilities.

To create OpenAIFive, the OpenAI team introduced a new class of reinforcement learning algorithms called Proximal Policy Optimization (PPO), which outperforms state-of-the-art techniques while being significantly easier to deploy and tweak [6].

Given an environment that delivers valuable and realistic observations for an agent, reinforcement learning produces excellent results. The environment design requires an easy and highly configurable tool to imitate real-world ideas and test researchers' theories. Unity, one of the most popular gaming engines globally, bills itself as an ecosystem that offers a global real-time platform with detailed physics and complete usability to meet research demands. Engineering, entertainment, customer service, and other fields use the research outputs, which subsequently appear in instructional simulators and mobile or VR applications with multi-platform compatibility [7].

In order to provide all the necessary information for agents and meet the needs of research and easy environment creation, Unity has published ML-Agents toolkits. It is open-source that allows researchers and developers to create an emulator environment on the Unity editor for interacting with them through a python API. The toolkit helps us define objects and events in the environment handled by C# scripts which then log and connect to the python algorithm. One of the critical components of the toolkit is Soft Actor-Critic (SAC) and PPO, which this research will utilize [8]. Although PPO is a state-of-the-art approach, in many cases, especially when the interaction in the environment becomes complex, it will be difficult for the agent to find the optimal solution. For example, in the very first learning stage, the agent exploration is represented by random actions, which may lead to sparse rewards. In numerous instances, the sparseness of the rewards can make the agent hardly improve its policy and get stuck in random actions loop. We can add more rewards to instruct the agent on such complex problems. Or we can start from a simpler environment and then gradually increase its complexity. This concept, called Curriculum Learning, has been shown to reduce training time and quality of local minima significantly [9]. In ML-Agents Toolkit, environment parameters may be added and changed during the training process. A curriculum is made of a sequence of lessons triggered by certain completion requirements. Each criterion should have a threshold to decide when the lesson ends for the chosen measure (e.g., cumulative reward or step progress). It is also possible to choose a minimum lesson duration and signal smoothing. Overall, a good curriculum lesson will result in less training time and better optimal behavior.

This paper aims to study how RL acts as a robot under Unity's ML-Agents. Specific tasks include target aim, collection of objects, and obstacle avoidance. We designed a new environment and made incremental improvements when we included DRL in the problem. Implementations include environment design, learning process and algorithm tuning for the best possible results. Then, we consider the possibility of trained intelligent agents as an alternative to hand-scripted bots for diverse interactions to player for a better commercial video game product.

2 Methodology

Self-play can be used with implementations of both Proximal Policy Optimization and Soft Actor-Critic. However, because the opponent is always changing, many scenarios appear to exhibit non-stationary dynamics from the viewpoint of a solitary Agent. Self-play has a high risk on causing serious problems with SAC's experience replay system. As a result, users are advised to utilize PPO [10].

2.1 *Environment Design*

Tank Battle plays out on a square map surrounded by four walls with two tanks shooting each other. Each tank has to move around the map to find the enemy, avoid rocks, take health packs, and align the cannon angle accurately; the game ends when one of them is eliminated or the time runs out. When the time runs out, that match is considered a draw. There are two main parts of the tank, the body and the turret (Fig. 1).

Fig. 1 Turret and body of the tank

- **Body.** The tank can move like a standard 4-wheel car, including actions: forward, backward, turn left, turn right. However, in this study, to reduce the complexity, the Agent will always move forward and cannot stop (can still turn left or right 20 degrees) and only automatically goes back for a fixed time after colliding with an obstacle.
- **Turret.** The turret is fixed on the vehicle's body and can rotate 360 degrees. (include two actions: rotate clockwise and counterclockwise). In addition, there is a cannon on the turret, from which the bullets are fired. Cannon can adjust the angle up and down to 5 and –5 degrees. Therefore, to accurately shoot the target, the Agent needs to skillfully align both the angle of the turret and the cannon. To aid in accurate aim, a ray cast from the cannon beams straight in the direction it is facing to the first object it hits, indicating the distance from the cannon to that object.

2.2 *Environment Learning*

Although the game is designed for humans to receive information through visual input (Fig. 2), the Agent observes the environment through numbers to minimize calculation and neural networks complexity. The game is designed for players to control the tank from a third-person perspective using input devices like mouses and keyboards. On the other hand, the Agent observes the environment through position, vector to the enemy, and distance provided by the Unity game engine at each time step (Table 1). It is considered to normalize all components of the agent's Vector Observations for a best practice when using neural networks, so all information is adjusted to range $[-1, +1]$. For a sequence of acts that lead to a match win, we give the Agent a reward (or a punishment). Table 2 lists all of the outcomes rewards that we identify. In experiment, we maximize the reward function that includes extra signals such as colliding with obstacles and collecting health packs. When computing the reward function, we also use a method to take advantage of the problem's zero-sum construction—for example, we symmetrize rewards by deducting the reward gained by the enemy.

For tracking obstacles and finding health packs, the Agent used RayPerception Sensor whose total size of: (Observation Stacks) $*$ $(1 + 2 *$ Rays Per Direction) $*$ (Num Detectable Tags $+ 2) = 1 * (1 + 2 * 5) * (2 + 2) = 44$ (Fig. 3).

During inference mode, the agent's policy will determine the actions that map the current situation based on the information gathered from Vector Observation and Ray Perception Sensor. The reward in reinforcement learning is an indication that the agent has made right series of actions. According to these rewards, the PPO algorithm optimizes the agent's decision to maximize the cumulative reward over time. The training is divided into Episodes, each Episode is a Tank Battle match. When a match ends, all environments and reward points will be reset and a new Episode begin.

Fig. 2 Tank battle's human "observation space"

Table 1 Vector observation

Current position (x, z)	2
Current health percent	1
Turret's vector direction (x, z)	2
Vector from itself to enemy (x, z)	2
Fire bullet cooldown	1
Distance from the cannon to the first object that raycast hits	1
Cannon angle	1
Enemy's current health percent	1
Enemy's velocity (x, z)	2
Distance to enemy	1
Total	14

Table 2 Shaped reward weights

Name	Reward	Description
Shooting accurately	0.1	Each bullet that hits the enemy will get a reward
Collect a health pack	3	
Collide with obstacle	– 1	Collide with walls or rocks
Turret direction	0.003	Every step if the turret's direction is facing the enemy
Penalty per step	– 0.0001	This penalty is applied every step for making the Agent kill the enemy faster
Win	2	

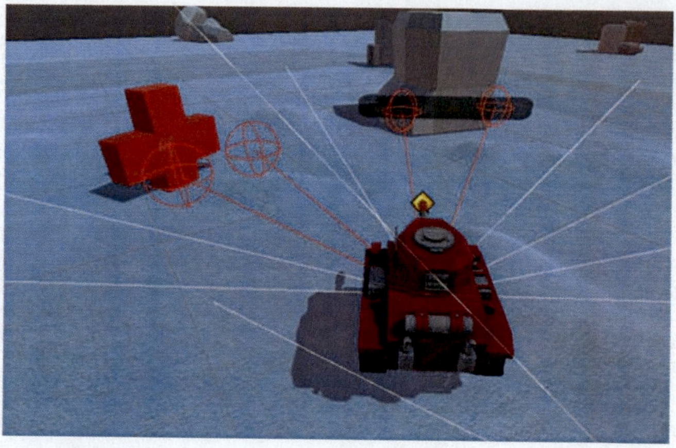

Fig. 3 Ray perception sensor

3 Experiments and Results

The statistics were saved by ML-Agents Toolkit and monitored via TensorBoard during the learning lesson. It gives us the ability to track and evaluate the learning process through data that has been visualized. Over the whole step count, a graph illustrates each separate training run with chosen metrics.

In the first lesson of Curriculum Learning, the environment will not contain rocks as obstacles for the agent to learn to shoot and not hit walls only. After about 3 million steps, the mean reward is at its peak. The environment starts to add some obstacles, increasing the amount gradually proportional to the mean reward. (Fig. 4).

In Fig. 4, the reward starts from 0, gradually increases to a peak of 4 in between steps 1 M and 2 M, then gradually stabilizes and maintains the oscillation amplitude from around 3. This result happens because there are not only the rewards received

Fig. 4 Tracking of environment metrics (cumulative reward)

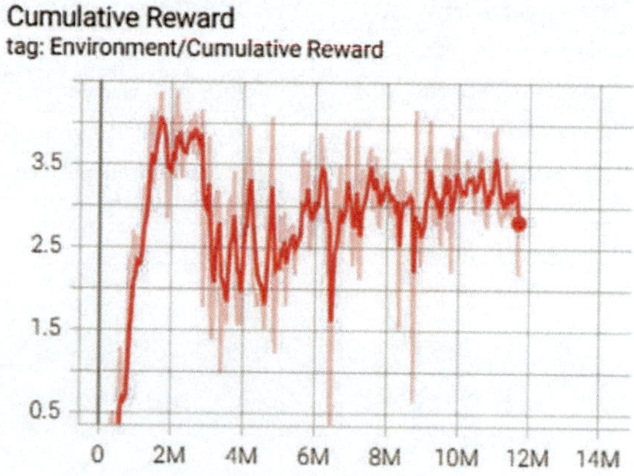

after each right action. The Agent also gets a + 2 reward for each game they win. With self-play, the opponent of the Agent will be the most recent version of itself (defined by play_against_latest_model_ratio = 0.65). When the policy improved, the Agent's opponents grew more assertive, making each episode ending in win/lose more pronounced.

Because of this reason that it is not reliable to evaluate policy improvement through the Cumulative Reward metric, the ML-Agent toolkit provides users with another metric to evaluate Agents in self-play called the ELO rating system. However, to use it, the Agent's reward must be designed in a zero-sum game, and the structure of winners with a positive reward, negative for losers, and 0 for a tie. This type of reward has been implemented by using 'SetReward()' to negative two if the Agent loses. Unfortunately, this implementation makes the learning unstable. Experiments show that after training the Agent to learn the game's basic rules in the first lesson of Curriculum Learning, the Agent knew to turn the cannon at the enemy and avoid the wall to optimize the reward. But later on, somehow the above reward shape made the Agent behavior become weird. They did not spin the turret in the right direction of the enemy anymore. They just roamed around in the environment and shot aimlessly. Agent evaluation becomes more difficult without the ELO metric because empirical observations must be applied more frequently. The mean length of the episode (Fig. 5) shows that Agents are killing each other much faster, meaning they are learning to shoot more precisely. However, after adding obstacles, projectiles are regularly blocked, causing the episode's length to increase dramatically and decrease over time.

Entropy, which measures the unpredictability of Agents' decisions, is another critical metric for evaluating the policy. As the training progresses, it steadily declines, indicating a well-selected beta hyperparameter. According to Fig. 6, the more training Agent has, the less random actions Agent will have.

One important note is the Normalize hyperparameter in the configuration file. This hyperparameter is recommended to use only when there are continuous actions. It is even said to be harmful with more straightforward discrete control problems.

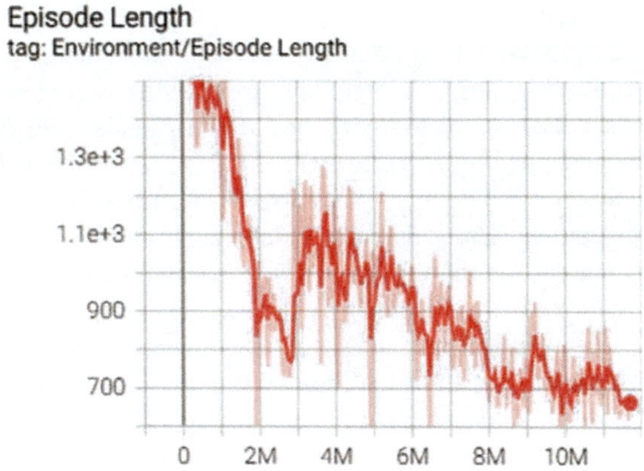

Fig. 5 Tracking of environment metrics (episode length)

Fig. 6 Entropy metrics

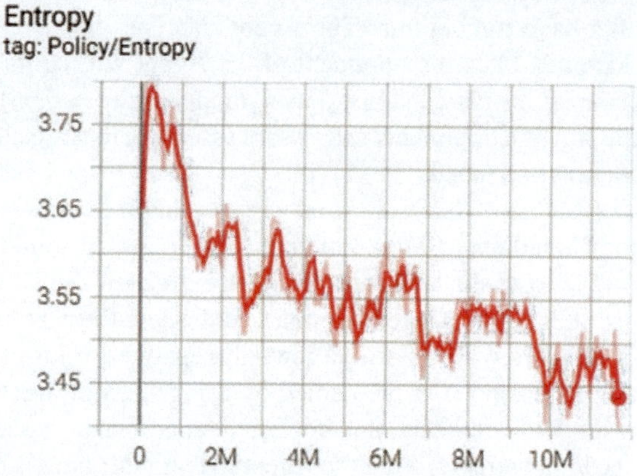

In comparison, all of the actions in this study are purely discrete actions. Experiments show that, after only about the first 250 k steps, the neuron network somehow converges fast to some weird local minimum. Making the Agent's behavior selects only one action in each action branch. Expressly, they only turn in one direction, go in a circle, and constantly rotate the turret clockwise. They do not even fire any bullets. This issue is entirely resolved after the hyperparameter switches to True.

4 Conclusion and Future Works

This study demonstrates the performance and possibilities of intelligent agent training by ML-Agents Toolkits. The Agent was able to learn the basic rules of the game quickly. It can avoid obstacles and walls, collect health packs, and face its turret toward the enemy. However, the way the Agent observes their surroundings is not visual observations, which is very costly, making shooting a complex problem. As humans play the game through a screen and control their tank by keyboard and mouse, they can effortlessly aim and shoot precisely to trounce the Agent. Although we can make the Agent to do even better if we increase the hidden units and improve its observations, it is quite hard for the Agent to play the game as good as human. The reason is due to limitations of ML-Agents itself. We can configure the training by changing Hyperparameters in the configuration file but interfering in the neural network too deeply is not allowed. Therefore, we can conclude that ML-Agents Toolkit and Unity engine still have high potential for commercial in video games. However, the more complex the environment is, the harder the agent to learn. So causal games are most likely the best suit for this commercial due to its simplicity.

We would like to add more agents and make Tank Battle a Cooperative game for further work. In addition to shooting each other and collecting health packs, agents on the same team can also fire special bullets to heal teammates and diversify interactions and tactics. Moreover, we will also alternate entirely current the vector observation to visual observation by adding a camera following the turret so that the Agent can learn the ability to aim more precisely, and apply the RL methods to machine learning problems such as [11–13].

References

1. Sutton RS, Barto AG, Williams RJ (1992) Reinforcement learning is direct adaptive optimal control. IEEE Control Syst Mag 12(2):19–22. https://doi.org/10.1109/37.126844
2. Li Y (2017) Deep reinforcement learning: an overview. arXiv:1701.07274
3. Hsu FH (2002) Behind deep blue: building the computer that defeated the world chess champion. Princeton University Press
4. Silver D, Hubert T, Schrittwieser J, Antonoglou I, Lai M, Guez A, Lanctot M, Sifre L, Kumaran D, Graepel T, Lillicrap T, Simonyan K, Hassabis D (2018) A general reinforcement learning algorithm that masters chess, shogi, and go through self-play. Science 362(6419):1140–1144
5. Open AI et al (2019) Dota 2 with large scale deep reinforcement learning. arXiv:1912.06680
6. Schulman J, Wolski F, Dhariwal P, Radford A, Klimov O (2017) Proximal policy optimization algorithms. arXiv:1707.06347
7. Xie J (2012) Research on key technologies base Unity3D game engine. In: Proceedings of the 7th International Conference on Computer Science & Education (ICCSE), pp 695–699. https://doi.org/10.1109/ICCSE.2012.6295169
8. Juliani A et al (2020) Unity: a general platform for intelligent agents. arXiv:1809.02627
9. Bengio Y, Louradour J, Collobert R, Weston J (2009) Curriculum learning. In: Proceedings of the 26th Annual International Conference on Machine Learning (ICML '09). Association for computing machinery, New York, NY, USA, pp 41–48. https://doi.org/10.1145/1553374.1553380
10. Foerster J, Nardelli N, Farquhar G et al (2017) Stabilising experience replay for deep multi-agent reinforcement learning. In: Proceedings of the 34th International Conference on Machine Learning, vol. 70 (ICML'17). JMLR.org, pp 1146–1155
11. Hung PD, Giang DT (2021) Traffic light control at isolated intersections in case of heterogeneous traffic. In: Kreinovich V, Hoang Phuong N (eds) Soft computing for biomedical applications and related topics. Studies in computational intelligence, vol 899. Springer, Cham. https://doi.org/10.1007/978-3-030-49536-7_23
12. Hung PD (2020) Early warning system for shock points on the road surface. In: Luo Y (eds) Cooperative design, visualization, and engineering. CDVE 2020. Lecture Notes in Computer Science, vol 12341. Springer, Cham. https://doi.org/10.1007/978-3-030-60816-3_33
13. Su NT, Hung PD, Vinh BT, Diep VT (2022) Rice leaf disease classification using deep learning and target for mobile devices. In: Al-Emran M, Al-Sharafi MA, Al-Kabi MN, Shaalan, K (eds) Proceedings of International Conference on Emerging Technologies and Intelligent Systems. ICETIS 2021. Lecture Notes in Networks and Systems, vol 299. Springer, Cham. https://doi.org/10.1007/978-3-030-82616-1_13

An Insight of the Nexus Between Psychological Distress and Social Network Site Needs

Mei Peng Low and Siew Yen Lau

Abstract The passage of time has brought mankind to a seamless communication universe with informational technologies and social network sites (SNS). This study examines the correlation between psychological distress and SNS among the general public. Five SNS needs were examined. Quantitative research design specifically a cross-sectional approach with a self-administered questionnaire was used to reach to the pool of respondents. Purposive sampling method was applied. A total of 210 responses were collected from Malaysians aged 18 and above. The findings reveal that overall psychological distress has led to the SNS needs with personal integrative needs ($\beta = 0.332$) emerged as the core needs followed by diversion need ($\beta = 0.241$), affective needs ($\beta = 0.239$), social interactive needs ($\beta = 0.210$) and cognitive needs ($\beta = 0.197$). While bulk of the studies examines the use of SNS leading to psychological distress, the current study empirically relates psychological distress as the antecedents of SNS usage. The findings offer insights to the respective authorities and mental associations for drawing up recouping strategies and programs to cope with mental health issues via SNS.

Keywords Social Network Sites (SNS) · SNS needs · Psychological distress

1 Introduction

The internet is a product of technological innovation that connects the global wide area network and computer systems worldwide [1]. The advent of new technological revolution has augmented the internet's functions to be more visible and influential. As a result, people are greatly impacted by technological innovation. Recently,

M. P. Low (✉) · S. Y. Lau
Faculty of Accountancy and Management, Universiti Tunku Abdul Rahman, Selangor Kajang, Malaysia
e-mail: lowmp@utar.edu.my

S. Y. Lau
Department of International Business, Universiti Tunku Abdul Rahman, Selangor Kajang, Malaysia

M. Al-Emran et al. (eds.), *International Conference on Information Systems and Intelligent Applications*, Lecture Notes in Networks and Systems 550,
https://doi.org/10.1007/978-3-031-16865-9_11

social network via the internet platforms, namely social network sites have become prevalent. In fact, the influences have cascaded to economic advancement, human life, and social development [2]. The evolution of internet and social network sites coupled with technological innovations are immersing into all aspects of human society, extending to international relations, and the international strategic pattern. The work by Al-Qaysi, Mohamad-Nordin, and Al-Emran [3] have highlighted the perverseness of SNS in particular the Facebook usage.

SNS was an internet-based service that allowed users to establish a public or semi-public profile within a limited system, articulate a list of other users with whom they share a connection, and get their list of connections within the system [4]. The dynamic nature of the internet has altered the definition of SNS over the last 25 years. Recently, Aichner et al. [4] defined SNS as a networked communication platform in which participants (1) have profiles that are uniquely recognizable based on user-supplied information, the information given by other users, and/or system information; (2) describe openly relationships that others can observe and explore; and (3) consume, create, and/or engage with user-generated content streams given by others. Currently, the SNS users worldwide has accounted for more than half of the world population of 7.9 billion [5]. In Malaysia, there are 27.43 million of SNS users which accounts for 86% of the Malaysia total population [5, 6].

Year 2020 was unthinkable that hit hard by COVID-19 pandemic. Various measures were implemented by the governments such as nationwide lockdown, cessations of public activities and social distancing practices. People and organizations worldwide have had to adjust to new norms of work and life. Accordingly, a new phenomenon is observed with an inevitable surge of digital technologies demand and internet usages [7]. These changes come along with numerous social challenges such as general public's mental health and internet addictions [8].

According to the World Health Organization (WHO) [9], mental health encompasses subjective well-being, self-perceived, freedom, competency, interpersonal relying, and self-actualization of one's mental and moral capacity, among others. WHO describes mental health as a condition of well-being in which the individual realizes his or her abilities, able to cope with the usual demands of life, able to work successfully and meaningfully, and ability to contribute to a particular group.

The COVID-19 attack have exacerbated to rising mental health issues such as suicide cases and self-harming acts. The Royal Malaysian Police have reported an astounding number of 468 suicides between January and May of 2021 [10]. The figure indicates that there is average three suicide cases each day which has tripled the number in 2020. The alarming statistics deserve some immediate attention.

Putting the pervasiveness of SNS and COVID-19 pandemic together, the development has inseminated many research interests. Against this background, we explore the possible correlation between psychological distress and SNS needs as part of the digital surge scenarios during the pandemic.

2 Literature Review

Past relevant studies were examined to develop current research. The concerns of psychological distress was referred through various medical journals such as *International Journal of Mental Health and Addiction* and journals from *US National Library of Medicine National Institutes of Health.* For SNS needs, *Cyberpsychology, Behavior, and Social Networking Journal, Computers in Human Behavior, Telematics and informatics Journals* were examined to build the research idea and variables.

2.1 *Psychological Distress*

Psychological distress is a widespread mental health issue in the population [11]. It is an emotional discomfort caused by daily pressures and obligations that are difficult to manage. Generally, emotional discomfort are typified by exhaustion, depression and anxiety symptoms [12]. These symptoms frequently cohabit with typical somatic complaints, chronic illnesses, and medically unexplained disorders. When an individual encounters excessive demands and inadequate support from external factors, and simultaneously experiences lack of internal control, psychological distress would occur.

World Health Organization [8] enlightened the five psychological distress features displayed by patients are perceived incapacity to cope, changes in an emotional state, suffering, communication of irritation, and self-harm. These features could be reflected in six fundamental daily idioms of low morale and pessimism about the future, suffering and pressure, self-depreciation, social retreat and isolation, somatization and self-back down [11]. Failure to properly identify and seek immediate treatment can lead to chronicity, attempt suicide and tragedy.

2.2 *Social Network Sites (SNS) Needs*

According to Chen [13], SNS has emerged as a need in everyday interpersonal interactions. People are increasingly concerned about the considerable impacts of SNS in numerous aspects of their lives including social difficulties, performance decline, interference with school, family, and job, and mental issues. In fact, Wang et al. [14] confirmed a reciprocal link between the passive use of SNS and subjective well-being. Passive SNS use may be harmful to subjective well-being since it lacks social support and may elicit envy and jealously.

Referring to Katz et al. [15]'s earlier work, there are five needs people acquired from mass media, specifically diversion, cognitive, personal integrative, social integrative, and affective needs. Lately, Ali et al. [16] and Sharif [17] adopted the same five needs to expound on SNS needs.

2.2.1 Diversion Needs

Diversion needs are also known as tension free needs. Cressey and McDermott [18] and McQuail [19] described diversion needs as "escape from boredom or challenges, as well as an emotional release." People listen to music and access social media to reduce tension or to pass time when they are bored. Also, people may have numerous pressures in their lives that they do not want to confront, therefore they use media to escape from them. As such, one of the SNS needs is diversion needs.

2.2.2 Cognitive Needs

Cognition refers to the mental processes involved in learning and comprehension [20]. Thinking, knowing, remembering, analyzing, and problem-solving are examples of cognitive processes. These are the higher-level brain processes that include language, imagination, perception, and planning [20]. Meanwhile, cognitive psychology is the set of behavioral individuals thinking mechanism and processes that occur during cognition. People utilize social media to obtain information and to satisfy their mental and intellectual requirements [16]. Often, people watch the news to satisfy this cognitive desire. Likewise, people join social groups in SNS to search for information. Hence, SNS is a mean to meet the needs for knowledge, understanding, curiosity, exploration, predictability, creativity, and discovery that represents the intellectual desire.

2.2.3 Personal Integrative Needs

Personal integrative needs include self-esteem and respect. People want reassurance to build their position, trustworthiness, strength, and authority, which is accomplished via the use of SNS. They utilize SNS to watch commercials and learn which items are in vogue, and they adapt appropriately to modify their lifestyle and fit in with others. Besides, gratifications acquired from SNS use also include the methods of reinforcing particular ideals [21]. In this vein, people rely on SNS to meet their desire for self-esteem [22] by rescuing their status, to gain respect, credibility, confidence, stability as 5well as power [23].

2.2.4 Affective Needs

Affective needs refer to the emotional fulfilment and pleasure that people obtain from SNS. Typically, affective needs focus on awareness and growth in attitudes emotions, and feelings [24]. The affective domain describes people's emotional reactions and their capacity to sense the delights or suffering of others [25]. Often, people are identified with the characters and the emotions they exhibit. If they experience sorrow, the audience will feel sad along with them, and if they are happy and joyful, the audience will share the similar mood with them.

2.2.5 Social Integrative Needs

Aristotle, the Greek philosopher once said that human beings are "social creatures" and naturally seek the companionship of others as part of their well-being. The sayings reinforced in the social integrative needs to interact and socialize with family, friends, and society. Social integrative needs are based on individual connection and interaction with the outside world [26]. People utilize SNS to connect, to interact and to improve their social connections with their friends, family and alliances by discussing various issues. SNS fulfils the social integrative needs by presenting a platform and avenue for individuals to connect, to discuss subjects, to contribute ideas and to give opinions among their networks [17].

2.3 Uses and Gratification Theory and Hypotheses Development

Uses and Gratification Theory (UGT) by Katz et al. [15] explains how and why people are actively seeking out specific types of media. The central focus of UGT is "What do people do with media?" and "Why do people use media?" [19, 27]. Following the scholarly research by Sundar and Limperos [28] and Gil de Zúñiga et al. [29], they unanimously informed that people receive gratifications through media that fulfil their social, informational and leisure needs. Applying to current psychological distress conditions as the consequence of lockdown and social distancing, UGT is used to examine the correlations between psychological distress and the five SNS needs.

From a therapeutic perspective, when people encounter a stressful state of mind, it is recommended to attempt a diverting activity to mitigate the stress level. According to Orchard et al. [30] social maintenance and freedom of expression are some of the motivations for SNS usage. With this, we hypothesize that people face with physiological distress are diverting the negative emotions toward SNS usage. H1 is developed.

H1: Psychological distress leads to SNS diversion needs.

Cognitive psychology describes the set of behaviors relate to the effort of understanding and exploring to fulfil our curiosity and predictability. This intellectual seeking effort is known as the cognitive needs. According to Phua et al. [31, 32], people increasingly embrace SNSs as tools for communication and information purposes. We are of interest to uncover the plausible relations between physiological distress and cognitive needs via the SNS usage in H2.

H2: Psychological distress leads to SNS cognitive needs.

Personal integrative needs are construed as the self-esteem need. People use media to reassure their status, gain confidence and credibility. Park et al. [32, 33] found that one of the reasons for users to participate in Facebook groups is self-status. Therefore, we hypothesize that people encounter psychological distress use SNS to regain their confidence and status. With this, H3 is formed.

H3: Psychological distress leads to SNS personal integrative needs.

Affective needs relate to sentiments, strengthening aesthetic, and emotional experience. It encompasses all kind of emotions and moods which sought for gratification through SNS. Likewise, study by Phua et al. [31] also informed that SNS is used to meet the emotional and social desires. H4 is developed to investigate the correlation between psychological distress and affective needs.

H4: Psychological distress leads to SNS affective needs.

Social interaction needs reflect the nature of humankinds that needs interaction and not isolation. Gil de Zúñiga et al. [29] explained that SNS usage led to enhanced social interaction, knowledge, diversion, escapism and civic participation. We hypothesize the social interaction needs is a natural mean when people encounter with psychological distress. H5 is produced.

H5: Psychological distress leads to SNS social interaction needs.

2.4 Research Framework

Against the backdrop set forth, the following framework is posited to proceed with current research (Fig. 1).

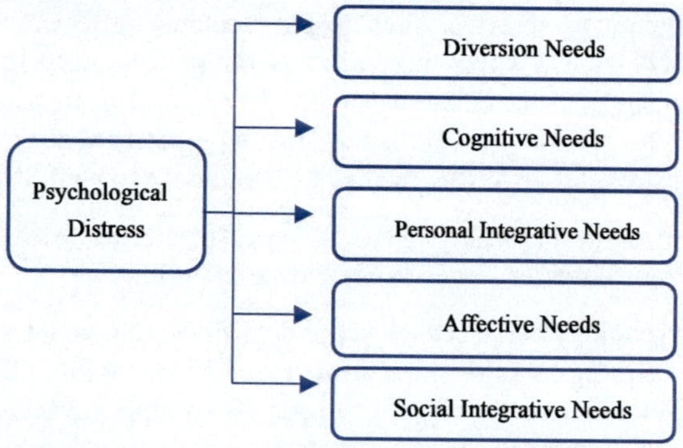

Fig. 1 Research framework

3 Research Methodology

Quantitative research design specifically cross-sectional approach through the use of self-administered questionnaire was operationalized in this study. As a symmetrical sampling was not the main concern in this study, purposive sampling method was applied. Targeted respondents were contacted and given the explanation of the research objectives before seeking for their voluntary participation. The data collection took three months and successfully collected a total of 210 responses from the Malaysians aged 18 and above.

The questionnaire was structured in three sections; respondents' demographic profile; experience of psychological distress and SNS needs. Hopkins Symptom Checklist (HSCL-10) from Yuan [34] was adopted to measure psychological distress while the five SNS needs were adopted from Ali et al. [16]. The respondents were required to rate their level of agreement based on Five-point Likert statements in the questionnaire. The complexity of the path modeling in SNS needs justified the use of Partial least square structural equation modeling (PLS-SEM) in performing the statistical analysis [35].

4 Research Findings

Table 1 provides an overview of the respondents' profiles. The majority of respondents are in the age groups of 18–39 years old (81.43%) with females made up 57.62% of the total polled. Most of respondents are with upper secondary school qualifications (32.38%) and degree (29.05%). The employed (41.43%) and self-employed (20.0%) dominated the responses.

Hair et al. [35] recommended that the analysis of PLS-SEM approach begins with the measurement model assessment before proceeding to structural mode assessment. Measurement model assessment entails reliability assessment that encompasses variables factor loadings, composite reliability (CR), and average variance extracted (AVE). In term of validity, discriminant validity was assessed using heterotrait–monotrait (HTMT) as suggested by Henseler et al. [36].

Table 2 shows that all the measurement items surpass the recommended threshold for factor loading, Cronbach's Alpha, CR and AVE. The HTMT in Table 3 informed that none of the HTMT values were greater than 0.90 [37, 38]. Henceforth, it concludes that measurement reliability and discriminant validity for the present study had been established.

Prior to assessing the structural model, the issue of collinearity was addressed using variance inflated factor (VIF) [38]. Table 5 indicates that all the VIF values below 3.3, informing the absence of collinearity in the model.

Thereon, bootstrapping procedure was performed using 1,000 resampling to generate the t-values to measure the statistical significance of the path coefficients.

Table 1 Respondents' profile

Demographic	Value	Frequency	Percentage (%)
Age	18–29 years old	91	43.33
	30–39 years old	80	38.10
	40–49 years old	24	11.43
	50–59 years old	13	6.19
	60 and above	2	0.95
Gender	Female	121	57.62
	Male	89	42.38
Educational level	Primary school	3	1.43
	Lower secondary	14	6.67
	Upper secondary	68	32.38
	Pre-university	17	8.10
	Diploma	40	19.05
	Bachelor degree	61	29.05
	Post graduate	2	0.95
	Others	5	2.38
Occupation	Student	63	30
	Employed	87	41.43
	Self-employed	42	20.00
	Unemployed	11	5.24
	Retired	7	3.33
Living area	Urban area	167	79.52
	Rural area	12	5.71
	Suburban area	31	14.76

The results of path co-efficient assessment is presented in Table 4 in which all the proposed hypotheses (H1 to H5) were found to be significant with p value < 0.05.

Subsequently, R^2, the variance explained in the dependent constructs, i.e., the five SNS needs, Q^2 predictive relevance and f^2 effect size were also being examined and the results are shown in Table 5. Overall, the R^2 for SNS needs are below 0.100 except personal interactive needs is 0.108, which indicates that 10.8% of the variance in personal interactive needs can be explained by psychological distress. Meanwhile, the overall Q^2 values are larger than 0 indicate that exogenous constructs possess predictive capacity over psychological distress. The results further show that among all the exogenous constructs, psychological distress has the medium effect on personal integrative needs ($f^2 = 0.121$) while others have low effect size (f^2 ranging from 0.004 to 0.064).

Table 2 Measurement model assessment

Construct/item	Factor loading > 0.7	Cronbach's Alpha > 0.8	CR > 0.7	AVE > 0.5
Diversion needs				
D1	0.788	0.860	0.902	0.697
D2	0.832			
D3	0.851			
D4	0.865			
Cognitive needs				
C1	0.902	0.920	0.943	0.805
C2	0.901			
C3	0.930			
C4	0.855			
Personal integrative needs				
PI1	0.875	0.913	0.938	0.790
PI2	0.914			
PI3	0.876			
PI4	0.891			
Affective needs				
A1	0.710	0.859	0.904	0.703
A2	0.907			
A3	0.854			
A4	0.869			
Social integrative needs				
SI1	0.875	0.891	0.924	0.754
SI2	0.912			
SI3	0.823			
SI4	0.861			
Psychological distress				
PD1	0.881	0.952	0.959	0.704
PD2	0.793			
PD3	0.894			
PD4	0.739			
PD5	0.897			
PD7	0.887			
PD8	0.873			
PD9	0.890			
PD10	0.879			

Table 3 HTMT discriminant validity

	1	2	3	4	5	6
Affective needs						
Cognitive needs	0.805					
Diversion needs	0.890	0.854				
Personal integrative needs	0.841	0.626	0.729			
Psycho distress	0.240	0.200	0.249	0.333		
Social integrative needs	0.890	0.880	0.889	0.732	0.226	

Table 4 Hypotheses testing

Hypothesis	Path coefficient	Standard deviation	T statistics	P values	Decision
H1: psychological distress -> diversion needs	0.241	0.056	4.281	0.000	Supported
H2: psychological distress -> cognitive needs	0.197	0.060	3.280	0.001	Supported
H3: psychological distress -> personal integrative needs	0.332	0.060	5.492	0.000	Supported
H4: psychological distress -> affective needs	0.239	0.055	4.354	0.000	Supported
H5: psychological distress -> social integrative needs	0.210	0.058	3.640	0.000	Supported

Table 5 Structural model assessment: collinearity, coefficient of determination, predictive relevance and effect size

Construct	VIF	R^2	R^2 Adj	Q^2	f^2
Affective needs	1.001	0.055	0.050	0.033	0.058
Cognitive needs	1.023	0.039	0.034	0.028	0.004
Diversion needs	1.069	0.060	0.055	0.033	0.064
Personal integrative needs	1.369	0.108	0.104	0.075	0.121
Social integrative needs	1.410	0.046	0.041	0.031	0.048

5 Discussion and Conclusion

The research was examined the correlations between psychological distress and SNS needs. The results demonstrate a positive relationship between psychological distress and the five SNS needs. The findings explicate that when people experience psychological distress, they use SNS to fulfil their needs. However, among the five SNS needs that we had examined, personal integrative needs is the strongest needs followed by diversion needs and affective needs. The findings show consistency with the UGT.

The fact behind the significant personal integration needs during psychological distress could be attributed to the speed of information dissemination [39]. By using SNS, it can reach a large number of audiences in a short period of time. Therefore, it was used as an avenue to meet personal integration needs. In term of the diversion needs, it is related to the concept of escapism. According to Wu et al. [40], diversion needs is also known as escapism by engaging in activities that are absorbing to the point of offering an escape from unpleasant realities, problems, and pressures. Hence, this offers an explanation to the correlation between psychological distress and SNS diversion needs. Meanwhile, recent research by Pang [41] highlighted the positive affective values of mobile social media. Drawing from the hedonic values, SNS users' affective responses underline emotional profits and self-sufficiency. Hence, a positive relationship is posited between psychological distress and affective needs.

The research findings produce two conclusions. First, there is a positive relationship between psychological distress and SNS needs. Second, psychological distress arouses the SNS usage as it enables the fulfilment of different types of SNS needs. With majority of the respondents were dominated by Gen Y and Z, it was observed that when psychological distress attack, they used SNS to meet the personal integrative needs, diversion needs and affective needs but less on cognitive needs and social integrative needs. These findings could serve good insights to mental health association and social network sites policy makers to cultivate a healthy mindset in the society as well as tackling the concern of rising suicide cases during the pandemic. Some of the past studies have indicated the dark side of SNS, however, current research enlightens that SNS could serve a practical platform for counselling too.

Despite that this research had provided some informative insights of the correlation between psychological distress and SNS needs, it suffers from a few shortcomings. The main flaw stem from the sample size in the context of societal well-being research. Notwithstanding that this research follow the guidelines of the recommended sample size, yet in order to generalize the findings, a larger pool of responses would be beneficial for social well-being context. In addition, current research does not embrace the uniqueness potential arise from diverse demographic profile. It will be of interest to conduct a multigroup analysis by segmenting various demographic such as age, race, income levels to obtain more comprehensive findings. To further validate the findings, it is also recommended to use weighted PLS (WPLS) algorithm to attain better average population evaluations when a set of appropriate weight is possible [42].

Future researchers may desire to address these shortcomings and further expand to scope of data collection from many sources to validate the information gained. In-depth interviews with respondents would be beneficial, particularly because the psychological distress component varies depending on the situation and background.

References

1. Attaran M (2017) The internet of things: limitless opportunities for business and society. J Strateg Innov Sustain 12(1):11
2. Ternes K, Iyengar V, Lavretsky H, Dawson WD, Booi L, Ibanez A, Eyre HA (2020) Brain health Innovation Diplomacy: a model binding diverse disciplines to manage the promise and perils of technological innovation. Int Psychogeriatr 32(8):955–979. https://doi.org/10.1017/S1041610219002266
3. Al-Qaysi N, Mohamad-Nordin N, Al-Emran M (2020) Employing the technology acceptance model in social media: a systematic review. Educ Inf Technol 25(6):4961–5002
4. Aichner T, Grünfelder M, Maurer O, Jegeni D (2021) Twenty-five years of social media: a review of social media applications and definitions from 1994 to 2019. Cyberpsychology Behav Soc Networking 24(4):215-222. https://doi.org/10.1089/cyber.2020.0134
5. Statista (2021) Active social network penetration in selected countries and territories as of January 2021. https://www.statista.com/statistics/282846/regular-social-networking-usage-penetration-worldwide-by-country/
6. Simon K (2021) Digital 2021: Malaysia. Data reportal, February 2021. https://datareportal.com/reports/digital-2021-malaysia
7. Pandey N, Pal A (2020) Impact of digital surge during Covid-19 pandemic: a viewpoint on research and practice. Int J Inf Manage 55:102171. https://doi.org/10.1016/j.ijinfomgt.2020.102171
8. Li YY, Sun Y, Meng SQ, Bao YP, Cheng JL, Chang XW, Shi J (2021) Internet addiction increases in the general population during COVID-19: evidence from China. Am J Addict 30:389–397. https://doi.org/10.1111/ajad.13156
9. World Health Statistics (2021) Monitoring health for the sustainable developments goals. World Health Organization. https://cdn.who.int/media/docs/default-source/gho-documents/world-health-statistic-reports/2021/whs-2021_20may.pdf?sfvrsn=55c7c6f2_8
10. Hani (2021) 468 suicide cases in the first five months of 2021, 1 July 2021. https://themalaysianreserve.com/2021/07/01/468-suicide-cases-in-the-first-five-months-of-2021/. Accessed 14 Aug 2021
11. Arvidsdotter T, Marklund B, Kylén S, Taft C, Ekman I (2016) Understanding persons with psychological distress in primary health care. Scand J Caring Sci 30(4):687–694. https://doi.org/10.1111/scs.12289
12. Costa DK, Moss M (2018) The cost of caring: emotion, burnout, and psychological distress in critical care clinicians. Ann Am Thorac Soc 15(7):787–790. https://doi.org/10.1513/AnnalsATS.201804-269PS
13. Chen A (2019) From attachment to addiction: the mediating role of need satisfaction on social networking sites. Comput Hum Behav 98:80–92. https://doi.org/10.1016/j.chb.2019.03.034
14. Wang JL, Gaskin J, Rost DH, Gentile DA (2018) The reciprocal relationship between passive social networking site (SNS) usage and users' subjective well-being. Soc Sci Comput Rev 36(5):511–522. https://doi.org/10.1177/0894439317721981
15. Katz E, Haas H, Gurevitch M (1973) On the use of the mass media for important things. Am Sociol Rev 164–181. https://doi.org/10.2307/2094393
16. Ali I, Danaee M, Firdaus A (2020) Social networking sites usage & needs scale (SNSUN): a new instrument for measuring social networking sites' usage patterns and needs. J Inf Telecommun 4(2):151–174. https://doi.org/10.1080/24751839.2019.1675461
17. Sharif EAM (2020) Uses and Gratifications (U&G) and UTAUT3: understanding the use of the Social Networking Site (SNS)-Facebook among Senior Citizens. Int J Adv Res Technol Innov 2(3):13–23. http://myjms.mohe.gov.my/index.php/ijarti/article/view/10815/5076
18. Cressey DE, McDermott RA (1973) Diversion: background and definition. University of Michigan Press, Ann Arbor
19. McQuail D (1972) The television audience: a revised perspective. In: Sociology of mass communications, pp 135–165

20. Alibali MW, Nathan MJ (2018) Embodied cognition in learning and teaching: action, observation, and imagination. In: International handbook of the learning sciences. Routledge, pp 75–85
21. Rauschnabel PA (2018) Virtually enhancing the real world with holograms: an exploration of expected gratifications of using augmented reality smart glasses. Psychol Mark 35(8):557–572. https://doi.org/10.1002/mar.21106
22. Xie Y, Qiao R, Shao G, Chen H (2017) Research on Chinese social media users' communication behaviors during public emergency events. Telematics Inform 34(3):740–754. https://doi.org/10.1016/j.tele.2016.05.023
23. Lin YH, Hsu CL, Chen MF, Fang CH (2017) New gratifications for social word-of-mouth spread via mobile SNSs: uses and gratifications approach with a perspective of media technology. Telematics Inform 34(4):382–397. https://doi.org/10.1016/j.tele.2016.08.019
24. Casey A, Fernandez-Rio J (2019) Cooperative learning and the affective domain. J Phys Educ Recreat Dance 90(3):12–17. https://doi.org/10.1080/07303084.2019.1559671
25. Meishar-Tal H, Pieterse E (2017) Why do academics use academic social networking sites? Int Rev Res Open Distrib Learn 18(1):1–22. https://doi.org/10.19173/irrodl.v18i1.2643
26. Gleeson DM, Craswell A, Jones CM (2019) Women's use of social networking sites related to childbearing: an integrative review. Women Birth 32(4):294–302. https://doi.org/10.1016/j.wombi.2018.10.010
27. Menon D, Meghana HR (2021) Unpacking the uses and gratifications of Facebook: a study among college teachers in India. Comput Hum Behav Rep 3:100066. https://doi.org/10.1016/j.chbr.2021.100066
28. Sundar SS, Limperos AM (2013) Uses and grats 2.0: new gratifications for new media. J Broadcast Electron Media 57(4):504–525
29. Gil de Zúñiga H, Jung N, Valenzuela S (2012) Social media use for news and individuals' social capital, civic engagement and political participation. J Comput Mediat Commun 17(3):319–336
30. Orchard LJ, Fullwood C, Galbraith N, Morris N (2014) Individual differences as predictors of social networking. J Comput Mediat Commun 19(3):388–402
31. Phua J, Jin SV, Kim JJ (2017) Uses and gratifications of social networking sites for bridging and bonding social capital: a comparison of Facebook, Twitter, Instagram, and Snapchat. Comput Hum Behav 72:115–122
32. Al-Qaysi N, Mohamad-Nordin N, Al-Emran M (2020) What leads to social learning? Students' attitudes towards using social media applications in Omani higher education. Educ Inf Technol 25(3):2157–2174
33. Park N, Kee KF, Valenzuela S (2009) Being immersed in social networking environment: Facebook groups, uses and gratifications, and social outcomes. Cyberpsychol Behav 12(6):729–733
34. Yuan H (2021) Internet use and mental health problems among older people in Shanghai, China: The moderating roles of chronic diseases and household income. Aging Ment Health 25(4):657–663. https://doi.org/10.1080/13607863.2020.1711858
35. Hair JF Jr, Matthews LM, Matthews RL, Sarstedt M (2017) PLS-SEM or CB-SEM: updated guidelines on which method to use. Int J Multivar Data Anal 1(2):107–123. https://doi.org/10.1504/IJMDA.2017.087624
36. Henseler J, Ringle CM, Sarstedt M (2015) A new criterion for assessing discriminant validity in variance-based structural equation modeling. J Acad Mark Sci 43(1):115–135
37. Gold AH, Malhotra A, Segars AH (2001) Knowledge management: an organizational capabilities perspective. J Manag Inf Syst 18(1):185–214
38. Diamantopoulos A, Siguaw JA (2006) Formative versus reflective indicators in organizational measure development: a comparison and empirical illustration. Br J Manag 17(4):263–282
39. Ayodele OS, Atanda AA (2020) Study of the use of website and social networking sites as public relations dialogic tools in universities in Kogi State Nigeria. Media Commun Curr 4(2):149–170. http://journals.unimaid.edu.ng/index.php/mcc/article/view/113
40. Wu J, Holsapple C (2014) Imaginal and emotional experiences in pleasure-oriented IT usage: a hedonic consumption perspective. Inf Manag 51(1):80–92

41. Pang H (2021) Identifying associations between mobile social media users' perceived values, attitude, satisfaction, and eWOM engagement: the moderating role of affective factors. Telematics Inform 59:101561
42. Low MP, Cham TH, Chang YS, Lim XJ (2021) Advancing on weighted PLS-SEM in examining the trust-based recommendation system in pioneering product promotion effectiveness. Qual Quant 1–30

Factors Influencing the Intention to Adopt Big Data in Small Medium Enterprises

Ahmed F. S. Abulehia, Norhaiza Khairudin, and Mohd Hisham Mohd Sharif

Abstract Making a data-driven decision is not just the forte of big business. Even small and medium businesses can benefit from big data. These days, companies are making adjustments to their business model to incorporate big data. Therefore, companies want to reap these fruits, big data set helps analyse and reveal trends, patterns, and correlations. as to whether the company is connected to the Internet or not, they need information that helps them to grow and prosper in their work, and here comes the role of big data. In the current research, the researcher discusses the factors that help to adopt big data in SMEs in Palestine. The researcher approached quantitative statistical analysis and (TOE) theory was adopted to build the study model. The measurement tool, which is the questionnaire, was built to collect data. The study consisted of 310. The SmartPLS program was used to test the hypotheses. The results indicated that there is significant relationship between technological, organizational, and environmental factors and the intention to adopt big data in SMEs in Palestine, except the governmental support.

Keywords Big data adoption · TOE · SMEs · Palestine

1 Introduction

During the fourth industry revolution, Big Data problems have become one of the most important issues and trend for years [1]. The use of big data analytics (BDA) has revolutionized the way businesses compete. To make better decisions, enhance

A. F. S. Abulehia (✉) · N. Khairudin · M. H. M. Sharif
School of Accountancy, Universiti Utara Malaysia (UUM), 06010 Sintok, Kedah, Malaysia
e-mail: ahmed.abulehia@gmail.com

N. Khairudin
e-mail: norhaiza@uum.edu.my

M. H. M. Sharif
e-mail: hisham79@uum.edu.my

M. Al-Emran et al. (eds.), *International Conference on Information Systems and Intelligent Applications*, Lecture Notes in Networks and Systems 550,
https://doi.org/10.1007/978-3-031-16865-9_12

output, produce knowledge, and improve ideas, it shows how to find hidden patterns in a set of raw data [2].

Big Data not only created for large companies, but business of all sizes have the opportunities to benefits from Big Data [3]. In most countries, SMEs make a significant contribution to the economy. Particularly in Palestine, 95% of economic enterprises are SMEs, which accounts for more than 55% of GDP and pro-vide more than half of private sector employment [4]. Nowadays more and more SME are seeking information technology (IT) solutions to big data management [1, 5].

Big Data is presented as a cutting-edge IT, marvel or a solution based on several technologies [6]. Big data is the information asset characterized by its Volume (much larger than traditional data sets), Velocity (the rapid speed with which it is produced and available), Variety (of formats in particular), Variability (over time and diversity of sources), and Volatility (inconsistent levels of production) [7]. This technology provides businesses with the tools and procedures they need to handle large amounts of unstructured and structured data for a variety of objectives [8].

According to a recent analysis by the Global Organization for Analytics on big data adoption, since 2014, more executives have started to grasp the potential commercial advantages of analytics and have expedited their adoption of big data initiatives [9]. The business climate in emerging economies had changed considerably, and there was fierce market rivalry. Consequently, the introduction of Big Data has altered the way business' function and compete, given this competitive environment, businesses have been compelled to use a variety of cutting-edge Information Technologies (IT) to better their company operations and performance [10].

Despite the importance of SMEs to a country's economy, the rate of Big Data adoption in SMEs is low [11], Since Big data adoption is particularly difficult for SMEs [12], As a result of technological, organizational, and environmental factors which includes the main indicators that might affect the intention to adopt Big Data. In addition, there aren't many studies looking at the factors that influence SMEs in their adoption of BDA [13, 14].

According to Dubey et al. (2020) companies need Big Data to enhance their organizational performance and SMEs without Big Data are hardly to stay competitive in the global market. Moreover, Big Data is gaining traction, reshaping business paradigms and revealing new paths from insights to value [6]. In addition, BDA, or the analysis of structured and unstructured data from customers and devices to do business, is a recognized area for IT innovation and investment [15].

As a result, the current study trying to investigate the determinants of BDA adoption among SMEs to encourage SMEs to adopt Big Data to sustain their performance and stay competitive in the local and global market. This study uses technological-organizational-environmental (TOE) paradigm. The TOE model's strength is its adaptability in describing the "technological factors (Relative Advantage, Compatibility, Complexity) and organizational factors (Top Management Support, & Organizational readiness), in addition to the Environmental Factors (Competitive Pressure and Government Support)" [16]. Based on the above, the research question is:

RQ1. To what degree may TOE contexts influence BDA uptake among Palestinian SMEs?

2 Literature Review

2.1 *Big Data Analytics (BDA)*

Although there is no universally acknowledged definition for "big data," it is useful to consider some of the most often accepted definitions. Business intelligence and analytics were the topics on which we decided to base our working definitions for this project (BDA). The optimization of your big data infrastructure is important from both a technical and a business standpoint. According to some experts, Big Data refers to a new generation of technologies and architectures that are aimed at economically utilizing large data volumes of information [17]. While technical components are concerned with the technology and how it might be used to accomplish the desired result, business components are concerned with the application of innovative approaches to assist corporate executives in gaining a competitive advantage. Big Data is a term that refers to massive Data sets or information movements that have been acquired from a variety of sources. However, integrating data from several sources into a single source is a difficult task [18]. Others say big data is a technical tool or kind of business analytics that helps organizations handle enormous amounts of data quickly [10]. A company's operations may be improved, new insights gained, activities accelerated, and economic value created with big data [19].

2.2 *Theoretical Background*

The technical factors examine a technology's endogenous and exogenous qualities that influence adoption. For example, a company's perceived value from new technology may influence its adoption intentions [20, 21]. In business, the relative advantage is the degree to which a company's use of technology is superior to other companies' use of current technology. According to Ghobakhloo et al. (2011), SMEs are willing to adopt new technologies if the advantages outweigh the present ones.

Compatibility assesses a new system's compatibility with the existing system. A company's culture and business operations are reflected in the use of technology, Verma and Bhattacharyya (2017) recognized compatibility as a primary driver of technology adoption. Compatibility is also a significant predictor of BDA adoption. Firms should improve their rule and process flexibility to improve BDA compatibility the results show that SMEs are more likely to adopt and utilize BDA if it aligns with existing organizational procedures and standards.

For example, new technology or system may not acquire momentum if seen as too ambitious or difficult to implement. Changing the way people collaborate is difficult, therefore the "new technology must be easy to use to get adoption" [22]. Employees must quickly understand new technologies since the adoption process is uncertain and challenging with modern technology. Complexity influences the adoption of an invention leaving decision-makers undecided Compared to other technical qualities of innovation, complexity has a negative association with adoption [23]. According to a recent study on big data adoption, complexity hinders adoption, SMEs are less likely to accept innovation if they perceive it would require a lot of time and effort.

The trialability of an IT innovation [24] is measured. It is crucial for early adopters, such as SMEs, who know the innovation's effectiveness from the outset [23]. As a consequence, it allows early innovators to decrease uncertainty and claim that the sooner an innovation is revealed, the better.

"The extent to which an invention's consequences are visible to others," says observability [25]. Describe observability as "the process by which firms perceive the success factor of other enterprises that have previously exploited big data." Observability's influence on technological adoption has been studied extensively, with mixed findings. Observability was recognized as a determinant of innovation adoption inside firms by Kapoor et al. (2014). Meanwhile, Siew et al. (2020), found a significant correlation between observability and techno-logical adoption. One of the factors influencing BDA adoption in grocery stores is observability. Research has shown a modest correlation between technology uptake and observability [26].

Organizational Factors

Environmental variables are components of the environment that organizations may confront [27]. Firms are more susceptible to the external dynamic ecology. The TOE model predicts that external variables such as competition, government regulations and support may affect SMEs' acceptance of BDA. Chen et al. (2015) defines competitive heaviness as "influences from the external environment that drives the business to use BDA." It's the pressure from customers, suppliers, and competitors. Observed that new technology is more effective when Firms are being pressed to compete on a global scale. Grandon and Pearson (2004) found that competition affects technology acceptance in SMEs in five out of ten cases. It is unclear if environmental law affects Egyptian SMEs' sustainable manufacturing practices or not, some studies think that growing BDA usage by competitors will force owners and managers to gather business information and analytics properly and professionally to stay competitive [28] (Fig. 1).

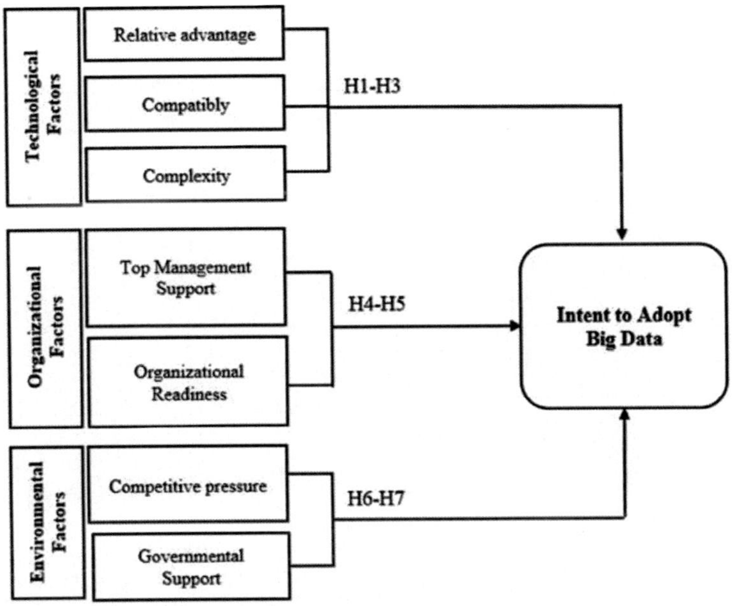

Fig. 1 The model of study

3 Research Method

The questionnaire is adopted in this research, as research is quantitative and was conducted to see the factors that affect the adoption of big data in medium and small companies in Palestine Where the number of small and medium-sized companies in Palestine, based on records, and the study sample was 310 from managers of small and medium-sized companies and CEOs.

4 Data Analysis

We used SmartPLS version 3.3.2 for partial least squares (PLS) Modelling. To analyze the statistical data by collecting the answers of company managers to the questionnaire, where the questionnaire was distributed using Google Forms. Convergent and Discriminant Validity models are used in the initial step of testing [29]. In order to verify that the model was valid and reliable, the research team moved on to testing the structural model.

If a given item properly assesses the latent construct it is designed to measure, it has convergent validity [30]. For assessing convergent validity, the item loadings were analyzed to see whether they were above or below the 0.7 thresholds. It was also necessary to look at the Avg. variance extracted (AVE) and Composite Reliability

(CR). It was determined that both the AVE and CR were above the acceptable levels of 0.5 and 0.7 for the corresponding metrics. The Results of all latent variables' item loadings are shown in Table 1 below. Convergent validity of latent constructs is thus confirmed.

Table 1 Cross loading analysis

Constructs	Items	Factor loadings	Cronbach's Alpha	CR	(AVE)
Intention to adopt big data	IABD-1	0.787	0.838	0.892	0.674
	IABD-2	0.814			
	IABD-3	0.843			
	IABD-4	0.837			
Relative advantage	RA-1	0.782	0.866	0.899	0.599
	RA-2	0.727			
	RA-3	0.776			
	RA-4	0.795			
	RA-5	0.796			
	RA-6	0.763			
Compatibility	CMP-1	0.837	0.898	0.929	0.766
	CMP-2	0.875			
	CMP-3	0.894			
	CMP-4	0.895			
Complexity	CPX-1	0.766	0.841	0.893	0.677
	CPX-2	0.827			
	CPX-3	0.857			
	CPX-4	0.839			
Top management support	TMS-1	0.804	0.813	0.877	0.64
	TMS-2	0.819			
	TMS-3	0.792			
	TMS-4	0.786			
Organizational readiness	OR-1	0.772	0.815	0.878	0.643
	OR-2	0.754			
	OR-3	0.863			
	OR-4	0.815			
Competitive pressure	CP-1	0.816	0.827	0.897	0.744
	CP-2	0.905			
	CP-3	0.865			
Governmental support	GS-1	0.885	0.876	0.924	0.801
	GS-2	0.893			
	GS-3	0.907			

Table 2 Discriminant validity HTMT

	Compatibility	Competitive pressure	Complexity	Governmental support	Intention to adopt big data	Organizational readiness	Relative advantage
Compatibility							
Competitive pressure	0.841						
Complexity	0.79	0.84					
Governmental Support	0.763	0.913	0.798				
Intention to adopt big data	0.851	0.953	0.885	0.872			
Organizational readiness	0.508	0.592	0.48	0.562	0.623		
Relative advantage	0.774	0.923	0.792	0.939	0.925	0.579	
Top management support	0.808	0.876	0.909	0.885	0.955	0.513	0.907

Table 3 Discriminant validity (Fornell-Larcker's test)

	Compatibility	Competitive pressure	Complexity	Governmental support	Intention to adopt big data	Organizational readiness	Relative advantage	Top management support
Compatibility	**0.875**							
Competitive pressure	0.728	**0.863**						
Complexity	0.69	0.703	**0.823**					
Governmental support	0.68	0.778	0.689	**0.895**				
Intention to adopt big data	0.741	0.795	0.747	0.752	**0.821**			
Organizational readiness	0.438	0.486	0.403	0.48	0.518	**0.802**		
Relative advantage	0.69	0.786	0.682	0.819	0.795	0.491	**0.774**	
Top management support	0.694	0.719	0.755	0.745	0.79	0.418	0.763	**0.8**

Table 4 Structural model estimates (path coefficients)

Hypo	Relationships	Std. beta	Std. error	T-value	P-values	Decision
H1	Relative advantage -> intention to adopt big data	0.203	0.065	3.117	0.002	Supported
H2	Compatibility -> intention to adopt big data	0.136	0.05	2.701	0.007	Supported
H3	Complexity -> intention to adopt big data	0.144	0.047	3.061	0.002	Supported
H4	Top management support -> intention to adopt big data	0.23	0.061	3.764	0.000	Supported
H5	Organizational readiness -> intention to adopt big data	0.095	0.037	2.6	0.010	Supported
H6	Competitive pressure -> intention to adopt big data	0.218	0.067	3.268	0.001	Supported
H7	Governmental support -> intention to adopt big data	0.007	0.06	0.109	0.914	Not supported

There are two ways to measure validity: discriminant and cross-validation. HTMT was investigated to make certain that it has discriminant validity. Henseler et al. (2015) first advocated the measure, which was then approved and revised by Franke and Sarstedt (2019). The maximum HTMT value that should be used is 0.90. Table 2 shows the HTMT findings, and it's clear that they all fall inside the range of acceptable values. Thus, each and every structure is different from the others. Using the measurement model, it was found that the constructs were both reliable and valid.

Hair et al. (2014) recommend that the skewness and kurtosis of the items be used to test for multivariate normality. It was observed that the data was not normal, following [31]. In the multivariate analysis, skewness and kurtosis are statistically significant at less than 0.05. As a result, the model's path coefficients, standard error, t values, and p values were all reported in accordance with [32]. 310 samples were used in the bootstrapping process. Path coefficients, p-values, and t-values were used to test the hypotheses. In addition, the magnitude of the impact was considered. Table 4 summarizes all of the criteria that were satisfied (Table 3).

5 Hypothesis Testing

Using a tenfold technique to evaluate predictive significance, Shmueli et al. (2019) proposed using PLS predict to generate case-level predictions. If there is a little difference between the items in PLS-SEM, the predictive significance is validated; on the other hand, if the difference is large, it is not. However, if most of the differences are low, the predictive power is weak, and the opposite is true if the majority of the differences are large (Fig. 2).

Fig. 2 Testing of hypotheses

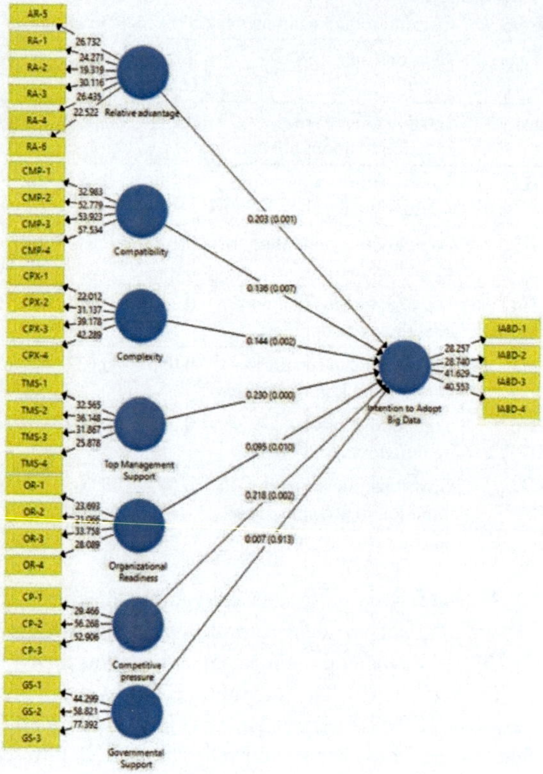

6 Discussion and Conclusion

This study adds to the body of knowledge on big data adoption by examining the several factors considered to be involved and exploring the correlations between these variables and organizational intent to adopt. The study has been contributed to the related causal paths, or configurations of antecedents, which can influence the intention to adopt big data, based on data has been collected from managers and owners of SMEs in Palestinian Territories. TOE models were used to develop the research's theoretical model. The findings show that elements from the technological, organizational, and environmental settings all influence organizational adoption of big data. Thus, our key findings contribute to a better understanding of the big data diffusion process.

In the technology context, relative advantage, compatibility, and complexity are the most significant factors influencing intention to adopt Big Data, although their roles in the big data adoption process differ considerably. For example, complexity may exert a greater influence on intention to adopt Big Data than compatibility. As previous studies have indicated [33–35]. Our findings demonstrate that executives have more confidence when making an adoption choice and embracing new information technology when they have a relative advantage. Top management support is a crucial factor influencing big data adoption intentions. This finding is consistent with Gangwar (2018) and L. Wang et al. (2018) of literature that defines the relationship between organizational factors and adoption.

In addition, regarding to environmental context the study found that competitive pressure significantly supports the intention to adopt Big Data. However, the results found no evidence to suggest that governments support is significant in line with Yadegaridehkordi et al. (2018) and [36]. Unlike previous studies on government support toward adopting Big Data [37], our results indicate that government support is not a key factor that significantly influences big data adoption.

From practical perspectives, current study would be useful for decision-makers within the manufacturing sector by offering a guideline for policymakers in developing countries such as Palestine. Further, the current study also supports the notion that practitioners must first initiate a coherent and unambiguous data-driven culture and infrastructure if they aim to benefit from BDA.

According to [38], developing countries face crucial issues such as insufficient IT resources, poor communication, and a scarcity of professionals. Governments must also provide favorable circumstances and adequate enforcement for SMEs to adopt new and innovative technologies. Due to economic instability and limitations imposed on Palestine, policymakers need to pay more attention to SMEs and provide adequate support. Therefore, still there is a need of more studies, guidelines, and models to assist SMEs to take advantage of new technologies such as Big Data. Decision-makers need also to enhance their understanding and knowledge about the effective adoption of BDA in SMEs environment.

References

1. Wang S, Wang H (2020) Big data for small and medium-sized enterprises (SME): a knowledge management model. J Knowl Manag 24(4):881–897. https://doi.org/10.1108/JKM-02-2020-0081/FULL/PDF
2. de Vasconcelos JB, Rocha Á (2019) Business analytics and big data. Int J Inf Manag 46:250–251. https://doi.org/10.1016/J.IJINFOMGT.2019.03.001
3. Mandal S (2018) An examination of the importance of big data analytics in supply chain agility development: a dynamic capability perspective. Manag Res Rev 41(10):1201–1219. https://doi.org/10.1108/MRR-11-2017-0400
4. Alfoqahaa S (2018) Critical success factors of small and medium-sized enterprises in Palestine. J Res Mark Entrep 20(2):170–188. https://doi.org/10.1108/JRME-05-2016-0014/FULL/PDF
5. O'Connor C, Kelly S (2017) Facilitating knowledge management through filtered big data: SME competitiveness in an agri-food sector. J Knowl Manag 21(1):156–179. https://doi.org/10.1108/JKM-08-2016-0357/FULL/PDF
6. Park JH, Kim MK, Paik JH (2015) The factors of technology, organization and environment influencing the adoption and usage of big data in Korean firms. In: 26th European regional conference of the international telecommunications society "What next for European telecommunications?"

7. Miah SJ, Vu HQ, Gammack J, McGrath M (2017) A big data analytics method for tourist behaviour analysis. Inf Manag 54(6):771–785. https://doi.org/10.1016/J.IM.2016.11.011
8. Kwon O, Lee N, Shin B (2014) Data quality management, data usage experience and acquisition intention of big data analytics. Int J Inf Manag 34(3):387–394. https://doi.org/10.1016/j.ijinfomgt.2014.02.002
9. Gandomi A, Haider M (2015) Beyond the hype: big data concepts, methods, and analytics. Int J Inf Manag 35(2):137–144. https://doi.org/10.1016/J.IJINFOMGT.2014.10.007
10. Faizi S, Sałabun W, Rashid T, Watróbski J, Zafar S (2017) Group decision-making for hesitant fuzzy sets based on characteristic objects method. Symmetry 9(8):136. https://doi.org/10.3390/SYM9080136
11. Coleman S, Göb R, Manco G, Pievatolo A, Tort-Martorell X, Reis MS (2016) How can SMEs benefit from big data? Challenges and a path forward. Qual Reliab Eng Int 32(6):2151–2164. https://doi.org/10.1002/QRE.2008
12. Sen D, Ozturk M, Vayvay O (2016) An overview of big data for growth in SMEs. Procedia - Soc. Behav. Sci. 235:159–167. https://doi.org/10.1016/J.SBSPRO.2016.11.011
13. Maroufkhani P, Wagner R, Wan Ismail WK, Baroto MB, Nourani M (2019) Big data analytics and firm performance: a systematic review. Information 10(7). https://doi.org/10.3390/INFO10070226.
14. Tien EL, Ali NM, Miskon S, Ahmad N, Abdullah NS (2020) Big data analytics adoption model for Malaysian SMEs, pp 45–53 (2020). https://doi.org/10.1007/978-3-030-33582-3_5.
15. Zomaya AY, Sakr S (2017) Handbook of big data technologies, pp 1–895. https://doi.org/10.1007/978-3-319-49340-4.
16. Grant D, Yeo B (2018) A global perspective on tech investment, financing, and ICT on manufacturing and service industry performance. Int J Inf Manag 43:130–145. https://doi.org/10.1016/J.IJINFOMGT.2018.06.007
17. Mikalef P, Pappas IO, Krogstie J, Giannakos M (2018) Big data analytics capabilities: a systematic literature review and research agenda. Inf Syst E-Bus Manag 16(3):547–578. https://doi.org/10.1007/s10257-017-0362-y
18. Li G, Yang X, Jun W, Tao Y (2018) A theoretical credit reporting system based on big data concept: a case study of humen textile garment enterprises. In: ACM international conference proceeding series, pp 22–26. https://doi.org/10.1145/3206157.3206163.
19. Davenport TH, Dyché J (2013) Big data in big companies. Baylor Bus Rev 32(1):20–21. http://search.proquest.com/docview/1467720121?accountid=10067%5Cnhttp://sfx.lib.nccu.edu.tw/sfxlcl41?url_ver=Z39.88-2004&rft_val_fmt=info:ofi/fmt:kev:mtx:journal&genre=article&sid=ProQ:ProQ:abiglobal&atitle=VIEW/REVIEW:+BIG+DATA+IN+BIG+COMPANIES&title=Bay
20. Kapoor KK, Dwivedi YK, Williams MD (2014) Examining the role of three sets of innovation attributes for determining adoption of the interbank mobile payment service. Inf Syst Front 17(5):1039–1056. https://doi.org/10.1007/S10796-014-9484-7
21. Gu VC, Cao Q, Duan W (2012) Unified Modeling Language (UML) IT adoption — a holistic model of organizational capabilities perspective. Decis Support Syst 54(1):257–269. https://doi.org/10.1016/J.DSS.2012.05.034
22. Kandil AM, Ragheb MA, Ragab AA, Farouk M (2018) Examining the effect of toe model on cloud computing adoption in Egypt. Bus Manag Rev 9(9):113–123
23. Alshamaila Y, Papagiannidis S, Li F (2013) Cloud computing adoption by SMEs in the north east of England: a multi-perspective framework. J Enterp Inf Manag 26(3):250–275. https://doi.org/10.1108/17410391311325225
24. Laurell C, Sandström C, Berthold A, Larsson D (2019) Exploring barriers to adoption of Virtual Reality through Social Media Analytics and Machine Learning – an assessment of technology, network, price and trialability. J Bus Res 100:469–474. https://doi.org/10.1016/J.JBUSRES.2019.01.017
25. Rogers EM, Singhal A, Quinlan MM (2014) Diffusion of innovations. In: An integrated approach to communication theory and research. Routledge, pp 432–448

26. Sun S, Cegielski CG, Jia L, Hall DJ (2018) Understanding the factors affecting the organizational adoption of big data. J Comput Inf Syst 58(3):193–203. https://doi.org/10.1080/08874417.2016.1222891
27. Xu W, Ou P, Fan W (2017) Antecedents of ERP assimilation and its impact on ERP value: a TOE-based model and empirical test. Inf Syst Front 19(1):13–30. https://doi.org/10.1007/S10796-015-9583-0/TABLES/6
28. Lautenbach P, Johnston K, Adeniran-Ogundipe T (2017) Factors influencing business intelligence and analytics usage extent in South African organisations. S Afr J Bus Manag 48(3):23–33
29. Anderson JC, Gerbing DW (1988) Structural equation modeling in practice: a review and recommended two-step approach. Psychol Bull 103(3):411
30. Hair M, Hult JF, Ringle GTM, Sarstedt CM (2017) A Primer on partial least squares structural equation modeling (PLS-SEM). Sage, Thousand Oaks, p 165
31. Ngah AH, Gabarre S, Eneizan B, Asri N (2021) Mediated and moderated model of the willingness to pay for halal transportation. J Islam Mark 12(8):1425–1445. https://doi.org/10.1108/JIMA-10-2019-0199
32. Hair JF, Hult GTM, Ringle CM, Sarstedt M (2014) A Primer on partial least squares structural equation modeling (PLS-SEM). Eur J Tour Res 6(2):211–213
33. Verma S, Chaurasia S (2019) Understanding the determinants of big data analytics adoption. Inf Resour Manag J 32(3):1–26. https://doi.org/10.4018/IRMJ.2019070101
34. Verma S, Bhattacharyya SS, Kumar S (2018) An extension of the technology acceptance model in the big data analytics system implementation environment. Inf Process Manag 54(5):791–806. https://doi.org/10.1016/j.ipm.2018.01.004
35. Sam KM, Chatwin CR (2019) Understanding adoption of big data analytics in china: from organizational users perspective. In: IEEE international conference on industrial engineering and engineering management, December 2019, vol 2019, pp 507–510. https://doi.org/10.1109/IEEM.2018.8607652.
36. Wang L, Yang M, Pathan ZH, Salam S, Shahzad K, Zeng J (2018) Analysis of influencing factors of big data adoption in Chinese enterprises using DANP technique. Sustainability 10(11). https://doi.org/10.3390/su10113956.
37. Jang W-J, Kim S-S, Jung S-W, Gim G-Y (2019) A study on the factors affecting intention to introduce big data from smart factory perspective, vol 786
38. Mangla SK, Raut R, Narwane VS, Zhang Z, Priyadarshinee P (2020) Mediating effect of big data analytics on project performance of small and medium enterprises. J Enterp Inf Manag. https://doi.org/10.1108/JEIM-12-2019-0394.

Examining Intentions to Use Mobile Check-In for Airlines Services: A View from East Malaysia Consumers

Ling Chai Wong, Poh Kiong Tee, Chia Keat Yap, and Tat-Huei Cham

Abstract The purpose of this study was to determine the factors that influence both attitudes and behavioural intentions toward airline services via mobile check-in in East Malaysia. The intention of consumers to use mobile check-in for airline services was examined, as well as the role of attitude as a mediator between perceived usefulness, perceived ease of use, perceived trust, and perceived enjoyment. The study sampled 256 respondents using the snowball method and analysed them using PLS-SEM 3.0. Except for perceived ease of use, three of the four independent variables were found to have a positive effect on attitudes toward mobile check-in services. The perceived usefulness of mobile check-in had no effect on behavioural intention to use airline services via mobile check-in. Additionally, perceived ease of use was found to be insignificant when it came to attitudes and behavioural intentions toward using mobile check-in for airline services. Meanwhile, it has been demonstrated that attitude serves as an ideal mediator between perceived enjoyment, perceived trust, perceived usefulness, and behavioural intention. The current study has several managerial implications for the airline industry, particularly for self-service operations. Limitations include the inability to generalise the findings of this study to other industries or country settings.

L. C. Wong (✉) · P. K. Tee · C. K. Yap
School of Marketing and Management, Asia Pacific University of Technology and Innovation, Kuala Lumpur, Malaysia
e-mail: lingchaiwong10@gmail.com

P. K. Tee
e-mail: seantee@live.com

C. K. Yap
e-mail: j.yap@hw.ac.uk

T.-H. Cham
UCSI Graduate Business School, UCSI University, Kuala Lumpur, Malaysia
e-mail: jaysoncham@gmail.com

C. K. Yap
School of Social Sciences, Heriot-Watt University Malaysia, Wilayah Persekutuan Putrajaya, Putrajaya, Malaysia

M. Al-Emran et al. (eds.), *International Conference on Information Systems and Intelligent Applications*, Lecture Notes in Networks and Systems 550,
https://doi.org/10.1007/978-3-031-16865-9_13

Keywords Mobile check-in · Technology acceptance modal · Perceived trust · Perceived enjoyment

1 Introduction

Technological advancement caused a necessitates change in service delivery style and process, from traditional face-to-face service delivery to using self-service technologies such as mobile, kiosk machines, and online, to help service providers lower operating costs and reduce waiting lines from a consumer standpoint [1]. The same goes for the airline industry, also influenced by advanced technology's wave of self-service appeal, particularly in mobile check-in [2]. Nevertheless, the number of people using self-check is expected to rise as a result of the elimination of the traditional check-in method at low-cost carrier terminals (KLIA2). Lee [3] stated that Mobile Check-In (MCI) acceptances were not specifically addressed, and mobile check-in for airline services adoption is still in its infancy, particularly in East Malaysia. Airline services are considered the most convenient transportation mode in Sabah and Sarawak due to the uneven shape of the earth's surface in Sabah and Sarawak makes infrastructures such as highways and Mass Rapid Transit difficult to be implemented [3]. Despite the fact that Sabah and Sarawak are Malaysia's two largest states, there has been a lack of studies that specifically address the preferences of East Malaysians, specifically studies related to technology adoption for airline services.

Individuals can now check in their airline services online using various digital gadgets such as mobile phones, computers, and kiosk system machines. Self-check-in technologies are expected to improve check-in efficiency and reduce operational costs [4, 5]. However, as reported by the media, mobile check-in services from AirAsia and Malaysia Airlines always encountered problems with their mobile applications [6]. Moreover, the mobile check-in option is rendered ineffective when more than one passenger travels at the same time [7]. Again, the preceding discussion did not focus exclusively on the East Malaysian market. To the extent of the researcher knowledge, prior studies only focused on examining mobile check-in services at the Kuala Lumpur International Airport 1 and 2 (KLIA1 & 2), there has been little research in East Malaysia's airports [4]. Clearly, there is a gap, and a study focused on these markets is required.

Nevertheless, airline check-in is critical for both airlines and airports to determine whether passengers intend to travel or not. Based on the evidence presented above, there are inconsistencies in the service delivery system and passengers' satisfaction, particularly for mobile check-in services for passengers in Malaysia, specifically East Malaysia. As a result, studying the factors influencing users' intentions to use mobile check-in services is critical for the use of modern technology in the airline industry. As the example, the current study will benefit stakeholders in Malaysia's service industry by providing pertinent user feedback on their expectations for the application of self-service technologies. This information is important for airline service providers and the government to develop an effective system for engaging with customers to maintain the quality and competitiveness of Malaysian airline services in the age of globalisation.

2 Literature Review

Prior to the actual behaviour, a person's intention is referred to behavioural intention (BI); the likelihood, pleasure, engagement, and consideration of accepting or rejecting a specific system [8]. Employment of Technology Acceptance Model (TAM) to underpin the present study, in concur with Theory of Planned Behaviour and Theory Reasoned Action, several important predictors have been identified and studied in predicting the intention to use mobile check in for airlines services in this study. These factors inclusive of perceived usefulness (PU), perceived ease of use (PEU), perceived trust (PT), perceived enjoyment (PE) and attitude (ATT) as the mediating variables.

2.1 Perceived Usefulness

PU refers to the degree to which a user believes that utilising a system will improve their job performance [8]. PU includes the user gaining the benefit or usefulness, such as being able to complete the task faster and more conveniently [9, 10]. People believe that technology will enable and assist them to perform better on a task. Similarly, perceived usefulness is conceptualized as convenient, efficient, effective, and useable for the consumer to check-in for the airline service. Indeed, most of the past studies suggested that perceived usefulness positively impacts attitude and behavioural intention in mobile marketing [11, 12]. However, there has been a limited focus on mobile check-in for airline services particularly for the East Malaysia market regarding to the relationship between PU, ATT, and BI. Therefore, to close the gap aforementioned, the following hypotheses were developed:

H1: PU has a positive impact on ATT to use MCI for airline services.
H2: PU has a positive impact on BI to use MCI for airline services.

2.2 Perceived Ease of Use

PEU refers to a consumer's belief that using a system will save them time and effort [8, 13]. PEU, according to Dahlberg, Mallat, and Öörni [14], includes ease to learn, control, understanding, use, clarity, and flexibility of use. According to the preceding discussion, the current study conceptualised PEU as being free of mental effort. The mobile check-in procedure is straightforward and simple to learn. In terms of relationship, PEU was found to be significantly related to attitude and actual use

[15]. Additionally, previous studies also confirmed the significant impact of PEU on ATT and BI [5, 15, 16] and proved that PEU is able to form a positive significant relationship with both ATT and BI in this context. However, little empirical study has been conducted in mobile check-in for airline services except for Wong [1] to confirm the direct relationship between PEU, ATT, and BI. As a result, the current study was conducted to identify such a relationship as hypotheses below:

H3: PEU has a positive impact toward ATT to use MCI for airline services.
H4: PEU has a positive impact toward BI to use MCI for airline services.

2.3 Perceived Enjoyment

TAM's construct of PE was added by Van der Heijden [17] in a study on the use of websites to the original TAM. Perceived enjoyment is defined as the user experiencing something fun, pleasurable, or enjoyable while interacting with a particular system [18, 19]. Pleasure or enjoyment was defined as the level of delight experienced by an individual in a preferred environment [20]. In addition to that, Holdack, Lurie-Stoyanov, and Fromme [21] accounted for a broad range of PE definitions including fun, felt good, exciting, enjoyable, and interesting. The current study adopts the conceptual definition of perceived enjoyment from Holdack et al. [21].

It was discovered that perceived enjoyment was significantly related to behavioural intention [20]. PE was also found to be significantly associated with attitude in a study of mobile social network game sustainable use intention [22]. However, in terms of the relationship between PE, ATT, and BI, there has been a limited focus on mobile check-in for airline services, particularly in the East Malaysia market. As a result, in order to bridge the aforementioned gap, the following hypotheses were developed:

H5: PE has a positive impact toward BI to use MCI for airline services.
H6: PE has a positive impact toward attitude to use MCI for airline services.

2.4 Perceived Trust

PT refers to a party's ability to earn the confidence or reliance of exchange partners, including the integrity and dependability of one party toward another [23]; trustworthy, reliable, and comfortable [24]. Singh and Sinha [25] define PT as an emotional state that compels one to trust another based on the other's acceptable behaviour. The concept of PT in this study is defined as the user trust the boarding pass showed in the system, comfortable with the system, and believe their information is protected [24]. PT was found to positively influence attitudes toward online shopping [26] and attitudes toward e-hailing services [27]. However, there is a dearth of empirical evidence establishing a relationship between PT, ATT, and BI in the field of mobile check-in for airline services. Hence, in order to bridge the aforementioned gap, the following hypotheses were developed:

H7: PT has a positive impact toward BI to use MCI for airline services.
H8: PT has a positive impact toward ATT to use MCI for airline services.

2.5 *Attitude*

Attitude is defined in the Theory of Reasoned Action (TRA) as "an individual's assessment of a system that has been used in the user's job. The positive or negative value that an individual associates with the fact of producing a behaviour is referred to as the individual's attitude toward the behaviour [28]. Similarly, Tee et al. [5] stated that when a consumer has a strong favourable ATT toward technology, it will undoubtedly be adopted. The concept of attitude in this study was adapted from Nagaraj's [29] definition of attitude as the consumer's feeling of whether something is good or bad, favourable or unfavourable, wise or foolish, positive or negative, and beneficial or detrimental. Again, little attention was focus on mobile check in context empirically. As a result, to fill the gap, mobile check-in for airline services is assumed to have a similar relationship to the following hypothesis:

H9: ATT has a positive impact toward BI to use MCI for airline services.

Numerous studies on consumer behavior found that attitude (ATT) is a main predictor on consumer behavior [8, 13, 28]. TAM model deemed that the user' adoption behavior is determined by their attitudes, and attitudes are jointly affected by perceived usefulness and perceived ease to use [8, 30]. Numerous studies applying TAM only tested the direct relationship between belief and attitude or behavioral intention, and as expected, belief variables were found significantly predict attitude. However, there was limited studies included attitude as a mediator. Indeed, Davis's [8] original work on TAM was not included attitude, and he did admit that people intention to use a technology can be influence by their attitude toward the technology. Questions remain about the mediating role of attitude toward the adoption of digital wallet in the TAM. Thus, the present study includes attitude as a mediator to further testify the direct and indirect relationships between the four independent variables (PU, PEOU, PE and TRU) and the dependent variable (adoption of digital wallet) via the mediator (ATT). As a result, mobile check-in for airline services is assumed to have a similar relationship to the following hypothesis:

H10: ATT mediates between PEU, PU, PE, PI and BI to use MCI for airlines services.

The research framework for this study is depicted in Fig. 1. It includes independent variables such as perceived usefulness, perceived ease of use, perceived enjoyment, and perceived trust, as well as mediators such as attitude and intention to use mobile check in. Perceived usefulness, perceived ease of use, attitude, and behavioural intention to use mobile check-in were all derived from TAM, whereas

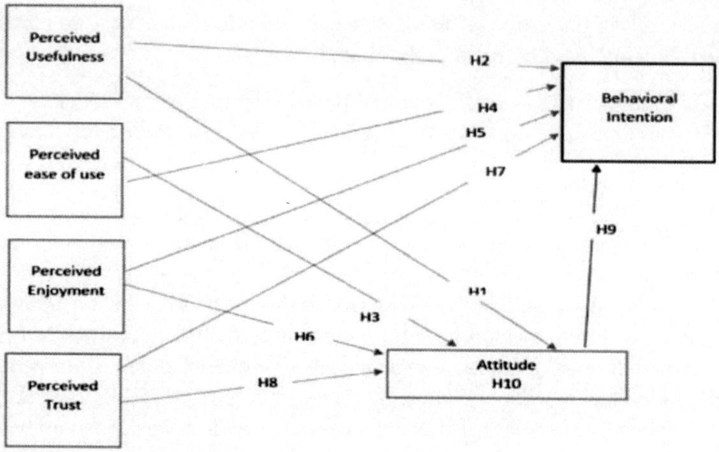

Fig. 1 Research model

perceived enjoyment and perceived trust were added to TAM for the purpose of assessing consumers' intentions to use mobile check-in for airline services in East Malaysia [1]. The framework was proposed to address the following questions:

1. Does perceived usefulness, perceived ease of use, perceived enjoyment, and perceived trust have a positive impact toward attitude?
2. Does perceived usefulness, perceived ease of use, perceived enjoyment, and perceived trust have a positive impact toward attitude?
3. Does attitude has a positive impact toward behavioral intention?
4. Does attitude mediates the relationship between perceived usefulness, perceived ease of use, perceived enjoyment, perceived trust and behavioral intention?

3 Research Method

The current study obtained 256 samples, all of which met the sample 129 minimum requirement. The sample are identified using snowball sampling method and self-administered questionnaires were adapted to gather the primary data. Snowball sampling is best method to reach unknown or rare populations and enables to identify respondent who meets the criteria for inclusion in this study [31]. In fact, there is a lack of statistical record about the respondents who have experienced the mobile check-in for airlines industry.

The questionnaire is divided into seven sections. The section A is about usage background that help the researcher to pre-screen the valid respondent and followed by section B (PU), section C (PEU), section D (PT), section E (PE), section F (BI), section G (ATT), and section G (demographic profile). For each variable, such as PEU, PU, ATT, and BI, five measurement items were adapted from [8]. In term

PE and PT, there were measurement items adapted from Holdack et al. [21] and Ghazizadeh et al. [24].

The collected data were analyzed using Smart-PLS 3.2's Partial Least Squares-Structural Equation Modelling (PLS-SEM). This technique was chosen to promote analytical rigour and more consistent estimations [32, 33]. Additionally, the model specification's characteristics, simplicity, and absence of strict distributional assumptions all contribute to the choice [32]. The analysis included evaluating the measurement model, the structural model, and the mediation model, as well as justifying the study's hypotheses.

4 Results

The final sample consists of 256 respondents. 121 out of 256 respondents (47.3%) are male, while 135 (52.7%) are female. 36% of total respondents are under the age of 20, followed by those age 21 to 30 years (176 respondents or 68.8%), those age 31 to 40 years (34 or 13.3%), and those age 41–50 years (8 or 3.1%). The remaining 2%, or 0.8% are over the age of 50. The respondents' educational attainment was deemed to be high; 178 (69.5%) respondents had a bachelor's degree, 52 (20.3%) respondents held a STPM or a college diploma, while 13 (5.1%) respondents held a master's degree. The remaining 12 (4.7%) respondents and 1 (0.4%) respondent were SPM and below, and PhD degree holders, respectively. Also, 71.7% of respondents had a monthly income of between RM3001 and RM4000, follows by respondents with incomes ranging from RM1001 to RM2000, RM2001 to RM3000, RM0 to RM1000, and RM5000 and above. For the ethnicity, majority of the respondents (113 respondents or 44.1%) are Chinese, followed by Bumiputera (66 respondents or 25.7%), Malay (56 respondents or 21.9%), and India (21 respondents or 8.2%).

The PLS-SEM technique was used to predict BI to use MCI for airline services and the role of ATT as a mediator. Numerous reliability and validity analyses were conducted in order to validate the measurement model, referred to as Convergent validity. Convergent validity is a term that refers to the degree to which a measure correlates with other measurements of the same phenomenon. Convergent validity was determined in accordance with the recommendations of Hair et al. [34], specifically by examining item loadings, average variance extracted (AVE), and composite reliability (CR). According to scholars, the loadings value must be greater than 0.708, the AVE must be greater than 0.50, and the CR must be greater than 0.70. As shown in Table 1, all loadings are range from 0.716 to 0.832, the AVE exceeded 0.50 and the CR exceeded 0.70, implying that convergent validity was achieved.

In addition, Discriminant validity analysis was used to quantify distinct concepts by examining the Heterotrait-Monotrait (HTMT) criterion measures of potentially overlapping concepts. The HTMT is a measure of latent variable similarity. Table 2 shows that all HTMT Criterion values were below 0.85 and 0.90 based on Henseler et al. [35]. Therefore, discriminant validity was established in this study.

Table 1 Result of convergent validity

	Items	Loadings	AVE	CR
Attitude	6	0.716–0.785	0.529	0.871
Behavioural intention	5	0.750–0.817	0.623	0.892
Perceived enjoyment	5	0.750–0.813	0.624	0.892
Perceived trust	5	0.766–0.832	0.581	0.873
Perceived usefulness	5	0.749–0.816	0.612	0.887
Perceived ease of use	5	0.698–0.769	0.528	0.847

Notes CR = Composite reliability; AVE = Average variance extracted

Table 2 Result of discriminant validity

	ATT	BI	PE	PT	PU	PEU
Attitude						
Behavioural intention	0.787					
Perceived enjoyment	0.581	0.589				
Perceived trust	0.603	0.657	0.662			
Perceived usefulness	0.657	0.562	0.558	0.594		
Perceived ease of use	0.595	0.523	0.697	0.709	0.717	

Notes ATT = Attitude, BI = Behavioural Intention, PE = Perceived Enjoyment, PU = Perceived Usefulness, PEU = Perceived ease of use

As for the structural model, Table 3 shows the results of the bootstrapping on the significance of the path estimates of the hypothesised relationships. The relative importance of the exogenous constructs in predicting ATT to use MCI for airline services revealed that PU ($\beta_1 = 0.315$, *t*-value = 5.019, $p < 0.01$) was the most important predictor, followed by PT ($\beta_8 = 0.204$, *t*-value = 2.730, $p < 0.01$), and PE ($\beta_6 = 0.187$, *t*-value = 2.522, $p < 0.01$), which supported H1, H6 and H8. However, PEU ($\beta_3 = 0.082$, t-value = 1.223, p > 0.01), which H3 is rejected. On the other hand, the relative importance of the exogenous constructs in predicting BI to use mobile check-in for airline services indicates that, PT ($\beta_7 = 0.233$, *t*-value = 3.506, $p < 0.01$) and PE ($\beta_5 = 0.141$, *t*-value = 2.210, $p < 0.05$) were found positive effect. Hence, H7 and H5 are supported. However, H2 ($\beta_7 = 0.073$, *t*-value = 1.120, $p >$ 0.01) and H4 ($\beta_4 = -0.042$, *t*-value = 0.613, $p > 0.01$) were found insignificant relationship toward BI to use MCI for airline services. The results also indicate a significant effect between ATT and BI to use MCI for airline services ($\beta_9 = 0.452$, *t*-value = 6.846, $p < 0.01$). Thus, H9 in this study is supported.

Table 4 shows that the mediation analysis confirms attitude as the significant mediator in the relationship between PE, PT, and PU toward BI to use MCI for airline services. The bootstrapping analysis showed that the indirect effect of $\beta = 0.084, \beta = 0.092, \beta = 0.142$ significant with a t-value of 2.737, 2.594, and 3.869. The indirect effect of 95 percent Boot CL: [LL = 0.037, UL = 0.152], [LL = 0.038, UL

Table 3 Result of hypotheses testing

Relationship	Std beta	Std error	t-value	p-value	Decision
H1: PU -> Attitude	0.315	0.063	5.019	0.000**	Supported
H2: PU -> BI	0.073	0.065	1.120	0.131	Rejected
H3: PEU -> Attitude	0.082	0.067	1.223	0.111	Rejected
H4: PEU -> BI	−0.042	0.068	0.613	0.270	Rejected
H5: PE -> BI	0.141	0.064	2.210	0.014*	Supported
H6: PE -> Attitude	0.187	0.074	2.522	0.006*	Supported
H7: PT -> BI	0.233	0.066	3.506	0.000**	Supported
H8: PT -> Attitude	0.204	0.075	2.730	0.003*	Supported
H9: Attitude -> BI	0.452	0.066	6.846	0.000**	Supported

Notes ** p-value < 0.001, * p-value < 0.05, ns = not significant

Table 4 Result of mediation analysis

Hypothesis	Std beta	Std error	*t*-value	P value	5%	95%	Decision
H10a: PE > ATT > BI	0.084	0.035	2.437	0.007	0.037	0.152	Supported
H10b: PT > ATT > BI	0.092	0.035	2.594	0.005	0.038	0.154	Supported
H10c: PU > ATT > BI	0.142	0.037	3.869	0.000	0.091	0.214	Supported
H10d: PEU > ATT > BI	0.037	0.031	1.183	0.118	0.013	0.089	Rejected

Notes ATT = Attitude, BI = Behavioural Intention, PE = Perceived Enjoyment, PU = Perceived Usefulness, PEU = Perceived ease of use

= 0.154], and [LL = 0.091, UL = 0.214], does not straddle a 0 in between, indicating the mediation effect. Hence, H10a, H10b, and H10c are supported. No mediation effect was found between PEU and BI to use MCI for airline services where the indirect effect of $\beta = 0.037$, and insignificant *t* value of 1.183. The indirect effect of 95 percent Boot CL: [LL = −0.013, UL = 0.089], straddle a 0 in between, indicating the mediation effect. Hence, H10d was rejected.

5 Discussion and Conclusion

The purpose of this study was to ascertain consumers' BI when employing MCI for airline services. The findings indicated that only three (i.e., PT, PE and ATT) out of the five predictors are significantly related to behavioural intention. Consumer attitude toward the mobile check-in system appears to be the most significant ($\beta = 0.452$) predictor amongst the five constructs. This indicates that if a person has a strong favourable attitude toward a new system or technology, it will undoubtedly be adopted. The important role of attitude in the adoption of new technology has been widely discussed and recognised in most of the past theories such as TAM [8], TRA

[28] and the Unified Theory of Acceptance and Use of Technology (UTAUT) [13]. Following this, perceived trust and perceived enjoyment were found significantly influenced attitude and behavioural intention. This means that the MCI users are very particular on reliability, safe (protection of their information), and comfortable and enjoyment to be used for MCI service. This result is analogous to substantiate studies on e-ticketing [5], mobile wallet [12], and e-hailing services [27].

It appears that PU and PEU are not significant predictors toward the BI to use the MCI for airline services. The results are contradicted with the previous studies claimed the positive relationship between PU and PEU on the BI toward new system [5, 15, 16]. Perhaps, most of the users in Malaysia still think that traditional counter-in is more useful and easier for them to complete the check-in process compared to using mobile check-in. Particularly, human interaction is still an important element in service organisation like airline industry [36]. Although PU has no significant direct impact on BI, it does affects user's BI indirectly via ATT. The result indicated that PU was fully mediated by attitude in which ATT absorb most of the PU impact on BI. Moreover, this finding suggests that most of users in Malaysia, especially the users in East Malaysia found that the new mobile check-in for airline services is still complication and troublesome. As reported, mobile check-in services from AirAsia and Malaysia Airlines always encountered problems with their mobile applications [6, 7]. These should be the main reasons that explained why people perception on the application of MCI was complicated and not easy to use, lead to the insignificant effect of toward BI to use the MCI services.

The study's findings have significant implications for theoretical and managerial practice in MCI for airlines service area. Theoretically, the current study adds PE and PT to the (TAM) in the context of mobile check-in for airline services, providing empirical support for the model [1]. Inconsistency between PEU and BI was confirmed in this study, as was the absence of a relationship between PEU and ATT. This finding closes a gap in the preview study. Additionally, the research contributes significantly to the body of knowledge by addressing a gap in MCI services. As previously stated, there has been little research on MCI for airline services in East Malaysia.

Practically, this study provides compelling evidence that may assist marketing managers and airlines service providers in better understanding the BI to use MCI for airlines service. Due to the changing technological environment in which services operate, the transformation of face-to-face service toward self-service technology, particularly in the mobile industry, has occurred [25]. Thus, airlines service providers should develop a MCI system that benefits the passenger, is convenient for the passenger, is preferred by the passenger, and is pleasant and desirable for the airline passenger. While developing an ATT toward using MCI for airlines does not guarantee BI to use MCI for airlines, it does play a critical role in developing BI to use mobile check-in for airlines. As a result, service providers constantly strive to ensure that their customers have a favourable ATT toward them [27].

Future research could look into the user experience of other smartphones from different manufacturers for comparison purposes. The differences in expected experience between students and non-students, as well as between users of different ages, are worth investigating in order to tailor specific designs for specific groups. Despite its limitations, this study has provided some insights to smartphone industry leaders on designing and marketing their products to maximise user satisfaction.

References

1. Wong LC (2013) The relationship of perceived usefulness, perceived ease of use, perceived enjoyment, and perceived trust toward behavioral intention to use mobile check-in for airlines service and the mediating role of attitude. Master dissertation, University Malaysia Sabah
2. Van NTT, Vrana V, Duy NT, Minh DXH, Dzung PT, Mondal SR, Das S (2020) The role of human–machine interactive devices for post-COVID-19 innovative sustainable tourism in Ho Chi Minh City, Vietnam. Sustainability 12(22):9523
3. Lee C (2020) Economic reforms in the aftermath of regime change in Malaysia. Asian Econ Policy Rev 15(2):239–257
4. Taufik N, Hanafiah MH (2019) Airport passengers' adoption behaviour towards self-check-in Kiosk Services: the roles of perceived ease of use, perceived usefulness and need for human interaction. Heliyon 5(12):e02960
5. Tee PK, Gharleghi B, Chan YF (2014) E-Ticketing in airline industries among Malaysian: the determinants. Int J Bus Soc Sci 5(9):168–174
6. The Edge Market (2019) Temporary systems disruption at KLIA, says Malaysia Airports. https://www.theedgemarkets.com/article/temporary-systems-disruption-klia-says-malaysia-airports
7. Apple Store (2021) AirAsia customer feedback on mobile apps. https://apps.apple.com/my/app/airasia/id565050268?mt=8
8. Davis FD (1989) Perceived usefulness, perceived ease of use, and user acceptance of information technology. MIS Q 13:319–340
9. Saadé R, Bahli B (2005) The impact of cognitive absorption on perceived usefulness and perceived ease of use in on-line learning: an extension of the technology acceptance model. Inf Manag 42(2):317–327
10. Yang KC, Shih PH (2020) Cognitive age in technology acceptance: at what age are people ready to adopt and continuously use fashionable products? Telematics Inform 51:101400
11. Chen CC, Tsai JL (2019) Determinants of behavioural intention to use the personalized location-based mobile tourism application: an empirical study by integrating TAM with ISSM. Future Gener Comput Syst 96:628–638
12. Al-Sharafi MA, Al-Qaysi N, Iahad NA, Al-Emran M (2021) Evaluating the sustainable use of mobile payment contactless technologies within and beyond the COVID-19 pandemic using a hybrid SEM-ANN approach. Int J Bank Mark 40:1071–1095
13. Venkatesh V, Davis FD (1996) A model of the antecedents of perceived ease of use: development and test. Decis Sci 27(3):451–481
14. Dahlberg T, Mallat N, Öörni A (2003) Trust enhanced technology acceptance model consumer acceptance of mobile payment solutions: tentative evidence. Stockh Mobil Roundtable 22(1):145–156
15. Prastiawan DI, Aisjah S, Rofiaty R (2021) The effect of perceived usefulness, perceived ease of use, and social influence on the use of mobile banking through the mediation of attitude toward use. APMBA (Asia Pac Manag Bus Appl) 9(3):245–266
16. Alhasan A, Audah L, Ibrahim I, Al-Sharaa A, Al-Ogaili AS, Mohammed JM (2020) A case-study to examine doctors' intentions to use IoT healthcare devices in Iraq during COVID-19 pandemic. Int J Pervasive Comput Commun 58–72

17. Van der Heijden H (2003) Factors influencing the usage of websites: the case of a generic portal in The Netherlands. Inf Manag 40(6):541–549
18. Ramayah T, Ignatius J (2005) Impact of perceived usefulness, perceived ease of use and perceived enjoyment on intention to shop online. ICFAI J Syst Manag (IJSM) 3(3):36–51
19. Sun H, Zhang P (2006) Causal relationships between perceived enjoyment and perceived ease of use: an alternative approach. J Assoc Inf Syst 7(1):24
20. So KKF, Kim H, Oh H (2021) What makes Airbnb experiences enjoyable? The effects of environmental stimuli on perceived enjoyment and repurchase intention. J Travel Res 60(5):1018–1038
21. Holdack E, Lurie-Stoyanov K, Fromme HF (2020) The role of perceived enjoyment and perceived informativeness in assessing the acceptance of AR wearables. J Retail Consum Serv 65:102259
22. Wang H, Lee K (2020) Getting in the flow together: the role of social presence, perceived enjoyment and concentration on sustainable use intention of mobile social network game. Sustainability 12(17):6853
23. Moorman C, Deshpande R, Zaltman G (1993) Factors affecting trust in market research relationships. J Mark 57(1):81–101
24. Ghazizadeh M, Peng Y, Lee JD, Boyle LN (2012) Augmenting the technology acceptance model with trust: commercial drivers' attitudes towards monitoring and feedback. In: Proceedings of the human factors and ergonomics society annual meeting, vol 56, no 1. Sage Publications, pp 2286–2290
25. Singh N, Sinha N (2020) How perceived trust mediates merchant's intention to use a mobile wallet technology. J Retail Consum Serv 52:101894
26. Tee PK, Lim KY, Ng CP, Wong LC (2022) Trust in green advertising: Mediating role of environmental involvement. Int J Acad Res Bus Soc Sci 12(1):1771–1786
27. Chia KM., Rohizan A, Tee PK, Tajuddin AR (2019) Evaluation of service quality dimensions toward customers' satisfaction of ride-hailing services in Kuala Lumpur Malaysia. Int J Recent Technol Eng 7(5S):102–109. ISSN 2278-3075
28. Ajzen I, Fishbein M (1977) Attitude-behavior relations: a theoretical analysis and review of empirical research. Psychol Bull 84(5):888–918
29. Nagaraj S (2021) Role of consumer health consciousness, food safety & attitude on organic food purchase in emerging market: a serial mediation model. J Retail Consum Serv 59:102423
30. Lien CH, Hsu MK, Shang JZ, Wang SW (2021) Self-service technology adoption by air passengers: a case study of fast air travel services in Taiwan. Serv Ind J 41(9–10):671–695
31. Marshall MN (1996) Sampling for qualitative research. Fam Pract 13(6):522–526
32. Sarstedt M, Hair JF, Ringle CM, Thiele KO, Gudergan SP (2016) Estimation issues with PLS and CBSEM: where the bias lies! J Bus Res 69(10):3998–4010
33. Tee PK, Cham TH, Low MP, Lau TC (2021) The role of organizational career management: comparing the academic staff' perception of internal and external employability in determining success in academia. Malays Online J Educ Manag 9(3):41–58
34. Hair JF, Risher JJ, Sarstedt M, Ringle CM (2019) When to use and how to report the results of PLS-SEM. Eur Bus Rev 31(1):2–24
35. Henseler J, Ringle CM, Sarstedt M (2015) A new criterion for assessing discriminant validity in variance-based structural equation modelling. J Acad Mark Sci 43(1):115–135
36. Awang Z, Yazid A (2017) Inflight service quality of Malaysia Airlines: validation using SEM and AMOS. Int J Acad Res Bus Soc Sci 7. https://doi.org/10.6007/IJARBSS/v7-i10/3395

Spreading Faster Than the Virus: Social Media in Spreading Panic Among Young Adults in Malaysia

Farah Waheeda Jalaludin, Fitriya Abdul Rahim, Lit Cheng Tai, and Tat-Huei Cham

Abstract The late-2019 Covid-19 outbreak has shifted global attention to social media. Governments used social media to raise public health awareness. However, the internet was flooded with disinformation and conspiracy theories, among other things. This condition may cause unwarranted alarm, compromising the health system and damaging the mental health of social media users, especially young people who dominate the internet population. The function of social media in spreading fear during this epidemic must be investigated. This study's goal is to see how social media might spread fear among Malaysian young people. A total of 400 university students took part in this online survey. The results show that fake news, mental health and anxiety, changes in public behaviour and sharing information are significantly related to panic behaviour. Implications of the research findings are discussed.

Keywords Social media · Panic · Covid-19 · Young adult

1 Introduction

COVID-19 was discovered in Malaysia on January 25, 2020, related to three Chinese people who had contact with an infected individual in Singapore [1]. Following that, a

F. W. Jalaludin (✉) · F. Abdul Rahim · L. C. Tai
Faculty of Accountancy and Management, Universiti Tunku Abdul Rahman, Kajang, Selangor, Malaysia
e-mail: farah@utar.edu.my

F. Abdul Rahim
e-mail: fitriya@utar.edu.my

L. C. Tai
e-mail: tailc@utar.edu.my

T.-H. Cham
UCSI Graduate Business School, UCSI University, Kuala Lumpur, Malaysia
e-mail: jaysoncham@gmail.com

M. Al-Emran et al. (eds.), *International Conference on Information Systems and Intelligent Applications*, Lecture Notes in Networks and Systems 550,
https://doi.org/10.1007/978-3-031-16865-9_14

cluster called tabligh was discovered in mid-March 2020, sponsored by an Islamic missionary organisation entitled Tablighi Jama'at, which recruited roughly 16 000 and 1500 members of religious groups from Malaysia and abroad, respectively. WHO declared the COVID-19 pandemic on March 11, 2020 [2] due to an unprecedented increase in cases, deaths, and disease transmission. Escalation prevention in public and medical settings is critical [3].

Malaysia has been virtually isolated since March 18, 2020. Malaysian authorities have taken severe steps, in addition to stringent emergency regulations. Travel restrictions were implemented [4]. Prolonged universities and schools closures as well as workplace distancing were implemented to mitigate COVID-19. Infectious disease identification rates are low due to lack of understanding. To raise public awareness, Ministry of Health launched a series of campaigns using posters, billboards, TV and radio ads [5]. As a dependable venue for rapid health communication, social media technologies are gaining traction [6]. Similar to how bogus news, prejudice, and racism are propagated by social media [7], so is public health information. It has been widely reported that the spread of COVID19 as well as other medical disinformation have been transmitted via unfiltered sources such as social media sites. Information overload (infodemic) is a severe public health issue [8]. In order to assess the effectiveness of government preventative measures and policies, social media platforms and websites must be evaluated for public awareness [9].

In March 2021, Malaysia passed a legislation against "false news." In addition to a daily penalty of RM1,000, anyone found guilty of distributing false news risk a RM100,000 fine and/or three years in jail [10]. Social media has been awash with complaints about the fast spread of COVID-19. Misinformation and false news propagate quicker than trustworthy information on social media, endangering health systems and individuals' mental health [11]. One week after learning that double coptis (shuang huanglian), a traditional Chinese herbal medicine commonly used to treat colds and flu, can effectively contain the multiplication of SARS-CoV-2, the virus responsible for COVID-19 in human bodies, [12] highlighted that this drug was sold out in China in less than 24 h. The realisation that this treatment was not the miracle cure swiftly dimmed the national enthusiasm. Because of this false information also, several 5G network towers have been dismantled or damaged [13] due to the belief that the virus can be transmitted through 5G network.

Studies show that social media may fuel the COVID-19 infodemic [14]. However, the degree to which social media influences people's health is unknown. This is critical since social media may have both positive and negative societal consequences. With the surge of numerous rumors, pieces of misinformation, and hoaxes appearing on several social media platforms, this paper aims to study the impact of social media in spreading panic towards young adults during the Covid-19 pandemic. This is due to the fact that the entire population are relying on social media especially during the first wave where majority were under lock down, causing them to turn to social media to communicate and searching for information that potentially affect their lives.

2 Review of Literature

2.1 *Uses and Gratification Theory (UGT)*

Users and gratification theory (UGT) [15] examines the benefits that pull and keep customers engaged with diverse media and content. The uses and gratifications hypothesis are relevant to social media since it is based on communication literature. People use specific media to meet their interests for a variety of reasons [16] such as to examine a medium's functions while also considering the audience's motives. [17] found that social characteristics greatly affected students' Facebook intents and [18] found that the UGT process is an important predictor of Facebook use intensity. According to [19], in addition to pleasure, social media is also used for information collection. UGT is linked to peer acceptance, information probing, and relationship conservation according to [20] which include amusement, prestige, knowledge, and sociability [21]. They now feature a greater understanding of information and news sharing behaviour. [22] connects news sharing to those seeking socialisation, status, and knowledge sharing which are more persuaded to disseminate news online. Prior study links information sharing to socialisation and communication [23]. Studies suggest that the inherent properties of social media, such as high interaction and unfettered information flow, may foster anxiety. This is based on earlier research and UGT theory. The UGT hypothesis has been used extensively in media studies, but not to study whether the pleasures people get from using social media make them more eager to spread fear through social media.

2.2 *Social Media and Pandemics*

In the H1N1 pandemic, social media had a significant influence on public health [24]. High frequency of misinformation, social media fear, public opinion, and linguistic misconceptions were discovered. The COVID-19 outbreak prompted many people to seek information online. In addition, the media must provide accurate information, dispel rumours and discrimination, and raise public awareness of health-related issues. To learn more about the COVID-19 epidemic, many people turned to social media. Social media became the main source of information [25–27] for the first time ever. Following the COVID-19 outbreak, social media platforms have seen a rise in false rumours, misinformation and conspiracy theories concerning the virus's origin, according to [28, 29], social media has been found to be one of the most efficient tools of spreading information about specific dangers [30]. However, disinformation and rumours spread faster on social media than factual information, possibly jeopardising the credibility and balance of the news media and, in particular, the healthcare industry [11]. As news regarding COVID-19 is widely disseminated on social media, some contentmay be incorrect or misleading. False information regarding the coronavirus might cause fear if it spreads faster than the virus [7], according to [31]. Because of this, misinformation and conspiracy theories abounded after the virus struck China. The rumours sparked global panic. Information systems enabled by social media [7] linked the data. [32] stated the importance to Reassure and advise against spreading sickness-related misinformation. [7]. In a pandemic, incorrect information published on social media is considered to cause broad public fear. Malaysian government gazetted the new Emergency (Essential Powers) (No. 2) Ordinance 2021 on fake news in March 2021 [33].

Decision-Making and Public Awareness and Panic

Modern communication relies heavily on social media. The internet has 4.54 billion users while social media has 3.8 billion [34]. Over 4 billion people use social media. Social media could help to raise public awareness [35]. When COVID-19 was found outside China, it became viral on social media. 19 million worldwide mentions for COVID-19 overnight [35]. They may be used to assess the success of government prevention programmes and legislation [36]. Social media expansion has created new routes for public communication and news distribution, say Merchant and Lurie (2020). This means they may propagate both right and false facts. Social media, according to [37], has facilitated stakeholder interaction. Hence, social media may change peoples' views. Social media has broadened its influence on decision-makers perspectives, as shown by [38]. Every political party uses social media to convey messages cheaply [39]. Thus, it is proposed that:

H1: There is a significant positive relationship between decision making and public awareness towards panic behaviour

Covid-19 Fake News and Panic

In recent months, the most concerning trend has been the propagation of false information during the COVID-19 outbreak. It's becoming tougher to distinguish fake news from real news [40]. As a consequence of disinformation on social media, the public is worried about the COVID-19 pandemic. As a result of this, many people believe they can be cured with seawater, bleach, and oregano [41]. Research by [42] indicated that the more people use social media to learn about COVID-19, the more fear. Thus, it is proposed that:

H2: There is a significant positive relationship between Covid 19 fake news towards panic behaviour

Mental Health and Anxiety and Panic

Mental health is defined as an individual's psychological, emotional, and social well-being. It impacts how we think, feel, behave, respond to stress, interact with others, and even make choices [43]. Social media may be used to connect, assist, and communicate [44] However, increased use of social media may lead to a continuous need to connect as well as negative experiences, affecting users' mental health [45]. Anxiety, stress, and depression have been associated to social media consumption in teenagers

[46]. [47] stress the need of protecting mental health during the COVID-19 epidemic. Because lockdowns require the use of technological equipment, cyberpsychology must be addressed alongside virus concern. Numerous studies on the consequences of social media have linked prolonged use of platforms like Facebook to negative symptoms of melancholy, anxiety, and stress [48–50]. Thus, it is proposed that:

H3: There is a significant positive relationship between mental health and anxiety towards panic behaviour.

Changes in Public Behaviour and Panic

Social media campaigns promoting healthy behaviours have been shown to promote good behavioural changes and even prevent bad ones. Utilizing social media to communicate about social and behavioural change is necessary in tackling large-scale difficulties [51]. Businesses have begun using these platforms to shift perspectives. They are becoming more popular for online marketing and purchasing [52]. Concerned about the pandemic, all levels of government turn to social media. Several networks provided expert about Covid-19. As part of the public awareness effort, many media outlets aired stories and safety suggestions. Facebook, Twitter, and other social media sites have the power to influence public safety [53]. Authorities could utilise Google Trends to forecast user behaviour and prevent panic-related searches. As social media health campaigns expand, fewer people become sick [54]. Thus, it is proposed that:

H4: There is a significant positive relationship between changes in public behaviour towards panic behaviour.

Sharing Information and Panic

Instead of requiring rigorous vetting and validation before being considered authoritative, social media allows anybody to be a source [52, 55], the drive to assist others motivates the desire to share knowledge. People utilise social media to keep family, friends, and others informed about major life events. The more individuals that trade news, the more likely they are to propagate false information [56]. According to [57], false health information has been deliberately circulated. False health information may harm public safety by persuading people into accepting needless health risks. Thus, it is proposed that:

H5: There is a significant positive relationship between sharing information towards panic behaviour

The following depicts the research model of this study (Fig. 1):

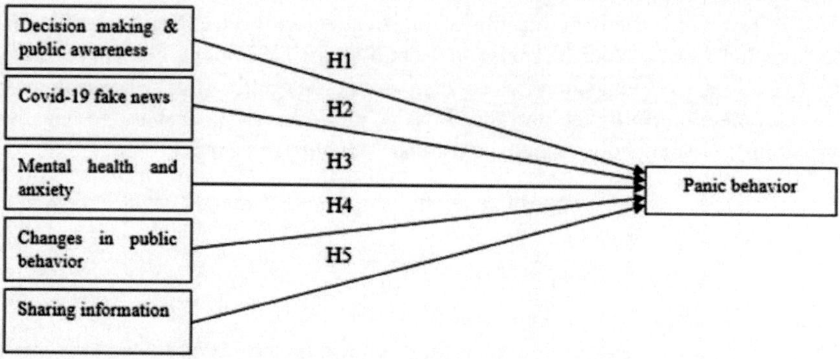

Fig. 1 Research model

3 Research Method

The study's target population was young people aged 20 to 40 from Klang Valley. Unlike conventional media, users of social media actively shape and create their own experience [48]. Social media usage is an essential part of the growth process for teenagers and young adults [48, 58]. The Google Forms questionnaires were delivered to students at a private institution in Malaysia through Facebook. The questionnaire has three stages, starting with three screenings. The responder must be a Malaysian millennial on social media. The first portion comprises demographic data such as gender, age, education level, and device used to connect social media. On a second segment, construct measures are utilised to quantify panic behaviours. The five factors' measurement items were taken from [38, 56], and [9]. The final portion of the questionnaire dealt with panic behaviour and included questions from [30]. A total of 400 responses were received from the target respondents and all questionnaires were usable. A sample size of 400 is considered large and sufficient for a multivariate research study [59–61].

Table 1 shows convergent validity is achieved as all constructs were above 0.7 for CR index and 0.5 for. All values were within the recommended threshold indicating the reliability of constructs is all considered to be good and acceptable [62].

Table 2 shows discriminant validity test by using HTMT. HTMT ratios for each construct are lower than 0.85 which was recommended by [63] except for 'MH-PBeh' which was 0.917. In summary, the constructs representing satisfactory discriminant validity and was not a serious threat in the study.

Figure 2 was presented with a direct path from decision making and public awareness, Covid 19 fake news, mental health and anxiety, public behaviour and sharing information. All variable show significant at the p-value, fake news (FN) (0.008), mental health and anxiety (MHA) (0.000), public behaviour (PB) (0.019), sharing information (SI) (0.005) respectively. All variables are contributed towards panic behaviour except for decision making (DM), where P value is (0.180) which exceeded 0.05.

Table 1 Reliability and validity assessments

Constructs	Items	Loadings	CA	CR	AVE
Decision making	4 items	0.811–0.864	0.869	0.909	0.715
Fake news	5 items	0.837–0.924	0.941	0.955	0.810
Mental health	5 items	0.926–0.941	0.963	0.971	0.870
Public behaviour	5 items	0.878–0.923	0.947	0.959	0.825
Sharing information	5 items	0.921–0.945	0.965	0.973	0.878
Panic behaviour	3 items	0.910–0.954	0.926	0.953	0.872

Notes CR = Compostite relaibility; AVE = Average variance extracted

Table 2 Discriminant validity: HTMT

Constructs	DM	FN	MH	PB	PBeh	SI
DM						
FN	0.454					
MH	0.481	0.849				
PB	0.572	0.724	0.793			
PBeh	0.440	0.848	0.917	0.796		
SI	0.459	0.801	0.792	0.814	0.829	

Note HTMT Values < 0.85

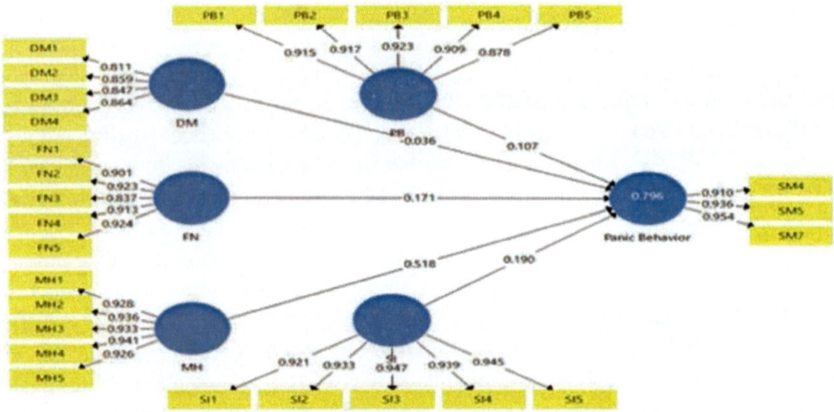

Fig. 2 PLS- SEM model with path coefficients

The models had:

1) a direct path from Covid 10 fake news towards panic behaviour
2) a direct path from mental health and anxiety towards panic behaviour
3) a direct path from public behaviour towards panic behaviour

Table 3 Hypothesis testing

		Beta	P values	VIF	Decision
H1	DM—PB	0.036	0.180	1.398	Not supported
H2	FN—PB	0.171	0.008**	3.400	Supported
H3	MH—PB	0.518	0.000**	3.858	Supported
H4	PB—PB	0.107	0.019**	3.291	Supported
H5	SI—PB	0.190	0.005**	3.514	Supported

Notes ** p-value < 0.001, * p-value < 0.05, ns = not significant

4) a direct path from sharing information towards panic behaviour

Table 3 indicates that the rest of the hypotheses proposed were supported with the exception to decision making to panic behaviour (DM-PB) where P value = 0.180, not significant. All the four factors in terms of FN, MHA, PB and SI are positively related to panic behaviour, supporting H2, H3, H4, and H5. DM (H1) did not find support, indicating a non-significant relationship between DM public behaviour.

The Variance Inflation Factor (VIF) was examined to identify multicollinearity issues. Table 3 shows that multicollinearity is not an issue among the exogenous latent constructs, since all VIF values were below 5. Thus, multicollinearity is not a threat in this study.

4 Discussion and Conclusion

The results of this study demonstrate that people are prompted to panic when they read falsified news on social media. Anxieties and doubts experienced by respondents during the COVID-19 pandemic may have been exacerbated by the dissemination of outdated and unconfirmed information about the disease, according to data collected from respondents. The way social media campaign in promoting healthy behaviours being run affect the public sentiment. A properly executed campaign put the public at ease. Information sharing among social media users could also triggers panic as misinformation occurs.

According to the findings of this study, panic was produced by young Malaysians' use of social media during the outbreak of COVID-19. Posting information on pandemics, such as COVID-19, on social media should be done with utmost caution. Additionally, anybody who distributed information on COVID-19 through social media was urged to check the content's legitimacy and reliability before to making it publicly available. This study is crucial for academics, as it modelled the elements that predict panic spreading on social media. The international community, health-care providers, legislators, in particular, the Malaysia government, can better navigate the delivery of pertinent information at this critical period of the pandemic. Due to time constraints, this research only concentrate on a single country; hence, only on one medium of communication. A future study will compare this to other media outlets and countries.

References

1. Mohd Radzi SF, Hassan MS, Mohd Radzi MAH (2020) How do Southeast Asia countries respond and mitigate to novel coronavirus pandemic? A lesson from Malaysia. Asia Pac J Public Health 32(8):453–455. https://doi.org/10.1177/1010539520962970
2. Mheidly N, Fares J (2020) Leveraging media and health communication strategies to overcome the COVID-19 infodemic. J Public Health Policy 41(4):410–420. https://doi.org/10.1057/s41271-020-00247-w
3. Sohrabi C, Alsafi Z, O'Neill N, Khan M, Kerwan A, Al-Jabir A, Iosifidis C, Agha R (2020) World Health Organization declares global emergency: a review of the 2019 novel coronavirus (COVID-19). Int J Surg 76:71–76. https://doi.org/10.1016/j.ijsu.2020.02.034
4. Tang KH (2020) Movement control as an effective measure against Covid-19 spread in Malaysia: an overview. J Public Health. https://doi.org/10.1007/s10389-020-01316-w
5. Holland & Knight homepage. https://www.hklaw.com/en/insights/publications/2020/04/the-impact-of-covid19-on-your-advertising-and-marketing-campaigns. Accessed 12 Apr 2022
6. Chen K, Luo Y, Hu A, Zhao J, Zhang L (2021) Characteristics of misinformation spreading on social media during the COVID-19 outbreak in China: a descriptive analysis. Risk Manag Healthc Policy 14:1869–1879. https://doi.org/10.2147/rmhp.s312327
7. Depoux A, Martin S, Karafillakis E, Preet R, Wilder-Smith A, Larson H (2020) The pandemic of social media panic travels faster than the COVID-19 outbreak. J Travel Med 27(3):taaa031
8. Naeem SB, Bhatti R (2020) The Covid-19 'infodemic': a new front for information professionals. Health Info Libr J 37(3):233–239. https://doi.org/10.1111/hir.12311
9. Al-Dmour H, Masa'deh R, Salman A, Abuhashesh M, Al-Dmour R (2020) Influence of social media platforms on public health protection against the COVID-19 pandemic via the mediating effects of Public Health Awareness and behavioral changes: Integrated model. J Med Internet Res 22(8). https://doi.org/10.2196/19996.
10. Al Jazeera homepage. https://www.aljazeera.com/news/2021/3/12/malaysia-cites-covid-19-misinformation-with-new-fake-news-law. Accessed 12 Mar 2022
11. Tasnim S, Hossain MM, Mazumder H (2020) Impact of rumors and misinformation on COVID-19 in social media. J Prev Med Public Health 53(3):171–174. https://doi.org/10.3961/jpmph.20.094
12. Zou W, Tang L (2020) What do we believe in? Rumors and processing strategies during the COVID-19 outbreak in China. Public Underst Sci 30(2):153–168. https://doi.org/10.1177/0963662520979459
13. Forbes homepage. https://www.forbes.com/sites/brucelee/2020/04/09/5g-networks-and-covid-19-coronavirus-here-are-the-latest-conspiracy-theories. Accessed 12 Mar 2022
14. Cato S, Iida T, Ishida K, Ito A, Katsumata H, McElwain KM, Shoji M (2021) The bright and dark sides of social media usage during the COVID-19 pandemic: survey evidence from Japan. Int J Disaster Risk Reduct 54:102034. https://doi.org/10.1016/j.ijdrr.2020.102034
15. Blumler JG, Katz E (1974) The uses of mass communications: current perspectives on gratifications research
16. Halpern D, Valenzuela S, Katz J, Miranda JP (2019) From belief in conspiracy theories to trust in others: which factors influence exposure, believing and sharing fake news. In: Social computing and social media. Design, human behavior and analytics, pp 217–232. https://doi.org/10.1007/978-3-030-21902-4_16
17. Cheung CM, Chiu PY, Lee MK (2011) Online social networks: why do students use Facebook? Comput Hum Behav 27(4):1337–1343

18. Dhir A, Tsai CC (2017) Understanding the relationship between intensity and gratifications of Facebook use among adolescents and young adults. Telematics Inform 34(4):350–364
19. Introne J, Gokce Yildirim I, Iandoli L, DeCook J, Elzeini S (2018) How people weave online information into pseudoknowledge. Social Media+Soc 4(3):205630511878563. https://doi.org/10.1177/2056305118785639
20. Dunne Á, Lawlor MA, Rowley J (2010) Young people's use of online social networking sites—a uses and gratifications perspective. J Res Interact Mark 4(1):46–58. https://doi.org/10.1108/17505931011033551
21. Park H, Blenkinsopp J (2008) Whistleblowing as planned behavior—a survey of south korean police officers. J Bus Ethics 85(4):545–556. https://doi.org/10.1007/s10551-008-9788-y
22. Thompson N, Wang X, Daya P (2019) Determinants of news sharing behavior on social media. J Comput Inf Syst 60(6):593–601. https://doi.org/10.1080/08874417.2019.1566803
23. Chiu CM, Hsu MH, Wang ETG (2006) Understanding knowledge sharing in virtual communities: an integration of social capital and social cognitive theories. Decis Support Syst 42(3):1872–1888. https://doi.org/10.1016/j.dss.2006.04.001
24. Chew C, Eysenbach G (2010) Pandemics in the age of Twitter: content analysis of tweets during the 2009 H1N1 outbreak. PLoS ONE 5(11). https://doi.org/10.1371/journal.pone.0014118
25. Gonzalez-Padilla DA, Tortolero-Blanco L (2020) Social media influence in the COVID-19 pandemic. Int Braz J Urol 46(Suppl 1):120–124
26. Al-Qaysi N, Mohamad-Nordin N, Al-Emran M (2021) Developing a comprehensive theoretical model for adopting social media in higher education. Interact Learn Environ 1–22. https://doi.org/10.1080/10494820.2021.1961809
27. Al-Qaysi N, Mohamad-Nordin N, Al-Emran M (2019) What leads to social learning? Students' attitudes towards using social media applications in Omani higher education. Educ Inf Technol 25(3):2157–2174. https://doi.org/10.1007/s10639-019-10074-6
28. Torales J, O'Higgins M, Castaldelli-Maia JM, Ventriglio A (2020) The outbreak of COVID-19 coronavirus and its impact on global mental health. Int J Soc Psychiatry 66(4):317–320. https://doi.org/10.1177/0020764020915212
29. Al-Qaysi N, Mohamad-Nordin N, Al-Emran M (2020). Factors affecting the adoption of social media in Higher Education: a systematic review of the technology acceptance model. Stud Syst Decis Control 571–584.https://doi.org/10.1007/978-3-030-47411-9_31
30. Demuyakor J (2020) Social media and COVID-19 pandemic: enhancing panic or preventing it? Int J Humanit Arts Soc Sci 6(5):211–222
31. Ahmad AR, Murad HR (2020) The impact of social media on panic during the COVID-19 pandemic in Iraqi Kurdistan: online questionnaire study. J Med Internet Res 22(5):19–56. https://doi.org/10.2196/19556
32. Hornmoen H, McInnes C (2018) Social media communication during disease outbreaks: findings and recommendations. In: Hornmoen H, Backholm K (eds) Social media use in crisis and risk communication, Bingley, UK.
33. The Star homepage. https://www.thestar.com.my/news/nation/2021/03/11/jail-rm100000-fine-for-those-who-spread-fake-news-on-covid-19-emergency-from-friday-march-12. Accessed 21 Mar 2022
34. Statista homepage. https://www.statista.com/statistics/617136/digital-population-worldwide/. Accessed 12 Mar 2022
35. Vox homepage. https://www.vox.com/recode/2020/3/12/21175570/coronavirus-covid-19-social-media-twitter-facebook-google. Accessed 12 Mar 2022
36. Merchant RM, Lurie N (2020) Social media and emergency preparedness in response to novel coronavirus. JAMA 323(20):2011. https://doi.org/10.1001/jama.2020.4469
37. Gökalp B, Karkın N, Calhan HS (2020) The political use of social networking sites in Turkey. In: Handbook of research on managing information systems in developing economies, pp 503–521. https://doi.org/10.4018/978-1-7998-2610-1.ch025
38. Kaya T, Sağsan M, Medeni T, Medeni T, Yıldız M (2020) Qualitative analysis to determine Decision-makers' attitudes towards E-government services in a defacto state. J Inf Commun Ethics Soc 18(4):609–629. https://doi.org/10.1108/jices-05-2019-0052

39. Nulty P, Theocharis Y, Popa SA, Parnet O, Benoit K (2016) Social media and political communication in the 2014 elections to the European Parliament. Elect Stud 44:429–444
40. Huynh TLD (2020) The COVID-19 risk perception: a survey on socioeconomics and media attention. Econ Bull 40(1):758–764
41. Lampos V, Moura S, Yom-Tov E, Cox IJ, McKendry R, Edelstein M (2020) Tracking COVID-19 using online search 93:4–9
42. Hou Z, Du F, Jiang H, Zhou X, Lin L (2020) Assessment of public attention, risk perception, emotional and behavioural responses to the COVID-19 outbreak: social media surveillance in China
43. Ulvi O, Karamehic-Muratovic A, Baghbanzadeh M, Bashir A, Smith J, Haque U (2022) Social media use and mental health: a global analysis. Epidemiologia 3(1):11–25. https://doi.org/10.3390/epidemiologia3010002
44. Cham TH, Cheng BL, Ng CKY (2020) Cruising down millennials' fashion runway: a cross-functional study beyond Pacific borders. Young Consum 22(1):28–67
45. Masedu F, Mazza M, Di Giovanni C, Calvarese A, Tiberti S, Sconci V, Valenti M (2014) Facebook, quality of life, and mental health outcomes in post-disaster urban environments: the L'Aquila earthquake experience. Front Public Health 2. https://doi.org/10.3389/fpubh.2014.00286
46. Glazzard J, Stones S (2019) Social media and young people's mental health. In: Selected topics in child and adolescent mental health. IntechOpen. https://doi.org/10.5772/intechopen.88569
47. Roy D, Tripathy S, Kar SK, Sharma N, Verma SK, Kaushal V (2020) Study of Knowledge, attitude, anxiety & perceived mental healthcare need in indian population During COVID-19 pandemic. Asian J Psychiatry 51:102083. https://doi.org/10.1016/j.ajp.2020.102083
48. Berryman C, Ferguson CJ, Negy C (2017) Social media use and mental health among young adults. Psychiatr Q 89(2):307–314. https://doi.org/10.1007/s11126-017-9535-6
49. Coyne SM, Stockdale L, Summers K (2019) Problematic cell phone use, depression, anxiety, AND self-regulation: evidence from a three year longitudinal study from adolescence to EMERGING adulthood. Comput Hum Behav 96:78–84. https://doi.org/10.1016/j.chb.2019.02.014
50. Escobar-Viera CG, Whitfield DL, Wessel CB, Shensa A, Sidani JE, Brown AL, Chandler CJ, Hoffman BL, Marshal MP, Primack BA (2018) For better or for worse? A systematic review of the evidence on social media use and depression among lesbian, gay, and bisexual minorities. JMIR Ment Health 5(3). https://doi.org/10.2196/10496
51. Cham TH, Lim YM, Aik NC, Tay AGM (2016) Antecedents of hospital brand image and the relationships with medical tourists' behavioral intention. Int J Pharm Healthc Mark 10(4):412–431
52. Wang Y, Dai Y, Li H, Song L (2021) Social media and attitude change: information booming promote or resist persuasion? Front Psychol 12. https://doi.org/10.3389/fpsyg.2021.596071
53. Rovetta A, Bhagavathula AS (2020) COVID-19-related web search behaviors and Infodemic attitudes in Italy: infodemiological study. https://doi.org/10.2196/preprints.19374
54. Misra AK, Sharma A, Shukla JB (2015) Stability analysis and optimal control of an epidemic model with awareness programs by media. Biosystems 138:53–62. https://doi.org/10.1016/j.biosystems.2015.11.002
55. Hayes JL, King KW (2014) The social exchange of viral ads: referral and coreferral of ads among college students. J Interact Advert 14:98–109. https://doi.org/10.1080/15252019.2014.942473
56. Apuke OD, Omar B (2021) Fake news and COVID-19: modelling the predictors of fake news sharing among social media users. Telematics Inform 56:101475
57. Pulido CM, Villarejo-Carballido B, Redondo-Sama G, Gómez A (2020) Covid-19 infodemic: more retweets for science-based information on coronavirus than for false information. Int Sociol 35(4):377–392. https://doi.org/10.1177/0268580920914755
58. Best P, Manktelow R, Taylor B (2014) Online communication, social media and adolescent wellbeing: a systematic narrative review. Child Youth Serv Rev 41:27–36. https://doi.org/10.1016/j.childyouth.2014.03.001

59. Hair JFJ, Black WC, Babin BJ, Anderson RE, Tatham RL (2010) Multivariate data analysis a global perspective. Pearson Education International, Englewood Cliffs
60. Saunders M, Lewis P, Thornhill A (2012) Research methods for business students, 6th edn. Pearson, Englewood Cliffs
61. Memon MA, Ting H, Cheah JH, Ramayah T, Chuah F, Cham TH (2020) Sample size for survey research: review and recommendations. J Appl Struct Equ Model 4(2):1–20
62. Hair JF Jr, Hult GTM, Ringle C, Sarstedt M (2014) A primer on partial least squares structural equations modeling (PLS-SEM). Sage, Thousand Oaks
63. Henseler J, Ringle CM, Sarstedt M (2015) A new criterion for assessing discriminant validity in variance-based structural equation modeling. J Acad Mark Sci 43(1):115–135

Social Media Co-creation Activities Among Elderly Consumers: An Innovation Resistance Perspective

Tat-Huei Cham, Eugene Cheng-Xi Aw, Garry Wei-Han Tan, and Keng-Boon Ooi

Abstract Since its inception, social media has disruptively transformed consumers' consumption patterns. Social media's unique attributes that allow consumers to voice their opinion and engage in multi-way social conversations with various stakeholders have encouraged them to engage in co-creation activities on social media. However, the participation among the elderlies in the social media co-creation activities remains minimal to date. This study aims to examine the influence of risk and functional barriers in explaining the resistance towards co-creation activities on social media among the elderly. The moderating role of perceived trust was also investigated in the proposed relationships. The data was gathered from 356 respondents using a self-administered online questionnaire. The study presents the importance of functional barriers (e.g., incompatibility and perceived complexity) and risk barriers (e.g., privacy risk and security risk) in influencing elderlies' resistance to social media co-creation activities. The research findings and implications are discussed.

Keywords Social media · Co-creation · Resistance · Risk barriers · Functional barriers · Perceived trust · Elderlies

1 Introduction

In the current era of connectivity, Information and communications technology (ICT) has disruptively transformed the way consumers consumed, behave, and experience product/services. This paradigm shift is driven by various enablers such as the availability of smart phone, computers, advance software applications, affordable internet broadband, and innovative Internet-based technologies. Among all the enablers, social networking service or better known as "social media" is reported to plays a substantial role in gearing digitalization efforts within the consumers market [1, 2]. According to [3], social media is a cluster of applications that is build based on technology that supports the formation and sharing of mass information. Since its

T.-H. Cham · E. C.-X. Aw (✉) · G. W.-H. Tan · K.-B. Ooi
UCSI Graduate Business School, UCSI University, Kuala Lumpur, Malaysia
e-mail: eugenecx.aw@gmail.com

M. Al-Emran et al. (eds.), *International Conference on Information Systems and Intelligent Applications*, Lecture Notes in Networks and Systems 550,
https://doi.org/10.1007/978-3-031-16865-9_15

inception, social media has emerged as a dominant digital communication channel in the consumer market whereby the system has transformed the way businesses interact with their clienteles [4, 5]. From the marketing perspective, social media has been regarded as a marketing communication strategy that could shape consumers' perceptions and behaviour by allowing them to involve directly in the product/service acquisition process in which they can learn, share information, purchase, and evaluate the products or brands [6].

In addition, the past literature has indicated that social media communication come in two forms, namely firm-created content that are developed by business entities and user-created social media that are established by the users [4, 6]. From the marketing point of view, the two-way communication nature of social media allows consumers to engage in multi-way social conversations with various stakeholders. This scenario has provided consumers with a sense of "empowerment" and encouraged them to be involved in co-creation activities on social media. Co-creation is built on ongoing collaboration and communications between businesses and consumers in creating new products or services [7]. Social media allows businesses to connect with vast numbers of consumers, strengthen their relationship with the customers, offer customers unique consumption experiences, and improve their product offerings [4, 6]. Given the importance of communication in co-creation, businesses and their customers are increasingly relying on social media platforms to facilitate the co-creation process [7].

Past studies have documented that the success of co-creation in social media can only be achieved with interactive and engaging communication between business and customers [8]. In other words, all parties must participate in the communication process in order to ensure that the co-creation activities take place effectively. However, there are limited attempts in understanding participation of consumers from different age group in the co-creation activities on social media. For example, most of the findings from the reviewed literature uncovered that co-creation activities on social media mostly involve younger generation group of consumers/users, neglecting the potential significance of the older cohort's involvement [9, 10]. Yet, an increasing number of older consumers has been reported as active social media users who involve in buying goods or services on social media [11]. Moreover, it was also found that elderlies actively used social media as the platform for information sharing [12, 13] and as a reference point for their purchase decision making [14]. Recognised as the potential segment who have more buying power and disposable income than other population segments, the participation of the elderlies is indeed important to ensure the success of co-creation activities on the social network sites (SNS). However, the review of literature found that the participation rate among the elderly in co-creation activities remains scant to date.

In view of the importance of co-creation, the deficiency in research as highlighted above represents a significant gap from the consumer behaviour standpoint that is worth to be investigated in view of the potential of this group of consumers. Drawing from the innovation resistance theory, the present study was set to examine the underlying reasons behind elderlies' resistance towards the co-creation activities on social media. The findings from the study are expected to provide an inclusive viewpoint by providing preliminary view to the rejection towards co-creation activities among the elderly.

2 Literature Review and Hypotheses Development

As highlighted in the information system (IS) literature, technology adoption is one of the important areas of research that seeks to understand how consumers react to a technological innovation. In the context of technology adoption, Innovation Resistance Theory (IRT) can be explained as the resistance-oriented behaviour of users towards the technology [15]. Users' technology resistance represents a common problem as consumers tend to resists the new and improved components/functions if they do not see any benefit and value from it. Hence, resistance from the users is often regarded as the barrier for the technological innovation to diffuse and sustain. Grounded on the foundation of IRT, the current study intends to explore the fundamental of resistance towards co-creation in social media through its antecedents (i.e. perceived complexity, perceived incompatibility, privacy risk, and security risk) and consequences (i.e. perceived trust and non-adoption intention) (Fig. 1).

2.1 Factors Influencing Resistance Towards Co-creation

According to [16], complexity in the present study can be explained as "the degree to an innovation is perceived as relatively difficult to understand and use". [17] highlighted that complexity for a subject can be view from two perspectives namely the (1) complexity of the idea of innovation and (2) the complexity in executing the idea. The past literature reported that perceived complexity associated with technological innovations has a significant impact on the acceptance and adoption rate among the users [18, 19]. Apart from that, complexity of the technological innovations was also found to have a direct impact on users' rejection towards the innovation [20]. Likewise, the complexity in social media content co-creation is reported to discourage the usage of such function [21]. As such, it is anticipated that complexity could create resistance among the older consumers to engage in social media co-creation activities as well.

Perceived incompatibility in the present study refers to the level to which innovation is professed to be inconsistent with the past experiences, needs, and existing values of the users [6]. It was reported that compatibility is vital for new technological innovation adoption as it could reduce the possible uncertainty associated with the technology [20]. Moreover, less compatible technological innovation as perceived by the individual would make him/her reject the adoption of the said technology [22]. Perceived compatibility is regarded as one the key determinant that promotes speedy adoption of co-creation among the users on social media [23]. Drawing from the evidence above, this study suggests that perceived incompatibility will have a direct impact on resistance towards co-creation activities on social media among the older consumers.

In the online environment, the issues of privacy has always been a concern for many of the internet users. Consistent with this argument, privacy concerns are reported to be the significant challenges for the acceptance of products and services that are related to technology [24]. According to [25], privacy risk can be defined as the potential exposure of a user's private information as a result of using a technological product or service. In the context of social media, privacy has been regarded as the major concern for its adoption due to the exposure of the users' private information (e.g., personal interests, name, geographic location, birthdate, etc.) and possible data usage by marketing and advertising companies. It was argued that such information could be maliciously used or violated by irresponsible parties [26]. To concur, the existing literature has reported that privacy risk has found to have negative impact on users' social media usage [27] and co-creation intention [28]. In view of this, there is a possibility that privacy risk can create rejection among the users towards co-creation activities on social media.

Security risk in the present study is defined as the possible loss of personal information or fraud, which exposes the security of an online user [29]. Since the nature of social media is operated entirely online, there would be a risk of computer security and information breaches that could compromise the confidentiality of data of the users [30]. Moreover, the issues related to social media's system malfunction, inadequate internal processes, and slow response by the admin are among the security hazards that could hinder the use of social media. Hence, it is undeniable that security risk remains as the primary concern for social media usage in view of the security issues associated with it, such as the possibility that personal information could be exposed and used for fraudulent activities [31]. The past studies have recorded that user are reluctant to adopt a technology if they perceived that the security risk is high [32]. Drawing from the evidence above, the following hypotheses are postulated:

H1: Perceived complexity has a direct influence on the resistance towards co-creation activities on social media.
H2: Perceived incompatibility has a direct influence on the resistance towards co-creation activities on social media.
H3: Privacy risk has a direct influence on the resistance towards co-creation activities on social media.
H4: Security risk has a direct influence on the resistance towards co-creation activities on social media.

2.2 *Linking Resistance, Perceived Trust and Non-adoption Intention*

Consumers' resistance has often been regarded as a major challenge for technological innovation adoption and usage. This is due to the fact that consumer resistance could

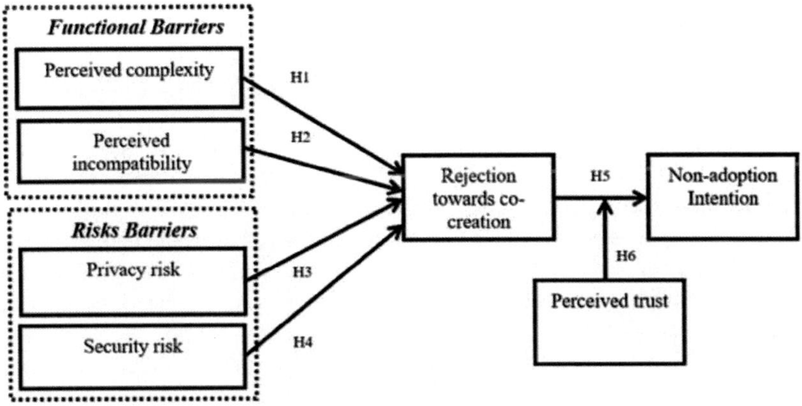

Fig. 1 Research model

determine whether the innovation is successful or vice-versa. Past studies have highlighted that user's resistance towards innovation has an impact on their non-adoption intention towards the innovation [11, 33]. In other words, this scenario shows that the users may choose not to adopt co-creation activities in social media if they have the sense of resistance towards it. Moreover, it was documented that lacking of trust among the users will also inhibit the adoption rate of a new technology or innovation [34]. To concur, the evidence from the recent studies also reported that users' trust greatly influences their resistance and non-adoption of certain innovations and technologies [33, 35]. Extending to this logic, it is anticipated that the relationship between user's resistance towards co-creation activities and their non-adoption intention could be moderated by the level of their trust. As such, the following hypotheses are postulated:

H5: Resistance has a direct influence on the non-adoption intention towards co-creation activities on social media.

H6: Perceived trust has a moderating effect on the link between elderlies' resistance and intention not to adopt co-creation activities on social media.

3 Research Methodology

The data in this study was collected through online self-administered questionnaire (via Qualtrics) from the elderly respondents as suggested by [11]. The online survey questionnaire was distributed to 400 respondents through Facebook, WhatsApp, and email. A purposive sampling approach with screening criteria was adopted in this study with the aim to obtain reliable response and ensure the respondents meet the qualifying criteria before they participated in the study. The criteria imposed for the screening purpose were (1) the respondents must be at least 60 years old of age, (2) they owned at least a social media account, and (3) they never participated in any co-creation activities on social media before. Following the data cleaning procedures, 356 responses were retained to be use for further analysis and hypothesis testing purposes.

In addition, all the measurement items for the variables included in this study were sourced from the prior literature [6, 11, 33]. The items included in the questionnaire were measured with the use of a six-point Likert scale whereby 6 implies strongly agree while 1 implies strongly disagree. The questionnaire was then pretested with the experts to ensure clarity of the questions, sequential arrangement, and the requirements of face validity are achieved. Moreover, Harman's single-factor analysis was conducted to examine the aspect of common method bias before proceed with the data analysis [36]. Since the highest single factor only contributed 29% of the variance (<40% threshold recommended), it can be assumed that common method bias is found not to be an issue for the present study.

4 Data Analysis

In term of demographic profile, the sample for the present study comprised of 53.1% men and 46.9% women, who were married (84.6%), single (11.2%), divorced (2.3%), and the rest are widowed (1.9%). More than half of the respondents (53.4%) held a diploma degree, while 25.6% held a Bachelor's degree, 10.7% held a primary/secondary school qualification, 8.7% held a Master's degree and 1.6% held a Doctorate degree. In terms of usage, majority (52.3%) of the respondents uses social media between 7–9 h per day.

The data analysis in this study was conducted with the use of AMOS statistical software based on the two-steps approach (measurement model and structural model assessment) as suggested by [37]. Confirmatory factor analysis was used to address the model fit of the measurement model before assessing the constructs' convergent and discriminant validity. According to [37], a research model can be regarded as fit if the value χ^2/df (Normed Chi-square) ≤ 3, RMSEA (Root Mean Square Error of Approximation) ≤ 0.08, GFI (Goodness of Fit) ≥ 0.90, PNI (Parsimony Normed Fit Index) ≥ 0.50 and TLI (Tucker-Lewis index) ≥ 0.90. The results of the measurement model indicated that the χ^2/df = 1.231, GFI = 0.935, RMSEA = 0.026, TLI = 0.982, and PFI = 0.782, suggesting the establishment of model fit. Moreover, [37] proposed that convergent validity for the measurement model is established if (1) the loadings for all the items exceed 0.60, (2) the constructs' average variance extracted (AVE) is larger than the recommended value of 0.50, and (3) the constructs' composite reliability (CR) is larger than the recommended value of 0.70. The findings from statistical analysis output in Table 1 indicated that all the loadings value for all the items are larger than 0.60 and the value of the AVE and CR is above 0.50 and 0.70 respectively, thus suggest that convergent validity was established in this study.

As for the discriminant validity, this aspect was addressed through the examination of the value of AVE (squared root) compared to the value of variance shared between

Table 1 The result of convergent validity

	Items	FL	AVE	CR
RESIST	5	0.709–0.817	0.594	0.879
SECURE	3	0.733–0.898	0.658	0.851
PRIVACY	4	0.751–0.786	0.581	0.847
TRUST	4	0.654–0.800	0.557	0.832
COMPLEX	3	0.713–0.757	0.536	0.776
INCOMPATIBILITY	3	0.702–0.786	0.543	0.781
NAI	3	0.669–0.785	0.578	0.802

Notes TRUST = Perceived trust, INCOMPATIBILITY = Perceived incompatibility; RESIST = Resistance, NAI = non-adoption intention; COMPLEX = Perceived complexity; SECURE = Security risk, PRIVACY = privacy risk, CR = Composite reliability, AVE = Average variance extracted, FL = Factor loadings

any two constructs. According to [38] discriminant validity is said to be established if the value of the variance shared between other constructs are lesser than the value of AVE (squared root). As highlighted in Table 2, it was found that the value of AVE that has been squared root (in italics) is larger than the value of variance shared with other constructs (in bold), thus suggesting that discriminant validity is established in this study. As for the structural model assessment, the model fit was assessed before proceeding to hypotheses testing. The analysis of the structural model revealed that the $\chi^2/df = 1.290$, GFI = 0.931, RMSEA = 0.029, TLI = 0.977, and PFI = 0.801, indicating the structural model is considered fit. Table 3 shows the analysis results for the causal paths related to the developed hypotheses.

Results showed that all the hypotheses were supported. For instance, both the functional (e.g. perceived complexity and perceived incompatibility) and risk (e.g. privacy risk and security risk) barriers were found to have positive direct effect on

Table 2 The result of discriminant validity

	1	2	3	4	5	6	7
RESIST	*0.771*[b]						
SECURE	**0.578**[a]	*0.811*					
PRIVACY	**0.503**	**0.427**	*0.762*				
TRUST	**0.418**	**0.372**	**0.342**	*0.746*			
COMPLEX	**0.433**	**0.405**	**0.275**	**0.083**	*0.732*		
INCOMPATIBILITY	**0.425**	**0.234**	**0.333**	**0.233**	**0.275**	*0.737*	
NAI	**0.199**	**0.229**	**0.225**	**0.264**	**0.083**	**0.033**	*0.760*

Notes TRUST = Perceived trust, INCOMPATIBILITY = Perceived incompatibility; RESIST = Resistance, NAI = non-adoption intention; COMPLEX = Perceived complexity; SECURE = Security risk, PRIVACY = privacy risk, [a] The off-diagonal values (in bold) signify the variance shared between constructs; [b] The diagonal values (in italics) signify the squared root average variance extracted by the construct

Table 3 Result of path analysis

	Standardized estimate (β)	Critical ratio	Hypothesis
H1: COMPLEX ⟶ RESIST	0.154	2.569*	Yes
H2: INCOMPATIBILITY ⟶ RESIST	0.206	3.816**	Yes
H3: PRIVACY ⟶ RESIST	0.216	4.026**	Yes
H4: SECURE ⟶ RESIST	0.342	5.848**	Yes
H5: RESIST ⟶ NAI	0.284	2.986*	Yes

Notes TRUST = Perceived trust, INCOMPATIBILITY = Perceived incompatibility; RESIST = Resistance, NAI = non-adoption intention; COMPLEX = Perceived complexity; SECURE = Security risk, PRIVACY = privacy risk, * and ** denote significant at 95% and 99% confidence level respectively

Table 4 Result of interaction analysis

			Con. Interval	
Variable	β	SE	lower bound	upper bound
Model: Perceived trust moderate the resistance - non-adoption intention link				
H6: Interaction (RESIST X TRUST)	0.186	0.044	0.073	0.213

Notes TRUST = Perceived trust, RESIST = Resistance, SE = Standard Error, β = Co-efficient Beta, Con.Interval = Confidence intervals at 95%, U.L = Upper Limit, L.L = Lower Limit, * p-value < 0.05, ** p-value < 0.001

elderlies resistance towards co-creation activities on social media, which in turn, influence their intention not to adopt it ($\beta = 0.284$, $p < 0.05$). For the moderating effect of perceived trust, the interaction analysis through 2,000 bootstrap samples indicated in Table 4 shows that the interaction analysis is significant grounded on the 95 per cent confidence interval measure with upper level of 0.213 and lower level of 0.073 . Additionally, the graph highlighted in the second figure indicated that elderlies who have less trust towards co-creation activities on social media is steeper than those who have higher trust. This scenario shows that the association between the resistance of elderlies and non-adoption intention towards the co-creation activities on social media is more substantial for users with a low level of trust (Fig. 2).

5 Discussion and Implications of Research Findings

The results of this study indicated that perceived complexity, perceived incompatibility, privacy risk, and security risk would positively influence elderlies' resistance towards co-creation activities on social media, which in turn, influence their non-adopt intention. This finding spelt out the importance of functional and risk aspects

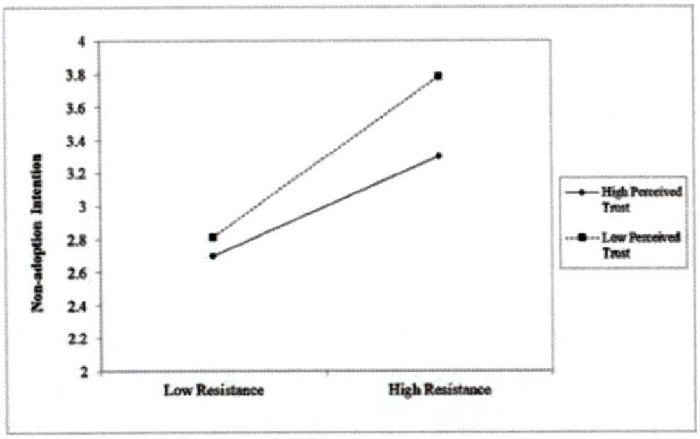

Fig. 2 Plot of interaction analysis

in influencing one's sense of resistance towards co-creation activities. Correspondingly exemplified through the studies by [21–23], it was argued that the aspects of complexity and compatibility would impact the elderlies' consideration towards the social media co-creation activities acceptance and adoption. This scenario is plausible as digital technicalities, uncertainties and complications associated with social media platforms may results in rejection from users due to the difficulties in dealing with it [23].

As for the context of risk barriers, the present study successfully put forward that both privacy and security risks were found to significantly impact the elderly's resistance to co-creation activities on social media which is in line with prior literature [27–30]. This outcome is possible due to the potential leakage of a user's private information, negligence among information handlers, fraud and uncertainties resulting from the social media platforms [30, 31]. Adhered to the findings by [32], consumers' privacy concern is regarded as a fundamental consideration between one's readiness and involvement with co-creation activities on social media. As such, it is anticipated that risk implications are intensified if elderlies are neither prepared for co-creation activities on social media nor intend for such endeavour. Moreover, the underperformance of the social media platforms operators in handling co-creation activities on their platform could also stir negative association to the security risk whilst intensifying elderlies' rejection towards the innovation as well [34, 35].

In addition to the above, the obtained findings in the present study have successfully highlighted the significant of perceived trust as the moderator in the resistance and intention to adopt relationship. Despite the importance of resistance, this study put forward the importance of perceived trust in explaining users' non-adopt intention towards co-creation activities. This finding thus highlights that trust is a vital element that could influence the likelihood of an individual's intention to adopt a certain technology as it has the capability to reduce one's worries and fears [33, 35]. Specifically, lacking of trust among the elderlies in this case will inhibit their participation in co-creation activities on social media, in which have direct effect on their non-adoption intention [32–34].

Consequently, social media companies should invest sufficient resources in system and risk management aspects when dealing with their platform design. As for the functional aspects, the social media operators should consider enhancing their platform functions such as graphic user interface, layout of the platform, and having responsive support team so that the social media platform will be perceived as user friendly. Besides, it is recommended for social media operators to constantly educate their users and collect feedback about the performance of their social media platform for improvement purposes. All the feedback received from the users from time to time should be considered in the organisation's strategic planning purposes to layout sustainable social networking and operating policy in the long term.

In terms of security and privacy aspects, the social media operators should consider using an advanced authentication mechanism, make configurable security and privacy setting easily accessible to users, make report users function available, make their platform privacy policy transparent and available to the public, train their staff on how to handle social media security issues, regularly review social media security issues, and hold awareness campaigns on issues how to deal with common social media security risks for users from time to time. The awareness campaign should include agendas that expose the users to the understanding of phishing, scams, malware attacks, hacks, social network privacy settings, monitoring and social streams, and other agendas that could help them protect themselves when using social media. In this case, such initiatives would make the users particularly the elderly feel safer to take part in social media activities and improve their overall perception toward their social media platforms.

In summary, the finding from the present study contributed to the consumer behaviour and technology management literature. Specifically, this study is one the few that focused on the effect of both risk and functional barriers on resistance among the older consumers towards co-creation activities on social media. Moreover, the study also highlights the moderating role of perceived trust in explaining the non-adoption intention. The findings hereby offer an inclusive view on how social media operators can overcome the concerns encountered by the elderlies for them to participate in social media co-creation activities in the long term.

References

1. Cham TH, Cheng BL, Ng CKY (2020) Cruising down millennials' fashion runway: a cross-functional study beyond Pacific borders. Young Consum 22(1):28–67
2. Wong CH, Tan GWH, Loke SP, Ooi KB (2015) Adoption of mobile social networking sites for learning? Online Inf Rev 39(6):762–778
3. Kaplan AM, Haenlein M (2010) Users of the world, unite! The challenges and opportunities of Social Media. Bus Horiz 53(1):59–68
4. Cham TH, Cheng BL, Low MP, Cheok JBC (2020) Brand Image as the competitive edge for Hospitals in Medical Tourism. Eur Bus Rev 31(1):31–59
5. Wong LW, Tan GWH, Hew JJ, Ooi KB, Leong LY (2020) Mobile social media marketing: a new marketing channel among digital natives in higher education? J Mark High Educ 32:113–137
6. Cham TH, Lim YM, Sigala M (2022) Marketing and social influences, hospital branding, and medical tourists' behavioural intention: before-and after-service consumption perspective. Int J Tour Res 24(1):140–157

7. Piller F, Vossen A, Ihl C (2012) From social media to social product development: the impact of social media on co-creation of innovation. Die Unternehmung Swiss J Bus Res Pract 66(1):7–27
8. Jouny-Rivier E, Reynoso J, Edvardsson B (2017) Determinants of services co-creation with business customers. J Serv Mark 21(2):85–103
9. Hsu LJ, Yueh HP, Hsu SH (2021) Subjective social capital and loneliness for the elderly: the moderator role of line and Facebook use. Soc Media+Soc 7(3):20563051211043906
10. Moore RC, Hancock JT (2020) Older adults, social technologies, and the coronavirus pandemic: challenges, strengths, and strategies for support. Soc Media+Soc 6(3):2056305120948162
11. Cham TH, Cheah JH, Cheng BL, Lim XJ (2021) I am too old for this! Barriers contributing to the non-adoption of mobile payment. Int J Bank Mark. https://doi.org/10.1108/IJBM-06-2021-0283
12. Ramírez-Correa PE, Rondán-Cataluña FJ, Arenas-Gaitán J, Grandón EE, Alfaro-Pérez JL, Ramírez-Santana M (2021) Segmentation of older adults in the acceptance of social networking sites using machine learning. Front Psychol 12:1–12
13. Cham TH, Cheng BL, Lee YH, Cheah JH (2022) Should I buy or not? Revisiting the concept and measurement of panic buying. Curr Psychol 1–21. https://doi.org/10.1007/s12144-022-03089-9
14. Toska A, Zeqiri J, Ramadani V, Ribeiro-Navarrete S (2022) Covid-19 and consumers' online purchase intention among an older-aged group of Kosovo. Int J Emerg Mark. https://doi.org/10.1108/IJOEM-12-2021-1875
15. Hew JJ, Leong LY, Tan GWH, Ooi KB, Lee VH (2019) The age of mobile social commerce: an Artificial Neural Network analysis on its resistances. Technol Forecast Soc Chang 144:311–324
16. Rogers EM (2003) Diffusion of innovations, 5th edn. Free Press, New York
17. Ram S (1987) A model of innovation resistance. In: Melanie W, Paul A (eds) Advances in consumer research, vol 14. ACR North American Advances
18. Adapa S, Fazal-e-Hasan SM, Makam SB, Azeem MM, Mortimer G (2020) Examining the antecedents and consequences of perceived shopping value through smart retail technology. J Retail Consum Serv 52:101901
19. Araujo T (2018) Living up to the chatbot hype: the influence of anthropomorphic design cues and communicative agency framing on conversational agent and company perceptions. Comput Hum Behav 85:183–189
20. Tsai JM, Cheng MJ, Tsai HH, Hung SW, Chen YL (2019) Acceptance and resistance of telehealth: the perspective of dual-factor concepts in technology adoption. Int J Inf Manag 49:34–44
21. Breidbach CF, Maglio PP (2016) Technology-enabled value co-creation: an empirical analysis of actors, resources, and practices. Ind Mark Manag 56:73–85
22. Turja T, Aaltonen I, Taipale S, Oksanen A (2020) Robot acceptance model for care (RAM-care): a principled approach to the intention to use care robots. Inf Manag 57(5):103220
23. Cheng JH, Yu CK, Chien FC (2021) Enhancing effects of value co-creation in social commerce: insights from network externalities, institution-based trust and resource-based perspectives. Behav Inf Technol 41:1755–1768
24. Hérault S, Belvaux B (2014) "Privacy paradox" et adoption de technologies intrusives Le cas de la géolocalisation mobile. Decis. Mark 67–82
25. Featherman MS, Pavlou PA (2003) Predicting e-services adoption: a perceived risk facets perspective. Int J Hum Comput Stud 59(4):451–474
26. Ayaburi EW, Treku DN (2020) Effect of penitence on social media trust and privacy concerns: the case of Facebook. Int J Inf Manag 50:171–181
27. Van Schaik P, Jansen J, Onibokun J, Camp J, Kusev P (2018) Security and privacy in online social networking: risk perceptions and precautionary behaviour. Comput Hum Behav 78:283–297
28. Tajvidi M, Richard MO, Wang Y, Hajli N (2020) Brand co-creation through social commerce information sharing: the role of social media. J Bus Res 121:476–486
29. Ariffin SK, Mohan T, Goh YN (2018) Influence of consumers' perceived risk on consumers' online purchase intention. J Res Interact Mark 12(3):309–327

30. Zhang Z, Gupta BB (2018) Social media security and trustworthiness: overview and new direction. Future Gener Comput Syst 86:914–925
31. Dwivedi YK, Kelly G, Janssen M, Rana NP, Slade EL, Clement M (2018) Social media: the good, the bad, and the ugly. Inf Syst Front 20(3):419–423
32. Rehman ZU, Baharun R, Salleh NZM (2020) Antecedents, consequences, and reducers of perceived risk in social media: a systematic literature review and directions for further research. Psychol Mark 37(1):74–86
33. Leong LY, Hew TS, Ooi KB, Wei J (2020) Predicting mobile wallet resistance: a two-staged structural equation modeling-artificial neural network approach. Int J Inf Manag 51:102047
34. Gutierrez A, Boukrami E, Lumsden R (2015) Technological, organisational and environmental factors influencing managers' decision to adopt cloud computing in the UK. J Enterp Inf Manag 28(6):788–807
35. Ullah F, Sepasgozar SM, Thaheem MJ, Al-Turjman F (2021) Barriers to the digitalisation and innovation of Australian Smart Real Estate: a managerial perspective on the technology non-adoption. Environ Technol Innov 22:101527
36. Low MP, Cham TH, Chang YS, Lim XJ (2021) Advancing on weighted PLS-SEM in examining the trust-based recommendation system in pioneering product promotion effectiveness. Qual Quant 1–30
37. Hair JFJ, Black WC, Babin BJ, Anderson RE, Tatham RL (2010) Multivariate data analysis a global perspective. Pearson Education International, Englewood Cliffs
38. Fornell C, Larcker DF (1981) Structural equation models with unobservable variables and measurement error: algebra and statistics. J Mark Res 18(3):382

Acceptance of IoT Technology for Smart Homes:A Systematic Literature Review

Siti Farah Hussin, Mohd Faizal Abdollah, and Ibrahim Bin Ahmad

Abstract The Internet of things for smart home (IoT SH) technology is the latest technology for homes that integrates sensors, functional software, and network connections. However, the acceptance of IoT SH technology remain low. Hence, Information System (IS) researchers have shown interest in determining the theories and models that influence the acceptance of this technology. This study can assist IoT SH practitioners in enhancing the functionality and shortcomings of IoT products or services for smart homes in order to attract more users. Apart from identifying theories and models, this study will suggest a direction for future research. A systematic literature review was conducted to explore IoT SH by re- viewing previous studies from 2018 to February 2022 with a total of 22 selected research papers. The results show that previous studies covered different technology acceptance theories related to IoT SH namely Technology Acceptance Model (TAM), Unified Theory of Acceptance and Use of Technology (UTAUT), Unified Theory of Acceptance and Use of Technology (UTAUT 2) and alternative theory to measure the factors. The findings of this review will aid academics, especially novice researchers in understanding the current trends and gaps, as well as future work for IoT SH research.

Keywords Internet of things (IoT) · Technology acceptance theories · Systematic literature review

S. F. Hussin (✉) · M. F. Abdollah · I. B. Ahmad
Faculty of Information Technology and Communication, Universiti Teknikal Malaysia Melaka, Melaka, Malaysia
e-mail: farah.hjhussin@gmail.com

M. F. Abdollah
e-mail: faizalabdollah@utem.edu.my

I. B. Ahmad
e-mail: ibrahim@utem.edu.my

M. Al-Emran et al. (eds.), *International Conference on Information Systems and Intelligent Applications*, Lecture Notes in Networks and Systems 550,
https://doi.org/10.1007/978-3-031-16865-9_16

1 Introduction

The term "Internet of things" (IoT) was introduced by Kevin Ashton, who was a digital innovation expert in 1999. Nowadays, IoT is one of the technologies used in smart homes. A smart home, according to Yang et al [1], is a home with sensors that are able to control equipment remotely, monitored home condition, equipped with high-tech home appliances and connected to the networks. According to [2], the global IoT SH technology market is expected to grow by 21.1% between 2021 and 2028, from USD 99.89 billion to USD 380.52 billion.

The primary purpose of IoT SH is to make living at home more comfortable with better security for the residents [3], making living safer and easier for older adults [4, 5], providing elderly better healthcare services [6], automation [1, 7], and energy reduction in residential sectors [8].

The IoT technology is still considered a new technology used in smart homes, and IoT SH will only be successful if users accept and subsequently adopt this technology [9]. Despite its advantages, IoT SH has a low level of acceptance [10]. Therefore, researchers and practitioners use theories and models to find out the factors that influence the acceptance of IoT SH among users. According to [1, 7, 11–15], the use of various theories and models is important because IoT SH potential users come from various age groups, backgrounds, and usage types of IoT SH services.

Although studies on the acceptance of IoT SH technology has been explored for more than five years, there is no up-to-date study on the theories and models that impact user's acceptance of this technology. In addition, acceptance studies had a significant impact on IoT SH technology, however only a few comprehensive studies have been conducted on this subject [16].

Studies on the acceptance of IoT SH enable researchers to identify gaps associated with the current knowledge. There is a need to conduct research on theories and models according to the latest IoT technology development because technology is constantly evolving [13]. Therefore, this study aims to examine about the theories and models used in the acceptance of IoT SH.

The systematic literature review technique is used to identify and evaluate studies related to IoT SH acceptance. This paper is divided into five sections. Section 2 focuses on the research methodology. Technology acceptance theories include TAM, UTAUT, UTAUT 2, and alternative theory that were applied in previous IoT SHs are discussed in Sect. 3. Then, Sect. 4 suggested the possible directions for future research. The conclusions of the current study are discussed in Sect. 5.

2 Research Methodology

The majority of important relevant research is published in prestigious international publications. Therefore, Scopus, Science Direct, and Google Scholar databases were used as the data source in this study. The search string technique was based on two research questions:

1. What are the theories or models applied in previous IoT SH studies?
2. What are the potential future research directions for IoT SH?

The search strings were as follows: ((“Aged” OR “Old people” OR “Aging population” OR “Senior citizens” OR “Older adult”) OR (“Healthcare”) OR (“Save Energy” OR “Energy Management”) OR (“Security” OR “Surveillance”) OR (“Home Management” OR “Home Automation”) OR (“Comfort” OR “Quality of Life”) AND (“Smart Home Technology” OR “Smart House”) AND (“Internet of Things” OR “IoT”) AND (“adoption” OR “behavioral AND intention”, OR “acceptance”) AND (“empirical” OR “quantitative” OR “qualitative” OR “mixed method”)). This study focused on articles from 2018 to the end of February 2022. The inclusion criteria were as follows: (1) English articles, (2) quantitative, qualitative, or mixed-method research, and (3) the theories or models of acceptance of IoT SH. Next, the exclusion criteria were as follows: (1) papers that are not in the English language, (2) papers that are lesser than four pages in length, (3) papers that are not related to IS scope, (4) technical paper and not related to acceptance/adoption (5) papers that were published before 2018, except papers that are related to the theory or model, and (6) Ph.D. or Master’s theses. The database searching had identified 331 articles. The number of articles was reduced to 224 after the screening process based on the inclusion criteria. Then, after screening the titles and abstracts, 70 articles were chosen. As a result, after reading the entire content, only 39 articles were chosen. Lastly, 22 articles that met the inclusion criteria were retained and reviewed.

Figure 1 shows the study of acceptance theories and models for IoT SH revealed an increase, with the most studies published in 2021 (8 papers), followed by 2020 (5 papers), 2019 (5 papers), and 1 paper published in the first two months of 2022. Figure 2 exhibits the most IoT SH acceptance investigations with a total of 3 publications in the USA, Germany, Malaysia, and South Korea.

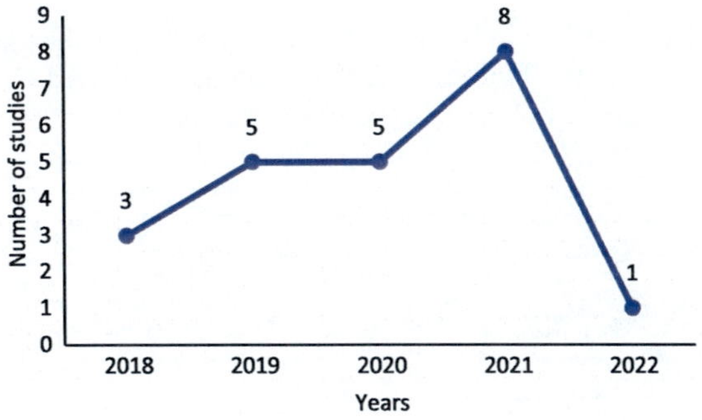

Fig. 1 Publication trends from 2018 to February 2022

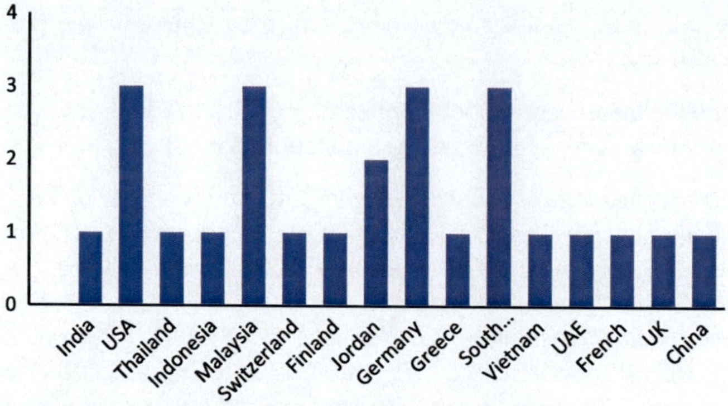

Fig. 2 Distribution of IoT SH acceptance studies based on countries

3 Technology Acceptance Theories

This study has reviewed the acceptance theories and models that have been used in prior works of literature, that include TAM (as shown in 3.1), UTAUT (as depicted in 3.2), UTAUT 2 (as shown in 3.3), and alternative acceptance theory for IoT SH (summarized in 3.4).

3.1 Technology Acceptance Model (TAM)

Davis et al. [23] created TAM to investigate the feasibility of new information system or technology adoption within an individual, and it can also be used to predict attitudes toward technology as well as behavioral intention, often known as the intention to use the technology, based on perceived ease of use and perceived usefulness. Perceived Usefulness (PU), Perceived Ease of Use (PEoU), Attitude Toward Using (A), and Behavioral Intention (BI) are the four primary factors found in TAM. Figure 3 shows the illustration of TAM.

Figure 4 indicates TAM was further simplified by removing the mediation of attitude toward using (A) and limiting it to three factors: Perceived Usefulness (PU), Perceived Ease of Use (PEoU), and Behavioral Intention (BI). This is because the

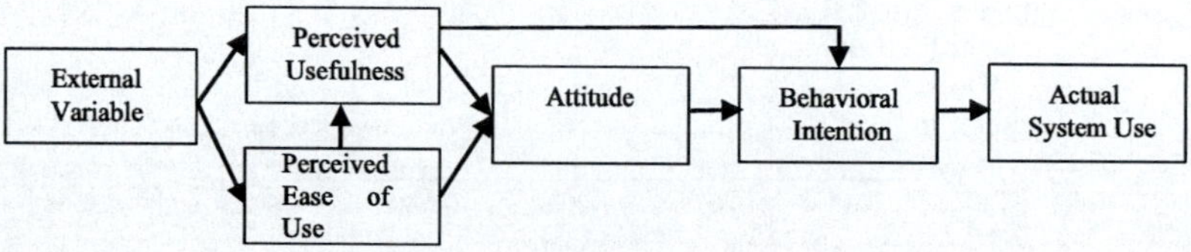

Fig. 3 The illustration of TAM (Davis et al. 1989)

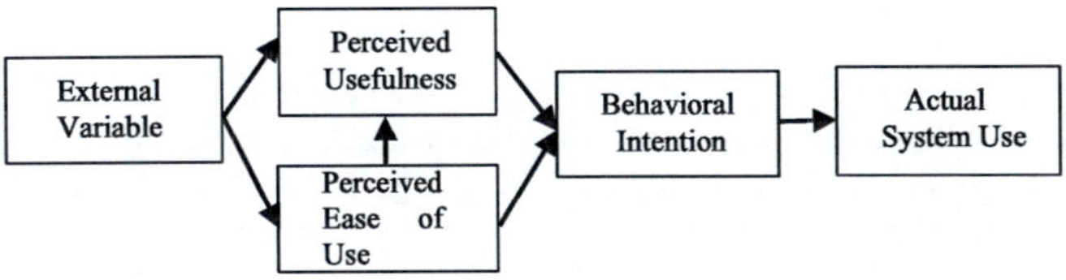

Fig. 4 TAM without the mediation of attitude toward using (A) (Venkatesh and Davis 1996)

mediation of attitude toward using (A) had a little significant impact on the coefficients of PU or PEOU. As a result, neither TAM 2 created by Venkatesh and Davis [17] nor the TAM 3 developed by Venkatesh and Bala [18] used the mediation of attitude toward using (A).

TAM has been frequently utilized to measure the intention to use various information system technologies. As a result, eleven previous researchers have used TAM as a theoretical model to investigate the acceptance of IoT SH technology. Park et al. [3] combined the original TAM with four value concepts, namely hedonic value, comfortable value, security value, and economic value. This study found that the total standardized effects for security consisting of perceived security, perceived system reliability, and compatibility gave the highest reading of 0.540 compared to economic (0.512), comfort (0.318), and hedonic (0.070). The most significant factors are perceived usefulness and compatibility, while perceived connectedness and control have a moderate impact. In addition, enjoyment and perceived system reliability have a weak influence on the behavioral intention and attitude toward IoT SH. Etemad-Sajadi and Gomes Dos Santos [4] added several factors to TAM to measure the acceptance of connected health technologies used at home by the elderly.

Nikou [9] combines the original TAM along with other factors and discovered PU, PEoU, compatibility and consumer perceived innovativeness (CPI), are important factors in influencing the intention to use IoT SH. Meanwhile, the perceived cost is a factor that contributes to the negative impact on the adoption of IoT SH. In addition, Al-Husamiyah and Al-Bashayreh [10] combined the original TAM with Theory of Planned Behavior (TPB) and Innovation Diffusion Theory (IDT). This is because TPB has factors related to the ability to fully control users' behavior and IDT focuses on technology related factors.

Guhr et al. [19] incorporated TAM with TPB and privacy theory. According to the finding of this study, the most crucial factor determining the intention to use IoT SH was privacy concerns. A study by researchers [20] related to smart home technologies in Greece found that Greek consumers had the following: 1) moderate level for the usefulness of new technologies and trust, 2) the compatibility of IoT SH technology ranges from moderate to high, and 3) social influence in the use of IoT SH ranges from moderate to low. Hubert et al. [21] combined TAM with IDT and risk theory. Findings from this study show that the most important factor in contributing to the intention to use was compatibility and usefulness, whereas the major barrier to using IoT SH was risk perception.

Researchers [22] combined the original TAM with four external factors. One of the study's contributions was to investigate the relationship between awareness and attitude. Researchers found that users' awareness had a significant impact on attitudes toward IoT SH. It is found that if users have a high awareness of IoT SH, then users will have a high attitude towards the use of IoT SH. Furthermore, the studies discovered that trust and perceived enjoyment have a positive significant impact on IoT SH attitudes, and perceived risk has a negative significant impact on users' trust. A study by researchers [24] in Danang City, Vietnam discovered that perceptions of connectivity, perceptions of ease of use, perceptions of affordability and compatibility have significant impacts on acceptance of new technology. In addition, the most important factors on the intention to use IoT SH are perceptions of ease of use, perceptions of usefulness and personal innovation.

Researchers [25] added one factor, which is quality of life, into the original TAM. The study population involved millennials aged between 19–35 years old. The findings of this study demonstrate that using IoT SH can increase happiness and well-being, resulting in a higher quality of life. In a study of smart homes systems in Malaysia, Wei et al. [26] examined three elements for perceived ease of use that consist of clear interface, attractiveness and consistency. In addition, two elements were used for perceived usefulness: information completeness and information accuracy. Perceived Privacy and Perceived security were also added as independent factors in this study. Results show that all factors were significant except information completeness. The TAM applied in previous IoT SH studies are summarized in Table 1. The (/) symbol represent mediation of attitude towards using (A) in the TAM.

3.2 *Unified Theory of Acceptance and Use of Technology (UTAUT)*

The UTAUT model was created by Venkatesh et al. [27] to measure users' intentions on information system. The Behavioral Intention (BI) of users to use technology is determined by four factors that consist of performance expectancy, effort expectancy, social influence, and facilitating conditions. Besides that, gender, age, experience, and voluntariness of use serve as moderating components that affect the four factors of usage intention and behavior. This theory is a unification of factors from the eight acceptance previous models, which are TAM, TPB, social cognitive theory, Theory of Reasoned Action (TRA), Diffusion of Innovation (DOI) Theory, Motivational Model, a combination of TPB and TAM and model of personal computer use. Figure 5 shows the UTAUT model.

A study that uses UTAUT theory related to users' needs, preferences, opinions, and intentions related to IoT SH was conducted by Arar et al. [5]. The study involved 110 respondents aged between the 40 s to 60 s. Findings from the study revealed that 67% of respondents suffered from chronic diseases. According to the researcher, the

Table 1 Summary of TAM

Authors	Sample size/Research strategies	Mediation of attitude toward using (A)	Additional factors	Consequences
3	799 respondents Survey	/	Enjoyment, Perceived connectedness, Perceived control, Perceived system reliability, Compatibility, Perceived cost, Perceived security,	Intention to use
4	213 respondents survey		Social presence, Trust and Degree of intrusiveness	Intention to accept
9	156 respondents survey		Compatibility, Trialability, Observability, Consumer perceived innovativeness, Perceived cost	Intention to use
10	750 respondents survey	/	Perceived compatibility, Perceived convenience, Perceived connectedness, Perceived cost, Perceived privacy risk, Perceived behavioral control	Intention to use
19	187 respondents survey		Perceived behavioral Control, subjective norms, Privacy concerns (Awareness of privacy practices, Secondary use of personal information, Perceived intrusion, Perceived surveillance)	Intention to use

(continued)

Table 1 (continued)

Authors	Sample size/Research strategies	Mediation of attitude toward using (A)	Additional factors	Consequences
20	108 respondents survey		Perceived enjoyment, Perceived compatibility, Trust, social influence, Perceived cost	Intention to use
21	409 respondents survey		Compatibility, Trialability, Result demonstrability, Visibility, Perceived risk Overall, Perceived risk Security, Perceived risk Performance, Perceived risk time	Intention to use
22	258 respondents survey	/	Trust, Awareness, enjoyment, Risk	Intention to use
24	287 respondents survey		Personal innovation, Perception of affordability, Perception of connectivity, Perception of compatibility, Perception of risk	Intention to use
25	206 respondents survey	/	Quality of life	Intention to use
26	102 respondents survey		Perceived security, Perceived privacy	Intention to use

increasing number of factors is due to the IoT SH necessitates the transmission and management of personal health data. Therefore, maintaining security and fostering trust are critical. As a result, the adoption of technology can be better understood by integrating predictive characteristics such as perceived security.

Researchers [6] have constructed a model by combining UTAUT with four other factors to test the acceptance of IoT SH healthcare services for the elderly in four countries. One of the contributions of the study was to examine the impact of expert opinion on the healthcare system. The results of this study found that the expert

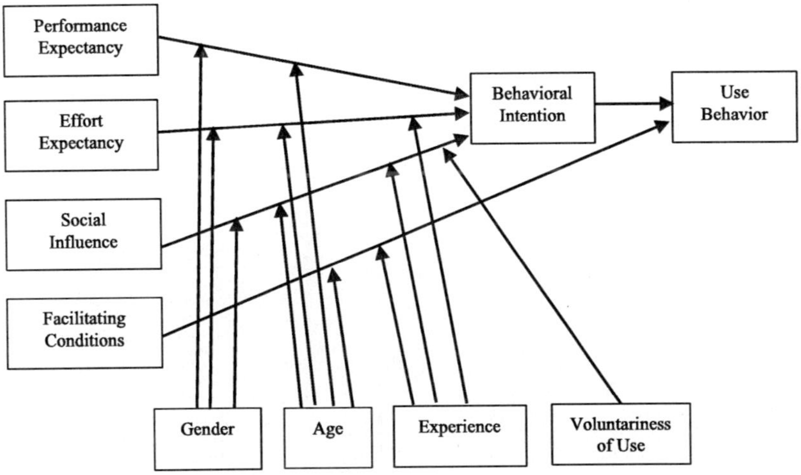

Fig. 5 UTAUT model (Venkatesh et al. 2003)

Table 2 Summary of UTAUT model

Authors	Sample size/ Research strategy	Additional factors	Consequences
5	110 respondents Survey for 55 respondents in their 40 s Survey and interview for 55 respondents in their 60 s	Perceived security, Anxiety about technology	Intention to use
6	239 respondents survey	Perceived trust, Expert advice, Technology anxiety, Perceived cost	Intention to use

advice factor has recorded a significant value. This is because the elderly depending on the advice of experts such as doctors and pharmacists when it comes to the use of technology in healthcare. However, social influence indicates not significant on the intention to use the healthcare system. The summary of UTAUT model applied in previous IoT SH studies are present in Table 2.

3.3 *Unified Theory of Acceptance and Use of Technology 2 (UTAUT 2)*

Venkatesh et al. [28] developed UTAUT 2 in 2012. It is an extension of UTAUT that was first introduced in 2003. Habit, hedonic motivation and price value are three additional factors in the UTAUT 2 model. Age, gender, and experience moderate the

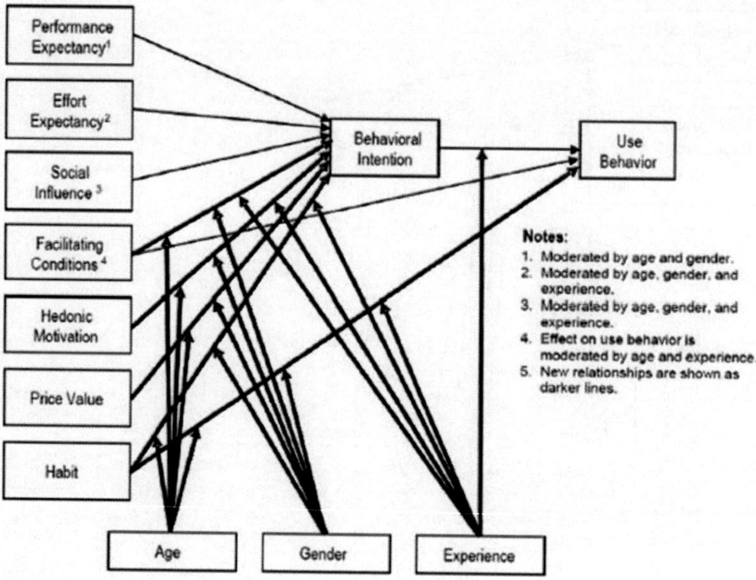

Fig. 6 UTAUT 2 model (Venkatesh et al. 2012)

effects of these factors on behavioral intention and technology use. Figure 6 shows the illustration of the UTAUT 2 model.

Aldossari and Sidorova [29] used the UTAUT 2 model with three additional factors. However, the habit factor was not included in this study model. According to the findings of this investigation, performance expectancy, effort expectancy, social influence, hedonic motivation, price value, trust, and security risk played significant roles in the acceptance of IoT SH. Nevertheless, it was found that facilitating condition had no significant impact on IoT SH acceptance. Furthermore, the attitude factor is used as a mediator in this model.

Researchers [30] combined the dimensions of the smart home (safety/security, health, comfort/convenience and sustainability) along with UTAUT 2, and personal innovation in the IT domain. In addition, types of education and gender were used as moderating effects between the factors of the model. This study focuses on the population of digital natives among highly educated young adults. The findings of the

Table 3 Summary of UTAUT 2 model

Authors	Sample size/ Research strategy	Additional factors	Consequences
29	424 respondents Survey respondents in their 60 s	Security risk, Privacy risk, Trust	Intention to use
30	206 respondents Survey	Safety security, Health, Convenience comfort, sustainability, Personal innovativeness	Intention to use

study found that comfort/convenience was the most significant primary motivator for the acceptance of IoT SH. The UTAUT 2 model applied in previous IoT SH studies are summarized in Table 3.

3.4 Alternative Acceptance Model for IoT SH

Seven researchers used the model that contained alternative theory and factors for measuring the acceptance of IoT SH. The researcher [1 and 7] built a model based on smart home features, smart home service preferences [11], and smart home trust that included general trust, privacy trust, and security trust [12]. Next, a researcher [13] created a model on salient beliefs that affected smart locks technology adoption for smart homes. Then, a model of smart thermostats technology adoption was developed by a researcher [14]. This study used a mixed-method methodology that combined TAM/UTAUT 2 with functional concerns, hedonic/symbolic benefits, and privacy concerns. Meanwhile, researcher [15] built a model on voice-enabled smart home systems. Table 4 shows the summary of the alternative acceptance model applied in previous IoT SH studies.

Table 4 Summary of the alternative acceptance model

Authors	Sample Size/ Research Strategy	Alternative Theory	Factors	Consequences
1	216 respondents survey	Automation	Perceived automation, Perceived controllability, Perceived interconnectedness, perceived reliability	Adoption intention

(continued)

Table 4 (continued)

Authors	Sample Size/ Research Strategy	Alternative Theory	Factors	Consequences
7	137 respondents survey	Characteristics of smart homes	Affinity for technology Interaction, technology Optimism, Privacy disposition, Trust Disposition, Experiences with smart homes, Perceived privacy risk, Trust in automation	Intention to use
11	400 respondents survey	Service preferences of smart homes	Convenience, Safety, Energy, Healthcare	Intention to use
12	2033 respondents survey	Trust of smart homes	Awareness, Ownership, experience of use, Trust in privacy, Trust in security, Trust in general, Satisfaction	Future intention to use and recommendation
13	531 respondents survey	The beliefs of key users in relation to the effect of smart locks	Perceived usefulness, Malfunction concerns, Perceived relative Advantage, Security Concerns, Negative effect, Novel benefits, Privacy Concerns	Adoption intention
14	612 respondents Interview and survey	The beliefs of key users in relation to the experiential and esthetic benefits of using smart thermostats	TAM/UTAUT 2: Cost concerns, Effort expectancy Performance expectancy, Hedonic/Symbolic benefits: techno-coolness Functional concerns: Compatibility concerns, Installation concerns, Reliability concerns, Privacy concerns	Adoption intention

(continued)

Table 4 (continued)

Authors	Sample Size/ Research Strategy	Alternative Theory	Factors	Consequences
15	475 respondents survey	Voice-enabled smart home systems	Technology optimism, Subjective norm, Perceived enjoyment, Familiarity, System quality	Perceived trust

4 Future Research Direction

The implementation of longitudinal study is one of the recommendations for future research. Researchers use longitudinal studies to evaluate the same individuals over time to detect changes that may occur over a short or lengthy period. For example, longitudinal research combined with established acceptance model or alternative model can be used to track changes in trust factors over time. It is believed that an individual can change from security concerns to data privacy concerns. This transition can happen to individuals due to policy changes, as well as the addition of knowledge and awareness of individuals.

According to several studies, consumers' acceptance of cutting-edge technology is intimately linked to their individual qualities. For that reason, personal factors such as gender, age, income, regions, education, resident type, cultural background, and internet availability should be included when building a research model. Future studies should look into how the price levels of a smart homes affects its acceptance. As a result, the existing acceptance theory can be broadened to incorporate price level factor, or new theories and models for price level factor can be developed by academicians.

In addition, research can also be conducted on the various sorts of services offered in smart homes. Due to the fast-paced evolution and the change of the smart home market, many services are established and dissolved at the same time. For this reason, future studies should focus on the acceptance of new types of IoT SH services in response to market developments.

5 Conclusions

IoT technology is evolving and provides many benefits to human life and the environment. As a result, IoT technology is widely used in the smart homes. Although reports reveal the increased demand for IoT SH in the coming years, the current acceptance of IoT SH are quite low.

Therefore, a systematic literature review was used to identify theories and models for assessing factors that influence IoT SH acceptance in order to address this gap. This review includes studies that were published from 2018 to February 2022. After applying the search string method, 22 papers were selected and the remaining papers were removed for not fulfilling the inclusion criteria. This study has reviewed four acceptance theories and models used in IoT SH previous studies, namely TAM, UTAUT, UTAUT 2, and alternative acceptance theory. Through this acceptance model, various factors that influenced the acceptance of IoT SH were identified.

TAM has been shown to be effective in explaining IoT SH technology acceptance. The results of this review are consistent with previous studies, demonstrating that TAM has been employed in the majority of studies to investigate the intention to use Internet of Things (IoT) technology in smart homes. However, more research in alternative IS theories and models related to the acceptance of IoT SH are required. This is because many previous studies used TAM, UTAUT, UTAUT 2, as well as the use of existing technologies model that ignored the unique characteristics of smart home technologies and their abilities to provide a wide range of functional, aesthetic, and sensory advantages.

The results also indicated that the majority of studies were carried out in the USA, Germany, Malaysia, and South Korea. Since cultural differences may have an impact on IoT SH acceptance, an empirical study should be conducted in countries that have only a few or no research on this matter. According to the findings, most previous studies used quantitative methodology except only one used mixed-method approach. is suggested that qualitative methodologies should be used in future studies to examine IoT SH acceptance. In terms of consequences of the acceptance model, most researchers used intention to use or also known as behavioral intention (16), followed by adoption intention (3), future intention to use and recommendation (1), perceived trust (1), and intention to accept (1).

A study of trust in smart homes among the population in the United Kingdom involved a very large population of 2033 respondents. The finding of this study revealed that elderly people and less educated people had less trust in smart home devices. Furthermore, unlawful data collecting will have an impact on people's willingness to utilize IoT SH. On the other hand, an empirical study on the context of system used on smart homes using TAM has the least number of respondents with a total of 102 potential user. This study found that a clear interface, consistency and attractiveness were the most important factors influencing the intention to adopt smart home.

In conclusion, this systematic review is useful for both IS academics and IoT SH practitioners. Furthermore, the findings of this study will benefit academics, particularly novice researchers, in identifying existing trends and gaps in IoT SH studies, as well as the future work in this research area.

References

1. Yang H, Lee W, Lee H (2018) IoT smart home adoption: the importance of proper level automation. J Sens
2. Fortune Business Insights Homepage. https://www.fortunebusinessinsights.com/industry-reports/smart-home-market-101900 Accessed 14 Mar 2022
3. Park E, Kim S, Kim YS, Kwon SJ (2018) Smart home services as the next mainstream of the ICT industry: determinants of the adoption of smart home services. Univ Access Inf Soc 17(1):175–190
4. Etemad-Sajadi R, Gomes Dos Santos G (2019) Senior citizens' acceptance of connected health technologies in their homes. Int J Health Care Qual Assur 32(8):1162–1174
5. Arar M, Jung C, Awad J, Chohan AH (2021) Analysis of smart home technology acceptance and preference for elderly in dubai, UAE. Designs 5(4):70
6. Pal D, Funilkul S, Charoenkitkarn N, Kanthamanon P (2018) Internet-of-things and smart homes for elderly healthcare: an end user perspective. IEEE Access 6:10483–10496
7. Schomakers EM, Biermann H, Ziefle M (2021) Users' preferences for smart home automation – investigating aspects of privacy and trust. Telematics Inform 64(July):101689
8. Ji W, Chan EHW (2019) Critical factors influencing the adoption of smart home energy technology in China: a Guangdong province case study. Energies 12(21):4180
9. Nikou S (2019) Factors driving the adoption of smart home technology: an empirical assessment. Telematics Inform 45:101283
10. Al-Husamiyah A, Al-Bashayreh M (2021) A comprehensive acceptance model for smart home services. Int J Data Netw Sci 6(1):45–58
11. Chang S, Nam K (2021) Smart home adoption: the impact of user characteristics and differences in perception of benefits. Buildings 11(9):393
12. Cannizzaro S, Procter R, Ma S, Maple C (2020) Trust in the smart home: findings from a nationally representative survey in the UK. PLoS One, 15(5):e0231615
13. Mamonov S, Benbunan-Fich R (2021) Unlocking the smart home: exploring key factors affecting the smart lock adoption intention. Inf Technol People 34(2):835–861
14. Mamonov S, Koufaris M (2020) Fulfillment of higher-order psychological needs through technology: the case of smart thermostats. Int J Inf Manag 52:102091
15. Liu Y, Gan Y, Song Y, Liu J (2021) What influences the perceived trust of a voice-enabled smart home system: an empirical study. Sensors 21(6):1–22
16. Marikyan D, Papagiannidis S, Alamanos E (2019) A systematic review of the smart home literature: a user perspective. Technol Forecast Soc Change 138:139–154
17. Venkatesh V, Davis FD (2000) A theoretical extension of the technology acceptance model: four longitudinal field studies. Manage Sci 46(2):186
18. Venkatesh V, Bala H (2008) Technology acceptance model 3 and a research agenda on interventions. Decis Sci 39(2):273–315
19. Guhr N, Werth O, Blacha PPH, Breitner MH (2020) privacy concerns in the smart home context. SN Appl Sci 2(2):1–12
20. Pliatsikas P, Economides AA (2022) Factors influencing intention of Greek consumers to use smart home technology. Appl Syst Innov 5(1):26
21. Hubert M, Blut M, Brock C, Zhang RW, Koch V, Riedl R (2019) The influence of acceptance and adoption drivers on smart home usage. Eur J Mark 53(6):1073–1098
22. Shuhaiber A, Mashal I (2019) Understanding users' acceptance of smart homes. Technol Soc 58:101110
23. Davis FD, Bagozzi RP, Warshaw PR (1992) Extrinsic and intrinsic motivation to use computers in the workplace. J Appl Soc Psychol 22(14):1111–1132
24. Van Hung T, Thao TNP, Kieu TNT, Hien DQ (2021) Research on factors influencing intention to use smart home devices in Danang. In: Proceedings - 2021 21st ACIS international semi-virtual winter conference on software engineering, artificial intelligence, networking and parallel/distributed computing, SNPD-Winter 2021, pp 208–212

25. Mohamad ZZ, Meor Musa SU, Abdul Razak R, Ganapathy T, Mansor NA (2021) Internet of things: the acceptance and its impact on well-being among millennials. Int J Serv Technol Manag 27(4/5/6):265
26. Wei NT, Baharudin ASA, Hussein L, Hilmi MF (2019) Factors affecting user's intention to adopt smart home in Malaysia. Int J Interact Mob Technol 13(12):39–54
27. Venkatesh V, Morris MG, Davis GB, Davis FD (2003) User acceptance of information technology: toward a unified view. MIS Q 27(3):425–478
28. Venkatesh V, Thong JY, Xu X (2012) Consumer acceptance and use of information technology: extending the unified theory of acceptance and use of technology. MIS Q 36(1):157–178
29. Aldossari MQ, Sidorova A (2020) Consumer acceptance of Internet of Things (IoT): smart home context. J Comput Inf Syst 60(6):507–517
30. Baudier P, Ammi C, Deboeuf-Rouchon M (2020) Smart home: highly-educated students' acceptance. Technol Forecast Soc Change 153:119355

Nautical Digital Platforms with Navigator-Generated Content: An Analysis of Human–Computer Interaction

Diogo Miguel Carvalho

Abstract The exponential growth of Information and Communication Technologies (ICT) has made it possible to distribute content in more ubiquitous and simple ways. ICTs have contributed to the rise of User-Generated Content (UGC) platforms, which have become essential in maritime navigation, supporting systems and applications. Nonetheless, it is important to understand how the interface's design is conceived in most of these systems, meeting usability heuristics to assist the navigator's decision-making and prevent human error.

This article aims to compare and analyze a set of technological systems and mobile applications that aim to promote the maritime information sharing and aid the navigation, based on a set of Usability Heuristics and guidelines of the International Maritime Organization (IMO). Furthermore, this analysis has provided the relevance of Human–Computer Interaction (HCI) as a field to aid decision making and prevent human error in maritime navigation systems.

Keywords Human–computer interaction · User interface design · Situational visual impairments · Nautical digital platforms · Nautical applications · Benchmarking

1 Introduction

With the evolution of the telecommunications industry and the mass production of electronic components, communication started to be more ubiquitous in the human's daily life. An example is the growth of digital networks [1, 2], which allows increasing human communication through online platforms. This expansion allows humans to quickly communicate and inform themselves concerning subjects of interest.

Access to information content has progressed thanks to the fusion of Information and Communication Technologies (ICTs) and traditional media. In this environment,

D. M. Carvalho (✉)
Departamento de Comunicação E Arte/DigiMedia, Universidade de Aveiro, Campus Universitário de Santiago, 3810-193 Aveiro, Portugal
e-mail: diogocarvalho28@ua.pt

M. Al-Emran et al. (eds.), *International Conference on Information Systems and Intelligent Applications*, Lecture Notes in Networks and Systems 550,
https://doi.org/10.1007/978-3-031-16865-9_17

the consumer also started to act as a creator and distributor of digital content into the global network, becoming not just a consumer but a "prosumer" [3]. Thus, in this prosumer's role, the user can entertain himself and consume self-made contents [4].

As a consequence, media convergence fomented easy access to content produced by their consumers – user-generated content (UGC) – mediated by platforms, usually online [3, 4]. According to some authors [3, 5], UGC refers to tailored multimedia contents, produced and developed by amateur users, aiming to disseminate their own contents, willingly made for the internet.

Regarding the content's sharing to the globally distributed network, relevant data concerning specific phenomena can be obtained through a user- or community-wise distribution: crowdsourcing [6]. With crowdsourcing [7], tailored and shared contents can be acquired collaboratively and collectively, just like in Wikipedia or OpenStreetMap, tools in which users – among a community – modify and edit information content.

In the context of meteorological conditions, the emergence of ICTs is a pillar that aids media communication since it can create, interact, and disseminate content regarding this subject onto the digital network, but also can be a mean to represent and alert for potential risks during maritime navigation tasks.

Maritime navigation aims to constantly find the safest and most efficient route, avoiding collisions [8], grounds, or any other adverse scenarios that endanger the vessels or their crew. Also due to ICTs emergence, the International Maritime Organization (IMO) introduces the term "e-navigation", which concerns the process of collecting and sharing maritime information on board or inland by electronic means to enhance peer to peer navigation.

Some studies [9, 10] already interconnect UGC with communities and maritime data, emphasizing the relevance to study the interfaces and Human–Computer Interaction (HCI) paradigms [11] that mediate maritime navigation-related content to fulfill the requirement to assist maritime navigation without compromising the users' cognitive load.

In this scope, this paper presents a comparative study that analyses a set of nautical platforms and maritime navigation support services, in the shape of technological systems, resorting to Nielsen's Usability Heuristics [12], psychological character criteria (i.e. decreasing cognitive load and aiding timely decision-making), situational visual limitations [13], and IMO guidelines [14, 15].

The paper is organized into four sections: following this (1) introduction, the second section presents a (2) theoretical framework about the relevance of human aspects to support the user interface design; followed by the (3) benchmarking (comparative) analysis, which includes description, and evaluation and its results of nautical platforms selected; and finally, the (4) conclusions are presented on section four.

2 The Relevance of HCI in the Navigator's Role

Interpreting and observing the human brain and behavior is crucial to understanding the information process, based on a sensorial register (i.e. visual, auditory, or tactile experience), in order to memorize the actions to aids humans in: problem-solving; stimulating reasoning; and making assertive decisions, taking into account the circumstances and the context [16].

For this reason, it is important to understand the human as a key element in maritime safety and apply strategies to promote safe navigation [17]. In fact, human error is the dominant factor in causing maritime accidents [18], hence it is crucial to understand the cognitive roots of these accidents.

In this sense, mental overload, decision-making, performance, stress, and the interactions that occur between navigator-machine are relevant topics to study since these can assist in the optimization of the technological integration into maritime environment, having in regard the human being's needs [18].

Decision-making is a relevant factor to promote safety at any navigation stage, that involves selecting an option among several possibilities. Nonetheless, the decision level of importance can differ in terms of complexity since it depends on a set of factors (i.e. emotional and social) that are inherent to this decision-making process [19].

Notwithstanding emotions being relevant to support decision-making, it is crucial to understand HCI and its constituent areas as an essential pillar of mediation to support maritime navigation's interfaces. The HCI field has a critical role in human beings since they have considerable plasticity and uncontrollability due to their physical, psychological, social and spiritual characteristics [20]. The HCI's role can negatively influence human error through [20, 21]: (i) decreasing situational awareness (SA), as navigation support system interfaces, may represent information that biases wrong decision-making; (ii) ambiguous interpretation of certain icons; (iii) high demand cognitive load due to interface interactions; and (iv) the presence of less agreeable weather conditions (e.g. sunlight reflections, wind, waves, temperature, and noise).

However, HCI aims to reduce problems (i.e. connected to emotional, technical and conceptual factors) that human beings may have when interacting with any type of electronic system mediating through an interface [11]. To solve a set of problems that can result from several scientific domains, the HCI discipline resorts to several other disciplines (e.g., psychology, anthropology, industrial design, and computer science), being interdisciplinary and capable of embedding contributions from several areas.

In this sense, usability [22, 23] should ensure that an interactive system follows the interaction aspects between the individual and the technology. The various interaction paradigms [11, 24], from graphical user interfaces (GUI) – which adopted windows, icons, menus, and pointers (WIMP) – to the most contemporaneous touch interfaces, have contributed to a shift in the way information is consumed and users communicate between themselves.

Mostly in maritime navigation, technological systems mediated by interfaces are used to display information to support navigation (e.g. nautical charts, depth measured by a sounder, radar, anemometer), that can be chart plotters or Electronic Chart Display and Information System (ECDIS) [25, 26]. According to some authors [26, 27], the most frequent interaction paradigms in navigational systems are buttons and touch screens – however, interactions can be affected by the environment: wet or damp fingers, sunlight reflection, wind, or rain.

Although some concerns must be taken regarding the interaction with technological systems, it is crucial to ensure an optimal usage of any User Interface (UI) in any context [28]. For this purpose, it is essential to resort to usability [22, 23], which allows us to identify the repeatability that a user has or has not with a given technological system. To avoid any usability or interactions problems with interfaces' design, during the UI conceptualization process, the design should be evaluated by a set of heuristics [12] and guidelines [29].

In addition to the interaction and usability aspects, it is relevant to understand the User Experience (UX), being a concept distinct from usability. In a product design point of view, Norman [30] emphasizes that products should not only focus on production and technologies but also should integrate experience, aesthetics and interaction quality. So, the UX design focuses on the quality, satisfaction and emotions that a given subject has when using a product [30, 31].

Experience, interaction, and usability are very important concepts to design a maritime technological system. However, the maritime context is very restricted, where it should study, not only users' emotions and interactions, but how the new technology usage affects the navigator's psychological factors (e.g., decision making, human error, situational awareness).

In the case of sailing vessels, in which technological screens are located outdoor [26], it is recognized that situational visual impairments (SVI) [13] occur more frequently, making it difficult to information access and representation mediated by technological interfaces. Moreover, UI design must mitigate these limitations to aid the individual's decision-making.

3 Benchmarking

The present study aims to (1) identify a set of technological solutions (i.e., applications available on the market and scientific projects/proofs-of-concept) that (2) support decision-making and prevent human error, specifically aspects related to the required steps to access information and allow their immediate access to information in an interface. Furthermore, it intends (3) to carefully analyze a set of interfaces, which fit within the heuristic principles and guidelines established for this research.

3.1 UI Design Usability Heuristics

To individually evaluate the interfaces that support navigation, some usability heuristics of interface design [12] are used, namely:

Consistency and Standards. This principle contributes to reducing the user's learning curve, being that the visual line applied to a system maintained throughout the various interfaces, especially during the process of maritime navigation, which requires greater attention from navigators.

Error Prevention. This heuristic prevents an action from being executed without having the user's confirmation: e.g., if fisher has wet hands and wants to visualize his/her route in an interface, he may not be able to click on the desirable button and lose his information about the voyage's process. In this case, the system must prompt a dialogue box asking the fisherman whether he wants or not to cancel an action.

Visibility of System Status. This heuristic is crucial for the interface to communicate any feedback information to the user since he needs to be informed about essential navigational information, such as position, direction, meteo-oceanographic conditions.

User Control and Freedom. An interface should allow the navigator to amend a wrong interaction since the UI design should be as unambiguous as possible and capable of decreasing the user's cognitive load (e.g., back button).

3.2 IMO Guidelines

Complementarily, this study's focus on maritime navigation aims to understand whether the nautical interfaces designed in the majority of the nautical digital platforms and systems analyzed, observing some criteria: aid decision-making, the cognitive load reduction, and the standards and guides established by the IMO [14, 15] – notwithstanding being idealized for the interfaces of ships' screens, these guides contain essential contributions to helping maritime navigation on diverse devices.

The IMO guidelines selected for this research complement the heuristic analysis based on usability principles. Thus, the chosen guidelines—which did not represent a redundancy concerning the selected usability heuristics—were: representation of relevant information to aid navigation; risks of over-reliance on nautical interfaces (e.g. malfunction of the system, which can lead to human error or representation of outdated hydrographic values); operational use of the interface (e.g. zooming, selecting and modifying content); setting of safety values according to the vessel's descriptions; reading all the chart's symbols and abbreviations; and representation of several types of screen orientation (e.g. north-up).

3.3 Description of the Applications: State-of-the-Art

The search for maritime technological systems resorted to Apple's mobile application store (App Store), the IEEExplore database and Google Scholar's search engine—searching the words maritime navigation, boating, data sharing and e-navigation.

In this sense, this study gathered four technological systems: two mobile applications [32, 33], one electronic data sharing system [34], and one functional prototype [35]. The selection reasons of these technological artefacts are due to the integration of a maritime navigation interface, preferably with graphical and image representation; the popularity of the application market; and the possibility for the user to create contents to share with a community. In this sense, a description is presented for each of these systems, synthesizing their main functionalities.

Navionics Boating App [32]: This service is a mobile application that allows displaying nautical maps developed by Navionics and the user to plot a route. In parallel, while the user navigates with Navionics Boating, he/she can share a range of information (e.g., maritime signals, obstacles, points of interest) with the Navionics community. Moreover, this application presents another feature, namely embedding the map in a chart plotter, in which it is possible to visualize the information acquired and shared by the mobile application.

NaAVIC [33]: is a mobile application that integrates features of an ECDIS and is developed for all types of vessels that aim to navigate in a professional, safe and reliable way with mobile devices. This mobile application presents a map with information – bathymetric data, wrecks and routes – that is shared and updated due to the user's collaboration. This service follows a crowdsourcing approach, in which users can provide and consume data represented in the application, being this one of the main advantages over an ECDIS.

DYNAMO [34]: is an environmental and marine system that collects data by using sensors installed on recreational boats. This system aims to share the maritime information collected with a community of sailors. In this innovative strategy, the system integrates the instruments that already exist in a boat and share this information via Internet of Things (IoT) with the community. In this project [34] it was perceived that DYNAMO is able to present additional and important information in an interface map, compared to Navionics and OpenSeatMap maps.

ESABALT [35]: Basalt is a scientific project that developed a maritime data sharing software between ships, leisure boats and coastal authorities. This project has the following objectives: improve maritime safety; enable intelligent navigation; and encourage citizens' participation in coastal environmental monitoring. In this perspective, a prototype, based on end-user's requirements, was developed to simulate scenarios related to maritime safety, data sharing, situational awareness, and communication between authorities, to validate the ESABALT project concept in a Baltic Sea environment.

Heuristics & Guidelines		Navionics Boating App	DYNAMO	NaAVIC	ESABALT
Decision-making	Direct access to nautical information (ex: speed, nautical miles)	✔	✔	✖	!
Decision-making	Less than 3 steps to view information on the interface	✔	✔	✖	✖
Heuristics	Consistency and standards	✔	✖	✖	✔
Heuristics	Error prevention	✔	✔	✖	!
Heuristics	Visibility of system status	✔	✔	✔	✔
Heuristics	User control and freedom	✔	!	!	!
Situational visual impairments	Content visibility in outdoor scenarios	!	!	!	✖
IMO Guidelines	Risks of over-reliance on nautical interfaces	!	!	✔	✔
IMO Guidelines	Operational use of electronic interfaces (ex.: zoom, selecting and modification of content)	✔	✔	✔	✔
IMO Guidelines	Setting of safety values according to the vessel descriptions;	✔	✔	!	✖
IMO Guidelines	Appropriate Use of Icons	✔	✖	✖	✖
IMO Guidelines	Presentation of several types of screen orientation (ex: north-up)	✔	✔	✖	✖

Fig. 1 Benchmarking analysis

3.4 *Comparative Analysis*

To proceed with the systems' analysis, the interface design of each service was assessed based on heuristics and guides. In the aim of this evaluation, a green tick was attributed to systems that answered the usability and IMO guidelines criteria, a yellow exclamation mark for the systems that are close to meeting the requirements, and a red cross to services that do not meet the specific requirement (Fig. 1).

The first parameter in Fig. 1, Decision-making, is fulfilled by Navionics Boating and DYNAMO since those systems summarized information and made it accessible, hence aiding the user's decision-making. The ESBALT prototype synthesizes information that supports navigation, however, it requires the user to click on buttons to display additional information (e.g. oil spill level). Lastly, the NaAVIC application does not fulfill this parameter since it requires users to constantly interact with the mobile phone to access information.

Concerning the (usability) Heuristics [12]: only the ESBALT prototype and the Navionics Boating application satisfied the "Consistency and standards" criteria by maintaining the coherence of the interface's graphical elements and their positioning. Moreover, regarding the "User control and freedom" it is evident in the Navionics Boating app since the user can easily cancel the navigation mode through the "Cancel" button, which, in the other solutions analyzed, is not so straightforward (i.e. it is not possible to identify the action to perform in order to end the route). The heuristic complied by all those systems was "Visibility of the system status", being represented by the identification (through a graphic symbol) of the users' position on the map.

Regarding the applications' readability in outdoor scenarios ("Situational visual impairments" parameter), it was perceived that in most applications, the UI's colors affect the map's interpretation, and the iconography and texts' understanding (Fig. 2).

✔ green tick - The platform complies with the analysed criteria

! yellow exclamation mark - The platform presents some constraints that do not comply with the analysed criteria

✘ red cross - The platform does not comply with the analysed criteria

Fig. 2 Benchmarking analysis – criteria symbols used

Fig. 3 NaAVIC and boating app evaluation in an outdoor environment

a) Mobile application: NaAVIC   b) Mobile application: Boating App

Nonetheless, by analyzing of the Navionics Boating app screen (Fig. 2 - b) it can be perceived that the upper part of the interface may be adequate to place relevant information, over a dark background and with a larger (than the standard) font size to ensure legibility.

Throughout this research, there is evidence of similarities between Navionics Boating app and DYNAMO. Specifically, concerning decision-making and human error prevention, those interfaces present univocal information and synthesize essential information that aids navigators' decision-making, as well as respond to the IMO guidelines that support navigators' cognitive aspects. Notwithstanding that NaAVIC and the ESBALT prototype represent a useful solution for the promotion of maritime safety, the design of nautical interfaces does not meet the criteria established in Fig. 1 that aims to overcome usability problems, provide an optimal UX design, and allow information access in maritime environments.

4 Discussion and Conclusions

Digital platforms are increasingly being used to support maritime navigation, allowing the creation and distribution of essential contents that aid nautical navigation. The contribution of multidisciplinary disciplines enhances the design of technological systems. Complementarily, given the prominence of understanding human behavior and how it affects human–computer interactions, it is relevant to study the human as the decision-maker and the interactions implied by human's decisions in order to mitigate maritime accidents.

This paper depicts a comparative study, which selected four technological systems that support maritime navigation: NaAVIC, Navionics Boating App, DYNAMO, ESBALT's prototype. To establish comparison criteria, Usability Heuristics and IMO guidelines were used to assess the technological systems' UI and interaction in order to understand how those aspects affect the navigators' decision-making when they resort to such systems whilst navigating. It was concluded that the Navionics Boating App presents an agreeable response to most criteria, namely to share content with other users, the interface's design; and improve decision-making process. The remaining three systems pose more disadvantages due to their complexity in assisting the navigator and preventing an accident since they require more clicks and the content's visibility is hindered.

Furthermore, from this analysis, a few considerations can be highlighted for further studies, namely: consider IMO and Usability guidelines, to reduce the interaction steps required to assist decision-making, provide consistency on the overall UI design, and symbolize the users' position on a map to inform them on the system's status; use darker backgrounds and larger font sizes to enhance outdoor content's readability; and benefit for UGC to update meteo-oceanographic data in real time, aiding navigators' decision-making.

Although it was not evidently considered for the comparative analysis, the underlying UGC integrated in the maritime technological systems, which allows the sharing of knowledge by users, may aid the anticipation of the navigators' decision, hence decreasing their cognitive load (related to the decision-making process). In fact, two of the analyzed apps (i.e. Boating App and NaAVIC) include this method of information acquisition, although it is more intuitive in the Boating App.

The strategy used for the benchmarking analysis, as well as the criteria established, can be applied in future studies to support the design and evaluation of nautical interfaces, aiding the humans' decision-making and UGC knowledge sharing in maritime navigation.

Notwithstanding, this research is limited to the author's analysis, it is pertinent to evaluate and analyze these interfaces with participants, and even experts, to achieve more conclusive and representative results of the target audience's needs.

Acknowledgements Thanks are due to FCT/MCTES for the financial support to DIGIMEDIA (UIDP/05460/2020+ UIBD/05460/2020), through national funds.

References

1. Castells M (2012) Sociedade em Rede. Gulbenkian
2. Ribeiro F, Silva AM (2019) da: Infocomunicação como projeto comum de diálogo e prática = Infocommunication as a common dialogue and practice project. In: Ciências da comunicação: vinte anos de investigação em Portugal/10a Congresso SOPCOM
3. Jenkins H (2006) Convergence culture. New York University Press
4. Dwyer T (2010) Media convergence. McGraw-Hill Education (UK)

5. Krumm J, Davies N, Narayanaswami C (2008) User-generated content. IEEE Pervasive Comput 7:10–11
6. Levina N, Arriaga M (2014) Distinction and status production on user-generated content platforms: Using Bourdieu's theory of cultural production to understand social dynamics in online fields. Inf Syst Res 25:468–488
7. Howe J (2006) The rise of crowdsourcing. Wired Mag 14:1–4
8. IALA: IALA NAVGUIDE 2018 DIGITAL COPY - IALA AISM. https://tinyurl.com/2eejtvz3 Accessed 18 Oct 2021
9. Thombre S, Kuusniemi H, Söderholm S, Chen L, Guinness R, Pietrzykowski Z, Wołejsza P (2016) Operational scenarios for maritime safety in the baltic sea. Navigation 63:521–531. https://doi.org/10.1002/navi.161
10. Wright RG (2017) Scientific data acquisition using navigation sonar. In: Oceans 2017 – Anchorage, pp 1–6
11. Dix A, Finlay J, Abowd GD, Beale R (2004) Human-computer interaction. Pearson
12. Nielsen J 10 Heuristics for User Interface Design. https://www.nngroup.com/articles/ten-usability-heuristics/ Accessed 21 Dec 2019
13. Vatavu RD (2017) Visual impairments and mobile touchscreen interaction: state-of-the-art, causes of visual impairment, and design guidelines. Int J Hum Comput Interact 33:486–509. https://doi.org/10.1080/10447318.2017.1279827
14. International Maritime Organization: MSC/Circ.982: Guidelines On Ergonomic Criteria For Bridge Equipment And Layout. , London (2000)
15. International Maritime Organization: MSC.1/Circ.1503/Rev.1: ECDIS – GUIDANCE FOR GOOD PRACTICE. https://www.classnk.or.jp/hp/pdf/activities/statutory/ism/imo/msc1-circ1503-rev1.pdf Accessed 22 Jan 2022
16. Atkinson RC, Shiffrin RM (1968) Human memory: a proposed system and its control processes. In: Psychology of learning and motivation Elsevier, pp 89–195
17. IMO: Human Element. https://www.imo.org/en/OurWork/HumanElement/Pages/Default.aspx Accessed 22 Jan 2022
18. Barnett ML, Pekcan CH (2017) The human element in shipping. Encycl Marit Offshore Eng:1–10
19. Eysenck MW, Keane MT (2020) Cognitive psychology: a student's handbook. Psychology press
20. Han S, Wang T, Chen J, Wang Y, Zhu B, Zhou Y (2021) Towards the human–machine interaction: strategies, design, and human reliability assessment of crews' response to daily cargo ship navigation tasks. Sustainability 13(15):8173. https://doi.org/10.3390/su13158173
21. Islam R, Khan F, Abbassi R, Garaniya V (2018) Human error probability assessment during maintenance activities of marine systems. Saf Health Work 9:42–52. https://doi.org/10.1016/j.shaw.2017.06.008
22. Nielsen J Usability 101: Introduction to Usability. https://www.nngroup.com/articles/usability-101-introduction-to-usability/ Accessed 22 Dec 2019
23. Nielsen J (1993) Usability engineering. Morgan Kaufmann, San Francisco
24. Rogers Y, Sharp H, Preece J (2019) Interaction design: beyond human-computer interaction. John Wiley & Sons
25. Weintrit A (2009) The electronic chart display and information system (ECDIS): an operational handbook. CRC Press
26. Müller-Plath G, Jung D, Müller M (2018) Based design and usability guidelines for electronic charting systems (ECS) in yachting and boating research-based design and usability guidelines for electronic charting systems (ECS) in yachting and boating ☆. Int J e-Navig Maritime Econ 10:32–48
27. Mills S (2005) Designing usable marine interfaces: some issues and constraints. J Navig 58:67–75. https://doi.org/10.1017/S0373463304003078
28. Shneiderman B, Plaisant C, Cohen M, Jacobs S, Elmqvist N, Diakopoulos N (2016) Designing the user interface: strategies for effective human-computer interaction. Pearson
29. Smith SL, Mosier JN (1986) Guidelines for designing user interface software. Citeseer

30. Norman D (2013) The design of everyday things. The Perseus Books Group
31. Obrist M, Tscheligi M, De Ruyter B, Schmidt A (2010) Contextual user experience: how to reflect it in interaction designs? In: Conference on Human Factors in Computing Systems - Proceedings. ACM Press, New York, New York, USA, pp 3197–3200. https://doi.org/10.1145/1753846.1753956
32. Navionics | Mobile App for Boating and Fishing. https://www.navionics.com/fin/apps/navionics-boating Accessed 11 Jan 2021
33. Peyton D, Kuwalek E, Alla A (2019) NaAVIC, a free downloadable ECS app that runs on ENC data streamed directly from a cloud-based infrastructure specifically designed for marine navigation. TransNav Int J Mar Navig Saf Sea Transp 13(1)
34. Di Luccio D, Riccio A, Galletti A, Laccetti G, Lapegna M, Marcellino L, Kosta S, Montella R (2020) Coastal marine data crowdsourcing using the Internet of floating things: improving the results of a water quality model. IEEE Access. 8:101209–101223. https://doi.org/10.1109/ACCESS.2020.2996778
35. Thombre S, et al (2017) Proof-of-concept demonstrator to improve safety of maritime navigation in the Baltic Sea. In: 2017 European navigation conference (ENC), pp 232–241 https://doi.org/10.1109/EURONAV.2017.7954213

Digital Sweetness: Perceived Authenticity, Premium Price, and Its Effects on User Behavior

F.-E. Ouboutaib, A. Aitheda, and S. Mekkaoui

Abstract Smartness is continually impacting every territory. It is a huge component of industry, business, and society. User-technology relationship seeks to promote the creation of value for society. Though, this shift has been accompanied with global health crises. While the previous waves have ignored humans by focusing on machines, recent literature calls for a user-oriented approach to preparing an authentic social environment.

The reality of local communities is also changing and little is known about the perception of authenticity which is a driving force to generate digital contents. To handle this challenge, through the partial-least-squares (PLS), this paper aims to understand the interaction between trust, premium price, and user behavior. Findings underline the importance of authenticity and the use of its determinants can create a worldwide community helping to encourage content sharing that is very useful for the user-oriented digital strategies.

Keywords User authenticity · Smart societies · Food consumption · PLS

1 Introduction

Technology domain has encouraged smartness in economic and social entities. It has transported a baseline of performance, competitiveness, and opportunities [1, 2]. Advancements are marked by the significant automation, digitization process, and massive use of information technologies that influence implicitly changes/inequalities in small and medium businesses [3]. Among this, local economies are being invaded by globalized products. Thus, the postmodern consumer enjoys wide choices that are increasingly accessible with the development of digital marketing solutions.

F.-E. Ouboutaib (✉) · A. Aitheda · S. Mekkaoui
National School of Commerce and Management, Agadir, Morocco
e-mail: fouboutaib@gmail.com; fatima-ezzahra.ouboutaib@edu.uiz.ac.ma

Research Team in Marketing Management and Territorial Communication, Agadir, Morocco

M. Al-Emran et al. (eds.), *International Conference on Information Systems and Intelligent Applications*, Lecture Notes in Networks and Systems 550,
https://doi.org/10.1007/978-3-031-16865-9_18

Overtime, the industrial movements in the food sector, intensive agricultural product, have been accompanied with worldwide health crisis [4, 5]. For nutritionists, the eating habits have become sources of health problems. Overweight and obesity are exploding as the hidden hunger which is presented as the lack of micronutrients (vitamins and minerals) [6]. Obesity among adults has increased from 11.7% in 2012 to 13.2% in 2016, and it is more important in North America than in Africa. What is alarming is this change is based on the overconsumption of ultra-processed products, it is not only related to income but also to the increased urbanization. Overconsumption of globalized products with high proportions of added sugar, fat, and important chemical preservatives are common ingredients of various industrial food [7, 8]. Studies have shown that poor nutrition is the main source of recent diseases. The number of overweight adolescents worldwide has increased from 11 million in 1975 to 213 million in 2017. Research addresses accusations to the highly industrialized food system [9, 10]. In this regard, smart society challenges aim to situate human and small local communities at the focus of business and ecosystem [11].

Namely, local distribution of agricultural products has declined in parallel with the rise of the globalized Agri-Food-System [12] which has distanced the producers from the consumers; regional culture has been perceived as an old heritage [13]. However, this concern has changed following the alarming health crises [14–16]. Food production should answer the melange of objective and subjective needs of consumers, and establish resilience and assurance [15].The link with food demands a basis of trust, credibility, and authenticity to meet the expectations of consumers [4, 5, 17, 18], and the digital dynamic is called to balance with different ecosystem's actors. Markets are becoming global but they still are constituted by small territories.

Although the ongoing marketing concern in studying authenticity, it is on an emergent phase. Digital interaction needs more examinations [17]. This research addresses this gap. It aims to understand the perceived authenticity and its effects on user behavior. It argues that a deep comprehension of the determinants of user authenticity in context of food consumption may help online managers to decide on the shared content. Users and local communities are calling to work together to shift toward smart society. Authors consider that human concerns should be above the greatest technology that is the baseline of the smartness. It responses the following questions: how does authenticity influence user? And what is the role of premium price and media for user? It designs a quantitative study with 300 Moroccan consumers. It has based on the complex modeling by using the PLS-SEM approach in Smartpls3.

2 Research Model and Design

2.1 *Hypothesis Development: From the Return to Local Production*

Scholars have highlighted the advantages of a proximity production system in order to mitigate the issues of poor nutrition and to respond to social, economic, and ecological requirements [19, 20, 21]. The reconfiguration of proximity agriculture is an awareness that announcing a return to a new relationship with the local [12, 22, 23, 24]. For example, the Slow Food movement in Italy denounces the harmonization of food products caused by globalization and shifts the focus to local specialties [25]. This global awareness of the current reality of the agri-food, productivism, is at the heart of the social transition of the agricultural sector. Local movements meet in the importance of revalorizing the products of small farmers. This return is in line with the aim of the concept of society 5.0 [2–26]. The harmonization of this concept with the realities of local communities can offer a positive impact and reduce the gap between giant groups and small-medium communities which often produce and commercialize the local production. This synchronization could be a baseline of the sustainable development which also aim the valorization of local resources and traditional life style [11]. In this vein, the determinants of authenticity are a strategic component in the digitalization era of business [17, 28, 29]. Postmodern paradigm has been greatly enriched by the technological revolutions. Main findings have underlined that the concerns of social research have progressed with the expansion of the industry [30]. Authenticity is very linked to the origin of product [31, 32] and its producer [33]. Also, the association with time remains an important determinant for consumer [34]. Consumer accepts its own definition of authenticity. It is based on the use of traditional manners during the production process [35]. Studies underline that authenticity is not an objective character, but it may be a subjective valuation [36]. Hence:

H.1. user perceived authenticity influences positively the user digital behavior.

2.2 *To the Perceived Authenticity in the Digital Era*

From the first use of the steam engine to the mass digitalization, the significant development made in the field of information technology has dramatically changed the everyday life of users. Society 5.0 is not a linear extension but it aims to place human wellbeing in the core worldwide agenda [26]. It relies on decreasing of economics and social gaps in societies [1], by taking advantages from industry 4.0 benefits [26]. This shift to digital life calls for a deep baseline to go with these changes. According to Aslam et al. [37], the focus of now is not on innovation, however, also on its execution which needs a user orientation financially viable and marketable objects. Researchers stress that this smartness occurs by making this concept more valuable, useful, practical, implementable and meaningful to society [1–26, 38, 39]. As commonly acknowledged, to get a positive effect on key performance indicators, business strategies should enhance the consumer needs; society 5.0 looks to develop the world in which socials and economics gaps will be reduced [1]. Enterprises can bring the maximum of technological progress and subsist with the globalization conflicts; as well the small or medium firms should take advantages.

For researchers in social and human sciences, the consumption of standardized products has caused an appetite for a new revival with the local. Globalization has set the pace of consumption according to the stock market seasons of international production and not to the natural seasons of farmers. We are witnessing as never before an increased demand for meaning, significance, credibility, trust, and authenticity to consume food [40, 41, 42, 43]. This awareness is also favored by the massif usage of media and its fast-growing platforms [44, 45], the implementation of tractability system in food sector [5], and the abundance of recommendations from nutritionists [46, 47]; studies have shown that media coverage and nutritional information overload amplify consumers' concerns: an uncontrolled information creates contradictions in the minds which negatively influence food behavior. Nagler [48] found that the consumers most exposed to media information are those who are increasingly moving away from healthy behaviors. He stressed that the general public is not always able to interpret the correct meaning of the nutritional message. Therefore, the informed public can admit the existence of contradictions between results: different conditions of the experiment (a beta-carotene control) often leads to different conclusions, but the general public does not necessarily focus on the scientific protocol. Moreover, the media propaganda will draw more attention to the contradiction than to the clarification. This fast-growing of social platforms is enabling the speedy exchange of virtual content, but also it is very worrying. According to [17] the mass utilization of digital in marketing practices seems to create a misapprehension of online authenticity. In a real context, consumers can evaluate the authenticity dimensions, but in digital context users are unable to contact the vendor directly. In this regard, the virtual relationship contributes to deception, less authenticity, and fraud [17–45]. Researchers accentuate the key role of trust in food consumption in the digital era [4–45, 49, 50], it is a requirement for digital interaction [28–45].Therefore

H.2. User perceived authenticity influences positively classical media trust
H.3. User Perceived authenticity influences positively social media trust
H.4. Trust in classical media influences positively user digital behavior
H.5. Trust in social media influences positively user digital behavior

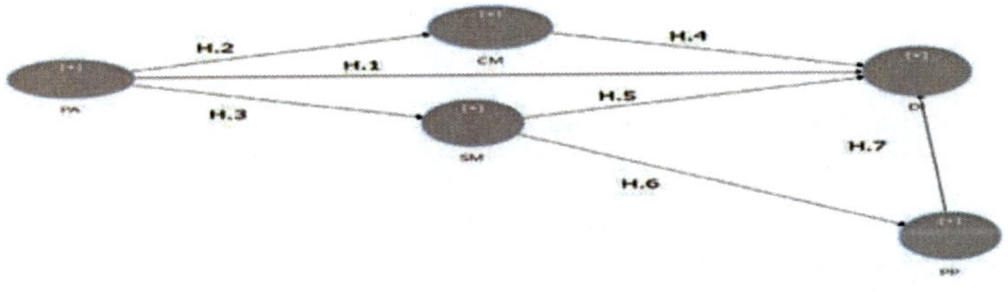

Fig. 1 Research's model

The literature has underline that perceived authenticity increase self-congruence [51] that meets the consumer's need for self-improvement and boosts their self-esteem [52]. In parallel, we can accept that, as a result, user perceived authenticity leads to a good connection with the product which has a positive impact on behavioral outcomes, as willingness to pay a premium price [53]. Hence:

H.6. Trust in social media influence positively the premium price
H.7. Premium price influences positively the user digital behavior

2.3 *Research Design*

The respondents were the consumer interacted with digital platforms in context of local food consumption in Morocco. The participation was voluntary and we have selected 300 users, who used social platforms, who have reported to all questions successfully. This sample number is ample in context of PLS-SEM approach [54, 55, 56, 57] for a higher order construct, which is recommended in marketing [54, 55] to test the hypothesis (Fig. 1). All items are obtained from the literature and adjusted to the research's setting (Table 5).

3 Findings

The research's model contains the higher order Perceived Authenticity construct. It has measured by three constructs: origin, tradition, and the self-representation. The statistical analysis has based in disjoint two-stage approaches in PLS-SEM [54]. In the first stage, we have assessed the three inferior constructs in light of the standard criteria. In the second stage, the perceived authenticity has been assessed with the latent variables score from the first phase [55, 56].

Table 1 Assessment's results of measurement model

Construct	Item	Loading	Alpha	CR	AVE
PA	At1/At2/At3	0.882/0.934/0.765	0.825	0.897	0.746
CM	Cm1/Cm2/Cm3	0.852/0.853/0.836	0.804	0.884	0.717
SM	Sm1/Sm2/Sm3	0.904/0.898/0.847	0.862	0.914	0.780
PP	PP1/PP2/PP3	0.907/0.927/0.872	0.886	0.929	0.814
UB	UB1/UB2/UB3	0.846/0.789/0.795	0.740	0.851	0.657

Table 2 Discriminant validity

	CM	UB	PA	SM	PP
CM	0.847				
UB	0.434	0.810			
PA	0.397	0.557	0.863		
SM	0.580	0.477	0.414	0.883	
PP	0.374	0.569	0.578	0.450	0.902*

It is accepted since the reliability and convergent validity were established [57]

3.1 *Measurement Model*

The assessment draws on the evaluation of the internal consistency with Cronbach's alpha and composite reliability (CR), convergent validity with average variance extracted (AVE) and discriminant validity [55, 56] (Table 1).

We observe that all reflectively constructs are above the critical values [56]: loading >0.7, Alpha >0.7, and AVE >0.5. Also, the square root of AVE is higher than the correlation for all constructs (Table 2).

3.2 *Structural Model*

The structural model assessment is based on the general criteria of PLS-SEM: the evaluation of the collinearity between constructs, significance and relevance of the path coefficient, and R^2 [54, 55]. Table 3 shows that the R^2 of the construct of user behavior is 0.448 which is considered as heigher in consumer behavior, also the Q^2 value is larger than 0 that suggests a good predictive relevance of the construct [55, 56].

Analyses underline that trust in classical and social media do not have a positive effect on the user (rejection of H.4, H.5). The important influence of perceived authenticity denotes the importance of the first real experience for users which is in line with [17]. The good effects of perceived authenticity on trust in Classical and Social Media demonstrate its importance. To enhance the user interaction, managers

Table 3 Evaluation of R^2 and Q^2 of constructs

Construct	R^2*	Q^2**
Authenticity	–	0.482
Trust in classic medias	0.158	0.419
Trust in social medias	0.171	0.533
Premium price	0.203	0.593
User behavior	0.448	0.315
* > 0.2; ** >> 0 [55]		

Table 4 Results of hypothesis testing (Second stage)

Hypothesis	Path	T value*	Pvalue**	Confidence intervals	Decision
H.4: CM-> UB	0.122	1.672	**0.095**	−0.020 –0.248	Rejected
H.2: PA-> CM	0.397	7.986	0.000	0.295–0.476	Accepted
H.1: PA -> UB	0.273	4.475	0.009	0.151–0.390	Accepted
H.3: PA -> SM	0.414	8.640	0.000	0.309–0.491	Accepted
H.5: SM -> UB	0.162	2.413	**0.016**	0.036–0.301	Rejected
H.6: SM -> PP	0.450	9.357	0.000	0.351–0.534	Accepted
H.7: PP -> UB	0.293	4.275	0.000	0.151–0.417	Accepted
* > 1.96; ** < 0.01 [55]; boostrapping 5000					

should communicate about the determinants of authenticity. The food sector still has difficulties, because users need the real contact to verify the authenticity. In this vein, entities should consider that community managers play a key role: users expressed the need for direct and human contact.

4 Conclusion

The first aim of this paper was to propose a lecture on the determinants of authenticity in the digital era and its association with behavioral variables. Therefore, this present consideration connects prior research on the emergence of society 5.0 [1–26] and the authenticity craze for postmodern consumers [36–49, 59].

The mass utilization of digital by postmodern consumer can present a key source for an effective and an efficient marketing practices. Findings offer different insights for managers and government. This research shows that user authenticity is determined by the definition of the origin, tradition, and also the self-definition of user. Thus, authors suggest that communication should focus on the realities of the small communities behind the production, because they affect positively users. The usage of its determinants can create and develop a worldwide community. This action can encourage content sharing between users which is very useful for the user-oriented digital strategies for manager (sharing economy).

However, the local food sector still suffers and needs more implementation of the technologies [39], this is arising especially when regarding the small- and medium-scaled firms and/or food processors that are associated with social skill, old know-how and technological limitations [39]; it requests a deep communication in order to increase credibility and limit the negative effect of the previous scandals [4, 5, 16]. The fast worldwide-changing in societies and the implementation of technologies around the everyday life call researchers to constitute a synergy between technological development and its effect on social and cultural dimensions. Once the aim of the technology meets the firm's need and fits within the social reality and the capability of humans, then the firm is capable of making a meaningful technology for itself and also for users.

This study has limitations. Authors advice to perform more investigation in food consumption in particular, and cultural consumption in general. It is also very important to study the linkage between the sharing economy and its implementation in this context. Authors stress that it is very urgent to deep understanding on how we can accompany our societies in this changing period to shift serenely to 5.0's level with our authenticity as humans.

Appendices: Items of construct

Table 5 Items and Constructs

Perceived authenticity of user	Refined from [28, 33, 36]
Trust in classical Media (no use of internet) and Digital Media (Social platforms)	Refined from [17, 59]
Premium price	Refined from [60]
User behavior	Refined from [28, 61]

References

1. Fukuda K (2020) Science, technology and innovation ecosystem transformation toward society 5.0. Int J Product Econ 220:107460
2. Xu LD, Xu EL, Li F (2018) Industry 4.0: state of the art and future trends. Int J Product Res 56(8):2941–2962
3. Roblek V, Meško M, Krapež A (2016) A complex view of industry 4.0. Sage open 6(2):2158244016653987
4. Lang B, Conroy DM (2021) Are trust and consumption values important for buyers of organic food? a comparison of regular buyers, occasional buyers, and non-buyers. Appetite 161:105123

5. Garaus M, Treiblmaier M (2021) The influence of blockchain-based food traceability on retailer choice: The mediating role of trust. Food Control 129:108082
6. FAO, IFAD, UNICEF, WFP & WHO (2017) The State of Food Security and Nutrition in the World 2017. Building resilience for peace and food security. Rome, FAO
7. Nglazi MD, Ataguba JE-O (2022) Overweight and obesity in non-pregnant women of child-bearing age in South Africa: subgroup regression analyses of survey data from 1998 to 2017. BMC Public Health 22(1):1–18
8. Allali F (2017) Evolution des pratiques alimentaires au Maroc. J Med Sur 4(1):70–73
9. Development Initiatives (2018): 2018 global nutrition report: shining a light to spur action on nutrition. Bristol, UK: Development Initiatives
10. Shekar M, Popkin B, (eds.) (2020) Obesity: health and economic consequences of an impending global challenge. World Bank Publications
11. Battino S, Lampreu S (2019) The role of the sharing economy for a sustainable and innovative development of rural areas: a case study in Sardinia (Italy). Sustainability 11(11):3004
12. Morgan K, Marsden T, Murdoch J (2008) Worlds of food: place, power, and provenance in the food chain. Oxford University Press on Demand
13. Bailly A (2002) Vers un nouvel ordre alimentaire local-global: le cas de la restauration. Revue dEconomie Regionale Urbaine 2:319–332
14. Harrison M (2013) Disease and the modern world: 1500 to the present day. John Wiley & Sons
15. Bakalis S, Gerogiorgis D, Argyropoulos D, Emmanoulidis C (2022) Food Industry 4.0: opportunities for a digital future. In Juliano P, Buckow R, Nguyen MH, Knoerzer K, Sellahewa J (eds.), Food engineering innovations across the food supply chain, Academic Press, pp 357–368
16. Yang Y et al (2019) Fraud vulnerability in the dutch milk supply chain: assessments of farmers, processors and retailers. Food Control 95:308–317
17. Davis R, Sheriff K, Owen K (2019) Conceptualising and measuring consumer authenticity online. J Retail Cons Serv 47:17–31
18. Zhou X, Van Tilburg WA, Mei D, Wildschut T, Sedikides C (2019) Hungering for the past: Nostalgic food labels increase purchase intentions and actual consumption. Appetite 140:151–158
19. Bowen S, Mutersbaugh T (2014) Local or localized? exploring the contributions of franco-mediterranean agrifood theory to alternative food research. Agric Hum Values 31(2):201–213
20. Pachoud C, Labeyrie V, Polge E (2019) Collective action in Localized Agrifood Systems: An analysis by the social networks and the proximities. study of a Serrano cheese producers' association in the Campos de Cima da Serra/Brazil. J Rural Stud 72:58–74
21. Lamine C, Garçon L, Brunori G (2019) Territorial agrifood systems: a Franco-Italian contribution to the debates over alternative food networks in rural areas. J Rural Stud 68:159–170
22. Wiskerke J (2009) On places lost and places regained: reflections on the alternative food geography and sustainable regional development. Int Plan Stud 14(4):369–387
23. Veenhuizen V (2014) René, ed.: Cities farming for the future: urban agriculture for green and productive cities. IDRC
24. Brinkley C, Manser GM, Pesci S (2021) Growing pains in local food systems: a longitudinal social network analysis on local food marketing in Baltimore County, Maryland and Chester County. Pennsylvania. Agric Hum Values 38(4):911–927
25. Pietrykowski B (2004) You are what you eat: The social economy of the slow food movement. Rev Soc Econ 62(3):307–321
26. Carayannis EG, Morawska-Jancelewicz J (2022) The Futures of Europe: society 5.0 and industry 5.0 as driving forces of future universities. J Knowl Econ 1–27
27. Balaban D, Szambolics JA (2022) Proposed model of self-perceived authenticity of social media influencers. Media Commun 10(1):235–246
28. Arya V, Verma H, Sethi D, Agarwal R (2019) Brand authenticity and brand attachment: how online communities built on social networking vehicles moderate the consumers' brand attachment. Iim Kozhikode Soc Manag Rev 8(2):87–103

29. Eigenraam AW, Eelen J, Verlegh PW (2021) Let me entertain you? the importance of authenticity in online customer engagement. J Interact Mark 54:53–68
30. Mumford L (1961) The city in history: its origins, its transformations, and its prospects. Sci Soc 27(1):106–109
31. Moulard J, Babin BJ, Griffin M (2015) How aspects of a wine's place affect consumers' authenticity perceptions and purchase intentions: the role of country of origin and technical terroir. Int J Wine Bus Res 27(1):61–78
32. Morhart F, Malär L, Guèvremont A, Girardin F, Grohmann B (2015) Brand authenticity: an integrative framework and measurement scale. J Consum Psychol 25(2):200–218
33. Camus S (2004) Proposition d'échelle de mesure de l'authenticité perçue d'un produit alimentaire. Recherche et Appl Mark (French Ed) 19(4):39–63
34. Beverland M (2005) Brand management and the challenge of authenticity. J Product Brand Manag 14(7):460–461
35. Cova B, Dalli D (2009) Working consumers: the next step in marketing theory? Mark Theory 9(3):315–339
36. Napoli J, Dickinson SJ, Beverland MB, Farrelly F (2014) Measuring consumer-based brand authenticity. J Bus Res 67(6):1090–1098
37. Aslam F, Aimin W, Li M, Ur Rehman K (2020) Innovation in the Era of IoT and industry 5.0: absolute innovation management (AIM) framework. Information 11(2):1–24
38. Fukuyama M (2018) Society 5.0: aiming for a new human-centered society. Jpn Spotlight 27:47–50
39. Hasnan NZN, Yusoff YM (2018) Short review: application areas of industry 4.0 technologies in food processing sector. In: 2018 IEEE student conference on research and development (SCOReD), pp 1–6
40. Akbar MM, Wymer W (2017) Refining the conceptualization of brand authenticity. J Brand Manag 24(1):14–32
41. Hu W, Batte MT, Woods T, Ernst S (2012) Consumer preferences for local production and other value-added label claims for a processed food product. Eur Rev Agric Econ 39(3):489–510
42. Cova B, Maffesoli M (2015) postmodernité et tribalisme. Regards croisés sur la consommation 2:167–183
43. Vita B, Deitiana T, Ruswidiono W (2021) The online marketing of Indonesian street food in Jakarta. Cogent Bus Manag 8(1):1–20
44. Gon M (2021) Local experiences on Instagram: Social media data as source of evidence for experience design. J Destin Mark Manag 19:100435
45. Martindale L (2021) I will know it when I taste it': trust, food materialities and social media in Chinese alternative food networks. Agric Hum Values 38(2):365–380
46. Marocolo M, Meireles A, de Souza HLR, Mota GR, Oranchuk DJ, Arriel RA, Leite LHR (2021) Is social media spreading misinformation on exercise and health in Brazil? Int J Environ Res Public Health 18(22):1–10
47. Kabata P, Winniczuk-Kabata D, Kabata PM, Jaśkiewicz J, Połom K (2022) Can social media profiles be a reliable source of information on nutrition and dietetics? Healthcare 10(2):1–8
48. Nagler RH (2014) Adverse outcomes associated with media exposure to contradictory nutrition messages. J Health Commun 19(1):24–40
49. Portal S, Abratt R, Bendixen M (2019) The role of brand authenticity in developing brand trust. J Strat Mark 27(8):714–729
50. Hernandez-Fernandez A, Lewis Mathieu C (2019) Brand authenticity leads to perceived value and brand trust. Euro J Manag Bus Econ 28(3):222–238
51. Beverland M, Farrelly F (2010) The quest for authenticity in consumption: consumers' purposive choice of authentic cues to shape experienced outcomes. J Consum Res 36(5):838–856
52. Kressmann F, Sirgy MJ, Herrmann A, Huber F, Huber S, Lee DJ (2006) Direct and indirect effects of self-image congruence on brand loyalty. J Bus Res 59(9):955–964
53. Fritz K, Schoenmueller V, Bruhn M (2017) Authenticity in branding–exploring antecedents and consequences of brand authenticity. Eur J Mark 51(2):324–348

54. Sarstedt M, Hair JF Jr, Cheah JH, Becker JM, Ringle CM (2019) How to specify, estimate, and validate higher-order constructs in PLS-SEM. Australas Mark J (AMJ) 27(3):197–211
55. Hair JF, Howard MC, Nitzl C (2020) Assessing measurement model quality in PLS-SEM using confirmatory composite analysis. J Bus Res 109:101–110
56. Sarstedt M, Cheah JH (2019) Partial least squares structural equation modeling using SmartPLS: a software review. J Mark Anal 7(3):196–202
57. Al-Emran M, Al-Maroof R, Al-Sharafi MA, Arpaci I (2020) What impacts learning with wearables? an integrated theoretical model. Interact Learn Environ 21
58. Lee TH, Arcodia C, Novais MA, Kralj A, Phan TC (2021) Exploring the multi-dimensionality of authenticity in dining experiences using online reviews. Tour Manage 85:104292
59. Benamour Y (2000) Confiance interpersonnelle et confiance institutionnelle dans la relation client-entreprise de service: Une application au secteur bancaire français. Doctoral dissertation Paris 9
60. Zeithaml V, Berry LL, Parasuraman A (1996) The behavioral consequences of service quality. J Mark Anal 60(2):31–46
61. Price L, Arnould E (1999) Commercial friendships: service provider–client relationships in context. J Mark Manag 63(4):38–56

Factors Affecting Students' Adoption of E-Learning Systems During COVID-19 Pandemic: A Structural Equation Modeling Approach

Tareq Obaid, Bilal Eneizan, Mohanad S. S. Abumandil, Ahmed Y. Mahmoud, Samy S. Abu-Naser, and Ahmed Ali Atieh Ali

Abstract The provision and usage of online and e-learning systems are becoming the main challenge for many universities during COVID-19 pandemic. E-learning system such as Moodle has several fantastic features that would be valuable for use during this COVID-19 pandemic. However, the successful usage of the e-learning system relies on understanding the adoption factors. There is a lack of agreement about the critical factors that shape the successful usage of e-learning systems during the COVID-19 pandemic; hence, a clear gap has been identified in the knowledge of the critical factors of e-learning usage during this pandemic. Therefore, an extended version of the Technology Acceptance Model (TAM) was developed to investigate the underlying factors that influence Students' decisions to use an e-learning system. The TAM was populated using data gathered from a survey of 389 undergraduate Students' who were using the based-Moodle e-learning system at Alazhar University. The model was estimated using Structural Equation Modelling (SEM). A path model was developed to analyze the relationships between the factors to explain students' adoption of the e-learning system. The findings indicated that Computer Anxiety,

T. Obaid (✉) · A. Y. Mahmoud · S. S. Abu-Naser
Faculty of Engineering and IT, Alazhar University, Gaza, Palestine
e-mail: tareq.obaid@alazhar.edu.ps

A. Y. Mahmoud
e-mail: ahmed@alazhar.edu.ps

S. S. Abu-Naser
e-mail: abunaser@alazhar.edu.ps

B. Eneizan
Business School, Jadara University, Irbid, Jordan

M. S. S. Abumandil
Faculty of Hospitality, Tourism and Wellness, Universiti Malaysia Kelantan, Kota Bharu, Malaysia
e-mail: nad.ssa@umk.edu.my

A. A. A. Ali
Candidate at School of Technology and Logistics Management, Universiti Utara, UUM Sintok, 06010 Kedah, Malaysia

M. Al-Emran et al. (eds.), *International Conference on Information Systems and Intelligent Applications*, Lecture Notes in Networks and Systems 550,
https://doi.org/10.1007/978-3-031-16865-9_19

Course Content, Hedonic Motivation, Perceived Environment, Subjective Norm, and Technical Support effect significantly on both ease of use and usefulness. Subjective Norm effect significantly on intention to use. Perceived Ease of Use and Perceived Usefulness effect significantly on intention to use.

Keywords E-Learning · TAM · SEM · Adoption · Palestine

1 Introduction

As we see now in the world, the COVID-19 pandemic is forcing educational institutions such as universities to shift rapidly to distance and online learning. COVID-19 has forced universities around the world to adopt online learning. We are now in a state of emergency and must react with different and available ways of learning such as eLearning systems and mobile learning applications. Online learning is not new to learners, nor is distance learning. However, COVID-19 is reviving the need to explore online teaching and learning opportunities [1].

In general, e-learning is implementable by utilizing a learning management system (LMS) [2, 3]. A majority of the universities in Palestine utilize open-source LMS such as the Moodle platform. As LMS provides significant advantages as particularly evident throughout the COVID-19 pandemic thus far, numerous studies have been conducted to look at e-learning system adoption and the main factors that contribute to it from various e-learning perspectives including the students [4] examined an interpretive case study based on the realistic and social perceptions of students that are positively affected by habitual activities. The study suggested e-learning approaches using social media platforms like Facebook and WhatsApp for e-learning. Meanwhile, [5] revealed that academic performance and organizational aspects are positively correlated in the context of remote teaching [6] proposed a new e-learning adoption framework that groups numerous features of ISS and diffusion of innovation (DOI), and found the features to be significantly related to e-learning adoption. Additionally [7] studied four factors including EOU and technical usage of LMS in affecting its usage intention.

Numerous approaches had been utilized to look at the success factors of e-learning activities [8–11]. Several theoretical frameworks have been employed to examine the hindrances to adoption including the (UTAUT) and (DOI) [12; 13]. The (TAM) has been extended to incorporate the factors of sharing of knowledge and acquisition to produce a new model for assessing e-learning system usage [2]. The significance of certain factors has also been identified using cutting-edge artificial intelligence (AI) techniques [14]. Yet, there are still very few studies on other factors such as technical support which can improve LMS productivity [15].

As usage willingness and acceptance are crucial in identifying the success of a given system [16], low usage of the system hinders the materialization of its full benefits [17]. This leads to the failure of the system and results in wasted money for the university [18]. Limited studies had looked at this topic from the student's perspective [19]. To gain better insight into the students' e-learning requirements, all facets of e-learning system adoption must be examined which would ultimately lead to the system's successful implementation [20]. Thus far, no study has examined the drivers and hindrances of e-learning adoption specifically in the COVID-19 pandemic setting although such systems had been implemented in numerous universities three years ago. Hence, this study aims to look at the key influencing factors to e-learning system adoption in the COVID-19 setting [21].

2 Literature Review

System usage determines an information system's success [17]. Hence, student acceptance is a critical component of any success. A number of studies had examined e-learning adoption issues worldwide. In Malaysia, for example, the TAM with IDT model was used to determine the important determinants influencing e-learning system usage among students [22]. The researchers discovered that relative advantages perceived compatibility, complexity, and reported enjoyment and some other factors all have a substantial impact on students' decision to use e-learning systems [23].

Several other existing studies had examined e-learning acceptance from the standpoints of technology, the organization, and the environment [10, 24]–[26]. Aldammagh et al.. [5], introduced the revised IS success model which identifies the six components of successful e-learning adoption. System usage has been identified to be more significant than user satisfaction in improving user satisfaction, thus leading to greater usage intention [27]. The revised IS success model is a highly prominent model for examining the success of e-learning adoption.

There are other IS success models in existence [28] including Davies' Technology Acceptance Model (TAM); Gable, Sedera, and Chan's (IS) success, and DeLone and McLean's Information System Success Model (ISSM). TAM was introduced in 1989 to address information technology usage acceptance and willingness rather than usage success. Meanwhile, ISSM focuses on the net benefits derived from the successful usage of IS. As such, none of the models can be considered a "one size fits all" solution. The adoption of these models must match the given study objective [29] introduced the TAM [30], which later on emerged as a prominent innovation adoption model employed by numerous researchers to examine the impact of novel technologies on users [31]. The time-based relationship between belief, attitude, intention, and behavior, according to [29], can aid in the forecast of new technology utilization. TAM is an alteration of TRA which determines a person's behavioral intention towards usage [32]. PEOU has been identified to directly affect PU [33; 34].

3 Theoretical Framework and Hypotheses

This part of the study explicates the revision and redefinition of the factors which affect e-learning system adoption among university students. Towards that end, this study adds six constructs i.e., Computer Anxiety, Course Content, Hedonic Motivation, Perceived Enjoyment, Subjective Norm, and Technical Support to the original TAM which already entails the constructs' ease of use and usefulness. The proposed framework is shown in the Figure below. The hypothesized relationships based on past findings from the literature are explicated in the subsequent sub-sections.

3.1 Subjective Norms

This construct describes an individual's perception of what other people believe that he/she should decide in relation to the performance of a certain behavior [35; 37] defined it as a person's sense of social pressures to engage in specific behaviors. Subjective norms can significantly predict the intention to use computer technology whether in a direct manner [37]. or indirect manner [38]. But the existing outcomes to this are inconsistent. Certain studies asserted that subjective norms have no significant effect [29], while others suggest that the construct declines over time and remains significant only in binding situations [39]. According to [40]. subjective norms significantly affect perceived usefulness. Hence, this current study forms the hypotheses:

H1: The intention to use e-learning systems is positively affected by subjective norms.

H2: The ease of use of e-learning systems is positively affected by subjective norms.

H3: The usefulness of e-learning systems is positively affected by subjective norms.

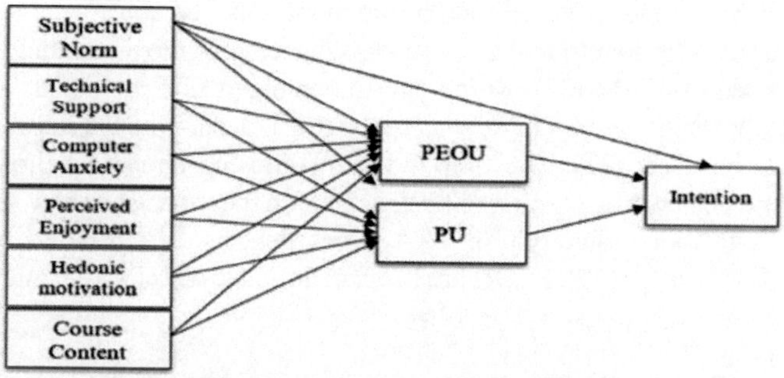

Fig. 1 The proposed research model

3.2 Technical Support

Technical support availability predominantly determines technology acceptance in the teaching setting [41], particularly in the early phase of technology adoption. According to [42], facilitating conditions i.e., technical support and external control are the main drivers of perceived ease of use in the context of information technology. Existing empirical findings revealed that unsuccessful e-learning projects are mainly those that lack technical support [43, 44] added the construct of technical support into the original TAM as an external variable to explicate WebCT usage. Hence, this current study forms the hypotheses:

H4: The ease of use of e-learning systems is positively affected by technical support.
H5: The usefulness of e-learning systems is positively affected by technical support.

3.3 Computer Anxiety

Computer anxiety entails the fear or nervousness related to computer system usage [41]. It is the overall negative perception towards computer usage [34]. Computer anxiety involving the usage of new interfaces and the performance of system tasks is a major hindrance. Many studies have confirmed the prevalence of computer anxiety towards e-learning system usage. Past research had shown that computer anxiety significantly affects PEOU [45]. In line with past evidences by [46; 47] this current study forms the hypotheses:

H6: The ease of use of e-learning systems is negatively affected by Computer anxiety.
H7: The usefulness of e-learning systems is negatively affected by Computer anxiety.

3.4 Perceived Enjoyment

According [55] enjoyment entails the extent to which a system's utilization is perceived to be enjoyable despite any repercussions. An e-learning system may attract higher user engagement if it has interactive entertainment functions [13]. Low enjoyment has been linked to greater usage effort [48, 30] had validated the causal link between enjoyment and PEOU. Several studies had provided evidence that integrating the construct of enjoyment into the original TAM as an external determinant could help in explaining e-learning system adoption and usage better [49].

H8: The ease of use of e-learning systems is positively affected by Perceived Enjoyment.

H9: The usefulness of e-learning systems is positively affected by Perceived Enjoyment.

3.5 Hedonic Motivation

Hedonic motivation is the measurement of a user's perceived delight and entertainment [49, 50] incorporated this construct into their revised model to identify the effect of intrinsic utilities. According to [50], the key effect of hedonic motivation is derived from the sense of newness and innovativeness in utilizing novel systems. Past research revealed that hedonic motivation significantly affects the adoption of certain technologies [47] and the utilization of e-learning [51]. This current study proposes that if users enjoy utilizing an e-learning system, their likelihood to continue using it becomes higher.

H10: The ease of use of e-learning systems is positively affected by Hedonic motivation.

H11: The usefulness of e-learning systems is positively affected by Hedonic motivation.

3.6 Course Content

[52] proved the significance of curriculum design in improving e-learning performance. Simultaneous presentations of texts and images in the e-learning system along with animation and narration improve the course's illustrations and hence allow the students to better comprehend the course. Meanwhile, [53] found that the nature of the course significantly determines students' decision to adopt an e-learning system. Course syllabuses with excessive practical work and the need for extreme technical expertise are rather unsuitable to be used on e-learning platforms. Hence, course content is hypothesized to have a positive impact on behavioral intention to use an e-learning system.

H12: The ease of use of e-learning systems is positively affected by Course content.

H13: The usefulness of e-learning systems is positively affected by Course content.

3.7 Perceived Ease of Use

This construct describes the degree to which a person believes that using a particular system will require minimal effort [54]. PEOU has been shown in previous studies to have a positive impact on behavioral intention to use the system [55]. Likewise, PEOU also influences the direct or indirect acceptance of a given system via PU.

H14: PEOU positively affects the intention to use e-learning system.

3.8 Perceived Usefulness

This construct describes how often an individual believes that using a particular system will enhance his or her work performance [56]. PU has been shown in a number of studies to be a strong predictor of behavioral intention to use an e-learning system [46]. Some other studies also found that PU positively affects behavioral intention (BI) whether directly or attitudinally [57]. PU also reveals the user's degree of belief that the usage of a new technology will provide future benefits.

H15: PU positively affects the intention to use e-learning system.

4 Methodology

This study is quantitative in nature. The needed data was collected using close-ended questionnaires distributed to 400 respondents, with questions regarding the previously discussed constructs. Relevant statistical tools were used to determine the questionnaire's reliability and validity. The measurement of the items in the questionnaire was done using a 5-point Likert scale whereby 1 = strongly disagree and 5 = strongly agree.

Random sampling was applied in this study. The questionnaire was distributed among 420 undergraduate students et al.-Azhar University in Palestine. 389 questionnaire was received and valid. The unit of analysis was the undergraduate students et al.-Azhar University who familiar with Moodle based e-learning system at Alazhar University.

Data analysis was performed using PLS-SEM, specifically Smart PLS 3 [58]. This is a variance-based structural equation modeling technique which aids the analysis of complex models with multiple relationships. Its aim is to predict and test the developed hypotheses, and eventually provide empirical proof of the findings.

Table 1 Full collinearity

CA	CC	HM	PE	SN	TS	PU	PEOU
1.02	1.59	1.46	1.32	1.36	1.16	2.36	2.78

Note: CA = Computer Anxiety, CC = Course Content, HM = Hedonic Motivation, PE = Perceived Environment, SN = Subjective Norm, TS = Technical Support, PU = Perceived Usefulness, PE = Perceived Ease of Use

5 Data Analysis

The current work employs variance-based SEM i.e., partial least square using Smart PLS version 3.3.2 [58]. for examining the results of the data analysis. According to and [43] if the study is done for the predictive purposes Partial least square is an appropriate technique for data analysis.

Since the research used a single source data there can be an issue of Common method variance (CMV) [49]. Hence the study followed [54] to avoid this issue. If the value of VIF is 3.3 or above there is a concern of common method variance. Table 1 shows the VIF and it can be noticed that all VIF values are under the threshold value i.e., less than 3.3 hence no serious concern of single source bias in our data.

For data analysis, we used a two-step procedure. We run the convergent and discriminant validity measurement model in the first phase [2]. The study moved on to the next level of structural model testing after establishing the model's validity and reliability.

Convergent validity indicates whether a particular item adequately measures a latent construct that it is supposed to measure [58]. The items loading was assessed for testing the convergent validity and we found that all the item loading were above the suggested value of 0.7. Moreover the (AVE) and (CR) were also examined. The values for the AVE and CR were found to be above the accepted values of 0.5 and 0.7 respectively. Table 2 depicts the results of item loadings, AVE and CR of all the latent variables. Hence confirming the convergent validity of the latent constructs.

Discriminant validity is the second type of validity assessment. For ensuring discriminant validity HTMT was examined. The measure was initially recommended by [58]. and later endorsed by [59]. The recommended HTMT values is maximum 0.90. Table 3 below presents the results of HTMT and it can be noticed that all the values are appropriate as per recommended values. Hence each construct is distinct from the others. The measuring model's results validate the constructs' reliability and validity.

Table 2 Convergent validity

Items	Loading	CR	AVE
CA1	0.85	0.90	0.69
CA2	0.89		
CA3	0.81		
CA4	0.76		
CC1	0.91	0.92	0.74
CC2	0.83		
CC3	0.85		
CC4	0.84		
HM1	0.84	0.89	0.68
HM2	0.84		
HM3	0.84		
HM4	0.78		
ITU1	0.81	0.92	0.74
ITU2	0.87		
ITU3	0.84		
ITU4	0.91		
PE1	0.76	0.91	0.66
PE2	0.81		
PE3	0.71		
PE4	0.88		
PE5	0.88		
PEOU1	0.71	0.88	0.59
PEOU2	0.75		
PEOU3	0.79		
PEOU4	0.76		
PEOU5	0.83		
PU1	0.75	0.90	0.70
PU2	0.84		
PU3	0.85		
PU4	0.89		
SN1	0.90	0.91	0.73
SN2	0.71		
SN3	0.90		
SN4	0.93		
TS1	0.90	0.92	0.73
TS2	0.85		
TS3	0.77		
TS4	0.90		

Table 3 Discriminant validity (HTMT)

	CA	CC	HM	ITU	PE	PEOU	PU	SN	TS
Computer anxiety									
Course content	0.11								
Hedonic motivation	0.09	0.56							
Intention to use	0.13	0.31	0.45						
Perceived environment	0.09	0.5	0.34	0.26					
Perceived ease of use	0.15	0.56	0.72	0.65	0.51				
Perceived usefulness	0.17	0.49	0.59	0.55	0.44	0.9			
Subjective norm	0.05	0.44	0.34	0.4	0.41	0.59	0.42		
Technical support	0.06	0.12	0.36	0.39	0.14	0.45	0.37	0.27	

6 Structural Model

[60] suggest to test the multivariate normality using skewness and kurtosis of the items. Following [49] we tested the skewness and kurtosis and found that the data was not normal. The multivariate skew-ness and kurtosis have p-values less than 0.05. Hence the suggestion of [59] were followed the path coefficients and the S.E along with t values and p values were reported for the model. The bootstrapping was per-formed for 5000 samples. The hypotheses were tested based on path coefficients, p-values and t-values. Moreover, the effect size has been taken into the account too. Table 5 provides the summarized form of the all the criterionmet.

With 6 predictors on PEOU, the R2 was 0.56, which demonstrates that all six predictors explained 56% of the variance in PEOU. The PU was also predicted by 6 predictors with R2 of 0.39 showing that all the 6 predictors explained 39% variation in PU. Moreover, ITU was predicted by 3 variables with R2 of 0.33 hence explaining 33% variance in ITU. The individual relationships of Computer Anxiety → PEOU (β = -0.09, p = 0.01), Course content → PEOU (β = 0.10, p = 0.02), Hedonic motivation → PEOU (β = 0.35, p = 0.00), Perceived Enjoyment → PEOU (β = 0.18, p = 0.00), Subjective norm → PEOU (β = 0.27, p = 0.00), Technical support → PEOU (β = 0.18, p = 0.00), Computer Anxiety → PU (β = –0.11, p = 0.01), Course content → PU (β = 0.13, p = 0.01), Hedonic motivation → PU (β = 0.28, p = 0.00), Perceived environment → PU (β = 0.19, p = 0.00), Subjective norm → PU (β = 0.13, p = 0.01), Technical support → PU (β = 0.17, p = 0.00), PEOU → Intention to use (β = 0.37, p = 0.00), PU → Intention to use (β = 0.16, p = 0.02), Subjective norm → Intention to use (β = 0.13, p = 0.01) were found to be significant hence the impact of computer anxiety on PEOU and PU found to be negative and rest all relationships were found to be positive. In a nutshell it can be concluded that all 15 (H1 to H15) hypotheses were supported.

For the effect size (f2), it shows the change in R2 when a certain construct is removed from the model. According to [7] 0.02, 0.15 and 0.35 for the f2 is representing the small, medium, and large effect size. Thus, the study found that all the supported hypothesis has a small effect size, except HM →PEOU having the medium effect size of the study. Table 4 and Fig. 1 illustrates the results for the hypothesis testing for the H1 to H15 of the study.

Table 4 .

Hyp	Relationship	Beta	Se	T value	P Value	F2	R2	VIF	Result
H1	**CA → PEOU**	– 0.09	0.04	2.58	0.01	0.02	0.56	1.02	Accepted
H2	**CC → PEOU**	0.10	0.04	2.35	0.02	0.02		1.59	Accepted
H3	**HM → PEOU**	0.35	0.04	8.78	0.00	0.19		1.46	Accepted
H4	**PE → PEOU**	0.18	0.04	4.70	0.00	0.06		1.32	Accepted
H5	**SN → PEOU**	0.27	0.04	7.08	0.00	0.13		1.30	Accepted
H6	**TS → PEOU**	0.18	0.04	4.63	0.00	0.06		1.16	Accepted
H7	**CA → PU**	– 0.11	0.04	2.68	0.01	0.02	0.39	1.02	Accepted
H8	**CC → PU**	0.13	0.05	2.56	0.01	0.02		1.59	Accepted
H9	**HM → PU**	0.28	0.05	5.96	0.00	0.09		1.46	Accepted
H10	**PE → PU**	0.19	0.05	3.71	0.00	0.04		1.32	Accepted
H11	**SN → PU**	0.13	0.05	2.80	0.01	0.02		1.30	Accepted
H12	**TS → PU**	0.17	0.04	4.01	0.00	0.04		1.16	Accepted
H13	**PEOU→ ITU**	0.37	0.07	4.97	0.00	0.07	0.33	2.78	Accepted
H14	**PU → ITU**	0.16	0.07	2.41	0.02	0.02		2.36	Accepted
H15	**SN → ITU**	0.13	0.05	2.51	0.01	0.02		1.36	Accepted

Table 5 PLS Predict

ITEMS	RMSE PLS	RMSE LM	PLS - LM	Q2_ PREDICT
ITU1	0.94	0.90	0.04	0.22
ITU2	0.84	0.79	0.05	0.16
ITU3	0.97	0.92	0.05	0.15
ITU4	0.90	0.89	0.01	0.19
PEOU1	0.83	0.71	0.12	0.37
PEOU2	1.28	1.17	0.11	0.22
PEOU3	1.04	1.01	0.03	0.34
PEOU4	0.90	0.88	0.02	0.40
PEOU5	1.10	1.13	–0.03	0.27
PU1	0.96	0.89	0.07	0.24
PU2	1.26	1.19	0.07	0.22
PU3	1.06	1.05	0.01	0.29
PU4	1.12	1.12	0.00	0.26

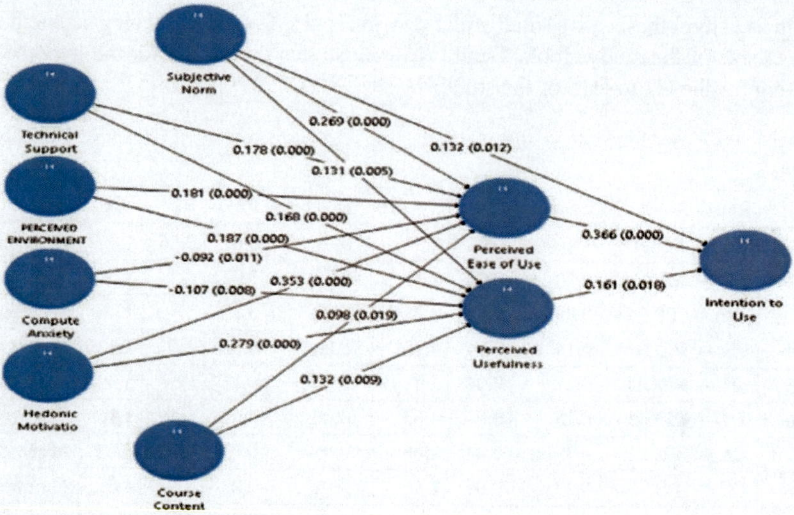

Fig. 2 Hypothesis Testing

7 Hypothesis Testing

PLS predict, a sample-based approach that provides case-level predictions with a tenfold procedure for assessing predictive significance, was proposed by [45]. The lower the differences in the items in PLS-LM, the stronger the predictive power, whereas the higher the difference, the predictive relevance is not confirmed. However, if most of the differences are low then there is moderate predictive power and contrast is the case where the majority is a high difference. Table 5 shows that almost all the errors of the model were less than the LM model suggesting indicating that the model is highly predictive.

8 Discussion and Implications

The primary goal of this work is to investigate the elements that may influence undergraduate students' intentions to use e-learning systems in Arab countries, particularly in the case of the COVID 19 pandemic. The study extended TAM theory by adding Computer Anxiety, Course Content, Hedonic Motivation, Perceived Environment, Subjective Norm, and Technical Support to the TAM model. The findings indicted that Computer Anxiety, Course Content, Hedonic Motivation, Perceived Environment, Subjective Norm, and Technical Support effect significantly on both ease of use and usefulness. Subjective Norm significantly effects on intention to use. PEOU and PU significantly affect the intention to use.

We contribute to the literature on e-learning system adoption by investigating the effect of Computer Anxiety, Course Content, Hedonic Motivation, Perceived Environment, Subjective Norm, and Technical Support on both ease of use and usefulness. Subjective Norm on intention to use. PEOU and PU on intention to use. Little studies have been carried out on the intention to use different types of e-learning system by undergraduate studies in developing countries especially after the effect of COVID 19 pandemic. Thus, researchers should be conduct other studies to help the both students and universities in using e-learning systems in the effective way.

The practical implications of the study help the universities to provide an effective e-learning system to undergraduate students, where the e-learning has become an important part in the education in the developing countries, for example 50% face to face learning and 50% the both lectures and students will use e-learning.

9 Limitations and Future Studies

The first limitation of the current work is that the study's sample includes of undergraduate students from Al-Azhar University, and future studies could apply the same approach to other universities. Second, because this study was conducted in Palestine, the findings may not be generalizable to other countries; therefore, we recommend future research to apply the same approach to other nations. Third, we used SMARTPLS 3 to analyze the data, and future research could be conducted using AMOS to test the study's model.

References

1. Al-Tahitah AN, Al-Sharafi MA, Abdulrab M (2021) How COVID-19 pandemic is accelerating the transformation of higher education institutes: a health belief model view. In: Arpaci I, Al-Emran M, A. Al-Sharafi M, Marques G (eds) Emerging technologies during the era of COVID-19 pandemic, vol 348. Studies in Systems, Decision and Control. Springer, Cham, pp 333–347. https://doi.org/10.1007/978-3-030-67716-9_21
2. Murad DF, Heryadi Y, Wijanarko BD, Isa SM, Budiharto W (2018) Rec-ommendation system for smart LMS using machine learning: a literature review. In: 2018 international conference on computing, engineering, and design (ICCED), pp 113–118
3. Aldheleai YM, Tasir Z, Al-Rahmi WM, Al-Sharafi MA, Mydin A (2020) Modeling of students online social presence on social networking sites with academic performance. Int J Emerg Technol Learn 15(12). https://doi.org/10.3991/ijet.v15i12.12599.
4. Makumane MA (2021) Students' perceptions on the use of LMS at a Lesotho uni-versity amidst the COVID-19 pandemic. Afr Identities 1–18
5. Aldammagh Z, Abdaljawad R, Obaid T (2021) Factors driving e-learning adoption in palestine: an integration of technology acceptance model and is success model. Financ Internet Q e-Finanse 17(1)
6. Al Zoubi SI, Alzoubi AI (2019) E-learning benchmarking adoption: a case study of sur university college. Int J Adv Comput Sci Appl 10(11)

7. Nurakun Kyzy Z, Ismailova R, Dündar H (2018) Learning management system implementation: a case study in the Kyrgyz Republic. Inter Learn Environ 26(8):1010–1022
8. Alajmi Q, Sadiq A, Kamaludin A, Al-Sharafi MA (2017) E-learning models: The effectiveness of the cloud-based E-learning model over the traditional E-learning model. https://doi.org/10.1109/ICITECH.2017.8079909.
9. AlAjmi Q, Al-Sharafi MA, Yassin AA (2021) Behavioral intention of students in higher education institutions towards online learning during COVID-19. In: Arpaci I, Al-Emran M, A. Al-Sharafi M, Marques G (eds) Emerging Technologies During the Era of COVID-19 Pandemic, vol 348. Studies in Systems, Decision and Control. Springer, Cham, pp 259–274. https://doi.org/10.1007/978-3-030-67716-9_16
10. Aldheleai YM, Al-Sharafi MA, Al-Kumaim NH, Al-Rahmi WM (2021) Investigating the impact of the sense of privacy on the correlation between online learning interaction and students' academic performance. In: Al-Emran M, Shaalan K (eds) recent advances in technology acceptance models and theories, vol 335. studies in systems, decision and control. Springer, Cham, pp 485–496. https://doi.org/10.1007/978-3-030-64987-6_28
11. Al-Emran M, Al-Maroof R, Al-Sharafi MA, Arpaci I (2020) What impacts learning with wearables? an integrated theoretical model. Int Learn Environ. https://doi.org/10.1080/10494820.2020.1753216
12. Obaid T (2018) Determine process training key factors and job performance in higher education sector Int J Eng Technol 7(4.15):477–480 (2018)
13. Gamede BT, Ajani OA, Afolabi OS (2021) Exploring the adoption and usage of learning management system as alternative for curriculum delivery in South African higher education institutions during COVID-19 lockdown Int J High Educ 11(1):71-84
14. Cavus N, Mohammed YB, Yakubu MN (2021) Determinants of learning management systems during COVID-19 pandemic for sustainable education. Sustainability 13(9):5189
15. Obaid T, Eneizan B, Naser SSA, Alsheikh G, Ali AAA, Abualrejal HME, Gazem NA (2022) Factors contributing to an effective e- government adoption in palestine. In: Saeed F, Mohammed F, Ghaleb F (eds) Advances on intelligent informatics and computing, vol 127. Lecture Notes on Data Engineering and Communications Technologies. Springer, Cham, pp 663–676. https://doi.org/10.1007/978-3-030-98741-1_55
16. Almaiah MA, Alismaiel OA (2019) Examination of factors influencing the use of mobile learning system: an empirical study. Educ Inf Technol 24(1):885–909
17. Almaiah MA, Al-Khasawneh A, Althunibat A (2020) Exploring the critical challenges and factors influencing the E-learning system usage during COVID-19 pandemic. Educ Inf Technol 25(6):5261–5280
18. Obaid T (2020) Factors driving e-learning adoption in palestine: an integration of technology acceptance model and IS success model. Available at SSRN 3686490
19. Al Mulhem A (2020) Investigating the effects of quality factors and organizational factors on university students' satisfaction of e-learning system quality. Cogent Educ 7(1):1787004
20. El-Masri M, Tarhini A (2017) Factors affecting the adoption of e-learning systems in Qatar and USA: extending the unified theory of acceptance and use of technology 2 (UTAUT2). Educ Technol Res Devlopment 65(3):743–763
21. Obaid T et al. (2022) Factors contributing to an effective e-government adoption in Palestine. In: International conference of reliable information and communication technology, pp 663–676
22. Almaiah MA, Jalil MA, Man M (2016) Extending the TAM to examine the effects of quality features on mobile learning acceptance. J Comput Educ 3(4):453–485
23. Eneizan BM, Abd Wahab K, Zainon MS (2016) Prior research on green marketing and green marketing strategy : critical analysis Singaporean J Bus Econ Manag Stud 5(5):1–19. https://doi.org/10.12816/0033265
24. Jaradat MRM (2014) Understanding individuals' perceptions, determinants and the moderating effects of age and gender on the adoption of mobile learning: developing country perspective. Int J Mob Learn Organ 8(3–4):253–275
25. Al-Sharafi MA, AlAjmi Q, Al-Emran M, Qasem YAM, Ald-heleai YM (2021) Cloud computing adoption in higher education: an integrated theoretical model 335 https://doi.org/10.1007/978-3-030-64987-6_12

26. Qasem YAM Abdullah R, Yah Y, Atan R, Al-Sharafi MA, Al-Emran M (2021) Towards the development of a comprehensive theoretical model for examining the cloud computing. Adopt Organ Level 295:63. https://doi.org/10.1007/978-3-030-47411-9_4
27. Mtebe JS Raphael C (2018) Key factors in learners' satisfaction with the e-learning system at the University of Dares Salaam, Tanzania. Australas J Educ Technol 34(4)
28. Adeyemi Abdulwahab Olanrewaju, IOI (2020) 2 Record and library Journal 6(1):69–79
29. Davis FD, Bagozzi RP, Warshaw PR (1989) User acceptance of computer technology: a comparison of two theoretical models. Manage Sci 35(8):982–1003
30. Lee B-C, Yoon J-O, Lee I (2009) Learners' acceptance of e-learning in South Korea: theories and results. Comput Educ 53(4):1320–1329
31. Abbad MM, Morris D, De Nahlik C (2009) Looking under the bonnet: factors affecting student adoption of e-learning systems in Jordan 1–25
32. Al-Fuqaha A, Guizani M, Mohammadi M, Aledhari M, Ayyash M (2015) Internet of things: a survey on enabling technologies, protocols, and applications. IEEE Comm Surv Tutorials 17(4):2347–2376
33. Obaid T, Abdaljawad R, Abumandil M (2020) COVID-19 and the digital trans-formation of higher education: What insights Palestinian institutes can share? IJAR 6(8):109–114
34. Talebian S, Mohammadi HM, Rezvanfar A (2014) Information and communication technology (ICT) in higher education: advantages, disadvantages, conveniences and limitations of applying e-learning to agricultural students in Iran. Procedia Soc Behav Sci 152:300–305
35. Ajzen I, Fishbein M (1975) A Bayesian analysis of attribution processes. Psychol Bull 82(2):261
36. Van Raaij EM, Schepers JJL (2008) The acceptance and use of a virtual learning environment in China. Comput Educ 50(3):838–852
37. Mathieson K (1991) Predicting user intentions: comparing the technology acceptance model with the theory of planned behavior. Inf Syst Res 2(3):173–191
38. Venkatesh V, Davis FD (2000) A theoretical extension of the technology acceptance model: four longitudinal field studies. Manage Sci 46(2):186–204
39. Venkatesh V, Thong JYL, Xu X (2012) Consumer acceptance and use of information technology: extending the unified theory of acceptance and use of technology. MIS Q: 157–178
40. Ku ECS, Wu WC, Chen YJ (2016) The relationships among supply chain partnerships, customer orientation, and operational performance: the effect of flexibility. IseB 14(2):415–441. https://doi.org/10.1007/s10257-015-0289-0
41. Saade R, Kira D (2009) 44 Proceedings of the 2009 InSITE Conference, vol 8. https://doi.org/10.28945/3386
42. Belhadi A, Mani V, Kamble SS, Khan SAR, Verma S (2021) Artificial intelligence-driven innovation for enhancing supply chain resilience and performance under the effect of supply chain dynamism: an empirical investigation. Ann Oper Res 0123456789.https://doi.org/10.1007/s10479-021-03956-x
43. Garg R (2017) Optimal selection of E-learning websites using multiattribute decision-making approaches. J Multi-Criteria Decis Anal 24(3–4):187–196
44. Talukder KI, Mubasshira T, Hasnat MA, Factors affecting student's perception and actual uses of lms in malaysian universities
45. Shenhar AJ, Wideman RM (1996) Improving PM: linking success criteria to project type. Proc Proj Manag 96:71–76
46. Fan C, Zhang C, Yahja A, Mostafavi A (2021) Disaster City digital twin: a vision for integrating artificial and human intelligence for disaster management. Int J Inf Manag 56:102049. https://doi.org/10.1016/j.ijinfomgt.2019.102049.
47. Teo T, Noyes J (2014) Explaining the intention to use technology among pre-service teachers: a multi-group analysis of the unified theory of acceptance and use of technology. Interact Learn Environ 22(1):51–66
48. Simonson MR, Maurer M, Montag-Torardi M, Whitaker M (1987) Development of a standardized test of computer literacy and a computer anxiety index. J Educ Comput Res 3(2):231–247

49. Rahi S, Ghani MA, Ngah AH (2019) Integration of unified theory of acceptance and use of technology in internet banking adoption setting: evidence from Pakistan. Technol Soc 58:101120
50. Rahi S, Ghani MA, Ngah AH (2020) Factors propelling the adoption of internet banking: the role of e-customer service, website design, brand image and customer satisfaction. Int J Bus Inf Syst 33(4):549–569
51. Al-Emran M, Teo T (2020) Do knowledge acquisition and knowledge sharing really affect e-learning adoption? an empirical study. Educ Inf Technol 25(3):1983–1998
52. Bhuasiri W, Xaymoungkhoun O, Zo H, Rho JJ, Ciganek AP (2012) Critical success factors for e-learning in developing countries: a comparative analysis between ICT experts and faculty. Comput Educ 58(2):843–855
53. Baumann-Birkbeck L et al (2015) Benefits of e-learning in chemotherapy pharmacology education. Curr Pharm Teach Learn 7(1):106–111
54. Kock N (2015) Common method bias in PLS-SEM: a full collinearity assessment approach. Int J e-Collab (ijec) 11(4):1–10
55. Hassanzadeh A, Kanaani F, Elahi S (2012) A model for measuring e-learning systems success in universities. Expert Syst Appl 39(12):10959–10966
56. Na S, Heo S, Han S, Shin Y, Roh Y (2022) Acceptance model of artificial intelligence (AI)-based technologies in construction firms: applying the technology acceptance model (TAM) in combination with the technology–organisation–environment (TOE) framework. Buildings 12(2). https://doi.org/10.3390/buildings12020090
57. Jan AU, Contreras V (2011) Technology acceptance model for the use of information technology in universities. Comput Hum Behav 27(2):845–851
58. Ringle CM, Wende S, Becker JM (2015) SmartPLS 3, Boenningstedt: SmartPLS GmbH 584
59. Shmueli G, et al. (2019) Predictive model assessment in PLS-SEM: guidelines for using PLSpredict. Euro J Mark
60. Hair JF, Hult GTM, Ringle CM, Sarstedt M (2014) A primer on partial least squares structural equation modeling (PLS-SEM). sage publications. Euro J Tour Res 6(2):211–213

Mining Educational Data to Improve Teachers' Performance

Abdelbaset Almasri, Tareq Obaid, Mohanad S. S. Abumandil, Bilal Eneizan, Ahmed Y. Mahmoud, and Samy S. Abu-Naser

Abstract Educational Data Mining (EDM) is a new paradigm aiming to mine and extract the knowledge necessary to optimize the effectiveness of the teaching process. With normal educational system work, it's often unlikely to accomplish fine system optimisation due to the large amount of data being collected and tangled throughout the system. EDM resolves this problem by its capability to mine and explore these raw data and as a consequence of extracting knowledge. This paper describes several experiments on real educational data wherein the effectiveness of Data Mining is explained in the migration of the educational data into knowledge. The's experiment goal at first was to identify important factors of teacher behaviors influencing student satisfaction. In addition to presenting experiences gained through the experiments, the paper aims to provide practical guidance on Data Mining solutions in a real application.

Keywords EDM · Knowledge · Survey · C4.5

A. Almasri (✉) · T. Obaid · A. Y. Mahmoud · S. S. Abu-Naser
Faculty of Engineering and IT, Alazhar University, Gaza, Palestine
e-mail: a.masrey@alazhar.edu.ps

T. Obaid
e-mail: tareq.obaid@alazhar.edu.ps

A. Y. Mahmoud
e-mail: ahmed@alazhar.edu.ps

S. S. Abu-Naser
e-mail: abunaser@alazhar.edu.ps

M. S. S. Abumandil
Faculty of Hospitality, Tourism and Wellness, Universiti Malaysia Kelantan, Bharu, Malaysia
e-mail: mohanad.ssa@umk.edu.my

B. Eneizan
Business School, Jadara University, Irbid, Jordan

M. Al-Emran et al. (eds.), *International Conference on Information Systems and Intelligent Applications*, Lecture Notes in Networks and Systems 550,
https://doi.org/10.1007/978-3-031-16865-9_20

1 Introduction

The quality of classroom instructions is primarily determined via the evaluation of teaching performance. Hence, the Teacher Assessment Survey (TAS) is used by a majority of educational institutions to measure the level of student satisfaction and to mine knowledge regarding teaching behaviours. Commonly, the TAS aims to answer the following questions [1, 2]: Which teaching constructs that students find satisfactory/unsatisfactory? Are the students' dissatisfactions related to certain attributes such as gender, course major, faculty? Is there a specific pattern to the findings such as a higher prevalence of student dissatisfaction in a certain faculty/major? Can the students be categorized into certain groups which share certain characteristics? Are there certain predictors to student dissatisfaction? Can certain attributes be identified within the groups of dissatisfied students?

The same queries can be applied to the group of satisfied students. This paper mainly aims to develop data mining models that can explicate teacher behaviors which are significantly linked to student satisfaction. This can be identified via the scores given by students on certain features such as their teachers' personality and scientific background. Data mining is effective in identifying significant student satisfaction determinants and the correlations between them. The field of data mining has been growing rapidly based on the accumulated data from various institutions [3]. Several novel data mining methods had been explored to describe data extraction processes such as data pre-processing, analysis and representation. Data mining mainly aims to initiate classification models [4], rules of association [5], evolution and deviation analysis as well as to cluster akin data objects [3]. The data should first be prepared for mining i.e., by cleansing and transforming the data into a mining-ready format [5]. According to Llorente & Morant, (2011), Educational Data Mining offers a firm platform for educational applications. EDM is capable of extracting learning-related knowledge. Figure 1 shows how data mining significantly provides educational knowledge that contributes to the making of proper decisions towards optimizing educational systems and how data mining usage in educational institutions leads to the formation of an interactive cycle towards improving learning.

This study mainly aims to utilize data mining towards improving student achievement via: i) a thorough comprehension of prevailing teaching behaviors, ii) identification of teaching behaviors which significantly affect student satisfaction and of which are good predictors of teacher performance, and iii) the designing of future plans for attaining set improvements according to the results observed.

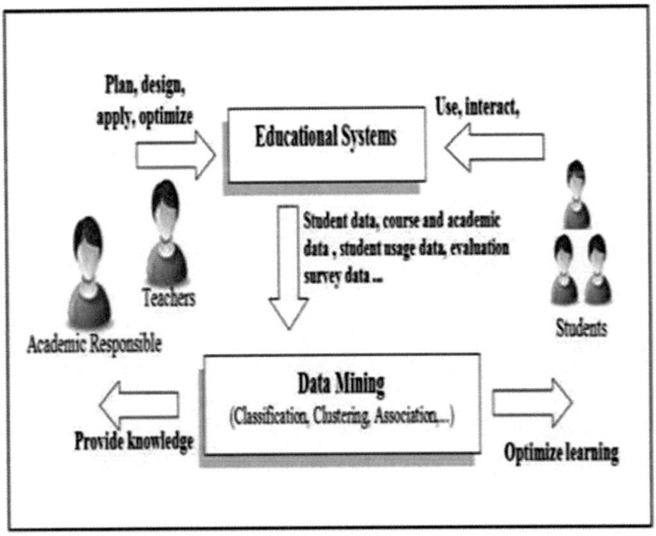

Fig. 1 Data mining cycle in educational institutions

2 Related Works

Data mining was used by Taherifar and Banirostam [7] on their data gathered via a survey on university students in Turkey. The authors utilized the principle component analyses for reducing the dataset followed by the "two-step and Kohonen clustering algorithms". Next, the Quest decision tree algorithm was utilized on the outcomes of the two-step clustering followed by the extraction of the key predictors of student satisfaction.

Hamada and Abadi [8] surveyed students' opinions regarding their teachers and mined the collected data. The authors then presented the results of the analysis utilizing the WEKA tool. Hemaid & El-Halees [9] also used data mining to investigate the factors affecting teaching performance. The authors proposed a teacher performance evaluation model using data mining methods (e.g., association and classification rules). The methods were also employed via WEKA on the real-world teacher data gathered in the context of Gaza City.

Likewise, Pal and Pal, [10] employed several data mining methods in evaluating university lecturers' performance including the Naive Bayes, ID3, CART, and LAD tree. The best algorithm with the lowest average errors was demonstrated by the Naïve Bayes classifier.

Palshikar et al. (2009) demonstrated the analysis and processing of survey responses by utilizing several data mining methods and the newly-introduced QUEST tool. Their real-world case study utilized QUEST for analyzing employee satisfaction survey responses [11] introduced a novel teacher performance prediction method by analyzing educational surveys. This method uses classification and sequential pattern mining for identifying and ascertaining meta-patterns that describe the typical behaviors of teachers.

Abu-Naser et al. [12] examined the performance of second-year university students using an Artificial Neural Network model. The authors' proved that the model can predict more than 80% of the surveyed students' performance.

3 Dataset Description

The current study examines real-world data derived from an educational database and a higher education institution's online Teacher Assessment Survey (TAS). The said institution carries out a survey on each available course every semester. The survey seeks the opinion of the students on teaching-related matters, specifically their assessment of their teacher's support during classes. A total of 20 structured question items were incorporated in the survey, which the students need to respond to as either "Excellent", "Good", "Average", "Poor", or "Very Poor". Table 1 shows that the TAS questions are divided into four categories that correspond to a certain aspect of the teacher's behaviour.

The SI (satisfaction index) for each TAS question is generated prior to doing data analysis in order to determine the overall student satisfaction for that particular item (teaching behavior aspect). On the range 0 to 4, 4 is excellent, 3 good, 2 average, 1

Table 1 TAS questions

Categories	Items
Personal characteristics	1. The teacher is strict and dominant 2. The teacher appears stylish and decent 3. The teacher commits to the scheduled lecture dates 4. The teacher respects the students
Scientific background	5. The teacher is well-versed in the scientific teaching materials 6. The teacher responds to the students' queries in a clear manner 7. The teacher is broadly knowledgeable in diverse areas 8. The teacher presents teaching materials in ways suitable to the students' level 9. The teacher presents teaching materials coherently and sequentially 10. The teacher teaches all the course topics throughout the semester
Professional skills	11. The teacher uses examples to enrich the materials 12. The teacher utilizes techniques that develop the students' thinking 13. The teacher instills positivity in the students with regards to the specialization 14. The teacher spends significant time in presenting lecture materials and conducting scientific activities 15. The teacher grows his/her research skills by performing numerous research activities 16. The teacher urges students to utilize various sources of knowledge

(continued)

Table 1 (continued)

Categories	Items
Assessment	17. The teacher employs various questions for exams 18.The teacher uses highly scientific topics for exams 19. The teacher uses a proportional number of questions to the set exam time 20. The teacher is objective in assessing students' work and activities

bad and 0 is very poor. The SI of the ith question (Qi) replied by N students (v) is established by mapping the response of Qi to the numerical value based on the scale from 0 to 4. Calculating the SI of Qi, which has a defined domain Di of plausible responses (0.. |Di|-1) may be done using Eq. 1 (niv= the number of students that picked answer v for Qi). S(Qi) = 0% if all answers to Qi are 0. S(Qi) = 100 percent if all of the replies to Qi are |Di| – 1.

$$S(Q_i) = 100 \times \frac{\sum_{v=0}^{|D_i|-1} v \times n_{iv}}{(|D_i| - 1) \times N} \ldots\ldots 0 \le S(Q_i) \le 100.0 \ldots\ldots \text{for each question } Q_i \quad (1)$$

Each category (a group of linked questions of a given concern) had its own overall SI, which we calculated as the average of the SI S (Cj) for the category (Cj) including N questions (see Eq. 2).

$$S(C_j) = \frac{\sum_{i=1}^{N} S(Q_{ij})}{N} \quad (2)$$

Subsequent to data processing, 608 records were derived in which each contain 29 attributes explicating a course and the student' overall satisfaction level with its teaching. Table 2 shows the attributes along with the respective descriptions as derived from the source database following calculations of the measures for satisfaction.

Table 2 Dataset attributes of experiment 1

Fields	Description	Values domain		Direction
Perschar	Total si for the	≥80	Good	Input
	Personal characteristics	79–65	Average	
	Category	<65	Poor	
	Total si for the	≥80	Good	Input
scbackground	Scientific background	79–65	Average	
	Category	<65	Poor	
	Total si for the	>80	Good	Input

(continued)

Table 2 (continued)

Fields	Description	Values domain		Direction
Profskills	Professional skills	79–65	Average	
	Category	<65	Poor	
	Total si for the	≥80	Good	Input
Assessment	Assessment category	79–65	Average	
		<65	Poor	
	Teacher performance	>80	Good	Output
Techperfavg	Overall average si for	79–65	Average	(Target)
		<65	Poor	

Table 3 Dataset attributes of the clustering process

Fields	Description	Values domain		Direction
Faculty	Faculty in which the course is being taught			Input
Question1_SI_Bin Question20_SI_Bi n	Categories based on the mean values and standard deviation of the field distribution	x < (Mean – Std. Dev) (Mean – Std. Dev) ≤x ≤ (Mean + Std. Dev) x > (Mean + Std. Dev	-1 (poor) 0(average) 1(good)	Input
No_Students_Bin	Number of student enrollment in the course	x < (Mean – Std. Dev) (Mean – Std. Dev) ≤ ≤ (Mean + Std. Dev) x > (Mean + Std. Dev	– 1(small) 0(average) 1(large)	Input

4 Experiment 1: Predicting Student Satisfaction Solely via the Responses

Here, the model for predicting student satisfaction towards teacher performance is built solely by using the responses to the survey items i.e. without considering any other data. This experiment utilized the SI attributes of the TAS categories by assigning a class label to the attribute values (poor, average, good), Firstly, the overall average SI for all the questions is discerned to determine the class label.

Following the preparation of the dataset for mining (see Table 2), the "c4.5 classification algorithm is employed" i.e., a tree-based classification and prediction approach which recursively partitions the training dataset into subgroups of equivalent target field values. The c4.5 investigates the input fields to discover the optimal split i.e., by assessing the impurity index decrease as a consequence of the split. The split results in many subgroups which are further divided into additional subgroups until one of the halting criteria is triggered [13, 14]. Figures 2 and 3 provide the categorization results of the TechPerfAvg (overall average SI for teacher performance) as a target class.

The classification results show that the best teacher performance predictor is the teacher's scientific background.

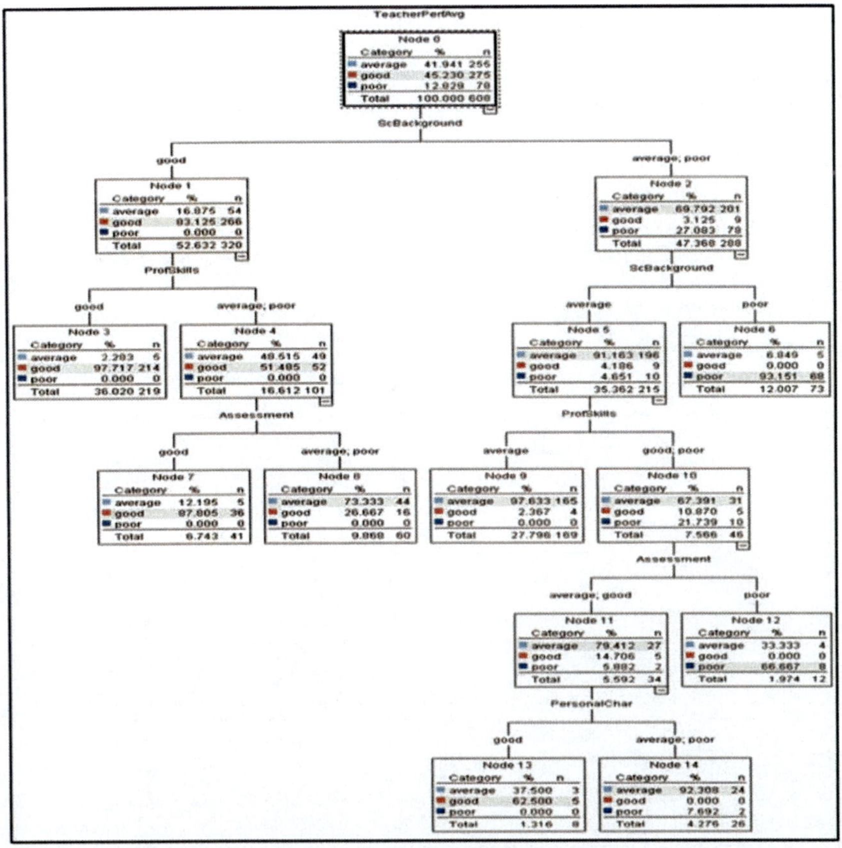

Fig. 2 Teaching behavior classification tree in experiment 1

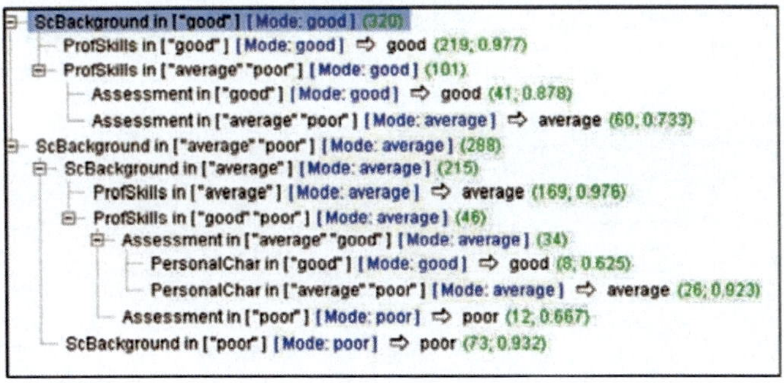

Fig. 3 Teaching behavior classification rules in experiment 1

5 Experiment 2: Using Responses and Achievement Average

Here, teacher performance is described using all the student responses and average course achievement. Figure 4 illustrates that an accuracy rate of 94.2% is achieved for the resulting classification tree. Figure 5 presents the classification's rule-based view. Below is an explanation of some to the rules:

Rule 1: If (Q12 is [poor, average] and Q9 is [poor]), then teacher performance is poor.

Rule 2: If (Q12 is [good] and Q18 is [good] and Q1 is [good]), then teacher performance is good.

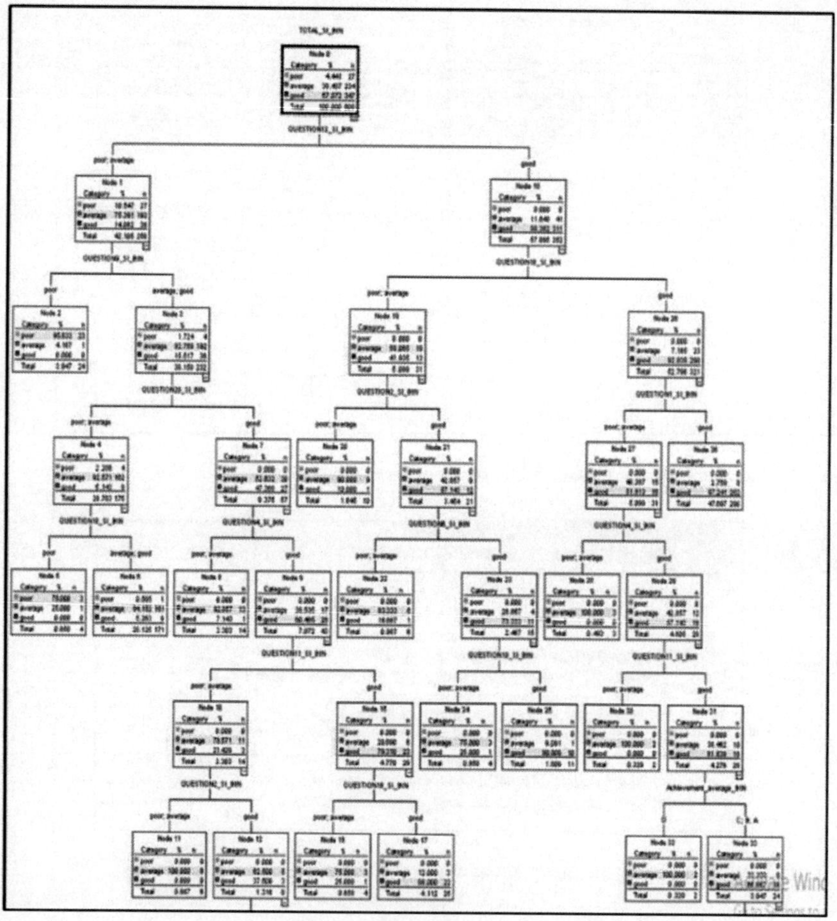

Fig. 4 Teacher performance classification tree in experiment 2

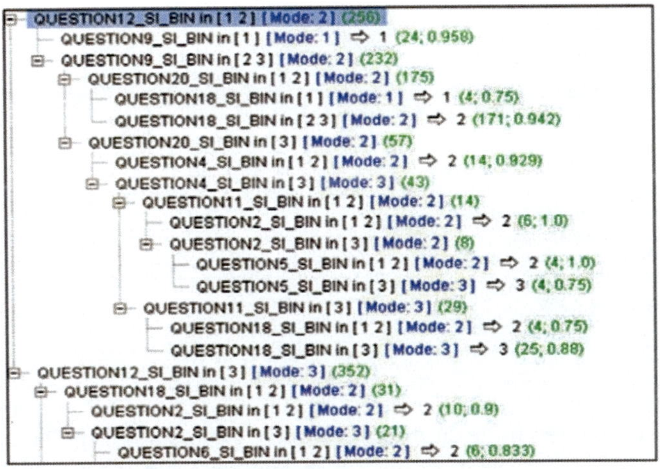

Fig. 5 Teacher performance classification rules in experiment 2

Based on the rules, it was found that attribute Q12 i.e. the teaching method contributing to the students' growth in thinking significantly determines the performance of the teacher.

6 Experiment 3: Predicting Student Satisfaction Based on Student Responses and Data

Experiment 3 aims to identify the factors affecting the students' satisfaction with their teacher's performance and to develop a student satisfaction prediction classification model. The next sections describe the data mining process beginning with the data preparation to the actual data mining procedure and finally data evaluation.

Data Preprocessing: The SI values for the TAS questions and student enrollment numbers were discerned into categories according to the mean values and standard deviation of the value distributions. The attributes and their respective description are presented in Table 4.

Table 4 Dataset attributes of the classification process

Fields	Description	Values domain		Direction
Faculty	Faculty in which the course is taught	Agriculture, Arts, Dental, Economics, Education, Engineering, Islamic, Law, Medical sciences, Pharmacy, Science		input
Question1_SI_Bin Question20_SI_Bin	Categories based on the mean values and standard deviation of the field distribution	x < (Mean – Std. Dev) (Mean – Std. Dev) < = x < = (Mean + Std Dev) x > (Mean + Std. Dev	– 1 (poor) 0 (average) 1 (good)	Input
Satisfaction	Student satisfaction	True False		Output

Data Mining Functionality: Firstly, the student satisfaction responses were grouped into 3 clusters using the k-means clustering algorithm [14]. This clustering process categorizes the data based on their similarities. Table 4 shows the input data for this process. The data for student satisfaction is grouped into Cluster 1, Cluster 2, and Cluster 3, and plotted and colored based on the overall SI percentage as shown in Figure 6. Cluster 3 is observed to present the highest data of student dissatisfaction.

Secondly, the output of the clustering process is used to establish a new category called 'Satisfaction'. This category is attributed to course data that does not fit into Cluster 3. Next, the classification algorithm is applied to develop a classification model and determine the key factors that drive student satisfaction with regards to their teacher's performance. Table 4 below presents the data fields for building the model. Figure 7 presents the resulting classification tree.

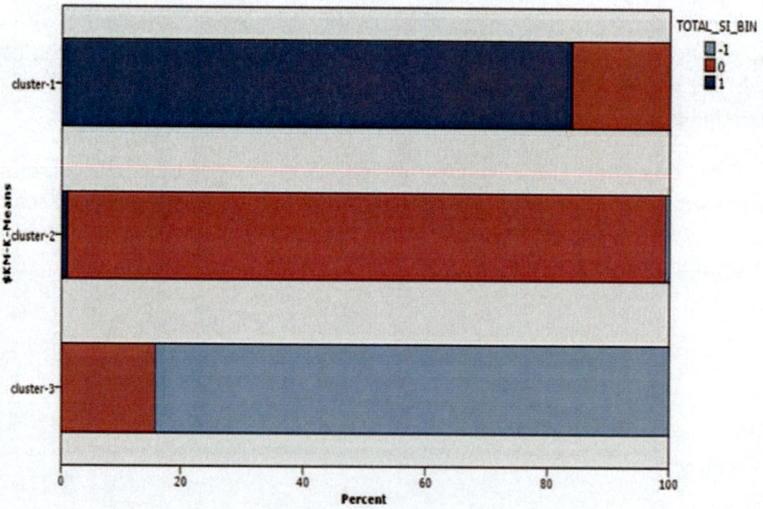

Fig. 6 Satisfaction clusters graph of experiment 3

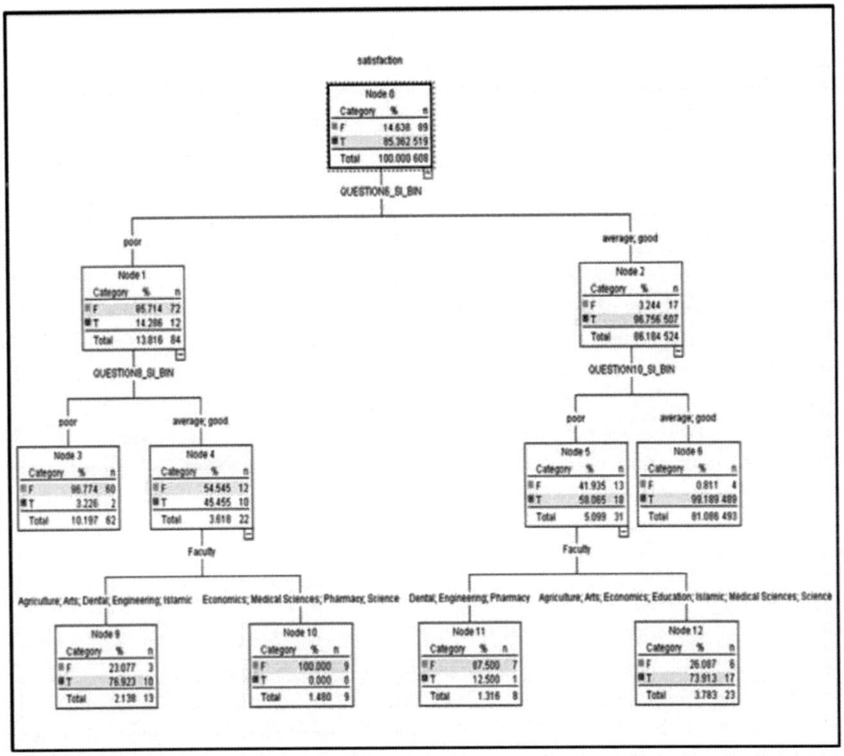

Fig. 7 Classification tree of experiment 4

Evaluation: The experiment employed "608 course teaching records", out of which 592 were correctly classified with 97.37% accuracy. Based on the data mining, several major factors contributing to teacher performance were identified. The first one was identified via Question 6 i.e. how the teacher answers the students' questions. The second factor was identified via Question 10 i.e. the course topics covered during the semester. The third factor was identified via Question 8 i.e. how the teacher presents teaching materials to suit the students' level. It was found that the attribute of 'faculty' contributes to the classification of student satisfaction in relation to their teacher's performance. It was also found that a majority of scientific colleges focus on the teaching construct i.e. how the teacher covers the curriculum throughout the semester.

7 Conclusion

The current paper demonstrates the significance of data mining methods in examining and discerning educational data. The study identifies the teaching constructs which influence student satisfaction as well as the key predictors of teacher performance. The data mining methods employed include data pre-processing, c4.5 classification algorithm, and K-means clustering algorithm. The study fundamentally shows the mining and processing of data gathered from survey responses, as well as the likely predictors of student satisfaction towards their teacher's performance. This study managed to fulfill its objective i.e. to explore data concerning student satisfaction using data mining methods. Numerous attributes were tested of which some were found to be effective for predicting student satisfaction. One key predictor identified was teaching constructs that help grow the students' thinking. Meanwhile, the classification of teacher performance was primarily attributed to the teacher's clear responses to questions, the coverage of course topics throughout the semester, and the proper presentation of course materials.

References

1. Berk RA (2005) Survey of 12 strategies to measure teaching effectiveness. Int J Teach Learn High Educ 17(1):48–62
2. Palshikar GK, Deshpande S, Bhat SS (2009) QUEST discovering insights from survey responses. In: AusDM, pp 83–92
3. Han J, Kamber M (2011) Pei. Data mining concepts and techniques MK
4. Kamber M, Winstone L, Gong W, Cheng S, Han J (1997) Generalization and decision tree induction: efficient classification in data mining. In: Proceedings seventh international workshop on research issues in data engineering. high performance database management for large-scale applications, pp 111–120
5. Agrawal R, Imieliński T, Swami A (1993) Mining association rules between sets of items in large databases. In: Proceedings of the 1993 ACM SIGMOD international conference on Management of data, pp 207–216
6. Liorente R, Morant M (2011) Data mining in higher education New Fundam Technol Data Min: 201–220
7. Taherifar E, Banirostam T (2016) Assessment of student feedback from the training course and instructor'performance through the combination of clustering methods and decision tree algorithms. Int J Adv Res Comput Sci Softw Eng 6(2):56–64
8. Ahmadi F, Ahmad S (2013) Data mining in teacher evaluation system using WEKA. Int J Comput Appl 63(10):14–18
9. Hemaid RK, El-Halees AM (2015) Improving teacher performance using data mining. Int J Adv Res Comput Commun Eng 4(2)
10. Pal AK, Pal S (2013) Evaluation of teacher's performance: a data mining approach. Int J Comput Sci Mob Comput 2(12):359–369
11. Barracosa J, Antunes C (2011) Anticipating teachers' performance. In: Proceedings of the KDD Workshop: Knowledge Discovery in Educational Data, pp 77–82

12. Abu-Naser SS, Zaqout IS, Abu Ghosh M, Atallah RR, Alajrami E (2015) Predicting student performance using artificial neural network. Fac Eng Inf Technol
13. Kohavi R, Quinlan JR (2022) Data mining tasks and methods: classification: decision-tree discovery. In: Handbook of data mining and knowledge discovery, pp 267–276
14. Ruggieri S (2002) Efficient C4. 5 [classification algorithm]. IEEE Trans Knowl Data Eng 14(2):438–444

Effectiveness of Face-to-Face Computer Assisted Cooperative Learning in Teaching Reading Skills to Yemeni EFL Learners: Linking Theory to Practice

Amr Abdullatif Yassin, Norizan Abdul Razak, Tg Nor Rizan Tg Mohamad Maasum, and Qasim AlAjmi

Abstract This paper aimed at investigating the effectiveness of face-to-face Computer Assisted Cooperative Learning (CACL) in teaching reading skills. It employed a mixed-method design as the data were collected through pre and post-test and semi-structured interviews. The pre and post-test of reading skills were analyzed through t-test, and the qualitative data were analyzed through thematic patterns. The findings showed a significant difference between the pre and post-test of reading skills, and the qualitative data analysis showed that face-to-face CACL has academic, social, and cognitive advantages. These findings showed that face-to-face CACL effectively teaches reading skills, which is attributed to the design of CALL and the implementation of cooperative learning principles. It is concluded that CALL and cooperative learning have a complementary advantage in teaching reading skills through face-to-face CACL. Therefore, teachers need to focus on the learning theories of CALL activities and implement the five principles of cooperative learning to make teaching reading skills through CACL more effective for EFL learners.

Keywords Computer assisted cooperative learning · CALL · Face-to-face interaction · Reading skills · Interactive reading model · Yemeni EFL learners

A. A. Yassin (✉)
Centre of Languages and Translation, Ibb University, Ibb, Yemen
e-mail: amryassin84@gmail.com

A. A. Yassin · N. A. Razak · T. N. R. T. M. Maasum
Faculty of Social Sciences and Humanities, Universiti Kebangsaan Malaysia, Bangil, Malaysia

Q. AlAjmi
Department of Education, College of Arts and Humanities, A' Sharqiyah University, Ibra, Oman

M. Al-Emran et al. (eds.), *International Conference on Information Systems and Intelligent Applications*, Lecture Notes in Networks and Systems 550,
https://doi.org/10.1007/978-3-031-16865-9_21

1 Introduction

Computer Assisted Cooperative Learning (CACL) links cooperative learning and CALL, and it is not a new method. CACL started in the field of education in the 1980s, like the studies [1–4].

Using CALL in the field of education started in the 50 s of the twentieth century [5], and its incorporation in the field of education is constantly increasing [6, 7]. Previous studies reported many advantages for technology, including making the process of teaching learner-centered [8–13], giving immediate feedback [14, 15], providing interesting features [16], and improving language skills and vocabulary [3, 17–20]. However, CALL still has several limitations because it dehumanizes the process of learning [21], marginalizes the role of the teacher [22], results in social communication distance [23], and requires updated teaching approaches [11, 24, 25]. Therefore, face-to-face CACL might help to solve such limitations, because cooperative learning increases the interaction between the students and the teachers. In other words, the limitation of technology, such as lacking interaction and facing learning difficulties might be solved with face-to-face CACL since it focuses more on the interaction among students to support each other. Another important point is that cooperative learning ensures that every group supports each student to understand the lesson through the exchange of ideas and constructive interaction [26].

Reviewing previous studies on CACL under different names and titles [1, 21, 27–29], there is no evidence for a framework for face-to-face CACL implementation, especially implementing the five principles of cooperative learning with CACL in teaching reading skills. Therefore, this study attempts to bridge this gap by providing a theoretical framework for teaching reading skills, focusing on reading models, the elements of CALL, and cooperative learning principles.

Besides, the focus of this study is on teaching reading to Yemeni EFL learners because their reading improvement has not been encouraging in the last two decades [30]. This might be attributed to the idea that the process of teaching in Yemen is teacher-centered, which makes the students passive participants in the classroom [31, 32]. Further, lecturers have less focus on utilizing technology in teaching English in Yemen [33], so students might not get enough practice to improve their reading skills. Therefore, there is a need to introduce new teaching methods to help Yemeni students overcome reading difficulties like face-to-face CACL [34]. Accordingly, the main objective of this study is to investigate the effectiveness of face-to-face CACL in teaching reading skills, and it aims at answering the following questions:

1. What is the effect of face-to-face CACL on reading skills among Yemeni university EFL learners?
2. What is the effect of face-to-face CACL on bottom-up reading skills among Yemeni university EFL learners?
3. What is the effect of face-to-face CACL on top-down reading skills among Yemeni university EFL learners?

2 Literature Review

2.1 *Teaching Through Computer Assisted Cooperative Learning*

Different studies have investigated teaching English through CACL. The two studies by AbuSeileek [27, 28] in using CACL in teaching communication skills. AbuSeileek [27] conducted a comprehensive study on the effectiveness of using cooperative and collective learning in teaching speaking and listening. The study concluded that the cooperative computer-mediated technique was more effective in teaching oral skills. The second study by AbuSeileek [28] focused on communication skills. This study investigated the effect of cooperative learning as well as positive interdependence and individual accountability on communication skills achievement among EFL undergraduate learners. The study findings showed that this method is effective in improving students' communication skills. The study also found that a small group of five students outperformed the two other groups which are composed of two and seven students.

Researchers have also focused on using CACL in teaching writing skills. The study [35] aimed to investigate the effect of computer cooperative learning on improving language skills among secondary school students in Hong Kong. The collaborative communication was done through email. The data in this study were collected through a pre- and post-survey and interviews. The study concluded that CACL helped the students to improve their writing skills, gain a positive attitude towards cooperative learning, and acquire high motivation. Similarly, the study [36] aimed to investigate the effect of cooperative learning on writing among Chinese college students. This study used a mixed-method design, and the findings showed that cooperative learning is effective with CALL instruction as it makes students more active during classes. Also, cooperative task-based activities are more effective than traditional instructional methods. This is also supported by [37] who showed that online cooperative learning helps to reduce the level of learning anxiety.

Other studies focused on cooperative online learning in different contexts, such as [38] among students of management and the study [39] in the area of education. The study [38] aimed at investigating the effectiveness of Computer-Supported Collaborative Learning (CSCL) in improving the students' academic achievement. The findings of the study showed that the role of the teacher in keeping cooperative learning is essential for the success of teaching through online cooperative learning. Another study [39] investigated the students' satisfaction with online cooperative learning. The findings of this quantitative study showed that the students were interested in studying cooperatively through the online platform since it helps them to support each other learning.

The above studies have used CACL in teaching different courses; however, they did not clarify the learning theories that underline CALL activities. Also, they did not explain how the teaching process implemented the five principles of cooperative learning since Johnson and Johnson [26] have asserted that CACL requires the implementation of these principles; otherwise, the learning process will be close to group activities instead of CACL.

2.2 *Teaching Reading Skills Through Computer Assisted Cooperative Learning*

Few studies have used CACL in teaching reading skills. One of the studies that focused on group work is [40]. This study used Computer-assisted reciprocal early English reading (CAREER) system to teach reading to early English learners. Students studied vocabulary and reading through the CAREER program. This study showed that the CAREER system helped early readers to improve their reading skills. Another study [41] used a Tag-based Collaborative reading learning System (TACO). This system was the tool for the students to create a learning environment with cooperative learning. One of the TACO system features is aiding teachers in accurately assessing literacy among students. The post-test findings showed a significant improvement in reading scores among participants who used the TACO system, which was due to collaborative tag sharing among the participants.

Two other studies have used the term CACL in teaching reading skills. Al-Salem [42] investigated the effectiveness of Computer Assisted Synchronous Learning in teaching reading skills to fresh female EFL students in KSA. The findings showed that cooperative learning and using technology helped students improve reading comprehension. Also, Sioofy and Ahangry [21] aimed to investigate the effectiveness of CACL in improving reading comprehension. The study's findings showed that students in the experimental group outperformed the students in the control group in the post-test as their reading comprehension was improved with CACL, which supports the findings of [43] that cooperative learning activities enhance students' reading skills.

To sum up, teaching through CACL is proved to be effective in teaching reading skills. However, some scholars have used 'cooperative learning' and 'collaborative learning', interchangeably while the difference between them is a need to implement the five principles of cooperative learning in CACL. Also, collaborative learning is based on socio-constructivism, particularly Zone of Proximal Development, so collaborative learning helps students to move from the lower end to the upper end. In comparison, cooperative learning is more systematic, as it depends on the Social Interdependence Theory. Besides, few studies have been carried out on CACL in teaching reading skills, and there is no clear evidence for the implementation of the five principles of cooperative learning with CACL. Furthermore, previous literature did not explain the role of CALL in CACL because most of them focused on cooperative learning without paying attention to the complementary value of CALL features to cooperative learning. This guides the researchers in the current study to pay more attention to the design of the web-based CALL and the design of reading activities according to learning theories, which are behaviorist CALL and cognitive CALL.

2.3 Theoretical Framework

The study's theoretical framework depends on the Social Interdependence Theory of cooperative learning and two learning theories for CALL activities (behaviorism and cognitivism). Besides, the interactive reading model is used for reading skills because it integrates both bottom-up and top-down reading skills. The whole process of development of CALL and implementation of face-to-face CACL was based on ADDIE instructional design model as discussed below.

This study adopted the five principles of cooperative learning by Johnson and Johnson [26], who stated that these five principles are based on the Social Interdependence Theory. So, the implementation of cooperative learning in the current study depended on implementing the five principles of cooperative learning as follows. First, positive interdependence was achieved in two ways. The first one is through the tutorials as the students read them and discuss them within the group. The second is the exercises which the students do them cooperatively. Second, in this study, promotive interaction was face-to-face. The students sat next to each other to discuss the materials and do group exercises. Third, the individual accountability principle was achieved through the students' exercises individually. Also, each student had a specific responsibility to help the group including the facilitator, summarizer, recorder, and reporter. Fourth, interpersonal and small group skills were essential as students must communicate only in the English language inside the classroom. Fifth, group processing in this study was achieved by making the students at the end of each class discuss the actions that helped them to achieve the learning goals as well as the difficulties they encountered during the lesson within the group (group processing) and with the whole class (whole class group processing).

The cooperative learning strategy used in this study was STAD, which included introducing the teacher's skills, doing group exercises, doing individual exercises, and rewarding the top team in every class. So, the teacher introduces the skill at the beginning of the class and then the students discuss the skill theoretically within each group, and they have to write their scores for each exercise. Then, the students do exercises as groups. After that, every student has to practice exercises individually, and s/he has to write his scores for each exercise. Finally, the teacher counts the scores of each group, both the group exercises and the individual exercises, to reward the top team in every class.

The theories that are used in the design of the website are behaviorism and cognitivism. So, behaviorist CALL activities refer to drills and practice, which can be done through the different exercises provided to the learners. These exercises are the stimuli that require responses or answers from the students [16, 44, 45]. Also, cognitive CALL activities refer to providing challenging exercises to learners that require the students to think and organize their learning [44]. In other words, cognitive CALL activities in this study refer to providing challenging materials for the students [16, 45]. Therefore, the content on the website of this study is constantly challenging for the students during the whole course.

Table 1 Matching TOEFL reading skills to reading taxonomy by [47]

Bottom-up reading skills	Top-down reading skills
1. Finding meaning from structural clues 2. Finding meaning from word parts 3. Meaning from the context of difficult words 4. Meaning from context for easy words 5. Pronoun reference 6. Stated detail question 7. Where specific information is found 8. Transition questions	Analyze 1. Main idea 2. Organization of ideas 3. Unstated details 4. Implied details Interpret 1. Purpose 2. Tone 3. Course

In terms of reading, the study adopted the interactive reading model by Rumelhart [46] which linked bottom-up and top-down approaches. In this model, learners are not passive participants as there is a kind of dialogue between the reader and the text. Both the text and the reader are important in creating the meaning, and the role of the reader is to decode and interpret the text. Based on this model, reading skills included micro (bottom-up) reading skills and macro (top-down) reading skills. Also, the revision exercises included questions from both approaches, which require the student to read the text interactively from parts to whole text and vice versa. The classification of reading skills into micro reading skills (bottom-up skills) and macro reading skills (top-down skills) was according to the taxonomy of Champeau, Marchi, and Arreaza-Coyle [47], as shown in Table 1 below.

Finally, this whole process of development and implementation was within the frame of the ADDIE instructional model, which includes five phases: analysis, design, development, implementation, and evaluation. In the analysis phase, the researchers carried out a needs analysis that covered 60 students to investigate the students' skills to study. The needs analysis results showed that the students need to study all the skills listed in the survey, which are 15 reading skills. In the design phase, the skills and the pre and post-test were adopted from [48], because there is a wide variety of skills that prepare the students for university study, and the skills can be divided into bottom-up and top-down skills. In the development phase, the researchers adopted the materials from [48] to design the website for the study, taking into consideration the features of behaviorist CALL as it provides different drills to the students and cognitive CALL since it provides the students with constant challenging materials [16, 45]. In terms of implementation, Social Interdependence Theory was used because the treatment depends on implementing the five principles of cooperative learning. Also, STAD was the cooperative learning strategy of teaching. The evaluation was in the form of pre and post-test and semi-structured interviews. Figure 1 below shows the theoretical framework of the study.

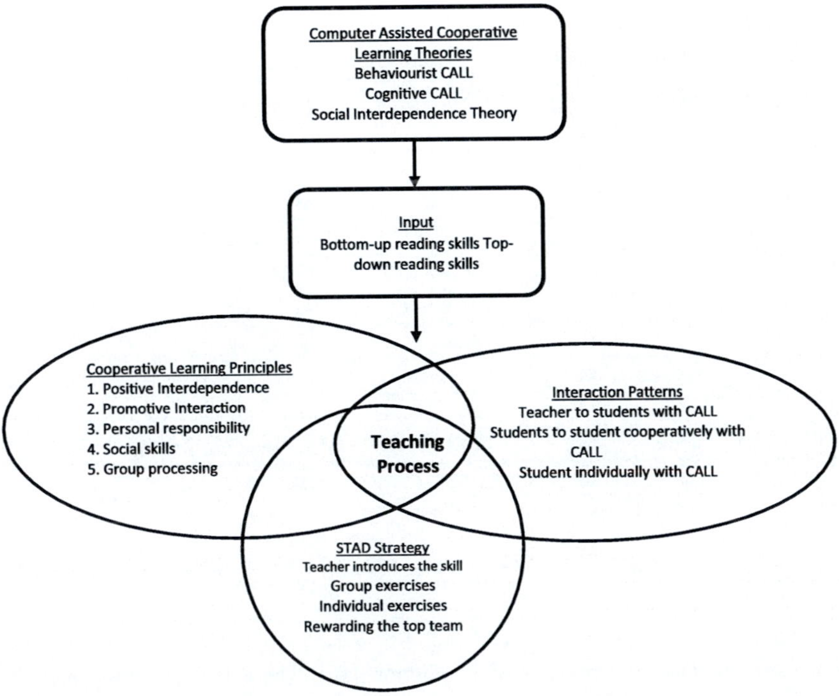

Fig. 1 Face-to-face Computer Assisted Cooperative Learning framework

3 Methods

3.1 *Research Design*

This study employed a mixed-method approach as the research data were collected through both quantitative and qualitative instruments. The quantitative data were collected through a pre and post-test, while the qualitative data were collected through semi-structured interviews. According to [49], quantitative data helps generalize the results, and qualitative data gives the researcher an in-depth investigation. Therefore, the triangulation of both data sources will help to understand better the effect of face-to-face CACL on reading skills.

3.2 *Sample*

This study used purposive sampling because the samples are Yemeni EFL students who should be students at one of the Malaysian universities and enrolled in a program

Table 2 Students' background information

No	Stage of study	Gender	University	Major
1	Postgraduates	M	UNIZA	Pharmacology
2		M	UM	Electrical Engineering
3		M	UM	Architecture
4		M	Limkokwing	MBA
5		M	UKM	Molecular Biology
6		M	UPM	MBA
7		F	UPM	Accounting
8		F	UPM	Sociology
9	Undergraduates	M	Help University	Financial Management
10		M	APU	Telecommunication Engineering
11		M	UTM	Software Engineering
12		M	APU	IT
13		M	Help University	Business
14		M	APU	IT
15		M	APU	IT

that uses English as a medium of instruction [50]. Therefore, the study participants were 15 Yemeni EFL students who are studying different majors in different public and private universities in Malaysia. Besides, interviews were carried out until reaching the saturation point where the participants' answers were repeated [51]. Accordingly, the interviews were made with five participants. According to [52], participants in intervention studies should not be less than 15 students, so the number of participants is enough to carry out this study. Even though the number of participants is limited to 15 learners, the data triangulation will help better understand the effect of face-to-face CACL on teaching reading skills. Besides, all the participants are at the same level of reading proficiency which was tested through the pre-test. The participants' background information is shown in Table 2.

3.3 Data Analysis

Before carrying out the paired sample t-test, the researcher analyzed the four assumptions of this test, namely skewness, kurtosis, normality, and homogeneity. The results of the analysis showed that the skewness is -0.092, and kurtosis is 0.254. This showed that the result is between +2 and −2, which is the accepted value to analyze the data using a paired-sample t-test [53]. Also, De Winter [54] proved that a t-test is feasible when dealing with a small sample size, making a paired sample t-test suitable for the study.

The pre and post-test were analyzed using SPSS (Version 22) using t-test inferential statistics. In terms of the interviews, they were analyzed in the form of thematic patterns. The interviews were transcribed and then sent back to the interviewees for member checking. Then, the researcher coded the interviews and categorized them in the form of different themes for triangulation with the quantitative data [51].

3.4 Validity and Reliability

The pre and post-test were adopted from Phillips [48]. Although experts designed it in teaching English to EFL learners, the researchers have ensured its validity and reliability. So, the researchers have sent the reading test to two academicians, who are experts in teaching reading to EFL learners. They stated that the test is suitable for university students. Also, researchers checked the reading test reliability using SPSS (Version 22) by distributing it to 16 students. Cronbach's alpha result was 0.759, which shows good internal consistency.

In terms of the validity and reliability of the qualitative data, the researchers have used different measures to increase the trustworthiness of the study findings. First, the researcher distributed the interview protocols to three academicians for validation. Second, the researcher used purposive sampling to choose the interviewees, considering their age and level of study, to avoid receiving one attitude or opinion [52]. Third, the researcher asked the interviewees to choose the language of the interview to avoid any kind of misconception [52]. Fourth, [52] stated that the trustworthiness of qualitative data is affected when there is a poor transcription for the interviews. Therefore, to avoid such problems, the researcher transcribed the interviews and sent them back to the participants to see if they wanted to add or modify anything in the transcription. The participants checked the transcription, and they stated that the transcription is identical to the interviews. Finally, the researcher triangulated the data collected from the interviews with the results of the questionnaires to link and support the results [55]. This helps to use the qualitative data to support the quantitative data, which is essential to investigate the themes related to teaching reading through face-to-face CACL, which helped the students to improve their reading skills.

3.5 Ethical Considerations

The study adopted reading tutorials and reading exercise materials from Phillips [48], and the researchers have obtained permission from Person to reproduce the materials in the web-based CALL of the study. Also, the researchers have explained to the participants that participation in the study is voluntary, and there is no risk for students. Also, the students have the right to withdraw from the intervention at any time. All the students volunteered to participate in the study and signed consent forms

Table 3 T –value and level of significance of Pre and Post-test total scores

Total scores	N	Mean	SD	t-value	Df	Sig. (2-tailed)
Pre-test	15	18.53	7.130	8.716	14	0.000**
Post-test	15	32.33	6.683			

**. Difference is significant at the 0.01 level (2-tailed)

that allowed the researchers to use the data for research purposes only. Moreover, the researchers have got verbal consent from the interviewees at the beginning of every interview to record the interviews. The researchers assured the participants that the data would be confidential and used for research purposes only.

4 Results

This section presents the analysis of the quantitative data in the form of pre and post-test and the qualitative data in the form of thematic patterns followed by triangulation.

What is the effect of face-to-face Computer Assisted Cooperative Learning on reading skills among Yemeni university EFL learners?
To find out if there is a significant difference in the students' performance after using CACL in studying reading skills, the researcher used a paired sample t-test to compare the pre-test and the post-test. The result of the whole pre-test was compared with the result of the whole post-test, and the result is shown in Table 3 above.

Table 3 above shows that the mean of the pre-test is 18.53, and the mean of the post-test is 32.33. The mean value of the pre-test is greater than the mean value of the pre-test, which indicates that the students' performance is better after CACL training than their performance before the CACL training. Also, it shows that there is a significant difference between reading comprehension pre-test and reading comprehension post-test (t-value = 8.716, $P = 0.000 > 0.05$). Therefore, it is concluded that CACL positively affects teaching reading skills to Yemeni EFL students.

What is the difference in micro reading skills between the pre-test and the post-test among Yemeni EFL learners?
To find out if there is a significant difference in the students' performance in micro reading skills (bottom-up skills) after using CACL in studying reading skills, the researcher used a paired sample t-test to compare the pre-test and the post-test. The result of the analysis is shown in Table 4 below.

Table 4 T –value and level of significance of Pre and Post-test scores of bottom-up reading skills

Micro skills scores	N	Mean	SD	t-value	Df	Sig. (2-tailed)
Pre-test	15	14.87	5.604	7.236	14	0.000**
Post-test	15	24.47	4.688			

**. Difference is significant at the 0.01 level (2-tailed)

Table 4 above shows that the mean of the bottom-up reading skills in the pre-test is 14.87, and the mean of the bottom-up reading skills in the post-test is 24.47. The mean value of bottom-up skills in the pre-test is greater than the mean value of bottom-up skills in the pre-test, which indicates that students' performance concerning micro reading skills is better after using CACL. Also, it shows a significant difference between micro reading skills in the pre-test and the micro reading skills in the post-test (t-value = 7.236, $P = 0.000 > 0.05$). Consequently, it is concluded that CACL positively affects teaching bottom-up reading skills to Yemeni EFL students.

What is the difference in macro reading skills between the pre-test and the post-test among Yemeni EFL learners?

To find out if there is a significant difference in the students' performance in top-down reading skills after using CACL in studying reading skills, the researcher used a paired sample t-test to compare the pre-test and the post-test. The result of the analysis is shown in Table 5 below.

Table 5 below shows that the mean of the top-down reading skills in the pre-test is 3.67 and the mean of the top-down reading skills in the post-test is 7.20. The mean value of the post-test is greater than the mean value of the pre-test. This indicates that students' performance in top-down reading skills is better after studying through face-to-face CACL. Also, it shows a significant difference between top-down reading skills in the pre-test and the top-down reading skills in the post-test (t-value = 6.046, $P = 0.000 > 0.05$). Accordingly, it is concluded that CACL positively affects teaching top-down reading skills to Yemeni EFL students.

The quantitative data is supported by the analysis of the interviews that led to different themes that made face-to-face CACL effective in teaching reading skills. These themes can be categorized into three main themes, namely academic, psychological, and social advantages.

Academic Advantages

The qualitative data analysis led to many academic themes that helped the students improve their reading skills. The first theme the students highlighted is that the

Table 5 T –value and level of significance of Pre and Post-test scores of top-down reading skills

Macro skills scores	N	Mean	SD	t-value	Df	Sig. (2-tailed)
Pre-test	15	3.67	1.759	6.046	14	0.000**
Post-test	15	7.20	2.210			

**. Difference is significant at the 0.01 level (2-tailed)

integration of CALL and face-to-face cooperative learning helped them improve their reading skills. This appeared in the statement of the students below.

> **S1**: "My performance in the post-test was better than my performance in the pre-test because of the group study inside the classroom. Also, it is due to using the computer which we used it during the study and when we have group activities or self-exercises, I mean individual exercises."

> **S2**: "I think they are integrated with each other. Computer and cooperative learning are integrated with each other. It is true that the student might use the computer alone, aaa but aaa he will not get the benefit which he came to get. For example, in reading, he will read normally as if he is reading a book, but cooperative learning gives you the information in a nice way as groups and as a group activity. The student might lack things, and this thing is available with his classmate. This makes it cooperative."

According to the statements of the students, there are different activities to be carried out inside the classroom. These activities are related to face-to-face cooperative learning with CALL. The dynamic interaction among the students when they work on computers is a key factor of face-to-face CACL that helps the students to improve their reading skills.

The second theme is that face-to-face CACL makes a balance between theory and practice in teaching reading skills. This can be found in the excerpt below.

> **S3:** "aaa I feel that the class was divided in an amazing way. The division was perfect. For example, if we study theoretically only, I think we would not be able to reach the expected benefit. Also, if we studied using the computer only, the learning process will be boring, and we would not be able to reach to the expected benefit."

> **S4**: "And, aaa Computer Assisted Cooperative Learning was a new skill for and a new learning method for me, to be in a group and aaa do several things such as doing many activities inside the classroom. And, aaa we start by doing tutorials then we having exercises in groups, then individual …"

The discussion of the participants above showed that the theoretical knowledge of the reading skills and the practice of the exercises on the website is one of the advantages of face-to-face CACL that helped them improve their reading skills. This clearly shows that the tutorials on the website were helpful for the students, as the students need to understand the skill theoretically before moving to practice. Also, face-to-face cooperative learning helps students since students help each other get feedback and reinforce their learning.

Another theme is that face-to-face CACL helps students improve their reading skills because of the feedback, which reinforces their understanding and practice. The face-to-face CACL helps the students negotiate and get feedback from CALL, teacher, and students, which reinforces their understanding of reading skills. Student 3 stated that:

> **S3:** "I think that the theoretical explanation of the skill at the beginning and the group discussion after that and the exercises reinforce the idea more, reinforces the understanding more. Sometimes, one of the students might misunderstand the idea, so the group members explain it more so that it is understood in a better way."

According to the expression above, face-to-face CACL activities guided the students to support each other, which helps the students minimize learning differences. If one of the students faces difficulty in understanding the materials, the other students help him in this regard. So, all the students will have the chance to understand the course materials, which is one factor that made face-to-face CACL helpful for the students to improve their reading skills. Another theme is that linking face-to-face CACL to the interactive reading approach helped them improve their reading skills. The skills taught to the students were both the bottom-up skills and the top-down skills, which helped the students understand a wide range of skills and practice them with face-to-face CACL.

> **S1**: "The improvement was good or we can say excellent for me. I got many skills like the linkage between the skills and the questions, and the way of reading now is better than the past."

> **S3**: "When you reach skill 10, this means that you have studied 10 skills. Therefore, in every passage you study, you should practice these skills or most of these skills. This makes reading challenging."

According to the students, the students practiced different exercises in line with behaviorist CALL, and the materials were challenging for the students in line with cognitive CALL. Such features of CALL helped the students to improve their reading skills further.

Another theme is learning autonomy, as students could depend on themselves more during their studies. This theme is shown in the statement of student 5 below.

> **S5:** "So, I studied how to answer depending on myself and finish the answers in the given time."

The above theme shows that face-to-face CACL helped students improve each reading skill. The main aim of cooperative learning is to strengthen the individual so that every student can do similar exercises successfully. This supports the role of face-to-face cooperative learning when using CALL to improve the students' language skills.

Social Advantages

This part discusses the social themes that helped students improve their reading when they studied reading skills using face-to-face CACL. Other students support the first theme. The students made it clear that they are socially active to support each other, and if one of the students feels bored, isolated, or distracted, they attract his attention and help him to understand the lesson. One of the students stated:

> **S5:** "if one of the students is distracted or did not understand, our duty as a group is to help him and attract his attention to the lesson. It happens sometimes in the middle of the lesson or at the end of the class."

The above expression also shows that face-to-face cooperative learning with computers helps students focus on the exercises. The cooperative nature of learning guided the learning process to be task-based oriented without being distracted by face-to-face communication during the classes.

Another theme is that in face-to-face CACL, the students provide emotional support to the other students to motivate them to participate and give their answers. Student 5 gave a situation with one of his groupmates.

S2: "amazing learning cordiality. You feel that you are close to your classmates and your teacher. It took away the learning phobia."

S5: "We also used to give him chances to participate and things to speak. When he becomes the reporter, sometimes we help him with some points to say. He takes the challenge seriously, and we noticed that he got a lot of benefit. Also, all of us got benefit in my group because we help each other."

An important theme is that in face-to-face CACL, the role of the teacher is essential during the process of learning. His role is not limited to gathering the students and supervising them; however, he is considered a main reference and source in learning using CACL as stated by students.

S2: "yes, the teacher is the main source. Secondly, after we finish working as groups and individually, the teacher used to explain aaa the passage; what is it and how to reach the answers."

To summarize, face-to-face CACL depends on social interaction among the students to facilitate learning for the whole class. The students' statements showed that effective interaction among the students is essential for improving their learning, especially that face-to-face interaction might go beyond academic support to provide emotional support so that weak students can perform better inside the classroom.

Psychological Advantages

This category discusses the psychological themes that helped the students to improve their reading skills. The first theme is motivation and self-confidence, as students gained this feeling towards reading and study in general because of face-to-face CACL, as shown in the comment below.

S1: "The course was a beginning for an essential learning motivation which the student will gain at the end of the course."

S3: "Now after the course, my confidence is increased, and the sense of boring is decreased. Now, I can read a passage or a book more comfortably than before, especially after gaining background about reading skills."

Another theme that students highlighted is that face-to-face CACL led them to be less bored inside the classroom and more excited about reading. This led to another theme: face-to-face CACL made learning interesting, motivating, and anxiety-free, as shown in the students' statements below.

S4: "there were rewards given to us as motivation and warming up activities at the beginning making us excited for the classes."

S5: "there was anxiety at the beginning of the course because I was not familiar with my classmates. However, the shyness was becoming less and less when I came to know my classmates more, and there are daily activities for every group, and the students contact the teacher every day and get feedback from him."

The students' comments above show that students could overcome different psychological learning barriers when they studied reading skills using face-to-face CACL. This is attributed to face-to-face cooperative learning since the students could build a learning community inside the classroom. So, they felt comfortable discussing and negotiating with each other. Effective communication among the students gives them a sense of "cordiality" inside the classroom, which helps them overcome any learning barrier. Such social advantages of face-to-face CACL are essential for the students to adapt psychologically to the classroom, in terms of having an interest in learning, feeling less shy and less anxious, and getting the motivation to improve their reading skills.

To sum up, the quantitative data analysis showed that the students could improve their reading skills. Also, the qualitative data analysis gave a clear picture of the elements of face-to-face CACL that helped the students improve their reading skills, which can be categorized under three general themes, namely academic, social, and cognitive themes. The discussion above in this section is directly linked to face-to-face cooperative learning and CALL elements. The implementation of the five principles of cooperative learning during teaching reading skills with the STAD strategy was essential for the success of face-to-face CACL implementation to teach reading skills. Also, the behaviorist and cognitive elements of CALL played an important role in helping the students improve their reading skills. Table 6 below shows the different themes of face-face CACL that helped the students improve their reading skills, as expressed in the interviews. Besides, the researchers linked these themes to the elements of face-to-face CACL according to the cooperative learning activities of the students inside the classroom and the features of the designed website.

5 Discussion

This study aimed at investigating the effect of face-to-face CACL on teaching reading skills, including both bottom-up and top-down reading skills. The study's findings showed that face-to-face CACL is an effective method of teaching reading skills, and this result is in line with the findings of [21, 42]. However, this study gives an in-depth investigation as it showed that face-to-face CACL helps the students to improve bottom-up reading skills and top-down reading skills. This confirms the argument of [56] that cooperative learning is more effective when used with CALL. Moreover, this confirms the idea of the interactive reading approach by [57]. The interactive reading approach helps the students improve both bottom-up and top-down reading skills because the two approaches are linked during reading.

Theoretically, the link between the behaviorist CALL, cognitive CALL, and cooperative learning showed that it is effective in teaching reading skills. Thus, drill-and-practice is a feature of behaviorist CALL associated with behaviorism and bottom-up reading skills, and the challenging materials are a feature of cognitive CALL associated with cognitive CALL [16, 45, 58]. Besides, the implementation of the principles of cooperative learning played an essential role in improving learners' reading

Table 6 Themes of teaching reading skills through face-to-face CACL

Themes	Subthemes
Academic	Improving bottom-up and top-down reading skills
	Practicing different CALL exercises
	Feedback from classmates, computer and teacher
	Challenging CALL exercises
	Effective communication
	Minimizing learning differences
	Improving learning autonomy
	Task-based activities of reading
	Balancing between teaching and practice
	Learning autonomy
Social	Emotional support
	Control isolation
	Control distraction
Psychological	Raising motivation
	Raising self-confidence
	Raising interest
	Reducing anxiety
	Reducing shyness

skills. Thus, positive interdependence motivates the students to help each other to understand the skill and practice the exercises; promotive interaction encourages the students to exchange ideas to understand the lesson and answer the exercises; individual accountability motivates every student to improve his skills for his benefit and the benefit of his group; social skills assures that the students use English only as a means of communication which improves other skills like communicative skills; and group processing helps the students reflect on their study and share experiences to get benefit from the group, the whole class, and the teacher. Hence, the elements of reading exercises, CALL features, and principles of cooperative learning are integrated, and the role of each theoretical component is important for the success of the implementation of CACL.

Furthermore, face-to-face cooperative learning played a vital role in helping learners to improve their reading skills. One of the main advantages of face-to-face interaction is improving communicative skills [19, 22, 27, 28]. Face-to-face interaction makes the students socially active inside the classroom since they have to discuss and debate with the other group members to achieve the required tasks [59]. Students provided academic support to their classmates and emotional support to the weak individuals in their groups, which is in line with the primary goal of cooperative learning, which is to strengthen every individual to do similar tasks individually. Again, this is also related to learning autonomy since learners become able to achieve similar tasks, which is one of the advantages of cooperative learning [26]. Therefore, using face-to-face cooperative learning in CACL increases learning autonomy because the students can practice different exercises as groups and then do individual drills.

Besides, another advantage of face-to-face interaction is that the students could overcome psychological language learning barriers such as anxiety and boredom, and they became more motivated and self-confident to improve their reading skills. Such psychological barriers were greatly minimized because of face-to-face interaction, and the students stated that they could feel a sense of "cordiality". This makes face-to-face interaction with CACL the main factor to help the students adapt to the learning environment and feel attached to the learning community inside the classroom.

Although face-to-face CACL is a student-centered approach, the teacher did not lose his role inside the classroom. The role of the teacher was essential as he was the reference for the students if they faced any difficulty in understanding the materials. Also, the role of the teacher did not affect student-centered learning as the students used to achieve the tasks depending on themselves. Hence, face-to-face CACL might solve the problems stated by previous literature concerning the marginalization of the role of the teacher when using technology in the process of language teaching and learning [22]. Also, the students' need for the guidance of the teacher from time to time supports that CALL cannot replace the role of the teacher totally, as stated by [18]; however, it is effective when used with other learning methods such as face-to-face cooperative learning.

Furthermore, using face-to-face CACL helps the students to get immediate feedback from students, feedback from their classmates, and feedback from the teacher. These three sources of feedback show that face-to-face CACL is a solution to the problem raised by [11] that CALL might not be suitable for students with different levels. Face-to-face CACL helps the students overcome learning differences in group work, especially when there is mixed-group cooperative learning.

Besides, STAD is an effective method with CACL for different reasons. First, it helps to balance theoretical knowledge and practice, which helps raise the benefit since some students are weak in reading due to lacking reading skills. Second, STAD is a cooperative learning strategy, but the reward creates a sense of competition. Therefore, cooperative learning among the students and the sense of competition with the other groups was a source of motivation for the students [9, 60]. Competition motivates each group to cooperate and improve their skills to be the top team, a healthy practice inside the classroom.

6 Implications

The study findings lead to many implications. First, face-to-face cooperative learning and CALL have complementary advantages for the students; hence, it is essential to pay attention to the design of CALL in teaching through CACL. Second, face-to-face CACL helps the students adapt socially and psychologically to the learning environment, which is essential for the students to improve their academic performance. Third, face-to-face interaction with CACL humanizes the use of technology in the form of social and emotional support. Such aspects proved to be important in language learning as it helps the students create a community of learning inside the classroom. Fourth, even though face-to-face CACL is a student-centered approach, the role of the teacher was not marginalized. He was available to help the students when they needed him, which was helpful for the students; this assures that technology is effective in the process of teaching, but it cannot replace the role of the teacher totally, at least until now. Fifth, although the students receive immediate feedback from CALL and necessary feedback from the teacher, the feedback from the other students is of equal importance to improving the students' performance and controlling learning differences inside the classroom.

7 Conclusion

The study investigated the effect of face-to-face CACL on improving reading skills. It proved that the students favored face-to-face interaction with CACL, especially because of its academic, social, and psychological advantages. The improvement of the students in reading skills in the post-test supports the importance of linking learning theories of CALL activities and cooperative learning principles with SATD strategy. The findings show that face-to-face cooperative learning humanizes the use of CALL during teaching so that the students can support each other academically and emotionally. Also, implementing the five principles of cooperative learning is essential to make CACL activities cooperative among students. The study's contribution is in the proposed framework for face-to-face CACL, and the implementation showed that it is effective in teaching reading skills. The proposed framework guides researchers from the stage of analysis to the evaluation stage, taking into consideration learning theories of CALL exercises and cooperative learning principles and strategies. The study might be replicated in other EFL settings in teaching reading skills, and the framework might be adapted to teach other language skills or courses.

References

1. Johnson RT, Johnson DW, Stanne MB (1986) Comparison of computer-assisted cooperative, competitive, and individualistic learning. Am Educ Res J 23(3):382–392
2. King A (1989) Verbal interaction and problem-solving within computer-assisted cooperative learning groups. J Educ Comput Res 5(1):1–15
3. Hooper S (1992) Cooperative learning and computer-based instruction. Educ Tech Res Dev 40(3):21–38

4. Neo M (2004) Cooperative learning on the web: a group based, student centred learning experience in the Malaysian classroom. Aust J Educ Technol 20(2):1–20
5. Yang Y (2010) Computer-assisted foreign language teaching: Theory and practice. J Lang Teach Res 1(6):909
6. Yang X, Kuo L-J, Ji X, McTigue E (2018) A critical examination of the relationship among research, theory, and practice: Technology and reading instruction. Comput Educ 125:62–73
7. Jarvis H (2013) Computer assisted language learning (CALL): Asian learners and users going beyond traditional frameworks. Asian EFL J 15(1):190–201
8. Intratat C (2007) Investigation on advantages and disadvantages in using English CALL according to the opinions of Thai university students and lecturers. KMUTT Res Dev J 30(1):3–20
9. Qasim A, Ali S, Adzhar K, Al-Sharafi MA (2017) E-learning models: the effectiveness of the cloud-based E-learning model over the traditional E-learning model. In: 2017 8th international conference on information technology (ICIT), Amman, Jordan, 17–18 May 2017. IEEE, pp 12–16. https://doi.org/10.1109/ICITECH.2017.8079909. http://ieeexplore.ieee.org/document/8079909/
10. Al-Ajmi Q, Al-Shaalan MK, Al- MA, Chellathurai GJ (2021) Fit-viability approach for E-learning based cloud computing adoption in higher education institutions: a conceptual model. In: Recent Advances in Technology Acceptance Models and Theories. Springer International Publishing, Cham, pp 331–348
11. Dina AT, Ciornei S-I (2013) The advantages and disadvantages of computer assisted language learning and teaching for foreign languages. Procedia Soc Behav Sci 76:248–252
12. Galavis B (1998) Computers and the EFL class: their advantages and a possible outcome, the autonomous learner Engl Teach Forum 36(4):27
13. Razak NA, Yassin AA, Maasum TNRTM (2020) Formalizing informal CALL in learning english language skills. In: Enhancements and limitations to ICT-based informal language learning: emerging research and opportunities. IGI Global, pp 161–182
14. Fang Y (2010) Perceptions of the computer-assisted writing program among EFL college learners. J Educ Technol Soc 13(3):246–256
15. Rahimi M, Tavakoli M (2015) The effectiveness of CALL in helping Persian L2 learners produce the English vowel/ɒ. GEMA Online J Lang Stud 15(3):17–30
16. Hammad ZM, Hussin S (2017) The implementation of computer assisted language learning in EFL classroom: Arab EFL learner's attitudes and perceptions. Int J Stud Engl Lang Literat 5(9):27–37
17. Levy M (1997) Computer-assisted language learning: context and conceptualization. Oxford University Press, London
18. Chapelle CA (2001) Computer applications in second language acquisition. Cambridge University Press, Cambridge
19. Lee K-W (2000) English teachers' barriers to the use of computer-assisted language learning. Internet TESL J 6(12):1–8
20. Lee G, Wallace A (2018) Flipped learning in the English as a foreign language classroom: outcomes and perceptions. TESOL Q 52(1):62–84
21. Sioofy M, Ahangari S (2013) The effect of computer assisted cooperative language learning on Iranian high school students' language anxiety and reading comprehension. Int J Fore Lang Teach Res 1(3):45–59
22. Lai C-C, Kritsonis WA (2006) The advantages and disadvantages of computer technology in second language acquisition. Online Sub 3(1):1–6
23. Jayachandran J (2007) Computer assisted language learning (CALL) as a method to develop study skills in students of engineering and technology at the tertiary level. Ind Rev World Literat Engl 3(2):1–7
24. Yassin AA, Razak NA, Maasum TNRTM (2021) Innovation attributes of F2F computer-assisted cooperative learning in teaching reading skills. Int J Web-Based Learn Teach Technol (IJWLTT) 17(3):1–17

25. Al-Emran M, Arpaci I, Salloum SA (2020) An empirical examination of continuous intention to use m-learning: an integrated model. Educ Inf Technol 25(4):2899–2918
26. Roger T, Johnson DW (1994) An overview of cooperative learning. Creat Collab Learn, 1–21
27. AbuSeileek AF (2007) Cooperative vs. individual learning of oral skills in a CALL environment. Comput Assist Lang Learn 20(5):493–514
28. AbuSeileek AF (2012) The effect of computer-assisted cooperative learning methods and group size on the EFL learners' achievement in communication skills. Comput Educ 58(1):231–239
29. Arpaci I, Al-Emran M, Al-Sharafi MA (2020) The impact of knowledge management practices on the acceptance of Massive Open Online Courses (MOOCs) by engineering students: a cross-cultural comparison. Telematics Inform 54:101468
30. Azman H, Bhooth AM, Ismail K (2013) Readers reading practices of EFL yemeni students: recommendations for the 21st century. GEMA Online J Lang Stud 13(3)
31. Yassin AA, Abdul Razak N (2017) Investigating the relationship between foreign language anxiety in the four skills and year of study among yemeni university EFL learners. 3L: Southeast Asian J Engl Lang Stud 23(4)
32. Al-Sohbani YAY (2018) Foreign language reading anxiety among Yemeni secondary school students. Int J Engl Lang Transl Stud 6(1):57–65
33. Al-kadi AMT (2013) English acquisition through unstructured internet use in Yemen. Internet J Lang Cult Soc 38:1–14
34. Yassin AA, Razak NA, Maasum NRM (2019) Investigating the need for computer assisted cooperative learning to improve reading skills among Yemeni university EFL students: a needs analysis study. Int J Virt Pers Learn Environ (IJVPLE) 9(2):15–31
35. Greenfield R (2003) Collaborative e-mail exchange for teaching secondary ESL: a case study in hong Kong. Lang Learn Technol 7(1):46–70
36. Li X, He J (2012) Study on cooperative learning of college ESL writing in network environment. Int J Knowl Lang Process 3(1):35–53
37. Altun H, Korkmaz Ö (2012) Computer, electrical & electronic engineering students' attitude towards cooperative learning. Online Sub 7(3):220–228
38. Chen Y-F, Cheng K-W (2009) Integrating computer-supported cooperative learning and creative problem solving into a single teaching strategy. Soc Behav Pers Int J 37(9):1283–1296
39. Kuo Y-C, Walker AE, Belland BR, Schroder KE (2013) A predictive study of student satisfaction in online education programs. Int Rev Res Open Distrib Learn 14(1):16–39
40. Lan Y-J, Sung Y-T, Chang K-E (2009) Let us read together: development and evaluation of a computer-assisted reciprocal early English reading system. Comput Educ 53(4):1188–1198
41. Chen J-M, Chen M-C, Sun YS (2010) A novel approach for enhancing student reading comprehension and assisting teacher assessment of literacy. Comput Educ 55(3):1367–1382
42. Al-Salem M (2016) The effectiveness of cooperative-online synchronous learning in promoting reading skills of freshman female students at the college of languages and translation. MA Thesis, King Saud University
43. Klingner JK, Vaughn S (2000) The helping behaviors of fifth graders while using collaborative strategic reading during ESL content classes. TESOL Q 34(1):69–98
44. Warschauer M, Healey D (1998) Computers and language learning: an overview. Lang Teach 31(2):57–71
45. Anderson T (2008) The theory and practice of online learning. Athabasca University Press, Athabasca
46. Rumelhart D, LaBerge D, Samuels S, Clark H (1977) Basic processes in reading: perception and comprehension
47. C. L. Champeau de Lopez, G. Marchi, and M. E. Arreaza-Coyle, "Taxonomy: Evaluating reading comprehension in EFL," in *Forum*, 1997, vol. 35, no. 2: ERIC, p. n2.
48. D. Phillips, *Longman complete course for the TOEFL test: Preparation for the computer and paper tests*. Longman London, 2001.
49. J. W. Creswell, *Qualitative, quantitative and mixed methods approaches*. Sage, 2014.
50. Etikan I, Musa SA, Alkassim RS (2016) Comparison of convenience sampling and purposive sampling. Am J Theor Appl Stat 5(1):1–4

51. Creswell JW, Creswell JD (2017) Research design: qualitative, quantitative, and mixed methods approaches. Sage publications, Thousand Oaks
52. Cohen L, Manion L, Morrison K (2002) Research methods in education. Routledge, London
53. George D (2011) SPSS for windows step by step: a simple study guide and reference, 17.0 update, 10/e. Pearson Education India
54. De Winter JC (2013) Using the Student's t-test with extremely small sample sizes. Pract Assess Res Eval 18(1):10
55. Long T, Johnson M (2000) Rigour, reliability and validity in qualitative research. Clin Eff Nurs 4(1):30–37
56. Lihong Z (2008) Cooperative learning models in advanced business english reading course. J Hunan Univ Commer
57. Rumelhart DE (1994) Toward an interactive model of reading. International Reading Association
58. Cooper JL (1995) Cooperative learning and critical thinking. Teach Psychol 22(1):7–9
59. Ahour T, Mukundan J, Rafik-Galea S (2012) Cooperative and individual reading: the effect on writing fluency and accuracty. Asian EFL J Q 14:46
60. Norman DG (2005) Using STAD in an EFL elementary school classroom in South Korea: effects on student achievement, motivation, and attitudes toward cooperative learning. Asian EFL J 35(3):419–454

The Effect of B-learning Adoption on the Evolution of Self-regulation Skills: A Longitudinal Study on a Group of Private Universities' Freshman Students

Mohammed Ali Al-Awlaqi, Maged Mohammed Barahma, Tawfiq Sarea Ali Basrda, and Ali AL-Tahitah

Abstract This study aims to study the evolution of self-regulation skills when adopting B-learning schemes among undergraduate students, using a sample of 68 students who use a blended learning strategy. This study was designed as a longitudinal study to grasp the evolution of self-regulation skills among learners over a one year of adopting blended learning classes. Repeated measures ANOVA design has been used to analyze the longitudinal data over three waves survey. Repeated measure ANOVA was used to test the change of the groups' mean over time. The data of the study has been collected through three waves from the undergraduate students. The three waves of data collection were spaced four months apart. The study found that help-seeking and self-evolution have evolved significantly while the environment structuring, goal setting, time management, or task strategies skills didn't evolve significantly. The study has come up with practical recommendations of how to improve the interaction between learners and the blended learning scheme.

Keywords B-learning · Self-regulation skills · Longitudinal study · Repeated measures design · Yemen

M. A. Al-Awlaqi (✉)
Lebanese International University-Yemen, Sana'a, Yemen
e-mail: alzooka@gmail.com

M. M. Barahma
Shabwah University, Ataq, Yemen

T. S. A. Basrda
University of Aden, Aden, Yemen

A. AL-Tahitah
Faculty of Leadership and Management, Universiti Sains Islam Malaysia USIM, Nilai, Malaysia

M. Al-Emran et al. (eds.), *International Conference on Information Systems and Intelligent Applications*, Lecture Notes in Networks and Systems 550,
https://doi.org/10.1007/978-3-031-16865-9_22

1 Introduction

1.1 B-learning

Blended learning (B-learning) is a hypered type of learning that combines traditional class or face-to-face learning and flipped classroom learning techniques. It helps learners benefit from the traditional learning technique and allows them to use new technologies or online learning to advance their learning abilities. Learning technologies freed learners from the lock of traditional classrooms [1]. B-learning can offer learners many advantages, such as flexibility, more ability to share ideas, and a higher level of interaction between learners, helping them develop more leadership skills [2]. B-learning creates an autonomous environment that replaces the traditional controlled environment, achieving positive outcomes [3].

The problem-based learning approach (PBL) favorites traditional learning methods such as the face-to-face method because it looks at learning from collaboration and contextual learning perspectives [4]. Learners can't forgo the temptation of face to face learning pedagogy easily [5]. Unless B-learning can improve collaboration and contextual learning, it would represent a problem in the learning process. Thus, it would be an essential point of B-learning to develop self-regulation skills among learners to be an effective method of learning.

Previous literature discussed many critical areas of the B-learning technique. B-learning literature discussed the benefits of applying this method to learners [2]. It was found to positively impact leadership skills, ability to share ideas, learning interaction, technological competency, self-development, self-discipline, and a higher level of motivation [2, 6]. It significantly affected learning behaviour, knowledge retention, and learning engagement [7]. It is a new solution that helps many learners receive necessary education even during worse scenarios such as the Covid-19 pandemic [8].

Another theme found in the literature discussed the learner's side. Some of the previous studies investigated the learners' satisfaction after experiencing one or more of the B-learning modules. These studies usually found a high level of satisfaction among B-learning modules students or learners [9]. It is found that B-learning had a positive and significant impact on the students' exam grades and academic achievement [3, 7], student engagement [10], self-motivation in learning [11], and flexibility in assessment [13]. Moreover, the viability of blended learning was discussed for disadvantaged students [12].

Earlier literature revealed other important themes. B-learning readiness was one of them. Generally, B-learning was not a strange learning method, and learners show a high level of readiness to enroll in a B-learning experience [13]. This readiness varies according to gender, ethnicity, age, or field of study [13]. Others discussed different B-learning delivery methods and their effectiveness [14]. The literature discussed important issues when implementing the B-learning pedagogical strategy from an education institution perspective. It further found that the B-learning strategy can help institutions operate within their limited budgets. Education institutions should avoid investing in expensive It infrastructure during the early implementation stage [15]. Moreover, education institutions should ensure that all stakeholders accept and adopt B-learning reform to get a successful experience [16].

1.2 Self-regulated Learning

Self-regulated learning is a self-managed learning strategy in which learners reflect their metacognition on their learning process to choose the best strategies that maximize their learning gain [17]. Self-regulated learning strategies are linked to the success of E-learning schemes as self-regulated learners can control their learning process.

Self-regulated learning skills were discussed in the literature. B-learning is a mid-point technique between the regulated face-to-face learning method and the unregulated E-learning method. B-learning needs learners to depend on themselves more to best benefit from their learning scheme. Thus, in previous literature, self-regulated learning skills were an essential part of B-learning [18]. It was found that time management and self-evaluation were among the weakest point in B-learning.

Although self-regulation learning has been investigated, no previous study has investigated the long-term longitudinal development of self-regulation among B-learning students. Studying the longitudinal effect of the B-learning method on developing self-regulation learning skills was recommended as a prominent scheme for future studies [18]. Thus, this study filled the gap in previous literature by investigating the longitudinal development of self-regulated skills among B-learning learners.

2 Methodology

The current study used repeated measures ANOVA design to test the evolution of self-regulation skills among a group of first-year students at private universities in Yemen. Repeated measures design measures the change of a variable over two different data collection waves [19]. It is the most suitable design for longitudinal studies. Repeated measures ANOVA has superior advantages over other means comparison techniques such as t-test or ANOVA. A t-test can compare cross sectional data between two groups, while ANOVA can be used to compare cross sectional data among more than two groups. On the other hand, repeated measures can compare means among more than two groups and over time.

2.1 Measurements

Self-regulation Measurement
Self-regulation variables were measured using a scale developed by Barnard et al. [20]. This scale is called Online Self-regulated Learning Questionnaire (OSLQ). Also, the scale contains six sub-dimensions: Environment structuring, goal setting, time management, help seeking, task strategies, and self-evaluation. Twenty four items were used to measure these sub-dimensions. 4 items for environment structuring, 5 items for goal setting, 3 items for time management, 4 items for help seeking, 4 items for task strategies, and four items for self-evaluation. The creator validated this scale in his later works [21].

B-Learning Measurement
B-learning was measured using a scale from 1 to 3. The number one represents the score before taking any B-learning course. Number 2 represents the second data point after taking the first B-learning course, while 3 represents the data point after finishing the second B-learning course.

2.2 Data Collection

This study targeted 154 students who enrolled in their first year in one of the private universities in Yemen. The sample was selected randomly from the students asking for private teaching sessions on accounting courses. All the students in this study were first-year students at the business colleges. The study targeted the student who took accounting courses 101 and 102 in two consequent semesters. We excluded any students who asked for different private teaching sessions on any two courses because we did not want the nature of the course to affect the credibility of our findings. Only 68 students responded to the three waves of data collection. Thus, only these students were included in the study. Given that this study is exploratory and according to the minimum sample size requirement of the repeated measure technique, this sample size is considered satisfactory [22].

The students took offline and online courses on accounting-related materials. The offline sessions were given based on three hours a week. These sessions lasted for two consequent semesters, of which each semester consisted of 16 weeks. At the beginning of the first semester, two sessions were conducted for each group of students to explain the predetermined instructions that should be followed to accomplish the offline and the online parts of the courses. The online course was conducted using the free platform of G-classroom. This platform had all the necessary materials and instructions to pass the accounting class. Also, the platform contents gave the participants tools to be self-dependent. Finally, participants were given full autonomy in their learning process.

The respondents of this study were 62% females, and 38% were males. The sample showed that 22% of the respondents were 19 years old, 59% were 20 years old, and only 19% were 21 years old. No student in the study sample had any experience with any B-learning schemes.

3 Data Analysis

Descriptive statistics are conducted to give a general overview of the data collected in this study. The descriptive statistics results are shown in Table 1.

In the next step, reliability tests were conducted on the three-wave data set. Reliability tests are shown in Table 2.

3.1 Environment Structuring

We started our tests by examining the effect of using B-learning on the environment structuring variable among the students. The analysis showed that students' average

Table 1 Descriptive statistics

Variable	Waves	Mean	Std. deviation	Skewness	Kurtosis
Environment structuring	Wave 1	2.13	0.771	−0.234	−1.267
	Wave 2	2.22	0.826	−0.436	−1.399
	Wave 3	4.06	0.879	−0.117	−1.713
Goal setting	Wave 1	2.43	0.498	0.304	−1.966
	Wave 2	2.49	0.503	0.060	−2.058
	Wave 3	2.29	0.670	−0.424	−0.746
Time management	Wave 1	1.41	0.496	0.367	−1.923
	Wave 2	1.50	0.504	0.000	−2.062
	Wave 3	1.60	0.493	−0.430	−1.871
Help seeking	Wave 1	1.54	0.502	−0.181	−2.028
	Wave 2	2.56	0.500	−0.242	−2.001
	Wave 3	4.50	0.504	0.000	−2.062
Task strategies	Wave 1	1.41	0.496	0.367	−1.923
	Wave 2	1.46	0.502	0.181	−2.028
	Wave 3	1.63	0.621	0.440	−0.621
Self−evaluation	Wave 1	1.76	0.794	0.452	−1.266
	Wave 2	2.07	0.654	0.587	1.235
	Wave 3	4.00	0.846	0.000	−1.613

Table 2 Reliability tests

Variable	Waves	No. items	Cronbach alpha
Environment structuring	Wave 1	4	0.64
	Wave 2		0.62
	Wave 3		0.59
Goal setting	Wave 1	5	0.78
	Wave 2		0.73
	Wave 3		0.71
Time management	Wave 1	3	0.86
	Wave 2		0.83
	Wave 3		0.87
Help seeking	Wave 1	4	0.70
	Wave 2		0.69
	Wave 3		0.74
Task strategies	Wave 1	4	0.85
	Wave 2		0.80
	Wave 3		0.74
Self-evaluation	Wave 1	4	0.75
	Wave 2		0.68
	Wave 3		0.60

level before using B-learning classes was 2.13. In the second wave of collection, this average increased to 2.22. Meanwhile, in the last wave, this average increased significantly to 4.05.

Mauchly's test of sphericity shows a nonsignificant value of 0.998 with a p-value of 0.950. This indicated clearly that the sphericity assumption was fulfilled, as shown in Table 3.

The test for change between the three different waves shows a significant effect as shown in Table 4.

Table 3 Mauchly's test of sphericity (environment structuring)

Within subjects effect	Mauchly's W	Approx. chi-square	df	Sig.	Epsilon		
					Greenhouse–Geisser	Huynh–Feldt	Lower-bound
ES	0.998	0.103	2	0.950	0.998	1.000	0.500

Table 4 Test of within-subject effect (environment structuring)

Source		Type III sum of squares	df	Mean square	F	Sig.	Partial eta squared
ES	Sphericity assumed	160.892	2	80.446	112.554	0.000	0.627
Error(ES)	Sphericity assumed	95.775	134	0.715			

Table 5 Mauchly's test of sphericity (goal setting)

Within subjects effect	Mauchly's W	Approx. chi-square	df	Sig.	Epsilon		
					Greenhouse–Geisser	Huynh–Feldt	Lower-bound
GS	0.921	5.428	2	0.066	0.927	0.952	0.500

Table 6 Test of within-subject effect (goal setting)

Source		Type III sum of squares	df	Mean square	F	Sig.
GS	Sphericity assumed	1.304	2	0.652	2.112	0.125
Error (GS)	Sphericity assumed	41.363	134	0.309		

3.2 Goal Setting

Conducting tests to find the effect of using B-learning on the goal setting variable among the students. The analysis showed that students' average level before using B-learning classes was 2.43. In the second wave of collection, this average increased to 2.49, while in the last wave, this average was almost the same to 2.29.

Mauchly's test of sphericity shows a nonsignificant value of 0.998 with a p-value of 0.066. This indicated clearly that the sphericity assumption was fulfilled, as shown in Table 5.

The test for change between the three different waves shows an insignificant effect as shown in Table 6.

3.3 Time Management

Examining the effect of using B-learning on the time management variable among the students. The analysis showed that students' average level before using B-learning classes was 1.41. In the second wave of collection, this average increased to 1.50. Meanwhile, in the last wave, this average stayed at the same level of 1.6.

Table 7 Mauchly's test of sphericity (time management)

Within subjects effect	Mauchly's W	Approx. chi-square	df	Sig.	Epsilon[b]		
					Greenhouse–Geisser	Huynh–Feldt	Lower-bound
TM	0.992	0.538	2	0.764	0.992	1.000	0.500

[b]May be used to adjust the degrees of freedom for the averaged tests of significance. Corrected tests are displayed in the Tests of Within-Subjects Effects table.

Table 8 Test of within-subject effect (time management)

Source		Type III sum of squares	df	Mean square	F	Sig.
TM	Sphericity assumed	1.245	2	0.623	2.355	0.099
Error (TM)	Sphericity assumed	95.775	134	0.715		

Mauchly's test of sphericity shows a nonsignificant value of 0.992 with a p-value of 0.764. This indicated clearly that the sphericity assumption was fulfilled, as shown in Table 7.

The test for change between the three different waves shows an insignificant effect as shown in Table 8.

3.4 Help Seeking

The effect of using B-learning on the help-seeking variable among the students was tested. The analysis showed that students' average level before taking the B-learning classes was 1.54. In the second wave of collection, this average increased to 2.56, while in the last wave, this average increased significantly to 4.50.

Mauchly's test of sphericity shows a nonsignificant value of 0.997 with a p-value of 0.918. This indicated clearly that the sphericity assumption was fulfilled, as shown in Table 9.

The test for change between the three different waves shows significant effects, as shown in Table 10.

Table 9 Mauchly's test of sphericity (help seeking)

Within subjects effect	Mauchly's W	Approx. chi-square	df	Sig.	Epsilon[b]		
					Greenhouse–Geisser	Huynh–Feldt	Lower-bound
HS	0.997	0.171	2	0.918	0.997	1.000	0.500

[b]May be used to adjust the degrees of freedom for the averaged tests of significance. Corrected tests are displayed in the Tests of Within-Subjects Effects table.

Table 10 Test of within-subject effect (help seeking)

Source		Type III sum of squares	df	Mean square	F	Sig.
HS	Sphericity assumed	360.794	2	153.397	552.472	0.000
Error (HS)	Sphericity assumed	37.206	134	0.278		

Table 11 Mauchly's test of sphericity (task strategies)

Within subjects effect	Mauchly's W	Approx. chi-square	df	Sig.	Epsilon[b]		
					Greenhouse–Geisser	Huynh–Feldt	Lower-bound
TS	0.937	4.288	2	0.117	0.941	0.967	0.500

Table 12 Test of within-subject effect (task strategies)

Source		Type III sum of squares	df	Mean square	F	Sig.
TS	Sphericity assumed	1.853	2	0.926	3.042	0.051
Error (TS)	Sphericity assumed	40.814	134	0.305		

3.5 *Task Strategies*

Moreover, we tested the effect of using B-learning on the task strategies variable among the students. The analysis showed that students' average level before taking the B-learning classes was 1.41. In the second wave of collection, this average increased to 1.46, while in the last wave, this average was not increased significantly to 1.63.

Mauchly's test of sphericity shows a nonsignificant value of 0.998 with a p-value of 0.950. This indicated clearly that the sphericity assumption was fulfilled, as shown in Table 11.

The test for change between the three different waves shows insignificant effects, as shown in Table 12.

3.6 *Self-evaluation*

Finally, we tested the effect of using B-learning on the self-evaluation variable among the students. The analysis showed that students' average level before taking the B-learning classes was 1.765. In the second wave of collection, this average increased to 2.07. Meanwhile, in the last wave, this average was increased significantly to 4.00.

Table 13 Mauchly's test of sphericity (self-evaluation)

Within subjects effect	Mauchly's W	Approx. chi-square	df	Sig.	Epsilon[b]		
					Greenhouse–Geisser	Huynh–Feldt	Lower-bound
SE	0.999	0.050	2	0.975	0.999	1.000	0.500

[b]May be used to adjust the degrees of freedom for the averaged tests of significance. Corrected tests are displayed in the Tests of Within-Subjects Effects table.

Table 14 Test of within-subject effect (self-evaluation)

Source		Type III sum of squares	df	Mean square	F	Sig.
SE	Sphericity assumed	199.539	2	99.77	170.392	0.000
Error (SE)	Sphericity assumed	78.461	134	0.586		

Mauchly's test of sphericity shows a nonsignificant value of 0.998 with a p-value of 0.950. This indicated clearly that the sphericity assumption was fulfilled, as shown in Table 13.

The test for change between the three different waves shows a significant effect, as shown in Table 14.

4 Conclusion and Discussion

This study tried to test the evolution of self-regulation skills after taking B-learning classes among first-year students at some private universities in Yemen. The study found that students significantly developed their help-seeking and self-evaluation skills over time. On the other hand, students did not significantly improve their environment structuring, goal setting, time management, or task strategies skills after taking B-learning classes over a longer period. This study developed the previous literature by examining B-learning's longitudinal effect on self-regulation skills. Previous studies investigated the cross-section effect of B-learning on self-regulation skills [18]. However, this study filled this gap by examining the evolution of self-regulation skills by taking more B-learning courses over time.

The context of the study showed unique results. Consequently, this shows the importance of conducting this study in a challenging context such as Yemen. Previous work showed a positive impact on B-learning on self-regulation skills, while this study showed unique and different results [18]. In the study, Yemeni students did not show significant improvements in their environment structuring. This shows that Yemeni students do not treat B-learning technologies as an official learning environment. Yemeni students also failed to develop goal-setting skills, which resulted from the face-to-face learning environment. Usually, students rely on their instructors to set learning goals for them. B-learning imposed new experiences they were not familiar with. The same results were shown when it comes to time management. Yemeni students did not see challenges in meeting B-learning schedules and deadlines. This indicates an amateur interaction between students and the online technology of the B-learning environment. Students did not face enough enforcement from the B-learning environment to develop their task strategies skills.

5 Study's Implications

Following more innovative pedagogical strategies is essential for students in one of the least developed countries such as Yemen. Because of the low level of education performance, students in Yemen need to follow more innovative strategies to advance their learning skills. This study showed that Yemeni undergraduate students who followed the B-learning strategy failed to develop their self-regulated skills, which could help them advance their learning skills. Thus, educational institutions could design B-learning programs that help students structure their learning environment and learn how to create more comfortable places to study. Educational institutions should promote B-learning classes with additional training courses that help students develop more skills on how to set their learning goals. More training should be offered on studying time management. Educational institutions should change their students' mentality toward their evaluation. This is because most students believe they should be guided and evaluated only by their teachers and institution managers. Thus, more educational orientation should focus on building self-evaluation techniques among these students. This would promote more successful B-learning strategies implementation.

5.1 Limitations and Future Research

This study has its own limitations, one of which is the small sample size. This study examines longitudinally 68 students. This sample size could be satisfactory, yet a bigger sample size is favorable for future studies. This study used a repeated measure design to test the ovulation of self-regulation skills due to the sample size limitation. Future research could use more sophisticated techniques such as the latent growth model, which can enrich understanding of the dynamic evolution of self-regulation skills after taking B-learning courses.

References

1. Norberg A, Dziuban CD, Moskal PD (2011) A time-based blended learning model. Horizon 19:207–216

2. Fearon C, Starr S, McLaughlin H (2011) Value of blended learning in university and the workplace: some experiences of university students. Ind Commer Train 43:446–450
3. Siddiqui S, Soomro NN, Thomas M (2020) Blended learning source of satisfaction of psychological needs: an empirical study conducted on O-levels chemistry students in metropolis city of Pakistan. Asian Assoc Open Univ J 15:49–67
4. Bohle CK, Dailey-Hebert A, Gerken M, Grohnert T (2013) Problem-based learning in hybrid, blended, or online courses: instructional and change management implications for supporting learner engagement. In: Wankel C, Blessinger P (eds) Increasing student engagement and retention in e-learning environments: web 20 and blended learning technologies. Emerald Group Publishing Limited, pp 359–86. https://doi.org/10.1108/S2044-9968(2013)000006G015
5. Weil S, De Silva T-A, Ward M (2014) Blended learning in accounting: a New Zealand case. Meditari Account Res 22:224–244
6. Yousef Jarrah H, Alhourani MI, Al-Srehan HS (2021) Blended learning: the amount of requisite professional competencies in faculty members of Al Ain University from viewpoint of students. J Appl Res High Educ. https://doi.org/10.1108/JARHE-06-2021-0206
7. Al-Tahitah AN, Al-Sharafi MA, Abdulrab M (2021) How COVID-19 pandemic is accelerating the transformation of higher education institutes: a health belief model view. In: Arpaci I, Al-Emran M, Al-Sharafi MA, Marques G (eds) Emerging technologies during the era of COVID-19 pandemic. Studies in systems, decision and control, vol 348. Springer, Cham. https://doi.org/10.1007/978-3-030-67716-9_21
8. Bordoloi R, Das P, Das K (2021) Perception towards online/blended learning at the time of Covid-19 pandemic: an academic analytics in the Indian context. Asian Assoc Open Univ J 16:41–60
9. Stanislaus I (2021) Forming digital shepherds of the Church: evaluating participation and satisfaction of blended learning course on communication theology. Interact Technol Smart Educ 19:58–74
10. Siraj KK, Maskari AA (2019) Student engagement in blended learning instructional design: an analytical study. Learn Teach High Educ Gulf Perspect 15:61–79
11. Lo C-M, Han J, Wong ESW, Tang C-C (2021) Flexible learning with multicomponent blended learning mode for undergraduate chemistry courses in the pandemic of COVID-19. Interact Technol Smart Educ 18:175–188
12. Garrett DA, Lewis S, Whiteside AL (2015) Blended learning for students with disabilities: the North Carolina virtual public school's co-teaching model. Explor Pedagogies Diverse Learn, 67–93. https://doi.org/10.1108/S1479-368720150000027013
13. Adams D, Tan MHJ, Sumintono B (2020) Students' readiness for blended learning in a leading Malaysian private higher education institution. Interact Technol Smart Educ 18:515–534
14. Mahmud MM, Freeman B, Abu Bakar MS (2021) Technology in education: efficacies and outcomes of different delivery methods. Interact Technol Smart Educ 19:20–38
15. Abusalim N, Rayyan M, Jarrah M, Sharab M (2020) Institutional adoption of blended learning on a budget. Int J Educ Manag 34:1203–1220
16. Chowdhury F (2019) Blended learning: how to flip the classroom at HEIs in Bangladesh? J Res Innov Teach Learn 13:228–242
17. Zimmerman BJ (2008) Investigating self-regulation and motivation: historical background, methodological developments, and future prospects. Am Educ Res J 45:166–183
18. Onah DFO, Pang ELL, Sinclair JE (2021) Investigating self-regulation in the context of a blended learning computing course. Int J Inf Learn Technol 39:50–69
19. Verma JP (2015) Repeated measures design for empirical researchers, 1st edn. Wiley, Hoboken
20. Barnard L, Lan WY, To YM, Paton VO, Lai S-L (2009) Measuring self-regulation in online and blended learning environments. Internet High Educ 12:1–6
21. Barnard L, Paton V, Lan W (2008) Online self-regulatory learning behaviors as a mediator in the relationship between online course perceptions with achievement. Int Rev Res Open Distrib Learn 9. http://www.irrodl.org/index.php/irrodl/article/view/516
22. Guo Y, Logan HL, Glueck DH, Muller KE (2013) Selecting a sample size for studies with repeated measures. BMC Med Res Methodol 13:100

Perception of Word-Initial and Word-Final Phonemic Contrasts Using an Online Simulation Computer Program by Yemeni Learners of English as a Foreign Language in Malaysia

Lubna Ali Mohammed and Musheer Abdulwahid Aljaberi

Abstract This study aimed to examine the influence of different contexts (word-initial and word-final phonemic contrasts) on the perception of the phonemic contrasts among Yemeni learners of English-as-a-Foreign Language (EFL). The study also sought to ascertain the effect of Length of Residence (LOR) in Malaysia on the perception of selected phonemic contrasts in English by Yemeni EFL learners, as these contracts are presented in different contexts (word-initial and word-final positions). A total of forty-two Yemeni speakers living in Malaysia, 22 men and 20 women participated in this study; they were divided into two groups according to their LORs in Malaysia: group A (four months, short length of residence) and group B (three years, long length of residence). The results revealed a significant effect ($P < 0.05$) for different contexts (word-initial and word-final) on the perception of all participants and between both groups; In the word-initial position, all participants performed much better than in the word-final position.

Keywords Phonemic contrast · Length of residence · EFL learners · Contrastive analysis · Flege's Speech learning model · Minimal pairs

1 Introduction

English-as-a-Foreign-Language (EFL) poses both speaking and writing problems to Arabic learners, as reported by numerous studies [1–9]. Arabic learners of English are mostly taught by native speakers of Arabic who mostly use the Arabic language, rather than English, in the classroom and focus on sentence structure rather than

L. A. Mohammed (✉)
Department of TESL, Faculty of Social Sciences, Arts, and Humanities, Lincoln University College (LUC), Petaling Jaya, Malaysia
e-mail: lubnaali@lincoln.edu.my; luby_luda@Yahoo.com

M. A. Aljaberi
Faculty of Nursing and Applied Sciences, Lincoln University College (LUC), Petaling Jaya, Malaysia

M. Al-Emran et al. (eds.), *International Conference on Information Systems and Intelligent Applications*, Lecture Notes in Networks and Systems 550,
https://doi.org/10.1007/978-3-031-16865-9_23

correct pronunciation and articulation gestures for English sounds. This creates challenges for Yemeni learners of EFL when they are communicating in English, the lack of opportunity to practice English pronunciation and the prior English pronunciation learning experience are prominent problems in the improvement of English pronunciation and perception. Other factors that challenge the pronunciation and the perception of the English language are the English language instruction, the living in a native speaking country, and the length of practicing the English language [10]. These factors and many others, according to the authors' opinion, increase the possibility for L1 to influence L2 in all fields, especially in phonology where there is a lack of practice in the pronunciation of English sounds and words.

The Contrastive Analysis Theory (CAT) Stockwell et al. [11] focuses on the differences and similarities between the intended aspects of the study of two different languages (L1 and L2). It hypothesizes that the similarities in any two languages will bring about the positive transfer (no errors), whereas the differences will cause negative transfer or interlanguage [12]. In other words, L2 learners will encounter difficulties in discriminating speech sounds that are nonexistent in their L1. Similarly, models such as Flege's Speech Learning Model (SLM) explained the relationship between L1 and L2 phonology and proposed that ability to process non-native speech can be influenced by the native phonetic gap [13]. For instance, adult Japanese listeners find it difficult to discriminate American English [l] from[ɹ] [14]. According to SLM, /r/ is the only liquid phoneme present in Japanese, and Japanese listeners assimilate both the English [l] and [ɹ].

However, for L2 learners, forming phonological categories in their L2 phonology depends on their ability to successfully perceive and produce L2 sound contrasts that do not occur in their L1 phonology. The L1 phonological system constrains the improvement of perceptual ability since it works as a filter, filtering out L2 sounds that are absent in the L1 phonology. In this way, Flege's speech learning model suggests that improving the perception of phonemic differences between L1 and L2 sounds is the best way for learning a phonemic category of L2 sounds [15].

This study aims to examine the perceptual ability of selected English phonemic contrasts by the Yemeni learners of EFL in relation to different contexts (word-initial and word-final position). Therefore, the current study seeks to address the following research questions in relation to the Yemeni EFL learners:

1. *What are the effects of context (e.g., word-initial and word-final position) on the perceptual ability of phonemic contrasts in English among the Yemeni learners of EFL?*
2. *What is the effect of length of residence in Malaysia on the perception of the selected phonemic contrasts in English by the Yemenis learners of EFL as they are presented in different contexts (e.g., word-initial and word-final position?*

2 Literature Review

Research in speech perception in second language learning has posited that the perception of the new L2 sounds that have no counterpart in L1 will be easier learned and discriminated than L2 sounds that are close to L1 sounds [16, 17]. Moreover, cross-language research supports the effects of L1 phonological knowledge on L2 phonological perception and production in relation to different factors such as length of residence (LOR) [16–19].

2.1 *The Consonant Inventory of Arabic Language and English Language:*

Many researchers classified Arabic into three different varieties [20–23]: a) Classical Arabic, also known as Standard Arabic—the language of the Qur'an, Islam's Holy Book; b) Modern Standard Arabic (MSA), which is the standard formal language among Arabs and is mostly written than spoken; and c) Colloquial Arabic, which includes the informally spoken dialects used as a medium of daily contact and are mostly employed in oral communication. Diab, Habash [20] mentioned the existence of numerous dialectal Arabic groups, which can differ within the same country and among countries. This study is restricted to the phonology of Yemeni EFL learners, especially those who speak the Adeni and Ta'aizzi dialects of southern Yemen due to their similarities. The phonemic inventory of MSA includes 28 consonants: eight stops in which the voiced velar stop [g] is not listed, possibly one affricate (in case Arabic has only voiced palatal affricate [ʤ], twelve or thirteen fricatives, two nasals, one trill, one liquid, and two glides. In contrast, English has 24 consonants; six stops, two affricates, nine fricatives, three nasals, two approximants, and two liquids.

The existence of the voiced palatal affricate [ʤ] and voiced alveolar fricative [ʒ] in MSA is somewhat controversial. For example, Amayreh Mousa [24] opined that MSA contains the voiced palatal affricate [ʤ] but not the voiceless palatal affricate [tʃ], the voiced palatal fricative [ʒ], and the voiced velar stop [g]. On the other hand, Huthaily (2003) explained that MSA contains the voiced palatal fricative [ʒ] but not the voiced palatal affricate [ʤ], the voiceless palatal affricate [tʃ], or the voiced velar stop [g]. The dialects of interest in this study (Adeni and Ta'aizzi) use the voiced velar stop consonant [g] always, instead of the palatal affricate [ʤ] or the palatal fricative [ʒ]. In other words, the phonemic inventory of Yemeni dialects in this study includes the voiced velar stop [g] but not the voiced palatal affricate [ʤ], the voiceless palatal affricate [tʃ], or the voiced palatal fricative [ʒ]. Tables 1 and 2 show the consonant inventories of both the English Language and the Arabic dialect of southern Yemen. The English phonemic inventory presented in Table 2 was derived from [22].

A contrastive analysis for the phonemic inventory of MSA and English language was analyzed and studied by Mohammed and Yap (2009). They examined the perception of the phonemic contrasts between /p/ and /b/, /f/ and /v/, and /ʧ/ and /ʤ/. The

Table 1 The consonant inventory of Arabic dialects of south Yemen

	Labial	Labio-dental	Inter-dental		Denti-alveolar		Palatal	Velar		Uvular		Pharyngeal		Glottal
Stops	b				d	t		g	k	q				ʔ
					d̲	t̲								
Fricatives		f	ð	θ	z	s	ʃ			ʁ	χ	ʕ	ħ	h
			ð̲		s̲									
Nasals	m				N									
Lateral					L									
Trill					R									
Glides	w						j							

Table 2 The consonant inventory of the English language

	Labial		Labio-dental		Inter-dental	Alveolar		Alveolar		Palatal	Velar		Glottal
Stops	b	p				d	t				g	k	
Fricatives			v	f	ð	z	s	ʒ	ʃ				h
					θ								
Affricates								ʤ	tʃ				
Nasals	m					n					ŋ		
Lateral liquid						l							
Retroflex liquid						ɹ							
Glide										j	w		

researchers observed that Yemeni EFL learners found it difficult to perceive the absent phonemic contrasts in their L1. The scores of the discrimination tasks ranged from 44 to 81 out of 96, with percentages of 45.83 to 84.38%. The mean percentage of the discrimination task was 62.94% and the standard deviation was 9.63 [25]. In addition, the researchers found that the perception of these sounds could be improved as the length of residence is increased in the native L2-speaking country. To the author's knowledge, no previous study has attempted into the perceptual abilities of Arab English learners (speakers of Yemeni dialects in particular) in relation to their context, and therefore this study was conducted.

2.2 The Role of Technology in Pronunciation

In the study conducted byBusa [26] it was found that Practicing pronunciation with the visualization and comparing it with the native speakers was proven favorable. This method was considered to be significant and powerful for working on their pronunciation in English and asserted that after a few reiterations their inflection patterns would in general look like those of the native speakers [26]. According to the study conducted on Iranian EFL instructors 'Pronunciation Power programming' put a greater obligation on students rather than educators. Further teachers developed their pronunciation job into a student-centered instructional method. Thereby, switching their roles as EFL teachers from being an allocator of knowledge to facilitators and guides making students active learners [27]. Furthermore, based on the study conducted in Taiwan regarding computer-assisted pronunciation learning, it was found that the educators could see the growing experience of their graduates based on their learning reflections'. Thus, instructors can additionally customize the course to address the issues of the learners. In doing so the instructors can acquaint different intervening devices to work with their learning at various learning stages, thereby actually helping them to move to the further advanced stage of learning [28].

3 Research Methodology

3.1 Research Design

A Static Group Comparison design was used for the current studyIt's associated with pre-experimental design because it doesn't allow for much control over uncontrollable variables (such as L1, age, place and years of studying English, and the education level of the learner's parents) [29]. The static group comparison design requires two or more pre-existing groups, only one of which is exposed to the experimental treatment. No pre-treatment measures are employed. The researcher assumes that the groups are equal in all relevant aspects prior to the beginning of the study, except in their exposure to the independent variable. Then, the dependent variables for the groups are compared to assess the effect of the X-treatment.

In this study, two groups (group A and group B) of native Yemeni learners of English with different lengths of residence in Malaysia (the independent variable) were exposed to the experimental treatment (discrimination task). The scores of the discrimination task (the dependent variable) for the participants in each group were measured and compared to determine the relationship between them (the scores) and the LOR by investigating the effects of treatment in the discrimination experiment.

3.2 Sample

Forty-two Yemeni learners of EFL live in Malaysia, 22 men and 20 women, participated in this study. All had begun studying the English language from the age of 13 in Yemen. Their social interaction in English, both at school and home, was extremely limited, and none had studied the English language at any private institute prior to their arrival in Malaysia. All were monolinguals, and none have had any chance to practice English with native speakers. A convenient sampling was selected, generally all of the sample are from two governorates in the south of Yemen (Aden and Ta'aiz) according to their dialectal similarities.

According to their LORs in Malaysia, the participants were divided into two groups, i.e., Long Length of Residence (LLOR) and Short Length of Residence (SLOR). The samples in both groups were convenience samples.

- **Group A (SLOR)**: consisted of participants ranging between18–35 years (11 men and 10 women). They had been in Malaysia for less than one year to learn the English language or to study for a degree in various fields.
- **Group B (LLOR):** consisted of participants ranging between18–35 years (11 men and 10 women). They had been in Malaysia for at least two years prior to this study, meaning they had received more exposure and underwent more practice of the English language than group A. All were students of various institutions in Malaysian universities.

3.3 Data Collection

The participants were tested on the perception of three phonemic contrasts in English (/f/, /v/), (/p/, /b/), and (/ʧ/, /ʤ/), which are absent in Arabic, using a discrimination task. A questionnaire survey of the subject's background and a discrimination task from previous L2 research was used to obtain the data [16, 30].

4 The Stimuli

Twenty-four items were used as stimuli in the discrimination task. These items included six minimal pairs comprising three each in word-initial and word-final position (pan–ban/lap–lab), (fan–van/leaf–leave), (choke–joke/rich–ridge). They examined the perception of voiceless/voiced phonemic contrasts for the following phonemes in English (/f/ vs. /v/), (/p/ vs. /b/), and (/ʧ/ vs. /ʤ/) by native Yemeni learners. An online computer program AT&T text-to-speech was used to generate the stimuli (L2 words) using two UK models of speech, one male and one female, available from the program. Twenty-four tokens were produced: 12 words with a male voice and 12 words with a female voice. These words, which constituted the aural stimuli for the discrimination task, were chosen according to the voiceless/voiced phonemic contrasts (/p/, /b/), (/f/, /v/), and (/ʧ/, /ʤ/). The voiceless bilabial stop /p/ and voiced labiodental fricative /v/ are both absent in Arabic. Although the occurrence of (/ʧ/, /ʤ/) in Arabic is still being debated, both sounds are undoubtedly absent

Table 3 The word order of the discrimination experiment

Items		Discrimination task	Expected results	
panM1 panF1	banM1 banF1	panM1,banM1/panF1,banF1 panM1,panF1/panF1,panM1	D/D S/S	
		banM1,panM1/banF1,panF1 banF1,banM1/banM1,banF1	D/D S/S	
labM1 labF1	lapM1 lapF1	labM1,lapM1/labF1,lapF1 labM1,labF1/labF1,labM1	D/D S/S	
		lapM1labM1/lapF1, labF1 lapF1, lapM1/lapM1, lapF1	D/D S/S	
fanM1 fanF1	vanM1 vanF1	fanM1,vanM1/fanF1,vanF1 fanM1,fanF1/fanF1,fanM1	D/D S/S	
		vanM1fanM1/vanF1,fanF1 vanF1,vanM1/vanM1,vanF1	D/D S/S	
leafM1 leafF1	leaveM1 leaveF1	leafM1, leaveM1/leafF1, leaveF1 leafM1, leafF1/leafF1, leafM1	D/D S/S	
		leaveM1, leafM1/leaveF1, leafF1 leaveF1, leaveM1/leaveM1, leaveF1	D/D S/S	
chokeM1 chokeF1	jokeM1 jokeF1	chokeM1, jokeM1/chokeF1, jokeF1 chokeM1, chokeF1/chokeF1, chokeM1	D/D S/S	
		jokeM1, chokeM1/jokeF1, chokeF1 jokeF1, jokeM1/jokeM1, jokeF1	D/D S/S	
richM1 richF1	ridgeM1 ridgeF1	richM1,ridgeM1/richF1,ridgeF1 richM1,richF1/richF1,richM1	D/D S/S	
		ridgeM1, richM1/ridgeF1, richF1 ridgeF1, ridgeM1/ridgeM1, ridgeF1	D/D S/S	
		Total: 48	**24** **24**	**S** **D**

in the Yemeni dialects of interest in this study. For each contrast, two minimal pairs were chosen: one in the word-initial position and the other in the word-final position.

After generating the aural stimuli, the word order of the experiment was generated manually by considering the four outcomes of the signal detection task, which was achieved by allowing four expected responses for each item: two different (D) and two same (S). Forty-eight stimuli pairs were thus created out of six items. Table 3shows the word order that was created for the discrimination task. Once the word order was created, the PRAAT software was used to create the perceptual discrimination experiment. The forty-eight different stimuli were presented twice, resulting in 96 trials: 48 test trials (different) and 48 control trials (same). The stimulus items were presented with a silence duration of 0.8 s used as an inter-stimulus interval during the experiment. A laptop computer and headset were used to conduct the experiment.

5 Procedures

The participants were examined individually for around 20 min in a quiet room. The data were elicited in two phases: training and experimentation. The training phase was conducted first to orient the subjects and train them on using the computer to perform the discrimination task. Each participant was asked to wear a headset and sit in front of a laptop computer for the experimentation phase. Next, the participant initiated the experiment by clicking on the *click to start* button. In all, there were 96 trials. In each trial, the participant listened to two stimuli, and on a laptop screen, two choices appeared: same and different. The participant responded to the trials by clicking on either the *same* or *different* buttons as they heard the stimuli. The stimuli were presented in four blocks of 24 trials each. The subjects could take a short break between blocks; thus, there were three short breaks for the whole experiment. This phase took about 15 min for each subject (Figs. 1, 2,3).

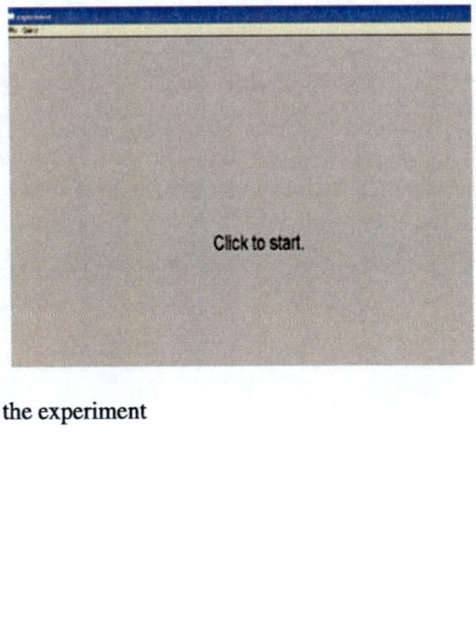

Fig. 1 Pre-start step of the experiment

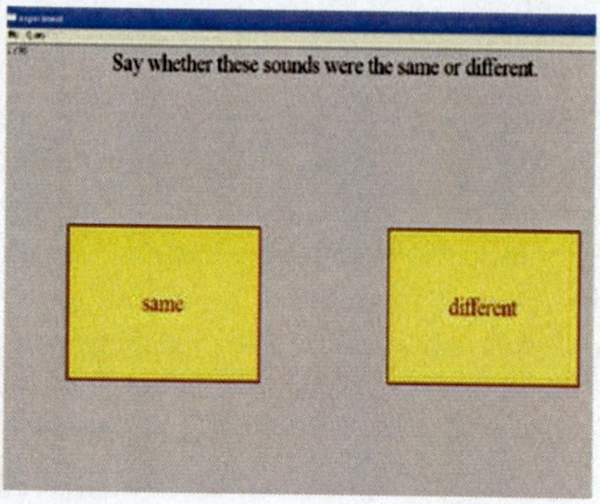

Fig. 2 The start step of the experiment

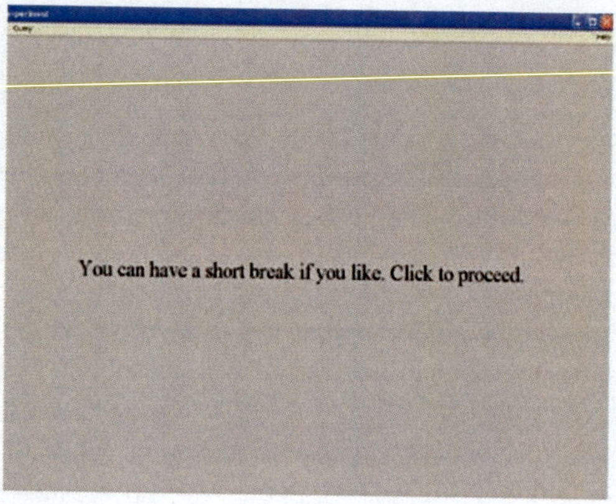

Fig. 3 Inter-experiment interval

6 Data Analysis

All the experiment results were extracted from the PRAAT software and transferred to Excel for scoring purposes. A score of either '0' or '1' was awarded for each trial: 0 for a wrong answer and1for a correct answer. The data was then analysed using the SPSS programme (Statistical Package for Social Science) to measure the perceptual ability of native Yemenis to differentiate English phonemic contrasts and to measure the differences between both groups (LLOR and SLOR Malaysia). To detect the effect of LOR on their perceptibility, we applied an independent sample T-test.

Table 4 Perception of the voicing contrasts by all participants in different contexts

Context	N	Mean	Minimum	Maximum	95% confidence interval for mean		Std. deviation
					Lower bound	Upper bound	
Word Initial	42	66.46	45.83	85.42	63.28	69.64	10.204
Word Final	42	59.42	33.33	83.33	55.69	63.15	11.983

7 Results

7.1 *The Effects of Context (Word-Initial and Word-Final Position) on the Perceptual Ability of Phonemic Contrasts in English*

The result shows clear differences in the mean scores of all participants' perceptions in different contexts. The perception of word-initial position (M = 66.46%, SD = 10.20) was higher than the perception of word-final position (M = 59.42%, SD = 11.98), as presented in Table 4.

7.2 *The Effect of Length of Residence in Malaysia on the Perception of the Selected Phonemic Contrasts in English by Yemenis Learners of English as They are Presented in Different Contexts (e.g., Word-Initial and Word-Final Position)*

The results showed a significant difference in P-value < 0.05 in the perceptual ability of the phonemic contrasts between the two groups of participants. In addition, both groups differed significantly with regard to word-initial and word-final positions (Tables 5 and 6). The independent sample T-test showed a significant effect for context on the perceptual ability of phonemic contrasts in English. The summary of the independent sample T-test is presented in Table 6.

Table 5 Descriptive statistics of participants' performance in different Contexts

Contexts	Groups	N	Mean	Std. deviation	Std. error mean
Word-initial	Group A	21	61.01	9.060	1.977
	Group B	21	71.92	8.2972	1.810
Word-final	Group A	21	53.37	10.384	2.266
	Group B	21	65.47	10.470	2.284

Table 6 Independent sample T-Test

Contexts	Mean differences	Std. error differences	Confidence interval of the difference 95%		t	df	Sig. (2 tailed)
			Lower	Upper			
Word-initial	−10.91	2.680	−16.33	−5.49	−4.07	40	0.001
Word-final	−13.00	3.218	−18.60	−5.59	−3.76	40	0.001

In summary, All participants in the two groups had significantly different perceptual abilities (LLOR and SLOR) in different contexts of phonemic contrasts (word-initial and word-final position). The participants in both groups performed better in the word-initial position than in the word-final position. The results are presented graphically in Fig. 4.

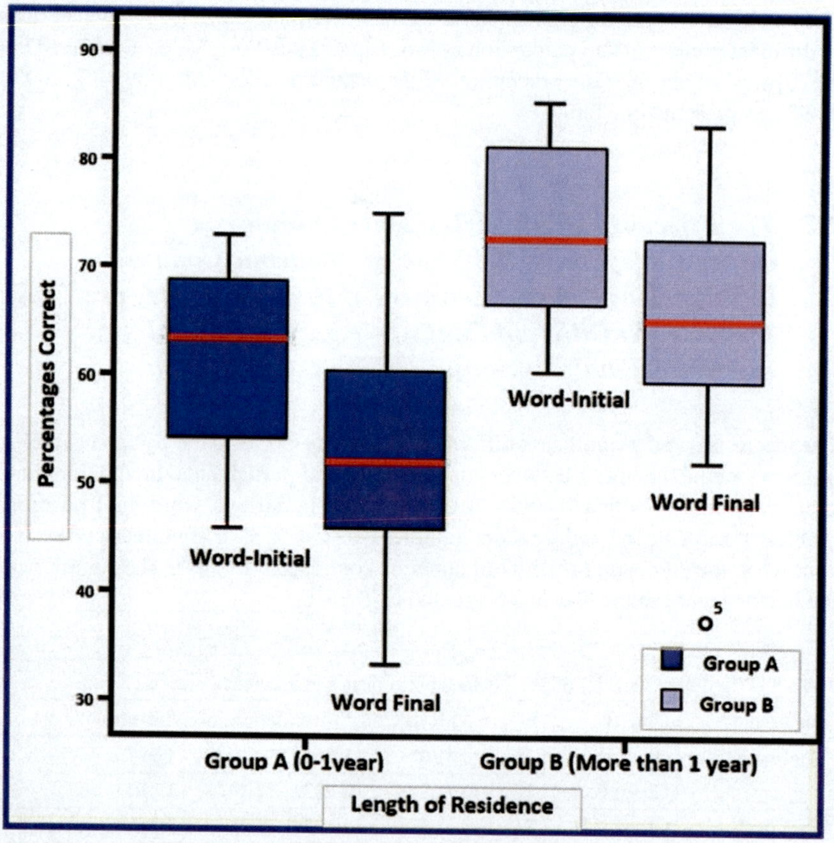

Fig. 4 The perception of phonemic contrasts in word-initial and word final position by all participants and between groups

8 Discussion

The goal of this study was to investigate the influence of different contexts of phonemic contrasts on the perception of the Yemeni EFL learners and to find out whether the LOR in Malaysia has an effect on the perception of the selected phonemic contrasts in English by Yemeni EFL learners as they are presented in different contexts. The results revealed a statistically significant difference in the scores of the two groups of participants on the discrimination test. Scores for the LLOR were higher than scores for the SLOR group. These results suggest that when an individual's LOR increases, their perceptual performance for voicing contrasts increases as well. In other words, increases in the LOR in Malaysia are directly proportional to increases in the participants' mean scores in the discrimination task. This could be because the higher the LOR, the more the exposure to a large amount of L2 input [31].

Clearly, native Yemeni EFL learners can improve their perceptual ability for the L2 sounds absent in their L1 and develop new phonetic categories over time. This finding supports the findings of previous studies [15–19, 32, 33]. Such results can be attributed to the level of integration achieved by Yemeni EFL learners in Malaysia, as was suggested by [32, 34].

Moreover, the results demonstrated that native Yemeni EFL learners faced more difficulties in the perception of English phonemic contrasts that are absent in their L1 in the word-final position compared to that in the word-initial position. This indicates that the context does have an effect on the perceptual level of the phonemic contrasts in English by native Yemeni EFL learners, as the performance of native Yemeni significantly differs when the phonemic contrasts are presented in different contexts. These results are in agreement with previous studies [35, 36]. Ding et al. [35] found that most mandarin students face challenges in perceiving and producing voicing contrasts of word-final stops in English; whereas, Maiunguwa [36] found that, in Hausa EFL learners, the production and perception of /v/, /θ/, and /ð/ in word-initial position were easier than it was in word-final position.

9 Implications for EFL Pedagogy

From the results of this study, Yemenis EFL learners will know that it is not impossible for the learners to acquire L2 new sounds. They will pay more attention to the mismatch between the two languages and try to seize any opportunity for greater exposure to the L2. The results explained that learners who lived in Malaysia for a long time performed better than those who lived in Malaysia for a shorter time. That means, as Yemenis learners were exposed to and practiced the second language; their phonemic categories developed and improved. in view of the results of the study,, the learners should help themselves by finding opportunities for exposure to the L2, and to use the L2 more often than their L1 even when communicating with fellow

L1 speakers. That is because the residence in an L2 country without seizing each opportunity to practice and expose to L2 will not cause any kind of improvement. A list of implications for EFL Pedagogy is stated below:

- L2 teachers should focus on teaching pronunciation and examine the perception of their students to be able to Perceive and communicate effectively in the L2. Teachers can highlight to their students the phonological differences between Arabic and the target language.
- Providing students with virtual interaction with native or native-like speakers of English can provide them exposure to the English language and increase the length of practicing the English language; so, by replacing the absence of the native speaker of English [37].
- Designing remedial activities and exercises concentrating on English pronunciation, listening, exercises of confusing words, and practicing voicing distinctions in the curriculum for the students to practice [8].
- Apply different methods when teaching English as a Foreign language. Furthermore, listening to native speakers on TV and the radio while watching English programs improves listening skills and improves appropriate pronunciation and phonemic perception.

10 Limitations

The limitations of this study are explained below:

- The results of this study will not be generalizable to all Yemenis ESL learners because the participants of this study were restricted to the participants who speak Arabic dialects in south Yemen only. In addition, the results of this study are also not generalizable because of the small sample size; only 42 participants took part in this study.
- The aural stimuli were prepared by using two models of native UK speakers available from the AT&T text-to-speech computer program. Then, the validity of the aural stimuli was tested with a near-native speaker; a Malaysian speaker who can perceive the relevant contrasts in the study. Nevertheless, this study lacks a control group (i.e. native speakers of the UK) to examine the validity of the aural stimuli.
- Studies that have examined the perceptibility of L2 sounds have claimed that age of learning a second language (AOL) and age of arrival (AOA) has a strong effect on the performance and the improvement of the perception ability of cross-language differences [16, 17]. However, this study looked only at the effect of length of residence on the perception of cross-language differences. The effect of AOA and AOL was not tested in this study. So, the difficulties that faced the Yemeni EFL learners in the perception of English phonemic contrasts may not be complete due to the differences in LOR in the two groups. The results would be more convincingly interpreted if the participants' AOA and AOL were also taken into consideration and matched in both groups that varied in LOR.

11 Conclusion and Recommendation

The main aim of this study was to examine the influence of different contexts (word-initial and word-final phonemic contrasts) on the perception of the phonemic contrasts among Yemeni EFL learners and to investigate the effect of LOR of Yemeni EFL learners in Malaysia on the perception of English phonemic contrasts that are absent in Yemeni dialects of interest in this study, i.e., (/p/, /b/), (/f/, /v/), and (/ʧ/, /ʤ/). The findings revealed a significant effect on the improvement of the perception of the phonemic contrasts for LOR in the L2 country: the LLOR participants living in Malaysia performed better than their SLOR counterparts. Moreover, the participants in both groups perceived English phonemic contrasts in word-initial position better than in word-final position.

This study could serve as a step in investigating the cross-language differences between Arabic and English and their effects on the perception–production ability. Additional research is needed to determine the influence of age of arrival and age of L2 acquisition on the development of Yemeni EFL learners' perceptual abilities in distinguishing English phonemic contrasts. Research is also required to look at how native Yemeni speakers perceive and produce English phonemic contrasts in connection to other parameters including language learning age and arrival age. Besides, the present study investigated the perception of English consonants that were absent in the selected Yemeni dialects. So, it will be interesting if further research examined the perception of both English consonants and vowels that do not occur in selected Yemeni dialects and other Yemeni dialects. Further research to be conducted with large sample size is also needed for generalizability.

References

1. Abbad AT (1988) An analysis of communicative competence features in English language texts in Yemen Arab Republic. University of Illinois at Urbana-Champaign, Ann Arbor, p 184
2. Abdul HF (1982) An analysis of syntactic errors in the composition of Jordanian secondary students. Yarmouk University, Jordan
3. Saeed MA, Ghazali K, Aljaberi MA (2018) A review of previous studies on ESL/EFL learners' interactional feedback exchanges in face-to-face and computer-assisted peer review of writing. Int J Educ Technol High Educ 15(1):6. https://doi.org/10.1186/s41239-017-0084-8
4. Al-Jaberi MA, Juni MH, Kadir Shahar H, Ismail SIF, Saeed MA, Ying LP (2020) Effectiveness of an educational intervention in reducing new international postgraduates' acculturative stress in malaysian public universities: protocol for a cluster randomized controlled trial. JMIR Res Protoc. 9(2):e12950. https://doi.org/10.2196/12950
5. Mohammed MAS, Al-Jaberi MA (2021) Google Docs or Microsoft Word? Master's students' engagement with instructor written feedback on academic writing in a cross-cultural setting. Comput Compos 62:102672. https://doi.org/10.1016/j.compcom.2021.102672
6. Musheer A-J, Juni MH, Shahar HK, Ismail SIJMJoM (2019) Acculturative stress and intention to dropout from the university among new postgraduate international student in publicuniversities, Malaysia. Malay J Med Health Sci 15(104)
7. Al HS (2014) Speaking difficulties encountered by young EFL learners. Int J Stud Engl Lang Literat 2(6):22–30

8. Rababah G (2003) Communication problems facing arab learners of English. J Lang Learn 3
9. Wahba EH (1998) Teaching pronunciation--why? In: Language teaching forum: ERIC, p 3
10. Agung A, Laksmi S, Yowani LD (2021) Common Pronunciation Problems of Learners of English. Institut Teknologi Sepuluh Nopember (ITS), Surabaya, Indonesia
11. Stockwell RP, Bowen JD, Martin JW (1965) The grammatical structures of English and Spanish. University of Chicago Press, Chicago
12. Isurin L (2005) Cross linguistic transfer in word order: evidence from L1 forgetting and L2 acquisition. In: Proceedings of the 4th international symposium on bilingualism, p 1130
13. Dufour S, Nguyen N, Frauenfelder UH (2007) The perception of phonemic contrasts in a non-native dialect. J Acoust Soc Am 121(4):EL131–EL6
14. Aoyama K, Flege JE, Guion SG, Akahane-Yamada R, Yamada T (2004) Perceived phonetic dissimilarity and L2 speech learning: the case of Japanese /r/ and English /l/ and /r. J Phon 32(2):233–250. https://doi.org/10.1016/S0095-4470(03)00036-6
15. Flege JE, MacKay IRA (2004) Perceiving vowels in a second language. Stud Second Lang Acquis 26(1):1–34. https://doi.org/10.1017/S0272263104026117
16. Flege JE, MacKay IRA, Meador D (1999) Native Italian speakers' perception and production of English vowels. J Acoust Soc Am 106(5):2973–2987. https://doi.org/10.1121/1.428116
17. Flege JE, Munro MJ, MacKay IRA (1995) Factors affecting strength of perceived foreign accent in a second language. J Acoust Soc Am 97(5):3125–3134. https://doi.org/10.1121/1.413041
18. Hammer Carol S, Komaroff E, Rodriguez Barbara L, Lopez Lisa M, Scarpino Shelley E, Goldstein B (2012) Predicting Spanish-English bilingual children's language abilities. J Speech Lang Hear Res 55(5):1251–1264. https://doi.org/10.1044/1092-4388(2012/11-0016)
19. Bedore LM, PeÑA ED, Griffin ZM, Hixon JG (2016) Effects of age of english exposure, current input/output, and grade on bilingual language performance. J Child Lang 43(3):687–706. https://doi.org/10.1017/S0305000915000811
20. Diab M, Habash N (2007) Arabic dialect processing tutorial. In: Proceedings of the human language technology conference of the NAACL, companion volume: tutorial abstracts, p 5–6
21. Holes C (2004) Modern Arabic: structures, functions, and varieties. Georgetown University Press, Washington
22. Huthaily K (2003) Contrastive phonological analysis of Arabic and English. University of Montana, Montana
23. Watson JC (2007) The phonology and morphology of Arabic. OUP, Oxford
24. Amayreh MM (2003) Completion of the consonant inventory of Arabic. J Speech Lang Hear Res 46(3):517–529. https://doi.org/10.1044/1092-4388(2003/042)
25. Mohammed LA, Yap NT (2010) The effect of length of residence on the perception of english consonants. In: Tan BH, Yong MF, Thai YN (eds) Language learning: challenges, approaches and collaboration. VDM Verlag Dr. Müller, pp 141–67
26. Busa MG (2008) New perspectives in teaching pronunciation
27. Gilakjani AP, Sabouri NB (2014) Role of Iranian EFL teachers about using '"pronunciation power software"' in the instruction of english pronunciation. Engl Lang Teach 7(1):139–148. https://doi.org/10.5539/elt.v7n1p139
28. Tsai P-h (2015) Computer-assisted pronunciation learning in a collaborative context: a case study in Taiwan. Turk Online J Educ Technol-TOJET 14(4):1–13
29. Ary D, Jacobs LC, Irvine CKS, Walker D (2018) Introduction to research in education. In: Cengage learning
30. Tsukada K, Birdsong D, Bialystok E, Mack M, Sung H, Flege J (2005) A developmental study of English vowel production and perception by native Korean adults and children. J Phon 33(3):263–290. https://doi.org/10.1016/j.wocn.2004.10.002
31. Larson-Hall J (2008) Weighing the benefits of studying a foreign language at a younger starting age in a minimal input situation. Second Lang Res 24(1):35–63. https://doi.org/10.1177/0267658307082981
32. Flege JE, Liu S (2001) The effect of experience on adults' acquisition of a second language. Stud Second Lang Acquis 23(4):527–552. https://doi.org/10.1017/S0272263101004041

33. Flege JE, Takagi N, Mann V (1996) Lexical familiarity and English-language experience affect Japanese adults' perception of /ɹ/ and /l/. J Acoust Soc Am 99(2):1161–1173. https://doi.org/10.1121/1.414884
34. Stevens G (2006) The age-length-onset problem in research on second language acquisition among immigrants. Lang Learn 56(4):671–692. https://doi.org/10.1111/j.1467-9922.2006.00392.x
35. Ding H, Zhan Y, Liao S, Yuan J (2015) Production of English stops by Mandarin Chinese learners. In: Proceedings 9th international conference on speech prosody 2018, pp 888–892
36. Maiunguwa A (2015) Perception and production of English fricatives by Hausa speakers. University of Malaya, Kuala Lumpur
37. Rintaningrum R (2016) Maintaining English speaking skill in their homeland through technology: personal experience. Asian EFL J

BMA Approach for University Students' Entrepreneurial Intention

Dam Tri Cuong

Abstract Entrepreneurs have viewed the foundation of the industries because they have given innovative business views that contributed to social and economic development. Besides, today with the expansion of entrepreneurship activities, more and more scholars have concentrated on the study of entrepreneurship. While undergraduate students have regularly considered potential entrepreneurs, entrepreneurial intentions have been the center variable to anticipate the university students' entrepreneurial behavior. The former studies in the literature proposed factors that affected students' entrepreneurial intention with various approaches. Yet, the traditional methods commonly disregarded the uncertainty associated with the selection of models. So, the outcome of model estimates might be biased and pointed to inaccurate inference in analyzing students' entrepreneurial intention. In opposite, the Bayesian model averaging (BMA) approach was also one of the thorough methods for solving model uncertainty, which enabled the assessment of the strength of results to alternative terms by estimating posterior distributions over coefficients and models. Therefore, this paper applied a Bayesian model averaging (BMA) approach to select the best models for university students' entrepreneurial intention. The finding through the BMA approach disclosed that there were the four best models for explaining the association between predictor variables and university students' entrepreneurial intention.

Keywords Bayesian model averaging · Entrepreneurial intention · University students · The entrepreneurial event model

1 Introduction

Entrepreneurship has thought the attention of researchers also policymakers for the latest of many years. The principal reason concerning this attention was the developing demand for entrepreneurs that speeded up economic growth by generating

D. T. Cuong (✉)
Industrial University of Ho Chi Minh City, Ho Chi Minh City, Vietnam
e-mail: damtricuong@iuh.edu.vn

M. Al-Emran et al. (eds.), *International Conference on Information Systems and Intelligent Applications*, Lecture Notes in Networks and Systems 550,
https://doi.org/10.1007/978-3-031-16865-9_24

novel thoughts and transforming them into beneficial investments [1]. Entrepreneurship was also considered the strategy for the development of a significant economy to improve the nation's economy and increased and sustained its competitiveness in challenging the growing tendencies of globalization [2]. Furthermore, entrepreneurs have viewed the foundation of the industries because they have given innovative business views that contributed to social and economic development [3]. Entrepreneurial intentions were considered that urge person to make business. On the other view, Hou et al. [4] said that undergraduate students were considered latent entrepreneurs, and entrepreneurial intentions have been the center predictor to anticipate the entrepreneurial behavior of university students.

The former studies in the literature suggested factors (such as perceived desirability [5], perceived feasibility [5, 6], propensity to act [5], entrepreneurial education [7–9], family background [8, 10], prior work experience [3, 11]) that affected students' entrepreneurial intention with various approaches. These ways generally selected only the best model among feasible choice models depending on a few model choice standards. Nonetheless, the traditional ways commonly disregarded the uncertainty associated with the models' selection. So, the models' estimates of outcomes might be biased and pointed to inaccurate conclusions in analyzing students' entrepreneurial intentions. Subsequently, it was critical to recognize the uncertainty between applicant models, particularly when the models were viewed as feasible despite differences in forecasts [12]. The Bayesian model averaging (BMA) approach was introduced by Draper [13] presented the statistical theoretical foundation concerning explaining the model uncertainty challenge in linear regression models. The BMA was also one of the thorough methods for solving model uncertainty, which enabled the assessment of the strength of results to option terms by estimating posterior distributions through coefficients and models [14]. Besides, the BMA approach had become popular in different fields, like management science, medicine, etc., since it could create more precise and dependable predictions than other methods [12].

Consequently, it was attractive for using the BMA approach to select the best models for university students' entrepreneurial intentions. Thus, this paper proposed a Bayesian model averaging (BMA) approach to select the optimal models for university students' entrepreneurial intention.

2 Literature Review

2.1 The Entrepreneurial Event Model

The entrepreneurial event model was recommended by Shapero and Sokol [15], it was implicitly an intention model, specific to the domain of entrepreneurship [16]. Shapero and Sokol [15] described the interaction of cultural and social factors that could lead to a business formulation by affecting an individual's thoughts. In this sense, the model viewed entrepreneurship as an alternative or possible choice that took place as the outcome of the external change [17]. Besides, among the current entrepreneurial intention models, the entrepreneurial event model by Shapero and Sokol [15] was one of the models that gained widespread consideration [18].

In the entrepreneurial event model, entrepreneurial behavior selection depended on three factors, perceived desirability, perceived feasibility, and propensity to act [18]. Moreover, the former studies suggested factors (such as entrepreneurial education [7–9], family background [8, 10], and prior work experience [3, 11]) that affected students' entrepreneurial intention.

2.2 Entrepreneurial Intention

Entrepreneurial intention originated from intention, which was a crucial notion of psychology. In the literature, there were many interpretations of entrepreneurial intention [18]. Krueger [19] described entrepreneurial intention as the commitment to beginning a new venture. Likewise, Engle et al. [20] argued that entrepreneurial intention was related to the person's intention to begin a new venture [20]. An entrepreneurial intention was also an entrepreneur's viewpoint that aimed to observe, experience, and performance on a particular purpose or method to reach the goal [18, 21]. Entrepreneurial intentions as the investigation and appraisal of information that was gainful to accomplish the target of business creation. The core of entrepreneurship was to must entrepreneurial intentions before beginning the real business since it decided the beginning stage of new business creation. An individual commitment that significantly affected shaping new ventures came from entrepreneurial intentions [3, 22].

3 Methodology

3.1 Analytical Method

In this research, the BMA method with RStudio software was applied to estimate the optimal model. With BMA not only selected the best model among reasonable choice models but could choose many models to explain the dependable variable. Besides, in this study, the independent variables were six factors (perceived desirability, perceived feasibility, propensity to act, entrepreneurship education, family background, and prior work experience), the dependent variable was the students' entrepreneurial intention.

As an analytical method to induce concurrent forecasts, the BMA approach estimated specific predictions depending on their posterior model probabilities, with the higher-performing estimates of models getting the larger weights than the lower performing models. Thus, the BMA approach could produce the averaged model, particularly in situations more than one model had a non-negligible posterior probability [12]. Therefore, let $M = \{M_1, \ldots, M_J\}$ denoted the collection of all models and let y signify the quantity of interest, as the future observed values, and then the posterior distribution of y, given the observed data D was

$$\Pr(y|D) = \sum_{j=1}^{J} \Pr(y|M_j, D)\, \Pr(M_j|D) \tag{1}$$

where:

$\Pr(y|M_j, D)$ was the mean of the posterior distribution of y based on the candidate model M_J, which was the result of the BMA method.

$\Pr(M_j|D)$ was the true prediction model (M_J)' probability which was related to the posterior model probability.

The posterior probability of the model M_J was given by

$$\Pr(M_j|D) = \frac{\Pr(D|M_j)\Pr(M_j)}{\sum_{l=1}^{J} \Pr(D|M_l)\Pr(M_l)}, \tag{2}$$

where

$$\Pr(D|M_j) = \int \Pr(D|\theta_j, M_j)\, \Pr(\theta_j|M_j)\, d\theta_j \tag{3}$$

was the marginal likelihood of the model M_J, θ_j was the vector of parameters of the model M_J, $\Pr(\theta_j|M_j)$ was the prior density of θ_j under model M_J, $\Pr(D|\theta_j, M_j)$ was the likelihood, and $\Pr(M_j)$ was the prior probability that M_j was the true model [23].

The posterior mean and variance of y were shown as follows [24]:

$$E(y|D) = \sum_{j=1}^{J} E(y|D, M_j)\Pr(M_j|D) \tag{4}$$

$$Var(y|D) = \sum_{j=1}^{J} Var(y,|D, M_j) + E(y|D, M_j)^2)\Pr(M_j|D) - E(y|D)^2 \tag{5}$$

3.2 Data and Sample

Data in this study were chosen from undergraduates in Ho Chi Minh City, Vietnam. A five-point Likert scale was employed to evaluate the factors (from 1 = entirely

Table 1 Demographic characteristics of undergraduates

Characteristics	Classifications	Frequency	Percent
Gender	Male	130	44.4
	Female	163	55.6
	Total	293	100
The school year	1st year	41	14.0
	2nd year	101	34.5
	3rd year	85	29.0
	Final year	66	22.5
	Total	293	100

object to 5 = entirely consent). The population sample was gathered by a convenient method through the online survey. The scale including of three indicators of perceived desirability from [25], four indicators of perceived feasibility from [5, 25], four indicators of propensity to act from [25], four items of entrepreneurship education from [26], four items of family background from [27], four items of previous work experience from [28], and three items of entrepreneurship intention from [28, 29].

4 Results

4.1 Descriptive Statistics

After discard of the questionnaire that did not complete information or answered with the same scales, 293 questionnaires were used for the final analysis. The analysis of gender and the school year of students was shown in Table 1.

As described in Table 1, about gender, the sample included 130 male students accounted for 44.4% and 163 female students accounted for 55.6%. Regarding the school year, the population sample consisted of 41 first-year students estimated at 14.0%, 101 s-year students estimated at 34.5%, 85 third-year students estimated at 29.0%, and 66 final-year students estimated at 22.5%.

4.2 Bayesian Model Averaging (BMA)

The select models result for the BMA method was demonstrated in Table 2.

Where

DES: Perceived desirability, FEA: Perceived feasibility, ACT: Propensity to act, EDU: Entrepreneurship education, FAM: Family background, EXP: Prior work experience, nVar: Number of variables, r2: Determination coefficient, BIC: Bayesian information criterion, post prob: posterior probability.

Table 2 Four selected models

	p! = 0	EV	SD	Model 1	Model 2	Model 3	Model 4
Intercept	100.0	−1.09541	0.26248	−1.1160	−0.7133	−1.1190	−1.1180
DES	94.9	0.20812	0.08001	0.2195		0.2190	0.2186
FEA	100.0	0.22476	0.05410	0.2239	0.2420	0.2235	0.2233
ACT	100.0	0.57640	0.07069	0.5753	0.6007	0.5732	0.5735
EDU	100.0	0.29857	0.06913	0.2954	0.3589	0.2936	0.2954
FAM	5.0	0.00029	0.01631			0.5963	
EXP	5.0	0.00024					0.0049
nVar				4	3	5	5
r2				0.606	0.590	0.606	0.606
BIC				−0.025	−0.024	−0.024	−0.024
post prob				0.849	0.051	0.050	0.050

Table 2 listed the posterior effect probabilities (p! = 0), expected values (EV) or posterior means, standard deviations (SD), and the best 4 models by using the BMA approach. The posterior effect probabilities ((p! = 0) implied the regression coefficient probabilities at differing zero (i.e. having an effect and related to the dependent variable). For instance, the posterior affect probability of perceived feasibility was 100 means the perceived feasibility variable appeared 100% in all models. Expected values (EV) or posterior means of the regression coefficient were estimated for all models. Standard deviations (SD) of the regression coefficient were estimated for all models. Model 1 in Table 2 signified the best model in four models; model 2 implied the second-best model after model 1, etc. The best model was evaluated by three indexes (r2, BIC, post prob), in which the post prob index was the most important.

As demonstrated in Table 2, the Bayesian model averaging outcomes of independent variables for university students' entrepreneurial intention. The posterior affect probabilities of three predictor variables (perceived feasibility, propensity to act, and entrepreneurship education) were 100%. This finding illustrated that these three predictor variables had in all the selected models. Besides, the posterior affect probability of the perceived desirability variable was 94.9%. This result revealed that the perceived feasibility variable had 94.9 times happening in all the selected models. Therefore, these four variables were the key factors influencing undergraduate students' entrepreneurial intention.

Moreover, in Table 2, model 1 suggested four variables (nVar = 4) including perceived desirability, perceived feasibility, propensity to act, and entrepreneurship education. These four antecedent variables explained the 60.6% variance of students' entrepreneurial intention. This model 1 had the lowest BIC index (-0.025) when compared with other models. Besides, the posterior probability that the model happened in the analysis was 84.9%, the highest index when compared with other models. Model 2 consisted of three variables (i.e. perceived feasibility, propensity to act, and entrepreneurship education). These three antecedent variables described the

59.0% variance of students' entrepreneurial intention. This model 2 had a low BIC index, but the posterior probability was very low (only 5.1%). Model 3 contained five variables (i.e. perceived desirability, perceived feasibility, propensity to act, entrepreneurship education, and family background). These five predictor variables explained the 60.6% variance of students' entrepreneurial intention. This model 3 also has a low BIC index, but the posterior probability was very low (only 5.0%) because the family background variable had a low affect probability (only 5.0%). Model 4 included five variables (i.e. perceived desirability, perceived feasibility, propensity to act, entrepreneurship education, and prior work experience). These five predictor variables also described the 60.6% variance of students' entrepreneurial intention. This model 4 also had a low BIC index, but the posterior probability was very low (only 5.0%) because the prior work experience variable had a low affect probability (only 5.0%).

When compared with other approaches (e.g. [5, 9, 10]) disclosed, these methods were given only one of the best models among reasonably selected models. Purwana [5] using SEM (structural equation modeling) showed that the best model comprised three predictor variables (perceived desirability, perceived feasibility, and perceived propensity to act) and one explained variable (students' entrepreneurial intention). Kabir et al. [9] by applying PLS-SEM suggested only the best model included four independents (attitude, subjective norm, entrepreneurial education, self-efficacy) and one explained variable (entrepreneurial intention). Joseph [10] conducted multiple regression analysis, and also identified only the accurate model, including four predictors (need for achievement, subjective norm, entrepreneurship education, economic situation) and one dependent (entrepreneurial intention). In contrast, as explained before, the BMA method identified the four best models.

5 Conclusions

This research applied the BMA approach to select the optimal models for undergraduate students' entrepreneurial intention. The results selected the best 4 models. Model 1 included four predictor variables (perceived desirability, perceived feasibility, propensity to act, and entrepreneurship education) and a dependent variable (students' entrepreneurial intention). Model 2 contained three independent variables (perceived feasibility, propensity to act, and entrepreneurship education) and a dependent variable (students' entrepreneurial intention). Model 3 consisted of five independent variables (perceived desirability, perceived feasibility, propensity to act, entrepreneurship education, and family background) and a dependent variable (students' entrepreneurial intention). Model 4 also consisted of five independent variables (perceived desirability, perceived feasibility, propensity to act, entrepreneurship education, and prior work experience) and a dependent variable (students' entrepreneurial intention). The findings also revealed that these results were in line with reality in real life. This was reasonable because we did not have only one option, practically we could have equivalent selections. The Bayesian approach could provide us with opportunities for thinking and evaluating the model's uncertainty. Besides, the findings also identified the four key determinants (perceived desirability, perceived feasibility, propensity to act, and entrepreneurship education) that influenced undergraduate students' entrepreneurial intention. Therefore, these findings also helped the education managers have some options for redesigning the curriculum and the content of the program entrepreneurship at universities. Likewise, entrepreneurship should be a compulsory subject at universities, especially in business schools, since entrepreneurship subjects will be the path to entrepreneurship behavior and becoming a future entrepreneur.

However, this research was conducted at one university in Ho Chi Minh city of Vietnam. So it was not representative of all universities and cities in Vietnam. Therefore, future research should continue to test these findings at other universities in Ho Chi Minh city as well as in other cities in Vietnam.

References

1. Turker D, Selcuk SS (2009) Which factors affect entrepreneurial intention of university students? J Eur Ind Train 33:142–159. https://doi.org/10.1108/03090590910939049
2. Shamsudin SFF, Mamun A, Nawi NB, Nasir NAB, Zakaria MN (2017) Factors affecting entrepreneurial intention among the Malaysian University students. J Dev Areas 51:423–431. https://doi.org/10.1353/jda.2017.0111
3. Israr M, Saleem M (2018) Entrepreneurial intentions among university students in Italy. J Glob Entrep Res 8:1–14. https://doi.org/10.1186/s40497-018-0107-5
4. Hou F, Su Y, Lu M, Qi M (2019) Model of the entrepreneurial intention of university students in the Pearl River Delta of China. Front Psychol 10:916. https://doi.org/10.3389/fpsyg.2019.00916
5. Purwana D (2018) Determinant factors of students' entrepreneurial intention: a comparative study. Din Pendidik 13:1–13. https://doi.org/10.15294/dp.v13i1.12971
6. Liñán F, Rodríguez-Cohard JC, Rueda-Cantuche JM (2011) Factors affecting entrepreneurial intention levels: a role for education. Int Entrep Manag J 7:195–218. https://doi.org/10.1007/s11365-010-0154-z
7. Hassan A, Saleem I, Anwar I, Hussain SA (2020) Entrepreneurial intention of Indian university students: the role of opportunity recognition and entrepreneurship education. Educ Train 62:843–861. https://doi.org/10.1108/ET-02-2020-0033
8. Looi KH, Khoo-Lattimore C (2015) Undergraduate students' entrepreneurial intention: born or made? Int J Entrep Small Bus 26:1–20. https://doi.org/10.1504/IJESB.2015.071317
9. Kabir S, Ahasanul H, Sarwar A (2017) Factors affecting the intention to become an entrepreneur: a study from bangladeshi business graduates perspective. Int J Eng Inf Syst 1:10–19
10. Joseph I (2017) Factors influencing international student entrepreneurial intention in Malaysia. Am J Ind Bus Manag 07:424–428. https://doi.org/10.4236/ajibm.2017.74030
11. Masoomi E, Zamani N, Bazrafkan K, Akbari M (2016) An investigation of the factors influencing entrepreneurial intention of senior agricultural students at Shiraz University. Int J Agric Manag Dev 6:431–437
12. Li G, Shi J (2010) Application of Bayesian model averaging in modeling long-term wind speed distributions. Renew Energy 35:1192–1202. https://doi.org/10.1016/j.renene.2009.09.003
13. Draper D (1995) Assessment and propagation of model uncertainty. J R Stat Soc Ser B 57:45–70. https://doi.org/10.1111/j.2517-6161.1995.tb02015.x

14. Montgomery JM, Nyhan B (2010) Bayesian model averaging: theoretical developments and practical applications. Polit Anal 18:245–270. https://doi.org/10.1093/pan/mpq001
15. Shapero A, Sokol L (1982) The social dimensions of entrepreneurship. In: Kent CA, Sexton DL, Vesper KH (eds) Encyclopedia of entrepreneurship. Prentice-Hall, Englewood Cliffs, pp 72–90
16. Krueger NF, Reilly MD, Carsrud AL (2000) Competing models of entrepreneurial intentions. J Bus Ventur 15:411–432. https://doi.org/10.1016/S0883-9026(98)00033-0
17. Miralles F, Riverola C, Giones F (2012) Analysing nascent entrepreneurs' behaviour through intention-based models. In: Proceedings of the 7th European conference on innovation and entrepreneurship, vols 1 and 2. https://doi.org/10.13140/2.1.4595.6161
18. Lu G, Song Y, Pan B (2021) How university entrepreneurship support affects college students' entrepreneurial intentions: an empirical analysis from China. Sustainability 13:3224. https://doi.org/10.3390/su13063224
19. Krueger N (1993) The impact of prior entrepreneurial exposure on perceptions of new venture feasibility and desirability. Entrep Theory Pract 18:5–21. https://doi.org/10.1177/104225879301800101
20. Engle RL, Dimitriadi N, Gavidia JV, Schlaegel C, Delanoe S, Alvarado I, He X, Buame S, Wolff B (2010) Entrepreneurial intent: a twelve-country evaluation of Ajzen's model of planned behavior. Team Perform Manag 16:35–57. https://doi.org/10.1108/13552551011020063
21. Brid B (1988) Implementing entrepreneurial ideas: the case for intention. Acad Manag Rev 13:442–453. https://doi.org/10.1177/0896920511399938
22. Choo S, Wong M (2006) Entrepreneurial intention: Triggers and barriers to new venture creations in Singapore. Singapore Manag Rev 28:47–64
23. Raftery AE, Madigan D, Hoeting JA (1997) Bayesian model averaging for linear regression models. J Am Stat Assoc 92:179–191. https://doi.org/10.1080/01621459.1997.10473615
24. Zou Y, Lin B, Yang X, Wu L, Muneeb Abid M, Tang J (2021) Application of the bayesian model averaging in analyzing freeway traffic incident clearance time for emergency management. J Adv Transp 2021:1–9. https://doi.org/10.1155/2021/6671983
25. Lepoutre J, Van den Berghe W, Tilleuil O, Crijns H (2011) A new approach to testing the effects of entrepreneurship education among secondary school pupils. In: Entrepreneurship, growth and economic development, pp 94–117. https://doi.org/10.4337/9780857934901.00010
26. Walter SG, Block JH (2016) Outcomes of entrepreneurship education: an institutional perspective. J Bus Ventur 31:216–233. https://doi.org/10.1016/j.jbusvent.2015.10.003
27. Van Auken H, Fry FL, Stephens P (2006) The influence of role models on entrepreneurial intentions. J Dev Entrep 11:157–167. https://doi.org/10.1108/et-09-2018-0194
28. Miralles F, Giones F, Riverola C (2016) Evaluating the impact of prior experience in entrepreneurial intention. Int Entrep Manag J 12:791–813. https://doi.org/10.1007/s11365-015-0365-4
29. Robledo JLR, Arán MV, Martin-Sanchez V, Molina MÁR (2015) The moderating role of gender on entrepreneurial intentions: a TPB perspective. Intang Cap 11:92–117. https://doi.org/10.3926/ic.557

A Systematic Review of Knowledge Management Integration in Higher Educational Institution with an Emphasis on a Blended Learning Environment

Samar Ibrahim and Khaled Shaalan

Abstract A knowledge management process is a collection of practices that can work effectively to benefit academicians and foster innovation at Higher Education Institutions (HEI). With the advancement in Information, Communication, and Technology (ICT) capabilities, these institutions are presented with opportunities as well as challenges to keep up with knowledge management. Together with the emergence of Learning Management Systems (LMS), institutions have an unprecedented opportunity to facilitate and improve the quality of teaching–learning resources. Many researchers have investigated the Knowledge management process integration and its implementation in HEIs. Some have also examined the benefits of LMS implementation, the barriers, and underutilization. In addition, researchers are interested in analyzing the best teaching and content delivery methods associated with LMS implementation, such as blended learning environments that integrate online and face-to-face delivery methods. This systematic review investigates knowledge management integration in HEIs and emphasizes the blended learning environment by examining the implementation of the learning management system. The review analyzes 16 studies collected from different databases between 2012 and 2021 dealing with knowledge sharing, and to a lesser extent, with knowledge creation and knowledge acquisition. A key finding of the review was that the knowledge management process could enhance an institution's ability to innovate. Through KM and LMS, an institution can transform the traditional face-to-face environment into a blended, innovative, convenient, flexible, and student-centric mode of delivery, which leads to organizational and stakeholder performance improvements. Unfortunately, the review also identified that implementation to date is not as effective as it should be.

K. Shaalan
British University in Dubai, Dubai, UAE
e-mail: khaled.shaalan@buid.ac.ae

S. Ibrahim (✉)
School of Art and Science, American University in Dubai, Dubai, UAE
e-mail: sibrahim@adjunct.aud.edu

M. Al-Emran et al. (eds.), *International Conference on Information Systems and Intelligent Applications*, Lecture Notes in Networks and Systems 550,
https://doi.org/10.1007/978-3-031-16865-9_25

Keywords Knowledge management · Higher education institutions · Knowledge management process · Knowledge management system · Learning management system

1 Introduction

Knowledge can be acquired or achieved through learning or practice [1]. Higher Educational Institutions (HEI) have an essential role to play in creating, transferring, sharing, and distributing knowledge and making it accessible to their communities and societies [2]. The process of acquiring, transferring, and managing knowledge creates opportunities as well as challenges for HEI to compete and keep pace with the continuous global changes and technological development in today's world. Such a process is critical to endure and ensure success [3]. Furthermore, for success to be sustained and maintained, knowledge needs to be continuously and adequately managed in an organization to achieve the vision, mission, and objectives [1]. The knowledge management process can be perceived as a collection of practices that can effectively benefit academicians and promote innovation for these institutions [4]. In the academic context, knowledge is generated by human resources, which are established through educational and research behaviors and practices. Hence, the HEI knowledge management process would be divided into academic and organization generated by faculties, students, administrators, and researchers. As such, for knowledge to be created, transferred, and shared effectively, this would depend on three factors: Human assets, management process, and well implemented and advanced technology [5, 27]

In addition to advanced technology, A successful knowledge management process requires HEI to be innovative and adopt clear policies and strategies. This would enhance the possibility for knowledge to be created through research, shared through teaching and learning, and transferred through communication [6]. Furthermore, the emergence of the digital-native generation and a broader internet have created an unprecedented foundation for further advancement that can be efficiently used as a knowledge resource implemented in research and education toward more competitive and innovative HEIs [7]. Information technology's progression enables knowledge sharing to become more feasible, which is further enhanced through collaborative research, where institutions develop a knowledge management system to acquire, create, share and organize knowledge. In essence, the advancing capabilities of ICT allow Learning Management Systems (LMS) to facilitate and improve the quality of teaching–learning resources [8].

Many researchers are interested in investigating knowledge management process integration and its implementation in HEIs, studying the benefit of implementing LMS, the barriers, and the underutilization in some cases [9]. In addition, researchers are also interested in exploring the various and best modes of teaching and different content delivery methods associated with LMS implementation, such as blended learning environments that integrate digital and face-to-face delivery methods [10].

They are studying different aspects of these environments and their association with the knowledge management practices and the LMS implementation towards a blended learning innovation environment; In which the educational approaches implemented in these HEI are more in line with creative and critical thinking approaches that adopt sharing and self-reflective educational methods [11].

This study aims to conduct a systematic review of Knowledge Management (KM) integration in HEIs, with an emphasis on a blended learning environment. Although many previous studies have developed various systematic reviews on similar areas, most previous reviews covered different scopes, locations, or objectives [12] or are in a different time range [2]. Below are the research questions that this study is intended to answer.

RQ1: What are the main KM processes implemented in HEI? What is the impact of this implementation?
RQ2: What is the distribution of the selected studies across the countries?
RQ3: What are the main research methods used in the selected studies, and which databases were involved in publishing these studies?
RQ4: What is the relation between the KM process and innovation in HEI?
RQ5: What is the impact of a blended learning environment with LMS implementation on academic performance in HEI?
RQ6: What framework can implement LMS in a more blended learning environment in HEI?

Section 2 captures the problem identification and its analysis. While Sect. 3 explains the methodological approach implemented in the systematic review that analyses Knowledge Management (KM) integration in HEIs with an emphasis on a Blended Learning (BL) environment. Section 4 summarizes the literature reviews' results. Finally, Sect. 5 illustrates the results, followed by Sect. 6, which covers discussion and conclusions review, and suggestions for future research.

2 Problem Identification

For the Knowledge management process, integration is a valuable and complex process that needs to be investigated and surveyed its implementation in higher educational institutions. Researchers noted that the most significant barriers to implementing KMP in HEI are the lack of a KM defined strategies and institutional approach to KM in general and in particular to LMS.

Due to the growing capabilities of ICT, learning management systems can be used to facilitate and enhance teaching–learning resources and create a more blended collaborative environment. However, the majority of these LMSs tools are mainly used as administrative and content distribution tools rather than effective systems for enhancing teaching and learning and creating an innovative blended environment. Therefore, a systematic review is conducted on KM integration in HEI and examines

LMS implementation in HEI to explore strategies and approaches to achieve a more blended innovative learning environment in HEI.

3 Literature Review

3.1 *Knowledge Management and Knowledge Management Process*

Knowledge is a synonym for information. It can take different forms such as ideas, opinions, values, facts and skills acquired through experience or education, and many other types [13]. In an organizational context, knowledge is the corporate assets owned by its member. It comprises the practical experience with critical and creative abilities for this organization to be innovative, competitive, and sustainable [13]. Knowledge can be represented in two different forms: tacit knowledge related to the human mind's perception versus explicit knowledge that can be seen [2, 14]. In order to gain new knowledge, individuals need to communicate and share their forms of knowledge with others [7]. In addition, knowledge management processes (KMP) are required to enable sharing. These KMPs include identifying, creating, transferring, processing, interpreting, storing, and sharing knowledge across an organization [13]. Today many organizations use well-managed knowledge to attain their goal and lead in their domain to achieve organizational innovation and compete globally [3].

3.2 *Knowledge Management Process in Higher Educational Institution*

HEIs have always been dealing with Knowledge Management, research, education, and service to their society inherent in their missions [15]. At the heart of the HEI mission are Knowledge Creation, Knowledge Dissemination, and Knowledge transfer [4]. Knowledge Creation is the elaboration of new knowledge, and as such, HEI focuses on expanding the boundaries of their knowledge through research activities, publications, and scientific discovery [6]. Knowledge Sharing occurs in HEI through seminars, conferences, and publications supported by culture and environment to foster knowledge sharing [13]. And Knowledge transfer is achieved through activities of teaching and learning, as well as sharing such knowledge with the public and organizations across different industries. Accordingly, HEI builds reputation and recognition through disseminating knowledge created by researchers to other stakeholders [4].

HEI has three objectives for the Knowledge process: to develop tasks with improved quality and efficiency, then to develop human resources at all levels of the organization, and finally, to develop knowledge bases in sectors to maximize their knowledge investment [13]. In these Institutions, the knowledge management process can be elaborated by performing various human tasks to improve teaching, evaluation, counseling, research, and all administration function [16]. Furthermore, the KM process is crucial for higher educational institutions' success, enabling them to perform more effectively and efficiently and improve their quality and competitiveness [17]. Therefore, HEI must develop strategies to transform tacit knowledge into explicit one to maximize the benefit of its intellectual assets. In addition, HEIs need to develop strategies and policies that encourage knowledge management practices [16].

In contrast, the absence of such KM strategies is one of the critical barriers to KMP implementation [16]. For example, a study by Hawkins shows that the KM process integration in HEIs is very limited, and it is only implemented by librarians [18]. Instead, what is required for an effective organizational KMP, is the efficient integration of all resources that incorporate human resources, management resources, and technological resources. Only then can HEI improve the existence of the KM processes and encourage and exchange information among all stakeholders [1].

3.3 *Towards a Blended Learning Environment*

Information and communication technologies (ICT) have developed rapidly in recent years, offering HEIs the opportunity to adapt to this advancement and benefit [19]. Blended learning combines face-to-face and online learning primarily conducted through a learning management system (LMS) and other web-based learning modes [20]. As part of a blended learning environment, LMSs can be seen as integrating collaborative and critical interactive platforms for various learning activities [21]. LMSs have gained popularity and have allowed the possibility to blend a learning environment supported by great learning and teaching resources, where lecturers can act more as facilitators or moderators and learners receive more feedback [9]. LMS is used as an optimization feedback-like process to improve blended learning effectiveness in such an environment. A standard LMS incorporates mediators within an interactive learning environment, enabling tools for managing inter/intra-action, coordination, and cooperation between learners [21]. More importantly, researchers advocate that LMSs become more adaptable and responsive and advance instructional and learning practices [22]. Unfortunately, some LMSs are used to distribute information and facilitate administration instead of ameliorating teaching and learning [21]. Many studies examine the different barriers related to technology or an institution that prevent reaching this objective, such as staff development, policy, and administrative support [11]. One such example is unfortunate evidence that faculty members underuse LMS tools for various reasons, including but not limited to resistance to change, time management, and training requirements [23]. Hence, the criticality to transform educators into "digitally literate" [11].

4 Research Methodology

For any progression in research development, a detailed critical review is needed to lay and create a foundation for any possible development or expansions, to foresee any issue, and reveal any hidden research areas and challenges. That would be in addition to presenting a complete view of certain research areas with all the latest and the critical updates in this field [24]. This systematic review follows the review guideline, general strategy, and general protocol suggested by [2] and [24]. In addition, the systematic review is conducted on the integration of Knowledge Management in Higher Educational Institutions with an emphasis on a Blended Learning environment.

This review was conducted in four different steps that include:

- An identification of the inclusion and exclusion criteria,
- Clear identification of the data sources
- Search strategies for selecting the articles, and
- Finally, data coding and analysis are used to analyze and summarize the results.

These steps are elaborated in the following sections:

4.1 Inclusion-Exclusion Criteria

The articles are analyzed and selected according to the inclusion-exclusion presented in (Table 1):

Table 1 Inclusion and exclusion criteria

Inclusion criteria	Exclusion criteria
Studies that discuss the KM, KM process	Studies that are not related to Knowledge management, or Learning Management system
Studies must be in English Available studies Limiting to Journals	Studies Not in English
Studies must be between 2012 and 2021	
Studies that discuss traditional or blended learning	

4.2 Data Source and Data Extraction

The articles were selected according to a vast and extensive range of searches that were done against various databases, such as Emerald, ACM Digital Library, Google Scholar, and Scopus.

Search Strategies/Search Keywords. Articles selected in this systematic review were chosen and narrowed down according to the following keywords or a combination of these keywords:

- "Knowledge management" and "Higher Educational Institutions"
- "Higher Educational Institutions" and "Knowledge management "
- "Learning management system" and "Higher Educational Institutions."
- "Higher Educational Institutions" and "Learning management system"
- "Learning management system" and "Blended learning environment."
- "Blended learning environment" and "learning Management system"
- "Blended learning environment" and "Knowledge management"
- "Blended learning environment" and "Higher Educational Institutions"

A total of 325 articles were retrieved by applying the above search keywords, of which 70 articles were duplicates. By applying the inclusion and exclusion criteria for each article, the analysis process ended with 16 articles. The search and analysis were performed according to the Preferred Reporting Items for Systematic Reviews and Meta-Analysis (PRISMA) [29]. Figure 1 is a presentation of the PRISMA flowchart.

4.3 *Quality* Assessment

Blackboard's The quality assessment is a critical and valuable type of appraisal implemented along with the inclusion–exclusion criteria. For this systematic review, A checklist of six quality assessment questions was designed to evaluate the quality of selected research articles, as shown in Table 3.

Each checklist question is given a score on a three-point scale: 1 for "yes," "0" for "No," and 0.5 for "partially." As such, each article will have total score between 0 and 6. Then the result of the assessment shows that all the articles passed and qualified for more assessment as shown in Table 2.

4.4 Data Analysis and Coding

The study will also analyze and code all features related to the research methodology and the method types used in the selected studies. In addition, the review will examine where and in which field the study is conducted. There was a formal approach to confirm the selected studies and exclude studies that do not clearly describe HEI

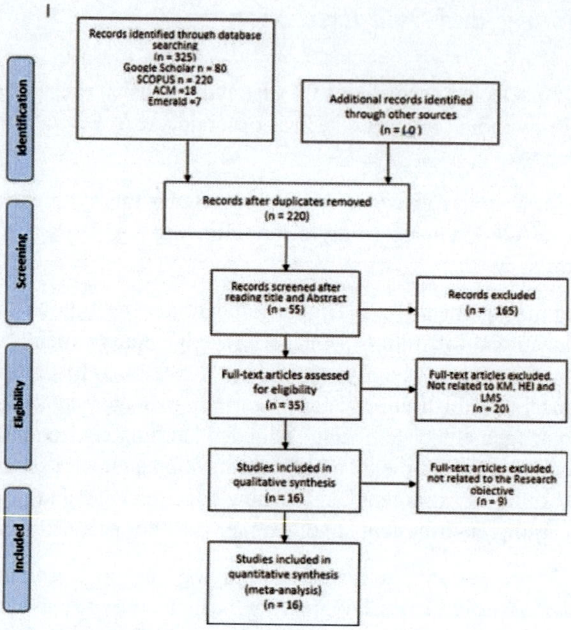

Fig. 1 Process of selected papers

Table 2 Quality appraisal checklist

1-Are the research aims clearly identified?
2-Are the KM processes integrated by the study clearly identified?
3-How suitable are the methods and the analysis?
4- How relevant is the main aim of the study to our Study?
5-Are the studies' results adding value to the literature?
6-Are the objective of LMS implementation clearly identified?

knowledge management integration nor emphasize a blended learning environment. An analysis of the selected studies is conducted in detail in the following sections. Appendix A provides a more comprehensive codebook, which includes all attributes along with the assessment coding; Appendix B presents the journal ranking, the number of citations, and the impact rate of the journal.

5 Results and Discussion

This systematic review analyzed sixteen studies between 2012 and 2021 that were selected and filtered according to the strategies mentioned above. As a result, it is

Table 3 Quality appraisal results

Study	Q1	Q2	Q3	Q4	Q5	Q6	Total	%
S1	1	0.5	1	0.5	1	1	5	83.33
S2	1	1	1	1	1	1	6	100
S3	1	1	1	0.5	0.5	1	5	83.33
S4	1	0	1	1	1	0.5	4.5	75
S5	1	1	1	0.5	1	1	5.5	91.66
S6	1	1	1	1	1	1	6	100
S7	1	1	1	1	1	1	6	100
S8	1	1	0.5	0.5	1	0.5	4	75
S9	1	1	1	1	1	0.5	5.5	91.66
S10	1	1	1	0.5	1	0	4.5	75
S11	1	1	1	1	1	0	5	83.33
S12	1	1	1	1	1	0	5	83.33
S13	1	1	0.5	1	1	0	4.5	75
S14	1	1	1	0.5	0.5	1	5	83.33
S15	1	1	1	0.5	1	0	4.5	75
S16	1	1	1	1	1	0.5	5.5	91.66

evident that research on knowledge management is prevalent and is progressing. Therefore, the findings are based on the research questions presented in this section.

Figure 2 below presents the number of studies per year, showing that most selected studies are between 2012 and 2021. While Fig. 3 depicts the vox views presenting the five different clusters/ keywords examined in these studies: knowledge management, higher educational institution, LMS, academic performance, and study. This confirms the strong relationship between Knowledge management and higher educational institution, as well as academic performance and LMS. Other predominant keywords used in these studies include innovation, cloud computing, employee empowerment, research, sharing, and effectiveness.

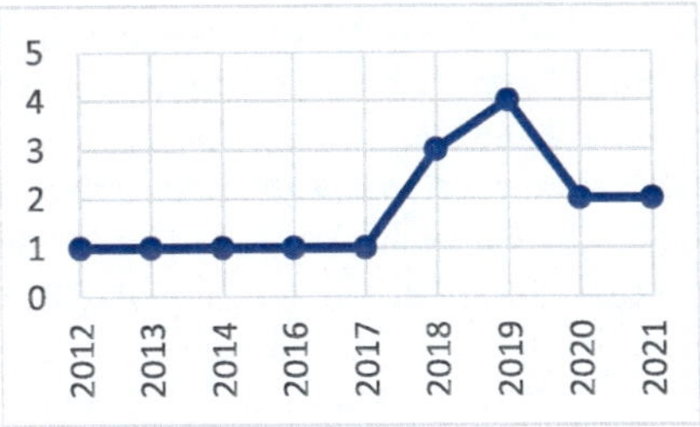

Fig. 2 Total Number of studies per year

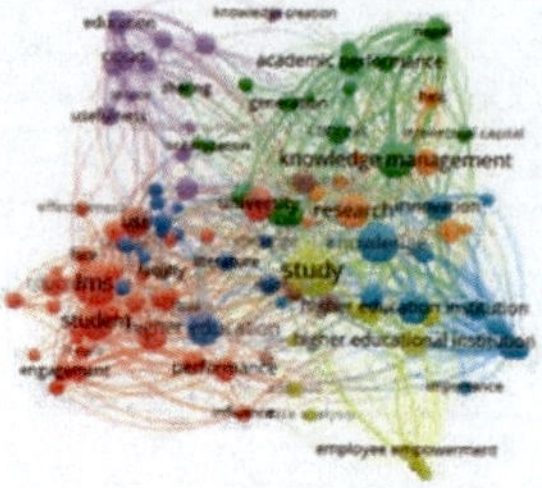

Fig. 3 Vos view visualization

5.1 RQ1: What Are the Main KM Processes Implemented in HEI, and What is the Impact of This Implementation

The main KM processes are presented in Table 4 below, which shows that not all studies discuss the various KM processes. Instead, the focus was mainly on sharing knowledge and less on other KM processes such as creation, acquisition and transfer, and application, and the least was on knowledge storage. In addition, many scholars investigate the impact of these processes on HEI. For example, Asiedu et al. (2020) debate that sharing knowledge activities when it expands between all institutional levels, i.e., departmental and faculty, would lead to a collaborative environment of sharing resources that enhance creativity. He emphasizes that Knowledge is only valuable if it is shared and integrated among all institutional levels [5]. Knowledge-sharing between faculty members allows institutions to exploit and capitalize on knowledge-based resources [25]. However, knowledge sharing can only happen within an institution with an open culture, nurturing teamwork, networking, and collaboration [16]. Other studies argue that a culture of implicit knowledge sharing exists, which could strengthen the research capacity in these institutions. And despite that it is more individualistic, there is a prevalent protective culture of knowledge assets [6]. The results indicated that institutions must develop policies to manage and share knowledge effectively [1].

Knowledge creation in HEI is the most critical factor in ensuring the survival of organizations and institutions. This, in return, would empower human capital [13]. According to Paudel et al. (2021), there is also a strong correlation between different aspects of the knowledge management process and faculty and academic performance in HEI. Besides, knowledge acquisition, utilization, and application can enhance innovation and performance within an organization [3].

Veer-Ramjeawon et al. (2019) emphasize the effect of transfer and application of knowledge occurring from universities to industry and the public sector, with all concepts of creating jobs, doing consulting, and the idea of continuing professional development. Furthermore, developing knowledge and creating knowledge are key aspects of academic excellence in the educational world, particularly in the areas of research and publishing [4]. In summary, the knowledge management process critically impacts HEI academic performance and organization performance.

Table 4 Main KMP applied in the selected Studies

Source	Knowledge creation	Knowledge acquisition	Knowledge sharing	Knowledge transfer	Knowledge storage	Knowledge application
S1			✓			✓
S2	✓	✓	✓			
S3						✓
S4						
S5	✓	✓	✓			
S6						
S7						
S8	✓	✓	✓		✓	✓
S9	✓	✓	✓	✓		✓
S10	✓	✓	✓			✓
S11	✓	✓	✓		✓	✓
S12	✓	✓	✓			
S13	✓	✓	✓	✓		✓
S14						
S15	✓	✓	✓			✓
S16	✓	✓	✓			

5.2 *RQ2: What is the Distribution of Studies Across the Countries?*

The following graph (Fig. 4) shows the distribution of the collected articles across the countries where these studies were conducted. As shown in the figure, Malaysia is the leading country in research on KM and HEI ($N = 3$), followed by Pakistan and Iran ($N = 2$). All the other countries are equally distributed: South Asia, Ghana, UK, US, Mauritius, Nepal, Pakistan, South Asia, UK, US, Vietnam. It is essential to investigate the location of the studies; it will provide us with a clear idea of where these studies originated and which countries have an interest in HEI research.

5.3 *RQ3: What Are the Main Research Methods Used in the Selected Studies, and Which Databases Were Involved in Publishing These Studies?*

Half of the selected studies were conducted using a survey method that used a questionnaire type of research. In contrast, the other half articles adopt different methods such as Survey/interview, Systematic review, and quantitative analysis study. This analysis shows that these studies are conducted using more qualitative analysis.

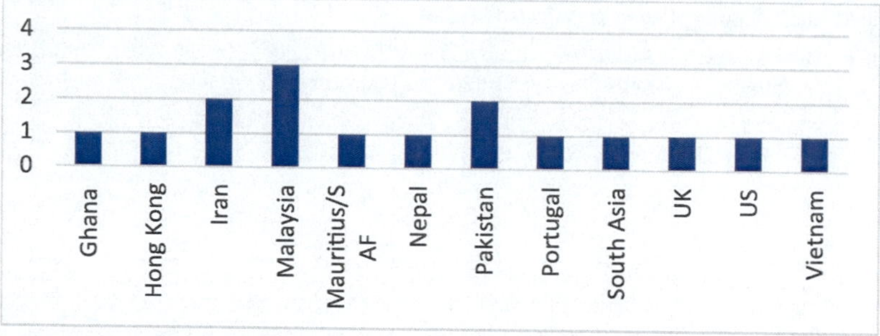

Fig. 4 Number of studies on HEI used in the SR by country

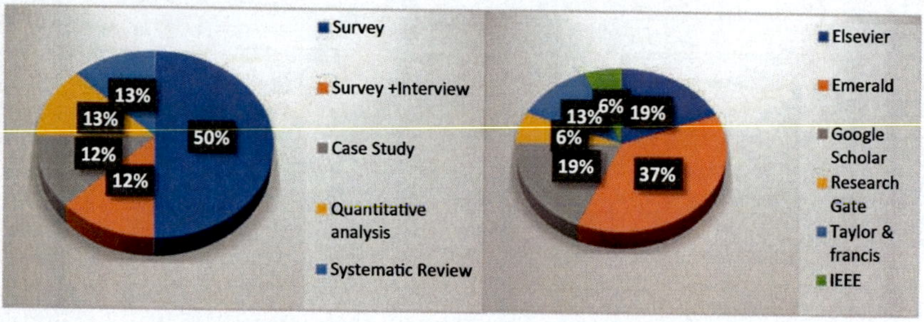

Fig. 5 2 pies display the different implementing research methods used by the studies covered in this review and the database to which these studies belong

Among the 16 papers that were reviewed, Emerald is the most research database used, where 8 of the 16 were sourced—followed by Elsevier and Google Scholar, Research Gate, and Taylor and Francis (Fig. 5).

5.4 *RQ4: What is the Relation Between the KM Process and Innovation in HEI?*

Innovation is necessary for organizations to improve their performance continuously. Several scholars have contemplated that the knowledge management process can enhance an institution's ability to innovate [13]. More so, many researchers show that KMP has emerged as a crucial trend for innovation in organizational practice [7]. It is a mechanism that addresses the complexities of innovation by helping in managing new and existing knowledge throughout the innovation process [5]. Individuals' innovative approaches are directly impacted by the knowledge creation of KMP [4]. The sustainability of higher education institutions depends on continuous improvement and innovation in curricula and services [5]. The innovation performance of these institutions can be represented by the way they continuously look for potential new ideas. It is mainly driven by their central functions of teaching and research [7]. In addition, HEI's organizational performance is strongly associated with innovation [7]. As part of HEI's continuous creation processes, knowledge management is similarly expected to enhance resources sharing.

Moreover, with improved innovation, knowledge-sharing has become a key contributor to helping HEI solve their problems through more innovative solutions [4]. Furthermore, Arpaci et al., [26] have investigated a cross-cultural analysis of the effects of knowledge management (KM) approaches on accepting Massive Open Online Courses (MOOCs). The study shows that KM practices, such as knowledge access, knowledge storage, knowledge application, and knowledge sharing can substantially affect the perceived usefulness of MOOCs.

5.5 *RQ5: What is the Impact of a Blended Learning Environment with LMS Implementation on Academic Performance in HEI?*

Few researchers have explored the blended learning environment and its substantial impact on academic performance in HEIs [11]. Evan et al. (2020) explains that achieving organizational performance can result from the unprecedented enhancements taking the traditional face-to-face environment through a learning management system in a convenient, flexible, and student-centered way to a whole new level of teaching and learning. Hence through LMS platforms, instructors can deliver a wide range of educational new innovative, and distinguished experiences. Thus, LMS can elevate the traditional setting to a more collaborative and interactive mode, creating a blended environment (Rhode et al., 2017). Another scholar emphasizes that LMS systems have been exhibited to offer faculties and students many tools and numerous methods for engaging in active teaching–learning, as well as improving the overall academic performance [20]. Furthermore Arpaci, (2017b) in his study explains that a Learning Management System is an integrated set of software that administers, tracks, reports, documents, and delivers online distance learning courses and blended learning. According to Rhode et al. (2017), a study conducted on student LMS practice, perseverance, and course achievement in a hybrid course indicated that the LMS data could provide a signal of students' academic performance. In addition, many studies suggested that it can be useful to implement IT-based KM intervention in HEIs to enhance the performance of areas of research and administrative services [1].

However, many researchers investigating the blended learning environment and LMS implementation have limited relation with the KMP. This is revealed in Table 4, which describes the KMP included in the studies. In Summary, if more strategies and planning and if KMP were critically better integrated into HEI, LMS would allow and achieve a more blended innovative learning environment.

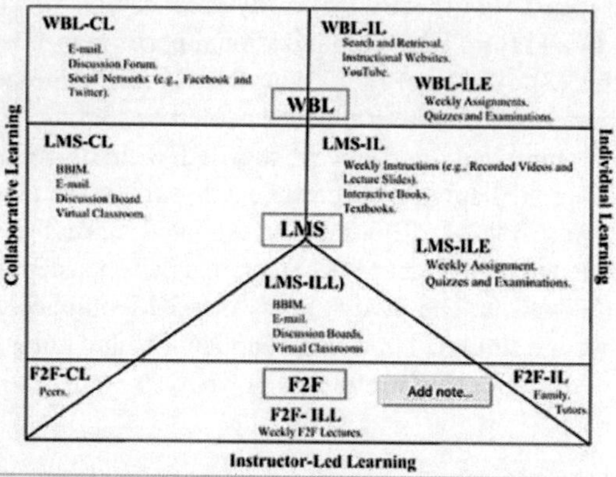

Fig. 6 Framework for LMS implementation

5.6 *RQ 6: What Framework is Needed to Implement LMS in a More Blended Learning Environment in HEI?*

A framework can be constructed to show the different modes of learning delivery that can be used in a blended learning environment, such as face-to-face learning, i.e., the traditional learning mode, the LMS mode, and the web mode, where learners are allowed online content using the web browser. In this framework, students can interact with the three modes of learning delivery to enable the teaching–learning process to be more interactive, collaborative, innovative, and blended to improve academic performance and student engagement. Several studies have indicated the crucial role of social media can play in enhancing classroom interactions and ensuring timely involvement in teaching and learning processes [9]. In addition, Cloud computing is another technological advancement that gives these experiences a new dimension where students and faculty can communicate in real-time [20]. As an example in the below framework used in this study, students can use the F2F mode of learning as direct interaction and acquiring knowledge, then extend through LMS mode for exchanging thoughts with others web mode for more alternate ideas and views [20] (Fig. 6).

6 Conclusion

The paradigm shift in ICT technology provides HEIs with a massive opportunity to manage their most valuable resource: knowledge. In HEI, this knowledge is built by researchers, shared through instruction and learning, and finally shared and transferred through communication and employment development [6]. A systematic review was developed to study Knowledge Management (KM) integration in

Higher Educational Institutions, with an emphasis on a Blended Learning environment. As per the methodology description, 16 papers were selected. These papers were assessed using a quality appraisal that selected only the high-quality studies. Research questions were discussed and analyzed, and the result concludes that knowledge management process integration is, to an extent, limited in HEI, where not all KM processes are presented entirely in the studies. This study concluded from the reviewed studies that Knowledge management practices and processes do contribute to innovation practices in HEI. And it also found that many studies emphasize that the knowledge management process can enhance an institution's ability to innovate. Also, the result of the study concluded that many studies identified that LMSs are used to distribute information and facilitate administration instead of ameliorating teaching and learning and achieving a more blended learning environment. Also, this study concludes that most of the reviewed studies describe LMS integration as a blended learning environment but didn't associate the blended learning environment with the KM processes. Finally, this study identified that Despite the various conclusions of the reviewed studies, all studies confirm that HEIs must invest in and develop knowledge management strategies, policies, and procedures enabled by innovative, collaborative learnings management systems (LMS) to differentiate the delivery of their mission around knowledge and learning to achieve the foreseen creative blended environment.

Acknowledgements This work is a part of a Knowledge Management course research paper at The British University in Dubai.

Appendix A

#	Source	Study purposes	Database	Method	Country	Year (publishing)	Study finding
S1	[9]	To evaluate various cloud computing tools in a blended learning environment	Elsevier	SR	Malaysia	2018	Benefits and limitations of utilizing literature these tools in a blended learning environment

(continued)

(continued)

#	Source	Study purposes	Database	Method	Country	Year (publishing)	Study finding
S2	[5]	To study the Influence of transformation leadership on KM processes and their impact on organizational learning and innovation in HEI	Emerald	Survey	Ghana	2020	Organizational learning and knowledge management positively affect innovation performance
S3	[20]	Effect of multiple learning modes, including face-to-face and LMS-based learning and web-based learning, on students' academic performance	Elsevier	Survey	Malaysia	2018	Multiple modes of learning delivery improve learning performance HEI
S4	[10]	the effect in promoting interactions between students, their teachers, and their learning materials (LMS)	Google Scholar	Survey	Vietnam	2020	Interactions, responses, and benefits of students vary towards blended learning
S5	[21]	To examine the use of LMSs in blended environments	ResearchGate	Survey-interview	Portugal	2014	Optimizes learning performance
S6	[11]	Utilizing the university's LMS is more effective through blended mode learning	Taylor and Francis	Case Study	Hong Kong	2019	blended learning enhanced learning management systems and made them more effective

(continued)

(continued)

#	Source	Study purposes	Database	Method	Country	Year (publishing)	Study finding
S7	[28]	Research the impact of essential achievement factors on students' experience with the LMS in a blended environment	IEEE	Survey	Malaysia	2018	Guidelines for universities to improve using LMS to facilitate the blended environment
S8	[13]	Examine the relationship between the KM Process and organizational development in HEI	Emerald	Survey	Iran	2016	The significant relationship between KM and professional development in HEI
S9	[16]	Determine if knowledge creation and sharing are related to cultural practices in higher education	Elsevier	Survey	UK	2013	Knowledge creation, transmission, and sharing in universities play a significant role in human development
S10	[1]	A systematic review of the knowledge management in HEI	Google Scholar	Systematic Review	South Asia	2019	Developing a framework for incorporating knowledge management in higher education
S11	[3]	explore the relationship between Knowledge Management and innovation and Intellectual Capital. and also examine the relationship between KM and organizational	Emerald	Survey	Pakistan	2019	KM affects OP by improving innovation and Intellectual Performance

(continued)

(continued)

#	Source	Study purposes	Database	Method	Country	Year (publishing)	Study finding
S12	[7]	The effect of knowledge management on innovation pace and quality and evaluating the facilitating aspect of the knowledge dissemination process	Emerald	Quantitative Study	Pakistan	2021	innovation speed and quality are affected by knowledge sharing and generation
S13	[4]	Finding the relationship between knowledge management and faculty performance in (HEIs)	Emerald	Survey-interview	Nepal	2021	A Modification in academic methods and in organizational arrangements would impact faculty members' performances and perspectives
S14	[22]	To understand what LMS do, faculty include their course in a different mode of study	Google Scholar	Quantitative Study	US Midwest	2017	An increase in the use of LMS in their course and learning
S15	[25]	Examine the relationship between (KM) and (OI) in higher education	Emerald	Survey	Iran	2019	The KM model predicted the aspects of organizational innovation in higher education
S16	[6]	This study aims to develop a model of KM that can be used as a basis for innovation while studying the similarities and differences between the two countries	Taylor and Francis-	Case Study	Mauritius and SA	2018	A profile that illustrates the similarities and differences was developed

(continued)

(continued)

#	Source	Study purposes	Database	Method	Country	Year (publishing)	Study finding

Appendix B

#	Source	Journal	Ranking	Citations	Impact ranking
S1	[9]	Computers and Education	Q1	165	10.88
S2	[5]	Learning Organization	Q2	11	3.01
S3	[20]	Telematics and Informatics	Q1	63	7.45
S4	[10]	Education and Information Technologies	Q1	26	2.917
S5	[21]	Educational Technology and Society	Q1	144	3.52
S6	[11]	Higher Education Research & Development	Q1	29	3.851
S7	[28]	IEEE access	Q1	48	3.367
S8	[13]	Kybernetes	Q2	85	1.75
S9	[16]	International Journal of Information Management	Q1	194	16.6
S10	(Kanwal, Nunes & Arif 2019)	IFLA Journal	Q1	6	1.667
S11	[3]	Journal of Enterprise Information Management	Q1	121	5.17
S12	[7]	Journal of knowledge management	Q1	0	4.745
S13	[4]	VINE Journal of Information and Knowledge Management Systems	Q2	0	2.75
S14	[22]	Online Learning Journal	Q1	98	2.46
S15	[25]	Kybernites	Q2	13	1.75

(continued)

(continued)

#	Source	Journal	Ranking	Citations	Impact ranking
S16	[6]	Studies in Higher Education	Q1	24	3

References

1. Kanwal S, Nunes MB, Arif M (2019) Knowledge management practice in South Asian higher education institutions. IFLA J 45(4):309–321. https://doi.org/10.1177/0340035219876958
2. Baptista Nunes JM, Kanwal S, Arif M (2017) Knowledge management practices in higher education institutions: a systematic literature review. In: IFLA WLIC 2017. http://library.ifla.org/view/conferences/2017/2017-08-18/731.html. http://www.ncbi.nlm.nih.gov/pubmed/20237272. http://www.pubmedcentral.nih.gov/articlerender.fcgi?artid=PMC2852648
3. Iqbal A, Latif F, Marimon F, Sahibzada UF, Hussain S (2019) From knowledge management to organizational performance: modelling the mediating role of innovation and intellectual capital in higher education. J Enterp Inf Manag 32(1):36–59. https://doi.org/10.1108/JEIM-04-2018-0083
4. Paudel KP, Bhattarai PC, Chalise M (2021) Interdependencies between knowledge management and academic performance in higher educational institutions. VINE J Inf Knowl Manag Syst. https://doi.org/10.1108/VJIKMS-01-2021-0005
5. Asiedu MA, Anyigba H, Ofori KS, Ampong GOA, Addae JA (2020) Factors influencing innovation performance in higher education institutions. Learn Organ 27(4):365–378. https://doi.org/10.1108/TLO-12-2018-0205
6. Veer-Ramjeawon P, Rowley J (2019) Embedding knowledge management in higher education institutions (HEIs): a comparison between two countries. Stud High Educ 45:2324–2340. https://doi.org/10.1080/03075079.2019.1608431
7. Iqbal A (2021) Innovation speed and quality in higher education institutions: the role of knowledge management enablers and knowledge sharing process. J Knowl Manag 25:1–27. https://doi.org/10.1108/JKM-07-2020-0546
8. Arpaci I (2017) Antecedents and consequences of cloud computing adoption in education to achieve knowledge management. Comput Human Behav 70:382–390. https://doi.org/10.1016/j.chb.2017.01.024
9. Al-Samarraie H, Saeed N (2018) A systematic review of cloud computing tools for collaborative learning: opportunities and challenges to the blended-learning environment. Comput Educ 124(March):77–91. https://doi.org/10.1016/j.compedu.2018.05.016
10. Bouilheres F, Le LTVH, McDonald S, Nkhoma C, Jandug-Montera L (2020) Defining student learning experience through blended learning. Educ Inf Technol 25(4):3049–3069. https://doi.org/10.1007/s10639-020-10100-y
11. Evans JC, Yip H, Chan K, Armatas C, Tse A (2020) Blended learning in higher education: professional development in a Hong Kong university. High Educ Res Dev 39(4):643–656. https://doi.org/10.1080/07294360.2019.1685943
12. Quint RUEC, Du B, Albert ROI (2013) Knowledge sharing in higher education institutions: a systematic review
13. Hasani K, Sheikhesmaeili S (2016) Knowledge management and employee empowerment: a study of higher education institutions. Kybernetes 45(2):337–355. https://doi.org/10.1108/K-04-2014-0077
14. Nahar R, Parvin S, Ullah KT, Parvez A (2020) Knowledge management: how the relationship works with organizational performance in the higher education sector? PalArch's J Archaeol

Egypt/Egyptol 17(7):9479–9501. https://www.archives.palarch.nl/index.php/jae/article/view/3904
15. Veer Ramjeawon P, Rowley J (2018) Knowledge management in higher education institutions in Mauritius. Int J Educ Manag 32(7):1319–1332. https://doi.org/10.1108/IJEM-05-2017-0129
16. Howell KE, Annansingh F (2013) Knowledge generation and sharing in UK Universities: a tale of two cultures? Int J Inf Manag 33(1):32–39. https://doi.org/10.1016/j.ijinfomgt.2012.05.003
17. Veer Ramjeawon P, Rowley J (2017) Knowledge management in higher education institutions: enablers and barriers in Mauritius. Learn. Organ 24(5):366–377. https://doi.org/10.1108/TLO-03-2017-0030
18. Wu M, Nurhadi D (2019) Continuous development of knowledge management for higher education institutions. J Teknol Kejuruan dan Pengajarannya 42(2):121–132
19. Yigzaw ST, Jormanainen I, Tukiainen M (2019) Trends in the role of ICT in higher education knowledge management systems: a systematic literature review. In: ACM international conference proceeding series, pp 473–480. https://doi.org/10.1145/3362789.3362805
20. Baragash RS, Al-Samarraie H (2018) Blended learning: investigating the influence of engagement in multiple learning delivery modes on students' performance. Telemat Inf 35(7):2082–2098. https://doi.org/10.1016/j.tele.2018.07.010
21. Dias SB, Diniz JA (2013) Towards an enhanced learning management system for blended learning in higher education incorporating distinct learners' profiles. Educ Technol Soc 17(1):307–319
22. Rhode J, Richter S, Gowen P, Miller T, Wills C (2017) Understanding faculty use of the learning management system. Online Learn. J. 21(3):68–86. https://doi.org/10.24059/olj.v%vi%i.1217
23. Asiri MJ, Mahmud RB, Abu Bakar K, Bin Mohd Ayub AF (2012) Factors influencing the use of learning management system in Saudi Arabian higher education: a theoretical framework. High Educ Stud 2(2), 125–137. https://doi.org/10.5539/hes.v2n2p125
24. Al-Emran M, Mezhuyev V, Kamaludin A, Shaalan K (2018) The impact of knowledge management processes on information systems: a systematic review. Int J Inf Manage 43(August):173–187. https://doi.org/10.1016/j.ijinfomgt.2018.08.001
25. S. Sadeghi Boroujerdi, K. Hasani, and V. Delshab, "Investigating the influence of knowledge management on organizational innovation in higher educational institutions," *Kybernetes*, vol. 49, no. 2, pp. 442–459, 2020, doi: https://doi.org/10.1108/K-09-2018-0492.
26. Arpaci Ibrahim, Al-Emran Mostafa, Al-Sharafi Mohammed A (2020) The impact of knowledge management practices on the acceptance of Massive Open Online Courses (MOOCs) by engineering students: a cross-cultural comparison. Telemat Inf 54:101468. https://doi.org/10.1016/j.tele.2020.101468
27. Arpaci I (2017) The role of self-efficacy in predicting use of distance education tools and learning management systems. Turk Online J Dist Educ 18(1):52–62. https://doi.org/10.17718/tojde.285715
28. Ghazal S, Al-Samarraie H, Aldowah H (2018) 'i am Still Learning': modeling LMS critical success factors for promoting students' experience and satisfaction in a blended learning environment. IEEE Access 6:77179–77201. https://doi.org/10.1109/ACCESS.2018.2879677
29. Moher D, Liberati A, Tetzlaff J, Altman DG, Altman D, Antes G, Tugwell P (2009) Preferred reporting items for systematic reviews and meta-analyses: the PRISMA statement. PLoS Med. https://doi.org/10.1371/journal.pmed.1000097

Undergraduate Students' Attitudes Towards Remote Learning During COVID-19 Pandemic: A Case Study from the UAE

Azza Alawadhi, Rawy A. Thabet, and Eman Gaad

Abstract The sudden closure of learning institutions due to the unprecedented COVID-19 pandemic has impacted education all over the world. With remote learning playing an increasingly important role in teaching during the pandemic, it is crucial to identify the variables that influence students' behaviors in using online education. Framed within the Technology Acceptance Model, this study examined undergraduate students' behavioral intention toward their remote learning experience at a federal higher education institution in the UAE. A random sample of 216 undergraduate students responded to an online survey. The results suggest that Perceived Ease of Use (PEU) and Perceived Usefulness (PU) positively impacted undergraduate students' acceptance of remote learning. In addition, data analysis revealed no significant difference between male and female students' attitudes towards remote learning. The results of this study are important to inform future efforts in facilitating institutional readiness for online education.

Keywords Remote learning · TAM · Higher education

1 Introduction

In December 2019, the world was struck with a pandemic that created massive damage to the educational system in history, affecting roughly 1.6 billion students in more than 190 countries all over the world [1]. As of March 2020, campus events, workshops, and face-to-face teaching were suspended to enforce social distancing. Meanwhile, higher education institutions had to react quickly and adapt to an unplanned, rapid and almost forced transition from a traditional classroom setting to

A. Alawadhi (✉)
Higher Colleges of Technology, Ras Al Khaimah, UAE
e-mail: aalawadhi@hct.ac.ae

R. A. Thabet · E. Gaad
The British University in Dubai, Dubai, UAE

M. Al-Emran et al. (eds.), *International Conference on Information Systems and Intelligent Applications*, Lecture Notes in Networks and Systems 550,
https://doi.org/10.1007/978-3-031-16865-9_26

online education [2]. Given the crucial role of online learning in facilitating education during crises, understanding how college students accept remote learning is important to enhance their academic success within an online setting. Furthermore, online learning is a relatively new teaching practice in UAE higher education, and it is still in its primary stage. Therefore, after one year of only online learning with no face-to-face teaching, it is essential to examine Emirati undergraduate students' remote and online learning experiences and identify the variables that significantly impact their attitudes to use online learning. The following main question guided the study: What are undergraduate students' attitudes towards remote learning during the pandemic?

In particular, the research questions the researchers intend to answer are as follows;

Q1: Do the factors of Perceived Ease of Use (PEU) and Perceived Usefulness (PU) affect students' Behavioral Intention (BI) towards remote learning?
Q2: Is there any significant difference among college students' behavioral intentions towards remote learning in terms of gender?

2 Conceptual and Theoretical Perspective

2.1 Remote and Online Learning During the COVID-19 Pandemic

Remote learning is a new concept that has been a focus of research in recent years [3–5]. Some researchers refer to teaching during the pandemic as Emergency e-Learning [6] or Emergency Remote Teaching [7]. Remote learning is referenced as a temporary sudden, and fast transition of face-to-face instructional delivery to distance and online education due to crises (e.g., the closure of academic institutions due to the lockdown), and it differs in meaning from pre-planned online learning [8]. There is often a lack of consistency in defining online learning in the literature. In their literature review, Moore, Dickson-Deane, and Gaylen [9] observed that the terms e-learning, online learning, and distance education are used interchangeably, though they encompass different meanings. Both Benson [10] and Conrad [11] believed that online learning is a newer and more recent version of distance education that offers flexibility and convenient learning, where students can learn at their own pace regardless of location and time. In this study, online and remote learning has been conceptualized as any learning that takes place in an entirely online environment using live video conferencing tools with no traditional face-to-face interaction. In this paper, the terms 'remote learning' and 'online learning' are used interchangeably to illustrate the education that took place during the closure of academic institutions in the UAE.

2.2 *Technology Acceptance Model*

This study uses the theoretical underpinnings of Davis (1989) Technology Acceptance Model (TAM) [12] as predictors of students´ attitudes towards remote learning. TAM asserts that there are two behavioral usability variables, perceived ease of use (PEU) and perceived usefulness (PU), which cause people to accept, reject or continue to use technologies [13]. PEU is defined as the degree to which an individual believes that using a particular technology will be free of effort, while PU refers to the degree to which an individual believes that using a particular system enhances his or her job performance.

According to TAM, attitude is another factor that affects users' acceptance. The attitude represents the individual personal evaluation of the use of technology [14]. In contrast, behavioral intention (BI) represents the actual use of a given information system which determines technology acceptance [15]. TAM explains that users' behavior and attitude are determined by their willingness to perform a specific task, driven by the system's perceived usefulness and ease of use. The following conceptual research model illustrates the relationships between the variables (Fig. 1).

3 Review of Related Literature

In recent years, the adoption, acceptance, and use of information technologies in education have been a critical research topic [16–18]. Researchers have investigated users' attitudes towards information technologies using several models to explain the variables that affect users' acceptance. One of these models is TAM, which examines users' attitudes towards technologies. A considerable amount of previous research employed TAM to investigate university students' attitudes and user behavior towards a variety of information technologies, including e-learning [19], mobile learning [20], Blackboard [21], Google classroom [22], and video-based learning [23]. These studies have found that TAM can predict and explain users' acceptance of information technology.

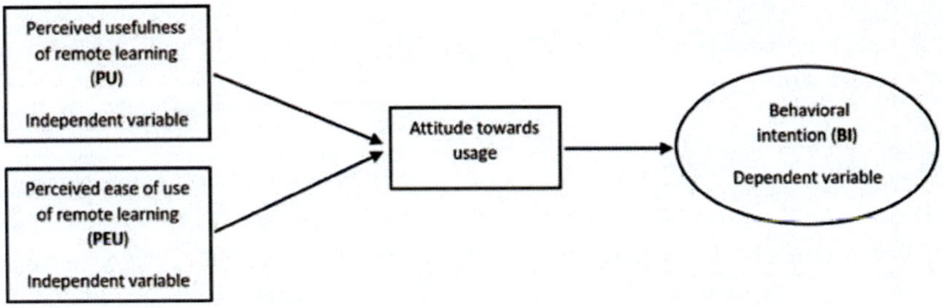

Fig. 1 Adapted from the technology acceptance model (TAM) developed by Davis (1989)

There is abundant evidence to show that perceived ease of use and usefulness impact students' behavior to use technology. For instance, Buabeng-Andoh [24] demonstrated that students' attitude greatly impacts their behavioral intent. However, the effect of ease-of-use and usefulness on behavioral intention was not reported. A similar pattern of results was obtained in Shroff, Deneen, and Ng's [25] study in which they surveyed university students' intent to use e-portfolios and reported that PEU and PU significantly influenced students' attitudes toward using e-portfolios. Previous work by Liu et al. [13] examined the relationship between online course design and different variables (e.g., perceived interaction; perceived usefulness, perceived ease of use). However, both Liu et al. [13] and Buabeng-Andoh's [24] experiments focused solely on the PEU and PU, but did not investigate users' BI. Comparable results were also found in Almekhlafy's study [26], who investigated Saudi students' perceptions of Blackboard during COVID-19 online learning and found that students' attitude plays a vital role in determining their intention to use Blackboard. There is considerable evidence from several studies of the impact of online learning on students' PU. For example, Jameel et al. [27] surveys show that perceived ease-of-use and usefulness are essential constructs that enhance the university students' behaviors towards e-learning. These findings are further supported by Alfadda and Mahdi [28], who studied the use of Zoom application for online learning during distance education. Their study confirms a positive relationship between the actual use of Zoom, PU, and students' attitudes. However, their study shows that there is no correlation between gender and other variables of TAM. Similarly, several researchers have studied the impact of gender on technology acceptance [15, 23]. However, the results of these studies were inconsistent. For instance, Al-Emran and Salloum [29] found that male students seem to have better perceptions of using mobile technologies for e-evaluation compared to female students. Nevertheless, Al-Emran, Elsherif, and Shaalan [30] study did not show any statistical gender differences. While the advantages and disadvantages of online learning have received the most attention in the literature [31, 32], very limited research has focused on the factors that impact students' behaviors toward remote learning during the pandemic. A review of the literature shows that apart from Jameel et al. [27], no previous research has sought to study Emirati students' attitudes towards remote learning and validate the technology acceptance model (TAM) in the UAE context during the pandemic.

4 Methodology

4.1 Research Design

Framed within a positivist paradigm, a quantitative cross-sectional design was used to address the purpose of the study [33]. This design enables the researchers to capture a snapshot of the current behaviors, thoughts, beliefs, and attitudes in a population [34]. The study was conducted at a medium-sized college in the UAE

Table 1 Demographic characteristics of the participants (n = 216)

Characteristics	Categories	Frequency (n)	Percentage (%)
Age group	17–18	54	25
	19–20	115	53.2
	21–22	27	12.5
	23 or above	20	9.3
Gender	Male	51	23.6
	Female	165	76.4
Major	Applied media	25	11.6
	Business	24	11.1
	CIS	65	30.1
	Education	5	2.3
	ETS	10	4.6
	Health science	32	14.8
	English	55	25.5
Competency of IT skills	Very good	82	38
	Moderate	128	59.3
	Low	6	2.8
Have you participated in online learning before COVID-19?	Yes	93	43.1
	No	123	56.9

during the academic year 2020–2021. The college hosts more than 1000 students with a separate male and female campus. Due to the pandemic, all programs were offered fully online, and some courses which required students to use the lab (e.g., Engineering) were delivered in a blended format. The online classes were delivered through Blackboard Collaborate Ultra and had the same academic rigor and quality as in face-to-face instruction pre-COVID-19. A simple random sample of 216 students (51 males and 165 females) responded to the online survey, with a response rate of 62%. All of the participants were UAE nationals who have attended a full year of online and remote education. Of the total sample, 23.6% were male, and 76.4% were female. Demographic profiles of the participants are provided in Table 1.

4.2 *Data Collection and Instrument Design*

Prior to data collection, ethical approval was obtained. A survey was developed using existing scales from prior TAM instruments [15, 35] to assess students' attitudes and behaviours toward the online learning. The survey was created via Google Forms, and it consisted of three sections. The first section was designed to elicit participants' demographic information, basic IT skills, and experience of online learning. In the

Table 2 Reliability analysis of the scale items

Variable	Cronbach Alpha (α)	No. of items
Perceived Ease of Use (PEU)	0.752	4
Perceived Usefulness (PU)	0.874	4
Behavioral Intention (BI)	0.921	4

second section, participants had to identify the advantages and disadvantages of remote learning. The last section consisted of 12 statements related to participants' attitudes towards remote learning, which were organized under three variables (PEU, PU, and BI). The 12-items were constructed with a 5-point Likert scale response format, ranging from Strongly Agree (5) to Strongly Disagree (1). A pilot test was administered to the target population. The pilot version of the survey was sent to random students by email and through WhatsApp groups. Appropriate revisions and modifications were made, including organizing the statements under three variables, reducing the number of statements, and adding more questions from the original scales.

4.3 Reliability and Descriptive Statistics

All of the statistical analysis was performed using SPSS version 23. Descriptive statistics were presented to provide an overview of the mean and standard deviation for key variables. Cronbach coefficient alpha (α) was calculated to estimate the internal consistency reliability (Table 2). All values were above 0.70, showing a good reliability level.

Table 3 shows the descriptive statistics. The mean of the items ranged from 3.92 to 3.32 and the standard of deviation shows that the distributions are narrow around the mean.

A Principle Component Analysis (PCA) with Varimax normalization rotation was performed with 12 survey items. The factorability of the matrix was determined using Kaiser–Meyer–Olkin (KMO) measure of sampling adequacy, and it was marvelous 0.939, which suggests that the items were suitable for factor analysis as it is above the commonly recommended value of 0.6 [36]. Furthermore, Bartlett's Test of Sphericity was significant ($p < 0.05$). When the total variance was calculated, it was found that two components with an eigenvalue greater than 0.1. The scree plot also indicated the existence of two components (Fig. 2).

All of the variables, which load under (component 1), indicate positive experience, and the only statement that loaded with (component 2) showed a feeling of frustration (Table 4). The researchers expected that the PEU4 statement to load differently because of the way it was phrased. The two extracted components together explain 71% of the variance.

Table 3 Descriptive statistics

Item	Mean	Std. deviation
PEU1. I find online learning easy to use	3.92	1.086
PEU2. I find it easy to contact my instructor in online learning	3.72	1.099
PEU3. Using online learning give me more control over my work	3.52	1.231
PEU4. I feel that online learning is often frustrating	3.32	1.156
PU1. I think online learning improve my learning	3.36	1.293
PU2. I think my classes are useful	3.56	1.106
PU3. I think online learning gives me greater control over my learning	3.51	1.137
PU4. I think online learning saves me time	3.77	1.247
BI1. I enjoy attending online classes	3.58	1.295
BI2. I find online learning interesting	3.58	1.251
BI3. I would like to use online learning in the future	3.56	1.410
BI4. Assuming that I have access to online learning, I intend to use it	3.53	1.099

Total items 12
Analysis (N) 216

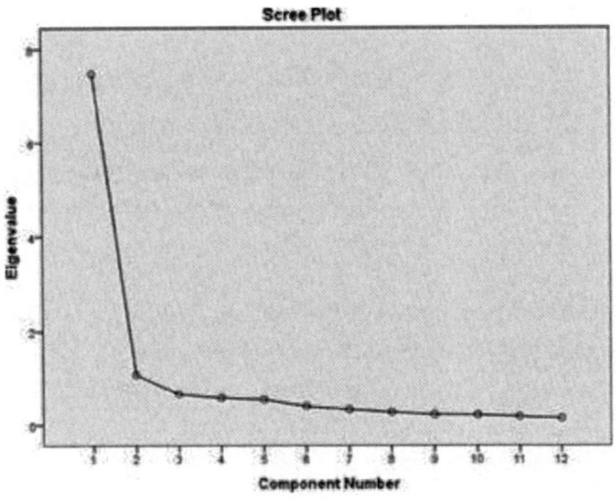

Fig. 2 Scree plot eigenvalue after PCA

5 Findings and Discussion

In this section, the results are reported based on the research questions. In total, 216 undergraduate students responded to the survey with a response rate of 62%.

> Q1: Do the factors of Perceived Ease of Use (PEU) and Perceived Usefulness (PU) affect students' Behavioral Intention (BI) towards remote learning?

Inferential statistics: A multiple linear regression was used to predict the impact of PEU and PU on students' BI towards remote learning. Before running regression, the normality assumption was calculated using Shapiro-Wilks test. The result shows that p value (0.000) is less than 0.05; therefore, the data is not normally distributed. Adjusted R square = 0.72, which means that the model accounts for 72% of variance in the behavioral intention.

Table 4 Rotated component matrix

Item	Component	
	1	2
BI2. I find online learning interesting	0.902	
PU2. I think my classes are useful	0.870	
BI1. I enjoy attending online classes	0.869	
BI3. I would like to use online learning in the future	0.856	
PU1. I think online learning improve my learning	0.841	
PU3. I think online learning gives me greater control over my learning	0.840	
BI4. Assuming that I have access to online learning, I intend to use it	0.836	
PEU3.Using online learning give me more control over my work	0.784	
PEU2. I find it easy to contact my instructor in online learning	0.747	
PEU1.I find online learning easy to use	0.724	
PU4. I think online learning saves me time	0.662	
PEU4. I feel that online learning is often frustrating		0.954

Extraction Method: Principal Component Analysis
Rotation Method: Varimax with Kaiser Normalization

To check whether there is a linear relationship between the independent and dependent variables, a scatterplot was generated by SPSS. A visual inspection of the plots as shown in Fig. 3 illustrates a proximate linear relationship between BI and PEU. Figure 4 shows the second scatterplot, which illustrates a proximate linear relationship between BI and PU. The analysis indicates that students have very positive intention towards remote learning. A similar positive influence of PU and PEU on BI has been reported by Alharbi and Drew [15] and Al Shammari [21].

Using the enter method, a significant model emerged at $(F = 279.629, p < 0.000)$. Significant variables are shown below:

Predictor Variable (independent variable) Beta p
PEU .152 $P = .006 < .05$
PU .732 $P = .006 < .05$

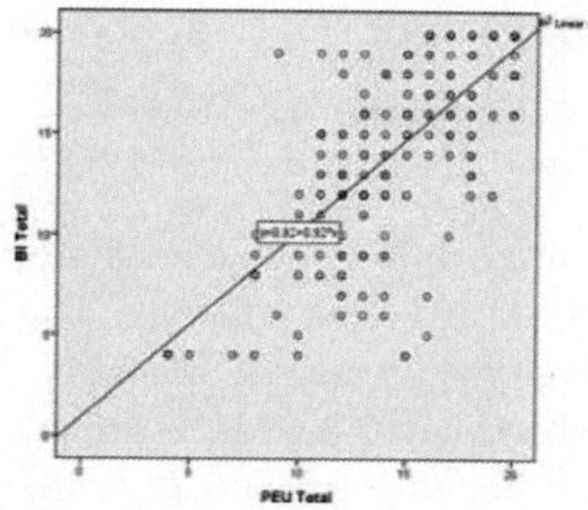

Fig. 3 Scatterplot showing a proximate linear relationship between BI and PEU

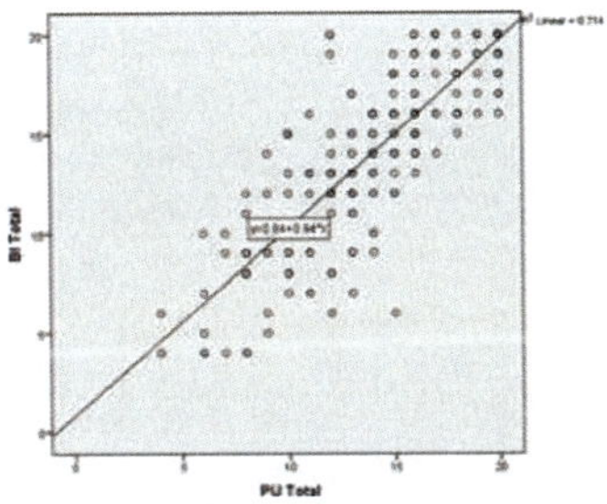

Fig. 4 Scatterplot showing a proximate linear relationship between BI and PU

As shown above, the PEU and PU are found to be predictors of undergraduate students' intention to use remote learning. These findings imply that students who found online learning easy to use and useful intend to use online learning in the future.

Q2: Is there any significant difference among college students' behavioral intentions towards remote learning in terms of gender?

Dependent variable: behavioral intention (BI)
Independent variable: gender

Descriptive statistics: As shown in Table 5, the mean of males is 15.12, and SD is 4.778 and the mean of females is 12.98 and SD is 4.478. These values imply that the mean of males is higher than the mean of females, but whether this difference is statistically significant or not, can be verified using the inferential statistics. Since the significance of Lavene's test ($p = 0.459$) is more than 0.05, there are equal variances, and the data is normally distributed.

Inferential statistics: An independent sample t-test was performed to examine whether there is a difference in means between male and female students' BI towards remote learning (Table 6). The analysis shows no significant differences in the means of male students and the mean of female students ($t = 1.558$ df 214, $p = 0.121 > 0.05$).

Meanwhile, it is noteworthy that the current findings are in line with Alfadda and Mahdi [28] and Al-Emran, ElShreif and Shaalan [30] findings, who revealed that gender did not predict a difference in terms of students' acceptance of technology information.

Table 5 The difference in means between male and female students

	Gender	N	Mean	Std. deviation	Std. error mean
BI total	Male	51	15.12	4.778	0.669
	Female	165	13.98	4.478	0.349

Table 6 An independent sample t-test

		Levene's test for equality of variances		t-test for equality of means				
		F	Sig.	t	df	Sig. (2-tailed)	Mean difference	Std. error difference
BI total	Equal variances assumed	0.55	0.45	1.55	214	0.121	1.136	0.729
	Equal variances not assumed			1.50	79.06	0.136	1.136	0.754

6 Conclusion and Future Research

This study examined undergraduate students' attitudes toward using remote learning during the pandemic in light of TAM model. A total of 216 undergraduate students from various departments replied to an online survey. The results of this study, which are consistent with prior research such as [35, 37], show that the students' intention to use online learning is impacted by their positive attitude, which is influenced by PEU and PU. Additionally, the study showed that mediating factors, such as gender, do not appear to be significant barriers to undergraduates' use of online learning. The generalizability of these results is subject to certain limitations as this study relied on self-reported data; therefore, there is a possibility of bias. Remote and online teaching has enabled higher education to continue during these extraordinary times. Moving forward from this pandemic, it is critical to assess this experience to increase the efficacy of higher education in the future. This cross-sectional study has sparked several lines for future research into both pedagogy and methodology. To begin with, remote and online learning in higher education is not a new concept. However, the COVID-19 crisis marks the first mass attempt of online and distance learning which needs to be further investigated. Future research may involve longitudinal designs that may yield valuable insight to capture changes in the students´ attitudes towards online learning over time. Based on TAM model, this study focused on two variables; PEU of remote learning and PU. Future research work should include additional variables such as self-efficacy, motivation, and enjoyment. In terms of the study contribution, this study extends the literature on the application of TAM model in higher education. It also bridges a gap in the literature as there is a scarcity of studies conducted in the UAE on the behavioral intention of Emirati undergraduate students (UAE nationals) towards the adoption of remote learning.

Acknowledgements This paper was submitted to the British University in Dubai as part of the assessment of one of the modules undertaken by the first author.

References

1. United Nations. Education during covid-19 and beyond. https://www.un.org/sites/un2.un.org/files/sg_policy_brief_covid-19_and_education_august_2020.pdf. Accessed 5 June 2021
2. Carrillo C, Flores MA (2020) COVID-19 and teacher education: a literature review of online teaching and learning practices. Eur J Teach Educ 43(4):466–487
3. Al-Tahitah AN, Al-Sharafi MA, Abdulrab M (2021) How COVID-19 pandemic is accelerating the transformation of higher education institutes: a health belief model view, vol 348. https://doi.org/10.1007/978-3-030-67716-9_21
4. Shin M, Hickey K (2020) Needs a little TLC: examining college students' emergency remote teaching and learning experiences during COVID-19. J Further High Educ 45:1–14
5. Lee K, Fanguy M, Lu XS, Bligh B (2021) Student learning during COVID-19: it was not as bad as we feared. Dist Educ 42(1):164–172
6. Murphy MP (2020) COVID-19 and emergency eLearning: consequences of the securitization of higher education for post-pandemic pedagogy. Contemp Secur Policy 41(3):492–505
7. Bozkurt A, Sharma RC (2020) Emergency remote teaching in a time of global crisis due to CoronaVirus pandemic. Asian J Dist Educ 15(1):i–vi
8. Hodges C, Moore S, Lockee B, Trust T, Bond A (2020) The difference between emergency remote teaching and online learning. Educause Rev 27:1–12
9. Moore JL, Dickson-Deane C, Galyen K (2011) e-Learning, online learning, and distance learning environments: are they the same? Internet High Educ 14(2):129–135
10. Benson AD (2002) Using online learning to meet workforce demand: a case study of stakeholder influence. Q Rev Dist Educ 3(4):443–452
11. Conrad, D (2006) E-Learning and social change: an apparent contradiction. In: Perspectives on higher education in the digital age, pp 21–33
12. Davis FD (1989) Perceived usefulness perceived ease of use and user acceptance of information technology. MIS Q 13:319–340
13. Liu IF, Chen MC, Sun YS, Wible D, Kuo CH (2010) Extending the TAM model to explore the factors that affect intention to use an online learning community. Comput Educ 54(2):600–610
14. Lee DY, Lehto MR (2013) User acceptance of YouTube for procedural learning: an extension of the Technology Acceptance Model. Comput Educ 61:193–208
15. Alharbi S, Drew S (2014) Using the technology acceptance model in understanding academics' behavioral intention to use learning management systems. Int J Adv Comput Sci Appl 5(1):143–155
16. Al Ajmi Q, Al-Sharafi MA, Yassin AA (2021) Behavioral intention of students in higher education institutions towards online learning during COVID-19. In: Emerging technologies during the era of COVID-19 pandemic. Springer, Cham, pp 259–274
17. Qasem YAM, Abdullah R, Yah Y, Atan R, Al-Sharafi MA, Al-Emran M (2021) Towards the development of a comprehensive theoretical model for examining the cloud computing adoption at the organizational level, vol 295. https://doi.org/10.1007/978-3-030-47411-9_4
18. Yassin AA, Razak NA, Saeed MA, Al-Maliki MAA, Al-Habies FA (2021) Psychological impact of the COVID-19 pandemic on local and international students in Malaysian universities. Asian Educ Dev Stud 10(4):574–586
19. Mohammadi H (2015) Investigating users' perspectives on e-learning: an integration of TAM and IS success model. Comput Hum Behav 45:359–374
20. Almaiah MA, Alismaiel OA (2019) Examination of factors influencing the use of mobile learning system: an empirical study. Educ Inf Technol 24(1):885–909
21. Al Shammari MH (2021) Devices and platforms used in emergency remote learning and teaching during Covid19: a case of English major students in Saudi Arabia. Arab World Engl J (AWEJ) Spec Issue Covid 19 Chall
22. Al-Maroof RAS, Al-Emran M (2018) Students acceptance of google classroom: an exploratory study using PLS-SEM approach. Int J Emerg Technol Learn 13(6):112
23. Pal D, Patra S (2020) University Students' perception of video-based learning in times of COVID-19: a TAM/TTF perspective. Int J Hum-Comput Interact 37:1–19

24. Buabeng-Andoh C (2021) Exploring University students' intention to use mobile learning: a research model approach. Educ Inf Technol 26(1):241–256
25. Shroff RH, Deneen CC, Ng EM (2011) Analysis of the technology acceptance model in examining students' behavioral intention to use an e-portfolio system. Aust J Educ Technol 27(4):600–618
26. Almekhlafy SSA (2020) Online learning of English language courses via blackboard at Saudi universities in the era of COVID-19: perception and use. PSU Res Rev 5:16–32
27. Jameel AS, Khald Hamzah A, Raad Al-Shaikhli T, Ihsan Alanssari A, Ibrahim MK (2021) System characteristics and behavioral intention to use E-learning. In: Learning, pp 7724–7733
28. Alfadda HA, Mahdi HS (2021) Measuring students' use of zoom application in language course based on the technology acceptance model (TAM). J Psycholinguist Res 50:1–18
29. Al-Emran M, Salloum SA (2017) Students' attitudes towards the use of mobile technologies in e-evaluation. Int J Interact Mob Technol 11(5):195–202
30. Al-Emran M, Elsherif HM, Shaalan K (2016) Investigating attitudes towards the use of mobile learning in higher education. Comput Hum Behav 56:93–102
31. Mukhtar K, Javed K, Arooj M, Sethi A (2020) Advantages, limitations and recommendations for online learning during COVID-19 pandemic era. Pak J Med Sci 36(COVID19-S4):S27
32. Rahiem MD (2020) The emergency remote learning experience of university students in Indonesia amidst the COVID-19 crisis. Int J Learn Teach Educ Res 19(6):1–26
33. Johnson B, Christensen L (2014) Educational research: quantitative, qualitative, and mixed approaches, 5th edn. SAGE publications, Los Angeles
34. Gay LR, Mills GE, Airasian P (2011) Educational research: competencies for analysis and applications, 10th edn. Pearson, Upper Saddle River
35. Bączek M, Zagańczyk-Bączek M, Szpringer M, Jaroszyński A, Wożakowska-Kapłon B (2021) Students' perception of online learning during the COVID-19 pandemic: a survey study of Polish medical students. Medicine 100(7):e24821
36. Kaiser HF, Rice J (1974) Little jiffy, mark IV. Educ Psychol Meas 34(1):111–117
37. Alfiras M, Bojiah J, Yassin AA (2020) COVID-19 pandemic and the changing paradigms of higher education: a gulf university perspective. Asian EFL J 27(5):1–9

Smart Campus Reliability Based on the Internet of Things

Khalid Adam, Mazlina Abdul Majid, and Younis Ibrahim

Abstract Nowadays, the Internet of Things (IoT) leads to efficient resource utilization and fosters the development of university campuses. The smart connected devices (things) help create smart campuses, which promises to transform into green campuses and achieve sustainable development. Therefore, designers will have to overcome significant implementation challenges to reach thousands or millions of devices to integrate the IoT on the university campus. Among these challenges, reliability has been identified as one of the critical issues for efficient IoT because unreliable sensing, processing, or transmission can cause false monitoring data reports, long delays, and even data loss, leading to vulnerabilities across smart campus applications. Unlike manufacturing or design faults, the worse behaviour of the unreliable smart campus, for example, transient faults that occur in IoT devices (also known as soft errors), do not happen consistently. External events, such as energetic particles striking the chip, cause these intermittent faults. These events do not result in permanent physical damage to IoT devices, but they can change signal transfers or stored values, resulting in incorrect smart campus application execution. This paper serves as a resource for smart campus reliability using the Internet of Things to understand smart campus sustainable development better.

Keywords Internet of Things · Smart campus · Reliability

K. Adam (✉)
Centre of Excellence for Artificial Intelligence and Data Science, University Malaysia Pahang, 26300 Gambang, Malaysia
e-mail: khalidwsn15@gmail.com

M. A. Majid
Faculty of Computing, University Malaysia Pahang, 26600 Pekan, Malaysia

Y. Ibrahim
Electrical and Computer Engineering, University of Saskatchewan, Saskatoon, Canada

M. Al-Emran et al. (eds.), *International Conference on Information Systems and Intelligent Applications*, Lecture Notes in Networks and Systems 550,
https://doi.org/10.1007/978-3-031-16865-9_27

1 Interdiction

At the moment, the integration of technology such as the IoT aims to meet the needs of a more demanding society that consumes more resources [1]. This technology's widespread adoption promotes the transformation of universities into smart campuses. Smart campuses improve people's comfort in education, construction, waste management, pollution, energy consumption, and so on [2–5]. However, reliability is a concern in IoT, where they are especially vulnerable to software faults, human faults, and transient faults that occur in IoT devices (also known as soft errors); it can assess the reliability of smart campus apps, leading to erroneous decision-making at this level, which might be deadly for campus end-users. [6–9]. The reliability of electronic devices is broadly estimated using MIL-HDBK-217, a military manual. Which has limits and hasn't been revised since 1995, despite its shortcomings, is nevertheless employed in reliability estimates by more than 80% of engineers. On another hand, there are other industrial and commercial criteria for measuring reliability. MIL-HDBK-217 has been replaced by the RIAC 217PlusTM methodology and software application; however, it is no longer free, it is more complex, and the approach is, at a minimum, the same [10].

Aside from that, there are several issues with calculating hardware reliability, and there is no standard method. As a result, methodological rigour, data quality, scope of analysis, uncertainty, and prediction process documentation vary greatly. For the reasons stated above, IEEE issued IEEE Std.1413 (Standard Framework for Hardware Reliability Prediction) in 2009 [11]. The Internet of Things includes a diverse variety of hardware of varying quality and reliability: much of this equipment is commercial, with no proven dependability and no data on failure rates, mean time to failure (MTTF), or mean time between failures (MTBF) [11]. However, these methods' calculated reliability is very difficult and often unexacted for smart campuses due to calculating the reliability of IoT devices only from the hardware perspective [12].

Smart campus reliability is controlled by IoT device (thing) failure rates and by software and human variables, making IoT deployment in smart campus sensitive applications difficult. [14–16]. As a result, the issue of reliability is frequently disregarded in smart campuses, prompting us to draw attention to the research gap as shown in Fig. 1. As a result, the purpose of this article is to examine smart campus reliability using the Internet of Things to acquire a deeper knowledge of smart campus long-term development. The following are the paper's main contributions:

- Presenting a thorough review of the available oriented towards reliability issues in the smart campus.
- Highlighting the future challenges for further research to enhance the smart campus' resilience.

The rest of this paper is organized as follows: Sect. 2 provides Motivation and Related Work, Reliability of Internet of things Caused by Soft Errors, Related Work. Section 3 Conclusion Remarks.

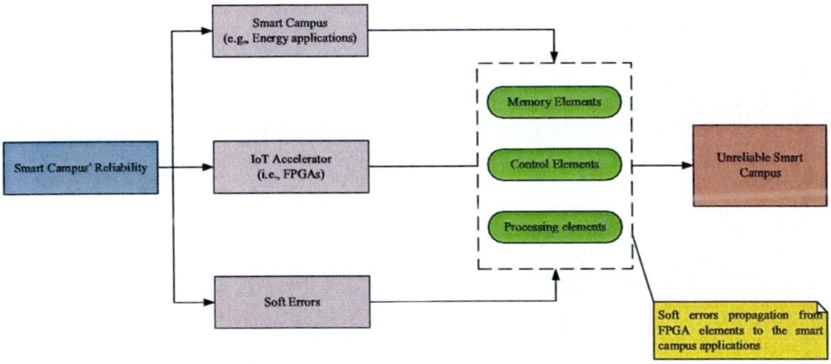

Fig. 1 Fault, error, and failure propagation in smart campus applications [11]

2 Motivation and Related Work

The Internet has revolutionized how individuals communicate with one another. The Internet of Things (IoT) aims to take this a step further by seamlessly connecting people and things, transforming colleges into smart campuses with huge economic and environmental benefits. Thus, the authors' motivation to analyze the smart campus reliability and the available solutions to mitigate the soft error problems.

2.1 Reliability of Internet of Things Caused by Soft Errors

With aggressive technology scaling, a soft error has become one of the most critical design issues in modern electronics as shown in Fig. 2. A soft error is temporary and cannot be replicated and becomes more frequent as feature sizes decrease along with chip supply voltages. As the semiconductor industry moves deeply into sub-micron technology, there is a rapid rise in chip susceptibility to soft errors. Such non-destructive events (soft errors) can cause IoT accelerators (i.e., FBGA) to generate an incorrect computational result or lose control of a disastrous device [11, 16, 17]. Thus, there are incorrect predictions in the smart campus applications. Soft errors are already a big problem in reliability smart campus applications, as well as in healthcare, aviation, and space.

When one or more bits flip from one value to another due to a soft error event, voltage fluctuation, source of electrical noise, or other reason, the data is corrupted (e.g., 0 to 1). Even if only one bit is changed, unintentional modification of data values might generate arbitrary undesirable system behaviour as shown in Fig. 2.

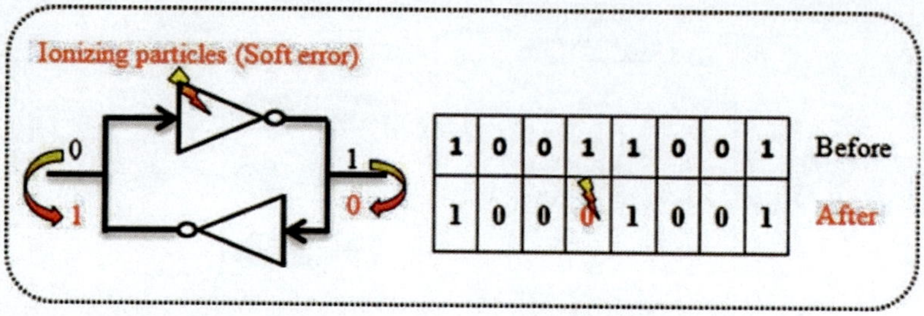

Fig. 2 (a) and (b) Example of Memory-element single particle strike [18]

2.2 *Related Work*

With the rapid development of the IoT, the construction of Smart Campus is the trend of universities. Implementing IoT lead to efficient resource utilization and foster the development of university campuses, where the smart connected devices (things) are helping to create smart campuses, which promises to transform to green campuses and achieve sustainable development [19, 20]. Therefore, this section provides a related work of Smart campus reliability behaviour based on IoT. The IoT devices (things) must perform reliably during the specified mission duration due to mission-critical nature of the internet of things applications. In other words, one of the most important prerequisites for IoT adoption in smart applications is reliability [3–5]. Therefore, in the IoT applications, transient faults (soft errors), failure to capture critical data, any network outage, data corruption, or loss during transmission or storage can all have disastrous consequences, such as mission failure, financial loss, and harm to people and the environment [8]. As a result, before IoT can be widely adopted on university campuses, academics, developers, and even customers must conduct dependability studies and design.

Several studies on fault tolerance in IoT have been offered to address the dependability issue. Previous research studies have used MIL-HDBK 217 as a classical approach to reliability assessment [10]. The Internet of Things (IoT) dependability is defined not only by the failure rate of IoT parts (things), but also by software and human factors in smart campuses. Many aims toward smart campuses and/or sustainable development to assist educational activities, energy, water, transportation, and all materials [3, 4, 19]. Nguyen et al. [1] demonstrate an IoT platform for monitoring environmental and human flows on a university campus. Fortes et al. [22] The University of Málaga adopted the smart campus (SmartUMA), which seeks to smart teaching, smart space, and smart parking and is powered by IoT. As a result, SmartUMA has developed UMA, a mobile application that allows students to access learning materials and engage in distance learning tasks while bravely watching videos created by teachers. The authors, however, do not assess the UMA smart campus's dependability.

Khajenasiri et al. [23] to benefit smart city applications, a survey on Internet of Things (IoT) technologies for smart energy control in smart city applications was undertaken. They stated that the Internet of Things is presently only being used in a few application areas to benefit both technology and humanity. IoT has a large range of applications, and in the not-too-distant future, it will be able to cover virtually all of them. They stated that energy conservation is an important part of civilization and that the Internet of Things might help in the construction of a smart energy control system that saves both energy and money. They talked about IoT architecture and how it relates to the smart city concept. One of the most difficult aspects of achieving this, according to the authors, is the immaturity of IoT hardware and software. They proposed that these issues be addressed in order to create an IoT system that is dependable, efficient, and user-friendly. Moore et al. [24] The impact of anomalous data on classification in an IoT application for recognising human activity was studied, and it was discovered that some classifiers were significantly more vulnerable to errors than others and that the data preparation method can also make the application more vulnerable to failure. In order to prevent major faults from entering the system, developers must make a concerted effort to construct and comprehend the dependability of applications hosted in internet of things infrastructure. The proclivity of sensors to "fail-dirty" is another source of concern in IoT device reliability [25–27]. When a sensor continues to provide inaccurate data after a failure, this phenomenon happens. This is a well-known issue, yet it is a little-understood one that hampers IoT settings. Because the sensor appears to be in good working order, this problem is particularly difficult to diagnose. Given that actuation typically has a tangible impact on people's lives [28–30], it's easy to see why. In an IoT setting, the influence of a false reading can be devastating. In comparison to this study paper, Table 1 summaries related studies in terms of topic and findings.

Table 1 The summary of the related work

Ref	Addressed Issues	Characteristics	Technology	Limitations
[10]	This study focused on reliability assessment	MIL-HDBK 217	Not mentioned	Software and human variables have a role in IoT dependability, as well as the failure rate of IoT pieces (things)
[22]	This study focused on implemented the smart campus (SmartUMA)	Smart education, smart space, smart parking, etc	Internet of Things	The authors don't consider the reliability of the UMA smart campus
[23]	This study focused on smart energy control	Deployed in a small number of application areas to benefit both technology and people	Internet of Things	The authors did not consider reliability

(continued)

Table 1 (continued)

Ref	Addressed Issues	Characteristics	Technology	Limitations
[24]	This study looked at the effect of anomalous data on categorization in an IoT application	Some classifiers were shown to be substantially more prone to mistakes than others, and the way data is prepared might also make the programme more sensitive to failure	Internet of Things	IoT reliability is determined not just by the failure rate of IoT parts (things), but also by software and human factors in smart campuses
[25]	This study aimed to propensity for sensors to "fail-dirty"	In an IoT world, the consequences of sending a false signal can be disastrous	Internet of Things	The authors did not take reliability into account Fault injection to assess the dependability

3 Conclusion Remarks

As the complexity and dynamics of IoT systems and applications grow, new characteristics of system complexity and dynamics may emerge, rendering existing dependability models and solutions ineffective or erroneous. New and efficient reliability models and methods are expected to capture the new qualities and behaviours, resulting in more effective and accurate IoT system reliability analysis, optimization, and design. Smart campuses require extremely reliable and efficient data storage and processing solutions due to the safety-critical or mission-critical nature of IoT applications, as well as the rapid growth of data produced. Additionally, IoT dependability is not usually the major issue in the IoT but understanding reliability might aid in the event of failure, i.e., where to seek a breakdown. This study serves as a resource for smart campus dependability researchers using the Internet of Things to acquire a better understanding of smart campus sustainable development.

Acknowledgements This research is supported by the UMP Green Technology Research Lab, University Malaysia Pahang (UMP) Research Grant (RDU190167) and Malaysia National Research Grant (FRGS/1/2018/ICT04/UMP/02/4)

References

1. Nguyen ST, Le BN, Dao QX (2021) AI and IoT-powered smart university campus: design of autonomous waste management. In: 2021 international symposium on electrical and electronics engineering (ISEE), pp 139–144
2. Jabbar WA, Wei CW, Azmi NA, Haironnazli, NA (2021) An IoT Raspberry Pi-based parking management system for smart campus. Internet Things 14(1):100387

3. Chagnon N, Gosselin L, Barnabe S (2021) Smart campuses: extensive review of the last decade of research and current challenges. IEEE Access 9:124200–124234
4. Valks B, Arkesteijn MH, Koutamanis A, den Heijer AC (2020) Towards a smart campus: supporting campus decisions with Internet of Things applications, Build Res Inf 49(1):1–20
5. Adenle YA, Chan EHW, Sun Y, Chau CK (2021) Assessing the relative importance of sustainability indicators for smart campuses: a case of higher education institutions in Nigeria. Environ Sustain Indicators 9:100092. https://doi.org/10.1016/j.indic.2020.100092
6. Ibrahim Y, Wang H, Bai M, Liu Z, Wang J, Yang Z, Chen Z (2020) Soft error resilience of deep residual networks for object recognition. IEEE Access 8:19490–19503
7. Ibrahim Y et al (2020) Soft errors in DNN accelerators: a comprehensive review. Microelectron Reliab 115:113969. https://doi.org/10.1016/j.microrel.2020.113969
8. Hammad KAI, Fakharaldien MAI, Zain J, Majid M (2015) Big data analysis and storage. In: International conference on operations excellence and service engineering, pp 10–11
9. Abich G, Gava J, Reis R, Ost L (2020) Soft error reliability assessment of neural networks on resource-constrained IoT devices. In: ICECS 2020 - 27th IEEE international conference on electronics, circuits and systems, proceedings. https://doi.org/10.1109/ICECS49266.2020.9294951
10. Pokorni S (2019) Reliability and availability of the Internet of things. Vojnotehnicki glasnik 67:588–600
11. Adam K, Mohamed II, Ibrahim Y (2021) A selective mitigation technique of soft errors for DNN models used in healthcare applications: DenseNet201 case study. IEEE Access 9:65803–65823
12. Azghiou K, Mouhib M, Koulali MA, Benali A (2020) An end-to-end reliability framework of the Internet of Things. Sensors (Switzerland) 20(4):2439. https://doi.org/10.3390/s20092439
13. Anagnostopoulos T et al (2021) Challenges and solutions of surveillance systems in IoT-enabled smart campus: a survey. IEEE Access 9:131926–131954
14. Imbar RV, Supangkat SH, Langi AZR (2020) Smart campus model: a literature review. In: 7th international conference on ICT for smart society: AIoT for smart society, ICISS 2020 - Proceeding. https://doi.org/10.1109/ICISS50791.2020.9307570
15. Nagowah SD, Ben H, Gobin B (2020) A systematic literature review on semantic models for IoT-enabled smart campus. Appl Ontol 16:27–53
16. Adam K, Mohd II, Younis YM (2021) The impact of the soft errors in convolutional neural network on GPUS: Alexnet as case study. Procedia Comput Sci 89–94
17. Adam K, Mohd II, Ibrahim Y (2021) Analyzing the soft error reliability of convolutional neural networks on graphics processing unit. J Phys: Conf Ser 1933(1):012045. https://doi.org/10.1088/1742-6596/1933/1/012045
18. Adam KI, Mohd I, Ibrahim Y (2021) Analyzing the resilience of convolutional neural networks implemented on gpus: Alexnet as a case study. Int J Electr Comput Eng Syst 12(2):91–103
19. Min-Allah N, Alrashed S (2020) Smart campus—A sketch. Sustain Cities Soc 59:102231. https://doi.org/10.1016/j.scs.2020.102231
20. Zaballos A, Briones A, Massa A, Centelles P, Caballero V (2020) A smart campus' digital twin for sustainable comfort monitoring. Sustainability (Switzerland) 12:1–33
21. Kempf J, Arkko J, Beheshti N, Yedavalli K (2011) Thoughts on reliability in the Internet of Things. In: Interconnecting smart objects with the Internet workshop, vol. 1, pp 1–4. Internet Architecture Board, Boston
22. Fortes S et al (2019) The campus as a smart city: University of málaga environmental, learning, and research approaches. Sensors (Switzerland) 19(6):1349. https://doi.org/10.3390/s19061349
23. Khajenasiri I, Estebsari A, Verhelst M, Gielen G (2017) A review on Internet of Things solutions for intelligent energy control in buildings for smart city applications. Energy Procedia 770–779
24. Moore SJ, Nugent CD, Zhang S, Cleland I (2020) IoT reliability: a review leading to 5 key research directions. CCF Trans Pervasive Comput Interact 2:147–163
25. Adam K, Mohd II, Ibrahim Y (2021) Analyzing the instructions vulnerability of dense convolutional network on GPUS. Int J Electric Comput Eng 11(5):2088–8708

26. Jacentha N, Maniam A, Dalbir S (2020) Towards data privacy and security framework in big data governance. Int J Softw Eng Comput Syst (IJSECS) 1(6):41–51
27. Jawad H, Aiman A, Anmar A (2020) An effective deep learning approach for improving off-line arabic handwritten character recognition. Int J Softw Eng Comput Syst (IJSECS) 6(2):53–61
28. Alsariera Y, Mazlina A, Zamli K (2015) SPLBA: an interaction strategy for testing software product lines using the bat-inspired algorithm. In: International conference on software engineering and computer systems (ICSECS), pp 148–153
29. Khalid A, Mazlina A, Jasni M (2016) Big Data prediction framework for weather Temperature based on MapReduce algorithm. In: 2016 IEEE conference on open systems (ICOS), pp 13–17
30. Alsariera A, Majid A, Zamli Z (2015) A bat-inspired strategy for pairwise testing. ARPN J Eng Appl Sci 10:8500–8506

Application and Exploration of Virtual Reality Technology in the Teaching of Sports Anatomy

Na Hou and Md. Safwan Samsir

Abstract Research Methodology: In this paper, the application of virtual reality technology in the teaching of motion anatomy is analyzed by means of literature method and logical analysis method. Research results: The application of virtual reality technology to the teaching of sports anatomy is conducive to cultivating students' three-dimensional thinking ability, making up for the lack of teaching conditions, saving costs, stimulating students' learning interest and initiative, and cultivating the ability to combine theory and practice. However, there are current deficiencies, such as imperfect technical hardware, insufficient capital investment, lack of software development of the virtual teaching system for sports anatomy, difficult development of teaching resources, and vr reality virtual helmets with a sense of vertigo, sluggish force feedback, and low screen resolution. This study proposes some specific solutions to this problem, such as further improving the hardware system of virtual reality technology, continuing to develop a more effective virtual teaching software system for motor anatomy, increasing the development of an effective monitoring and evaluation system for the teaching and learning of sports anatomy, and establishing a diversified evaluation and feedback mechanism. This study provides an idea for the teaching reform of sports anatomy, which can effectively improve the teaching effect of motor anatomy. In order to reform the teaching of sports anatomy and improve the teaching effect, it is intended to provide theoretical basis.

Keyword Sports anatomy · Virtual reality technology · Teaching

N. Hou · Md. Safwan Samsir (✉)
Faculty of Psychology and Education, University Malaysia Sabah, 88400 Kota Kinabalu, Sabah, Malaysia
e-mail: safwan.samsir@ums.edu.m

N. Hou
e-mail: houna0918@sina.com

N. Hou
Department of Physical Education, Xianyang Normal University, Xianyang 712000, China

M. Al-Emran et al. (eds.), *International Conference on Information Systems and Intelligent Applications*, Lecture Notes in Networks and Systems 550,
https://doi.org/10.1007/978-3-031-16865-9_28

1 Introduction

With the progress of science and technology and the rapid development of information technology, the education model should also keep up with the pace of the development of the times, make full use of modern information technology for teaching, and improve the learning effect of students. In recent years, in order to continue to promote the comprehensive development of education and information technology and realize the modernization of education, the state has issued a number of measures such as the Ten-Year Development Plan for Education Informatization (2011–2020) and the Action Plan for Education Informatization 2.0. [1] With the development of Internet technology and digital technology, the emergence of virtual reality technology has also created opportunities for the reform of the higher education model. Virtual reality technology is mainly through the computer simulation of the virtual environment, so that users can observe and operate the things in the virtual environment according to their own feelings, from the visual, auditory, tactile and other aspects [2], to the experiencer with an immersive sense of experience [3]. In March 2020, the Association for Information technology in College Education released the 2020 EDUCAUSE Horizon Report: Teaching and Learning Edition, which includes XR (AR, VR, MR, HAPTIC) as an emerging technology and practice [4], and predicts that its application to distance learners will become a future development trend [5]. In China, the "Thirteenth Five-Year Plan" for Education Informatization points out that we should unremittingly promote education informatization and strive to expand the coverage of high-quality educational resources by means of informatization [6]. This has greatly promoted the application of VR technology in the field of teaching. The application of virtual reality technology in the field of education and teaching will greatly improve the effect of education and teaching, and make it fun to teach [7].

Sports anatomy is the main compulsory course for physical education students in various colleges and universities, which occupies an important position in the curriculum system of physical education, providing the knowledge base of human morphology and structure for subsequent courses such as exercise physiology and sports medicine, and also providing theoretical support for sports teaching and sports training. Therefore, the teaching effect of sports anatomy is good or bad, which directly affects the teaching quality of the entire physical education major and the quality of the training of sports professionals [8]. Sports anatomy belongs to the category of morphology, the teaching content mainly includes the morphological structure of important motor organs of the human body and the components of various tissues and organs closely related to movement, the theoretical basis and principle of the analysis of motor action anatomy, etc., the amount of content is large, the proportion of knowledge is large, the concept of terminology is many, the noun is more, the student memory is called boring and difficult, and the traditional teaching method is mainly the combination of theory class in which the teacher teaches the teaching and the experimental class is observed and operated by many people in a group. In the theory class [9], the teacher mainly shows the human body structure through the pictures on the PPT, although there is a teacher's explanation, it is still difficult for students to form a clear understanding of the size, adjacency, walking, etc. of different structures through the two-dimensional vision composed of pieces and words; in the experimental class, due to the large number of members of the same group and can not repeatedly dissect the corpse, and even some colleges and universities lack of corpse specimens due to insufficient investment in laboratory funds, too few experimental class hours, etc. Over time, students lose interest and initiative in the study of this course, and the teaching effect is poor. Virtual Reality (VR) technology is used in the teaching of motor anatomy, which can make static anatomical pictures into three-dimensional animations, and form a complete picture of tissues, organs, systems, and human body structure, display the human body structure in three dimensions, and attract students' attention through two stimuli: sight and hearing [10]. Promoting the reform of teaching mode through the application of virtual reality technology teaching methods will greatly improve the interest of physical education students in learning motor anatomy, return to the starting point of attaching equal importance to theory and practice of this discipline, and contribute to the improvement of teaching quality.

In view of the above background, this study adopts the literature method and logical analysis method to read and sort out the relevant literature collected, analyzes the advantages and dilemmas of virtual reality technology applied to sports anatomy teaching, discusses the value of virtual reality technology applied to sports anatomy teaching, and proposes some specific solutions in order to provide a theoretical basis for the reform of sports anatomy teaching in sports majors.

1.1 Overview of Virtual Reality Technology (VR)

VR is a high-tech developed at the end of the twentieth century, including microelectronics technology, sensing technology, computer technology, simulation technology, etc. It uses computer hardware and software resources to create and experience virtual world integration technology, which also includes network technology, language recognition, computer graphics, computer vision, computer simulation, parallel processing and human–computer interaction and other technologies, which can realize dynamic simulation of the real world. The simulation environment is very realistic, the resulting dynamic environment can make real-time responses to the user's posture, language commands, etc., so that the user is immersed, and can break the limitations of time and space to experience the things in the simulation space, so that the user and the simulation environment can have multi-dimensional information interaction, so that the user can get the most real feedback on the objective world in the process of operation, and then produce a realistic sense of immersion, thereby causing thinking resonance, deepening the concept and reform and innovation [11].

VR has the following basic characteristics:

(i) Immersion, the visual, tactile, and auditory senses that the user feels in the virtual environment are the same as those felt in the real environment, so that the user feels that things have a high degree of authenticity, so that the user has a mental resonance, resulting in psychological immersion.

(ii) Interactivity, interactivity refers to the characteristics of the VR system that is different from the traditional three-dimensional animation, the user is no longer passively accepting the information given by the computer, but can use the interactive device to manipulate the virtual object, such as when the user through their own language and body functions to the virtual environment of people or things to perform a certain operation, the surrounding environment will also send a certain corresponding feedback.

(iii) Conceptual, conceivable means that users can use VR systems to obtain perceptual and rational understanding from the environment of qualitative and quantitative integration, thereby deepening concepts and germinating consciousness [12]. The VR system is mainly composed of professional graphics processing computers, application software systems, input devices and demonstration equipment, of which the input equipment mainly consists of helmet-mounted displays, stereoscopic headphones, head tracking systems and data gloves [13]. At present, VR technology is applied to film and television, design, medicine, military and other different fields, with the needs of the development of education informatization, VR technology has gradually been applied to the education industry, driving the reform of education mode. The application and research of VR technology in physical education teaching are mainly concentrated in competitive sports and sports training, and there are fewer theoretical research and applied research on sports majors. In view of the current teaching status of sports anatomy courses in colleges and universities, this paper focuses on the application value of VR technology in the teaching of sports anatomy.

2 The Value of VR Technology Applied to the Teaching of Motor Anatomy

2.1 VR Technology Can Display the Anatomy of the Human Body in Three Dimensions, Which Helps Students to Cultivate Three-Dimensional Thinking

In the traditional teaching of anatomy, it is impossible to completely and comprehensively show the structure of various organs and the adjacent relationship of each organ through the oral explanation of the teacher and the display of PPT pictures. VR technology is applied in the teaching of motor anatomy, which can build a three-dimensional digital model of human body structure in a computer to fully display the three-dimensional spatial structure of human body. After the generation of the human digital model, students can operate the highly imitation specimen in the virtual reality system, separate and assemble each structure and carefully observe it, to help students understand and master the structure and adjacent relationship of each organ. Moreover, VR technology helps students understand how organ systems coordinate their functions. Students can observe the animation of each organ system through stereoscopic virtual images, such as the whole process of blood vessel branches and flow through the body, and the composition of the urinary system, such as the subtle structure of the kidney, ureter, bladder and urethra. Therefore, VR technology makes abstract knowledge concrete and highly interactive, which helps students learn through thinking, makes students deeply realize that the human form and structure is the material basis for realizing physiological function, and improves students' interest in learning and learning effect.

2.2 The Immersive Teaching Features of VR Technology Help Students Apply the Theory of Sports Anatomy to Sports Practice

Sports anatomy is a branch of human anatomy, but different from human anatomy, it highlights the characteristics of sports major, focus on the impact of sports on human morphological structure and growth and development, anatomical analysis of sports technical movements, reveal the law of human movement. Movement technology action anatomy analysis is both the focus of learning and difficult, because the action analysis process is too complex, involved more knowledge, students are difficult to master, at the same time this part of the human anatomy knowledge and sports practice of important bridge, the cultivation of sports students' practical ability is very important. Through VR technology, can put the sports action through dynamic video demonstration, in each stage can show joint movement form, the direction of the human body, the movement stage of active muscle and against muscle, muscle contraction form and so on details, facilitate students to movement analysis theory and practice. Students can also wear a certain equipment, including virtual 3D helmet, infrared sensor, tracker, force feedback system and so on, and then observe the analysis of each movement when doing the movement through video playback. Adjusting the Angle direction of the movement has achieved the effect of exercising the target muscle. Through feedback, the purpose of physical movement anatomy analysis can also be mastered through the process of "simulation-observation-feedback-re-simulation", and apply the theory of movement analysis to practice.

2.3 The Application of VR Technology Can Make up for the Lack of Teaching Conditions and Save Costs

The structure of the human body is very complex, and there are certain differences between individuals. It is difficult to achieve the ideal results solely by relying on books and classroom explanation. In practical teaching, due to the lack of investment in physical education laboratory, the loss of anatomical laboratory specimens and models, and the lack of sources of dead bodies, and less experimental class, students can not get enough opportunities to exercise in the experimental class. The application of virtual reality in motor anatomy teaching is a good remedy for the above problems, especially in the developed network technology today, a large number of scenes and equipment in teaching can be built through virtual reality technology. Students can enter the VR anatomy operating system at any time to watch and dissect various parts of the human body. And most importantly, they can conduct repeated operations, so as to effectively solve the problem of insufficient specimens and irreversible anatomy of the body. Students only need to use VR anatomy teaching system in mobile phones and other mobile devices to complete teaching and learning, which can effectively save human specimens, reduce the utilization rate and loss rate of models and specimens, save teaching costs, improve teaching quality, and fully meet the teaching needs.

2.4 The Immersive and Interactive Characteristics of VR Technology Can Fully Stimulate Students' Interest in Learning

Virtual reality technology applied to movement anatomy teaching, not only the text, graphics, images, sound, animation organically, and is all-directional and perspective to students, constantly stimulate learners 'senses, greatly enhance students' learning experience, help students through active observation, improve students' learning focus, and improve the learning ability and learning efficiency. Virtual reality can effectively realize the visualization of teaching content and knowledge, enhance the immersion sense of learning, and increase the interaction between teachers, students, students and students and the environment. By creating high simulation teaching situations, VR technology provides rich perceptual clues and multi-channel feedback (such as auditory, vision, touch, etc.), and helps learners to transfer the anatomical knowledge of virtual situations to real sports to meet the needs of situational learning. Learners can also directly communicate with the surrounding virtual environment to realize human–computer interaction, so as to enrich perceptual knowledge and deepen the understanding of teaching content, which greatly stimulates students' interest and enthusiasm in learning.

3 The Dilemma and Countermeasures of VR Technology Application of Sports Anatomy Teaching

Using virtual reality and sports anatomy as keywords to search for relevant literature, analyze the literature, and combine the query to visit the teaching platform and experimental platform of various colleges and universities with sports majors, the research and analysis found that the main problems in the application of virtual reality technology in the teaching of sports anatomy in China's sports professional courses include the following points.

3.1 The School's Virtual Reality Technology Hardware is not Perfect, and the Capital Investment is Insufficient

According to the investigation and analysis of literature data, few teaching equipment in physical education is equipped with virtual reality technology in China. In addition to the traditional multimedia classroom, new devices such as tablets, high-speed networks and virtual simulation devices. Problems such as the high cost of the display equipment and the clarity of the display have not been well solved, and the price of complete virtual reality equipment is still very high for [14]. Due to expensive equipment and insufficient funds, schools restrict the application of virtual reality technology in education. The way to solve this problem is to actively reduce hardware costs and develop more software systems, reduce hardware prices and make more schools buy VR hardware, and actively develop more software systems to better support sports science classroom teaching.

3.2 Lack of Software Development of the Virtual Teaching System for Sports Anatomy

In terms of software, virtual reality education companies have not done enough work on software development, especially in sports human science courses, virtual reality resources are very scarce. The Sports Anatomy teaching system integrating virtual reality is not only a technical implementation problem, but also a series of "soft" issues such as online teaching design, learning methods, teaching methods, interaction methods, and development standards that require the in-depth promotion of theoretical research and practical research. In order to promote the development of software systems and save development costs, it is recommended to build a virtual reality teaching module on the original online teaching platform, study the traditional system platform and the maximum utilization of carrying resources in the upgrade process, and promote the online teaching system platform to play the aggregation effect of resources, realize resource sharing, and save costs.

3.3 It is Difficult to Develop Teaching Resources for Sports Anatomy

The development process of VR teaching resources includes: learning situation analysis, scene script design, interaction mode design, evaluation design, operation structure design, idea design and courseware development, etc. The development tools that need to be used include graphic image processing tools, 3D model construction tools and virtual reality resource development tools. At present, the development cost and technical difficulty have become the primary factors restricting its wide application. Experts who master computer technology do not have a deep understanding of movement technology and anatomy knowledge, and teachers who understand the theory of motor anatomy and motor technology analysis have poor computer level, so it is difficult to develop sports anatomy teaching resources. From the current results, there are more teaching resources for human anatomy, but in addition to the need to master the basic structure of the human body, sports anatomy also has its outstanding characteristics of sports, that is, the law of the influence of sports on the human organ system and growth and development, the anatomical analysis of human movement technology, etc. At this stage, China's teaching resources in this regard are still relatively scarce.

How to solve this dilemma? On the one hand, strengthen the close cooperation between computer experts and sports anatomy experts to tackle key problems, encourage school-enterprise cooperation and joint development, fully learn from the technical advantages of enterprises, support colleges and universities to set up special engineering centers, and timely apply the most cutting-edge technology to resource development, in order to ensure the use of sports anatomy teaching resources in the teaching effect; on the other hand, to strengthen the cultivation of professional talents, in order to change this situation, the department of Physical Education, Tsinghua University, Shanghai University Of Sport and other universities have taken the lead in setting up similar to "sports three-dimensional simulation", "the use of modern computer information technology in sports", "video cutting and recognition", and other professional master's or doctoral programs, has begun to cultivate for China's sports community both proficient in sports training and competition and proficient in computers master's, doctoral and other high-level talents [15].

3.4 VR Virtual Helmets Produce a Sense of Vertigo, Sluggish Force Feedback, and Low Screen Resolution

VR technology support needs to be further improved, VR system is a multi-sensory interactive system, but for now the most used is vision. Studies have shown that a more comfortable experience is only possible when the resolution reaches 4K or even higher, and most current VR monitors are far from sufficient resolution [16]. For young students, long-term wear may cause adverse visual effects. And according to the feedback of many adult users, wearing VR glasses for a long time will produce a sense of vertigo, and the sensing equipment and control equipment also have the phenomenon of slow feedback. All of the above aspects affect the user's sense of experience. However, with the further development of information technology, these problems will gradually be solved.

4 Conclusion

In short, the application of virtual reality technology to the teaching of sports anatomy, greatly expanding the learning space of students, and creating an immersive and personalized learning experience for them, can fully stimulate students' interest in learning, improve learning autonomy, make up for the shortcomings of the lack of experimental teaching specimen models, greatly improve the effect of sports anatomy teaching, and respond to the requirements of the Ministry of Education on the teaching reform of colleges and universities.

VR technology also has some shortcomings, due to technical limitations, expensive prices, high maintenance costs, VR virtual helmets produce a sense of vertigo, low force feedback sensitivity, somatosensory interaction is not fine enough, etc., resulting in the promotion and popularization of virtual reality technology is more difficult, which is also a bottleneck in the development of virtual reality technology. With the development of VR technology, I believe that these problems will be satisfactorily solved, making VR technology more and more perfect, and playing a greater role in teaching.

At the same time, VR technology applied to the teaching of sports anatomy should also be supported by the development of an effective learning monitoring and evaluation system to effectively evaluate the learning behavior in the virtual reality teaching environment."Virtual Reality + Sports Anatomy Teaching" is a systematic project, which not only requires all participants to have high information technology literacy, optimize teaching design and teaching content, but also strengthen the network supervision of virtual learning space and establish a diversified evaluation and feedback mechanism.

This study provides an idea for the teaching reform of sports anatomy, which can effectively improve the teaching effect of motor anatomy.

References

1. Notice of the Ministry of Education on printing and distributing the action plan for informatization of education 2.0. Teach Skills 6 (2018)
2. Gao Y, Liu J, Huang Z, Huang R (2016) The core elements of virtual reality technology to promote learning and its challenges. Res Electrochem Educ 37(10):77–87, 103
3. Li X, Zhang L, Zhao F, Chen J (2017) Research on teaching design of hybrid form under virtual reality/augmented reality. Res Electrochem Educ 38(7):20–25

4. EDUCAUSE.2020ECUCAUS Horizon report [EB/OL], 02 March 2020. http://libarary.Educause.edu/
5. Educause.edu/resou-rces/2020/3/2020-educause-horizon-report-teaching-and-learn-ing-edition
6. Liu G, Wang X (2020) Virtual reality reshapes online education: learning resources, teaching organization and system platform. China Electrochem Educ (11):87–96
7. Ministry of Education (2016) 13th five-year plan for informatization in education [EB/OL], 16 June 2016. http://www.ict.edu.cn/laws/new/n20160617_34574.shtml
8. Cao Y (2017) Applied research on virtual reality technology in teacher education in the United States: a case study of the University of Central Florida. Comp Educ Res 39(6):93–102
9. Zhu H, Ma X (2007) Experiment course reformation of sports anatomy. Lab Sci 2(1):52–53
10. Huang H (2002) The application of modern multimedia teaching in the teaching of the motion system of "Motion Anatomy". Sports Sci Res 6(3):65–67
11. Liu H (2020) To explore the application of virtual reality technology in the teaching of anatomy experiments. Int Infect Dis 9(2):243
12. Blood EB (1990) Device for quantitatively measuring the relative position and orientation of two bodies
13. Presence metals utilizing direct current magnetic fields 18:235 (1989)
14. Hightower J (2003) Location systems for ubiquitous computing. Computer 8:563
15. Wei Y, Yang X, Wang F (2004) Virtual reality and simulation. National Defense Industry Press, Beijing. (to be published)
16. Dou Y (2014) The application of VR technology in middle school biology classroom teaching. Middle School Biol Teach 12:30

Research on the Application of Virtual Reality Technology in Physical Education in Colleges and Universities

Shengqi Wang and Mohamad Nizam Bin Nazarudin

Abstract The rapid development of the global information industry and the iterative update of cutting-edge high-tech technologies are gradually leading countries around the world to an intelligent development path based on 5G technology, artificial intelligence and other high-tech technologies. With the widespread intelligentization of terminals in the education system, virtual reality technology is gradually being widely used in the field of education and teaching. Physical education in colleges and universities is a highly specialized subject, which requires teachers to demonstrate in person in the process of imparting knowledge, and puts forward very high requirements for teachers' technical teaching. Based on the technical difficulties in college sports technology courses and the actual demand for auxiliary teaching tools, this article uses the method of literature materials, questionnaires, expert interviews, mathematical statistics and other methods to carry out a study on 7 colleges and universities with strong sports majors in Shandong, China. The research and analysis aims to explore the auxiliary teaching of sports technology through virtual reality, so that students can quickly grasp the essentials and experience of sports technology, and can accurately see the complete method of technology, deduct points for technology, and make sports technology teaching more comfortable. Finally, in view of the demand for virtual reality technology in sports technology courses in colleges and universities and the problems existing in the integration of virtual reality and sports technology courses, key combing and suggestions are made, in order to enable students to better grasp the technical essence of sports technology.

Keywords Virtual reality · Colleges and universities · Sports technology · Teaching

S. Wang · M. N. Bin Nazarudin
Faculty of Psychology and Education, University Malaysia Sabah, 88400 Kota Kinabalu, Sabah, Malaysia
e-mail: mnizam@ums.edu.my

S. Wang (✉)
Department of Physical Education, Xianyang Normal University, Xianyang 712000, China
e-mail: shengqiwang1986@sina.com

M. Al-Emran et al. (eds.), *International Conference on Information Systems and Intelligent Applications*, Lecture Notes in Networks and Systems 550,
https://doi.org/10.1007/978-3-031-16865-9_29

1 Virtual Reality Overview

1.1 Virtual Reality Concept

Virtual reality is Virtual Reality, or VR for short. Virtual reality technology is mainly to create simulation technology through vision, hearing and feeling in three-dimensional space scene [1]. It can simulate real-world scenarios through computers and related supporting equipment. For example: to obtain the experience of a real sense of space, to truly maximize the user's immersion, to feel the feeling of being there, and to observe and understand the objects in the three-dimensional space according to their own needs. This kind of virtual reality Although there is a certain gap between the feeling of the real object and the real object, it has brought people a lot of convenience and novelty experience, and can also solve many problems that are difficult to solve in real life. Some virtual reality technologies can be viewed directly with the naked eye, but others need to be viewed with interactive devices [2]. With the help of interactive devices, a stronger sense of visual immersion can be produced, and physical perception will also be enhanced.

1.2 Features of Virtual Reality

With the development of information technology, with the support of computers and their auxiliary equipment, the authenticity and experience of virtual reality scenes are getting stronger and stronger. Its characteristics are the three characteristics we often say, interactivity, immersion and intelligence [3] aspect. First of all, the application of computer virtual reality technology is interactive. With the support of interactive functions, users can interact with the virtual reality scene created by the computer according to their needs. In this way, some conditional virtual operations can be performed by using the operating handle, operating gloves, etc. connected to the computer, and the input of voice in the scene can be realized, and the response of the virtual scene can also be obtained after the operation, thus forming the effect of virtual reality interaction. Secondly, immersion is in the process of virtual reality application, which can allow users to obtain a good sense of immersion [4]. This is a personal subjective feeling. When users are immersed in virtual reality in operation, they may ignore the real time and space, so that the whole body Feel the virtual space–time field. Finally, the intelligent characteristics of computer virtual reality technology can prompt users to make intelligent responses, such as automatic calculation, intelligent reply, automatic operation, etc., according to the requirements of users.

2 The Practical Demands of Physical Technology Courses in Colleges and Universities

There are many movements that are difficult to understand immediately in the technology of physical education in colleges and universities. It takes a long time of training, thinking, and experience to get the experience and understand the movements correctly [5]. For example: kicking the ball on the inside of the instep and the outside of the instep in football technology, the kicker's stance, the angle of kicking the ball, and the timing of kicking the ball must be accurately understood to make the action in place. The tactical application of badminton, the movement of steps, and the technique of putting the ball before the net, the strength, angle of the wrist and the specific position of the action should be done, and the action is also quite fine. In competitive gymnastics, the number of rotations of the body, the force-producing part, the force-producing point, the body posture, and how to effectively cooperate with the hands and feet to complete the movement. Another example is the vaulting movement, which ends in a few seconds. Approaching movements, upper plate movements, first flying, top shoulder push, second flying, aerial movements and landing movements, due to the rapid completion of the movements, the teacher's explanation and demonstration alone are not enough for students to effectively understand the movements, let alone Master the action. Therefore, some movements in sports technology are abstract and difficult to understand. At this time, virtual reality technology has played a huge advantage. It can make it difficult for teachers to explain and demonstrate the movements in place. Through the restoration and simulation of the movements, it can directly interact with the students. Immersive experience of the difficulties of sports technical movements and difficult and difficult movements such as exertion, angle, stance, movement, manipulation, and footwork can improve the efficiency of learning technical movements and assist teachers in learning sports skills more intuitively. Complete the analysis and study of difficult technical movements. Students have high enthusiasm for the use of virtual reality in sports technology, and many sports projects have a high degree of feeling after using them.

3 Research Objects and Methods

3.1 Research Object

For the sports technology courses in colleges and universities, a total of 7 colleges and universities, Shandong University, Shandong Normal University, Shandong Institute of Physical Education, Qufu Normal University, Liaocheng University, Qingdao University and Linyi University with relatively good sports majors in Shandong, China, were selected as the research objects. A total of 800 questionnaires were distributed, of which Shandong Institute of Physical Education is a professional sports college, 200 questionnaires were distributed, and 100 questionnaires were distributed to other universities. Finally, 739 questionnaires were recovered, of which 95 were recovered from Shandong University, 189 from Shandong Institute of Physical Education, and 93 from Shandong University. Qufu Normal University 89, Liaocheng University 88, Qingdao University 91, Linyi University 94.

3.2 Research Method

This research mainly adopts the research methods such as literature data method, questionnaire survey method, expert interview method and mathematical statistics method.

Literature Research
The preliminary review refers to a large number of domestic and foreign literature on information technology, virtual reality, physical education and other related aspects, focusing on the application of high-tech auxiliary teaching cases in physical education, and the integration of virtual reality technology and teaching. This paper summarizes and organizes the key points, and deeply analyzes the practical teaching effect of the application of virtual reality technology in the physical education course. This allows for accurate refinement and in-depth analysis.

Questionnaire Survey Method
Through investigation and analysis, the "Questionnaire on the Application of Virtual Reality in Sports Technology in Colleges and Universities" was formulated. In order to ensure the reliability of this questionnaire, the Cronbach's coefficient method was used to test the combined reliability of the questionnaire. The test results showed that the Cronbach's coefficient was greater than 0.5, which met the reliability requirements. The questionnaire was sent to 6 professors and experts in education and physical education in Shandong Institute of Physical Education for identification and evaluation. Based on their feedback, the questionnaire was supplemented and improved.

Expert Interview Method
Selecting physical education professors from 7 colleges and universities in Shandong Province, China as the interview subjects, and selecting 1 physical education professor from each college. Through the interviews, we learned that with the development of high-tech technology, the application of information-based auxiliary teaching in the teaching of sports technology is gradually becoming more and more extensive. In particular, the application of virtual reality technology can really help the rapid mastery of sports technology.

Mathematical Statistics
Statistical analysis software such as SPSS was used for relevant data analysis.

4 Results and Analysis

4.1 Questionnaire Survey Analysis

Basic Information About the Schools Visited

According to the questionnaire, 7 colleges and universities with outstanding sports in Shandong Province have accepted the questionnaire. Among them, Shandong Institute of Physical Education is the most professional sports institution of higher learning in Shandong Province. The number of students majoring in physical education is more representative, so the distribution of the questionnaires accounted for It can be seen from Table 1 that there is not much difference in the proportion of effective questionnaires returned, which proves that college sports students have a relatively high tendency and enthusiasm for virtual reality.

The Motivation of the Interviewed Students

The motivation of college students to use virtual reality technology can be roughly divided into the following categories, as shown in Table 2. Most students use it for the convenience of learning. Among them, assisting in mastering sports skills, improving interest in learning skills, improving learning confidence, and self-learning are all positive assisted sports technology learning. A very small number of students choose to use virtual reality technology for entertainment. First of all, 98.9% of students are used to assist students in mastering physical skills and movements, and 98.1% are used to improve students' confidence in learning. The essentials cannot be grasped immediately, and auxiliary teaching tools are needed to analyze the learning action, and the analysis of the technology through virtual reality can improve the learning confidence. Secondly, the use of virtual reality technology to improve the interest in skills learning skills accounted for 87.9%, which also reflects the common problems of some sports technical skills. Generally, technical exercises and training are in a boring state. With the help of virtual reality technology, it can stimulate the senses of students., stimulate students' interest in learning, and carry out effective learning. Self-directed learning accounts for 95.4%. Virtual reality technology simulates the

Table 1 Proportion of the number of schools interviewed by virtual reality technology

Interviewed school	Recycled copies	Proportion/%
Shan Dong University	95	12.8
Shandong Institute of Physical Education	189	25.6
Shandong Normal University	93	12.6
Qufu Normal University	89	12.1
Qingdao University	88	11.9
Liaocheng University	91	12.3
Linyi University	94	12.7

Table 2 Motivation of college students using virtual reality skills

Motivation to use	Number of people	Proportion/%
Assist in mastering physical skills	731	98.9
Increase interest in study skills	650	87.9
Improve learning confidence	725	98.1
Self-learning	705	95.4
Entertainment	16	2.2

real teaching environment, and simulates students' classrooms through action explanation and voice interaction. Students can learn independently, and this kind of learning ideology is high. The proportion of entertainment is only 2.2%. It can be seen that there are not many people who use virtual reality technology for pure entertainment. They are all to facilitate the rapid learning of sports skills.

College Students' Expectations for the Application of Virtual Reality Technology to Sports

Through the questionnaire survey, in the actual teaching environment, students feel that some sports projects should use virtual reality technology, as shown in Table 3. Gymnastics, sports dance, tennis, boxing, shooting, badminton, football, basketball, table tennis, accounting for 98.1%, 95.5%, 94.5%, 92.4%, 87.4%, 85.8%, 80.1%, 78.5% respectively %, 75.9%, gymnastics and sports dance are relatively difficult, and belong to the difficult technical movements in the classification of item groups, while tennis is slow to get started, and requires a certain amount of time to imitate the movements. The more difficult part of boxing is the ground movements. According to the development of the movement, the movement unlocking changes are reasonably carried out. The entanglement and locking of the ground are relatively complicated, and the teaching using virtual reality technology is more intuitive and clear. Next, the movement techniques, essentials of movement, movement angles, and psychological adjustment of the shooting items all require fine motor learning, and the movement, force, angle, and techniques of badminton also require fine motor learning and adjustment. In the end, the psychological expectation of table tennis is relatively low, which is related to China's national conditions. Table tennis is known as China's national ball. Chinese students learn table tennis quickly and understand the movements more deeply.

Table 3 Psychological expectation table of sports events using virtual reality technology

Category	Number of people	Proportion/%
Gymnastics	725	98.1
Sport dancing	706	95.5
Basketball	580	78.5
Football	592	80.1
Badminton	634	85.8
Tennis	698	94.5
Pingpong	561	75.9
Shooting class	646	87.4
Boxing	683	92.4

4.2 *Difficulties of Virtual Reality Technology in the Teaching of Sports Technology in Colleges and Universities*

At this stage, the introduction of virtual reality technology in college sports technology courses has achieved certain results, but on the whole, there are still many problems to be solved in the integration of virtual reality technology and sports technology.

Insufficient Development of Virtual Reality Technology Terminals
There has been a certain development and application of virtual reality technology in physical education terminals, but there are still certain obstacles in the development of virtual reality technology. Virtual reality technology developers and trained teachers and coaches are two skins. Lack of communication, unable to effectively embed the technical essentials and technical experience into the virtual reality technology, resulting in the virtual reality technology in the terminal side can only simulate the complete movement of sports technology, lacking the specific experience in the real teacher's teaching, the integration of essentials [6]. In the expert interviews, the professors of physical education also focused on the analysis of the inter-integration of sports technology teaching and virtual reality technology.

The Transition Between Virtual Reality Technology and Traditional Teaching Mode is Difficult
The teaching of sports technology has always used the traditional mode of teaching. The introduction of virtual reality technology in college sports teaching is relatively avant-garde [7]. It allows students to experience the complete practice of technology in a realistic environment, carefully observe the key points and difficulties of technology, and stimulate students to practice. However, due to the influence of many factors, the use of virtual reality technology as an auxiliary teaching in technical teaching in an imperfect virtual teaching environment cannot be applied on a large scale, nor can it replace the teaching of teachers. In the process of collision between

technology and traditional physical education mode, we should find an effective integration point of the two to better create a perfect virtual reality learning and operation platform for students, so that students can truly appreciate the charm of science and technology and experience it personally. The authenticity of virtual reality can better improve the ability of sports practice.

The Special Cost of Using New Technology in Teaching in Colleges and Universities is Limited

At present, most of the sports technology in colleges and universities is mainly taught by teachers and coaches on the spot. Although the application of new technologies is gradually increasing, it is only a very small part of some key colleges and universities. There are generally limited special funds for sports-assisted teaching in colleges and universities. Therefore, in virtual reality technology There is insufficient allocation of teaching application funds, and a large number of terminal equipment cannot be purchased well [8]. There are certain problems in the development of virtual reality technology and subsequent maintenance and upgrades. Therefore, the use of virtual reality in the field of sports technology will also be limited, which seriously hinders. The implementation of virtual reality technology in college physical education.

5 Conclusions and Recommendations

5.1 Conclusion

(i) Through the investigation of students and sports experts, it is known that the sports technology class really needs to be upgraded in teaching application, and it must be in line with the general environment of the era of science and technology. Virtual reality technology has a great teaching help in sports technology teaching. Can improve students' confidence in learning sports skills.
(ii) In the allocation of sports projects in colleges and universities, students have great difficulty in gymnastics, sports dance, tennis, and boxing, and need the assistance of virtual reality technology in teaching. The proportion of psychological expectations in research is more than 90%.
(iii) The promotion and application of virtual reality technology in colleges and universities is still subject to certain limitations, such as: terminal developers have limited development capabilities, and cannot develop rationally based on effective teaching experience. The funds allocated to PE teaching in colleges and universities are indeed limited, and the uniqueness of students' effective use of virtual reality technology cannot be achieved. In the teaching mode, it is not possible to transform from the traditional teaching mode to the high-tech-assisted teaching mode as soon as possible to form a more efficient sports technology classroom.

5.2 *Recommendations*

(i) The reform of colleges and universities requires the cooperation of various departments, continuously exerting the strength of the school and society, increasing capital investment, purchasing

Buy advanced virtual reality technology equipment to create a perfect virtual reality technology practice teaching platform for students.

(ii) Change the concept, actively explore new technologies with the attitude of scientific development, establish a high-reduction virtual practice environment for students, and realize the deep integration of virtual reality technology and traditional teaching mode.

(iii) Improve the construction of virtual reality technology teachers in colleges and universities. Teachers should establish correct educational concepts, actively face new technologies, On the premise of updating their professional knowledge, they should arm themselves with virtual reality technology to make themselves more combat-effective, so as to better serve students.

References

1. Chen J, Yao S (2006) Application of virtual reality technology in sports technology simulation. Sports Sci 9(3):42–46
2. Qi L, Liu Z, Liu Z (2010) Development and application of network-based multimedia courseware for basketball teaching in colleges and universities. Sci Chinese Acad Sci 2(8):86–95
3. Wang J (2016) Comparative experimental research on multimedia teaching in basketball teaching in ordinary colleges and universities. Heilongjiang High Educ Res 8(6):88–96
4. He Q (2020) A review of the application of virtual reality technology in physical education. Neijiang Sci Technol 8(3):55–62
5. Han J (2020) The application of virtual reality technology to basketball fixed-point shooting training - a review of "The foundation and application of virtual reality technology." Res Sci Technol Manage 10(5):22–29
6. Arpaci I (2017) The role of self-efficacy in predicting use of distance education tools and learning management systems. Turk Online J Dist Educ 18(1):52–62. https://doi.org/10.17718/tojde.285715
7. Arpaci I, Al-Emran, Al-Sharafi MA (2020) The impact of knowledge management practices on the acceptance of Massive Open Online Courses (MOOCs) by engineering students: a cross-cultural comparison. Telemat Inform 54:101468. https://doi.org/10.1016/j.tele.2020.101468
8. Arpaci I (2017) Antecedents and consequences of cloud computing adoption in education to achieve knowledge management. Comput Hum Behav 70:382–390

The Effectiveness of Tynker Platform in Helping Early Ages Students to Acquire the Coding Skills Necessary for 21st Century

Wafaa Elsawah and Rawy A. Thabet

Abstract Learning to programme is not easy. And so, for the last few years, many online environments have been developed to help kids acquire the coding skills needed in the 21st century in a fun and interactive way. This paper uses a mixed approach to investigate elementary students' performance in programming after engaging in a 2-week online programme using the Tynker platform. The data was collected through observations and surveys. Children used Blockly programming (Python-based) to create animated stories, collages, and games. At the end of the program, the learners were assessed by a multiple-choice quiz. Additionally, they created a project that covered all the concepts covered during the program.

Successful examples from classroom observations are drawn to illustrate how students make practical use of the Tynker platform. Additionally, 117 closed-question surveys were analysed to determine the students' accurate perceptions about the coding and online platform. The interpretation of the findings implies that the students' programming knowledge acquisition follows a progressive path. However, the findings show that while all students learned the basics of coding, there were some differences in performance and understanding. This paper bridges the gap related to the insufficient attention in educational research towards teaching coding to primary students. The findings would help stakeholders to develop more capacity-building training programmes for young learners.

Keywords Coding · Programming · Early ages students · Tynker · Technology · Online platforms · Constructivism

1 Introduction

With the massive spread of technology in recent years, coding is considered an essential skill for all students in the 21st century [1–6]; However, insufficient attention in educational research has been done towards teaching coding to primary students

W. Elsawah (✉) · R. A. Thabet
The British University in Dubai, Dubai, UAE
e-mail: 20000587@student.buid.ac.ae

M. Al-Emran et al. (eds.), *International Conference on Information Systems and Intelligent Applications*, Lecture Notes in Networks and Systems 550,
https://doi.org/10.1007/978-3-031-16865-9_30

[2, 6, 7]. Nevertheless, learning to code is not an easy task for young students; it requires abstract concepts and complex skills for kids in the early stages of development. In the trial, many coding platforms succeeded in teaching the kids these skills in a suitable manner. This is emphasised by the findings of a study conducted by Lekan & Abiodun [1] and revealed that online coding platforms and age-appropriate development environments allow an easy entry into this field. These platforms play an essential role in learning the computational thinking skills needed for children and beginners; they also enable the reuse of the learning contexts [8].

In terms of educational practice, many schools do not pay enough attention to this skill; additionally, it is challenging to bring those skills in a concise school's time frame. Many online coding platforms, like Tynker, play a vital role in supporting students with these skills. Therefore, a comprehensive study that examines the effectiveness of online coding platforms in teaching early-age students the computational thinking needed in the 21st century should be conducted. This paper examines the effects of Tynker, one of these popular platforms used by 100,000 schools and over 60 million children in 150 countries around the world; additionally, it is widely used in the UAE in schools and many governmental initiatives. This paper also reveals the importance of online coding platforms for young learners' coding skills development, which are essential to cope with 21st century requirements in light of the literature and online classroom observations for primary students. The paper investigates the Tynker coding platform and determines its learning techniques, considering students' levels of understanding and ability.

The purpose of the present paper, therefore, can be broken down into the following research questions:

How effective is the Tynker coding platform in helping early ages students acquire the coding skills required for the 21st century?
What is the impact of the Tynker coding platform on early ages students?

2 Literature Review

2.1 Introduction

The literature review provides a theoretical framework underpinning the current study; what follows is an account of the answer to the research questions, which are synthesised into three main themes: (1) Importance of Coding skills for students in 21st century, (2) The role of online coding platforms on developing coding skills to early age students, and (3) The effectiveness of Tynker coding platform.

2.2 *Theoretical Framework*

Among the many theories on education, the common underpinnings of Piaget's Theory of Cognitive Development, Rogers' Diffusion of Innovation Theory and Jerome Bruner's Discovery Learning Theory could be, quite understandably, the most intriguing apropos the purpose of this study.

According to Piaget, children begin to engage in symbolic play and learn to manipulate symbols during the preoperational stage (from two to seven); they do not understand concrete reasoning and struggle with logic and mentally transforming knowledge. Teachers must consider the progression of cognitive development when teaching coding to younger students. Children start organising their thoughts, applying logical thinking skills, and relying less directly on physical representations of concepts during the concrete operations stage, which lasts for about seven to twelve years [9].

Rogers' adoption of innovation theory entails a person doing something different than they previously did. According to LaMorte [10], people who adopt an innovation early have distinct traits from those who embrace it later. When targeting the students' inventions, it's critical to identify the skills of those students that will aid or hinder adoption. Numerous studies have also found that learning coding early improves student skills like engagement, motivation, confidence, problem-solving, communication, and learning performance [11].

Coming to Jerome Bruner in his discovery learning theory, he believed that discovery should be used in the classroom to help students acquire problem-solving skills and that students should learn by using their intuition, imagination, and creativity to discover facts, correlations, and new truths. Instructors who apply discovery theory to their students should utilise stories, games, visual aids, and other attention-getting strategies to pique students' interests and encourage them to think, act, and reflect in new ways [12].

These theoretical perspectives go well together to serve the purpose of this review, which is set to consider coding skills as critical for students' future success.

2.3 *Importance of Coding Skills for Students in 21st Century*

Coding has been defined as the primary means of teaching students Computer technology in schools, it enables them to understand better how computers work [9, 13–15]. Coding evolved due to enormous intellectual and humanitarian development that cannot be ignored [16]. There is general agreement on the importance of coding as a core and vital competence in the twenty-first century, and all students should start learning it at an early age. Some studies have reported five years [17, 18]. Other studies have reported that students as young as 3–4 years of age can learn to code [19, 20]. In the same vein, Manches & Plowman [4] stated that coding is equally important as reading and writing; it also develops various skills such as science and mathematics [11]; additionally, it gives young students lifelong skills for the future [21].

In the past few years, many countries have proposed policies to promote coding skills in education. In the United Arab Emirates, the ministry of education [22] highlighted the importance of the coding role in the UAE Vision 2021, which highlights science, technology, and innovation as the main drivers of growth and progress. In 2016, the previous President of the United States, Barack Obama, launched the Computer Science for All initiative in response to the need for coding in education [18]. This policy sought to prepare students to be technological innovators and engaged citizens in a technologically advanced world. The policy strengthened the need for studying computational thinking from kindergarten to high school. Similarly, the new curricula in England have been devised to incorporate computer science learning topics for five-year-old students. However, coding is not an easy skill for young children to acquire in an abstract way. Some challenges in implementation will necessitate further effort, so it is critical to explore convenient learning content suitable for young children in their early developmental stage and link coding to everyday reasoning [23]. In recent years, studies have started to consider online coding platforms for developing younger students' computational thinking skills. What follows is an account of several authors' perceptions about online coding platforms and their role in teaching coding skills to young students. The role of online coding platforms on developing coding skills to early age students.

According to Piaget's cognitive development theory, young children's thinking is mainly categorised by symbolic functions and intuitive thoughts; they cannot absorb abstract concepts and logic. This view is supported by several studies that state that children at their early ages are frequently unable to create sound logical thinking when they are confronted with unfamiliar issues, too much information, or facts that they cannot reconcile [23–25]. These developmental considerations must be considered when designing educational programmes to teach coding to young students [9].

Children should be exposed to coding with appropriate pedagogical approaches. Bruner and Roger, in their theories, subscribe to the view that play and discovery are important ways of early learning and tend to adopt a cross-curriculum approach that recognises the physical, cognitive, and innovative aspects of education. Therefore, the strategies must be built in a way that respects early age students' pedagogy [4]. Recently, researchers have shown an increased interest in using well-designed online coding platforms, which may be particularly well-suited for this learning in this developmental period [14, 26, 27]. These platforms make coding easy to use for young children to understand; they depend on visual coding called block-based coding, which gamifies the activities, uses goals, tales, and discoveries, and provides a more visually graphical environment [3, 25]. A recent study by Gray & Thomsen [28] reports that young students who learn coding through digital play and playful approaches readily immerse themselves in the problem-solving process and make worthwhile discoveries. However, the online learning platforms have not escaped issues like cheating, lack of communication, splitting the participants into groups, and technological challenges due to internet and power outages that hamper the class activities [29–33]. Therefore, it is recommended to integrate online coding platforms into different didactical approaches, such as blended learning scenarios or flipped classroom settings, to fully benefit from them.

2.4 The Effectiveness of Tynker Coding Platform

Tynker is a coding platform launched in April 2013 that teaches youngsters the fundamentals of coding and game creation and how to create apps and complete outstanding projects in a fun and interactive way [16]. Tynker utilises visual code blocks to introduce logic concepts to children by providing free activities, mainly games and stories, to learn code during the popular hour of code [21]. Many authors, including [3], argue that most teachers in this field get stuck choosing the appropriate course for their students and have little insight into online coding materials. Tynker overcame this challenge by offering courses designed for each specific age group for students from 5 to 17 years old. Additionally, it allows teachers to create virtual classrooms with their students and track their progress on different courses [14].

Through their "Hour of Code" courses, Tynker provides free activities for kids to learn to code and be creative at the same time. Schools can also benefit from including Tynker in their curriculum to allow students to learn the coding fundamentals found in all object-oriented programming languages [21]. It is also compatible with many operating systems like Windows and Mac, and it has an application installed on a device with a mobile operating system [25].

Despite the numerous benefits offered by Tynker, some authors [7, 34, 35] question the usefulness of the block-based coding used by it in learning the actual code; they argue that when the students move to text-based programming, they feel overwhelmed with the structure of the text programming language. Yet, Tynker tried to overcome this challenge by providing a "toggle" feature where students can see their actual text code while working in block format. The following sections give a more detailed account of Tynker and its effectiveness in teaching the youngest Emirati students the coding fundamentals in online classroom settings.

3 Methodology

3.1 Study Design

This paper employs a mixed-method approach to investigate UAE primary students' performance in coding after engaging in a 2-week online coding class through the Tynker coding platform. The observation method and surveys are used for the conduct of the present study. The study had four dimensions: coding, the importance of coding to early-age students, online coding platforms, and Tynker. Observations allow the

Table 1 Demographics information of the participants

	Students	Trainers
Number of participants	319	7
Age	7–10 years old	27–35 years old
Male	183	1
Female	136	6

researcher to describe current situations, giving the researcher a "written snapshot" of the situation. Observation is an appropriate method for this study purpose, which aims to explore the detailed learning experience of early-age students in coding through an online platform. The purpose of the survey is to have more reliable data about students' perspectives on learning coding in online settings using the Tynker platform.

3.2 Participants

Participants in this study were UAE primary students, trainers to facilitate the learning process, and parents who were also active participants in the program, as they were assisting their children in setting up the Zoom platform and submitting assignments and following up via WhatsApp.

The sample size consists of 319 students, all of them from the western region and ranging in age from 7 to 10, from the same racial and socio-economic backgrounds. The students have been distributed among seven trainers aged 27 to 35 years old to help and guide them during the program. The trainers were from different Arab nationalities and had been fully trained on the curriculum and how to deal with young students in an online context before the programme started (Table 1).

3.3 Data Collecting Tools

In the data collection, the paper includes two instruments: observations and surveys. Observations of students' progress and achievement are to assess the impact of the Tynker platform on early-age students' achievement. The researcher attended all the sessions and recorded the students' progress extracted from the Tynker dashboard. Additionally, she took the feedback from the trainers at the end of every day, and finally added the results together in the observation notes. A total of 117 anonymous surveys containing closed-ended questions were created by SurveyMonkey. The students are asked to answer on a Likert-Scale from strongly agreeing to strongly disagreeing with the aid of their parents, as the parents have the main role in supporting the students and communicating with the trainers during the learning process. The answers measure the participants' insights and conceptions about the programme and ensure the validity of the study. In this case, only the students who answer with a trust level of "strongly agree" or "agree" are counted.

3.4 Procedures

Zoom and WhatsApp applications were used as ways of communication between students and teachers. The program was a non-profit governmental initiative funded and organized by a governmental organization in UAE to teach the Emirati young students how to code. It has been designed following the constructionist approach and grounded theories like discovery, learning-by-doing, and innovation.

Students were divided into seven classes and attended ten daily induction zoom sessions 30-min each led by multiple instructors to introduce the main lesson concepts before directing the students to Tynker to start applying. After the zoom session, the instructors sent a pre-made video to the students in the WhatsApp group to serve as a reference in their practice phase. The students spent approximately 60–90 min finishing each lesson.

3.5 Materials

Materials included the Programming 201 course in Tynker; the course contains eight lessons with unplugged activities to solve Python programming puzzles to create animated stories, animations, and games. The children were free to choose to learn in the application or directly work in the browser. Paid Tynker accounts have been sent to the students two days before the program, with an explanatory video and student information sheet to explain how to use the platform effectively (Table 2).

3.6 Data Analysis and Results

Students' achievement in the Tynker platform was investigated by observing their engagement and performance in the online classroom, then compared to quiz scores and project grades. At the end of each lesson, there is a quiz the students should solve to complete the lessons. Each quiz has multiple choice questions that cover all the new programming concepts of the lesson. The researcher put the quiz mark as a progress indicator for the students on the observation sheet and compared it with their comprehension of programming concepts. Some students were confused about activating their Tynker account and accessing their class to start working on the first day. The language was also an obstacle for some students from fully understanding some programming concepts. The language barrier has a negative influence

Table 2 Practical sessions involved into the program

Module	Practical session	Title
Programming 201	P1	Introduction to programming 201
	P2	Loops and Animation
	P3	Creating a Scene
	P4	Jumping over Obstacles
	P5	Rotation
	P6	Broadcasting Messages
	P7	Time Limits
	P8	Pop the Balloon

on students' academic achievement [36]. Still, with the help of their teacher and following the guided instructions, they started immersing and working by the end of the day. The use of Tynker tools provided an interactive environment to motivate students to engage in the tasks. The instructors encouraged the students to progress and get confidence during their activities; moreover, they sent a daily honour board to all the students who finished their daily lesson efficiently to encourage them to continue working.

Regarding the zoom sessions, the students interacted well with them to ask about the new concepts and activities. Additionally, the students sought help when they had misunderstandings. The students rarely interacted or cooperated with each other; instead, they only interacted with their instructor.

Students' results in quizzes and various activities indicate that most students complete their assigned lessons effectively, doing well and making significant progress in learning to code; on the other hand, a few students fall behind and do not complete their tasks nor score a good grade in the quizzes and project. After the investigation and communication with both parents and students, it became clear that some parents forced these students to participate in the program. As a result, they did not watch the explanatory videos or follow the daily instructions sent by the instructor, nor did they try to progress. Few others apologized because of illness conditions.

Although all the trainers made a great effort to be present with the students, the lack of live interaction and communication posed a threat in the online learning environment and hampered some students from fully participating in the learning process. Consequently, some students withdrew after a few days because of their inability to deal with the platform and understand the coding concepts.

The results also show that the number of withdrawn girls is more significant than boys; nevertheless, attendance and participation percentages are convergent (Table 3).

Table 3 Percentage of students' attendance, participation, and withdrawal

	Boys (%)	*Girls*(%)
Withdrawal %	6	10
Attendance %	92	93
Participation %	96	95

The survey results show that the students were happy with the coding learning experience and were eager to learn more advanced programmes in the future. Furthermore, they are satisfied with the online environment and the learning process. The data collected from surveys was analysed by PIVOT tables and charts as follows (Table 4):

The weighted mean and standard of deviation for all questionnaire items were calculated as shown below (Table 5).

The first part of the survey was to investigate the effectiveness of the Tynker platform, the explanatory videos, and daily zoom sessions. The following charts report the students' answer frequency for this part (Fig. 1):

The second part of the survey questioned trainer support and the utility of daily instructional films. The majority of students agreed that their trainers were constantly available to assist them and that the explanation videos aided in the material's elucidation.

It is clear from students' answers that the trainers' support is very much correlated to the students' enjoyment of their learning experience and their ability to work flexibly in Tynker. The answers showed that the videos were very helpful in the curriculum comprehension. Furthermore, the trainers provided the needed support all the time (Fig. 2).

The last part of the survey asked the students about their evaluation of their level after completing this programme and their eagerness to learn more programming in an online context. The results showed that 111 students agreed that this programme advanced their programming skills, while only 6 students reported neutral answers. This is supported by their answer when they asked if they were eager to join more programming courses, 115 students showed their eagerness to join and only 2 students disagreed to join any other future programming course. However, when they asked whether they preferred to participate in future programmes online, 65 students strongly agreed to participate in online programs, while 51 students showed their unwillingness to participate in the programme again in the online context. This may be because of the technical difficulties that have been faced by some students, especially the younger ones (Figs. 3 and 4).

Table 4 Students' responses frequencies of the survey questions

	Strongly agree	Agree	Neutral	Disagree	Strongly disagree
I enjoyed online learning programming experience	97	15	5	0	0
Online learning through the Tynker platform is easy	79	31	5	2	0
I like accessing the Tynker platform daily and completing the lessons	107	8	2	0	0
The trainers provide support on the different communication tools	104	11	1	0	1
Explanatory videos contributed to more clarification of the material	79	31	7	0	0
The zoom explanation sessions provided by the trainers are helpful	86	28	3	0	0
My programming skills advanced after completing this program	89	22	6	0	0
I am eager to learn more about programming after this program	93	22	0	2	0
I'd like to learn more about online coding programs	65	0	1	51	0
Grand Total	799	168	30	55	1

4 Discussion of the Results

This study provides empirical evidence from online classroom observations that teach 7–10 years old students coding through the Tynker platform. It also seeks to investigate the effectiveness of online coding platforms to develop the acquisition of coding skills. The observations were conducted on 319 children participating in coding activities in an online programme to address the above proposition. Students' surveys were used to measure their satisfaction with the programme and their learning experience in online coding classes. By investigating the impact of the online settings on learning to code among early-age students, the results show that online coding environments have presented new opportunities and promoted the need to design

Table 5 Mean and SD

	Strongly agree	Agree	Neutral	Disagree	Strongly disagree	Total	Weighted Mean	Mean^2	SD
I enjoyed the online learning programming experience	97	15	5	0	0	117	4.786324786	23.16239	4.286732
Online learning through the Tynker platform is easy	79	31	5	2	0	117	4.598290598	21.57265	4.119995
I like accessing the Tynker platform daily and completing the lessons through it	107	8	2	0	0	117	4.897435897	24.11111	4.383341
The trainers provide support on the different communication platforms	104	11	1	0	1	117	4.854700855	23.81197	4.353994
Explanatory videos contributed to more clarification of the material	79	31	7	0	0	117	4.615384615	21.65812	4.128285
The zoom explanation sessions provided by the trainers are helpful	86	28	3	0	0	117	4.709401709	22.4359	4.210285

(continued)

Table 5 (continued)

	Strongly agree	Agree	Neutral	Disagree	Strongly disagree	Total	Weighted Mean	$Mean^2$	SD
My programming skills advanced after completing this program	89	22	6	0	0	117	4.709401709	22.48718	4.21637
I am eager to learn more about programming after this program	93	22	0	2	0	117	4.760683761	22.94872	4.264743
I'd like to learn more about online coding programs	65	0	1	51	0	117	3.675213675	15.7094	3.469033

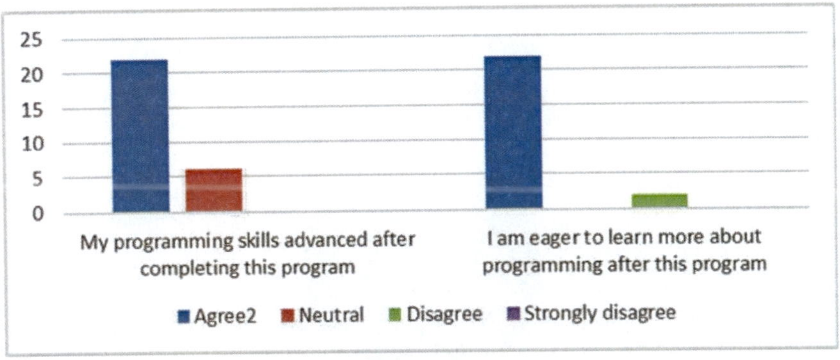

Fig. 1 Students' perspectives about the Tynker coding platform and the explanatory videos usefulness

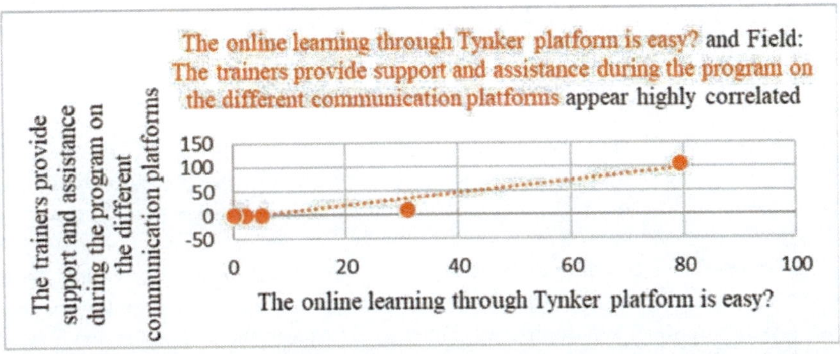

Fig. 2 Trainers' support effect on students' learning easiness through Tynker

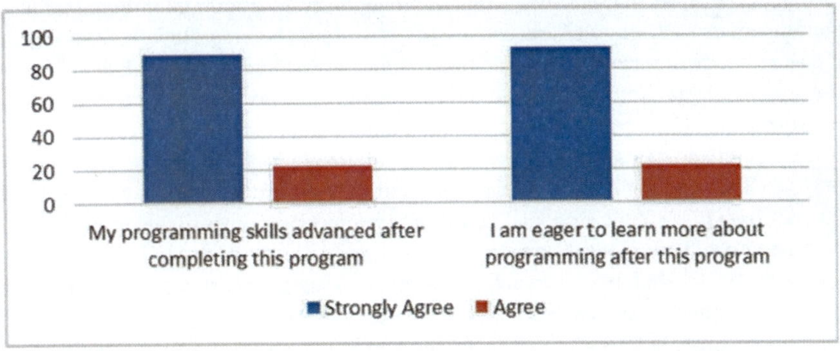

Fig. 3 Students' programming skills level and eagerness in learning more programming

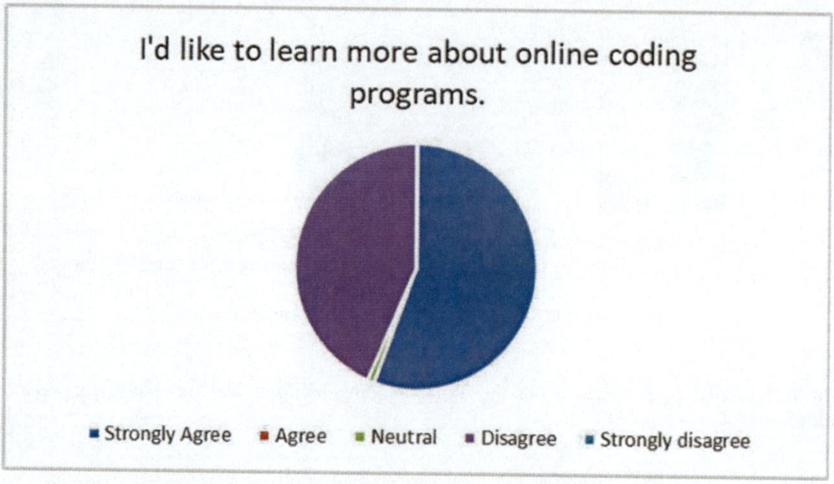

Fig. 4 Students' desire in joining more online programming courses

more coding experiences for learning. These results are consistent with most of the past literature. However, they are inconsistent with a few pieces of literature that question the usefulness of block-based coding used by online coding platforms [7, 34].

The course aims to help young students gain first-hand experience with Python programming and boost their interest in different topics regarding computer science. The programme had 345 registered students; 319 students completed the whole course, and 26 withdrew. The analysis of the activities and quiz results confirmed the success of the Tynker platform in teaching coding skills to early-age students; also, their performance in coding can improve in online coding environments as a consequence of learning and practise in a fun and interactive way. However, the lack of physical interaction between course instructors and participants is a frequent reason for some students' not finishing a course. Cheating also posed a threat to measuring the true success of some students, as we found that few mothers or older brothers of very young students who are seven years old are working on their behalf on their projects. The cheating issue was raised in the literature by many authors who reported that it is hard to track the students' actual programming progress in the online coding environments [30–33]. Still, most of the rest were determined to work and learn on their own. By the end of day 3, each class had at least ten highest achievers, finishing their lessons and activities early and eager to work more. Still, unfortunately, the inability to create subgroups in Tynker hampered the instructors from assigning new activities to them and applying the differentiation strategy between students in the same class.

The observations also show that, despite a few students getting stuck in the middle, they overcame that and understood the most foundational programming concepts after the teacher intervention. More importantly, most students in the sample were able to observe a programmed animation and deduce, through logic, the programming instructions necessary to demonstrate a range of mastery and creativity through coding.

By comparing these findings with the literature, it can be concluded that despite the vast growth of online coding platforms for kids, and their success in developing appropriate courses for them, how primary-aged children learn to code using these platforms still needs further investigation.

5 Conclusion

This study suggests that further consideration of coding education for children in the early years is needed. Moreover, there is still work that needs to be done to determine how easily these coding skills can be integrated into early childhood pedagogy. According to the findings, young students can follow a sequential programme on the growing number of coding platforms, which plays an essential role in teaching them the coding concepts in a fun, interactive, and appropriate way. The findings also imply that education ministries and decision-makers should pay more attention to engaging coding skills in early-age students' curricula.

With its interactive tools, fun activities, and puzzles, the Tynker Learning Platform is an excellent solution for teaching programming to early students. Tynker represents a game approach to critical thinking education for young children, allowing them to learn complex programming ideas through engaging and relevant methods for their ages and interests. Because of the nature of online educational resources within Tynker, it is possible to create new learning scenarios upon it. Furthermore, the content might be disseminated in a variety of ways that can suit the students.

6 Implications, Limitations, and Future Recommendations

The evidence from this study holds implications for the importance of introducing programming and coding in a fun and interactive way appropriate for early-age students. The study also suggests a framework that allows the teachers to capture the diversity of students, implementation, evaluation, and what exactly needs to be done on the online platforms to acquire the coding skills in each age phase. It also recommends using coding platforms in different educational approaches, such as blended and flipped classroom settings, to benefit from them wherein the teachers are present and can facilitate the learning process; moreover, they can overcome such cheating and lack of interaction issues.

This study has some limitations that need to be addressed. The study took place only in the western region; the participants included were from the same racial, cultural, and social backgrounds, limiting the ability to generalise results. Future studies are needed to conduct a similar investigation in different cultural contexts.

The study takes into account only the learners from grades 1 to 3. Further studies to adjudicate the efficacy of coding platforms at different school levels need to be conducted. The study does not consider the teachers' experience, training, and effectiveness that could affect this learning process. Some studies reveal that many teachers lack the training and knowledge of the discipline of coding. Therefore, it is crucial to shed light on the teacher's skills and how to cope with 21st century aspirations in future studies. Further research might also explore the usefulness of integrating coding with other curricula like math and science.

Acknowledgements This research article was a project submitted to the British University in Dubai during the master's studies of the first author.

References

1. Abiodun OS, Lekan AJ (2020) Children perceptions of the effectiveness of online coding as a supplement to in-person boot camps. Int J Sci Adv 1(3):187–191
2. Kanbul S, Uzunboylu H (2017) Importance of coding education and robotic applications for achieving 21st-century skills in north Cyprus. Int J Emerg Technol Learn 12(1):130–140. Kassel University Press GmbH
3. Kim AS, Ko AJ (2017) A pedagogical analysis of online coding tutorials. In: Proceedings of the Conference on Integrating Technology into Computer Science Education, ITiCSE, pp 321–326. Association for Computing Machinery
4. Manches A, Plowman L (2017) Computing education in children's early years: a call for debate. Br J Educ Technol 48(1):191–201. Blackwell Publishing Ltd
5. Mokhtar FA (2016) Recognizing possible limitations of e-learning through Edmodo. In: Proceedings of the ICECRS, vol 1, no 1. Universitas Muhammadiyah Sidoarjo
6. Vico F, Masa J, Garcia R (2019) ToolboX. Academy: coding & artificial intelligence made easy for kids, big data for educators. In: Proceedings of the 11th Annual International Conference on Education and New Learning Technologies (Edulearn19)
7. Lewis S (2020) Analysis of how primary-aged children learn to code: a year 5 case study using Ev3 LEGO® robotics and stimulated recall. MEd. Thesis. University of Central Queensland
8. Grandl M, Ebner M, Slany W, Janisch S (2018) It' s in your pocket: a MOOC about programming for kids and the role of OER It' s in your pocket: a MOOC about programming for kids and the role of OER in teaching and learning contexts. Graz University of Technology
9. Relkin E, De Ruiter LE, Bers MU (2021) Learning to code and the acquisition of computational thinking by young children. Comput Educ 169:104222. Elsevier Ltd
10. LaMorte W (2019) Diffusion of innovation theory. https://sphweb.bumc.bu.edu/otlt/MPH-Modules/SB/BehavioralChangeTheories/BehavioralChangeTheories4.html. Accessed 31 Oct 2021. /11/31
11. Turan S, Aydoğdu F (2020) Effect of coding and robotic education on pre-school children's skills of scientific process
12. Pappas C (2014) Instructional design models and theories: the discovery learning model. https://elearningindustry.com/discovery-learning-model. Accessed 31 Nov 2021
13. Arfé B, Vardanega T, Montuori C, Lavanga M (2019) Coding in primary grades boosts children's executive functions. Front Psychol 10:2713. Front Media S.A.
14. Manita F, Durão S, Aguiar A (2021) Faculdade De Engenharia Da Universidade Do Porto towards a live programming platform for K-12

15. Román-González M, Pérez-González JC, Jiménez-Fernández C (2017) Which cognitive abilities underlie computational thinking? Criterion validity of the computational thinking test. Comput Hum Behav 72:678–691 Elsevier Ltd
16. Cornella A: Education for humans in a world of smart machines. Barcelona: Institute of next Barcelona (n.d.)
17. Department for education: national curriculum (2014). https://www.gov.uk/government/collections/national-curriculum. Accessed 31 Oct 2021. /11/31
18. Smith M (2016) Computer science for all. https://obamawhitehouse.archives.gov/blog/2016/01/30/computer-science-all. Accessed 31 Nov 2021
19. Bers MU (2018) Coding as a Playground: Programming and Computational Thinking in the Early Childhood Classroom, 2nd edn. https://doi.org/10.4324/9781003022602. Accessed 1 Nov 2021
20. Strawhacker A, Bers MU (2019) What they learn when they learn coding: investigating cognitive domains and computer programming knowledge in young children. Educ Technol Res Dev 67(3):541–575. Springer New York LLC
21. Kaplancali UT, Demirkol Z (2017) teaching coding to children: a methodology for Kids 5+. Int J Element Educ 6(4):32. Science Publishing Group
22. Ministry of education: Science, Technology & Innovation Policy in the United Arab Emirates (2015)
23. Chen G, Shen J, Barth-Cohen L, Jiang S, Huang X, Eltoukhy M (2017) Assessing elementary students' computational thinking in everyday reasoning and robotics programming. Comput Educ 109:162–175 Elsevier Ltd
24. Berk LE, Meyers AB (2016) Infants and Children: Prenatal Through Middle Childhood. Pearson, London, UK
25. Kraleva R, Kralev V, Kostadinova D (2019) A methodology for the analysis of block-based programming languages appropriate for children. J Comput Sci Eng 13(1):1–10. Korean Institute of Information Scientists and Engineers
26. Sheehan KJ, Pila S, Lauricella AR, Wartella EA (2019) Parent-child interaction and children's learning from a coding application. Comput Educ 140:103601. Elsevier Ltd
27. Stephany F, Braesemann F, Graham M (2021) Coding together–coding alone: the role of trust in collaborative programming. Inf Commun Soc Routledge 24(13):1944–1961
28. Gray JH, Thomsen BS (2021) Learning through digital play: the educational power of children making and sharing digital creations
29. AlAjmi Q, Al-Sharafi MA, Yassin AA (2021) Behavioral intention of students in higher education institutions towards online learning during COVID-19. In: Arpaci I, Al-Emran M, Al-Sharafi MA, Marques G (eds) Emerging Technologies During the Era of COVID-19 Pandemic. Studies in Systems, Decision and Control, vol 348, pp 259–274. Springer, Cham. https://doi.org/10.1007/978-3-030-67716-9_16
30. Bozkurt A et al (2020) A global outlook to the interruption of education due to COVID-19 pandemic: navigating in a time of uncertainty and crisis. Asian J Distance Educ 15(1):1–126
31. De Jesus MA, Estrela VV, Mamani WDH, Razmjooy N, Plaza P, Peixoto A (2020) Using transmedia approaches in STEM. In: IEEE Global Engineering Education Conference. EDUCON. IEEE
32. Falco E, Kleinhans R (2018) Beyond technology: identifying local government challenges for using digital platforms for citizen engagement. Int J Inf Manag 40:17–20
33. Golden J, Kohlbeck M (2020) Addressing cheating when using test bank questions in online classes. J Account Educ 52:100671
34. De A, Do N (2021) Towards a live programming platform for K-12
35. Powers K, Ecott S, Hirshfield LM (2007) Through the looking glass: teaching CS0 with alice. In: SIGCSE 2007: 38th SIGCSE Technical Symposium on Computer Science Education, vol 1, pp 213–217
36. Yassin AA, Abdul Razak N, Qasem YA, Saeed Mohammed MA (2020) Intercultural learning challenges affecting international students' sustainable learning in Malaysian higher education institutions. Sustainability 12(18):7490

The Adoption of Cloud-Based E-Learning in HEIs Using DOI and FVM with the Moderation of Information Culture: A Conceptual Framework

Qasim AlAjmi, Amr Abdullatif Yassin, and Ahmed Said Alhadhrami

Abstract Cloud computing has led to the paradigm shift in information technology. However, its integration with the higher educational institutes remains a novel area to explore. The study aims to assess the adoption of Cloud-Based E-Learning in HEIs using DOI & FVM with the moderation of Information culture: a Conceptual Framework. A conceptual framework assimilates the Diffusion of Innovation theory & Fit-Viability model to fulfil the educational needs. A cross-sectional study design was used undertaking 33 institutions, where a close-ended questionnaire was used for collecting primary data. The gathered data were analyzed using Statistical Package for Social Sciences (SPSS). A significant impact of Relative Advantage ($p = 0.04$), Complexity ($p = 0.00$), Compatibility ($p = 0.00$), Trialability ($p = 0.01$), Observability ($p = 0.01$), Task ($p = 0.00$), Technology ($p = 0.00$), and IT infrastructure ($p = 0.02$) were found on student's performance. Moreover, the impact of Economic ($p = 0.60$) and Organization ($p = 0.70$) was found to be insignificant. Also, information culture significantly moderated the relationship between the adoption factors of Cloud-Based E-Learning in HEIs and Student's Performance ($p = 0.00$). The study proves beneficial for the decision makers concerning their focus on the factors that can help in yielding better academic outcomes linked to the adoption of cloud computing for e-learning. Therefore, the study has concluded that Could-Computing factors influences the value and the student's performance in HEIs in Oman. Also, the outcomes of the study highlighted the significance of the developed conceptual framework which serves as an introductory model for establishing an information culture within HEIs.

Q. AlAjmi (✉) · A. S. Alhadhrami
Department of Education, College of Arts and Humanities, A'Sharqiyah University, Ibra, Oman
e-mail: Alajmi.qasim@gmail.com

A. S. Alhadhrami
e-mail: Ahmed.alhadhrami@asu.edu.om

A. A. Yassin
English Department, Centre of Languages and Translation, Ibb University, Ibb, Yemen

M. Al-Emran et al. (eds.), *International Conference on Information Systems and Intelligent Applications*, Lecture Notes in Networks and Systems 550,
https://doi.org/10.1007/978-3-031-16865-9_31

Keywords Cloud computing · Diffusion of innovation · E-learning · Higher education institutes · Fit-viability model · Information culture

1 Introduction

Through the advent of modern technology and the Internet, the most common practice of teaching and learning in a classroom on blackboards or whiteboards is almost diminished in real time. This setup of education is taking its new shape called as 'e-Learning'. Ansong et al. [1] defined it as state-of-the-art educational technology adapted to support learning and teaching practices through instructions delivered on a digital device like computers and mobile phones. Cloud computing has innovated the structure of doing business. It was developed a decade before and it creates a paradigm shifts in Information Technology [2]. Cloud computing is a dynamic innovation platform that gives a wide range of digital framework to broaden a storage capacity of information. Also cloud computing gives an easy access to programming and equipment without any capital cost and also gives an easy path to administration and applications that can acknowledge insufficiency of interaction [3, 4]. Organization and administration have found thc solutions in cloud computing what the seek for minimizing the cost efficiently. The cost advantages cover through virtualization, scalability, and on demand hardware and software [4, 5].

An innovative concept of e-learning exists in the massive field of IT offering a number of services, which is purposeful in multidimensional directions of software, infrastructure, and platform of an organization. Technological factors comprise both internal and external technologies relevant to the organization as claimed by Ansong et al. [1]. In this case, technology not only denotes the features of software or hardware but also give insights on how well cloud-based e-learning can be adopted in teaching and learning practices [6]. Various technology adaptation models have been introduced by previous studies [7, 8]. The study posits a unified theory of acceptance and use of technology (UTAUT) (Venkatesh, Thong & Xu [9], and Technology-organization-environment (TOE) Baker [10] as primary models for the adoption of the technology at the individual level. DOI and FVM are also recognized as the models promoting the technology adaptation at the organizational level.

E-learning is classified as synchronous or asynchronous. Synchronous technology allows for live interaction between the instructor and the students (e.g., audioconferencing, videoconferencing, web chats etc.) while asynchronous technology involves significant delays in time between instruction and its receipt (e.g., E-mail, earlier video recording, discussion forums etc. Rogers [11] developed the theory of diffusion of Innovation, where he blended more than 500 research articles related to diffusion. Rogers has pointed out four main elements to boost the spread of origination including social systems, time interval, communication channels, and innovations. Tjan [12] inspiring from task-technology fit (TTF) model derived a model of Fit-Viability to address the adoption of new technology in an organization. Fit in this model measures the consistency of core competence, values, structure, and culture of an organization by adopting new technology, whereas, Viability calculates the value-added perspective of capital needs, infrastructure of economics, necessities of human resources, etc.

Therefore, this study uses a comprehensive framework to give better insight and institutional powers from the perspective of stakeholders in the adoption of cloud-based e-learning among higher educational intuitions in developing countries. It is believed that the theory of the Diffusion of Innovation and the model of Fit-Viability will assist HEIs in fulfilling the educational needs leading to the augmentation of the efficiency and production in different academics.

In addition, results of this study will assist in the professional development of the teachers by establishing a standardized framework to administer overall manual work which is reduced through the adoption of cloud-based e-learning technologies. The escalated dynamics of information technology are the outcomes of the advancement in cloud computing. It is one of the scientific breakthroughs, which has advanced the storage capacity, communication as well as accessibility of data from various locations [13]. The magnitude of cloud computing has enclosed the educational sector within its range providing it with various models as well as features for accessing and securing data through various devices. Along with it, its utilization also amplifies the connectivity as well as storage, which substantially contribute to the students' academic endeavour [14]. However, a comparison of its benefit and adaptation highlights that despite its increased advantages, its adaptation among the institutes and organization is slow [15]. The study of Pluzhnik & Nikulchev [16] further highlighted the significance of the cloud computing system considering its changing dynamics and study criteria, which require withholding of large data. Rindos, Vouk, & Jararweh [17] indicated that the utilization of cloud computing is substantially dependent on the institute environment where its' associated certain factors can impact the cloud computing adoption. Ssekakubo, Suleman & Marsden [18] illustrated that the illiteracy rate of ICT hinders the adaptation of the e-learning system among the institutes. Computer anxiety is also recognized as the hindering block towards its adaptation [19].

A recent study by Bulla, Hunshal & Mehta [20] has demonstrated that the adoption of cloud computing reduced the expenditure of computational operations as well as data storage. Al-Ajmi et al. [21] highlighted that the inclusion of dynamic scalability, virtualization technology, disaster recovery, and optimal server utilization, as well as on-demand cloud services, improve the educational performance of the institutes. Bibi & Ahmed [22] illustrate that the dynamic stability of cloud computing improves the institute's additional buffer processing without making the additional investment. Durairaj & Manimaran [23] highlighted the disaster recovery component of cloud computing stating its escalated performance for retrieving information from various sites. One drawback of these components is that it reduces the institutes planning for the disaster recovery [21]. Gai and Steenkamp [24] indicated that the virtualization of cloud computing acts as a simplification stimulator for the reconfiguration process. Various models are provided for assessing the adaption of the technology among the students, some of which are detailed below.

1.1 Diffusion of Innovations and E-Learning

In 1965, Everett Rogers developed a theory called diffusion of innovations (DOI) which assesses the occurrence of the social pressures as a result of innovation, a novel idea or the dispersion of new idea across community, organization or institution [25]. Rogers [26] in his theory highlighted that the spread of innovation is impacted by three elements encompassing innovation, communication channels, and necessary time, which impacts the adaptation of innovation and a social system, assimilating both internal and external factors.

In the context of e-learning, all learning-based technologies are blended together which results in the emergence of massive open online courses (MOOCs) and personalized online learning. This has been recognized as an essential innovation in the discipline of education. In this context, the study of Zhang et al. [27] can be considered which demonstrates that the perception and the attitude of the individual towards adaptation of the technology varies among the individuals based on their perceptions and attitudes concerning e-learning. Adoption remains confined based on its cost, quality, agility, and certification of degree, schedule control, and personal demands. The theory of diffusion assists in developing an understanding related to the short- and long-term innovative implications [28]. This model integrates five intrinsic characteristics such as compatibility, relative advantage, trialability, complexity, and observability, which significantly impact the adaptations of the technology. Mkhize, Mtsweni & Buthelezi [13] have revealed the effectiveness of this model on the adoption of computing technology by educational institutes. The adoption of e-learning suffers from limited access to the material [29]. Despite the adaptation, sustainability remains low given the poor implementation [30]. This indicates that concerns related to the adoption of the e-learning have not been comprehensively recognized.

1.2 DOI and Information Culture

The theory of diffusion of innovation assesses the dissemination of ideas among individuals. It expands from the two-flow theory, focusing on the conditions which accelerate or decelerate the possibility of adoption of an innovation, or a new idea or practice. It is found that the opinion leader substantially impacts the behavior of the individuals towards the adaptation of a particular idea or practices. The information culture asserts towards the notion that the adoption is not simultaneously practiced by everyone at a similar time instead it varies on a time sequence basis and the duration of being exposed to it [31]. The development of technology is based on the individual capability to adopt a reflective approach of the action instead of actually performing it. This reflection assists people in realizing the path where the world is heading and the influence of their behavior to adopt or not. Reflecting upon the development of e-learning, its increase is found to be widely spread across various disciplines [32]. ICT holds the potential to assist students in their learning endeavors as their effectiveness is dependent upon its acceptance [33]. Al-Gahtani [34] has indicated that the perception of the individual towards the technology is based on various factors such as their knowledge and skills. The adaptation of e-learning technology is based on human cultures in which they operate [35]. This emphasizes towards the formation of a culture where information regarding the development circulates around the adaptation of new technology. Schein's (2010) theory of organizational culture also stated that the organizational information culture and its various levels should be considered for stimulating the adoption of e-learning technologies.

1.3 FVM and E-Learning

The adaptation of the technology is particularly catered by Fit-Viability model. This model has been derived from the task-technology fit (TTF) model which was formulated by Goodhue & Thompson [36] Given the widespread importance of e-learning, its adaptation depends on two aspects, including its outcomes and satisfaction [37]. The adaptation of the computing technology in e-learning using Fit-Viability Model has been demonstrated by various studies. For instance, Liang et al. [38] showed its effectiveness in the adoption of mobile technology. Mohammed, Ibrahim, & Ithnin [39] have also supported the fit and viability dimensions for the adaptation of cloud computing in the context of developing countries.

1.4 FVM and Information Culture

Various studies have indicated the impact of information culture to adopt the computing technology [40, 41]. An association has been demonstrated between the IS system development and information culture in the study of Mukred, Singh, & Safie [41]. The culture of the institute and organization is reported to significantly impact the computing technology adoption [42]. Choo [43] has stated that the information culture of the organization significantly impacts its performance, though, the relation between the culture and organization lacks evidence.

2 Hypothesis and Conceptual Framework

The model proposed in this study is inclusive of three dimensions, including Fit-Viability Model, diffusion theory model, and information culture. The fit-viability model is adopted as it is found consistent with the requirements of the higher education institutions in terms of the institute structure, core competence, value, and culture whereas the factors derived from the DOI theory in the model include complexity, relative advantage, trialability, observability, and compatibility (Fig. 1).

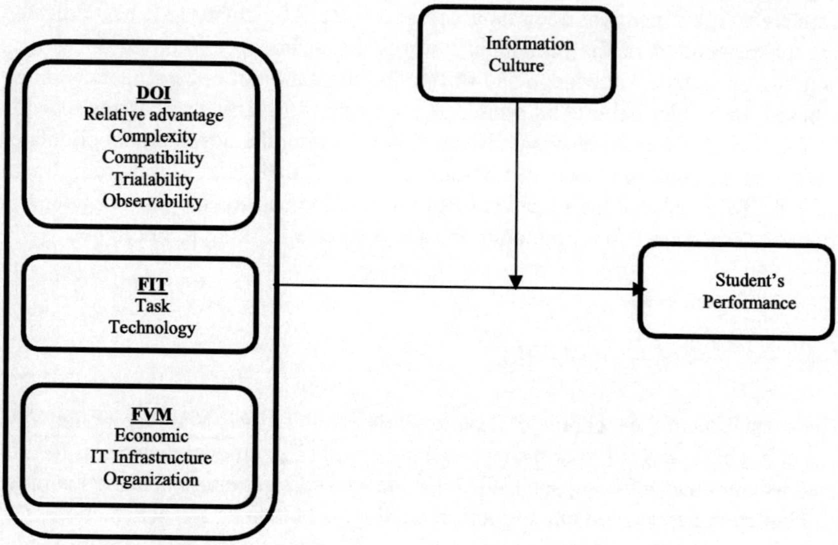

Fig. 1 DOI and FVM model

3 Research Hypothesis

The model proposed in this study is inclusive of three dimensions, including Fit-Viability Model, diffusion theory model, and information culture. The fit-viability model is adopted as it is found consistent with the requirements of the higher education institutions in terms of the institute structure, core competence, value, and culture whereas the factors derived from the DOI theory in the model include complexity, relative advantage, trialability, observability, and compatibility (Fig. 1). The aim of this study is to analyse adoption of cloud-based e-learning to identify patterns in the studied themes, on higher educational institution from the perspective of stakeholders of institutions in developing countries. The contribution of this study is to identify The FVM and ODI model fulfilling the educational need efficiently both in production and education. The approaches used to standardize the e learning educational phenomena, and their impact on the digital transformation of education and their impact on students in HEIs.

Hypotheses

H1: Relative advantage positively and significantly impacts student's performance in HEIs.

H2: Compatibility positively and significantly impacts student's performance in HEIs.

H3: Complexity positively and significantly impacts student's performance in HEIs.

H4: Trailability positively and significantly impacts student's performance in HEIs.
H5: Observability positively and significantly impacts student's performance in HEIs.
H6: Task characteristics positively and significantly impact student's performance in HEIs.
H7: Technology characteristics positively and significantly impact student's performance in HEIs.
H8: Economic feasibility positively and significantly impacts student's performance in HEIs.
H9: IT infrastructure positively and significantly impacts student's performance in HEIs.
H10: Organization support positively and significantly impacts student's performance in HEIs.
H11: Information culture positively and significantly moderates student's performance in HEIs.

4 Research Methods

4.1 Study Design

A quantitative cross-sectional design is used for investigating the adoption factors of cloud-based e-learning technologies using DOI and FVM undertaking the higher education institutions context.

4.2 Population and Sample

The targeted population of this study is higher educational institutions of Oman. In addition, the study has specifically targeted non-military colleges and universities offering 4 years graduate program or advanced degrees. The rationale behind the selection of these institutes is based on their offering of the e-learning courses. Moreover, 321 students have been selected from 33 institutions based on the sample size formula. In the formula, the confidence level takes 95%, confidence interval 5% and a population of 1950 students. These participants have been selected based on their knowledge and awareness with the e-learning systems adopted in the selected HEIs.

4.3 Instruments

A close-ended questionnaire has been structured based on the two adopted theories; DOI and FVM. The questionnaire has been divided into two major sections. The first section presents information about the demographic variables including age, gender, and qualification. In the second section, questions related to DOI and FVM are presented with respect to the adoption of cloud-based e-learning. Moreover, information culture has been introduced as a moderating variable in the questionnaire. A total of 19 sub-items are included in the 5 items of the FVM. These items include task characteristics (4 items), technology characteristics (4 items), economic feasibility (5 items), IT infrastructure (3 items), and organization support (3 items).

DOI theory has been used in the questionnaire to show the perceptions of the participants towards adopting the cloud-based e-learning. A total of 5 items are included in the DOI model comprising 24 sub-items as a whole. These items include relative advantage (5 items), compatibility (5 items), complexity (5 items), trialability (5 items), and observability (4 items).

4.4 Data Collection

SurveyMonkey website was used for collecting the data from the students studying in different years, programs, and degrees of the selected HEIs. 540 students have been provided with a structured questionnaire. The researcher has guided and administered the questionnaire to the students in case of any ambiguity and lack of understanding. In response, a total of 321 questionnaires were completed, accounting a response rate of 59.5%.

4.5 Validity and Reliability

A pilot test has been performed on the data collected to measure the validity of the questionnaire for further analysis. In this regard, data of 28 participants have been extracted from the final sample to perform the pilot test (Whitehead et al., 2016). Cronbach alpha has been used to measure the inter-reliability of the questionnaire for further assessment and examination.

4.6 Data Analysis

A structural equation modelling (SEM) has been used via Smart-PLS version 3.2.8 for analysing the data collected. Descriptive statistics have been presented for the demographic variables whereas confirmatory factor analysis and path analysis has been performed for DOI, FVM, and information cultural variables.

5 Results

Table 1 provides construct validity for the current study. The below Table 1 showed that all the measures were loaded into their respective construct with factor loadings greater than 0.60. The cross loadings criterion basically emphasized that all the measures of particular construct should have higher factor loading that is 0.96 in their respective construct rather than in any other construct.

Table 1 Construct validity

	IA	IMC	LPERF	SE	SF	SR	TA	VAS
RA1	0.81	– 0.23	0.01	– 0.12	– 0.18	0.04	0.22	– 0.26
RA3	0.86	– 0.33	0.44	– 0.07	0.03	– 0.15	0.29	0.09
COMPL1	– 0.22	0.66	– 0.03	0.15	– 0.13	0.06	– 0.09	0.03
COMPL2	– 0.12	0.80	– 0.13	0.29	– 0.09	0.21	0.15	– 0.09
COMPL3	– 0.39	0.83	– 0.14	0.22	0.03	0.40	0.02	0.07
COMPAT1	0.35	– 0.18	0.96	0.02	0.12	0.19	0.02	– 0.02
COMPAT2	– 0.01	– 0.01	0.67	– 0.04	0.11	0.12	0.14	0.21
TR1	– 0.10	0.32	0.02	0.88	0.23	0.55	0.17	0.08
TR2	– 0.03	0.27	– 0.01	0.87	0.16	0.52	0.29	– 0.05
OB3	– 0.15	0.16	– 0.01	0.83	0.05	0.34	0.05	0.19
OB4	– 0.10	0.26	0.03	0.86	0.27	0.53	0.12	0.01
OB2	0.02	– 0.13	0.27	0.05	0.81	0.11	0.14	0.17
TA3	– 0.20	0.01	0.06	0.26	0.84	0.25	0.14	0.09
TA4	0.06	– 0.04	0.03	0.16	0.80	0.11	0.17	– 0.01
TE1	– 0.01	0.34	0.22	0.45	0.19	0.86	0.20	0.12
TE2	0.11	0.26	0.29	0.40	0.06	0.79	0.12	– 0.04
EC3	– 0.22	0.36	0.06	0.43	0.22	0.84	0.08	0.14
EC4	– 0.10	0.23	0.11	0.61	0.20	0.86	0.08	0.07
ITI2	0.08	0.07	0.03	0.04	0.06	0.08	0.67	– 0.05
ITI3	0.39	0.02	0.11	0.18	0.12	0.14	0.89	– 0.10
ITI4	0.19	0.09	0.00	0.19	0.23	0.12	0.80	– 0.09
ORG1	– 0.07	– 0.04	0.01	0.05	0.08	0.07	– 0.16	0.95
ORG2	– 0.09	0.01	0.09	0.06	0.11	0.13	– 0.05	0.96
ORG3	– 0.08	0.12	0.01	0.07	0.04	– 0.12	– 0.08	0.70

Table 2 Convergent validity

Constructs	Composite Reliability	Average Variance Extracted (AVE)
Relative Advantage	0.823	0.700
Complexity	0.808	0.585
Compatibility	0.808	0.685
Trialability	0.917	0.734
Observability	0.860	0.673
Task	0.903	0.699
Technology	0.830	0.622
Economic	0.908	0.770
IT infrastructure	0.808	0.734
Organizations	0.917	0.673

Convergent validity basically aims to highlight the extent of convergence within the measures of particular variable. This helps to understand that either the measures are well-linked with each other or their convergence represents variable in adequate manner. Table 2 illustrates estimates of convergent validity.

The above table showed that all the variables have greater AVE coefficient than recommended 0.50, higher composite reliability 0.917 than the threshold of 0.734. Therefore, the study has achieved convergent validity in accordance with the recommended thresholds. Following table 3 shows Fornell & Larcker [44] criterion for assessing discriminant validity amongst the study variables.

The above table showed that all the variables have achieved discriminant validity using Fornell & Larcker [44] criterion. Following table 4 provides result of HTMT ratio for discriminant validity. All the constructs have less than 0.85 coefficients of

Table 3 Fornell & Larcker [44] criterion

Constructs	RA	COM	COMP	TR	OB	TA	TE	ECO	ITI	ORG
Relative Advantage	**0.84**									
Complexity	– 0.34	**0.77**								
Compatibility	0.29	– 0.15	**0.83**							
Trialability	– 0.11	0.30	0.01	**0.86**						
Observability	– 0.08	– 0.05	0.13	0.21	**0.82**					
Task	– 0.07	0.36	0.20	0.57	0.21	**0.84**				
Technology	0.30	0.07	0.06	0.19	0.18	0.14	**0.79**			
Economic	– 0.09	0.00	0.05	0.06	0.10	0.09	– 0.10	**0.88**		
IT infrastructure	– 0.09	0.01	0.09	0.06	0.11	0.13	– 0.05	0.96	**0.09**	
Organizations	– 0.07	– 0.04	0.01	0.05	0.08	0.07	– 0.16	0.95	0.06	**0.07**

HTMT ratio; hence, discriminant validity has been achieved. Table 5 provides path analysis of the variables according to structural model.

Relative advantage (0.13, $p < 0.10$) and complexity (0.35, $p < 0.10$) have significant relationship with student's performance. Compatibility (-0.20, $p < 0.10$) has statistically significant but negative relationship with student's performance. Furthermore, trialability (0.15, $p < 0.10$) and information culture (0.19, $p < 0.10$) have statistically significant and positive moderating impact on student's performance. Table 6 provides statistics related to predictive relevancy of the endogenous variables of the structural model. The below table showed that two reflective constructs of intra-firm resources namely tangible assets and intangible assets have 77% and 52% predictive relevance. However, four reflective constructs including trialability, observability, task and economic have the predictive relevancy of 76%, 17%, 75% and 3%, respectively. Lastly, compatibility and complexity have 13% and 9% predictive relevance respectively.

Table 4 Heterotrait-monotrait (HTMT) ratio

Constructs	IA	IMC	LPERF	SE	SF	SR	TA	VAS
Relative Advantage								
Complexity	0.50							
Compatibility	0.50	0.17						
Trialability	0.18	0.35	0.05					
Observability	0.25	0.17	0.26	0.26				
Task	0.21	0.36	0.27	0.65	0.23			
Technology	0.44	0.20	0.18	0.22	0.24	0.21		
Economic	0.32	0.16	0.25	0.13	0.14	0.15	0.14	

Table 5 Path analysis

	Estimate	S.D	T-Stats	Prob
Relative advantage → Student's performance	0.13	0.07	1.82	0.04
Complexity → Student's performance	0.35	0.05	6.74	0.00
Compatibility → Student's performance	– 0.20	0.07	2.80	0.00
Trialability → Student's performance	0.15	0.06	2.38	0.01
Observability → Student's performance	0.19	0.07	2.56	0.01
Task → Student's performance	0.87	0.02	37.16	0.00
Technology → Student's performance	0.41	0.08	5.08	0.00
Economic → Student's performance	0.87	0.02	44.53	0.60
IT Infrastructure → Student's performance	0.18	0.09	1.97	0.02
Organization → Student's performance	0.72	0.04	16.24	0.70
Adoption factors → Information culture → Student's Performance	0.88	0.02	39.76	0.00

Table 6 Predictive relevancy

Endogenous Variables	R Square	R Square Adjusted	Q Square
Relative Advantage	0.52	0.52	0.34
Complexity	0.09	0.07	0.04
Compatibility	0.13	0.12	0.04
Trialability	0.76	0.76	0.55
Observability	0.17	0.16	0.10
Task	0.75	0.75	0.52
Technology	0.77	0.77	0.46
Economic	0.03	0.03	0.01
IT infrastructure			
Organizations			

6 Discussion

The results of the study provide a firm base for the decision makers to determine the incorporation of cloud computing in the HEIs. Considering the relative advantage among the students, the study demonstrates its significant impact on cloud computing adaptation. The findings are corroborated by Almajalid [45] who stated that various educational institutes had assimilated their learning curriculum with the cloud computing technology, which aids teachers and students in enhancing their knowledge. This model helps in aligning the study goals with individual's goals promoting the engagement in educational cloud-based initiatives.

The current study also reveals the substantial impact of the computability, trialability, and complexity upon the learning endeavours of the students. It states that compatibility of the cloud system urges them to adopt indulge in the cloud-based education system. These results are consistent with the findings of Yatigammana, Johar, & Gunawardhana, who highlights that learner perceived compatibility also positively influences their intention to use the e-learning system. Duan et al. [46] also fall in-line with the findings of the present study stating that compatibility and trialability are positively associated with the adaptation of the e-learning system in the Chinese higher education.

Considering the DOI theory as a whole, the study has observed a significant impact on cloud computing among the HEIs. Buc & Divjak [25] have concluded same results stating that higher education culture and adaptation of the cloud computing facilitates the institutes in aligning its three-fold objective, i.e., education, research, and outreach. In the same context, endorsing the present study findings, White [47] writes that cloud computing model helps in exploring the organizational role and interrelation among the students, teachers, and the education system enhancing the teaching practices used. The present study findings have revealed a positive impact of the task and technology, along with IT infrastructure, whereas an insignificant impact is derived by the economic factors and organization of the system. Earlier studies supplement the research results such as Ellis & Loveless [48] show that in the higher education pedagogy, academic achievement cannot be isolated from technology, teaching process or innovation. Chan et al. [49] also reported same observation and demonstrated substantial significance of the cloud computing in the HEIs, positioning it as a stimulator for democratizing the educational goals and practices and meeting the changing dynamic demands of the leaners.

Similarly, Al-Ajmi et al. [21] stated that this system adds to the convenience of the educational network which improves its academic services and optimizes its reach to a wider area, network performance as well as support to the application of the system. The negative association of the economic factors and organizational support has been corroborated by earlier research [38, 50]. The conceptual framework of the study was found effective; however, there is certain limitation which still prevails in the present study. One limitation is its inclusion of the institutes in the region of Oman only. The restriction to a particular region institute impacts the generalizability of the study given the variant socio-economic dynamics in different countries. Another limitation is related to the design of the research which follows a quantitative design; however, a qualitative design can also be adopted for gathering valuable insights from the professionals who are associated with the implementation of cloud computing. Moreover, future researches can also improve the sample size for providing more valid results.

7 Conclusion

The study concluded that technology adoption factors, including Relative advantage, complexity, Compatibility, trialability, task, technology, and IT infrastructure, have a significant influence on students' performance in HIEs in Oman. Also, information culture has statistically significant and positive moderating impact on student's performance, which shows that information culture can strengthen the relationship between the adoption factors and student's performance. In other words, the value and the use of cloud computing in the HEIs in Oman can be enhanced through the above-mentioned factors. The outcomes of the study highlighted the significance of the developed conceptual framework which serves as an introductory model for establishing a cloud computing culture within HEIs.

The results of the study suggest the validation of the developed framework as well as its adaptation among the HEIs in the region for testing its descriptive and analytical stance. It implies that the educational institute policymakers can assess the factors which align with their institute objectives and evaluates whether cloud computing is suitable for their institutes or not. This study provides the foundation which helps institutes in determining their engagement with the educational cloud-based initiatives. Future studies are recommended to critically examine the relationship established in this study which will help expand the study horizon by providing the necessary explanation. Correspondingly, the present study shed light on the possible opportunities for the future research, emphasizing on evaluating the constructs proposed in the study considering the relationships that prevail between the dependents and the independent variables. Acceptance of selective software and applications such as Dropbox and Google Docs can also be studied in the academic environments for promoting cloud-based e-learning culture among students in HEIs.

References

1. Ansong E, Lovia Boateng S, Boateng R (2017) Determinants of e-learning adoption in universities: evidence from a developing country. J Educ Technol Syst 46(1):30–60. https://doi.org/10.1177/0047239516671520
2. Alajmi Q, Sadiq AS, Kamaludin A, Al-Sharafi MA (2018) Cloud computing delivery and delivery models: opportunity and challenges. Adv Sci Lett 24(6):4040–4044
3. Kayali M, Alaaraj S (2020) Adoption of cloud based e-learning in developing countries: a combination a of DOI, TAM and UTAUT. Int J Contemp Manag Inf Technol 1(1):1–7
4. Kaiiali M, Iliyasu A, Wazan AS, Habbal A, Muhammad YI (2019) A cloud-based architecture for mitigating privacy. Int Arab J Inf Technol 16(5):879–888
5. Kaiiali M, Sezer S, Khalid A (2019) Cloud computing in the quantum era. In 2019 IEEE conference on communications
6. Sharma K, Pandit P, Pandit P (2011) Critical success factors in crafting strategic architecture for e-learning at HP University. Int J Educ Manag 25(5):423–452. https://doi.org/10.1108/09513541111146350
7. Awa HO, Ukoha O, Emecheta BC (2016) Using TOE theoretical framework to study the adoption of ERP solution. Cogent Bus Manag 3(1):1196571. https://doi.org/10.1080/23311975.2016.1196571
8. Oliveira T, Martins MF (2011) Literature review of information technology adoption models at firm level. Electronic J Inform Syst Eval 14(1):110
9. Venkatesh V, Thong JY, Xu X (2016) Unified theory of acceptance and use of technology: a synthesis and the road ahead. J Assoc Inf Syst 17(5):328–376. https://doi.org/10.17705/1jais.00428
10. JT Baker (2012) The technology–organization–environment framework. Information systems theory, pp. 231–245. Springer, NY. https://doi.org/10.1007/978-1-4419-6108-2_12
11. Rogers EM (2010) Diffusion of innovations. Simon and Schuster
12. Tjan AK (2011) Finally, a way to put your Internet portfolio in order. Harv Bus Rev 79(2):76–85
13. Mkhize P, Mtsweni ES, Buthelezi P (2016) Diffusion of innovations approach to the evaluation of the learning management system used in an open distance learning institution. Int Rev Res Open Distrib Learn 17(3). https://doi.org/10.19173/irrodl.v17i3.2191
14. Ali S, Uppal MA, Gulliver SR (2018) A conceptual framework highlighting e-learning implementation barriers. Inf Technol People 31(1):156–180. https://doi.org/10.1108/itp-10-2016-0246
15. Raza MH, Adenola AF, Nafarieh A, Robertson W (2015) The slow adoption of cloud computing and IT workforce. Procedia Comput Sci 52:1114–1119. https://doi.org/10.1016/j.procs.2015.05.128
16. Pluzhnik E, Nikulchev E (2014) Virtual laboratories in cloud infrastructure of educational institutions. In emission electronics (ICEE), 2014 2nd international conference, pp. 1–3, IEEE https://doi.org/10.1109/emission.2014.6893974
17. Rindos A, Vouk M, Jararweh Y (2014) The virtual computing lab (VCL): an open-source cloud computing solution designed specifically for education and research. Int J Serv Sci Manag Eng Technol (IJSSMET), 5(2), 51–63. https://doi.org/10.4018/ijssmet.2014040104

18. Ssekakubo G, Suleman H, Marsden G (2011) Issues of adoption: have e-learning management systems fulfilled their potential in developing countries? In proceedings of the South African institute of computer scientists and information technologists conference on knowledge, innovation and leadership in a diverse, multidisciplinary environment, pp. 231–238. ACM, New York, NY, USA. https://doi.org/10.1145/2072221.2072248
19. Gutirrez-Santiuste E, Gallego-Arrufat MJ, Simone A (2016) Barriers in computer-mediated communication: typology and evolution over time. J e-Learn Knowl Soc 12(1):108–119
20. Bulla C, Hunshal B, Mehta S (2016) Adoption of cloud computing in education system: a survey. Int J Eng Sci 6375
21. Qasim AA, Arshah RA, Kamaludin A, Sadiq AS, Al-Sharafi MA (2017) A conceptual model of e-learning based on cloud computing adoption in higher education institutions. In electrical and computing technologies and applications (icecta), international conference, pp. 1–6. IEEE. https://doi.org/10.1109/icecta.2017.8252013
22. Bibi G, Ahmed IS (2017) A comprehensive survey on e-learning system in a cloud computing environment. Eng Sci Technol Int Res J 1(1):43–50
23. Durairaj M, Manimaran A (2015) A study on security issues in cloud-based e-learning. Indian J Sci Technol 8(8):757–765. https://doi.org/10.17485/ijst/2015/v8i8/69307
24. Gai K, Steenkamp A (2014) A feasibility study of platform-as-a-service using cloud computing for a global service organization. J Inf Syst Appl Res 7(3):28
25. Buc S, Divjak B (2015) Innovation diffusion model in higher education: case study of e-learning diffusion. Int Assoc Develop Info Soc
26. Rogers EM (2003) Elements of diffusion. Diffusion of innovations, 5(1.38)
27. Zhang L, Wen H, Li D, Fu Z, Cui S (2010) E-learning adoption intention and its key influence factors based on innovation adoption theory. Math Comput Model 51(11–12):1428–1432
28. Murphy J, Kalbaska N, Williams A, Ryan P, Cantoni L, Horton-Tognazzini LC (2014) Massive open online courses: strategies and research areas. J Hosp Tour Educ 26(1):39–43
29. AlAjmi Q, Arshah RA, Kamaludin A, Al-Sharafi MA (2021) Developing an Instrument for Cloud-Based E-Learning Adoption: Higher Education Institutions Perspective. In: Bhatia SK, Tiwari S, Ruidan S, Trivedi MC, Mishra KK (eds) Advances in Computer, Communication and Computational Sciences, vol 1158. Advances in Intelligent Systems and Computing. Springer, Singapore, pp 671–681. https://doi.org/10.1007/978-981-15-4409-5_60
30. Uden L, Corchado ES, Rodríguez JF, Santana DP, De la Prieta F (eds) (2012) Workshop on Learning Technology for Education in Cloud (LTEC'12). Springer Berlin Heidelberg, Berlin, Heidelberg https://doi.org/10.1007/978-3-642-30859-8
31. Kabunga NS, Dubois T, Qaim M (2012) Heterogeneous information exposure and technology adoption: the case of tissue culture bananas in Kenya. Agric Econ 43(5):473–486. https://doi.org/10.1111/j.1574-0862.2012.00597.x
32. Jenkins M, Browne T, Walker R, Hewitt R (2011) The development of technology enhanced learning: findings from a 2008 survey of UK higher education institutions. Interact Learn Environ 19(5):447–465. https://doi.org/10.1080/10494820903484429
33. Al-Tahitah AN, Al-Sharafi MA, Abdulrab M (2021) How COVID-19 Pandemic Is Accelerating the Transformation of Higher Education Institutes: A Health Belief Model View. In: Arpaci I, Al-Emran M, A. Al-Sharafi M, Marques G (eds) Emerging Technologies During the Era of COVID-19 Pandemic, vol 348. Studies in Systems, Decision and Control. Springer, Cham, pp 333–347. https://doi.org/10.1007/978-3-030-67716-9_21
34. Al-Gahtani SS (2016) An empirical investigation of e-learning acceptance and assimilation: a structural equation model. Appl Comput Infom 12(1):27–50. https://doi.org/10.1016/j.aci.2014.09.001
35. Kinasevych (2010) Effect of culture on online learning. In: Sudweeks F, Hrachovec H,Ess C (eds) Proceedings cultural attitudes towards communication and technology, Murdoch University, Australia, 420-427
36. Goodhue DL, Thompson RL (1995) Task-technology fit and individual performance. MIS Q: 213–236. https://doi.org/10.2307/249689

37. Eom SB, Ashill NJ (2018) A system's view of e-learning success model. J Innov Educ 16(1):42–76. https://doi.org/10.1111/dsji.12144
38. Liang TP, Huang CW, Yeh YH, Lin B (2007) Adoption of mobile technology in business: a fit-viability model. Ind Manag Data Syst 107(8):1154–1169. https://doi.org/10.1108/02635570710822796
39. Mohammed F, Ibrahim O, Ithnin N (2016) Factors influencing cloud computing adoption for e-government implementation in developing countries: instrument development. J Syst Inf Technol 18(3):297–327. https://doi.org/10.1108/jsit-01-2016-0001
40. Ejiaku SA (2014) Technology adoption: issues and challenges in information technology adoption in emerging economies. J Int Technol Inf Manag 23(2):5
41. Mukred A, Singh D, Safie N (2017) investigating the impact of information culture on the adoption of an information system in public health sector of developing countries. Int J Bus Inf Syst 24(3):261–284. https://doi.org/10.1504/ijbis.2017.10002805
42. John SP (2015) The integration of information technology in higher education: a study of faculty's attitude towards IT adoption in the teaching process. Contaduría y Administración 60:230–252. https://doi.org/10.2139/ssrn.2550007
43. Choo CW (2013) Information culture and organizational effectiveness. Int J Inf Manage 33(5):775–779. https://doi.org/10.1016/j.ijinfomgt.2013.05.009
44. Fornell C, Larcker DF (1981) Structural equation models with unobservable variables and measurement error: algebra and statistics. https://doi.org/10.2307/3150980
45. Almajalid R (2017) A survey on the adoption of cloud computing in education sector. arXiv preprint arXiv:1706.01136. https://doi.org/10.14445/22312803/ijctt-v60p102
46. Yatigammana MRKN, Johar MDGMD, Gunawardhana C (2013) Impact of Innovations attributes on e-learning acceptance among Sri Lankan postgraduate students. https://doi.org/10.4038/kjm.v2i1.6541
47. White I (2017) Modelling the complexity of technology adoption in higher education teaching practice
48. Ellis V, Loveless A (2013) ICT, pedagogy and the curriculum: subject to change. Routledge. https://doi.org/10.4324/9780203468258
49. Chan D, Bernal A, Camacho A (2013) Integration of ICT in higher education: experiences and best practices in the case of the university of Baja California in Mexico. In edulearn13 proceedings, pp. 1040–1049
50. Al-dheleai Y, Tasir Z, Al-Rahmi W, Al-Sharafi M, Mydin A (2020) Modeling of students online social presence on social networking sites and academic performance. Int J Emerg Technol Learn (iJET) 15(12):56–71

Online Learning During Covid-19 Pandemic: A View of Undergraduate Student Perspective in Malaysia

Ling Chai Wong, Poh Kiong Tee, Tat-Huei Cham, and Ming Fook Lim

Abstract The advancement of technology changes the mode of operation worldwide education industry, where educational services can be delivered either in face-to-face or online teaching. The outbreak of COVID-19 forced higher education institutions to shift from face-to-face teaching to fully online learning, even though online learning is yet to be fully implemented in many institutions. This trend has prompted us to study this interesting topic and gather information about undergraduate students' satisfaction with online learning from home due to the limited study focus on Malaysia. The operation of the study is based on the user satisfaction theories. A total of 156 questionnaires were distributed via judgement sampling guidelines. PLS-SEM was used for the data analysis. The results confirmed that the online learning system is useful but not user-friendly. Technical system quality is up to the satisfactory level from students' perception. Furthermore, the attitude was confirmed significantly impact undergraduate students' satisfaction with online learning. Discussion of the findings, implications, and direction for future research are also presented in the final section of the study.

Keywords Online learning · Perceived ease of use · Perceived usefulness · Perceived technical system quality · Students' satisfaction

L. C. Wong (✉) · P. K. Tee
School of Marketing and Management, Asia Pacific University of Technology and Innovation, Kuala Lumpur, Malaysia
e-mail: lingchaiwong10@gmail.com

P. K. Tee
e-mail: seantee@live.com

T.-H. Cham
UCSI Graduate Business School, UCSI University, Kuala Lumpur, Malaysia
e-mail: jaysoncham@gmail.com

M. F. Lim
Faculty of Business, Economics and Accountancy, Universiti Malaysia Sabah, Kota Kinabalu, Malaysia
e-mail: mfook89@gmail.com

M. Al-Emran et al. (eds.), *International Conference on Information Systems and Intelligent Applications*, Lecture Notes in Networks and Systems 550,
https://doi.org/10.1007/978-3-031-16865-9_32

1 Introduction

Online learning is a type of internet-based education [1], an innovative method that provides time- and space-free learning for the learners' convenience to study in time frame and location simultaneously [2]. Obviously, the online learning trend has brought many exciting features to teaching and learning and has created many incredible opportunities [3]. Some researchers claim online learning provides more opportunities for the students to interact with their lecturer than traditional classes [4, 5]. In addition, online discussions in asynchronous educational experiences facilitate a comprehensive review of knowledge by enabling students to prepare their ideas while posting a message to virtual conventions [6]. Asynchronous learning is preferred to facilitate in-depth online student discussions and connections [7]. Besides, a considerable body of evidence has also shown that e-learning can lead to substantial cost savings, often as much as 50% relative to conventional learning. The savings were attributed to reducing preparation time, decreasing institutional capacity and the prospect of extending programs with emerging educational technologies [8, 9]. Although many studies have been done on online learning, traditional face-to-face class remains the domain in the Malaysian education system. The traditional learning model can still generate a lucrative income for the institution. Hence, some institutions choose to put little effort into online learning before January 2020.

Unfortunately, the trend of learning was forced to be changed due to the outbreak of the novel COVID-19 when the Malaysian government imposed the movement control order (MCO) on March 18, 2020 [10]. The MCO caused a nationwide shutdown and forced all education to go online. To ensure smooth delivery of teaching and learning throughout the MCO period, some guidelines detailing the responsibilities of teachers, parents, students, and administrators were provided by the Ministry of Education [11]. However, both education service providers and users are not ready with the online learning system. The advancement of online learning from home evolved as a catalyst for today's educational institutions [12].

Moreover, education service providers are unsure how their customers perceive the system. Limited studies have covered how to use online learning platforms effectively during the COVID-19 pandemic. To fill the gap, conducting a study on students' satisfaction with online learning from home during the Covid-19 pandemic will give the direction on how to improve their online learning system. Accordingly, the current study is carried out to examine the relationship between online learning systems and teaching quality.

As students' satisfaction is progressively seen as a key factor in determining online learning services in market competition [13, 14], this study checks existing dimensions while establishing new dimensions to close existing gaps by utilizing Technology Acceptance Model (TAM) to analyze the behavioural trends of online learning users. TAM explains the importance of perceived ease of use (PEOU) and perceived usefulness (PU) in determining perceived satisfaction and user attitudes toward a technology [15–17]. Therefore, the attitude will influence the behavioural intention of using the system [18, 19]. This study inserts a new variable, technical system quality (TSQ), to enhance the application of TAM in the context of online learning.

2 Literature Review and Hypotheses Development

The shift from traditional face-to-face to online learning promotes a closer examination of the quality of instruction and course technology [19]. Some researchers [9, 20] define online learning as a more accessible form of distance learning, allowing students who are deemed unorthodox and unsatisfied, to have access to education services. Tuan and Tram [13] defined online learning as a form of teaching in which the multiple incorporations of technology are pursued, and it is a substitute for distance learning. In short, online learning represents the application of technology in education, and it is increasingly being studied. The domains of learning and teaching in the higher education institutions in Malaysia are undergoing major changes due to the COVID-19 pandemic [21, 22]. Many universities are starting to offer web-based courses that support classroom-based courses. Online learning is attractive to many students because it provides flexibility in engagement, ease of access, and accessibility, has been found to be perfectly suited to the current scenario. However, limited studies have covered how to effectively use the online learning platform to enhance student satisfaction with learning online from home.

2.1 Students' Satisfaction

Students are the primary customers of educational institutions [23]. Student satisfaction is defined as "the favorability of a student's subjective evaluation of the various educational outcomes and experiences". Student satisfaction has a significant consistency in predicting learning experience [24]. Their satisfaction is critical for universities seeking to promote prospective students. Many studies concluded that student satisfaction is critical in determining service consistency and efficiency [13, 14, 25]. Indeed, student satisfaction is very important because it is the only success measure of higher education service providers [26]. Due to the COVID-19 lockdown, there is a growing demand for online learning; student satisfaction is critical in this situation. Previous research found a link between PEOU, PU, attitude and user satisfaction with online learning [14, 27, 28]. Furthermore, a substantial body of literature indicates that TSQ is the primary factor influencing student satisfaction and IS utilization in the educational environment [29]. [30] discovered that the strongest relationship between TSQ and satisfaction is essential. Thus, PEOU, PU, and TSQ are theorized as the main predictors of student satisfaction toward online learning during the COVID-19 pandemic, and their relationships are discussed in the following sections.

2.2 Perceived Ease of Use (PEOU)

PEOU is characterized as the degree to which a person considers it would be effortless to use a particular system [17], which is an imminent driver of acceptance of new technology-based applications. PEOU is a variable that influences the behavioural intent of using the system [15, 17], particularly the adoption of new technologies by users who are looking for an easier way to accomplish a task [18]. Easiness is an essential element for an online learning system because easy to use can encourage students to use and accept online learning. Several studies have attempted TAM to study online learning and found that PEOU has a major effect on individuals' intention of using online learning systems [14, 31]. In addition, the PEOU has been used as a prerequisite for e-satisfaction in various studies [13–15]. Consequently, the greater the PEOU of online learning, the more optimistic the attitude and intention toward its use are. Thus, the likelihood of it being used and satisfied is greater. Hence, it is expected similar relationship may occur in this case, as hypothesized below:

H1: Perceived ease of use positively affects undergraduate students' attitude toward online learning from home during the COVID-19 pandemic.

H2: Perceived ease of use positively affects undergraduate students' satisfaction with online learning from home during the COVID-19 pandemic.

2.3 Perceived Usefulness (PU)

Perceived Usefulness (PU) is the degree to which the user believes that using a system would improve their job performance and help the user perform better in an organization [17]. PU is described as how a person considers their work performance enhanced by using a specific method. Studies have shown that PU has an important effect on the acceptance of technology, which can explain user behavioural intention [17, 27]. When coming to online learning, PU reflects the degree of reliability, effectiveness, and cost-efficiency from using technology, which significantly impacts the online users' satisfaction [27, 31]. Students are more satisfied and have favourable attitudes toward online learning systems once they believe it can help them accomplish their educational goals. Many researchers have applied TAM to online learning research and found that PU has a major effect on individuals' attitudes and intention of using online learning systems [17, 28, 32]. Consequently, the greater the PU of the online learning platform, the more optimistic the attitude and intention toward its use. Thus, the likelihood of it being used and satisfied is greater. We, therefore, hypothesized:

H3: Perceived usefulness positively affects undergraduate students' attitude toward online learning from home during the COVID-19 pandemic.

H4: Perceived usefulness positively affects undergraduate students' satisfaction with online learning from home during the COVID-19 pandemic.

2.4 Technical System Quality (TSQ)

In the success model of information system (IS) proposed by DeLone et al. [29], technical system quality (TSQ) refers to technological progress as well as the accuracy and efficiency of the information-producing communication system. System quality is linked to system reliability, user-friendliness, software quality, and program code consistency and maintenance [33]. For instance, Lin and Lu [34] claimed that many people still avoid using the internet because they want to avoid the slow response time, heavy Internet traffic, and lack of network connectivity. In addition, if existing users experience security issues or curriculum interruptions when using the system, this can lead to a decrease in the perception of user-friendliness of the machine, in effect influencing attitudes and behavioural intent to use the platform, as well as the satisfaction of the system user [28, 30]. In this case, the quality of the technical system is considered vital in influencing the beliefs of users of the website. TSQ may have a significant impact on undergraduate students' attitudes toward using online learning from home and their satisfaction during the COVID-19 pandemic. Nonetheless, none of the prior studies were conducted to investigate the impact of TSQ on students' satisfaction with online learning. To close the gap, a study on students' satisfaction with online learning from home during the COVID-19 pandemic as hypotheses below:

H5: Technical System Quality positively affects undergraduate students' attitude toward online learning from home during the COVID-19 pandemic.

H6: Technical System Quality positively affects undergraduate students' satisfaction with online learning from home during the COVID-19 pandemic.

2.5 Attitude

Attitude is defined as an individual's positive or negative feelings about engaging in the desired behaviour [17, 35]. The two are inextricably linked, and a positive attitude toward ICT is commonly regarded as a necessary condition for successful implementation [35]. Studies on the formation of attitudes show that beliefs and attitudes are linked, as are attitudes and behaviours. Several studies have found that the effectiveness and ease of use of online learning programs, perceived usefulness of online learning, and students' technical level and skills all impact students' attitudes [36]. After all, positive student attitudes and online learning behaviours are critical for student satisfaction and adoption of online learning [27, 31]. Based on the preceding discussion, attitude may directly influence undergraduate students' satisfaction with online learning and mediate the relationship between the independent variables (i.e., PEOU, PU & TSQ) and the dependent variable (i.e., student satisfaction). Hence, this study assumes that:

H7: Attitude positively affects undergraduate students' satisfaction with online learning from home during the COVID-19 pandemic.

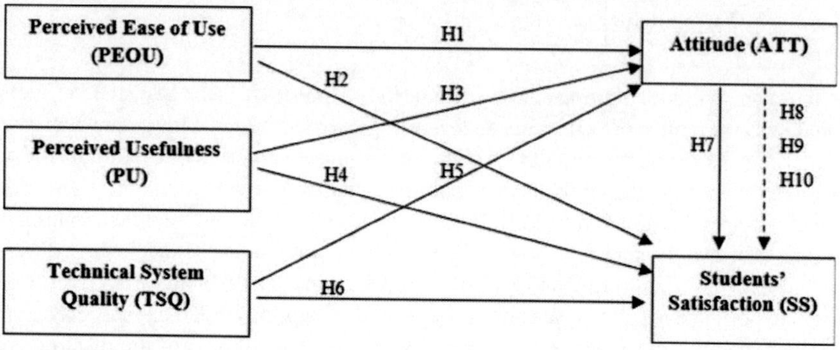

Fig. 1 Research model

H8: Attitude mediates the relationship between perceived ease of use and students' satisfaction with online learning from home during the COVID-19 pandemic.

H9: Attitude mediates the relationship between perceived usefulness and students' satisfaction with online learning from home during the COVID-19 pandemic.

H10: Attitude mediates the relationship between technical system quality and students' satisfaction with online learning from home during the COVID-19 pandemic.

Figure 1 presents this study's research framework, including perceived ease of use (PEOU), perceived usefulness (PU) and technical quality system (IQS) as independent variables, attitude (ATT) as mediators, and students' satisfaction (SS) as the dependent variable.

3 Research Method

A survey questionnaire via Google form was distributed to collect primary data from current undergraduate students. 156 responses were collected and utilized in data analysis. The questionnaire was divided into six sections. Section one presented the respondents' demographic questions. Section two has five questions related to PEOU, section three has five items to measure PU, section four consists of five items to measure TSQ, section five has five questions for ATT and section six has five items to measure SS. The five-point Likert scale from 1 (strongly disagree) to 5 (strongly agree) was used as the scale of measurement. Data collected were analyzed using PLS-SEM to assess the significance of the assumed relationship.

4 Results

Among 156 respondents, women accounted for 63.5% (n = 99), among the 156 interviewees, while men accounted for 36.5% (n = 57). The majority of respondents

Table 1 Full collinearity testing

ATT	PEU	PU	TSQ	SS
2.300	2.103	1.883	1.786	2.588

ATT = attitude; PEU = perceived ease of use; PU = perceived usefulness; TSQ = technical system quality; SS = students' satisfaction

were aged between 22 to 25 years (n = 88), followed by 18 to 21 years (n = 53) and 32 to 40 years (n = 8). The remaining seven persons were aged 26 to 30 years. Also, the majority (n = 109) have bachelor's degree programs, and the remaining (n = 47) have a Diploma. In terms of ethnicity, 68% (n = 106) of the respondents were Malay, followed by Chinese (n = 36) and Indians (n = 14).

This study uses the SmartPLS v3.3.8 as the analysis tool to examine the measurement and structural model. According to [37], if the data was collected using a single source, the Common Method Bias should be tested. As shown in Table 1, no bias exists in the single-source data since the single source in the current study did not have a serious bias (i.e., VIF < 3.3).

ATT = attitude; PEU = perceived ease of use; PU = perceived usefulness; TSQ = technical system quality; SS = students' satisfaction.

In the measurement model assessment (see Table 2), the validity and reliability of the instruments were tested. All the loadings value ranged from 0.727 to 0.846 (>0.708), the average variance extracted (AVE) values between 0.565 to 0.640 (>0.50) and composite reliability (CR) values between 0.866 to 0.899 (>0.70). All the threshold criteria for reliability and validity were met.

In addition, Heterotrait-Monotrait (HTMT) criterion was used to assess the discriminant validity. Table 3 shows that all HTMT Criterion values were below 0.85 [38]. Thus, discriminant validity was established in this study.

Table 4 reports the findings on the path coefficients and justifies the hypothesized relationships. PEU ($\beta = 0.175$, $p < 0.05$) and PU ($\beta = 0.271$, $p < 0.05$) have a positive relationship with ATT but are not associated with SS; thus, H1 and H3 were supported, but H2 and H4 were rejected. On the other hand, TSQ ($\beta = 0.227$, $p < 0.05$) was found to be positively related to SS but not to ATT, and ATT ($\beta = 0.390$, $p < 0.05$) has a positive relationship with SS. Hence, H6 and H7 were supported, but H5 was rejected. Overall, the model explained 47.2% ($R^2 = 0.472$) of the variance

Table 2 Convergent validity

	Items	Loadings	AVE	CR
Attitude	5	0.738–0.788	0.565	0.866
Students satisfaction	5	0.727–0.811	0.600	0.882
Perceived ease of use	5	0.737–0.771	0.570	0.868
Perceived usefulness	5	0.765–0.846	0.640	0.899
Technical system quality	5	0.726–0.819	0.606	0.885

CR = Compostite relaibility; AVE = Average variance extracted

Table 3 Discriminant validity (HTMT)

	ATT	PEOU	PU	SS	TSQ
Attitude					
Perceived ease of use	0.676				
Perceived usefulness	0.683	0.719			
Students' satisfaction	0.838	0.622	0.602		
Technical system quality	0.560	0.679	0.570	0.645	

ATT = Attitude; PEOU = Perceived ease of use; PU = Perceived usefulness; TSQ = Technical system quality; SS = Students' satisfaction

in ATT and 61.4% ($R^2 = 0.614$) in SS. The model displayed acceptable predictive relevance since all Q^2 values ($Q^2 = 0.256$ for ATT and 0.346 for SS) were > 0.

For the mediation analysis, the bootstrap indirect effect reported in Table 5 shows that only the indirect effect of PEU  ATT  SS ($\beta = 0.069$, $p < 0.01$) and PU  ATT  SS ($\beta = 0.108$, $p < 0.01$) were significant. Also, the confidence intervals bias-corrected 95% do not straddle a zero in between, indicating the mediation effect in these relationships. Hence, H8 and H9 were supported, but H10 was rejected.

Table 4 Result of structural model assessment

Relationship	Std Beta	Std Error	t-value	p-value	Decision
H1: PEOU - > ATT	0.175	0.075	2.171	0.015*	Supported
H2: PEOU - > SS	0.044	0.078	0.605	0.273	Rejected
H3: PU - > ATT	0.271	0.079	3.457	0.000**	Supported
H4: PU - > SS	0.042	0.068	0.523	0.301	Rejected
H5: TSQ - > ATT	0.078	0.090	0.874	0.191	Rejected
H6: TSQ - > SS	0.227	0.074	3.080	0.001**	Supported
H7: ATT - > SS	0.390	0.072	5.466	0.000**	Supported

ATT = Attitude; PEU = Perceived ease of use; PU = Perceived usefulness; TSQ = Technical system quality; SS = Students' satisfaction
*Notes **p-value < 0.001, * p-value < 0.05*

Table 5 Result of mediation analysis

Hypothesis	Std beta	Std error	*t*-value	P value	5%	95%	Decision
H8: PEOU > ATT > SS	0.069	0.035	1.928	0.052	0.006	0.140	Supported
H9: PU > ATT > SS	0.108	0.041	2.628	0.009	0.038	0.193	Supported
H10: TSQ > ATT > SS	0.036	0.037	0.831	0.406	−0.027	0.122	Rejected

ATT = Attitude; PEU = Perceived ease of use; PU = Perceived usefulness; TSQ = Technical system quality; SS = Students' satisfaction
*Notes **p-value < 0.001, * p-value < 0.05*

5 Discussion and Conclusion

This study was undertaken to understand students' satisfaction with online learning during the COVID-19 pandemic. The findings indicated that only two (i.e., perceived ease of use and perceived usefulness) out of the three predictors are significantly related to attitude. Hence, it implied that the undergrads prefer online learning systems that are easy to use, easy to navigate, and do not require much mental effort. The significant roles of perceived usefulness in online learning systems were analogies to the studies by [27, 31, 39]. Undergraduate students like online study systems if the learning systems are useful for their learning; the output of online learning is the same/or better than in the physical classroom. The student is expected to learn and get support from the instructor effectively via an online learning system. Surprisingly, technical system quality was found insignificant with attitude toward online learning systems, which contrasts with the previous studies [40, 41]. This finding is mainly due to the internet infrastructure in Malaysia, where the students expected some lags and were somehow disconnected from the online class. However, TSQ was significantly related to students' satisfaction with online learning. Although students expected some technical issues such as disconnections and response time delays during their online learning process, they dislike these issues. They cannot tolerate them in the long term [30, 42]. PEU and PU were irrelevant to students' satisfaction with online learning, which indicates that shifting to online classes during the pandemic made students feel stressed and uncomfortable [42]. In addition, facing the internet connection problem was the main issue affecting students' satisfaction, regardless of the system's ease of use and usefulness.

Furthermore, the attitude strengthened the link between perceived ease of use, perceived usefulness and students' satisfaction. An online learning system that is easy to use and navigate, while useful in assisting students' learning, doing coursework and preparing for ease, can generate a positive attitude toward the online system and lead to a higher level of satisfaction toward online learning systems. Thus, the university must create an online learning system that is easy to use and help the students learn effectively to increase student satisfaction with online learning systems. However, the attitude was not mediating the relationship between technical system quality and students' satisfaction with online learning systems. Students might see those technical issues such as bad internet connection are common online; nevertheless, if the technical problems persist for a longer time, they will not satisfy. As a result, students' attitudes toward online learning will not be changed in this manner.

In this study, TSQ was integrated with the constructs from TAM to predict the students' satisfaction with online learning. The findings indicate that TSQ was successfully incorporated and extended with TAM since TSQ was proven as one of the predictors of students' satisfaction with online learning. Practically, the findings of this study help the higher education institutions (HEIs) and the developers of online learning applications to understand better the factors influencing students' attitude satisfaction in online learning. The online systems need to be easy to use, navigate, and useful to affect students' attitudes. Also, technical issues such as lagging, the display and the internet connection need to be improved to increase students' satisfaction. With the inclusion of attitude, PU and PEU established a high level of student satisfaction, particularly to reduce their avoidance behaviour [43] in using online learning from home during the COVID-19 pandemic.

In conclusion, online learning has become a new normal in education [44]. Therefore, managing students' satisfaction, particularly in online learning, is vital for the institutions' performance [45]. To be effective, both educators and students must have a positive attitude toward the online learning platform. The student activities and behaviours (i.e., satisfaction) must be systematically monitored, especially when there are few or no opportunities for face-to-face encounters. Indeed, like any other business, the service provider must always ensure that the learning platform benefits both the educator and the student and remains sustainable in the long run [46–50].

References

1. Joshua Stern P (2020) Introduction to online teaching and learning. http://www.wlac.edu/online/documents/otl.pdf
2. Vanve A, Gaikwad R, Shelar K (2016) A new trend e-learning in education system. Int Res J Eng Technol 4(3):2395–2456. https://www.irjet.net/archives/V3/i4/IRJET-V3I457.pdf
3. Upadhyay S (2020) What will COVID-19 mean for teaching and learning? https://www.policyforum.net/the-opportunities-and-risks-of-taking-education-online/
4. Norman S (2016) 5 advantages of online learning: education without leaving home. https://elearningindustry.com/5-advantages-of-online-learning-education-without-leaving-home
5. Dumbauld B (2017) 13 great benefits of online learning. https://www.straighterline.com/blog/34-top-secret-benefits-of-studying-online/
6. Duffy TM, Dueber B, Hawley CL (1998) Critical thinking in a distributed environment: a pedagogical base for the design of conferencing systems. Electronic collaborators: learner-centered technologies for literacy, apprenticeship, and discourse, pp. 51–78
7. Bonk CJ, Hansen EJ, Grabner-Hagen MM, Lazar SA, Mirabelli C (1998) Time to "connect": synchronous and asynchronous case-based dialogue among preservice teachers. Electronic collaborators: learner-centered technologies for literacy, apprenticeship, and discourse, pp. 289–314
8. Ruiz JG, Mintzer MJ, Leipzig RM (2006) The impact of e-learning in medical education. Acad Med 81(3):207–212
9. Nagy A (2005) The Impact of e-learning. E-Content, pp. 79–96. https://link.springer.com/chapter/10.1007%2F3-540-26387-X_4
10. The Straight Time (2020) 14-day movement control order begins nationwide on wednesday. https://www.nst.com.my/news/nation/2020/03/575180/14-day-movement-control-order-begins-nationwide-wednesday
11. Karim KN (2020) Edu ministry introduces guidelines on online teaching, learning platforms. https://www.nst.com.my/news/nation/2020/03/578945/edu-ministry-introduces-guidelines-online-teaching-learning-platforms
12. Alsabawy AY, Cater-Steel A, Soar J (2013) IT infrastructure services as a requirement for e-learning system success. Comput Educ 69:431–451
13. Tuan L, Tram N (2022) Examining student satisfaction with online learning. Int J Data Netw Sci 6(1):273–280
14. Tran QH, Nguyen TM (2021) Determinants in student satisfaction with online learning: a survey study of second-year students at private universities in HCMC. Int J TESOL Educ 2(1):63–80

15. Tee PK, Gharleghi B, Chan YF (2014) E-Ticketing in airline industries among Malaysian: the determinants. Int J Bus Soc Sci 5(9):168–174
16. Davis FD (1989) Perceived usefulness, perceived ease of use, and user acceptance of information technology. MIS Q 13(3):319–340
17. Arpaci I, Al-Emran M, Al-Sharafi MA, Shaalan K (2021) A novel approach for predicting the adoption of smartwatches using machine learning algorithms. Recent advances in intelligent systems and smart applications, 185–195
18. Chia KM, Rohizan A, Tee PK, Tajuddin AR (2019) Evaluation of service quality dimensions toward customers' satisfaction of ride-hailing services in Kuala Lumpur Malaysia. Int J Recent Technol Eng 7(5S):102–109
19. Al-Emran M, Granić A, Al-Sharafi MA, Ameen N, Sarrab M (2020) Examining the roles of students' beliefs and security concerns for using smartwatches in higher education. J Enterp Inf Manag 34(4):1229–1251
20. Benson AD (2002) Using online learning to meet workforce demand: A case study of stakeholder influence. Q Rev Distance Educ 3(4):443–452
21. Tee PK, Cham TH, Low MP, Lau TC (2021) The role of organizational career management: comparing the academic staff' perception of internal and external employability in determining success in academia. Malaysian Online J Educ Manag 9(3):41–58
22. Lim TL, Omar R, Ho TCF, Tee PK (2021) The roles of work–family conflict and family–work conflict linking job satisfaction and turnover intention of academic staff. Aust J Career Dev 30(3):177–188
23. Tee PK, Eaw HC Oh SP, Han KS (2019) The employability of Chinese graduate in Malaysia upon returning to China employment market. Int J Recent Technol Eng, ISSN: 2277–3878, 8(2S): 358–365
24. Kadirova N, Lim LC, Benjamin CYF, Tee PK (2015) Service quality and postgraduate student satisfaction: a pilot study. Eur Acad Res 2(12):15483–15505
25. Sahin I, Shelley M (2008) Considering students' perceptions: the distance education student satisfaction model. Int Forum Educ Technol Soc 11(3):216–223
26. Barnett R (2011) The marketised university: defending the indefensible. The marketisation of higher education and the student as consumer, 53–65
27. Lin W-S, Wang C-H (2012) Antecedences to continued intentions of adopting e-learning system in blended learning instruction: a contingency framework based on models of information system success and task-technology fit. Comput Educ 58(1):88–99
28. Stoel L, Lee KH (2003) Modeling the effect of experience on student acceptance of web-based courseware. Internet Res 13(5):364–374
29. DeLone WH, McLean ER (1992) Information systems success: the quest for the dependent variable. Inf Syst Res 3(1):60–95
30. Bigne E, Moliner MA, Sanchez J (2003) Perceived quality and satisfaction in multiservice organisations: The case of Spanish public services. J Serv Mark 17(4):420–442
31. Al-Emran M, Al-Maroof R, Al-Sharafi MA, Arpaci I (2020) What impacts learning with wearables? An integrated theoretical model. Interact Learn Environ: 1–21
32. Park SY (2009) An analysis of the technology acceptance model in understanding university students' behavioral intention to use e-learning. Educ Technol Soc 12(3):150–162
33. Seddon PB (1997) A respecification and extension of the DeLone and McLean model of IS success. Proceedings of the IEEE international conference on information reuse and integration, vol. 8, issue 3, pp. 240–253
34. Lin JCC, Lu H (2000) Towards an understanding of the behavioural intention to use a web site. Int J Inf Manag 20(3):197–208
35. Ajzen I, Fishbein M (1977) Attitude-behavior relations: a theoretical analysis and review of empirical research. Psychol Bull 84(5):888–918
36. Aixia D, Wang D (2011) Factors influencing learner attitudes toward e-learning and development of e-learning environment based on the integrated e-learning platform. Int J E-Educ E-Bus E-Manag E-Learn 1(3):264–268

37. Kock N (2015) Common method bias in PLS-SEM: a full collinearity assessment approach. Int J e-Collab 11(4):1–10
38. Henseler J, Ringle C, Sarstedt M (2015) A new criterion for assessing discriminant validity in variance-based structural equation modeling. J Acad Mark Sci 43(1):115–135
39. Mailizar M, Almanthari A, Maulina S (2021) Examining teachers' behavioral intention to use e-learning in teaching of mathematics: an extended TAM model. Contemp Educ Technol 13(2):298–314
40. Maqableh M, Alia M (2021) Evaluation online learning of undergraduate students under lockdown amidst COVID-19 pandemic: the online learning experience and students' satisfaction. Child Youth Serv Rev 128:106160
41. Dhawan S (2020) Online learning: a panacea in the time of COVID-19 crisis. J Educ Technol Syst 49(1):5–22
42. Lee J-K, Lee WK (2008) The relationship of e-learner's self-regulatory efficacy and perception of e-learning environmental quality. Comput Hum Behav 24(1):32–47
43. Arpaci I, Karatas K, Kiran F, Kusci I, Topcu A (2021) Mediating role of positivity in the relationship between state anxiety and problematic social media use during the COVID-19 pandemic. Death Stud 46(4):1–11
44. The Star (2020) Online learning the new normal in education. https://www.thestar.com.my/opinion/letters/2020/04/16/online-learning-the-new-normal-in-education
45. Lee TH, Ching LC, Lim YM, Cham T (2019) H: University education and employment challenges: An evaluation of fresh accounting graduates in Malaysia. Int J Acad Res Bus Soc Sci 9(9):1061–1076
46. Cham TH, Cheng BL, Low MP, Cheok JBC (2020) Brand Image as the competitive edge for hospitals in medical tourism. Eur Bus Rev 31(1):31–59
47. Tee PK, Lim KY, Ng CP, Wong LC (2022) Trust in green advertising: mediating role of environmental involvement. Int J Acad Res Bus Soc Sci 12(1):1771–1786
48. Tee PK, Chan YF (2016) Exploring factors towards career success in Malaysia. Int Bus Manag 10(17):3936–3943
49. Cham TH, Cheah JH, Ting H, Memon MA (2021) Will destination image drive the intention to revisit and recommend? Empirical evidence from golf tourism. Int J Sports Mark Spons. https://doi.org/10.1108/IJSMS-02-2021-0040
50. Cheng BL, Shaheen M, Cham TH, Dent MM, Yacob Y (2021) Building sustainable relationships: service innovation at the pinnacle of touristic achievement. Asian J Bus Res 11(1):80–90

Dropout Early Warning System (DEWS) in Malaysia's Primary and Secondary Education: A Conceptual Paper

Wong Mikkay Ei Leen, Nasir Abdul Jalil, Narishah Mohamed Salleh, and Izian Idris

Abstract School dropout is an issue that plagues almost every nation globally, and Malaysia is not an exception. According to Malaysia's Education Ministry, dropouts are defined as Malaysian students in the school system who choose to leave before completing their education. Student dropout is a grave issue as it impacts the students and negatively implicates society and policymakers. At present, machine learning is the talk of the town as the world has a wealth of data that can be used freely. Machine learning is a method of data analysis that digitizes the development of analytical models. It is a technique that is based on the insights derived from data, recognize patterns, and make decisions with very little human intervention. From the Malaysian education perspective, no study thus far has looked into primary and secondary public-school dropouts while simultaneously using supervised machine learning. Consequently, this study intends to contribute to the literature, especially from Malaysia's perspective by proposing a dropout early warning system for primary and secondary students using supervised machine learning algorithms. The predictive model with machine learning has an enormous potential to develop early warning systems to identify and help students who are likely to drop out.

Keywords Dropout · Early warning system · Machine learning · Decision tree · Random forest · Naïve-Bayes

1 Introduction

Over the years, the number of students who dropped out of school has always been a concern. However, education experts have recently raised their concern that there will

W. Mikkay Ei Leen (✉) · N. A. Jalil · N. M. Salleh
Department of Business Analytics, Sunway University Business School, Sunway University, Subang Jaya, Malaysia
e-mail: mikkayw@sunway.edu.my

I. Idris
Department of Marketing and Innovation, Sunway University Business School, Sunway University, Subang Jaya, Malaysia

M. Al-Emran et al. (eds.), *International Conference on Information Systems and Intelligent Applications*, Lecture Notes in Networks and Systems 550,
https://doi.org/10.1007/978-3-031-16865-9_33

be a significant increase in school dropouts in the post-pandemic era. According to a recent report published by UNICEF and UNFPA, B40 households in urban areas have slashed an approximately 84% of their children's education expenditure [1]. Consequently, many children from B40 families had to quit school during the pandemic. This is a severe predicament for students and both society and policymakers [1].

Due to the current era of data, regardless of nation, the application of predictive analytics is the cutting edge in the education sector, along with banking, marketing, healthcare, and fraud detection. Predictive analytics has been extensively researched over the years due to its ability as an early warning system for forecasting potential academic results utilizing various student-related data [2–5]. In addition to that, predictive analytics would be able to forecast future events. The selection of B40 primary and secondary public-school students as the primary focus group parallels The Twelfth Malaysia Plan (RMK-12). In an attempt to reduce poverty among them, the Malaysian government identified education as one of the keys to alleviating poverty.

School dropouts are a grave concern as it affects the future generation of Malaysians—they are more inclined to encounter future economic adversity, social stigma, scarcer job opportunities, and lower income [6]. Despite the severity of the issue, there is no mechanism available thus far to predict primary and secondary public-school students from B40 families who are likely to drop out. Therefore, this research proposes a predictive analytics model by employing various supervised learning models to forecast at-risk students dropping out in advanced and take proactive measures to help these students. It is anticipated that this work will put forward a predictive analytics model that would accurately forecast students who are at risk of dropping out, and the Malaysian government and relevant stakeholders would be able to take pre-emptive measures to reduce the likelihood of their dropout.

The general society would be equally affected as they would face scarcity in the skilled workforce which inevitably, impacts the overall economic development of a nation. Due to these adverse consequences, students' dropouts have long been viewed as a grave educational obstruction by educators, government, and policymakers. In light of this conundrum, it is imperative for a dropout early warning system be established to ensure that the Ministry of Education can pinpoint at-risk students and take preemptive measures. Through the establishment of a dropout early warning system, it enables early intervention to ensure at-risk students would be able to complete their studies and have a better future. Hence, this paper proposes a conceptual framework for the dropout early warning system which would be able to predict Malaysia's primary and secondary students' intention to drop out.

2 Background

2.1 Machine Learning

The past years have witnessed a sudden surge in the utilization of the term Big Data—increasing the demand for advanced data analytics such as machine learning. Generally, machine learning is defined as a data analysis method that automates the development of analytical models [7]. Machine learning is based on the idea that systems can learn from data, recognize patterns, and make decisions with no or little human intervention [7]. Additionally, it can be classified into two learning algorithms, specifically, supervised and unsupervised learning.

Machine learning algorithms in supervised learning discover the connection between descriptive features (or predictors) and target features (or outcomes) in a dataset. We want to use the trained model from supervised learning to accurately predict future observations (prediction) or to understand better the relationship between the outcome and predictors (inference) [8]. The development of the prediction model for student dropout can be accomplished by utilizing supervised learning—machine learning algorithms ascertains the relationship between students' dropouts and different predictors. The dataset for supervised learning is identified as a labelled dataset—the dataset consists of a label (or target) that supervises the learning process [9]. It is crucial to note that the dataset must consist of both the target feature (or outcome) and descriptive features (or predictors) for supervised learning.

Conversely, in unsupervised learning, the machine-learning algorithms study the dataset's structure without the involvement of a target characteristic. In unsupervised learning, all variables involved in the analysis are utilized as inputs—hence, appropriate for clustering and association mining methods. Generally, unsupervised learning is apt for creating the labels in the data which are eventually utilized to employ supervised learning tasks [10]. For unsupervised learning, the main focus is to uncover the fundamental structure of the data instead of predicting the target feature. Therefore, in line with the objective of this study, supervised learning is selected as the algorithm of choice.

2.2 Predictive Learning Models

The utilization of machine learning in developing predictive models is ubiquitous in the literature. In this research, several classic Machine Learning techniques have been selected.

1. **Decision tree:** A decision support tool that is characterized by a tree-like model of decisions along with their possible outcomes. The tree can be characterized into two separate entities, specifically, decision nodes, branches, and tree leaves.

Typically, each node represents a test on an attribute value while the branches denote an outcome of the test, and lastly, tree leaves signify classes or class distributions [9].

2. **Random forest:** A classifier that consists of a number of decision trees on different subsets of a training sample and utilizes the average to optimize the dataset's predictive accuracy and control overfitting [8].
3. **Naïve-Bayes:** A measurable classifier stands that uses Bayes' theorem with independent theories. This classifier presumes that the occurrence of a particular feature in a class is distinct to the presence of any other feature [8].

3 Related Works

Typically, machine learning is employed to perform the following tasks, regression, classification, clustering, and dimension reduction. Regression is usually used in the estimation of numerical or continuous values. Next, clustering is generally employed to discover the natural grouping of data. The dimension reduction—to reduce variables in the analysis to improve the interpretability of the outcome and the algorithm's effectiveness. The classification technique is in parallel with the objective of this study, where it is used to predict students' dropouts. Indubitably, the classification technique has been utilized in the Malaysian education sector, but the emphasis is mainly on students' performance rather than dropout [11].

Undeniably, classification techniques have been employed in Malaysia's education system; nevertheless, it only centers around B40 university students [11]. Moreover, the literature demonstrates that dropout studies generally revolve around university students [6, 8, 11–15].

Chung and Lee [8] conducted research involving high school students in Korea where they utilized predictive modelling using machine learning, specifically random forest. It was reported that the developed predictive model showed a good performance in forecasting students' dropouts.

From Malaysia's perspective, there was a study conducted by [16], where the researcher looked into dropout trends and patterns among secondary school students in Perak. Nevertheless, this study did not predict students at risk of dropping out of school. The research aimed to understand the underlying reasons which lead to students dropping out. In another study by [11], they investigated B40 bachelor's degree students' dropout by constructing a predictive model using Decision Tree, Random Forest, and Artificial Neural Network. Their study showed that the Random Forest model is the best model for predicting B40 bachelor's degree students. [6] looked into Malaysia's private university students' dropout by employing two classifiers: decision tree and logistic regression model. The classifier execution is measured based on machine-learning performance's accuracy and misclassification rate. The study found that the decision tree has a better classification performance than the logistic regression model.

[13] proposed a predictive model to forecast student dropout employing two machine-learning approaches, logistic regressions and decision trees. The models are established using examination data—readily available in all universities. This work concluded that decision trees could produce better results than logistic regressions.

[15] adapted five machine-learning methods, decision trees, logistic regression, random forest, K-nearest neighbor and neural network algorithm to predict student dropout of Computer Science undergraduates. This work established that the logistic regression model is the best learners to forecast students' dropouts.

[14] researched to predict undergraduate and diploma students' dropout from ABC Faculty of XYZ University. In doing so, the authors applied the synthetic minority oversampling technique (SMOTE) and random forest algorithm to predict students' dropouts. The research found that the random forest algorithm accompanied by SMOTE offers the best accuracy result with 93.43%.

[17] investigated student dropout in India's university and proposed a predictive model using three computational techniques, Naïve-Bayes, decision tree and info gain. Students' dropout in this study is best predicted using the Naïve-Bayes technique.

Lastly, in a study by [18], the authors proposed a computational approach using educational data mining and various supervised learning techniques: decision tree, K-nearest neighbour, neural networks, Naïve Bayes, support vector machines, and random forest. These supervised learning techniques are used to evaluate the behavior different prediction models to determine students at risk of dropping out of university. The authors concluded that random forest and decision trees could present solid results compared to the rest of the proposed supervised learning techniques.

Thus, it is evident that there is a lack of an early warning system in predicting dropout among primary and secondary public-school students in Malaysia. Hence, this study will be utilizing three of the most promising supervised learning techniques, decision tree, random forest and Naïve-Bayes, to establish the predictive model to predict primary and secondary public-school students in Malaysia.

4 Methodology

Python will be used as the primary programming language to develop the predictive model in this study. It will also act as a database to keep and pre-process data and utilize it for characteristics selection and statistical tests. In this research, there are three critical phases involved.

Figure 1 depicts the phases of the projected predictive analytics model—forecast at-risk students of dropping out of school.

Phase 1

In Phase 1, raw data will be obtained from Bahagian Pembangunan dan Perancangan Dasar (BPPD), Kementerian Pendidikan Malaysia (Pendidikan Rendah). As denoted earlier, this study focuses on primary and secondary public-school students

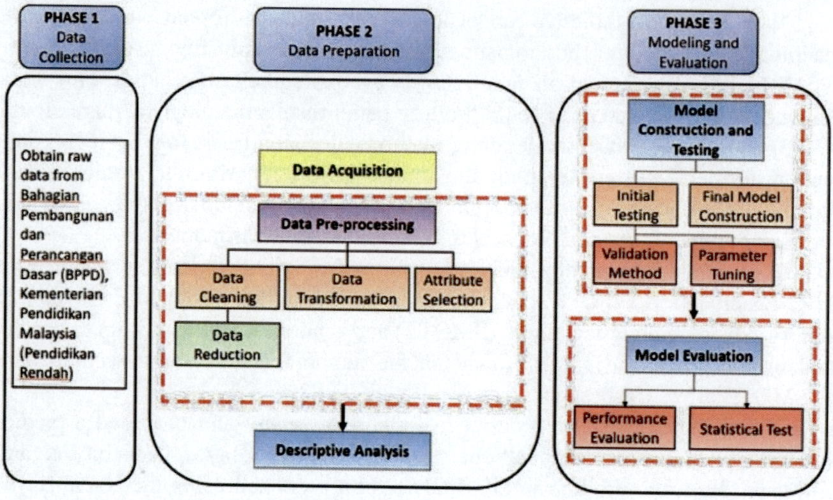

Fig. 1 Projected predictive analytics model of supervised learnin (adopted from (Sani et al., 2020b))

because it is pertinent to identify low-performance students early. The following raw data attributes are expected to be available upon request: student name, date of birth, gender, marital status, place of birth, postcode, family income, student status, sponsorship, and examination results. These are subject to change.

Phase 2

Upon the data acquisition, the dataset will undergo data pre-processing to prepare data for primary processing or further analysis. Often, actual data is inadequate, incomprehensible, as well as contains noise [11]. Therefore, raw data must undergo pre-processing to ensure that data is in a proper format for a selected miner tool and should be sufficient for a selected method. Furthermore, the data cleaning process will be done through the dimension reduction process. Incomplete, redundant, outliers, or obsolete data will be removed from the dataset. Data pre-processing is vital to ensure that only student records from B40 household income are included. Next, data must be transformed into an understandable structure to obtain appropriate information. Moreover, attributes with varieties of data will be accumulated by utilizing the hierarchical theory. Only significant characteristics will be chosen and employed to develop the dropout prediction model.

Phase 3

Phase 3 is the crucial phase in this study, and it comprises of several stages, specifically, model construction and testing as well as model evaluation. Three supervised learning algorithms are selected: decision tree, random forest, and Naïve-Bayes—determined upon the literature review process.

Model Construction and Testing

All prediction models (decision tree, random forest, and Naïve-Bayes) will be tested to ascertain the validation method and algorithm parameters used to generate a high-performance prediction model. The prediction model validation will be tested using the holdout and 10-folds cross-validation method. Next, all prediction models will be fabricated repetitively, utilizing different parameters to attain the highest accuracy.

Model Evaluation

In order to assess the performance of the prediction model, F measure, value of accuracy, precision, and recall will be used. Determination of the best prediction model will be accomplished through the statistical test.

In order to determine factors that lead to dropout, the researcher will be conducting an interview session among selected students who have been identified to be at risk. It is also planned that during the interview session, at-risk students will be given the opportunity to suggest ways to retain them in school.

5 Conclusion

It is important to be able to determine students who are at risk of dropping out of school. Establishing a dropout early warning system in the Malaysian education system would be highly beneficial, especially for educational administrators and students. Through this study, various parties will benefit from the outcome. Upon successful development of a predictive model using machine learning algorithms, it would predict dropout among primary and secondary public-school students. Apart from that, this study would be able to determine the common factors driving dropout from school as this study includes all Malaysian students from primary and secondary public schools Additionally, this study would contribute toward identifying the most suitable machine-learning algorithm to forecast students at risk. It is pertinent to determine the best algorithm to predict dropouts to ensure that the Ministry of Education, policymakers, and other stakeholders can propose data-based diagnosis as well as support systems for at-risk students.

References

1. UNICEF (202AD) Families on the Edge: mixed methods longitudinal research on the impact of the COVID-19 crisis on women and children in lower income families, United Nations Childrens' Fund, Malaysia and the United Nations Population Fund, no 1, pp 1527–1557
2. Guzmán-Castillo S et al (2022) Implementation of a predictive information system for university dropout prevention. Procedia Comput Sci 198(2020):566–571. https://doi.org/10.1016/j.procs.2021.12.287
3. Abdul Bujang SD, Selamat A, Krejcar O (2021) A predictive analytics model for students grade prediction by supervised machine learning. IOP Conf Ser Mater Sci Eng 1051(1):012005. https://doi.org/10.1088/1757-899x/1051/1/012005

4. Neves F, Campos F, Ströele V, Dantas M, David JMN, Braga R (2021) Assisted education: using predictive model to avoid school dropout in e-learning systems. Intell Syst Learn Data Anal Online Educ 2020:153–178. https://doi.org/10.1016/b978-0-12-823410-5.00002-4
5. Herodotou C, Rienties B, Boroowa A, Zdrahal Z, Hlosta M (2019) A large-scale implementation of predictive learning analytics in higher education: the teachers' role and perspective, vol 67, no 5. Springer. https://doi.org/10.1007/s11423-019-09685-0
6. Roslan N, Jamil JM, Shaharanee INM (2021) Prediction of student dropout in Malaysian's private higher education institute using data mining application. Turk J Comput Math Educ (TURCOMAT) 12(3):2326–2334. https://doi.org/10.17762/turcomat.v12i3.1219
7. Virvou M, Alepis E, Tsihrintzis GA, Jain LC (2020) Machine learning paradigms: advances in learning analytics, vol 158. Springer. https://doi.org/10.1007/978-3-030-13743-4_1
8. Chung JY, Lee S (2019) Dropout early warning systems for high school students using machine learning. Child Youth Serv Rev 96:346–353. https://doi.org/10.1016/j.childyouth.2018.11.030
9. Lee S, Chung JY (2019) The machine learning-based dropout early warning system for improving the performance of dropout prediction. Appl Sci (Switzerland) 9(15). https://doi.org/10.3390/app9153093
10. Hofmann T (2001) Unsupervised learning by probabilistic Latent Semantic Analysis. Mach Learn 42(1–2):177–196. https://doi.org/10.1023/A:1007617005950
11. Sani NS, Nafuri AFM, Othman ZA, Nazri MZA, Nadiyah Mohamad K (2020) Drop-out prediction in higher education among B40 students. Int J Adv Comput Sci Appl 11(11):550–559. https://doi.org/10.14569/IJACSA.2020.0111169
12. Jin C (2020) MOOC student dropout prediction model based on learning behavior features and parameter optimization. Interact Learn Environ. https://doi.org/10.1080/10494820.2020.1802300
13. Kemper L, Vorhoff G, Wigger BU (2020) Predicting student dropout: a machine learning approach. Eur J Higher Educ 10(1):28–47. https://doi.org/10.1080/21568235.2020.1718520
14. Utari M, Warsito B, Kusumaningrum R (2020) Implementation of data mining for drop-out prediction using random forest method. In: 2020 8th international conference on information and communication technology, ICoICT 2020. https://doi.org/10.1109/ICoICT49345.2020.9166276
15. Wan Yaacob WF, Mohd Sobri N, Nasir SAM, Wan Yaacob WF, Norshahidi ND, Wan Husin WZ (2020) Predicting student drop-out in higher institution using data mining techniques. J Phys Conf Ser 1496(1). https://doi.org/10.1088/1742-6596/1496/1/012005
16. Eshah Mokshein S, Teck Wong K, Ibrahim H (2016) Trends and factors for dropout among secondary school students in Perak. Policy Pract Teachers Teacher Educ 6(1):5–15
17. Hegde V, Prageeth PP (2018) Higher education student dropout prediction and analysis through educational data mining. In: Proceedings of the 2nd international conference on inventive systems and control, ICISC 2018, no ICISC, pp 694–699. https://doi.org/10.1109/ICISC.2018.8398887
18. De Santos KJO, Menezes AG, De Carvalho AB, Montesco CAE (2019) Supervised learning in the context of educational data mining to avoid university students dropout. In: Proceedings - IEEE 19th international conference on advanced learning technologies, ICALT 2019, pp 207–208. https://doi.org/10.1109/ICALT.2019.00068

Development of a Mobile Application for Room Booking and Indoor Navigation

Syahier Aqif bin Sabri, Mazlina Abdul Majid, Ali Shehadeh, and Abdul Rehman Gilal

Abstract Signboards and static maps placed around the buildings are mostly used to navigate the people. The use of signboards inside the building sometimes can cause confusions for people who are unfamiliar with the internal building architecture. It gets more difficult when the building consists of multiple level with several wings and junctions. Unlike outdoor navigations, indoor navigation applications are helpful to navigate users within the buildings. Several network devices are configured to collect the position of the users or objects. Precision of positioning in the indoor navigation is still an open research area. In this study, we develop an application to precisely navigate the users in the closed environments. We have used Universiti Malaysia Pahang (UMP) campus as a case for developing our application. The application also adds an additional feature for enabling users to book the university rooms (e.g., conference rooms). Once the users book the rooms, this application will navigate them to the booked locations within the building. The application has been developed using RAD methodology which allows the project to be divided into smaller parts based on modules involved which at the end was compiled together and can be completed within shorter time due to limited time constraint to complete the project. As final result, a mobile application has been successfully developed which able to elevate the project to next phase which carry out user acceptance test allowing the identification of target user insights of using an indoor navigation application in the environment selected for this project.

S. A. bin Sabri · M. A. Majid
Faculty of Computing, Universiti Malaysia Pahang, Lebuhraya Tun Razak, 26300 Gambang, Kuantan, Pahang, Malaysia

A. Shehadeh
Department of Civil Engineering, Hijjawi Faculty for Engineering Technology, Yarmouk University, Irbid 21163, Jordan

A. R. Gilal (✉)
Department of Computer and Information Sciences, Universiti Teknologi PETRONAS, 32610, Seri Iskandar, Perak, Malaysia
e-mail: rehman.gilal@utp.edu.my

M. Al-Emran et al. (eds.), *International Conference on Information Systems and Intelligent Applications*, Lecture Notes in Networks and Systems 550,
https://doi.org/10.1007/978-3-031-16865-9_34

Keywords Indoor navigation · Room booking · Mobile application · Smart campus

1 Introduction

A building is just a pile of bricks arranged to build rooms to accommodate people for multiple purposes e.g., living, shelter, working, studying or many other functions. Some buildings are big and confusing which can cause issue for newcomers to find out the specific location. The use of signboards and maps are sometime confusing for certain people with the increasing number of corridors and junctions inside the building [1]. For instance, usually universities campuses are big buildings. Almost thousand people visit campus every day. Outdoor navigation application, such as Google maps, can help the users to reach to the building. However, it is still time taking and confusing for newcomers to find out specific room number within the buildings. For example, sometimes teachers want to arrange urgent academic actives. The program coordinator may allocate the room to a building that may not be familiar to teachers or students. Sometimes, teachers and students attend the class activities back-to-back. In this situation, teachers or students may be late to the venue. Therefore, they need special guide to navigate within the building to reach to the venue on time.

With the introduction of technology of indoor navigation, issue mentioned above can be overcome. Nowadays, Wi-Fi based positioning method can be used to locate the users and objects. For example, Wi-Fi systems use access points placed inside the building. These access points can be used to determine the location of a device with precision of less than 15 m [2]. Other technology can be used to track the position of the user inside the building such as Bluetooth low energy (BLE) beacons. BLE beacons need to be placed around the building with beacons' coverage overlap with at least other two beacons' coverage which capable to track position of a device with accuracy of less than eight meter [4]. Therefore, for the development of this application for indoor navigation, BLE beacons technology is used as it is precise.

This study uses space of Faculty of Computing (FK) of Universiti Malaysia Pahang (UMP) to develop the application. First, a survey of FK building needs to be done to determine the position the BLE beacons need to be placed to ensure the coverage of the beacons combined to locate the position of the user's device. This is because more beacons used will result to better accuracy of the positioning. The best practice is to place between eight to 15 beacons for 1000 square meters [3]. After the survey, we have identified the number of beacons needed for this application. The next process is to make drawing of the layout of FK building. The rooms will be defined, obstacles and entrance of each rooms need to be assigned on the map. Paths need to be set inside the map as it will be used to set waypoint between the device's current position with the selected room. Once the beacons are placed inside the building, the position of each BLE beacons is needed to be marked inside the map as the input from each beacon will determine the position of the user according to Navigine user manual [4].

This application provides a feature to its users to book FK resources. Currently, we enable FK staff to book faculty room for any purposes e.g., meeting, replacement of classes, tests, or an event with instant confirmation. This feature also utilizes the navigation function to navigate them to the booked room inside FK building. Staff can manage the rooms under their supervision either to update the equipment available, room capacity and to change the status of the room either available to be booked or not. Recently, FK has moved to new faculty building consist of five level designed in E-shape which most of the students and staff are still not used to the localisation of the rooms inside the building. Even though with the signboards and static map are available inside the building, it still become an issue when someone is rushing to a room inside the building leading to possibility of someone to get lost. With the size of the building so big and multiple junctions, it will consume time for new students and staff to get used with the location of rooms inside the building even though there exist a static signboard, which sometimes it can be confusing for some people.

Moreover, academic and admin staff also need a platform to book a room in FK building. An on-the-spot confirmation system is required as some of the events may be urgent. Currently, there is a system in place provided by the Universiti Malaysia Pahang (UMP) named resource booking under e-community system. The system does not have updated the database of the room inside new FK faculty building. Therefore, it causes problem for staff to make a booking of any rooms inside the FK building. Additionally, resource booking through e-community require confirmation from admin which takes time to be approved. For instance, once staff makes a booking in the system, the status of the booking will be set as pending for approval. At mean time, another staff can also do the booking for the same room at same date and time. This problem can lead to a clash between two parties as the status of the booking for two parties still under pending even on the date the room is required. This clash incident shows how late admin approval issue of the resource booking system in e-community system cause the system to become less efficient.

Therefore, this study presents an application for FK staff to book the resources (e.g., room) in the faculty and get the indoor navigation to the room. The application is called FKguide in this paper. The project is only developed for android mobile application only using cloud database system that allows changes such as adding a new room booking into the database made by various user can be updated instantly and in real time as this can avoid clash room booking issue to occur when using developed application in this project [14]. The users of this application are FK staff and guest consist of FK students and visitors. The application will use Navigine Tracking Platform as third-party service to provide the indoor navigation function of the application. Indoor navigation technology used in this project is Bluetooth technology by deploying Bluetooth Low Energy (BLE) beacons.

2 Related Work

This section discusses the existing system and related applications. Three existing applications such as Indoors, Infsoft and Concierge Go are presented in the sub-sections below.

2.1 *Indoors*

Indoors [6] is a real time indoor navigation system with. Additional feature of this system for its room booking features is colleagues can be invited to the meeting once the booking is confirmed by the system. This system may be developed with various features to be added. For instance, it can be implemented by office building, public building such as airport, hospital, museum, retail store and industrial building such as warehouse, laboratory, and factory. The technology used for indoor tracking of this system are Wi-Fi positioning method or using Bluetooth Low Energy (BLE) beacons. The system is developed for web based and mobile application.

2.2 *Infsoft*

Second system is Infsoft [7] which focuses indoor positioning and navigation system with additional product for room utilization features. For room booking availability, this system implements the use of infrared sensors inside the rooms to provide real time status of its availability. This system can be developed with additional features that is suitable to be used at office buildings, public buildings such as hospital, retail store and industrial buildings such as factory and warehouses. The technology can be used by the system for indoor tracking are BLE beacons, Wi-Fi, Ultra-wideband, RFID and camera system. The system is developed for web based and mobile application.

2.3 *Concierge Go*

The third system is Concierge Go is developed by Fischer and Kerrn [8]. The application mainly focuses on room and desk booking system of a building which utilize the indoor navigation for better optimization of the system. For the room booking function, this system allows the user to make a booking and he/she need to check in once he/she arrived at the room during the booking time. This allows the system to remove ghost booking when a person who book the room not coming to the room within first few minutes of the booking to maximize the room usage. The suitability

Table 1 Specifications of the three existing systems

Specification	Indoors	Infsoft	Concierge Go
Room booking features	Colleague can be invited to the meeting once a booking has been made	Using infrared sensors for real time occupancy status	Require check-in to the booked room to remove ghost bookings
Type of building suitability	Office, public and industrial buildings	Office, public and industrial buildings	Office buildings
Indoor tracking technology	Wi-Fi, BLE beacons	BLE beacons, Wi-Fi, Ultra-wideband, RFID, Camera system	BLE beacons, Wi-Fi, Light source
Deployment	Web-based and mobile application	Web-based and mobile application	Web-based and mobile application

of this system is for office building only. The technology used for indoor tracking are BLE beacons, Wi-Fi, and light source. This system is deployed for web based and mobile applications. Table 1 above compares and summarises the three discussed systems.

3 Methodology

This project follows Rapid Application Development (RAD) methodology. RAD helps us to divide the project into smaller parts. RAD also allows changes to be made at any time to ensure all any new requirements required by the stakeholder during any phase of the development process can be added into the application. The disadvantages of RAD is the application is broken into modules which can lead to code integration issue during the compilation of the modules developed as most modules of the application are related to another module. This issue can lead to delay of the development process as more time is consumed to solve the code integration issue to ensure the modules combined enable the application to work as a whole system that relying to other module to fully functional as planned [11]. Figure 1 below shows the lifecycle of RAD.

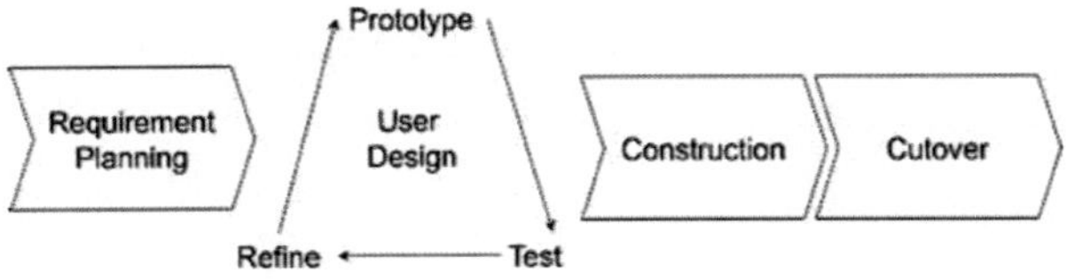

Fig. 1 A rapid application development

3.1 Requirement Planning

In the first phase, the problem is researched to allow the stakeholder to understand the problem that led to the development of this application. For instance, in this study, the main problems identified that some staff and students who are still unfamiliar with the location of rooms inside the building Second stage is defining the requirements needed to be added into system to fulfill the needs of the end user. The third stage is finalize the requirements and Software Requirement Document (SRS) is produced. The use case diagram presented in the Fig. 2 below.

Figure 2 shows the use case diagram of the FKguide consist of three actors which are staff, guest consist of student and visitors and Navigine as the third-party system used for indoor navigation function. The main modules of this system are manage login, manage room booking, manage navigation, manage building layout, manage room, and manage profile. The function of each module is elaborated in Table 2 below.

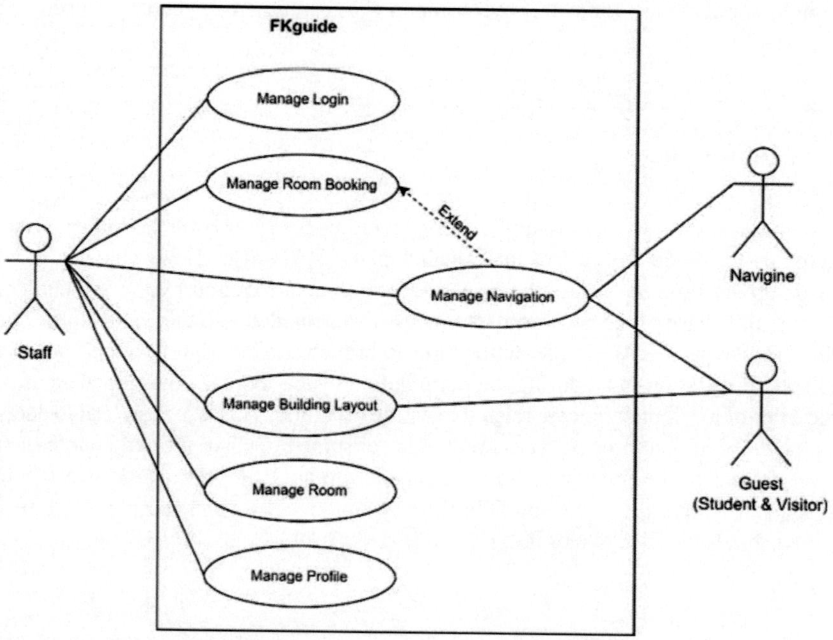

Fig. 2 Use case diagram for FKguide

Table 2 FKguide functions for each module

Module	Function
Manage login	To allow staff to login to their profile
Manage navigation	To allow the user (staff, student, and visitor) to use the indoor navigation feature inside the building to be navigated to the selected room
Manage room booking	To allow staff to make a room booking of room inside FK building and manage booking made. Staff also can use the indoor navigation feature to be navigated to the room booked
Manage building layout	To allow the user (staff, student, and visitor) to view the building layout of the building which also available offline
Manage room	To allow staff to manage room under their supervision. Staff can update the capacity and equipment provided in the room and can remove a booking made by other staff
Manage profile	To allow staff to view and edit their profile information

3.2 *User Design*

Once the scope of the project is defined and requirements have been finalized, initial models and prototypes are developed to allow the clients to evaluate whether it fulfill the end user requirements and changes to the requirement can be made earlier before the development is started to ensure the product able to satisfy the client. The prototype of the application to be developed is designed using Adobe XD. Software Design Document (SDD) is developed within this phase.

3.3 *Construction*

Consequently, the prototype designed and tested in the above phase is converted into working model or actual codes since most of the issues and improvements have been resolved during the iterative design phase. The application has been developed for Android mobile environment which allows the application to be portable to navigate user inside the building and can utilize the Bluetooth functionality of Android device to ensure the navigation features using BLE beacons implemented in the application functioned as planned to allow the implementation of Internet of Things (IOT) to be established on the campus in providing indoor navigation service to the dedicated user [13, 15]. The development environment, involving work areas and workspace for the developers, was finalized. In addition, the database was built based on the preliminary data structure designed at the previous phase. The database is using Firebase Firestore Cloud Database to store the details of the rooms inside the building, staff profile information and the room booking made by the staff. The features of real-time database from Firebase Firestore Cloud Database allows any changes made on the database can be updated to all users accessing the applications instantly. The application underwent a series of tests to make sure the developed application is working as expected and according to the initial design and fulfill the requirements [17].

3.4 Cutover

In this phase the application has been fully developed and ready to be deployed to real environment to be used by the end user once all modules have been integrated together and the testing in previous stage is satisfied. The user acceptance test is conducted to allow users to evaluate their experience while using the application while the developer continue to observe the application's behavior to identify bugs available to be solved to improve the reliability of the application. The results and feedback from the tester of user acceptance test carried out allows the identification of parts of the application that can be improvised to increase the functionality and usability of the application and fulfil the expectation of targeted user with the enhancement made in the future.

3.5 Used Software and Hardware

This sub section elaborates the system requirement in development of FKguide mobile application. System requirements are split into software and hardware requirement. It also explains the purpose of each hardware and software used in development of this project. Following Table 3 discusses the software and hardware used in this project development.

4 Implementation of Indoor Navigation System

For indoor tracking function, Navigine Tracking platform is used as third-party system in this application. Navigine is an indoor tracking system that allows developer to create an indoor map of a building, declare the rooms, obstacles, and path inside the building. Navigine also enables the integration of BLE beacons installed inside the building to function as tracking device to determine user's location. Once the process mentioned is complete, the indoor navigation system can be integrated with FKguide mobile application and can be accessed directly with the application.

When the user uses the indoor navigation function of the application, the BLE beacons emit the signal to user's device which the inputs from the signal of the BLE beacons are processed by Navigine to determine the position of the user inside the building when navigating the user from initial position to selected room.

Table 3 Software and hardware used for the application development

Software	Purpose
AbobeXD	To design the interface of the application and create storyboard of the flow of the functions of the application
Android studio	To code the application specifically for android smartphone
AutoCAD	To draw the layout of the FK building to be digitalized
Navigine developer	To create the indoor navigation system and integrate the system into the application developed
Drawio	To draw the diagrams to represent the details of the application
Microsoft word	To write the documentations of the application
Firebase firestore cloud database	To store and provide database required by the applications on a cloud database platform
Hardware	**Description**
Laptop	To develop the application and documentation preparation
Android smartphone with Android 8.0 or newer	To install the application, for run and testing purposes
Bluetooth Low Energy (BLE) beacons	Work to identify the position of the user's device inside the building

For this project, six BLE beacons is used. The BLE beacons is placed inside the building based on the following specifications and requirements by Navigine [16]:

1) Place eight to 15 BLE beacons per 1000 square meters which higher number of beacons will provide higher accuracy in tracking the user position.
2) BLE beacons need to be placed evenly inside the building area.
3) Avoid placing beacons in one straight line.
4) BLE beacons need to be attached on ceiling or at wall with height is inaccessible to avoid the beacons is moved by unauthorized person.
5) Ensure the user's device is in range or in zone of visibility of at least three BLE beacons at a time.

5 Results and Discussion

The system is developed using Java in Android Studio 11.0.10 for mobile application development. By using Android Studio, the interface of the application is designed, and the functions of the application are coded to ensure the application worked as required. For the database of the application, Firebase Database is used to store and

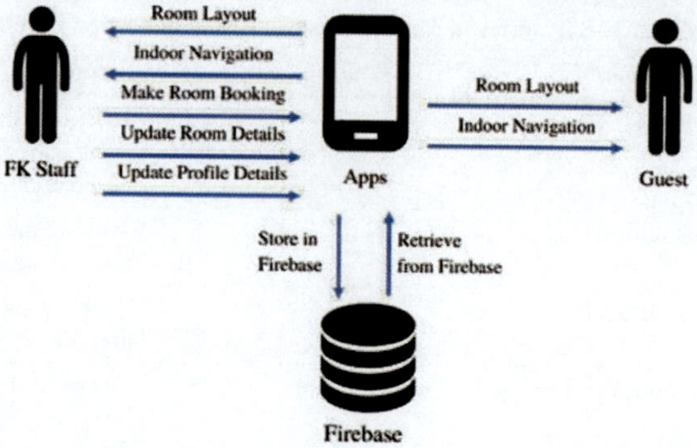

Fig. 3 FKguide architecture

provide the database required by the applications to run. Firebase allows the database to be stored on cloud and can be accessed by all the users when using the application. Figure 3 shows the process architecture of FKguide system.

In this project, the user acceptance test has been conducted and tested by 10 users. The UAT test participants were consist of five FK staff and five FK students. The UAT is performed using Google Form consist of 13 questions divided into three sections which are interface design, functionality, and overall user experience. Based on the result gathered for this section, for overall usability, four users were extremely satisfied, and six users were satisfied with the usability of this application. For graphic user interface, five users were extremely satisfied, and five users were satisfied with the design and user friendliness of the graphic user interface of the application. For overall functionality of the application, seven users were extremely satisfied, two users were satisfied, and one user rated the overall functionality as neutral. The results for overall user experience are depicted in the Fig. 4 below.

Based on the result of user acceptance test on overall user experience in Fig. 4, most of the testers are satisfied with the overall usability and graphic user interface of the application. The functionality received a neutral experience from user due to the accuracy issue of the indoor navigation features and other modules need to be improved as mentioned in system functionality above which can enhance the user experience and functionality of the application [9, 10, 25].

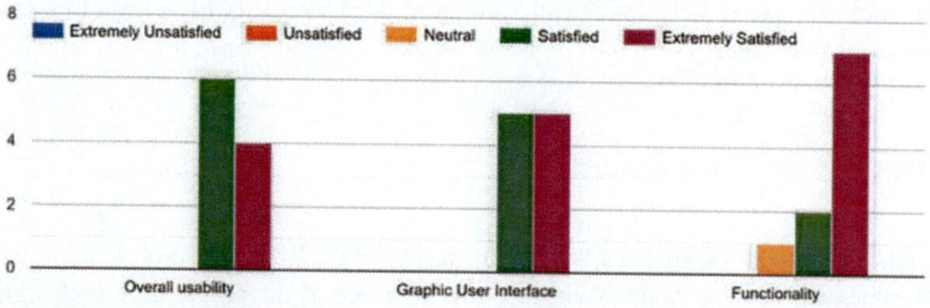

Fig. 4 Overall user experience results

One of the major concerns based on the result of user acceptance test carried out is the indoor navigation features of the application developed. As the indoor navigation system is still an open area which constantly in the process of improvement to enhance the quality and reliability of the indoor navigation system. Based on the results of a research study for indoor navigation using Bluetooth, it was stated that the issue and error while using the indoor navigation system using Bluetooth is the system sometimes misplaced the position of the device to somewhere near or far from actual position and caused by the low signal coverage of the devices or beacons to track the location of the user when moving around the building [12, 24].

Overall, of the feedbacks received from the user acceptance test can be referred to Jakob's ten usability heuristics. For the visibility of system status, some part of the system did not show the feedback such as notifications from an action made by user which user unable to realise the changes made based on their action. For the match between system and real world, the application only uses common phrases used by FK staff and students to ensure the familiarity of the users with the application without using any specific jargon. Third is user control and freedom, some part of the application unable to provide proper exit path for user as they will redirected back to previous interface without any exact path to redirect user back to respective home page. Most the forms in the application have been added checking to make sure no blank form is submitted to avoid any error either for login, booking or room details to make sure sufficient information is gathered to allow application to perform the action [18–23]. Submitting a blank form will trigger a message to notify the user that the form is not fully filled and unable to proceed the process [5].

This review has been further addressed to be put as focus when conducting maintenance of the application to increase the application's service quality. Some of the suggestion received from the testers to improve the user satisfaction and functionality of the application are as follows:

1) Person in charge of the room should be able to view the bookings made for their room and able to approve or cancel a booking made with reasons.
2) Room details need to have more attributes such as software installed inside the room's computer and the seating arrangement to ensure the staff can select a correct room for the purpose of the booking.
3) The maps of the building in the navigation modules need to be reoriented based on the user's heading to which direction to ease the user in identifying the path need to be followed.

6 Conclusion and Future Directions

This study presents a mobile application for resource booking and indoor navigation. Due to limitation of BLE beacons available to cover the whole building, only certain part of the building has been managed to be covered. The initial planning of the selecting the covered area and identify the rooms to be added into the application are successfully implemented. The mobile application is developed by implementing the functions of indoor navigation, room booking system and related functions such as staff can manage the room they oversee and manage their profile information and providing layout diagram of the building. To test the acceptance of the application among FK staff and students. Based on the user experience results, overall usability of the application reach satisfactory level with some of the module is at neutral due to some weakness of the functionality such as inaccuracy and time to taken to refresh the user location in indoor navigation function. The overall user interface design is satisfactory with further focus need to be set on increasing the efficiency and functionality of the application on every module.

During the development of this project, limitations and constraints are identified which limits the development process. The limitations of this project are time constraint, Number of BLE beacons used, Accuracy of indoor navigation function. Therefore, in the future, we will increase the coverage of the application by adding more BLE beacons inside the building allowing the application to provide the indoor navigation feature inside the whole FK building which increase the usability of the application. With fully coverage of indoor navigation, all rooms inside the building can be added to be selected for room booking feature. In the future, we shall also enable person in charge of the room should view the bookings made for their room and able to approve or cancel any booking made with reasons.

Acknowledgements This research is financially supported by the Green Technology Research Lab (GreenTech) University Malaysia Pahang. We would also like to acknowledge Universiti Teknologi PETRONAS and Yarmouk University, Irbid for their support to conduct this project.

References

1. Osman A et al (2020) Interactive virtual campus tour using panoramic video: a heuristic evaluation. J Comput Res Innov (JCRINN) 5(4):1–7
2. Thuong NT et al (2016) Android application for WiFi based indoor position: system design and performance analysis. In: 2016 international conference on information networking (ICOIN). IEEE
3. Huh J-H, Seo K (2017) An indoor location-based control system using bluetooth beacons for IoT systems. Sensors 17(12):2917
4. Torstensson D (2017) Indoor positioning system using bluetooth beacon technology
5. Nielsen J (2020) 10 usability heuristics for user interface design. Nielsen Norman Group. https://www.nngroup.com/articles/ten-usability-heuristics/. Accessed 16 Jan 2022
6. Indoor Positioning System for Enterprises | powered by indoo.rs, indoo.rs. https://indoo.rs/. Accessed 22 Mar 2022
7. RTLS Solutions (Real-Time Locating Systems) by infsoft. Infsoft.com. https://www.infsoft.com/. Accessed 22 Mar 2022
8. Room and desk booking software with workplace analytics | Fischer Kerrn. Fischer & Kerrn. https://fischerkerrn.com/. Accessed 22 Mar 2022
9. Gilal AR et al (2018) Finding an effective classification technique to develop a software team composition model. J Softw Evol Process 30(1):e1920
10. Tunio MZ et al (2018) Task assignment model for crowdsourcing software development: TAM. J Inf Process Syst 14(3):621–630

11. Despa ML (2014) Comparative study on software development methodologies. Database Syst J 5(3):37–56
12. Satan A (2018) Bluetooth-based indoor navigation mobile system. In: 2018 19th international Carpathian control conference (ICCC). IEEE
13. Alfiras M, Yassin AA, Bojiah J (2022) Present and the future role of the internet of things in higher education institutions. J Posit Psychol Wellbeing 6(1):167–175
14. Qasem YAM, Asadi S, Abdullah R, Yah Y, Atan R, Al-Sharafi MA, Yassin AA (2020) A multi-analytical approach to predict the determinants of cloud computing adoption in higher education institutions. Appl Sci 10(14):4905
15. Arpaci I (2017) Design and development of educational multimedia: the software development process for mobile learning. In: Khosrow-Pour M (ed) Blended learning: concepts, methodologies, tools, and applications, 2nd edn. IGI Global, Information Science Reference, Hershey, pp 366–384. https://doi.org/10.4018/978-1-5225-0783-3.ch018
16. "Home Page." Navigine Docs, docs.navigine.com. Accessed 22 Mar 2022
17. Ismail KA et al (2016) Big Data prediction framework for weather Temperature based on MapReduce algorithm. In: 2016 IEEE conference on open systems (ICOS). IEEE
18. Alsariera YA, Majid MA, Zamli KZ (2015) A bat-inspired strategy for pairwise testing. ARPN J Eng Appl Sci 10.8500-6
19. Masitry AK et al (2013) An investigation on learning performance among disabled people using educational multimedia software: a case study for deaf people. Int J Bio-Sci Bio-Technol 5(6):9–20
20. Gilal AR et al (2016) Balancing the personality of programmer: software development team composition. Malays J Comput Sci 29(2):145–155
21. Amin A et al (2020) The impact of personality traits and knowledge collection behavior on programmer creativity. Inf Softw Technol 128:106405
22. Gilal AR et al (2017) Effective personality preferences of software programmer: a systematic review. J Inf Sci Eng 33(6):1399–1416
23. Alshanqiti A et al (2021) Leveraging DistilBERT for summarizing Arabic text: an extractive dual-stage approach. IEEE Access 9:135594–135607
24. Basri S et al (2019) The organisational factors of software process improvement in small software industry: comparative study. In: International conference of reliable information and communication technology. Springer, Cham, pp 1132–1143
25. Alsariera YA, Majid MA, Zamli KZ (2015) SPLBA: an interaction strategy for testing software product lines using the Bat-inspired algorithm. In: International conference on software engineering and computer systems (ICSECS), pp 148–153

Determining Factors Affecting Nurses' Acceptance of a Hospital Information System Using a Modified Technology Acceptance Model 3

Saeed Barzegari, Ibrahim Arpaci, and Zohreh Hosseini Marznaki

Abstract The aim of this study is to investigate the factors influencing nurses' acceptance of a hospital information system. The study applied Technology Acceptance Model 3 (TAM-3) to explain behavioral intention to use the hospital information system. The research model was empirically tested on 302 nurses by using structural equation modelling (SEM). The results indicated that the subjective norm (SN), perceived ease of use (PEOU), and perceived usefulness (PU) were significant determinants of behavioral intention (BI). SN, image (IMG), job relevance (REL), output quality (OUT), results demonstrability (RES), and PEOU had significant effects on PU. Also, the computer self-efficacy (CSE), perception of external control (PEC), computer playfulness (PLAY), and enjoyment (ENJ) and computer anxiety (CANX) were significant determinants of PEOU. The research model explained 62% variance in the BI.

Keywords Hospital information system · Nurses · TAM-3

1 Introduction

Health Information Technology (HIT) delivers modern and sophisticated healthcare [1]. Main advantages of information technology are increasing the quality of health care, reducing medical errors and reducing the cost of health care [2]. However,

S. Barzegari (✉)
Department of Paramedicine, Amol Faculty of Paramedical Sciences, Mazandaran University of Medical Sciences, Sari, Iran
e-mail: barz_saeed@yahoo.com

I. Arpaci
Faculty of Engineering and Natural Sciences, Department of Software Engineering, Bandirma Onyedi Eylul University, 10200 Balikesir, Turkey
e-mail: iarpaci@bandirma.edu.tr

Z. H. Marznaki
Department of Nursing, Amol Faculty of Nursing and Midwifery, Mazandaran University of Medical Sciences, Sari, Iran

M. Al-Emran et al. (eds.), *International Conference on Information Systems and Intelligent Applications*, Lecture Notes in Networks and Systems 550,
https://doi.org/10.1007/978-3-031-16865-9_35

obstacles and challenges still exist in the application of information technology in the health sector. Investigating and identifying the prominent factors and trying to strengthen the positive factors to provide solutions may play a major role in the technology adoption [3]. Acceptance and use of technology by users is an important factor in the success of information technology implementation [4, 5]. Moreover, it is important to measure nurses' acceptance of HIS since they use these systems to carry out their daily routines [6].

In the field of technology acceptance, various models and theories have been proposed, including Theory of Reasoned Action, Technology Acceptance Model (TAM), Motivational Theory, Theory of Planned Behavior (TPB), Diffusion of Innovation (DOI) and Social Cognitive Theory [7]. The Technology Acceptance Model (TAM) was first proposed by Davis in 1985 [8], this model has been widely used as a conceptual concept in most experimental studies with different societies and technologies [9, 10]. According to the initial model, two types of beliefs play a key role in the acceptance of technology. These two beliefs include Perceived Usefulness (PU) and Perceived Ease of Use (PEOU), which together predict the attitudes towards the acceptance of a system and in turn affect the intention to use and ultimately determine the level of actual system use [11]. Venkatesh and Davis (2000) developed TAM-2 to explain how social impact processes (subjective norms, volunteering, and external reflection) and cognitive instrumental processes affect perceived usefulness and attitude [12]. Venkatesh and Davis added in the area of cognitive instrumental processes of job communication (the extent to which one feels the system is applicable to one's job), the quality of output (people's assessment of how tasks are performed by the technology in question), and the ability to explain results [13]. Then, Venkatesh and Bala (2008) combined the TAM 2 and the determinants of perceived ease of use and proposed the TAM-3.

In TAM-3, factors that affect PU are Subjective Norm (SN), Image (IMG), job relevance (REL), output quality (OUT), and results demonstrability (RES) (Venkatesh, & Bala, 2008). PEOU is determined by anchor variables (computer self-efficacy (CSE), perception of external control (PEC), computer anxiety (CANX), computer playfulness (PLAY)) and adjustment variable (Perceived Enjoyment (ENJ)) [13]. The aim of this study is to investigate the factors influencing nurses' acceptance of a hospital information system. The study used TAM-3 as a theoretical framework to explain nurses' acceptance of a hospital information system.

2 Theoretical Background and Hypotheses

Technology Acceptance Model was updated as TAM-3, focusing on expanding the number of determinants that affect PU and PEOU [14]. According to TAM-3, the subjective norm (SN), perceived ease of use (PEOU), and perceived usefulness (PU) are significant determinants of behavioral intention (BI). SN, image (IMG), job relevance (REL), output quality (OUT), results demonstrability (RES), and PEOU are significant determinants of PU. Further, the computer self-efficacy (CSE), perception of external control (PEC), computer playfulness (PLAY), and enjoyment (ENJ), and computer anxiety (CANX) are significant determinants of PEOU. Therefore:

H1. PU has a significant effect on BI
H2. PEOU has a significant effect on BI
H3. SN has a significant effect on BI
H4. SN has a significant effect on IMG
H5. SN has a significant effect on PU
H6. IMG has a significant effect on PU
H7. REL has a significant effect on PU
H8. RES has a significant effect on PU
H9. PEOU has a significant effect on PU
H10. PEC has a significant effect on PEOU
H11. CSE has a significant effect on PEOU
H12. CANX has a significant effect on PEOU
H13. CPLAY has a significant effect on PEOU
H14. ENJ has a significant effect on PEOU

3 Method

This cross-sectional study proposed TAM-3 to express user acceptance of Hospital Information Systems. This study was approved by the Medical Ethics Committee of the Mazandaran University of Medical Sciences. Using convenience sampling, 350 nurses of eight educational hospitals affiliated with Mazandaran university of Medical Sciences who had been using a hospital information system for longer than 1 year participated. Informed consent was obtained from all participants. 302 questionnaires (86.29%) completed by nurses were used for data analysis.

3.1 Data Analysis

Descriptive statistics were employed using SPSS version 20.0 to analyze demographic characteristics of participants. Path analysis was employed through Amos software version 24. We evaluated reliability, convergent, and discriminant validity of the constructs. Reliability was ensured if the scores of Cronbach alpha and composite reliability (CR) for all constructs were higher than 0.7. Convergent validity was confirmed if the average variance extracted (AVE) scores of all constructs were higher than 0.5. Also, to confirm discriminant validity we have two conditions: the square root of the AVE of each construct must be higher than the correlation between the items of that construct, and the factor loading of each item on its own construct must be higher than its cross-loadings on other constructs.

Table 1 Demographic characteristics

Variable	Frequency	Percent (%)
Gender		
Male	269	89.1
Female	33	10.9
Age		
Less than 30 years	120	39.7
Between 30 and 40 years	145	48
Between 40 and 50 years	33	10.9
50 years or above	4	1.3
Qualification		
Bachelors	292	96.7
Masters	10	3.3
Employment		
Less than 5 years	114	37.7
Between 5 and 10 years	125	41.4
10 years or above	63	20.9

4 Results

4.1 Demographics

In total 350 questionnaires were distributed among nurses. 302 of them were selected for data analysis, and the rest were deleted due to incomplete data. Table 1 shows the demographic characteristics of the participants. As shown in Table 1, 10.9% of the participants were male and 89.1% were female. The age of the respondents was in the range of 22–54 and their average age was 32.2. The academic qualification of participants was mostly bachelor (96.7%).

4.2 Normality, Reliability and Validity

To perform statistical analyzes, first the normality of the data was examined. The skewness of the items (except q14 $= -1.13$) lie in the range of $+1$ to -1, and the kurtosis lie in the range of -2.58 to $+2.58$, indicating that distribution of data was normal. After confirming the normality of the data, the internal reliability, convergent validity and discriminant validity of the model were examined. As shown in Table 2, Cronbach's alpha of all factors is in the acceptable range of 0.70–0.76 and composite reliability is in the acceptable range of 0.74–0.85. Therefore, the internal reliability of the data is supported by these results. According to the results, AVE is in the

Table 2 Factor loadings, Cronbach alpha, CR and AVE

	Factor loadings	Cronbach alpha	CR	AVE
VOL	0.75–0.79	0.71	0.81	0.59
SN	0.50–0.83	0.70	0.74	0.50
IMG	0.61–0.84	0.71	0.80	0.57
REL	0.71–0.81	0.72	0.81	0.60
OUT	0.79–0.83	0.76	0.85	0.65
RES	0.70–0.76	0.70	0.83	0.55
PEC	0.66–0.77	0.73	0.80	0.50
CSE	0.53–0.79	0.73	0.81	0.52
CANX	0.48–0.78	0.71	0.79	0.50
PLAY	0.65–0.77	0.76	0.80	0.50
ENJ	0.73–0.83	0.71	0.82	0.60
PEOU	0.71–0.78	0.74	0.83	0.55
PU	0.66–0.78	0.72	0.82	0.53
BI	0.74–0.80	0.71	0.82	0.60

acceptable range of 0.50–0.65 and also the factor loading of all items is higher than 0.48, which confirms the convergent validity.

As shown in Table 2, the square root values of the AVE (table diameter) are higher than the correlation values and confirm the discriminant validity of the constructs. Diagonal is the square root of AVE scores. Discriminant validity was confirmed because the factor loading of items on their own constructs is higher than their cross loading on other constructs. Also, according to the results of Table 2, the correlations between all constructs were lower than 0.85 and therefore, multicollinearity was not a concern. Based on the results, internal consistency, convergent, and discriminant validity were confirmed.

4.3 Model Fit

Confirmatory Factor analysis was used to test the model fit. The mean square error of approximation (RMSEA) < 0.08, comparative fit index (CFI) > 0.90, Tucker Lewis index (TLI) > 0.90, and goodness of fit index (GFI) > 0.90 used to examine goodness of fit in the model. Based on the results, the model was considered acceptable and the goodness of fit indices (CFI = 0.94, GFI = 0.95, and RMSEA = 0.06) indicated good model fit.

4.4 Structural Model

SEM approach using SPSS AMOS was employed to test the hypothesized relationships. The hypotheses testing results were presented in Table 3 and Fig. 1. According to results, subjective norm (SN), image (IMG), job relevance (REL), output quality (OUT), results demonstrability (RES), and perceived ease of use (PEOU) yielded approximately 55% of the variance for perceived usefulness (PU). Also, computer self-efficacy (CSE), perception of external control (PEC), computer anxiety (CANX), computer playfulness (PLAY), and enjoyment (ENJ) yielded approximately 30% of the variance for PEOU. Based on the results, all of the proposed hypotheses were supported.

Results indicated that the subjective norm (SN), perceived ease of use (PEOU), and perceived usefulness (PU) (path coefficients ranged = 0.26–0.42) were significant determinants of the behavioral intention (BI) ($R^2 = 0.62$). SN, image (IMG), job relevance (REL), output quality (OUT), results demonstrability (RES), and PEOU (path coefficients range = 0.09–0.14) were significant determinants the PU ($R^2 = 0.55$). Further, the computer self-efficacy (CSE), perception of external control (PEC), computer playfulness (PLAY), and enjoyment (ENJ) (path coefficients range = 0.14–0.44) and computer anxiety (CANX) (path coefficient = −0.11) were significant determinants of the PEOU ($R^2 = 0.30$).

Table 3 Path coefficients and hypotheses testing results

H	Dependent variable	Independent variable	Path coefficient	t-value	*p*-value	Result
H_1	PU	BI	0.38	7.37	<=0.001	Supported
H_2	PEOU	BI	0.42	8.69	<=0.001	Supported
H_3	SN	BI	0.20	7.16	<=0.001	Supported
H_4	SN	IMG	0.47	10.31	<=0.001	Supported
H_5	SN	PU	0.09	2.60	0.009	Supported
H_6	IMG	PU	0.09	2.49	0.013	Supported
H_7	REL	PU	0.10	3.71	<=0.001	Supported
H_8	RES	PU	0.08	2.33	0.020	Supported
H_9	PEOU	PU	0.65	17.42	<=0.001	Supported
H_{10}	PEC	PEOU	0.42	9.04	<=0.001	Supported
H_{11}	CSE	PEOU	0.13	2.98	0.003	Supported
H_{12}	CANX	PEOU	−0.11	2.32	0.020	Supported
H_{13}	CPLAY	PEOU	0.16	3.61	<=0.001	Supported
H_{14}	ENJ	PEOU	0.19	4.18	<=0.001	Supported

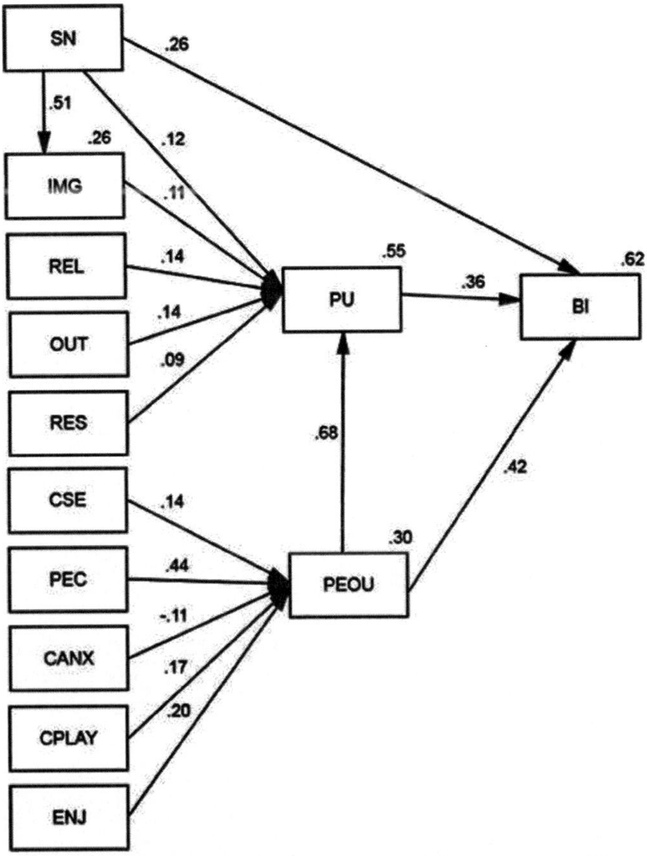

Fig. 1 Conceptual model with results

5 Discussion and Conclusion

The aim of this study is to investigate the factors influencing nurses' acceptance of a hospital information system. The study applied Technology Acceptance Model 3 (TAM-3) to explain behavioral intention to use the hospital information system. The research model was empirically tested on 302 nurses by using structural equation modelling (SEM). The results indicated that the subjective norm (SN), perceived ease of use (PEOU), and perceived usefulness (PU) constructs were significant determinants of behavioral intention (BI) ($R^2 = 0.62$). SN, image (IMG), job relevance (REL), output quality (OUT), results demonstrability (RES), and PEOU had significant effects on the PU. Moreover, the computer self-efficacy (CSE), perception of external control (PEC), computer playfulness (PLAY), and enjoyment (ENJ), and computer anxiety (CANX) (path coefficient $= -0.11$) were significant determinants of the PEOU.

The finding indicated that subjective norm (SN), image (IMG), job relevance (REL), output quality (OUT), results demonstrability (RES), and perceived ease of use (PEOU) constructs yielded approximately 55% of the variance in perceived usefulness (PU). Also, computer self-efficacy (CSE), perception of external control (PEC), computer anxiety (CANX), computer playfulness (PLAY), and enjoyment (ENJ) yielded approximately 30% of the variance in the PEOU.

References

1. Coiera E, Ash J, Berg M (2016) The unintended consequences of health information technology revisited. Yearb Med Inform 25(01):163–169
2. Wager KA, Lee FW, Glaser JP (2021) Health care information systems: a practical approach for health care management. Wiley, Hoboken
3. Ratwani RM, Reider J, Singh H (2019) A decade of health information technology usability challenges and the path forward. JAMA 321(8):743–744
4. Arpaci I (2017) The role of self-efficacy in predicting use of distance education tools and learning management systems. Turk Online J Distance Educ 18(1):52–62
5. Al-Emran M, Al-Maroof R, Al-Sharafi MA, Arpaci I (2020) What impacts learning with wearables? An integrated theoretical model. Interact Learn Environ 1–21. https://doi.org/10.1080/10494820.2020.1753216
6. Arpaci I, Al-Emran M, Al-Sharafi MA (2020) The impact of knowledge management practices on the acceptance of Massive Open Online Courses (MOOCs) by engineering students: a cross-cultural comparison. Telematics Inform 54:101468
7. Taherdoost H (2018) A review of technology acceptance and adoption models and theories. Procedia Manuf 22:960–967
8. Davis FD (1985) A technology acceptance model for empirically testing new end-user information systems: theory and results (Doctoral dissertation, Massachusetts Institute of Technology)
9. Arpaci I, Cetin Yardimci Y, Turetken O (2015) The impact of perceived security on organizational adoption of smartphones. Cyberpsychol Behav Soc Netw 18(10):602–608. https://doi.org/10.1089/cyber.2015.0243
10. Arpaci I (2017) The role of self-efficacy in predicting use of distance education tools and learning management systems. Turk Online J Distance Educ 18(1):52–62. https://doi.org/10.17718/tojde.285715
11. Scherer R, Siddiq F, Tondeur J (2019) The technology acceptance model (TAM): a meta-analytic structural equation modeling approach to explaining teachers' adoption of digital technology in education. Comput Educ 128:13–35
12. Venkatesh V, Davis FD (2000) A theoretical extension of the technology acceptance model: four longitudinal field studies. Manage Sci 46(2):186–204
13. Venkatesh V, Bala H (2008) Technology acceptance model 3 and a research agenda on interventions. Decis Sci 39(2):273–315
14. Al-Tahitah AN, Al-Sharafi MA, Abdulrab M (2021) How COVID-19 pandemic is accelerating the transformation of higher education institutes: a health belief model view. In: Arpaci I, Al-Emran M, Al-Sharafi AM, Marques G (eds) Emerging technologies during the era of COVID-19 pandemic. Studies in systems, decision and control, vol 348. Springer, Cham. https://doi.org/10.1007/978-3-030-67716-9_21

Psychometric Properties and Validation of the Persian Version of the Health Information Technology Usability Evaluation Scale

Hasti Mehdi Nezhad Doughikola, Ibrahim Arpaci, Meisam Rahmani, Toomaj VahidAfshar, and Saeed Barzegari

Abstract The quality of healthcare is highly dependent on nurses' performance and efficiency. In recent years, hospital information systems (HIS), healthcare software and applications have been introduced to enhance their ability and nurses have a key role in accepting and evaluating the HIS. This study tested psychometric properties and validation of a Persian version of the "Health Information Technology Usability Evaluation Scale" (Health-ITUES). Expert panel, which consisted of 10 nursing professors, assessed the content and face validity of the Persian Health-ITUES. Sample of the was 229 nurses employed in hospitals. Cronbach's alpha was used to test reliability along with "confirmatory factor analysis" (CFA) was used to test construct validity. Further, both discriminant and convergent validity were assessed. Content validity index (CVI) and content validity ratio (CVR) were .93 and .75 respectively. Scale content validity index (S-CVI) and item content validity index (I-CVI) were more than thresholds (.76 and .90, respectively). Goodness of fit indexes revealed the measurement model was fitted to data well ($\chi 2/DF = 2.49$, IFI = .939, CFI = .938, GFI = .903, RMSEA = .076). Cronbach's alpha values of the factors and total scale were ranged between .75 and .94. The results indicated that Persian version of the health-ITUES is a reliable and valid tool to measure information technology usability in nursing field. We recommend to researchers, health and nursing application developers to use iterative usability evaluation with HITUES to identify problems in early developing steps and address user-centered design.

H. M. N. Doughikola · S. Barzegari (✉)
Department of Paramedicine, Amol Faculty of Paramedical Sciences, Mazandaran University of Medical Sciences, Sari, Iran
e-mail: barz_saeed@yahoo.com

I. Arpaci
Faculty of Engineering and Natural Sciences, Department of Software Engineering, Bandirma Onyedi Eylul University, 10200 Balikesir, Turkey

M. Rahmani
Department of Health Information Management, School of Allied Medical Sciences, Tehran University of Medical Sciences (TUMS), Tehran, Iran

T. VahidAfshar
Iran University of Industries and Mines (IUIM), Tehran, Iran

M. Al-Emran et al. (eds.), *International Conference on Information Systems and Intelligent Applications*, Lecture Notes in Networks and Systems 550,
https://doi.org/10.1007/978-3-031-16865-9_36

Keywords Health-ITUES · Validation · Nurse

1 Introduction

The quality of healthcare is highly dependent on nurses' performance and efficiency [1, 2]. They need to collect and handle a lot of data every day. Volume of data may disrupt nursing process and reduce nurses' speed of reaction. Therefore, information systems, healthcare software and applications have been introduced to enhance their ability [3]. In recent years, hospital information systems (HIS) have been implemented to help clinicians, especially nurses, in their routines. Since their work encounters several challenges, policymakers must understand them and correctly assess their difficulties [4]. Given that they are the largest group of healthcare staff in hospitals and have a significant role in the treatment of patients, nurses have a key role in accepting and evaluating the HIS [5].

The acceptance of the HIS and related technologies by nurses may improve nursing services, and thereby, may have a positive impact on their performance [6]. However, factors related to the effectiveness of health information technology (HIT) are among the main concerns of nurses and executive managers [7]. Efficiency related concerns are substantial barriers to the adoption of the HIT [8]. Several survey instruments have been developed to assess users' perceptions about the effectiveness of the HIT, such as "Technology Acceptance Model" (TAM), "Unified Theory of Acceptance and Use of Technology" (UTAUT), and "Calculation of End-User Satisfaction". Although the achieved effectiveness through the interactions between system, users, and task is important in assessing the effectiveness of the HIT [9], most of the survey instruments overlooked "task" as a key element. Whereas Health-ITUES developed by Yen et al. considered the "task" as a key variable and focused on different levels of expectations. This instrument includes "quality of work-life", "perceived usefulness", "perceived ease of use", and "user control" factors [10].

The "quality of work-life" as a factor of this instrument, is essential in efficiency which includes the physical, social, psychological, and environmental aspects of employee behavior [11]. This factor consists of three items that are related to the organizational processes and efficiency [10]. The two concepts of "perceived usefulness" and "perceived ease of use" are described by the TAM. "Perceived usefulness" is the tendency of people to use or not using a system to the extent that they believe the system will help them to improve their tasks. "Perceived ease of use" implies that if a system is easy to use, users will use the system [12]. In the Health-ITUES, users' perceived usefulness was evaluated by nine items that assess the system's usefulness for a specific task. Five of these items have been adapted from the "perceived usefulness" in the TAM. The other four are related to effectiveness, information needs, system satisfaction, and ease of work. Also, the "perceived ease of use" factor consists of five items that focus on evaluating the user-system interaction [11]. The "User Control over the Information System" factor consists of 3 items, which are related to users' ability to control the information system. This factor includes error prevention, and information needs to reduce the performance problems of information systems [10]. The present study tested psychometric properties of the Persian version of the "Health Information Technology Usability Evaluation Scale" (Health-ITUES) among nurses.

2 Material and Methods

2.1 Translation Process and Face Validity

First, the Health-ITUES items were translated by a translator, who is specialist in the digital health and nursing from English to Persian. Translations took into account conceptual and cross-cultural equivalence rather than linguistic for phrases and words to ensure that translated instrument is simple, concise, and fit with Persian culture. The experts panel consisted of 10 nursing professors, modified the initial translation.

The expert panel evaluated the qualitative face validity by providing suggestions about suitability, relevancy, difficulty, ambiguity, and also the time needed to respond to the scale. Scores were rated on a five-point scale ranging from "1 = not important" to "5 = completely important". The formula of frequency × importance was use used to calculate impact score for each item and the items scored less than 1.5 were deleted.

2.2 Content Validity

The experts were requested to carefully read the scale and identify the errors in wording, item allocation, and grammar. Further, content validity ratio (CVR) and content validity index (CVI) were used to test content validity. The CVR indicates the degree of necessity of the item in the scale. The minimum acceptable level was 0.62 according to the Lawshe's table [13]. The item and scale level CVI, with a minimum threshold of 0.76 and 0.90 [14], were calculated to examine the relevancy of each item as well as the total scale, respectively.

2.3 Construct Validity and Reliability

SPSS (ver. 21) and Amos (ver. 20) were used to perform data analysis. Psychometric properties of the Persian Health-ITUES were tested by a "confirmatory factor analysis" (CFA), discriminant and convergent validity. The original Health-ITUES has 20 items and 10 samples are required for each item [15], therefore, the sample size (n = 229) is considered enough for a valid factor analysis. Ethical approval was granted by the affiliated university. Informed consents were obtained, and the aim of the research was explained to the participants before distributing the measurement.

"Chi-square, Chi-square/degree of freedom" (CMIN/DF $\leq$ 3), "Comparative Fit Index" (CFI $\geq$ 0.90), "Goodness-of-fit Index" (GFI $\geq$ 0.90), "Root Mean Square Error of Approximation" (RMSEA $\leq$ 0.06), and "Incremental Fit Index" (IFI $\geq$ 0.90) were used to check the fit between the data and measurement model. Convergent validity was evaluated by using "average variance extracted" (AVE). Further, discriminant validity was evaluated by using "maximal shared squared variance" (MSV) and "average shared square variance" (ASV). The "composite reliability" (CR) and AVE values for each factor should exceed 0.70 and 0.50, respectively. Moreover, the AVE values should be greater than the MSV and ASV. Cronbach's alpha values were checked to test internal reliability.

3 Results

3.1 Demographics

A total of 250 nurses were invited to participate the study, however, 229 nurses returned with informed consent and valid responses. Thereby, the response rate was 91.6%. The mean age and mean work experience of the respondents were 41.26 ± 51 and 10.09 ± 6.96 years respectively. The majority were females (59.4%, n = 136), Bachelor of Science (86.5%, n = 198), and most had no International Computer Driving License certificate (90.8%, n = 208), but most had used digital devices such as computer and smartphones every day (96.9%, n = 222).

3.2 Validity and Reliability

The mean age of the participants was 35.27 ± 8.16 years (ranged from 23 to 58 years). The results indicated that the CVR and CVI were 0.75. and 93, respectively. This indicated that the Persian Health-ITUES have an adequate content validity. The I-CVI and S-CVI of each item were more than the thresholds (0.76 and 0.90, respectively). The CFI, IFI, and GFI indices were above the threshold value of 0.90 (CFI = 0.938, GFI = 0.903; IFI = 0.939,), RMSEA was lower than 0.08 and on the favorable threshold (RMSEA = 0.076), and $\chi 2$/DF value was lower than 3, thus acceptable ($\chi 2$/DF = 2.49, P-value $\leq$ 0.001). Figure 1 shows the measurement model.

Factor loadings of the scale items (except for the Q4, Q5, Q11, Q12, and Q13) were above the threshold value of 0.40. Whereas five items having a low factor loading were deleted. Cronbach's alpha of the total scale was 0.79 and Cronbach's α values of the factors were ranged between 0.75 and 0.94. Table 1 shows the items with internal reliability coefficients.

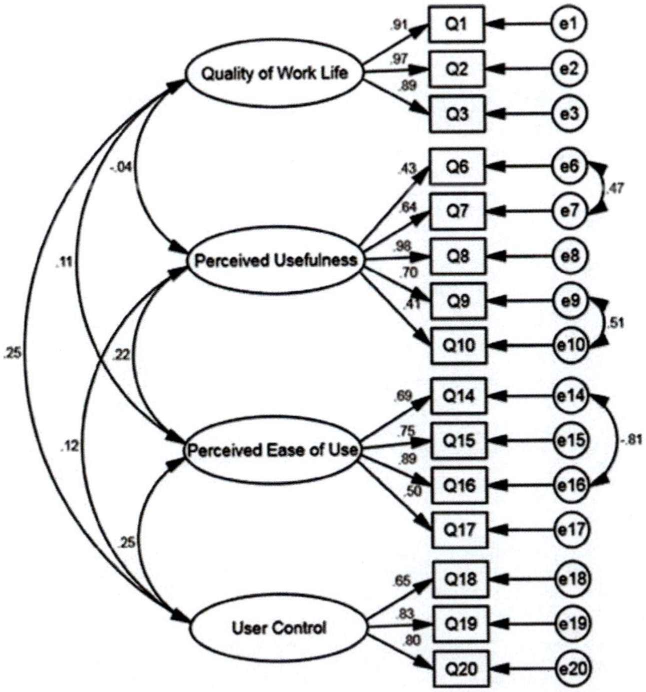

Fig. 1 Measurement model

According to the results, the AVE values (except PU) were greater than the threshold of 0.50. Further, the composite reliability (CR) of the factors were greater than the threshold of 0.70 (ranged between 0.72 and 0.95). Table 2 indicated that the health-ITUES has an adequate convergent, construct and discriminant validity along with internal reliability.

4 Discussion and Conclusion

The present study tested psychometric properties of the Persian Health-ITUES with a sample of nurses. The content validity, construct validity, face validity, along with internal reliability were investigated throughout the study. Prior research has been tested the psychometric properties of the Health-ITUES in different languages and cultural settings [9, 16]. Further, the Health-ITUES was used to measure usability of various information technologies. Househ et al. (2015) used the scale to determine the efficiency of mobile health software for diabetics [17]. In another study, Velez et al. (2014) used the scale to evaluate the efficiency of mobile health software for rural midwives in Ghana [18].

Table 1 Items with internal reliability coefficients

Construct	α	Item	α if item deleted
QWL	0.944	"I think HIS has been a positive addition to Nursing."	0.920
		"I think HIS has been a positive addition to our organization."	0.893
		"HIS is an important part of our staffing process."	0.938
PU	0.808	"Using HIS makes it more likely that I will be awarded a request."	0.800
		"Using HIS is useful for my work."	0.767
		"I think HIS presents a more equitable process for my requests."	0.730
		"I am satisfied with HIS."	0.734
PEU	0.751	"I do my works in a timely manner because of HIS."	0.810
		"Learning to operate HIS is easy for me."	0.740
		"It is easy for me to become skillful at using HIS."	0.621
		"I find HIS easy to use."	0.634
		"I can always remember how to log on to and use HIS."	0.763
UC	0.800	"HIS gives error messages that clearly tell me how to fix problems."	0.794
		"Whenever I make a mistake using HIS, I recover easily and quickly."	0.681
		"The information (such as on-line help, on-screen messages and other documentation) provided with HIS is clear."	0.702

QWL: Quality of Work Life; PU: Perceived Usefulness; PEU: Perceived Ease of Use; UC: User Control

Table 2 Discriminant and convergent validity

	CR	AVE	MSV	ASV	PEU	QWL	PU	UC
PEU	0.81	0.518	0.062	0.041	0.72			
QWL	0.95	0.852	0.062	0.025	0.11*	0.92		
PU	0.78	0.443	0.047	0.021	0.22*	0.14*	0.67	
UC	0.80	0.580	0.062	0.046	0.25*	0.25*	0.12*	0.76

* $p < 0.01$

The original scale (Health-ITUES) was developed and validated by Yen et al. (2010). They conducted an EFA and CFA by using data collected from 541 nurses in two healthcare organizations. They extracted four factors with 20 items. They found a good model fit (RMSEA = 0.064, SRMR = 0.085, CFI = 0.986, and TLI = 0.947) [9]. Our findings indicated that the content and face validity of the adapted scale were satisfactory. The measurement model of the Persian scale was assessed by the

CFA and the results showed the measurement model has a good fit with the data. Five items having a factor loading above the threshold value of 0.40 were eliminated. Thereby, Persian version of the scale had four dimensions with 15 items.

The internal reliability consistency of the total scale was satisfactory ($\alpha = 0.792$). Cronbach's α coefficients of the four factors were ranged from 0.75 to 0.94. The AVE and CR values were greater than 0.70 and 0.50, respectively. Pervan et al., suggested to check composite reliability if the AVE value was less than 0.50. The composite reliability was higher than 0.60, suggesting the convergent validity is adequate [19]. Further, the AVE values were higher than the MSV and ASV, indicating a good divergent validity [20–23]. This suggests that each item and its related factor are highly correlated. Overall, the Persian Health-ITUES has a good convergent and divergent validity.

In conclusion, the Health-ITUES, with its promising validity and reliability, can be used in to measure the efficiency of information systems and new technologies in nursing field. We recommend to researchers and healthcare application developers to use iterative usability evaluation with HITUES to identify problems in early developing steps and address user-centered design.

References

1. Tang C et al (2019) The influence of cultural competence of nurses on patient satisfaction and the mediating effect of patient trust. J Adv Nurs 75(4):749–759
2. Ammenwerth E et al (2011) Effect of a nursing information system on the quality of information processing in nursing: an evaluation study using the HIS-monitor instrument. Int J Med Informatics 80(1):25–38
3. Hao A et al (2006) Apply creative thinking of decision support in electrical nursing record. Stud Health Technol Inform 124:313–319
4. Rathert C et al (2019) Seven years after Meaningful Use: Physicians' and nurses' experiences with electronic health records. Health Care Manage Rev 44(1):30–40
5. Hsiao J-L, Chang H-C, Chen R-F (2011) A study of factors affecting acceptance of hospital information systems: a nursing perspective. J Nurs Res 19(2):150–160
6. Sharifian R et al (2014) Factors influencing nurses' acceptance of hospital information systems in Iran: application of the Unified Theory of Acceptance and Use of Technology. Health Inf Manage J 43(3):23–28
7. Staggers N et al (2018) The imperative of solving nurses' usability problems with health information technology. JONA J Nurs Admin 48(4):191–196
8. Yen P-Y, Bakken S (2012) Review of health information technology usability study methodologies. J Am Med Inform Assoc 19(3):413–422
9. Yen P-Y, Wantland D, Bakken S (2010) Development of a customizable health IT usability evaluation scale. In: AMIA annual symposium proceedings. American Medical Informatics Association
10. Kelbiso L, Belay A, Woldie M (2017) Determinants of quality of work life among nurses working in Hawassa town public health facilities, South Ethiopia: a cross-sectional study. Nurs Res Pract 2017
11. Nayak T, Sahoo CK (2015) Quality of work life and organizational performance: the mediating role of employee commitment. J Health Manag 17(3):263–273
12. Karahanna E, Straub DW (1999) The psychological origins of perceived usefulness and ease-of-use. Inf Manage 35(4):237–250

13. Lawshe CH (1975) A quantitative approach to content validity. Pers Psychol 28(4):563–575
14. Polit DF, Beck CT (2006) The content validity index: are you sure you know what's being reported? Critique and recommendations. Res Nurs Health 29(5):489–497
15. Plichta SB, Kelvin EA, Munro BH (2013) Munro s statistical methods for health care research. Wolters Kluwer Health/Lippincott Williams & Wilkins
16. Schnall R, Cho H, Liu J (2018) Health Information Technology Usability Evaluation Scale (Health-ITUES) for usability assessment of mobile health technology: validation study. JMIR Mhealth Uhealth 6(1):e4
17. Househ MS et al (2015) The Use of an Adapted Health IT Usability Evaluation Model (Health-ITUEM) for evaluating consumer reported ratings of diabetes mHealth applications: implications for diabetes care and management. Acta Informatica Medica 23(5):290
18. Vélez O et al (2014) A usability study of a mobile health application for rural Ghanaian midwives. J Midwifery Womens Health 59(2):184–191
19. Pervan M, Curak M, Pavic Kramaric T (2018) The influence of industry characteristics and dynamic capabilities on firms' profitability. Int J Fin Stud 6(1):4
20. Arpaci I, Sevinc K (2021) Development of the Cybersecurity Scale (CS-S): evidence of validity and reliability. Inf Dev. https://doi.org/10.1177/0266666921997512
21. Arpaci I, Karataş K, Baloğlu M, Haktanir A (2022) COVID-19 phobia in the United States: validation of the COVID-19 phobia scale (C19P-SE). Death Stud 46(3):553–559. https://doi.org/10.1080/07481187.2020.1848945
22. Arpaci I, Seong M, Karataş K (2021) Pandemic Awareness Scale (PAS): evidence of validity and reliability in a Turkish sample during the COVID-19 Pandemic. Trends Psychol. https://doi.org/10.1007/s43076-021-00113-y
23. Al-Tahitah AN, Al-Sharafi MA, Abdulrab M (2021) How COVID-19 pandemic is accelerating the transformation of higher education institutes: a health belief model view. In: Arpaci I, Al-Emran M, Al-Sharafi MA, Marques G (eds) Emerging technologies during the era of COVID-19 pandemic. Studies in systems, decision and control, vol 348. Springer, Cham. https://doi.org/10.1007/978-3-030-67716-9_21

The Influence of Social Media Use on Social Connectedness Among University Students

Balan Rathakrishnan, Soon Singh Bikar Singh, Azizi Yahaya, Mohammad Rahim Kamaluddin, Noor Hassline Mohamed, Anath Rau Krishnan, and Zaizul Ab Rahman

Abstract The pandemic of Covid-19 has changed the lifestyle of people nowadays. Students has to adapt to the new norms in which they need to rely on the digital mediums to interact with others. The main purpose of this study is to investigate the relationship between social media use and the connectedness among the university students in Malaysia during this pandemic of Covid-19. It also aims to investigate connection between the purposes of social media use (academic, socialization, entertainment and informativeness) and the level of social connectedness. Thirdly, the genders difference between social media use and social connectedness are investigated. The measurement used include the online social networking usage questionnaire and the social connectedness scale, and were distributed through snowball sampling method via the online platforms. A total of 300 respondents were recruited in this study with the mean age of 22.26. The results indicate that no significant relationship between social media usage and social connectedness. However, there was significant relationship between the purposes of using social media and social connectedness. Thirdly, no difference was found between females and males on the social media usage and social connectedness. Finally, this study highlights that the purpose of using social media could enhance the social relationship.

Keywords Social media usage · Social connectedness · University students

B. Rathakrishnan (✉) · S. S. B. Singh · A. Yahaya · N. H. Mohamed
Faculty of Psychology and Education, Universiti Malaysia Sabah, 88400 Kota Kinabalu, Sabah, Malaysia
e-mail: rbhalan@ums.edu.my

M. R. Kamaluddin
Faculty of Social Sciences and Humanities, Universiti Kebangsaan Malaysia, 43600 Bangi, Selangor, Malaysia

A. R. Krishnan
Labuan Faculty of International Finance, Universiti Malaysia Sabah, Labuan International Campus, Jalan Sungai Pagar, 87000 Labuan F.T., Sabah, Malaysia

Z. A. Rahman
Research Centre for Theology and Philosophy, Faculty of Islamic Studies, Universiti Kebangsaan Malaysia, UKM, 43600 Bangi, Malaysia

M. Al-Emran et al. (eds.), *International Conference on Information Systems and Intelligent Applications*, Lecture Notes in Networks and Systems 550,
https://doi.org/10.1007/978-3-031-16865-9_37

1 Introduction

Ever since the declaration of a pandemic due to the outbreak of Covid-19, the government of Malaysia has begun to implement the culture of "work from home" and "school from home". It can be "extremely difficult" to adjust rapidly associated with the movement restrictions, as well as confront uncertainty in their financial, jobs, education, and even daily needs. With the implementation of Movement Control Order (MCO), people could not go out and have their social life like those times before the outbreak of Covid-19. Hence, they could only stay at home. With that being said, people change the way they stay connected with others despite the social distancing and isolation at home during the pandemic and MCO in which everyone has to stay at home.

People interact with friends and family or even co-workers using social networking sites nowadays to get the sense of belonging [1]. With the sense of belonging, people can feel that they are a part of one or more social groups and it is sensible with digital platforms like social media platforms. By using the digital mediums, it mobilized the social interaction between people. Thus, it enhanced the social relationship among the people. On the other hand, social relationships are closely tied with social connectedness of how people feel that they are connected to the surroundings or the peers or family members [2].

The main purpose of this study is to investigate the influence of social media usage and the social relationship among the youth. Hence the objectives of the study are: 1) To investigate the correlation between social media usage patterns and social connectedness; 2) To investigate the relationship between the purposes of using social media and social connectedness among students; 3) To investigate the differences of social media usage among gender; 4) To investigate the differences of social connectedness among gender.

The contribution of this research is that it will have a considerable impact on academic and applied research by offering a causal explanation for how social media use can affect social relationships. Generally, social media directly influence the level of social interaction and psychological well-being such as depression and anxiety demonstrated in the past research, especially focused on loneliness and psychological distress in the workplace or certain specific social networking site such as Facebook. Besides, the findings could possess productive outcome to indicate that social media usage patterns have important implications to the social relationship among university students.

Nevertheless, there are insufficient studies about the relationship between social media usage patterns and social relationship related to social connectedness. Hence, it is essential to understanding and expands knowledge to discover different perspectives and growing evidence on this specific are in a Malaysian context.

The structure of this study utilised a cross-sectional, quantitative, and correlational research design. It was a self-reported survey method to collect data from the respondents. It involves the application of questionnaires to measure the data of the respondents. It was designed to investigate the influence of social media usage patterns on the social relationship of youth based on their experiences. In short, how the youths use the social media platform and whether certain functions influence social connectedness and peers' relationship were assessed. The questions used in the research were only close-ended questions. In addition, snowball sampling method was used to collect the data for this research. The questionnaire was prepared in Google Form and it was distributed to the respondents through online social media platforms and social networking sites with the link provided in order to reduce the physical contact in the midst of the pandemic of Covid-19.

2 Literature Review

2.1 Social Media Usage and Social Connectedness

In recent years, social media has grown in popularity as a tool for social interaction [3–5]. As it can bring opportunities like as connectivity to others, social media consumption might result in augmentation [6]. As social media is the platform for people to interact with others, it brings connectivity to others where people could have their sense of belonging towards their family, friends, classmates, or even co-workers. The significance of social media in creating social connections, implying that young people may have both positive and negative psychological impacts and it implies that social media produce a social connectivity paradox [7]. Consequently, by using social media, people might feel connected to one another.

People spend quite a number of hours on social media to keep connected with their friends and family. According to the statistics of Statista Research Department [8], internet users worldwide spent an average of 145 min per day on social media, up from 142 min the previous year in 2019 and 2020 and the global penetration rate of social networks is currently at 54%. Based on research, almost half of those polled said they used social media to communicate with friends and family, whereas filling free time and reading news stories proved to be two of the most common motivations for using social media, with over 21% of respondents using it to follow celebrities or influencers [8]. With these statistics, it could be said that people tend to use social media as a platform to pass their time and to interact with others. Thus, they feel connected to their surroundings and to what was happening globally with their interest. Nevertheless, the social connectedness in an online context could be different from the traditional social connectedness [9]. Thus, it is feasible to introspect the relationship between the usage of social media and social connectedness.

2.2 Purposes of Using Social Media

Technology advances the use of social media to keep in touch with friends. People could make new friends using social media. According to Aiello et al. [10], social networking using social media platform could capture actual friendship accurately which shares similar interests. The online social media support the friendship through shared life [11].

Besides communication purposes, social media platforms nowadays also serve as a tool for education purposes. Sutherland et al. [12] demonstrates that 52.8% of the university students said that university social media profiles made them feel more connected to their peers. It is consistent with the recent study stated that WhatsApp discussions are helpful and productive, and they increase motivation to actively participate in the lecture's topic in terms of promoting active learning and improve collaborative learning before and after lectures [13]. Hence, they would possibly feel connected and being comfortable with the online learning method using social media platforms like WhatsApp. Indirectly, it would influence the social relationship among the students.

2.3 Social Media Use and Gender

Genders could be the key variable in understanding the social media use. Different genders would have different purposes of using social media. Based on past research, both genders indicate interest in receiving information from social media [14]. Females use Facebook at a higher rate than males for sustaining current relationships, academic objectives, and following agendas, while males use it at a larger rate for forming new relationships [15].

When it comes to social media, males have a higher assessment of satisfaction and information quality than females [16]. In addition, men view social networking sites as a pragmatic communication medium but not as a meaningful platform for self-portrayal, whereas women appear to be motivated by a more hedonistic purpose of self-presentation, which causes them to be more worried about how others perceive them [17]. Within the context, males might have a different perspective of using social media in terms of perceiving the social relationship through social connectedness compared to females.

2.4 Social Connectedness and Gender

In terms of social connectedness, in comparison to female students, male students always have more social relationships [18]. As mentioned, there might be a difference between females and males towards perception of social relationship in terms of

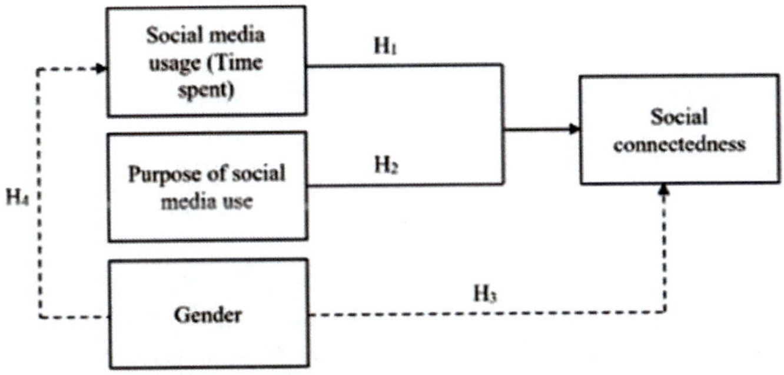

Fig. 1 Research framework

social connectedness in an online setting such as the social media use. However, there is less study reviews on the gender differences in social connectedness. Therefore, further research on this issue is needed to clarify the gender differences and the possible impact of it.

2.5 Research Framework

Grieve et al. [9] indicated that the usage of social media could be the predictor of social connectedness. Thus, it is plausible to hypothesized a positive association between the usage of social media and social connectedness (Fig. 1).

3 Methodology

3.1 Population and Study Context

The sample from this study covers the population of full-time undergraduate students in the universities in Malaysia. The targeted number of respondents was approximately between 250 to 350 participants. The inclusion criteria include university students with active status across Malaysia, aged between 19 and 24 years-old. Notwithstanding, 300 respondents were recruited in this study.

3.2 Instrument and Data Analysis

The Social Connectedness Scale (SCS) was originally developed by Lee and Robbins [2]. The scale consists of eight items whereby each item is scored on a 5-point Likert scale ranging from 1 ("*Strongly agree*") to 6 ("*Strongly disagree*"). The total sum value ranged between 8 and 48 points. The greater the score, the higher perceived social connectedness by the participant. The examples of items consist in the Social Connectedness Scale are the statements like "I feel disconnected from the world around me" and "I have no sense of togetherness with my peers". SCS indicated a good internal consistency (Cronbach's Alpha = 0.91) and with a goodness of fit index of 0.90 (2).

The Social Networking Usage Questionnaire (SNUQ) was adapted from previous measures assessing the social networking usage and social media usage [1, 19, 20]. The items were modified to suit this study. The items measure the frequency, duration, types of social media platforms (e.g., Twitter, Facebook, LinkedIn, Instagram), and the purposes of social networking usage. The purposes of social networking were split into 4 different subscales: Academic, Socialization, Entertainment, and Informativeness. The questions adapted from Gupta and Bashir [1] is a 5-point Likert scale, with each statement rated on five scale, (Always = 5, Often = 4, Sometimes = 3, Rarely = 2 and Never = 1). The examples of items consist in the Social Networking Usage questionnaire are the statements like "*I use social networking sites to keep in touch with my relatives.*" and "*I use social networking sites to learn about my curricular aspect*". It indicates a good internal consistency (Cronbach's Alpha = 0.83) and the convergent validity was found to be 0.593 to 0.894 [1, 21]. Moreover, a confirmatory factor analysis recorded $\chi 2/df = 2.348$, GFI = 0.929, CFI = 0.910, RMSEA = 0.061, indicating an adequate model fit; and reported a high construct reliability (CR = 0.759) and average variance extracted (AVE = 0.512), showing a high convergent validity [22].

All the data collected were analysed using IBM SPSS Statistics Version 27.0. Before conducting the analysis, the data collected were screened for missing values and normality. Descriptive analysis and inferential analysis were used (in Table 2 –4).

Table 1 Summary of measurement

Variables	Measurement	Source
Social connectedness	Social connectedness scale	Lee & Robbins 1995
Social media usage	Time spent (hours)	–
Purpose of using social media	The social networking usage questionnaire	Gupta & Bashir 2018

Table 2 Descriptive statistics for frequencies and percentages of the demographics (N = 300)

Characteristics	Frequency *(n)*	Percentage (%)
Male	100	30
Female	200	70
STPM/Matriculation/Diploma/A-level	78	23.4
Bachelor's degree	185	55.5
Master's degree	37	11.1
Social media use		
Little or no time	8	2.4
Between 1–3 h	78	23.4
Between 4–7 h	118	35.4
8 h or more	96	28.8
	Mean	Standard deviation
Age	22.6	1.35

4 Results

4.1 Respondent's Background

See Table 2.

4.2 Social Media Usage and Social Connectedness

See Table 3.

Table 3 Correlation between social media usage, purposes and social connectedness among university students

Variables	1	2	3	4	5	6
1. Time spent	–					
2. Academic	0.13	–				
3. Socialization	0.21**	0.62**	–			
4. Entertainment	0.21**	0.66**	0.56**	–		
5. Informativeness	0.12	0.66*	0.65**	0.53**	–	
6. Social connectedness	–0.002	0.26**	0.19**	0.15**	0.28**	–

Note ** $p < 0.01$: * $p < 0.05$

Table 4 Independent t-test for social media usage and social connectedness between male and female

Gender	*M*	*SD*	*t*	*P* value
Social media usage				
Male	3.03	0.77	−0.699	0.746
Female	3.11	0.78		
Social connectedness				
Male	34.02	8.77	−1.173	0.543
Female	35.58	8.73		

Table 5 Summary of results

Objective	Statistical analysis	Outcome
1	Pearson correlation	Insignificant
2	Pearson correlation	Significant
3	Independent-T test	Insignificant
4	Independent-T test	Insignificant

4.3 *Gender Differences of Social Media Usage and Social Connectedness*

See Table 4.

5 Discussion

Findings showed no significant relationship between social media usage and social connectedness. The emotional distance between oneself, others, and society which is measured by social connectedness [2, 23], was not associated with how many hours spent on social media. These findings are consistent with Ryan et al. [24] research, which suggested that interacting or communicating using social media does not mean that an individual is trying to establish the sense of belonging instead they might have been ignored by the other on social media and get a negative impact from the social media. Conversely, the present results are inconsistent with the findings from Taylor-Jackson et al. [25] study that found that time spent on social media could foster a greater sense of social connectedness when they used social media platforms to communicate with existing friends and experience positive socioemotional changes.

The findings showed a significant relationship between purposes of social media use and social connectedness. This finding is consistent with past research [12, 13, 26, 27], which indicated different purposes of using social media are directly related with the level of social connectedness. The purposes of social media usage positively correlated with the level of social connectedness in social relationship. It means that the more often the youth use social media with purpose, the greater the level of social connectedness. This implies that when a person has a purpose to engage in social media, they would feel more connected with their social network. For instance, if the person uses social media for the purpose of socialization, they might perceive a greater social connectedness. This is consistent with the study of Davis [26], which suggested that youth communicate through online social media could foster their connectedness with their peers.

The results show that no significant differences between female and male on the social media usage. Thus, the hypothesis was accepted. On the contrary of this results, it was inconsistent with the past researches stating that there are differences between females and males in using social media [28, 29]. Based on the findings in the present study, the female and males shared the same perception in using social media [15, 16, 30].

6 Conclusion

It can be concluded that the purpose of social media use such as academic, socialization, entertainment, and informativeness reveals the significant relationship on social connectedness among youth in Malaysia. The more often they use the social media for the four main purposes the greater the social connectedness perceived. Interestingly, the booster of social connectedness is when people use social media for informativeness purpose such as sharing new ideas. Besides, the time spent on social media could not be the predictor of social connectedness. Finally, the purpose of using social media and social connectedness is the same for both genders.

The results of this study could fill the knowledge gaps as most of the past researches investigated on the social media use with the impact of mental health problems, yet it did not imply the socioemotional aspects among the youth in Malaysia. Besides, it also opened a new insight on the gender differences on the social media use and social connectedness. The inconsistent findings with past studies reflect the equality of gender in perceiving their level of social connectedness through online social media platform. In addition, it provides some information regarding the purposes of social media usage social connectedness among youth in Malaysia. Finally, the findings of this study could help the clinicians, counsellors, and mental health practitioners to be aware and appreciate how the social media influences youth's social connections [31, 32]. Thus, appropriate strategies and interventions could develop to facilitate them establish a healthier social relationship and usage of social media.

This study demonstrated a few limitations. Firstly, the results of this study are inadequate to generalized to the population due to the small sample size and over-dominated distribution of respondent's age. Therefore, future research could attain a sufficient number of samples with a diverse range of participants in order to achieve a greater generalization for the results of the study. Despite having only students as the sample, future research could have considered youths age ranged at 18–30 years old no matter they are students or working adolescents. Therefore, it is recommended to investigate the differences between the students and working adolescents or unemployed youths on the perception of social media use and level of social connectedness on their social relationship.

Secondly, the cross-sectional method used in this study provide a limited causal explanation of the relationship between social media use and social connectedness. The study conducted during the pandemic of Covid-19 might have slight difference on the results with those studies conducted before and after the pandemic. Therefore, there might be discrepancies on the results of the similar studies. Thus, future study could consider a mix research design such as implement both qualitative and quantitative measures to conduct future studies. This is because it may provide a more detailed insights on the explanation of the reason behind people feel more or less connected with others using social media. Hence, it could provide a more accurate causal explanation to the readers and to fill the knowledge gaps in the field of psychology.

Acknowledgements This work is a part of a project submitted to Universiti Malaysia Sabah.

References

1. Gupta S, Bashir L (2018) Social networking usage questionnaire: development and validation in an Indian higher education context. Turk Online J Distance Educ 19(4):214–227
2. Lee RM, Robbins SB (1995) Measuring belongingness: the social connectedness and the social assurance scales. J Couns Psychol 42(2):232
3. Hart MJ (2010) A study on the motives of high school and undergraduate college students for using the social network site Facebook. Liberty University
4. Spiliotopoulos T, Oakley I (2013) Understanding motivations for Facebook use: usage metrics, network structure, and privacy. In: Proceedings of the SIGCHI conference on human factors in computing systems, pp 3287–3296
5. Parasuraman B, Satrya A, Rathakrishnan B, Muniapan B (2009) Analysing the relationship between unions and joint consultation committee: case studies of Malaysian and Indonesian postal industries. Int J Bus Soc 10(1):41–58
6. Ahn D, Shin D (2013) Is the social use of media for seeking connectedness or foe avoiding social isolation? Mechanisms unverlying media use and subjective well-being. Comput Hum Behav 29(6):2453–2462. https://doi.org/10.1016/j.chb.2012.12.022
7. Allen KA, Ryan T, Gray DL, McInerney DM, Waters L (2014) Social media use and social connectedness in adolescents: the positives and the potential pitfalls. Educ Dev. Psychol. 31(1):18–31
8. Statista Research Department (2021) Daily time spent on social networking by internet users worldwide 2012 to 2021. https://www.statista.com/statistics/433871/daily-social-media-usage-worldwide/
9. Grieve R, Indian M, Witteveen K, Tolan GA, Marrington J (2013) Face-to-face or Facebook: can social connectedness be derived online? Comput Hum Behav 29(3):604–609
10. Aiello LM, Barrat A, Schifanella R, Cattuto C, Markines B, Menczer F (2012) Friendship prediction and homophily in social media. ACM Trans Web (TWEB) 6(2):1–33
11. Vallor S (2012) Flourishing on facebook: virtue friendship & new social media. Ethics Inf Technol 14(3):185–199

12. Sutherland K, Davis C, Terton U, Visser I (2018) University student social media use and its influence on offline engagement in higher educational communities. Student Success 9(2):13–24
13. Dahdal S (2020) Using the WhatsApp social media application for active learning. J Educ Technol Syst 49(2):239–249
14. Karatsoli M, Nathanail E (2020) Examining gender differences of social media use for activity planning and travel choices. Eur Transp Res Rev 12(1):1–9
15. Mazman SG, Usluel YK (2011) Gender differences in using social networks. Turk Online J Educ Technol-TOJET 10(2):133–139
16. Idemudia EC, Raisinghani MS, Adeola O, Achebo N (2017) The effects of gender on the adoption of social media: an empirical investigation. AMCIS
17. Haferkamp N, Eimler SC, Papadakis AM, Kruck JV (2012) Men are from Mars, women are from Venus? Examining gender differences in self-presentation on social networking sites. Cyberpsychol Behav Soc Netw 15(2):91–98
18. Sultan S, Hussain I, Fatima S (2020) Social connectedness, life contentment, and learning achievement of undergraduate university students-does the use of internet matter? Bull Educ Res 42(1):111–125
19. Guo H (2015) Linking loneliness and use of social media [Master's thesis, University of Helsinki]. Helda. https://core.ac.uk/download/pdf/158607337.pdf
20. Kettle P, Gilmartin N, Corcoran MP, Byrne D, Sun T (2016) Time Well Spent? A survey of student online media usage. Maynooth University
21. Koay TY, Ayub N (2020) Social media usage, perceived social support, and loneliness among university students during the COVID-19 pandemic. In: Proceedings of the international seminar on counselling and well-being (ISCWB 2020), Microsoft Teams, Kuala Lumpur, Malaysia, 19 November 2020
22. Koay TY (2021) Perceived social support as the mediator between social media use and loneliness among the university students during the COVID-19 pandemic [Unpublished bachelor's thesis]. Universiti Malaysia Sabah
23. Rathakrishnan B, Bikar Singh SS, Kamaluddin MR, Ghazali MF, Yahaya A, Mohamed NH, Krishnan AR (2021a) Homesickness and socio-cultural adaptation towards perceived stress among international students of a public university in Sabah: an exploration study for social sustainability. Sustainability 13:4924. https://doi.org/10.3390/su13094924
24. Ryan T, Allen KA, Gray DL, McInerney DM (2017) How social are social media? A review of online social behaviour and connectedness. J Relationsh Res 8
25. Taylor-Jackson J, Abba I, Baradel A, Lay J, Herewini J, Taylor A (2021) Social media use, experiences of social connectedness and wellbeing during COVID-19. Mental Health Effects COVID-19 283–300
26. Davis K (2012) Friendship 2.0: adolescents' experiences of belonging and self disclosure online. J Adolesc 35(6):1527–1536
27. Rathakrishnan B, George S (2020) Gambling in Malaysia: an overview. BJPsych Int 2021(18):32–34. https://doi.org/10.1192/bji.2020.55
28. Lin KY, Lu HP (2011) Why people use social networking sites: an empirical study integrating network externalities and motivation theory. Comput Hum Behav 27(3):1152–1161
29. Rathakrishnan B, Singh SSB, Ghazali MF, Mohammed NH, Kamaluddin MR (2022) Motivation factors attributed to engaging in online studies amongst public university students. In: Lecture notes in networks and systems, vol 299, pp 217–226
30. Rathakrishnan B, Singh SSB, Kamaluddin MR, Yahaya A, Mohd Nasir MA, Ibrahim F, Rahman ZA (2021b) Smartphone addiction and sleep quality on academic performance of university students: an exploratory research. Int J Environ Res Public Health 18(16):8291. https://doi.org/10.3390/ijerph18168291

31. Beckstein A, Rathakrishnan B, Hutchings PB, Hassline Mohamed N (2021) The COVID-19 pandemic and mental health in Malaysia: current treatment and future recommendations. Malaysian J Public Health Med 21(1):260–267. https://doi.org/10.37268/mjphm/vol.21/no.1/art.826
32. Rathakrishnan B, Samsudin AR, Singh S, Juliana J (2017) Job preferences among marginalised and non-marginalised youths: A multi-ethnic study in sabah. Pertanika J Soc Sci Hum 25:55–66

Moderating Effect of Managerial Ownership on the Association Between Intellectual Capital and Firm Performance: A Conceptual Framework

Syed Quaid Ali Shah, Fong-Woon Lai, and Muhammad Kashif Shad

Abstract This document conceptualizes the intertwined nexus of intellectual capital and firm performance. The work also focuses on managerial ownership for moderating the effect between intellectual capital and performance. The alluded conceptual framework is valid for overall industries. This work uses a population of the Malaysian oil and gas industry. The census sampling technique is used. Based on prior studies and resource-based view theory, the study argues that the rise in organization value is directly related to the increased investment in intellectual capital. Besides, the document uses agency theory for conceptualizing managerial ownership for its multiplying effect between intellectual capital and performance. The study proposes a renowned "VAIC" model for computing intellectual capital. The document uses three performance indicators from management, shareholders, and the market perspective. The study provides essential intuitions to policymakers and practitioners on the crucial application of intellectual capital in value creation and providing a competitive advantage to the firms.

Keywords Intangible assets · Managerial ownership · Agency theory · Resource-based view theory

1 Introduction

Since the start, human beings have faced four main socio-economic phases. The first phase includes the primitive society; the second phase is the agricultural, followed by the industrial society in the third. The fourth phase we currently live in is the information-based society. In each socio-economic period, the firm's survival relied on different factors. For instance, land, infrastructure, and equipment were considered necessary for the growth of the business [1]. But with rising technological innovation, global challenges, and intense market competition, intellectual capital (Hereafter, IC)

S. Q. A. Shah (✉) · F.-W. Lai · M. K. Shad
Department of Management and Humanities, Universiti Teknologi PETRONAS, 32610 Seri Iskandar, Perak, Malaysia
e-mail: syed_18003337@utp.edu.my

M. Al-Emran et al. (eds.), *International Conference on Information Systems and Intelligent Applications*, Lecture Notes in Networks and Systems 550,
https://doi.org/10.1007/978-3-031-16865-9_38

is the crucial element for the success of the firms [2] because companies require liberal strategies, policies, and the application of intellectual capabilities regarding market development [3]. In this esteem, investors are pressurizing the management to give extra attention to the IC of the firm because it works as a cornerstone in a company's financial growth. The research shows that intellectual capital has contributed about 80% to the organization's value [4]. The authors of a related study [5] also revealed that the efficient utilization of IC obtains 50 to 90% of the firm's value.

Presently, concerned stakeholders are more knowledgeable regarding the significance of IC in sustainability and firm value addition. For this rationale, organizations are focusing on the management of IC to diffuse the opposing concerns of broad stakeholder groups. A research study revealed that renowned firms like Google and Microsoft invest more in IC than tangible and financial assets [6]. Hence, this justifies the argument that a company's success in the knowledge-based economy has a connection with IC. Nevertheless, academicians argue that the role of IC is more like a black box because of the indirect relationship with the firm performance [7]. In other words, it rewards the company after incurring substantial costs. Dalwai and Salehi [8] also highlight the indirect connection between IC and performance. Asif, Ting [9] argue that IC is an extra cost to the organization. It is reported as an expense and can put enterprise value at risk [10]. In such cases, the rolc of managers becomes essential to identify the threshold of investment in IC for a substantial outcome.

Prior studies have attempted to explore the effect of IC on firm performance, but this research area still seems under research and debatable due to controversial revelations by researchers [11]. The empirical testing suggests that IC plays a pivotal role in enhancing the firm's efficiency, shareholder, and profitability [12–14]. The authors have concluded positive and significant connections, respectively [12] and [2]. Some empirical tests fail to demonstrate a substantial relationship between IC and performance [15, 16]. Besides, the literature evidences the negative influence of IC on sales and market value [17]. Such mixed revelation can be a possible impact of the internal factors. Hence, it is pivotal to know the unveiled factors that can influence the nexus of IC and firm performance. In this esteem, the current document introduces a novel notion of managerial ownership as a moderating factor on the relationship of IC and firm performance. This moderating variable is critical due to its role in decision making and utilization of firm's resources.

Timely decisions are essential for business growth and economic success. In organizations, the investment decision is always in the hand of the managers because they are the one who monitors the day-to-day operations and has all the company's information. Given that, managers make decisions in the best interest of the business. Still, due to the opportunistic nature of humans, they might prefer their personal goals at the expense of shareholders. Perhaps, investment decisions related to the company's IC might be affected. In such cases, giving ownership to managers might be effective for a smooth business. Prior studies highlighted two concepts regarding managerial behavior, i.e., the interest-alignment hypothesis and the entrenchment hypothesis [18]. The first concept applies when overcoming the concerns between managers and shareholders by offering ownership to managers. On the contrary, the entrenchment hypothesis implies that when managerial ownership increases, the market value lessens due to less effectiveness of market discipline against shareholding managers. Despite this, literature shows that the firm value increases with the manager's ownership and vice versa [19]. Thus, in line with the interest-alignment hypothesis, this study attempts to moderate managerial ownership on the nexus of IC and performance. By owning shares in the business, managers might focus on the long-term value and make good decisions related to investment in IC which ultimately will enhance the firm value.

The alluded conceptualization is significant for every industry. The current study focuses on oil and gas firms due to their prominent role in the world's economy. Besides, this sector is of utmost importance due to its extensive exposure to broader risks, such as economic, social, governance, and ecological risks. IC is necessary for the oil and gas industry. The downstream workers vis-à-vis top stream employees will effectively execute the company operations by improving the IC. Moreover, the minimal cost and resource utilization resulting from IC will foster the company's returns. Increasing IC will accelerate the development of workers, which will result in sustainable economic, societal and environmental growth. In such a way, the industry can contribute to the United Nations 17 goals.

This document is crucial for academicians and practitioners. First, it provides insights on the IC in the firm's value creation. Secondly, it gives a fundamental understanding for researchers of how IC brings changes in enterprise value. Moreover, researchers can empirically validate the proposed framework with this basic understanding. Thirdly, it unveils managerial ownership as a factor that might affect IC and performance's nexus. Lastly, the document gives general insights for practitioners on the effective utilization of IC through managerial ownership, which might foster the firm's accounting performance.

The paper is split into several sections. Section 2 debates the literature review, conceptual framework, theoretical framework, and hypothesis development. Section 3 explains the methodology and operationalization of variables. Section 4 presents practical and methodological insights. Finally, Sect. 5 concludes the study.

2 Literature Review

This section reviews prior studies and explains IC's conceptualization, theoretical framework, and hypotheses development. The subsequent section sheds light on the IC.

2.1 Intellectual Capital

The hype about IC in businesses has highly drawn the researcher's attention. But researchers lack mutual consensus on a sole definition of IC. Consequently, IC has been defined differently. Edvinsson [20] defines IC as an intangible asset that positively influences the firm's performance but is not shown on a firm's balance sheets. In a related study, it is given that IC is not displayed on the balance sheet as it carries no actual value but is used for reporting purposes [2]. According to [12], IC is the asset absent on the balance sheet but significantly contributes to the firm value. Nevertheless, the importance of IC is highlighted by various researchers. Chen, Cheng [21] advocate on likage of the economic success with manufactured products vis-à-vis intangible assets. Additionally, the study urges investing in IC and managing it correctly to gain significant firm value [22]. Given that, companies enhance IC in employee training, education, business structure, and customers to compete and sustain success [11].

The researchers have reported different components of IC, but the most commonly acknowledged in the previous studies encompasses human, structural, and relational capital [11]. Skills, expertise, capabilities, knowledge, education, and experience fall under the category of human capital, which does not remain with the organization when the employees leave permanently. Structural capital is lifetime capital which includes intangible components such as system, structure, databases, management, and business strategies. Relational capital is the control and manages the relationship of the company. It includes the organization's relationship with external entities such as customers, suppliers, media groups, government, shareholders, and other stakeholders [2, 12, 23, 24].

Previous studies have measured intellectual capital through questionnaires, content analysis, and financial validation methods. For instance, Tobin's Q for measuring the intangible value [25], Skandia IC Navigator [20], IC-index [26], monitoring of intangible assets [27] and Value-added intellectual coefficient (VAIC) model [28, 29]. This document proposes the VAIC model demonstrated by Ante Pulic [28, 29]. This model computes IC and identifies whether it as a resource is efficiently utilized or not by the companies. Moreover, it is an easy method for measuring the organization's intangible assets using the balance sheet data. In other words, it shows how much the company creates a new value against a one-unit monetary investment in each source. The firm's value increase when the value of VAIC is larger [29]. VAIC is the most well-known technique to measure the company's IC [30].

2.2 *Conceptual Framework*

This study establishes a conceptual framework with literature and theoretical support, as depicted in Fig. 1. Prior studies have presented a linear causal relationship between IC and firm performance. Pulic [29], Ozkan, Cakan [2], and Tahir, Shah [12] have determined IC as a predictor variable against the performance. Despite the sustained relationship between IC and performance, limited attention is given to the factors that might compound the relationship between the two variables. This document proposes the missing aspect of managerial ownership on the nexus of IC and firm performance. Managerial ownership contributes to firm value in terms of sound

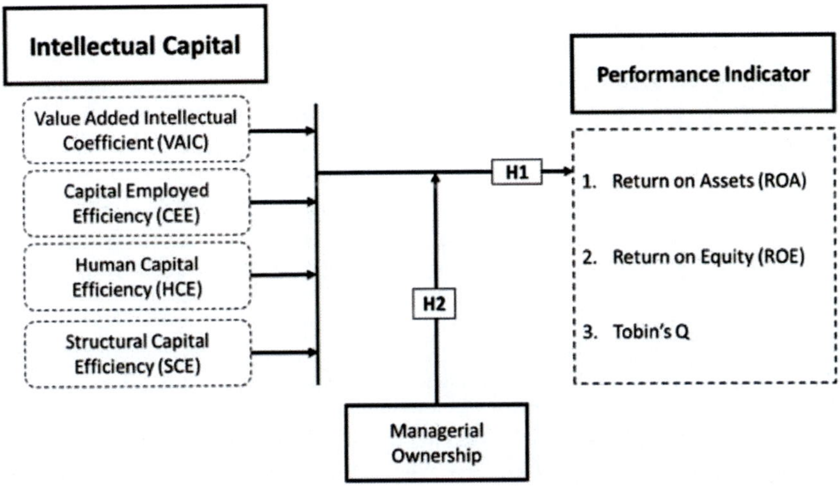

Fig. 1 Conceptual framework of the study

investment decisions and reduction of debts [31]. Managers are opportunistic because they easily recognize good and bad for them. Therefore, managers might ignore the investment in IC and count it as an extra cost to the company vis-à-vis might utilize it for their self-interest. Hence, issuing shares to managers will give them a sense of ownership of the company. Consequently, managers will put extra effort into the firm's long-term survival and invest in intangible assets that create firm value. IC is a long-term asset; therefore, managers will focus more on investing in IC and properly managing it. Ultimately, IC will give more output to the firm financial value [29].

The explanatory variable demonstrated in Fig. 1 is the IC proxied with VAIC. The dependent variable is ROA, ROE, and Tobin's Q. Managerial ownership is used as a moderator between IC and performance.

2.3 Theoretical Framework

Two corporate governance theories are used to support the alluded framework. Resource-based view theory establishes a direct association between IC and firm performance. At the same time, agency theory supports managerial ownership to strengthen the nexus of IC and firm performance.

Resource-Based View Theory (RBV)

RBV theory emerged in the 1980s and 1990s from the work of renowned researchers [32–35]. RBV educates on a firm's competitive advantage in the market by utilizing its resources. In other words, the competitive advantage lies in tangible and intangible resources at the firm's disposal. Such resources shall be preferred within the organizational strategy development, which eventually improves the long-term value. Given that, an organization's performance can be enhanced by utilizing its strategic resources, particularly intangible ones [12]. But the resources ought to be more precious, unique, limited, untransferable, and irreplaceable in order to ensure increased firm performance. All these qualities lie in the company's IC [36]. Hence, as an organization's strategic resource, IC can obtain a competitive edge and superior performance [37, 38]. The organization needs to focus on developing an efficient utilization of IC [38]. The greater the firm's IC, the high the firm's value will be [29]. Therefore, in light of RBV theory, IC as a resource will significantly positively impact the firm's performance.

Agency Theory
Agency theory emerged from the joint disciplines of economics and institutional theory in the late 1970s. It is invented by theorists Stephen Ross and Barry Mitnick [39]. But Jensen and Meckling [40] extend the theory concept from economics and institutional studies to various contexts, including information asymmetry, uncertainty, and risk. Its basis on the relationship between the principal (shareholders) and agents (management). Agents work on behalf of the principals and are expected to work in the principal's best interest. Deviation from the principal interest might lead to a conflict which causes inefficiencies and financial loss. This conflict remains minimal in the presence of a robust corporate governance structure. Managerial ownership is one of the corporate governance elements, among others. The interest of managers and the principal align when the managers are given right in the business shares. Hence, as per agency theory, managerial ownership aids in the mitigation of information asymmetry and agency costs which ultimately lead to higher financial performance. In line with the view of agency theory, managers as owners of the firm will focus on the investment in intellectual capital to increase the firm's performance. Additionally, monitoring of IC will be prudent to increase the company's return. Thus, the moderation of managerial ownership in the relationship between IC and firm performance might be significant.

2.4 *Hypothesis Development*

VAIC and Firm Performance
This nexus is determined in various studies within different contexts (Banking /finance, pharmaceutical, manufacturing, IT, etc.) and countries (Malaysia, Pakistan, India, UAE, Saudi Arabia, Australia, England, Germany, France, etc.). The primary focus had been on the effect of IC and financial performance.

The studies have proposed a significant association between the IC and firm performance based on the previous research. However, the results are different regarding the significance level and the sign of the coefficient. Many empirical studies assert a significant positive bonding between VAIC and performance. For instance, Chen Goh [41] examined a significant positive association between components of IC and firms' return on assets and market value. In the manufacturing industry of Thailand, Phusavat, Comepa [42] determined that IC significantly influences revenue growth, return on assets and return on equity. Besides, Riahi-Belkaoui [43] observed a significant positive relationship between IC and return on asset in US multinational firms. Likewise, the IC of Indian firms positively impacts profitability [44]. Tahir, Shah [12] showed a significant impact of IC on financial performance. Similar findings were evident by Ozkan, Cakan [2] in the banking industry of Turkey. However, another part of the literature found an insignificant impact of IC on firm performance [45, 46]. With the support of the resource-based view theory and discussion above, we suppose the below hypothesis:

Hypothesis 1: Intellectual capital has a significant positive impact on firm performance.

Moderating Role of Managerial Ownership

Notably, managerial ownership is recognized as an influencing corporate governance mechanism that aligns managers' and shareholders' interests [47]. Managerial ownership means that managers own the shares of the company. By having shares, managers monitor the company's operations with more care to get higher returns [47]. Besides, the managers diminish agency problems and, ultimately agency costs. Research studies have revealed that higher firm performance and value are associated with a high level of managerial ownership [48, 49]. Mangers put high interest in the firm's decision-making when they own shares in the business. Otherwise, they will affect by lousy decisions due to their shareholdings. Thus, it is expected that the managers having equity shares will be more careful in the decision-making related to IC. It has been proven that IC improves the firm value [29] and financial performance of the organization [12]. The managers will be enthusiastic about focusing on the IC to enhance firm performance and get a competitive advantage. Academicians have explored that the managers put more effort into the firm's long-term value when they own the shares. In this esteem, they might put all their efforts into strengthening intellectual capabilities and concentrating on IC for the firm's long-term survival [50]. Our idea of integrating managerial ownership is also supported by Anis [51], that IC and firm performance can be significantly moderated with the integration of corporate governance elements. In this esteem, it is assumed that the higher the managers' ownership, the more prominent the company's performance value will be. Thus, our study postulates the second hypothesis as,

Hypothesis 2: Managerial ownership significantly moderates the relationship between intellectual capital and firm performance.

3 Methodology

The alluded conceptual framework is valid for all industries. This document focuses on the population Malaysian oil and gas industry. The sampling technique is census sampling, in which all the firms are considered for analysis. The embedded reports, annual reports, and websites of the concerned firms will be sourced for the data collection of managerial ownership. Besides, Thomson Reuter DataStream will be used for the data collection of the financial ratios and IC. The study proposes panel data econometric techniques for exploring the impact of IC on firm performance. Hausman Test is suggested to be applied to decide between fixed effect and random effect [52, 53]. Moreover, endogeneity issues shall be covered using simultaneous equation models. The subsequent section explains the operationalization of the variables.

3.1 Measurement of Variables

Independent Variable

The study uses intellectual capital as an independent variable proxied by Pulic's model of value-added intellectual coefficient (VAIC) [28, 29]. As per Pulic's concept, IC or VAIC combines the three components given below.

$$VAIC = CEE + HCE + SCE \tag{1}$$

According to Ozkan, Cakan [2], the components of VAIC require the value addition (VA) of the firm.

$$VA = OP + EC + A/D \tag{2}$$

In Eq. 2, OP is operating profit, EC is employment cost, and A/D is amortization/depreciation of the firm. Now the first component of VAIC is CEE, calculated as follows:

$$CEE = VA/CE \tag{3}$$

In Eq. 3, VA is value added and CE is capital employed. The computation of HCE and SCE is given below:

$$HCE = VA/HC \tag{4}$$

$$SC = VA - CE \tag{5}$$

$$SCE = VA/CE \tag{6}$$

In Eqs. 4, 5, and 6, HC is personnel expenses while SC is the difference between VA and HC. The overall intellectual capital efficiency of the firm is calculated by putting Eqs. 3, 4, and 6 in Eq. 1.

Dependent Variables

Previous studies have used various performance indicators [54–61]. Here, firm performance is taken from three perspectives: Return on assets from a management perspective [12], return on equity from shareholders' perspective [55], and Tobin's Q from a market perspective [3]. Each dependent variable is calculated as follows:

$$Return\,on\,Assets = Net\,income\,/\,Total\,assets \tag{7}$$

$$Return\,on\,Equity = Net\,income\,/\,Shareholder's\,equity \tag{8}$$

$$Tobin's\,Q = Equity\,market\,value\,/\,Equity\,book\,value \tag{9}$$

Moderating Variable

The moderating variable is managerial ownership which is the ratio of shares with directors and overall total share [62]. It is calculated as follows:

$$Managerial\,ownership = No\,of\,shares\,held\,by\,managers\,/\,Total\,ordinary\,shares \tag{10}$$

4 Significance of the Study

The practical and theoretical significance of the study is given in the below subsections.

4.1 Practical Implication

In the present era of the knowledge economy, IC performs the role of the driver's seat by driving the business to success. Corporations should equivalate IC with tangible assets because buildings and machines cannot produce any idea, product, or innovative strategy. IC innovates the product or services and improves the organization's process to create a new source of value. To get a competitive advantage, the firms must invest more in IC. Our study gives direction to practitioners and policymakers about the importance of IC and its role in the firm's value creation.

4.2 Theoretical Significance

This study extends the literature on IC and its impact on firm performance. It increases the knowledge of academicians, researchers, and managers about the IC and how it enhances firm productivity. Our study integrates two theories to support the moderating role of managerial ownership and a direct link between IC and a firm's performance. This document is the first which conceptualizes the moderating effect of managerial ownership on the nexus of IC and the firm's performance in the Malaysian oil and gas industry. It adds to the literature by integrating managerial ownership with IC and the firm's performance.

5 Conclusion and Future Direction

This paper aims to produce a conceptual framework by integrating managerial ownership with IC for better financial performance. We have used the notion of resource-based view and agency theory to support the direct link and moderating role of director ownership between IC and performance. It is postulated that if the firm invests more in the IC, its value will be higher. Similarly, the interest-alignment hypothesis holds when managers are granted business shares.

This document is purely a conceptual study. Therefore, future studies can empirically validate our proposed model. Moreover, studies can be conducted by considering other exciting elements of corporate governance, such as board diversity, board committees, board independence, and institutional investors, for integration with IC.

Acknowledgements "The researchers would like to acknowledge Yayasan Universiti Teknologi PETRONAS (YUTP) for funding this research grant under Cost Center: 015LC0-188, Management and Humanities Department, Universiti Teknologi PETRONAS and Center of Social Innovation (CoSI) for the support to conduct this research"

References

1. Nuryaman TI (2015) The influence of intellectual capital on the firm's value with the financial performance as intervening variable. Procedia Soc Behav Sci 211:292–298
2. Ozkan N, Cakan S, Kayacan M (2017) Intellectual capital and financial performance: a study of the Turkish Banking Sector. Borsa Istanbul Rev 17(3):190–198
3. Hejazi R, Ghanbari M, Alipour M (2016) Intellectual, human and structural capital effects on firm performance as measured by Tobin's Q. Knowl Process Manag 23(4):259–273
4. Ahmed A, Khurshid MK, Yousaf MU (2019) Impact of intellectual capital on firm value: the moderating role of managerial ownership. Preprints
5. Noradiva H, Parastou A, Azlina A (2016) The effects of managerial ownership on the relationship between intellectual capital performance and firm value. Int J Soc Sci Humanit 6(7):514

6. Ong T, Yeoh L, Teh B (2011) Intellectual capital efficiency in Malaysian food and beverage industry. Int J Bus Behav Sci 1(1):16–31
7. Li Y, Zhao Z (2018) The dynamic impact of intellectual capital on firm value: evidence from China. Appl Econ Lett 25(1):19–23
8. Dalwai T, Salehi M (2021) Business strategy, intellectual capital, firm performance, and bankruptcy risk: evidence from Oman's non-financial sector companies. Asian Rev Account 29(3):474–504
9. Asif J, Ting IWK, Kweh QL (2020) Intellectual capital investment and firm performance of the Malaysian energy sector: a new perspective from a nonlinearity test. Energy Res Lett 1(3):13622
10. Kweh QL et al (2019) Intellectual capital, governmental presence, and firm performance of publicly listed companies in Malaysia. Int J Learn Intellect Cap 16(2):193–211
11. Shah SQA et al (2021) The inclusion of intellectual capital into the green board committee to enhance firm performance. Sustainability 13(19):1–21
12. Tahir M et al (2018) Intellectual capital and financial performance of banks in Pakistan. Dialogue (Pakistan) 13(1):105–118
13. Makki MAM, Lodhi S, Rohra C (2009) Impact of intellectual capital on shareholders earning. Aust J Basic Appl Sci 3(4):3386–3398
14. Pew TH, Plowman D, Hancock P (2007) Intellectual capital and financial returns of companies. J Intellect Cap 8(1):76–95
15. Iranmahd M et al (2014) The effect of intellectual capital on cost of finance and firm value. Int J Acad Res Account Fin Manage Sci 4(2):1–8
16. Mehralian G et al (2012) Intellectual capital and corporate performance in Iranian pharmaceutical industry. J Intellect Cap 13(1):138–158
17. Ge F, Xu J (2021) Does intellectual capital investment enhance firm performance? Evidence from pharmaceutical sector in China. Technol Anal Strateg Manage 33(9):1006–1021
18. Chen Y-R, Chuang W-T (2009) Alignment or entrenchment? Corporate governance and cash holdings in growing firms. J Bus Res 62(11):1200–1206
19. Clay DG (2002) Institutional ownership and firm value. SSRN 485922
20. Edvinsson L (1997) Developing intellectual capital at Skandia. Long Range Plan 30(3):366–373
21. Chen MC, Cheng SJ, Hwang Y (2005) An empirical investigation of the relationship between intellectual capital and firms' market value and financial performance. J Intellect Cap 6(2):159–176
22. Maditinos D et al (2011) The impact of intellectual capital on firms' market value and financial performance. J Intellect Cap 12(1):132–151
23. Joshi M et al (2013) Intellectual capital and financial performance: an evaluation of the Australian financial sector. J Intellect Cap 14(2):264–285
24. Mondal A, Ghosh SK (2012) Intellectual capital and financial performance of Indian banks. J Intellect Cap 13(4):515–530
25. Stewart TA (1997) Intellectual Capital: the new wealth of nations. Doubleday Dell Publishing Group Inc., New York
26. Roos J, Edvinsson L, Dragonetti NC (1997) Intellectual capital: navigating the new business landscape. Springer
27. Sveiby KE (1997) The new organizational wealth: managing & measuring knowledge-based assets. Berrett-koehler Series. Berrett-Koehler Publishers
28. Pulic A (1998) Measuring the performance of intellectual potential in knowledge economy. In: 2nd McMaster word congress on measuring and managing intellectual capital by the Austrian team for intellectual potential. McMaster University, Hamilton
29. Pulic A (2004) Intellectual capital – does it create or destroy value? Meas Bus Excell 8(1):62–68
30. Vishnu S, Gupta VK (2015) Performance of intellectual capital in Indian healthcare sector. Int J Learn Intellect Cap 12(1):47–60
31. Sun J et al (2016) Ownership, capital structure and financing decision: evidence from the UK. Br Account Rev 48(4):448–463
32. Wernerfelt B (1984) A resource-based view of the firm. Strateg Manag J 5(2):171–180

33. Barney J (1991) Firm resources and sustained competitive advantage. J Manag 17(1):99–120
34. Grant RM (1996) Toward a knowledge-based theory of the firm. Strateg Manag J 17(S2):109–122
35. Spender J-C, Grant RM (1996) Knowledge and the firm: overview. Strateg Manag J 17(S2):5–9
36. Molodchik M, Shakina E, Bykova A (2012) Intellectual capital transformation evaluating model. J Intellect Cap 13(4):444–461
37. Clarke M, Seng D, Whiting RH (2011) Intellectual capital and firm performance in Australia. J Intellect Cap 12(4):505–530
38. Marr B, Gray D, Neely A (2003) Why do firms measure their intellectual capital? J Intellect Cap 4(4):441–464
39. Mitnick BM (2019) Origin of the theory of agency: an account by one of the theory's originators. SSRN 1020378
40. Jensen MC, Meckling WH (1976) Theory of the firm: managerial behavior, agency costs and ownership structure. J Financ Econ 3(4):305–360
41. Chen Goh P (2005) Intellectual capital performance of commercial banks in Malaysia. J Intellect Cap 6(3):385–396
42. Phusavat K et al (2011) Interrelationships between intellectual capital and performance. Ind Manag Data Syst 111(6):810–829
43. Riahi-Belkaoui A (2003) Intellectual capital and firm performance of US multinational firms. J Intellect Cap 4(2):215–226
44. Pal K, Soriya S (2012) IC performance of Indian pharmaceutical and textile industry. J Intellect Cap 13(1):120–137
45. Pitelli Britto D, Monetti E, da Rocha Lima Jr J (2014) Intellectual capital in tangible intensive firms: the case of Brazilian real estate companies. J Intellect Cap 15(2):333–348
46. Celenza D, Rossi F (2014) Intellectual capital and performance of listed companies: empirical evidence from Italy. Meas Bus Excell 18(1):22–35
47. Brickley JA, Lease RC, Smith CW (1988) Ownership structure and voting on antitakeover amendments. J Financ Econ 20:267–291
48. Li X, Sun ST, Yannelis C (2018) Managerial ownership and firm performance: evidence from the 2003 Tax Cut. SSRN 2285638
49. Hanson RC, Song MH (2000) Managerial ownership, board structure, and the division of gains in divestitures. J Corp Finan 6(1):55–70
50. Mohd-Saleh N, Che Abdul Rahman MR (2009) Ownership structure and intellectual capital performance in Malaysia. Asian Acad Manage J Account Fin 5(1):1–29
51. Anis I (2013) Corporate governance-driven to intellectual capital and corporate performance: empirical study in Indonesian banking industry. In: International conference on business, economics, and accounting
52. Shah SQA et al (2022) Developing a green governance framework for the performance enhancement of the oil and gas industry. Sustainability 14(7):3735
53. Shah SAA, Shah SQA, Tahir M (2022) Determinants of CO2 emissions: exploring the unexplored in low-income countries. Environ Sci Pollution Res
54. Lai F-W, Shad MK, Shah SQA (2021) Conceptualizing corporate sustainability reporting and risk management towards green growth in the Malaysian oil and gas industry. SHS Web Conf 124:04001
55. Shah SQA et al (2018) Factors affecting liquidity of banks: empirical evidence from the banking sector of Pakistan. Colombo Bus J 9(1):1–18
56. Jan AA et al (2019) Bankruptcy profile of the Islamic banking industry: evidence from Pakistan. Bus Manage Strateg 10(2):265–284
57. Jan AA, Lai F-W, Tahir M (2021) Developing an Islamic Corporate Governance framework to examine sustainability performance in Islamic Banks and Financial Institutions. J Clean Prod 315:128099
58. Jan AA et al (2021) Integrating sustainability practices into Islamic corporate governance for sustainable firm performance: from the lens of agency and stakeholder theories. Qual Quant

59. Shad MK, Lai FW (2015) A conceptual framework for enterprise risk management performance measure through economic value added. Global Bus Manage Res 7(2):1–11
60. Shad MK et al (2020) The efficacy of sustainability reporting towards cost of debt and equity reduction. Environ Sci Pollution Res 2020:1–12
61. Shad MK et al (2019) Integrating sustainability reporting into enterprise risk management and its relationship with business performance: a conceptual framework. J Clean Prod 208:415–425
62. Moudud-Ul-Huq S, Biswas T, Proshad Dola S (2020) Effect of managerial ownership on bank value: insights of an emerging economy. Asian J Account Res 5(2):241–256

Motivational Elements of Online Knowledge Sharing Among Employees: Evidence from the Banking Sector

Alaa S. Jameel, Aram Hanna Massoudi, and Abd Rahman Ahmad

Abstract This study aims to examine the impact of self-efficacy, reputation, reciprocity, altruism, and enjoyment on the online knowledge sharing among employees. The study was conducted in the banking sector. The data were collected from four private banks. Smart-PLS., was applied to analyzed 187 valid questionnaires. The results indicated that self-efficacy, reputation, reciprocity, altruism, and enjoyment have a positive and significant impact on online knowledge sharing among bank employees. Therefore, banks should establish a conducive online knowledge-sharing environment to encourage reciprocal connections and interpersonal interactions among employees. Employees that actively participate in knowledge exchange are encouraged, which help in stimulating the reciprocal online knowledge sharing behavior. The comprehensive model of this study proposes to measure online knowledge sharing in the Banking Sector. The present literature does not take into account such a broad perspective.

Keywords Online knowledge sharing · Motivation · Reputation

1 Introduction

At the present time, working individuals are increasingly turning to the internet for new information and skills. As a result, employees must acquire information and increase their technical capabilities via the internet sources in order to utilize

A. S. Jameel (✉)
Department of Public Administration, Cihan University-Erbil, Kurdistan Region, Iraq
e-mail: alaa.salam@cihanuniversity.edu.iq

A. H. Massoudi
Department of Business Administration, Cihan University-Erbil, Kurdistan Region, Iraq
e-mail: aram.massoudi@cihanuniversity.edu.iq

A. R. Ahmad
Faculty of Technology Management and Business, Universiti Tun Hussein Onn Malaysia, Batu Pahat, Johor, Malaysia
e-mail: arahman@uthm.edu.my

M. Al-Emran et al. (eds.), *International Conference on Information Systems and Intelligent Applications*, Lecture Notes in Networks and Systems 550,
https://doi.org/10.1007/978-3-031-16865-9_39

knowledge-intensive activities. The Internet has surpassed printed media as the primary source of information, and realizing how to get information and knowledge via the internet has become a critical area of various studies [1].

The Internet has provided individuals with unparalleled access to information resources; online reading and retrieval have become the primary method of obtaining information. Furthermore, online information reduces the cost of seeking both time and efficiency [2]. Bharati et al. [3] emphasize the importance of internet platforms in facilitating online knowledge exchange by expanding individuals' reach beyond face-to-face contact. Knowledge is not an item that can be easily acquired, transported, shared, or traded based on its location [4]. Nonetheless, one of the most challenging difficulties confronting businesses is motivating knowledge sharing (KS) [5]. KS is critical for organization because it allows people to transfer their knowledge into organizational knowledge, resulting in new knowledge [6]. In most cases, the reason for doing anything is related to the individual's motivations. The current study will examine five elements that are considered the most critical motivational factors behind online knowledge sharing, these elements are: (reputation, self-efficacy, reciprocity, altruism and enjoyment). A few studies have been conducted to measure these elements in Iraq, particularly with online knowledge sharing. In spite of this, our expertise in online information-sharing remains limited [7]. Previous studies, which were conducted in different countries with different cultures and banking systems, cannot be compared to the Iraqi context. Especially, investigating the antecedents of online knowledge sharing in the banking industry. This study aims to test the impact of reputation, self-efficacy, reciprocity, altruism, and enjoyment on the online KS among employees in the banking sector.

2 Literature Review and Hypotheses Development

2.1 Online Knowledge Sharing

The pandemic of COVID-19 has opened the road for online KS among individuals in an organization through online platforms such as Google Meet, Zoom, Microsoft Teams, and others. Thus becoming the new trend in working and collaborating among individuals. The importance of online knowledge exchange in the organization has received a substantial attention since organizations' interest in shifting work settings to an online or virtual platform is growing. The online KS means to transfer and exchange information and knowledge among individuals through online platforms [15].

The importance of online KS inside the organization and workgroups has been recognized as critical factor in increasing productivity [16]. Individuals share knowledge online, which is crucial for organization when individuals operate from different locations, particularly during the COVID-19 pandemic. Through information exchange among individuals and knowledge recording for reuse, online KS helps firms gain a competitive advantage [17]. The transmission of knowledge online among individuals in a company is called online knowledge sharing behavior [18]. Individuals can contribute to generating organizational knowledge by exchanging ideas and knowledge assets through active online KS.

Online knowledge sharing is highly related to technological dimensions. IT enables new ways of working and cooperating among individuals in the workplace, and they are usually viewed as helpful in knowledge sharing [16]. An online platform must offer appropriate features and attributes, such as usability and user-friendliness to drive knowledge sharing behavior. When the online knowledge platform is of good quality, it is expected that more individuals would utilize it to exchange knowledge [19]. Traditional library-based information-seeking is being replaced by online information, opening new frontiers for knowledge management [7]. Employees in an organization can integrate and share their knowledge face-to-face and online, leading to better performance, more productivity, and innovative skills, which is considered a keys in giving the organization a long-term competitive advantage [20]. The underpinning theory of the current study consisted of expectancy theory, which applied to KS, and social cognitive theory, which applied to IS.

2.2 *Hypotheses Development*

Reputation

Reputation is considered a motivational factor that enhances online knowledge sharing among individuals. Usually, individuals tend to share their knowledge online if this KS is recognized [16]. To increase their reputation as a professional in coworkers' eyes, individuals may share information to brag about or let colleagues know that they are informed and hold valuable expertise. Individuals will provide information if they believe it will help them improve their reputation. Nguyen et al. [16] reported that individuals tend to share their knowledge online to enhance their reputation, and empirically reported reputation had a significant impact on online KS among banks employees. However, Hosen et al. [24] indicated in their study conducted among students in 10 private universities in Malaysia that reputation had a substantial effect on knowledge sharing, and reputation can increase the intention of knowledge sharing. Therefore, reputation significantly impacts KS intention [25], and the quantity of KS [26]. Thus, the researchers postulate the following hypothesis:

H1: Reputation has a positive and significant impact on online KS among banks employee.

Self-efficacy

According to Olatokun and Nwafor [27], the employees will not share their knowledge without self-efficacy and indicated that self-efficacy is the main condition for knowledge sharing. Self-efficacy is the belief in one's ability to provide helpful knowledge to others. Employees are more likely to share their knowledge when they have a sense of self-efficacy about their profession [12]. Individuals with high levels of self-efficacy are more willing to share their knowledge, leading to KS. Self-efficacy is able to enhance and improve online knowledge sharing [7] and reported self-efficacy has a significant impact on online KS. Nguyen et al. [16] reported online KS significantly impacted by self-efficacy in the context of the banks' sector. However, self-efficacy had a considerable effect on KS [4] and the quantity of KS [26].

H2: Self-efficacy has a positive and significant impact on online KS among banks employee.

Reciprocal

Individuals' perceptions of reciprocity include the belief that the present of KS behavior will lead to future KS by others. Therefore, when individuals offer their information to others, they may assume that others will reciprocate their knowledge [16]. As a result, information givers frequently expect to be compensated for their efforts [13]. Therefore, individuals tend to have a high level of reciprocity when they offer information and receive KS by others in return [12]. Empirically, Nguyen et al. [16] reported that reciprocity has a significant impact on online KS among employees. Reciprocity enhanced and improved the online KS, and statistically, reciprocity had a significant effect on online KS [30]. Similarly, Hoseini et al. [25] indicated that reciprocity significantly impacted the KS intention. Al Hawamdeh and AL-edenat [4] reported that reciprocity has a considerable impact on KS and the quantity of KS [26].

H3: Reciprocity has a positive and significant impact on online KS among banks employee.

Altruism

Altruism refers to the selfless act of helping others without expecting anything in return. In an online community, altruism is critical to knowledge sharing. Altruism is a personality trait that motivates people to actively assist others in attaining a set of goals while improving their learning performance [24]. Altruism is the extent to which an individual is prepared to help others without expecting anything in return [25]. Empirically there is no broad agreement on the impact of altruism on KS. According to Hoseini et al. [25], altruism has a significant impact on KS, and altruism is able to increase the intention of KS among individuals. In addition, Sedighi et al. [26] indicated that altruism had a statistical effect on KS quantity. On the other hand, Hosen et al. [24] reported that altruism had an insignificant impact on KS among students.

H4: Altruism has a positive and significant impact on online KS among banks employee.

Enjoyment

The level to which individuals believe that sharing information will result in the sense of enjoyment is known as enjoyment [15]. People usually visit a website if it entertains them. The experience of happiness when using mobile applications in both deliberate and unconscious phases contributes to the users' participation [32].

Individuals who prefer sharing their knowledge have an internal incentive that stems from a sense of moral duty, which typically outweighs the urge to maximize self-interest [33]. Enjoyment statistically has a significant impact on the intention of KS [25]. In addition, Al Hawamdeh and AL-edenat [4] reported that enjoyment could enhance the KS and enjoyment has a significant impact on KS.

H5: Enjoyment has a positive and significant impact on online KS among banks employee.

3 Methodology

The data for this study were gathered through a self-administered questionnaire, and a quantitative method was used to do it. The quantitative approach is widely used in business research [34].

And the questionnaires ensure to collect the data in a short time, less effort and with a high number of respondents [35]. Additionally, this study used the convenience sampling method. Three hundred questionnaires were distributed among employees in four private banks in Erbil, Kurdistan Region, Iraq. A 198 questionnaires were returned, and the response rate was 66%. Therefore, after checking the missing values and outliers, 187 questionnaires were valid for analysis. Of the total sample, most of the respondents were male, with 63%, and females, with 37%. Additionally, most of the employees held bachelor's degrees with 88.2% and were between 20 and 39 years old with 88%. Additionally, the data were analyzed by Smart-PLS 3.33 and all the instruments adopted from previous studies are depicted in Table 1.

4 Results

In this section, Smart-PLS will use two steps to analyze the data: Measurement model and structural model. The first measurement model, the purpose of this step is to measure the validity, reliability, convergent and discriminant validity.

The factor loading cutoff level as recommended by Hair et al. [36] is 0.7. All the factor loading showed greater than 0.7 as depicted in Table 1, except ALT1 and ATL5 showed poor loading thus they were removed. Additionally, the reliability is measured by two indices, the Cronbach's alpha (CA) and Composite Reliability (CR); the cutoff level for both mentioned indices are 0.7 [36], and as depicted in Table 1, both CA and CR values are greater than 0.7 thus the reliability has been achieved. Finally, the convergent validity is measured by the average of variance extracted (AVE), and the cutoff level of AVE should be 0.5 or greater [36]. As illustrated in Table 1, all the AVE values are <0.5. Thus, Convergent validity has been achieved.

The Heterotrait-Monotrait Ratio (HTMT) should be <0.90 [36]. Table 2 illustrates all the HTMT values are less than 0.90. Thus, the discriminant validity has been achieved.

Table 1 Construct Reliability and Validity

	Items	Outer loadings	CA	CR	AVE	Sources
Altruism	ALT2 ALT3 ALT4 ALT6	0.726 0.846 0.821 0.764	0.801	0.869	0.625	[24]
Enjoyment	ENJ1 ENJ2 ENJ3 ENJ4 ENJ5 ENJ6 ENJ7	0.790 0.897 0.923 0.842 0.885 0.897 0.788	0.944	0.953	0.743	[15, 37]
OKS	OKS1 OKS2 OKS3 OKS4 OKS5	0.802 0.850 0.890 0.879 0.837	0.905	0.930	0.726	[7, 24]
Reciprocal	REC1 REC2 REC3 REC4 REC5	0.816 0.853 0.887 0.824 0.768	0.887	0.917	0.690	[15, 37]
Reputation	REP1 REP2 REP3 REP4 REP5	0.812 0.876 0.855 0.743 0.794	0.875	0.909	0.668	[24]
Self-Efficacy	SE1 SE2 SE3 SE4 SE5	0.832 0.894 0.842 0.849 0.833	0.904	0.929	0.723	[7, 37]

Table 2 Heterotrait-Monotrait Ratio (HTMT)

	Altruism	Enjoyment	OKS	Reciprocal	Reputation	Self-Efficacy
Altruism						
Enjoyment	0.183					
OKS	0.458	0.126				
Reciprocal	0.369	0.073	0.512			
Reputation	0.425	0.087	0.606	0.389		
Self-Efficacy	0.299	0.064	0.523	0.479	0.577	

The second step is the structural model, in this step, the researchers will test the proposed hypotheses. This step showed the R^2 is 0.44 which means that the independent variables explained the dependent variable by 44%, which is considered moderate [36]. The hypotheses results showed that all the proposed hypotheses were accepted, as illustrated in Table 3 and Fig. 1. The H1, H2, H3, H4 and H5 showed the P-value < 0.05 and the T-value > 1.96; thus, all the hypotheses are supported.

Table 3 Hypotheses results

Hypotheses paths	Original sample	Sample mean	Standard deviation	T values	P values	Result
H1: Reputation -> OKS	0.322	0.324	0.058	5.556	**0.000**	Supported
H2:Self-Efficac -> OKS	0.180	0.179	0.063	2.872	**0.004**	Supported
H3: Reciprocal -> OKS	0.231	0.230	0.055	4.219	**0.000**	Supported
H4: Altruism -> OKS	0.143	0.145	0.045	3.198	**0.001**	Supported
H5: Enjoyment -> OKS	0.111	0.115	0.048	2.326	**0.020**	Supported

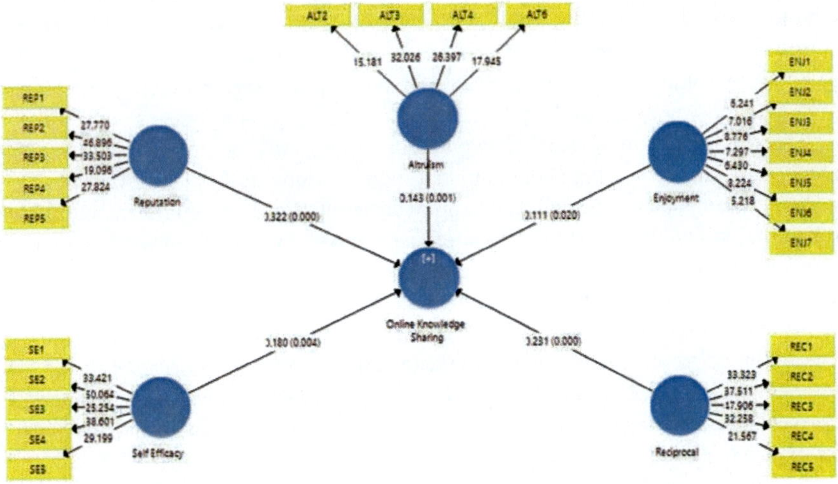

Fig. 1 Structural model

5 Discussion

The results of this study indicated that self-efficacy significantly improve the online KS among the employee; this result is in line with previous studies [7, 16]. When the employees have a sense of self-efficacy, they tend to share their knowledge among peers on online platforms. However, individuals with strong self-efficacy are more likely to share their expertise online among peers, according to the current findings. Individuals with a high level of self-efficacy are more inclined to participate in organizational activities and wish to contribute. Furthermore, because they are obedient and perform well in their tasks, they tend to share expertise to guarantee that they operate successfully and prevent errors. Since online knowledge sharing is typically voluntary, self-efficacy is critical. Employees who mistrust their ability to share knowledge are less likely to engage in online knowledge sharing behaviors.

The results indicated the reciprocity can significantly improve the online KS among the employee, this result in line with previous studies [4, 16, 30]. Reciprocity benefits can heavily influence individual attitudes toward online KS. As a result, when individuals have strong reciprocity, they are more inclined to share knowledge online throughout the organization and among peers. Yet, increased reciprocity in the workplace leads to information sharing online and resources exchange, resulting in joint gains such as maintaining capital and improving performance.

The results also indicated that enjoyment significantly improved the online KS among the employee; this result is in line with previous studies [4, 25]. Individuals tend to share the knowledge online when they feel this action is enjoyable. Therefore, managers should improve employees emotional state during online KS in order to boost self-enjoyment. Nevertheless, enhancing job design by giving employees greater autonomy may also help them build a sense of self-enjoyment.

The results also indicated that reputation significantly improved the online KS among the employee. This result is in line with previous studies [16, 24, 25].

The findings also revealed that employee reputation is a significant motivator for online knowledge sharing. Thus, efficient usage of online knowledge sharing is advantageous for job efficiency because of its communication visible properties, such as message transparency and network translucence. Moreover, individuals may determine "who knows what" and "who knows whom" by sharing knowledge online, this action helps to develop meta-knowledge and reduce repetition at work. Therefore, enhancing the employee's reputation is the primary motivator for online KS. Furthermore, establishing a favorable image and reputation is beneficial to the banks' sector since it builds trust among its employees.

Finally, the results indicated that altruism significantly improved the online KS among the employee this result in line with previous studies [24, 26]. Motivated individuals by altruism seek pleasure in assisting others without expecting anything in return. Individuals assist for many reasons. Some may be altruistic, while others may not. The importance of altruism is that it seems to be an exception to the widely held belief that behavior is governed by rewards and punishments and the implication that individuals are fundamentally selfish.

6 Conclusion

This study examined the motivational elements that lead to online knowledge sharing among employees in the banking sector. The results indicated that self-efficacy, reciprocity, enjoyment, reputation, and altruism are significantly impact online knowledge sharing among the employees in the banking sector. However, reputation showed the most critical element that encourages the employees to share their knowledge; this might be due to the country's culture, and the people pay more attention to their reputation than other elements. The findings showed that several factors could motivate the employee to share the knowledge online. Banks may create a variety of incentive schemes to encourage employees to use their internet search skills in knowledge sharing with colleagues.

References

1. Jameel AS, Karem MA, Ahmad AR (2022) Behavioral intention to use E-Learning among academic staff during COVID-19 pandemic based on UTAUT model. In: Al-Emran M, Al-Sharafi MA, Al-Kabi MN, Shaalan K (eds) Proceedings of International Conference on Emerging Technologies and Intelligent Systems. ICETIS 2021. LNNS, vol 299, pp 187–196. Springer, Cham. https://doi.org/10.1007/978-3-030-82616-1_17
2. Jameel AS, Karem MA, Aldulaimi SH, Muttar AK, Ahmad AR (2022) The acceptance of E-Learning service in a higher education context. In: Al-Emran M, Al-Sharafi MA, Al-Kabi MN, Shaalan K (eds) Proceedings of International Conference on Emerging Technologies and Intelligent Systems. ICETIS 2021. LNNS, vol 299, pp 255–264. Springer, Cham. https://doi.org/10.1007/978-3-030-82616-1_23
3. Bharati P, Zhang W, Chaudhury A (2015) Better knowledge with social media? Exploring the roles of social capital and organizational knowledge management. J Knowl Manag 19(3):456–475
4. Al Hawamdeh N, AL-edenat M (2022) Investigating the moderating effect of humble leadership behaviour on motivational factors and knowledge-sharing intentions: evidence from Jordanian public organisations. VINE J Inf Knowl Manag Syst
5. Hong D, Suh E, Koo C (2011) Developing strategies for overcoming barriers to knowledge sharing based on conversational knowledge management: a case study of a financial company. Expert Syst Appl 38(12):14417–14427
6. Ipe M (2003) Knowledge sharing in organizations: a conceptual framework. Hum Resour Dev Rev 2(4):337–359
7. Zhang W, Jiang Y, Zhang W (2021) Antecedents of online knowledge seeking of employees in technical R&D team: an empirical study in China. IEEE Trans Eng Manag 1–10
8. Jameel AS, Abdalla SN, Karem MA, Ahmad AR (2020) Behavioural intention to use E-Learning from student's perspective during COVID-19 pandemic. In: Proceedings - 2020 2nd annual international conference on information and sciences, AiCIS 2020, pp 165–171
9. Akhavan P, Jafari M, Fathian M (2005) Exploring the failure factors of implementing knowledge management system in the organizations. J Knowl Manag Pract 6
10. Titi Amayah A (2013) Determinants of knowledge sharing in a public sector organization. J Knowl Manag 17(3):454–471
11. Jameel AS, Ahmad AR (2020) The role of information and communication technology on knowledge sharing among the academic staff during COVID-19 pandemic. In: Proceedings - 2020 2nd annual international conference on information and sciences, AiCIS 2020, pp 141–147

12. Nguyen T-M, Nham TP, Froese FJ, Malik A (2019) Motivation and knowledge sharing: a meta-analysis of main and moderating effects. J Knowl Manag 23(5):998–1016
13. Kwahk K-Y, Park D-H (2016) The effects of network sharing on knowledge-sharing activities and job performance in enterprise social media environments. Comput Hum Behav 55:826–839
14. Wasko, Faraj (2005) Why should I share? Examining social capital and knowledge contribution in electronic networks of practice. MIS Q 29(1):35
15. Nguyen M, Malik A, Sharma P (2021) How to motivate employees to engage in online knowledge sharing? Differences between posters and lurkers. J Knowl Manag 25(7):1811–1831
16. Nguyen T-M, Ngo LV, Gregory G (2022) Motivation in organisational online knowledge sharing. J Knowl Manag 26(1):102–125
17. Khan NA, Khan AN (2019) What followers are saying about transformational leaders fostering employee innovation via organisational learning, knowledge sharing and social media use in public organisations? Gov Inf Q 36(4):101391
18. Lin H (2007) Knowledge sharing and firm innovation capability: an empirical study. Int J Manpow 28(3/4):315–332
19. Tan CN-L (2016) Enhancing knowledge sharing and research collaboration among academics: the role of knowledge management. High Educ 71(4):525–556
20. Li Z, Liu X, Wang W.M, Vatankhah Barenji A, Huang GQ (2019) CKshare: secured cloud-based knowledge-sharing blockchain for injection mold redesign. Enterp Inf Syst 13(1):1–33
21. Iglesias-Pradas S, Hernández-García Á, Fernández-Cardador P (2017) Acceptance of corporate blogs for collaboration and knowledge sharing. Inf Syst Manag 34(3):220–237
22. Ba S, Stallaert J, Whinston AB (2001) Research commentary: introducing a third dimension in information systems design—the case for incentive alignment. Inf Syst Res 12(3):225–239
23. Choi JH, Ramirez R, Gregg DG, Scott JE, Lee K-H (2020) Influencing knowledge sharing on social media: a gender perspective. Asia Pacific J Inf Syst 30(3):513–531
24. Hosen M, Ogbeibu S, Giridharan B, Cham T-H, Lim WM, Paul J (2021) Individual motivation and social media influence on student knowledge sharing and learning performance: evidence from an emerging economy. Comput Educ 172:104262
25. Hoseini M, Saghafi F, Aghayi E (2019) A multidimensional model of knowledge sharing behavior in mobile social networks. Kybernetes 48(5):906–929
26. Sedighi M, Lukosch S, Brazier F, Hamedi M, van Beers C (2018) Multi-level knowledge sharing: the role of perceived benefits in different visibility levels of knowledge exchange. J Knowl Manag 22(6):1264–1287
27. Olatokun W, Nwafor CI (2012) The effect of extrinsic and intrinsic motivation on knowledge sharing intentions of civil servants in Ebonyi State, Nigeria. Inf Dev 28(3):216–234
28. Lin H-F (2007) Effects of extrinsic and intrinsic motivation on employee knowledge sharing intentions. J Inf Sci 33(2):135–149
29. Jameel AS, Hamdi SS, Karem MA, Raewf MB, Ahmad AR (2021) E-Satisfaction based on E-service quality among university students. J Phys Conf Ser 1804(1):012039
30. Li C, Li H, Suomi R, Liu Y (2021) Knowledge sharing in online smoking cessation communities: a social capital perspective. Internet Res 32 (7):111–138
31. Fang Y-H, Chiu C-M (2010) In justice we trust: Exploring knowledge-sharing continuance intentions in virtual communities of practice. Comput Human Behav 26(2):235–246
32. Hsiao C-H, Chang J-J, Tang K-Y (2016) Exploring the influential factors in continuance usage of mobile social apps: satisfaction, habit, and customer value perspectives. Telemat. Inform 33(2):342–355
33. McLure Wasko M, Faraj S (2000) 'It is what one does': why people participate and help others in electronic communities of practice. J Strateg Inf Syst 9(2–3):155–173
34. Collis J, Hussey R (2013) Business research: A practical guide for undergraduate and postgraduate students. Macmillan International Higher Education
35. Sekaran U, Bougie R (2016) Research methods for business: A skill building approach, 7th edn. John Wiley & Sons, New York

36. Hair JF, Risher JJ, Sarstedt M, Ringle CM (2019) When to use and how to report the results of PLS-SEM. Eur Bus Rev 31(1):2–24
37. Singh JB, Chandwani R, Kumar M (2018) Factors affecting Web 2.0 adoption: exploring the knowledge sharing and knowledge seeking aspects in health care professionals. J Knowl Manag 22(1):21–43

Big Data and Business Analytics: Evidence from Egypt

Ahmed Elmashtawy and Mohamed Salaheldeen

Abstract Big data is one of the most valuable assets for businesses seeking to reach the broadest possible customer base. Big data offers significant benefits to corporate financial reporting, increasing its reliability and objectivity and transitioning from periodic to real-time reporting. The purpose of this research is to reveal the effect of big data on profit prediction. The research also investigated the effect of innovative business intelligence techniques and blockchain, as to dimensions of big data, on the disclosure quality of financial reports. A case study was conducted on the data of HSBC bank, which comes from the social networking website (Facebook) as one of the big data sources, to investigate the extent of the ability of big data in the preparation of predictive financial reports accurate. In addition to that, A total of 121 valid questionnaires were tested statistically to investigate the relationship between big data dimensions on the Disclosure quality of financial reports. The research concluded that big data gives businesses a competitive advantage in terms of operational efficiency, risk reduction, cost reduction, and technical and nontechnical innovation. Firms that are able to utilize innovative business intelligence technologies and blockchain database solutions can meet the challenges of big data applications to collect and analyze data in real-time.

Keywords Big data · Business intelligence · Blockchain · Financial reports disclosure quality

A. Elmashtawy
Universiti Malaysia Terengganu, Kuala Terengganu, Malaysia

A. Elmashtawy · M. Salaheldeen (✉)
Faculty of Commerce, Menoufia University, Shebin El-Kom, Egypt
e-mail: m_salah6000@yahoo.com

M. Salaheldeen
Universiti Sains Islam Malaysia, Nilai, Malaysia

M. Al-Emran et al. (eds.), *International Conference on Information Systems and Intelligent Applications*, Lecture Notes in Networks and Systems 550,
https://doi.org/10.1007/978-3-031-16865-9_40

1 Introduction

The world is now witnessing the fourth industrial revolution (Industry 4.0) led by big data and enabled by the advancement of technology, prevalence of the internet, and transition to automation. Organizations that embrace these enablers may achieve a competitive advantage [1, 2]. Big data technologies enable real-time monitoring of every change, resulting in the so-called "mirror world", which accurately reflects the physical world. As companies shift from products to knowledge, their competition increasingly revolves around innovation rather than product [3, 4]. This shift has created the knowledge-based economy, whose main commodity is information [4–7]. Conclusively, Big data offers significant benefits to corporate financial reporting, increasing its reliability and objectivity and transitioning from periodic to real-time reporting.

The use of big data helps organizations to achieve a competitive advantage, improve the quality of financial reports and disclosure, and support accounting practices. This study investigates the integration of business intelligence tools and blockchain with big data and how they may improve the disclosure quality of financial reports. The objective of this study is to examine how big data affected HSBC's quarterly profits. In addition, the impact of big data on the predictive ability of bank profit-ability is investigated. The study also investigated the impact of innovative business intelligence techniques and blockchain in the context of a big data environment on the quality of financial reports disclosure.

2 Literature Review

The emergence of big data is accompanied by many supportive techniques to extract knowledge from it, manage it efficiently, and verify its reliability. The outcomes of such techniques are reflected in the quality of financial reports and their transformation from periodic to real-time reports, increasing its reliability and objectivity and supporting the disclosure process. These are made possible by improving the organization's ability to evaluate elements that were previously not included in the budget due to the difficulty of evaluating them and information asymmetry [8]. Big data has also increased the ability of organizations to develop flexible and more accurate budget plans and support the audit process, the accuracy of which is ultimately reflected in the accuracy of financial reports [9].

Several studies [5, 10–12] have found a discrepancy between the use of big data and Generally Accepted Accounting Principles (GAAP). Furthermore, big data offers multi-faceted benefits to the accounting profession. For example, it replaces traditional skills and develops new skills for accountants, allowing them to play a more effective role within organizations. However, some studies [13, 14] highlight the challenges associated with big data, the most important of which are data quality, the availability of skills and infrastructure capable of analyzing and integrating it with corporate reports, and maintenance of data privacy and confidentiality.

Studies by [15–17] are amongst earlier studies, which were conducted to investigate the importance of big data in providing companies with competitive advantages, such as improving decision-making processes, forecasting, planning, risk management, increasing profit margins. According to [18], big data have an impact on changing accounting practices, changing the role of accountants and auditors, and supporting many fields of accounting, including financial accounting, management accounting, and auditing. Typically, big data will improve transparency and disclosure processes by transforming corporate reporting from periodic to real-time reports, as well as reducing the potential for information asymmetry [19].

Empirical evidence currently available has underlined that there is a gap between the application of big data and generally accepted accounting standards. Also, big data is a double-edged sword for the accounting profession, as it can replace traditional skills and has the ability to develop new skills for accountants, and thus they will have a more effective role within companies [20]. Meanwhile, the literatures also highlighted the challenges associated with the application of big data, the most important of which is ensuring data quality, as well as the availability of skills and infrastructure capable of analyzing and integrating it with corporate reports, as well as maintaining data privacy and confidentiality.

In Egypt, there is currently a lack of sufficient evidence about big data and its effect on a competitive advantage and disclosure quality of financial reports. Additionally, there is little evidence on the impact of big data technologies on the development of accounting practices. Furthermore, there is urgency for the International Financial Reporting Standards (IFRS) to establish a standard for new digital technologies, which is critical in today's business organizations.

3 Methodology

The importance of this research stems from the role that big data can play in improving the disclosure quality of financial reports and organizational operations, automating operational processes, and assisting managerial decisions to achieve a competitive advantage. The study's problem can be formulated in the following question "Is there a role for big data in increasing the quality of financial reporting?". We have conducted two separated studies. Our first study used archival data to examine how big data relates to quarterly profits and profit prediction. In the second study, we used the survey to investigate the effect of big data advantages, business intelligence, and blockchain on the disclosure quality of financial reports.

4 First Study

Previous studies have used big data as an element of budgetary planning to predict future earnings and as a measure to forecast the ability and flexibility of a firm to adapt to future changes. To provide practical evidence of how big data can be utilized to assure quality financial reporting, we adopt a case study approach based on an HSBC case study and analysis of its findings. That is, by investigating the existence of a relationship between the big data collected about the bank and the same bank's quarterly profits.

The bank's big data was also utilized in anticipating future profits as one of the aspects of the planning budget and as an indication that reflects the extent of the possibility of big data for forecasting to help in the bank's flexibility and ability to respond to future developments. The objectives of this study were to examine how big data affected HSBC's quarterly profits. In addition, the impact of big data on the predictive ability of bank profitability is being investigated. Figure 1 illustrates the research framework. The archival dataset aims to provide practical evidence on the extent to which big data can be used in a case-based experimental study. Here, the case study was HSBC.

The independent variable was the contents of the official Facebook page of HSBC in Egypt https://www.facebook.com/HSBCEgypt, while the dependent variable was the bank's quarterly profits.

Data Collection Sources: The data for the independent variable was collected via the Facebook Graph API, which is the primary method for entering and exiting data from the Facebook platform. It is a comprehensive HTTP-based API that apps utilize to programmatically query data, publish new events, and execute a wide range of operations. HSBC's quarterly financial statements were used to obtain dependent variable data. The Simple linear regression analysis tool was used. It is a method for estimating the value of one variable by knowing the value of the other variable via the regression equation.

The analysis was divided into two stages. First, the study analyzed the historical effect of big data on HSBC's quarterly profits from 2019 to 2021. Second, it forecasted ex-ante the impact of big data on the bank's future profits from 2021 to 2023. The

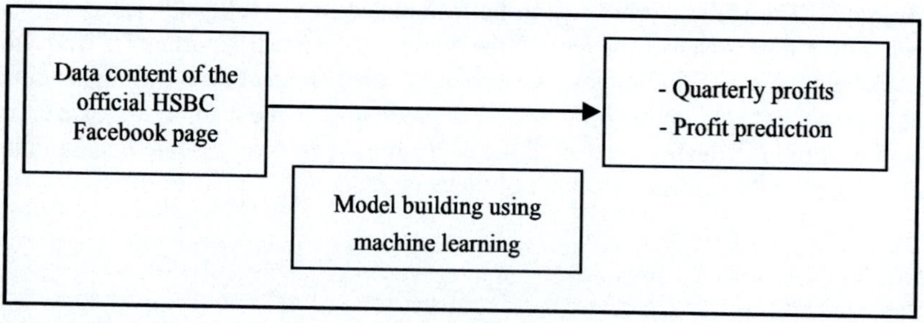

Fig. 1 First research framework

Table 1 Building of profits forecasting model

No	Year	Quarter	No. posts	Profit	Status	Number
1	2019	First	116	810,190.7	Number_likes	1,686,925
2	2019	Fourth	131	—567,299,800.0	Number_likes	1,651,146
3	2019	Second	184	459,003,658.0	Number_likes	2,696,512
4	2019	Third	148	1,355,186,691.0	Number_likes	2,175,620
5	2020	First	190	812,837.4	Number_likes	2,711,350
6	2020	Fourth	103	—1,530,420,497.0	Number_likes	921,476
7	2020	Second	177	582,050,652.0	Number_likes	2,180,025
8	2020	Third	97	532,712,924.0	Number_likes	934,000
9	2021	First	131	—387,637.7	Number_likes	798,439
10	2019	First	116	810,190.7	Number_comments	74,619
11	2019	Fourth	131	—567,299,800.0	Number_comments	61,723
12	2019	Second	184	459,003,658.0	Number_comments	70,589
13	2019	Third	148	1,355,186,691.0	Number_comments	60,082
14	2020	First	190	812,837.4	Number_comments	91,669

analysis began with the importing, cleaning, transformation, and visualization of data. The model was then constructed using machine learning.

The first step of the analysis was to import the data. Secondly, the data was cleaned. The profit dataset was created and renamed as such. The data were then summarized. Empty rows of data and duplicate data were deleted. Thirdly, the data was transformed by separating the year and month of each post. The dataset began with post id/year/month and converted to string every 3 months to a quarter. The data was grouped by year and quarter posts status. The Facebook data was combined with cash flow data.

Fourthly, the data was visualized by linking between posts and profits, plotting the relation between year and profit according to each quarter and between year and posts according to each quarter, and clarifying the relation between posts and profit in each year. Fifthly, post interactions were analyzed by examining the relation between comments and profits and between likes and profits. A profit forecasting model was constructed by compiling the customers' interactions. Table 1 shows the building of profits forecasting model based on interactions with posts analyses.

Our findings showed that profits increase as the number of posts increases, and profits increase in the first quarter of each year, decline by a small proportion in the second quarter, and decrease by a considerable percentage in the third and fourth quarters. The data also revealed that the first and third quarters of 2019 and 2020 saw the largest percentage of posts.

We concluded that the profits fall as the number of comments increases, implying that there is an inverse relationship between them. Which may represent client behavior and be an indication of their discontent with the bank's services and the bank's capacity to receive input to enhance its services. Profits rise as likes data rises, implying that there is a positive relationship between the two. Which may show client behavior in terms of expressing their happiness with the services given by the bank.

Finally, future profits were forecasted based on the number of posts using linear regression and machine learning (Cash flows = intercept + slope* no. Posts). Training and test data were first created. A model was then constructed and trained on the data. A final model was then constructed based on the results and was used to predict future profits using the test data to ensure its validity. Based on the findings of the previous investigation, big data had a statistically significant effect on HSBC's quarterly profits and its future profits.

5 Second Study

The second study was a survey to examine the relationship between big data dimensions (measured as big data advantages, supporting business intelligence tools for big data, and contribution of blockchain databases to the quality of big data) and the disclosure quality of financial reports. The measures of big data were adopted from [10, 21, 22] and the indicators of the disclosure quality of financial reports were adopted from [23, 24].

The objectives of this study were to examine the variances in the study sample's perceptions of big data, the quality of financial reporting disclosure, and the impact of big data on the quality of financial reporting disclosure. Our sample consists of the managers and data analysts in industrial companies listed in the EGX100 Index. The respondents of the survey were managers and data analysts in industrial companies from Egypt. A total of 121 questionnaires were electronically distributed to the respondents. Figure 2 below illustrates the second research framework.

The study used a Reliability analysis of Cronbach's Alpha Coefficient to assess the level of stability of the survey in order to assure the availability of reliability and confidence in the phrases included within, as well as their validity for the subsequent phases of analysis. Descriptive Statistics (Mean/Standard Deviation/Frequencies) used to describe the study sample. Pearson Correlation Coefficients used to measure the strength and direction of the relationship between the study variables. Multiple Regression Analysis used to measure the direct effects of the dimensions of the independent variable on the dimensions of the dependent variable separately. Table 2 shows the validity and reliability test for the study variables.

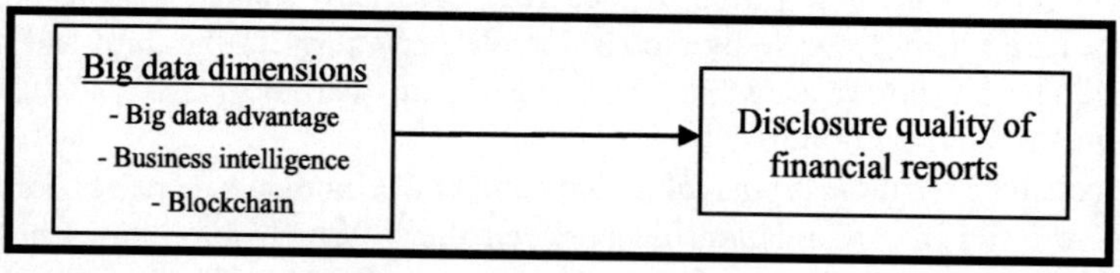

Fig. 2 Second research framework

The Table 2 shows that the Cronbach's alpha for the variables ranged between (0.739–0.822), while the Cronbach's alpha for all survey items was 0.920. These values are deemed acceptable in the sense that they reflect the availability of reliability and confidence in the study variables and confirm their validity for the subsequent phases of analysis. Cronbach's alpha and its square root, respectively, proved the survey list's reliability and validity.

Our findings showed that big data can provide a more complete picture of asset performance, more evidence to justify the values in which transactions are recorded, and a rich historical perspective for decision-making processes in measuring asset values and the basis for reaching fair value. Furthermore, big data aids the discovery of previously unseen elements in the budget. In addition, it enhances the accountant's decision-making confidence, accuracy, objectivity regarding the remaining elements of the budget, and ability to prepare more accurate forecasting reports.

We also noted that financial reports can be converted from periodic to real-time reports by integrating big data with Extensible Business Reporting Language (XBRL), enterprise resource planning (ERP) systems, data visualization, and cloud computing. More robust analytical models can be created using big data, which can improve the processes of control. Table 3 shows the effect of big data on the disclosure quality of financial reports.

Our results indicated that big data dimensions were significantly related to the quality of disclosure of financial reports. Blockchain had the strongest relationship with disclosure quality ($\beta = 0.416, p < 0.01$), followed by big data advantages ($\beta = 0.218, p < 0.05$) and business intelligence tools ($\beta = 0.184, p < 0.05$). The R^2 was 0.423, suggesting that the three dimensions of big data explained 42.3 percent of the variance in the disclosure quality of financial reports. The remainder of the variance is explained by other factors not included in the study. The results suggest that the application of big data and supporting technologies, such as business intelligence tools and blockchain, can improve the disclosure quality of financial reports.

Table 2 The value of the Cronbach's alpha coefficient and the validity

Variables	Reliability Coefficients/ Cronbach's alpha	Validity Coefficients
Big Data Advantages	0.739	0.860
Supporting business intelligence tools for big data	0.822	0.907
Contribution of blockchain databases to the quality of big data	0.783	0.885
The quality of financial reporting disclosure	0.804	0.897
All variables/Dimensions in the survey list	0.920	0.959

Table 3 Multiple linear regression results

Independent variable	Dependent variable	Unstandardized transactions		Standardized coefficients	t	p
		b	SE	β		
Big data advantages	Disclosure quality of financial reports	0.254	0.097	0.218	2.624*	0.010
Business intelligence		0.105	0.079	0.184	2.338*	0.045
Blockchain		0.356	0.082	0.416	4.320**	0.000
$F = 28.371$** *$p = 0.05$ **$p = 0.01$						
$R^2 = 0.423$ Adjusted $R^2 = 0.408$ $R = 0.651$						

6 Conclusion

The paper concluded that by utilizing innovative business intelligence technologies and blockchain database solutions, companies will be able to meet the challenges of big data applications to collect and analyze data in real-time They can also increase their ability to address privacy, security, and data management challenges. The use of blockchain databases as accounting ledgers is expected to eliminate many of the long-standing issues in auditing, accounting, and corporate governance, such as poor financial data quality, high auditing costs, in terms of both money and time, and inadequate corporate financial data security.

Big data can provide a more comprehensive picture of asset performance, provide additional evidence to justify the values in which transactions are recorded [18]. And provide a rich historical perspective for decision-making processes in measuring asset values and the basis for reaching fair value. Big data helps in the emergence of some elements that were not shown in the budget before. In addition to the confidence, accuracy, and objectivity of the accountant's decision regarding the remaining elements of the budget and the ability to prepare more accurate forecasting reports. Big data technologies lead to more robust analytical models, which helps to improve control processes.

This research showed that big data gives businesses a competitive advantage in terms of operational efficiency, risk reduction, cost reduction, technical and nontechnical innovation, increased sales volume, asset control, operational automation, and pre-crash preventive maintenance. In addition to assisting businesses to adapt to market changes. It improves supply chain management efficiency, enhances customer relationship management, and forecasts future changes. The research also concluded that the integration of business intelligence tools and blockchain databases with big data is critical to support accounting practices and develop the roles of accountants and auditors. We recommend the importance of adoption of big data, the development of corporate performance indicators, financial reporting standards, and the creation of new measurement tools commensurate with the application of big data.

References

1. Kagermann H (2015) Change through digitization—Value creation in the age of Industry 4.0. In: Albach H, Meffert H, Pinkwart A, Reichwald R (eds) Management of permanent change. Springer, Wiesbaden, pp 23–45. https://doi.org/10.1007/978-3-658-05014-6_2
2. Salaheldeen, M., Artificial Intelligence in Business Research: trends and future, in Emerging Issues and Challenges in Management Conference (2017) Faculty of Commerce. Menoufia University, Egypt
3. Salaheldeen M, Battour M, Nazri MA, Ahmad Bustamam US, Hashim AJCM (2022) The perception of success in the halal market: developing a halal entrepreneurship success scale, J Islamic Mark. ahead-of-print, no. ahead-of-print. https://doi.org/10.1108/JIMA-10-2021-0341
4. Salaheldeen M, Battour M, Nazri MA, Bustamam USA (2021) Prospects for achieving the sustainable development goals 2030 through a proposed halal entrepreneurship success index (HESI). SHS Web Conf. 124:08001. https://doi.org/10.1051/shsconf/2021124080
5. Ardito L et al (2019) Towards Industry 4.0: mapping digital technologies for supply chain management-marketing integration. Bus Process Manag J 25(2):323–346
6. Salaheldeen M (2015) Management control systems as a package: an application to science & technology parks: UPTEC case study. In: 8th conference on performance measurement and management control: nice, France
7. Salaheldeen M (2022) Opportunities for halal entrepreneurs in the Islamic digital economy: future and trends from a cultural entrepreneurship perspective. In: Ratten V (ed) Cultural entrepreneurship: new societal trends. Springer Nature Singapore, Singapore, pp 95–107
8. Tabesh P, Mousavidin E, Hasani S (2019) Implementing big data strategies: a managerial perspective. Bus Horiz 62(3):347–358
9. Sadasivam GS et al (2016) Corporate governance fraud detection from annual reports using big data analytics. Int J Big Data Intell 3(1):51–60
10. Vasarhelyi MA, Kogan A, Tuttle BM (2015) Big data in accounting: an overview. Account Horiz 29(2):381–396
11. Schneider GP et al (2015) Infer, predict, and assure: accounting opportunities in data analytics. Account Horiz 29(3):719–742
12. Bhimani A, Willcocks L (2014) Digitisation, 'Big Data' and the transformation of accounting information. Account Bus Res 44(4):469–490
13. Cai L, zhu Y (2015) The challenges of data quality and data quality assessment in the big data era. Data Sci J **14**:2
14. Sledgianowski D, Gomaa M, Tan C (2017) Toward integration of big data, technology and information systems competencies into the accounting curriculum. J Account Educ 38:81–93
15. Matthias O et al (2017) Making sense of Big Data–can it transform operations management? Int J Oper Prod Manag 37(1):37–55
16. Poleto T, de Carvalho VDH, Costa APCS (2015) The roles of big data in the decision-support process: an empirical investigation. In: Delibašić B, Hernández JE, Papathanasiou J, Dargam F, Zaraté P, Ribeiro R, Liu S, Linden I (eds) Decision Support Systems V – Big Data Analytics for Decision Making, vol 216. Lecture Notes in Business Information Processing. Springer, Cham, pp 10–21. https://doi.org/10.1007/978-3-319-18533-0_2
17. Noureldeen A, Salaheldeen M, Battour M (2022) Critical success factors for ERP implementation: a study on mobile telecommunication companies in Egypt. In: Al-Emran M et al (eds) Lecture Notes in Networks and Systems. Springer International Publishing, Cham, pp 691–701
18. Ianni M, Masciari E, Sperlí G (2021) A survey of big data dimensions vs social networks analysis. J Intell Info Syst 57(1):73–100
19. Basuony MA et al (2020) Big data analytics of corporate internet disclosures. Account Res J
20. Siano F, Wysocki P (2021) Transfer learning and textual analysis of accounting disclosures: applying big data methods to small (ER) datasets. Account Horiz 35(3):217–244
21. Deepa N et al (2022) A survey on blockchain for big data: approaches, opportunities, and future directions. Future Gener Comput Syst 131:209–226

22. Rabah K (2018) Convergence of AI, IoT, big data and blockchain: a review. Lake Institut J 1(1):1–18
23. Jha A (2019) Financial reports and social capital. J Bus Ethics 155(2):567–596
24. Chen S, Miao B, Shevlin T (2015) A new measure of disclosure quality: the level of disaggregation of accounting data in annual reports. J Account Res 53(5):1017–1054

Factors Affecting the BIM Adoption in the Yemeni Construction Industry

A. H. Al-Sarafi, A. H. Alias, H. Z. M. Shafri, and F. M. Jakarni

Abstract Performance of building construction was pointed over the past years to low standards of information management, which depend on project complexity. BIM application in building projects is generally seen as a sophisticated environment in the Yemeni construction industry, leading to cost and time overrun, labor productivity and poor design. This study aims to appraise the factors affecting BIM adoption in the Yemeni construction industry. In Yemen, a questionnaire survey of construction professionals and industry experts is being undertaken. The responses retrieved from the questionnaire were analyzed using descriptive statistics factor analysis and ranked accordingly. Findings show that the visualization of construction sequences is the most significant technological factor affecting BIM adoption. Greater collaboration with consultants and other project team members was ranked first as a process factor, whereas construction code is the most significant policy factor that hinders BIM adoption. Lack of top management support with a mean value of 3.61 is the most critical people factor and BIM readiness by project consultants as the significant environmental factors having a mean value of 3.9.

Keywords BIM adoption · Building information modelling · Construction · Factors

A. H. Al-Sarafi (✉) · A. H. Alias (✉) · H. Z. M. Shafri · F. M. Jakarni
Department of Civil Engineering, Faculty of Engineering, Universiti Putra Malaysia, UPM, 43400 Serdang, Selangor, Malaysia
e-mail: alsarafiali@gmail.com; gs51839@student.upm.edu.my

A. H. Alias
e-mail: aidihizami@upm.edu.my

H. Z. M. Shafri
e-mail: helmi@upm.edu.my

F. M. Jakarni
e-mail: fauzan.mj@upm.edu.my

M. Al-Emran et al. (eds.), *International Conference on Information Systems and Intelligent Applications*, Lecture Notes in Networks and Systems 550,
https://doi.org/10.1007/978-3-031-16865-9_41

1 Introduction

Building projects have a complex cycle and are fragmented in nature. Building projects go through phases from the beginning to the end, requiring a massive number of documents and data to complete the project scope. It also requires the collaboration and integration of multiple professionals from diverse organizations (Architecture, Engineering, and Construction (AEC)). Similarly, Babatunde et al. [1] identified and empirically investigated the characteristics that promote effective BIM adoption in the construction industry. Based on their height and number of stories, construction projects are divided into five categories: skyscrapers, high-rises, mid-rises, low-rises, houses, and others [2]. Low standards of information management, which are dependent on project complexity, have been blamed for poor building construction performance during the last 100 years. In the 1970s, Eastman attributed this to the inadequacy of construction drawings and the inability to visualize the project [3–5].

According to Yemen's Ministry of Public Works and Roads, infrastructure destruction cost the country more than $14 billion in the first seven months of the civil conflict. More challenge was the shortage of service availability, Ineffective regulation and legislation, low usage of local construction technologies, ineffective financial structure and improper use of local building resources [6]. Early in 2011, the Yemeni building industry saw a significant increase. The building industry suffered a significant drop once the war began in 2015, and several projects were halted. People were accustomed to the instability and conflict environment in early 2016 and began adopting new lifestyles and building their own homes despite a lack of resources and machinery. Other ongoing projects were financed by various voluntary organizations and the Social Fund for Development [7, 8].

Because of frequent natural environmental disasters, such as earthquakes and tsunamis, and other activities induced by human-made factors, such as conflicts and wars, post-disaster reconstruction (PDR) has gained increased attention worldwide. As the frequency of severe disasters rises, stakeholders are increasingly launching reconstruction efforts to mitigate the effects of those disasters on the built environment [9]. Yemen's construction industry is confronted with numerous obstacles that are causing it to deteriorate. Almost all projects in Yemen experience difficulties in achieving their objectives. It allows academics to dig deeper into the possibility of improving collaboration between diverse project participants in BIM-assisted construction projects.

2 Overview

A critical review of the literature on Building Information Modelling (BIM) for developing countries, including Yemen, was the first phase of background to the research subject area and various related factors that affect the BIM adoption for the construction industry in developing countries [10]. Thus, an overview of the construction industry in Yemen then reviews the factors affecting BIM adoption and its implementation, such as people, process, and technology. Gamil [11] Many difficulties and problems in Yemen's construction industry have been identified, indicating that the government's policies and strategies need to be improved. The construction industry in Yemen lacks a proper accrediting scheme for construction stakeholders. Inadequate cost and time preparation, a lack of efficient communication and collaboration platforms, and a lack of modern technology are all factors.

Moreover, Alaghbari and Wael [12] determine the most significant factors causing construction project delays in Sana'a, Yemen. Among the five classes, financial considerations came in first. Delays in implementing construction projects, especially public projects, have become common in Yemen. Furthermore, Kassem [13] states that the country's economy is inextricably linked to the oil and gas industry. Any construction project currently underway has its collection of risk factors. Again, Alaghbari [14] Construction project cost and time overruns are caused by various factors, including poor labour productivity. It is proposed that government policy emphasize technical education and apprenticeship programs. According to the study, the construction industry should comprehensively develop administrative and human capital. Since the 1970s, a variety of systems and models have been developed to improve the visualization of building components, and Building Information Modelling (BIM) in the early 2000s as the integration of Information Technology (I.T.) Information and Communication Technology (ICT) in the AEC industry [15].

Building information modelling (BIM) can revolutionize the Architecture, engineering, and construction (AEC) industry worldwide. Most importantly, BIM continues to be the most attractive innovation in the AEC industry Chan [12]; furthermore, BIM is more than simply a drawing and documentation tool; it is also more than just software, as it offers a more collaborative working technique. Another notion closer to BIMMI is that a firm's BIM maturity can help develop a BIM adoption model based on the Capability Maturity Model (CMM) [13]. Similarly, Rosli et al. [14], also from Malaysia, looked at the relationship between several factors influencing BIM adoption in the region. This is especially useful for the government, with limited financial resources to subsidize BIM adoption. The government subsidy can speed up the joining process and increase BIM adoption performance [16]. Several research studies have categorized the factors impacting BIM implementation in the construction industry into multiple categories and classes [17].

3 Methodology

This study uses a quantitative approach to collect data on the factors that affect BIM adoption in the Yemeni construction industry. In this research, the BIM adoption factors of the construction industry are grouped into their respective classes through exploratory factor analysis. For factor analysis, SPSS is used to carry out factor analysis. In SPSS, this method requires several steps, and these steps are derived from [18]. The reliability test is one of the essential tests in this study.

3.1 Determining the Characteristics that Influence the BIM Adoption

This research aims to determine what factors are driving Yemen's construction industry to adopt BIM. The first research approach involves finding all of the components linked to the relevant ideas using various resources such as scientific journal articles, conference proceedings, books, and documentaries [19]. To determine all associated factors, a complete literature review is conducted. For the primary search, a total of 248 publications were examined. The factors are then inspected and approved by a selection of qualified experts in Yemen's construction industry, including professionals from both the public and private sectors who use or don't use BIM. The next step will begin once all of the listed variables have been approved by experts.

The papers are first evaluated, and any unrelated documents are removed. The remaining studies are then submitted to a thorough examination to extract all BIM-related parameters. This approach yielded 62 papers that passed all of the previous phases. Several potential factors influencing BIM adoption have been identified in the 62 studies released between (2014 to 2021). Drivers, benefits, challenges, limits, essential success factors, initiatives, and other BIM-related issues are among these factors. The result is 125 factors as the primary list. The notion that each component describes is then reviewed qualitatively, and factors revolving around the same ideas are combined. This method resulted in a list of 89 factors being reduced. The next step is to organize these variables into categories. Grouping facilitates conceptual groups of components that share common qualities [20].

A quantitative survey approach is used with specialists from several engineering disciplines in the building business in Yemen. A review of BIM demonstrates its potential for adoption in the construction industry constant, shown in Fig. 1. as a conceptual model. Finally, the conceptual research model utilizes SEM in the data analysis phase's fourth phase. As a result, the research hypothesis is evaluated using the conceptual model.

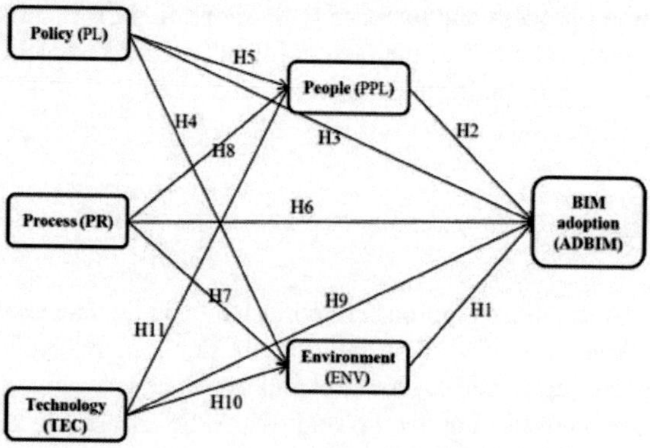

Fig. 1 Conceptual research model

According to the literature review, there is no apparent pattern for grouping factors that can be evaluated. Five groups emerge from the literature review's classification study, each labelled differently by various researchers. The final list of characteristics is initially divided into five categories. "Environment," "process," "technology," "people," and "Environment".

3.2 Data Gathering

To collect data, a questionnaire survey was employed, which is a quantitative tool for proving existing concepts and bolstering study findings with hypotheses and conclusions from previous investigations. The respondents' attitudes and understanding of BIM adoption determinants in the Yemeni construction industry were assessed using a Likert scale of 1 to 5 (1: strongly disagree; 2: disagree; 3: uncertain; 4: agree; and 5: strongly agree). The survey's respondents include architects, civil engineers, quantity surveyors, M&E, and others from the commercial and governmental sectors, as shown in Table 1.

The questionnaires were made available to the public over the internet. The Ministry of Public Works and Highways and Yemeni Engineers Syndicates (YES) was contacted on more than one visit to distribute the questionnaire online to all registered engineers since this is the best way of connecting during the Covid-19 pandemic. The survey was available for four months. The study's intended survey sample size was 475 people; however, 235 completed responses have been received. The response rate is expected to be around 49%.

Table 1 Demographic characteristics

		Frequency	%			Frequency	%
Qualification	High School	1	0.4	**Profession**	Architecture	33	14
	Diploma	5	2.1		Civil/Structural Engineering	147	62.6
	Bachelor	137	58.3		Electrical Engineering	13	5.5
	Masters	58	24.7		Mechanical Engineering	2	0.9
	PhD	34	14.5		Project Management	14	6
Specialization	Designer or Consultant	160	68.1		Construction Management	11	4.7
	Contractor/Construction	64	27.2		Quantity Surveying	3	1.3
	Client	11	4.7		Technical in panning team	5	2.1
Organization	Public	35	14.9		Others	7	3
	Private	94	40				
	Public and Private (Mix)	106	45.1				

3.3 *The Targeted Group for the Questionnaire*

Engineers from various disciplines connected to the construction sector were sought out for this study's targeted range of industry professionals. They had to be employed in governmental or engineering agencies to qualify. Furthermore, the survey does cover professionals who use BIM and those who do not utilize it yet work in management departments.

3.4 *Pre-test*

Pre-testing the questionnaire was necessary to guarantee that responders understood the questions. Pretesting was done by discussing with colleagues about the questionnaire. It also included a questionnaire review by experts in the same field to ensure that the questions were relevant, and that the questionnaire was correct in terms of simplicity and eligibility. Before the field survey, the pre-test helps to determine the data's reliability and validity [4].

4 Results and Discussion

4.1 *Reliability Test*

The Cronbach alpha value for variables impacting BIM adoption in the Yemeni construction sector is 0.971, indicating that the general dependability of the data acquired in this research is satisfactory.

4.2 *Descriptive Statistics*

To get at the paper's results, data from the survey were analyzed using descriptive statistics and factors analysis.

Technology Factors

The preliminary result was further explored by examining the basis of the technology factors and decisions of the respondent. The respondents were asked to rate the level of understanding with the technology factors that can assist BIM adoption in the Yemeni construction industry. The results are shown in Table 2.

The results are presented in descending order of mean values. The responses indicate that the visualization of construction sequences (TEC03) is the most significant technology factor, with a mean value of 3.94. The usefulness of digital transfer of data

Table 2 Technology factors mean ranking

I.D.	Mean	Std. deviation	Skewness		Kurtosis		Rank
			Statistic	Std. error	Statistic	Std. error	
TEC03	3.94	1.050	−1.256	0.159	1.369	0.316	1
TEC05	3.91	1.060	−1.100	0.159	0.830	0.316	2
TEC02	3.86	1.053	−1.229	0.159	1.254	0.316	3
TEC04	3.78	1.038	−1.152	0.159	1.016	0.316	4
TEC01	3.67	1.105	−0.970	0.159	0.389	0.316	5

(TEC05) BIM knowledge within the project (TEC02) with values of 3.91 and 3.86, respectively, are the second and third significant technology factors in the Yemeni construction industry. These results were supported by research conducted by [21]. Trialability Possibility of risk reduction with the try-out before adopting BIM in practice; and trying out various BIM features in my works to verify its effects (TEC04)] and Full automation in the construction industry (TEC01) are the least considered technology factors as suggested by the respondent with mean values of 3.78 and 3.67 respectively. This is also supported by the study [22]. From the kurtosis and skewness obtained, the result indicates the data to be normally distributed as its ranges of ±2 comply with the requirement as stated by [23].

Process Factors

The Preliminary result was further explored by examining the process factors that can assist in adopting BIM in the Yemen construction industry. Table 3 shows that greater collaboration with consultants and other project team members (PR13) was ranked first with a mean and standard deviation values of 4.13 and 1.015. This is followed by the Production of drawings and schedules (PR07) and Developing data exchange standards (PR12), ranked second and third with a mean value of 4.09 and 4.00, respectively. This finding is supported by the study [24].

Standard and rules (PR10), The leadership of senior management (PR03), and the Contractual sharing norm (PR04) with mean values of 3.81, 3.79, and 3.72 are the least considered process factors to assist the adoption of BIM in the Yemen construction industry. Skewness and Kurtosis values obtained also show that the data are normally distributed.

Policy Factors

Table 4 shows the result of policy factors to assist Yemen construction professionals in adopting BIM in the construction industry. Construction codes (PL08) with a mean value of 3.99 are ranked first and considered the most significant policy factors. Guidance on the use of BIM (PL05) and Organizational readiness (PL03) are ranked second and third with mean values of 3.97 and 3.85, respectively. This is following the findings of [25]. The data distributions show that it follows a normal distribution curve with all values ranging between ±2.

Regulation and policy (PL02), Financial resources of the organization (PL01) and Strong law legal institutions (PL04). It's the least significant policy factor aimed at

Table 3 Process factors mean ranking

I.D.	Mean	Std. deviation	Skewness		Kurtosis		Rank
			Statistic	Std. error	Statistic	Std. error	
PR13	4.13	1.015	−1.456	0.159	2.016	0.316	1
PR07	4.09	1.013	−1.514	0.159	2.233	0.316	2
PR12	4.00	0.934	−1.303	0.159	2.222	0.316	3
PR01	4.00	1.036	−1.388	0.159	1.827	0.316	4
PR09	3.90	0.955	−1.218	0.159	1.794	0.316	5
PR06	3.89	1.002	−1.085	0.159	1.028	0.316	6
PR02	3.89	1.060	−1.209	0.159	1.215	0.316	7
PR08	3.88	1.045	−1.020	0.159	0.605	0.316	8
PR11	3.83	0.982	−0.980	0.159	1.052	0.316	9
PR05	3.82	0.977	−0.931	0.159	0.909	0.316	10
PR10	3.81	1.045	−1.089	0.159	0.946	0.316	11
PR03	3.79	1.023	−0.950	0.159	0.640	0.316	12
PR04	3.72	1.044	−0.835	0.159	0.264	0.316	13

Table 4 Policy factors mean ranking

I.D.	Mean	Std. deviation	Skewness		Kurtosis		Rank
			Statistic	Std. error	Statistic	Std. error	
PL08	3.99	1.128	−1.261	0.159	1.005	0.316	1
PL05	3.97	0.995	−1.129	0.159	1.145	0.316	2
PL03	3.85	1.033	−1.128	0.159	1.077	0.316	3
PL07	3.80	1.169	−0.949	0.159	0.080	0.316	4
PL06	3.80	1.065	−1.068	0.159	0.824	0.316	5
PL02	3.75	1.037	−0.905	0.159	0.607	0.316	6
PL01	3.71	1.078	−1.001	0.159	0.640	0.316	7
PL04	3.68	1.080	−0.793	0.159	0.090	0.316	8

assisting BIM adoption in the Yemen construction industry, with mean values of 3.75, 3.71 and 3.68, respectively.

People Factors

People related factors towards assisting the adoption of BIM in the Yemen construction industry, as shown in Table 5, indicates that Lack of top management support (PPL04) is the most significant factor with a mean value of 3.61 among the variables ranked by the respondent. It is followed by a Lack of BIM expertise (PPL03) and Weak supervision and control (PPL06) with mean values of 3.57 and 3.53, respectively and ranked second and third. This finding follows the study conducted by [26]. Lack of demand by clients (PPL07), Errors by a design team in construction projects

Table 5 People factors mean ranking

I.D.	Mean	Std. deviation	Skewness		Kurtosis		Rank
			Statistic	Std. error	Statistic	Std. error	
PPL04	3.61	1.268	−0.729	0.159	−0.509	0.316	1
PPL03	3.57	1.229	−0.595	0.159	−0.734	0.316	2
PPL06	3.53	1.188	−0.585	0.159	−0.556	0.316	3
PPL01	3.53	1.156	−0.612	0.159	−0.561	0.316	4
PPL02	3.50	1.167	−0.597	0.159	−0.509	0.316	5
PPL07	3.48	1.171	−0.564	0.159	−0.514	0.316	6
PPL05	3.46	1.144	−0.511	0.159	−0.584	0.316	7
PPL04	3.61	1.268	−0.729	0.159	−0.509	0.316	8

(PPL05), and Lack of top management support (PPL04) are the least considered people-related factors in assisting the adoption of BIM in the Yemen construction industry. Data pattern shows that they are all normally distributed with kurtosis and skewness ranges between ±2.

Environments Factors
Environmental factors' contribution to assisting construction professionals in Yemen in adopting BIM indicates that BIM readiness by project consultants (ENV04) is ranked first as the most important factor, with a mean value of 3.9 and a standard deviation of 1.043. Method of communication between the team (ENV06) and Market demand, size and competition increase (ENV07) with mean values of 3.81 and 3.80 are ranked second and third environment factors, respectively. The three less significant environmental factors as shown by the respondent are Poor economic condition (ENV05), Poor Internet connectivity (ENV02) and Security of information on project data (ENV01), having mean values of 3.67, 3.52 and 3.49, respectively, as shown in Table 6. Also, kurtosis and skewness data obtained indicate that the data are normally distributed; hence the parametric analysis will be suitable for the data.

4.3 Exploratory Factor Analysis (EFA)

Factor analysis is a method for condensing many variables into a smaller number of factors. This method takes the most common variance from all variables and converts it to a single score.

Exploratory Factor Analysis (EFA) for factors influencing BIM adoption in the Yemeni construction industry
The KMO and spherical tests were performed before completing EFA to aid date factorability. The Kaiser–Meyer–Olkin (KMO) and Barlett's test of sphericity were 0.95 and significant (sign = 0.001), respectively. Principal component analysis was

Table 6 Environment factors mean ranking

I.D.	Mean	Std. deviation	Skewness		Kurtosis		Rank
			Statistic	Std. error	Statistic	Std. error	
ENV04	3.90	1.043	−1.191	0.159	1.265	0.316	1
ENV06	3.81	0.933	−1.018	0.159	1.269	0.316	2
ENV07	3.80	1.046	−0.982	0.159	0.605	0.316	3
ENV09	3.76	1.010	−1.039	0.159	0.980	0.316	4
ENV03	3.75	1.094	−1.027	0.159	0.614	0.316	5
ENV08	3.71	1.052	−0.790	0.159	0.150	0.316	6
ENV05	3.67	1.250	−0.814	0.159	−0.317	0.316	7
ENV02	3.52	1.214	−0.771	0.159	−0.341	0.316	8
ENV01	3.49	1.080	−0.568	0.159	−0.290	0.316	9

used to extract the 42 BIM influence adoption factors. According to the eigenvalue criterion larger than 1, Nineteen (19) factor was found after the maximum variance of Promax rotation. It has an eigenvalue of 47.242, 8.470, 4.093, 2.884 and 2.692% with pattern matrix analysis, as shown in Table 7. Component 1 comprised 13 items (PR11, PR12, PR07, PR13, PR09, PR05, PR06, PR08, PR10, PR02, PR04, PR01, PR03). Component 2 comprised eight items (PPL03, PPL04, PPL02, PPL06, PPL01, PPL05, PPL07, ENV02); Component 3 included eight items (ENV01, ENV08, ENV04, ENV07, ENV03, ENV06, ENV09, ENV05); Component 4 has eight items (PL03, PL04, PL07, PL02, PL08, PL01, PL05, PL06) and Component 5 comprises of 5 items (TEC01, TEC02, TEC03, TEC05, TEC04).

The result of 25 iterations of exploratory factor analysis with the principal component analysis extraction method and the Promax and Kaiser Normalization rotation methods. Variables discovered include:

- *Component 1:* This group of influencing factors accounted for 47.282% of the total variance, indicating its degree of importance. Besides, it was observed that the industry players acknowledge the process factors. Stakeholders need to understand the process factors that can influence BIM adoption, such as Companies' collaboration experience with project partners.
- *Component 2:* This BIM adoption influencing factors accounted for 8.47% of the total variance, the second crucial factor. Lack of BIM expertise is the most significant people related factor among all factors covered in the second component.
- *Component 3:* This influencing factor accounted for 4.093% of the total variance, making it the third most important factor. This factor is more about environmental influencing factors, with Security of information on project data being the most significant factor within the observed variables.
- *Component 4:* This accounted for 2.884% of the total variance, making it the fourth most important factor for adopting BIM in the Yemeni construction industry.

- *Component 5:* This is the minor significant factor with 2.692% of the total variance. It is technology-based mainly, with full automation in the construction industry being the most significant within the observed variables.

The component correlation matrix is shown in Table 8. The results revealed that the correlation between people and the industrial process is 0.497, and process with both environments, policy and technology 0.665, 0.717 and 0.659, respectively. Also, the correlation between environment and people is 0.510. The correlation between policy and people technology is 0.470 and 0.350, respectively. There is also a strong correlation between technology and policy, with a value of 0.618.

Table 7 Exploratory factor analysis for the assessment of the factors that influence BIM adoption in the Yemeni construction industry

Factors that influence BIM adoption	CODE	Component					Rank
		1	2	3	4	5	
Collaboration experience of companies with project partners	PR11	0.954					1
Developing data exchange standards.	PR12	0.918					2
Production of drawings and schedules	PR07	0.853					3
Greater collaboration with consultants and other project team members.	PR13	0.852					4
Collaboration (project) management tools.	PR09	0.785					5
Shared norms and collective expectations are diffused through information exchange activities.	PR05	0.776					6
Shared liability between project participants	PR06	0.755					7
Desire to have the design process go faster.	PR08	0.755					8
Standard and rules	PR10	0.744					9
Providing information on how to use BIM	PR02	0.732					10
Contractual sharing norm	PR04	0.657					11
Information availability and sharing	PR01	0.573					12
The leadership of senior management	PR03	0.485					13
Lack of BIM expertise	PPL03		0.888				1
Lack of top management support	PPL04		0.872				2
Lack of cooperative concept	PPL02		0.853				3
Weak supervision and control	PPL06		0.853				4
Lack of skills and knowledge of one of the partners	PPL01		0.812				5
Errors by a design team in construction projects	PPL05		0.811				6
Lack of demand by clients	PPL07		0.740				7
Poor Internet connectivity	ENV02		0.505				8

(continued)

Table 7 (continued)

Security of information on project data	ENV01	0.803			1
Risk management	ENV08	0.726			2
BIM readiness by project consultants.	ENV04	0.723			3
Market demand, size and competition increase	ENV07	0.720			4
Allows coordination and collaboration between disciplines	ENV03	0.717			5
Method of communication between the team	ENV06	0.640			6
Facility Management and buildings operation	ENV09	0.619			7
Poor economic condition	ENV05	0.498			8
Organizational readiness	PL03		0.800		1
Strong law legal institutions	PL04		0.783		2
Government incentives	PL07		0.699		3
Regulation and policy	PL02		0.622		4
Construction codes	PL08		0.609		5
Financial resources of the organization	PL01		0.451		6
Guidance on the use of BIM	PL05		0.404		7
The increased demand for design and build	PL06		0.394		8
Full automation in the construction industry	TEC01			0.890	1
BIM knowledge within the projects	TEC02			0.770	2
Visualization of construction sequences	TEC03			0.700	3
The usefulness of digital transfer of data	TEC05			0.609	4
Trialability Possibility of risk reduction with the try-out before adopting BIM in practice; and try out various BIM features in my works to verify its effects]	TEC04			0.498	5

Table 8 BIM influencing factors component correlation matrix

Component	Process	People	Environment	Policy	Technology
Process	1.000				
People	0.497	1.000			
Environment	0.665	0.510	1.000		
Policy	0.717	0.470	0.571	1.000	
Technology	0.659	0.350	0.545	0.618	1.000

The correlation matrix shows a good positive correlation between components that explain the related component's influence on each other.

5 Conclusion

This study explored the factors affecting Building Information Modelling (BIM) adoption in the Yemeni construction industry by reviewing literature and conducting a survey. The findings show good reliability of the data obtained. Factors loading was well tabulated as obtained from the factor analysis, indicating the variables' significance. However, this study demonstrates that BIM is a helpful decision tool for increasing construction productivity. Factors such as visualization of construction sequences, greater collaboration with consultants and other project team members, Construction codes, Lack of top management support, and BIM readiness by project consultants are the most significant factors affecting BIM adoption.

The Following Recommendations for This Study: i. During the examination, specific findings indicated the need for more research beyond the scope of the study's aims. However, due to the nature of this study, an in-depth analysis was not possible for several of the research issues. In the same way, more research is needed to expand and strengthen the study's findings. ii. The purpose of this thesis was to develop a framework for BIM adoption. More research on the parameters and their impact on various forms of infrastructure may be conducted. The scope of the study might be broadened to include the operational and destruction stages of the structures and inquiries into countries other than Yemen, with the potential for some fascinating international benchmarks. iii. Similar to other developed countries, the government should design or adopt construction policies to promote the use of BIM on every construction project. These policies would stimulate the implementation of BIM in Yemen.

References

1. Babatunde SO, Ekundayo D, Adekunle AO, Bello W (2020) Comparative analysis of drivers to BIM adoption among AEC firms in developing countries. J Eng Des Technol 23. https://doi.org/10.1108/JEDT-08-2019-0217
2. Imperiale R (2006) Getting started in real estate investment trusts. Wiley
3. Crotty R (2013) The impact of building information modelling: transforming construction. Routledge
4. Bahamid RA, Doh SI, Al-Sharafi MA, Rahimi AR (2020) Risk factors influencing the construction projects in Yemen from expert's perspective. In: IOP conference series: materials science and engineering, vol 712, no 1. https://doi.org/10.1088/1757-899X/712/1/012007
5. Elghdban MG, Azmy NB, Zulkiple AB, Al-Sharafi MA (2021) A systematic review of the technological factors affecting the adoption of advanced IT with specific emphasis on building information modeling, vol 295. https://doi.org/10.1007/978-3-030-47411-9_2

6. Sultan B, Alaghbari W (2014) Incompetent construction technologies and resources in the construction industry of Yemen. Labour 19:27
7. Gamil Y, Rahman IA, Nagapan S, Nasaruddin NAN (2020) Exploring the failure factors of Yemen construction industry using PLS-SEM approach. Asian J Civil Eng 21(6):967–975
8. Bahamid RA, Doh SI, Al-Sharaf MA (2019) Risk factors affecting the construction projects in the developing countries. IOP Conf Ser Earth Environ Sci 244:012040. https://doi.org/10.1088/1755-1315/244/1/012040
9. Baarimah AO et al (2022) A bibliometric analysis and review of building information modelling for post-disaster reconstruction. Sustainability (Switzerland) 14(1). https://doi.org/10.3390/su14010393
10. Baarimah AO et al (2021) A bibliometric analysis and review of building information modelling for post-disaster reconstruction. Sustainability 14(1):393. https://doi.org/10.3390/su14010393
11. Gamil Y et al (2017) Qualitative approach on investigating failure factors of Yemeni mega construction projects. In: MATEC web of conferences, vol 103. https://doi.org/10.1051/matecconf/201710303002
12. Alaghbari W, Sultan B (2018) Delay factors impacting construction projects in Sana' a -Yemen. PM World J VII:1–28
13. Kassem MA, Khoiry MA, Hamzah N (2019) Risk factors in oil and gas construction projects in developing countries: a case study. Int J Energy Sect Manage 13(4):846–861. https://doi.org/10.1108/IJESM-11-2018-0002
14. Alaghbari W, Al-Sakkaf AAA, Sultan B (2019) Factors affecting construction labour productivity in Yemen. Int J Constr Manage 19(1):79–91. https://doi.org/10.1080/15623599.2017.1382091
15. Latiffi AA, Mohd S, Kasim N, Fathi MS (2013) Building information modeling (BIM) application in Malaysian construction industry. Int J Constr Eng Manage 2(4A):1–6
16. Hosseini MR et al (2016) BIM adoption within Australian Small and Medium-sized Enterprises (SMEs): an innovation diffusion model. Constr Econ Build 16(3):71–86
17. Venkatesh V, Bala H (2008) Technology acceptance model 3 and a research agenda on interventions. Decis Sci 39(2):273–315
18. Williams B, Onsman A, Brown T (2010) Exploratory factor analysis: a five-step guide for novices. Australas J Paramed 8(3)
19. Bahamid RARA et al (2020) Risk factors influencing the construction projects in Yemen from expert's perspective. IOP Conf Ser Mater Sci Eng 712(1):12007. https://doi.org/10.1088/1757-899X/712/1/012007
20. Succar B (2009) Building information modelling framework: a research and delivery foundation for industry stakeholders. Autom Constr 18(3):357–375. https://doi.org/10.1016/j.autcon.2008.10.003
21. Li N, Becerik-Gerber B, Krishnamachari B, Soibelman L (2014) A BIM centered indoor localization algorithm to support building fire emergency response operations. Autom Constr 42:78–89. https://doi.org/10.1016/j.autcon.2014.02.019
22. Kumar VSS, Prasanthi I, Leena A (2008) Robotics and automation in construction industry. In: AEI 2008: building integration solutions, pp 1–9
23. Hair JF, Black WC, Babin BJ, Anderson RE, Tatham R (2006) Multivariate data analysis. Pearson Prentice Hall, Uppersaddle River
24. Ko C-H (2011) Production control in precast fabrication: considering demand variability in production schedules. Can J Civ Eng 38(2):191–199
25. Chan CT (2014) Barriers of implementing BIM in construction industry from the designers' perspective: a Hong Kong experience. J Syst Manage Sci 4(2):24–40
26. Ding Z, Zuo J, Wu J, Wang JY (2015) Key factors for the BIM adoption by architects: a China study. Eng Constr Archit Manage

Predicting the Effect of Environment, Social and Governance Practices on Green Innovation: An Artificial Neural Network Approach

Bilal Mukhtar, Muhammad Kashif Shad, and Lai Fong Woon

Abstract Few studies have been conducted to investigate whether the Environment, Social and Governance (ESG) practices could influence green innovation in small and medium enterprises (SMEs). Therefore, the purpose of this study is to predict the effect of Environment, Social, and Governance (ESG) practices on green innovation in SMEs. In this study, green innovation is segmented into two dimensions which are sustainable product innovation and sustainable process innovation. The data was collected through a questionnaire from medium-level IT firms and was analyzed using the Artificial Neural Network (ANN) approach. The findings indicated the different impactful factors of ESG practices to enhance green innovation. The results indicate that social and political contribution is the most impactful factor to enhance sustainable product innovation followed by pollution & waste and emission reduction. In addition, the findings of this study shows that pollution & waste is the most impactful factor to enhance sustainable process innovation followed by anti-competitive behavior and emission reduction. This study will provide insights on ESG practices as an important consideration to enhance green innovation among business, operations especially in SMEs. The findings of this paper are useful for regulators, legislators, shareholders, creditors, and practitioners in pursuing ESG practices that will not only improve financial performance but will also enhance green innovation.

Keywords ESG practices · Green innovation · SMEs · Sustainable Product innovation · Sustainable Process innovation

1 Introduction

Green innovation has become a popular concern in recent years as environmental issues such as resource depletion, energy consumption, and pollution have become

B. Mukhtar (✉) · M. K. Shad · L. F. Woon
Department of Management and Humanities, University Teknologi PETRONAS, 32610 Seri Iskandar, Perak, Malaysia
e-mail: bilal_20000996@utp.edu.my

M. Al-Emran et al. (eds.), *International Conference on Information Systems and Intelligent Applications*, Lecture Notes in Networks and Systems 550,
https://doi.org/10.1007/978-3-031-16865-9_42

the major global concerns [1]. In this regard, from the micro firm level to the macro national level, green innovation has been evolving vertically and horizontally in every dimension of human and organizational activities. Green innovation, from a technological standpoint, is a technological innovation activity that complies with eco-economic development standards to achieve resource conservation and environmental protection.

Over the last two decades, the need for green innovation has become more urgent as sustainable development has gained much attention in the scholarly world [2]. Green innovation can enhance sustainable development by improving environmental performance and reducing the negative effects of human activities. Furthermore, it has been argued that organizations that adopt green practices in their production processes have gained more sustainable development which further increased their overall productivity.

The competition in the market is severe and the pursuit of green innovation alone cannot meet up the requirements of multiple stakeholders [3]. For instance, the customers prefer to purchase environmentally friendly products and investors are looking for those organizations that is employing sustainable processes to design sustainable products [4, 5]. Hence, there is a need to enhance green innovation to fulfill the requirements of stakeholders.

Although prior research has shown the impact of several factors on green innovation performance. Like, Song et al. [6] argued that creativity climate has a positive impact on green innovation. Furthermore, Xu et al. [7] found the positive impact of independent research and development (R&D) on green innovation performance. However, few studies have focused on the influential drivers of green innovation. Therefore, it is important to acquire ESG practices for enhancing green innovation in organizations.

Furthermore, the relationship between distinct perspectives of ESG practices and green innovation has been thoroughly investigated in the prior literature. Suganthi [8] investigated the impact of corporate social responsibility (CSR) on the adoption of green innovation. Nevertheless, limited attention has been given to study the impact of ESG practices on green innovation [3]. In this respect, to fill this gap, the purpose of this study is to investigate the importance of ESG practices on green innovation. The study will not only improve the understanding of people on ESG practices and green innovation, but it will also show the factors of ESG practices that are most important and influential to enhance green innovation.

The concept of ESG practices is an extension of socially responsible investing (SRI) and a key indicator for sustainable development [9]. According to Morgan Stanley Capital International (MSCI), investment in ESG practices started in the 1960s as socially responsible investing (SRI) and is still gaining significance among institutional and individual investors.

Environmental, social, and governance are the three main pillars of ESG practices. The environmental practices are referred to handle the environmental issues, for example, pollution, waste, deforestation, carbon dioxide emissions, and climate change. The social practices of enterprises encompass all their interactions with diverse stakeholders such as employees, consumers, suppliers, the government, the local community, and so on. Finally, the governance practices where the investors have always been more interested than environmental and social practices referred to the corporation's managerial responsibilities, organizational transparency, and the quality of the organization's investment strategies.

In this regard, a growing number of enterprises have recognized the importance of ESG practices, and it also become the most focused area of academic attention. The influence of ESG practices on green innovation is beneficial in terms of meeting stakeholders' needs as well as ensuring sustainable development. As a result, businesses should not only pursue green innovation but also consider environmental, social, and governance (ESG) practices, which can send positive signals to all sectors of society.

Based on the above discussion, the purpose of this study is to evaluate the influence of ESG practices on green innovation. This study adopts the artificial neural network (ANN) method to look at how independent variables affect the dependent variables which may better reflect the most important factors of ESG practices in enhancing green innovation. The remaining article includes a critical review of relevant literature, followed by the methodology. The article ends with findings, conclusions, and practical implications.

2 Literature Review

The integration of Environment, Social, and Governance (ESG) practices provide a positive influence on the co-existence of green innovation [3]. Therefore, it is important to study the influence of ESG practices on green innovation [3]. From this perspective, some studies have been performed in academia and industries in recent years. No doubt, ESG practices can bring environmental, social, and governance benefits, but whether they can enhance green innovation in enterprises has always been strongly debated in academia.

The empirical studies on the relationship between ESG practices and green innovation are still emerging. Moreover, studying the relationship between corporate social responsibility (CSR) accounting environment, social and governance (ESG), Zhang et al. [3] indicated that the environmental and social perspective of CSR has a positive influence on green innovation. in addition, the social perspective of CSR has a substitution effect with green innovation which will gradually weaken as the firm value increases. Moreover, the governance perspective exhibits the adverse effects on green innovation with the increase of firm value.

Furthermore, Xu et al. [7] performed a study by collecting the data of Chinese listed companies between 2015–2018. They identified the positive impact of ESG practices on green innovation performance and further examined the positive moderating effect of ESG practices on the relationship between research and development (R&D) and green innovation as well. They indicated, that companies pollute the environment heavily should invest more in environmental protection and implement green technologies to modernize their production pattern as ESG activities communicate a positive messages and transmit green signals to their employees to obtain green outcomes.

2.1 Environmental Practices and Green Innovation

Several studies have been performed in the context of environmental practices's impact on green innovation performance. For instance, by analyzing the data from 2008 to 2012 from the top 100 listed companies in China, Li et al. [10] showed that the adoption of an environment management system fosters corporate green innovation performance. Furthermore, environmental regulations positively moderated and strengthened the relationship between the environmental management system (EMS) and green innovation. Studying the Chinese firms between 2011 to 2015, Pan et al. [11] segmented green innovation into pollution prevention and sustainable environmental innovation. Then, they discovered that environmental CSR is both linearly and curvilinearly connected to pollution prevention and sustainable environmental innovation respectively. Moreover, high environmental CSR showed that firms are more devoted to investing in pollution prevention innovation, which entertains the available resources to invest in sustainable environmental innovation.

2.2 Social Practices and Green Innovation

In the respect of the growing importance of social practices on green innovation Shahzad et al. [12] demonstrated that the prominence of CSR practices not only increases the environmental sustainability development but also helps to boost green innovation in the organization. Moreover, Abbas [13] noticed that CSR can improve corporate green practices (CGP) and its findings were supporting the results of [12]. Further, it has been revealed that CSR also positively influences the relationship between total quality management (TQM) and corporate green performance (CGP).

In their study on the relationship between CSR and green innovation of 121 Spanish wineries, Guerrero-Villegas et al. [14] found that increasing the CSR practices in the organization boosts the green innovation performance which further increases the sustainable development of the organization in terms of firm performance.

2.3 Governance Practices and Green Innovation

The mixed results have been seen while reviewing the literature on the impact of governance practices on green innovation. Using the data of 202 Taiwanese service and manufacturing companies Weng et al. [4] came to different conclusions. They noticed pressure from competitors and the government greater emphasis on green products and services and following the existing regulations respectively to achieve greater green innovation performance. They further suggested that the pressure of the customers is not the concern of managers, but the pressure of qualified suppliers is necessary to drive green innovation practices. Similarly, Zhaofang [15] observed that customer pressure is an important influence for driving in green innovation practices of Chinese third-party logistics (3PL) provider companies. The customer pressure enhances the organization's adaptability of green innovation because the customer wants to purchase green products that have not a harmful impact on the environment. In the context of moderating effect, flexibility-orientation improves the influence of customer pressure on green innovation practices of 3PL provider companies. The relative studies have shown in the literature below in Table 1.

3 Conceptual Framework

The impact of each factor of environment, social and governance is represented through conceptual model in Fig. 1.

4 Research Methodology

This section discusses the data collection, sampling methods and research instruments used in this study.

4.1 Data Collection and Sampling Method

To achieve the objectives of this study, an online survey technique was used to collect data from Malaysian medium IT firms operating in four states. Because according to the department of statistics Malaysia (DOSM), the contribution of these states in GDP is more than 40%. The Federation of Malaysian Manufacturers (FMM) list was used to draw the study sample. Based on the FMM directory (2019), there are more than 3300 SME firms operating across all states of Malaysia. In this study, the focus industry was medium IT firms. According to SME Corp Malaysia (2008), a firm having a number of employees from 75 to not more than 200 is categorized as medium firm. As recommended by [16], the calculation of sample size through G*Power is more reliable as compared to other techniques. Thus, based on G*power the minimum sample size is 98 with a power of 0.80 and effect size of 0.15. This sample size may fulfill the minimum sample size requirement under the ten times rule (Hair et al. 2014). Due to its simplicity and less complexity, researchers recommend

Table 1 Literature on ESG and green innovation

Author & Year	Purpose	Results	Relation
Li et al. (2019)	Analyzed the impact of environment management system (EMS) on green innovation.	Findings showed that EMS enhance the adoption of green innovation practices.	Positive
Pan et al. (2020)	Investigated the influence of environmental CSR on green innovation.	Findings showed that environmental CSR is both linearly and curvilinearly connected to pollution prevention and sustainable environmental innovation respectively.	Mixed
Shahzad et al. (2020)	Explored the effect of CSR practices on environmental sustainability and green innovation.	Results indicated that CSR practices not only enhance the environmental sustainability development but also encourage the green innovation in the organization.	Positive
Abbas (2020)	Examined the impact of CSR on green innovation	Efficiently capitalizing on CSR, enhance the adoption of corporate green practices	Positive
Guerrero-Villegas et al. (2018)	Investigated the relationship between CSR on green innovation and its effect on firm performance.	Findings showed the positive influence of CSR on the green innovation which further increase the firm performance.	Positive
Weng et al. (2015)	Investigated the impacts of stakeholders on green innovation.	Pressure from government, competitors and employees conduct has positive effects whereas, supplier and customer have no impact on green innovation.	Mixed
Zhaofang et al. (2018)	Investigated the impact of customer pressure on green innovation among Chinese third-party logistics 3PL providers companies.	Customer pressure is an important driver of green innovation as it has positive influence to enhance the green innovation.	Positive

it (Li et al., 2020; Leong et al. 2019). Moreover, Alwosheel et al. (2018) argued that if the research aim is to evaluate the model performance in terms of correctly classified exogenous indicators or hit rate or based metrics, then smaller data set may be used.

Before actual data collection, pretesting and pilot testing procedures were followed. The results of pilot testing highlight that Cronbach's Alpha value of all study variables achieved the minimum threshold level ($\alpha \geq 0.70$). The stratified sampling technique was used to collect the data. The data was collected in two months, period. Before the distribution of the questionnaire, the content of all items

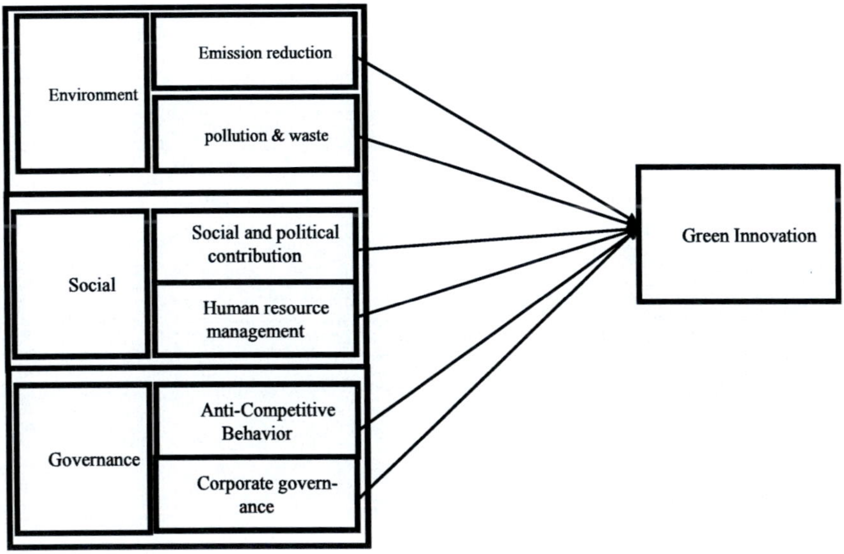

Fig. 1 Conceptual model

was validated by practitioners and academicians in the ESG field. A total of 125 completed questionnaires were collected and confirmed. However, 16 questionnaires were found incomplete, leaving a final of 109 functional samples. Based on G*power the study requires minimum of 92 respondents and since the completed questionnaires were obtained from 109 respondents. Thus, this sample size is significantly acceptable.

4.2 Operationalization of Measurement Items

In the study, the independent variable is ESG, which is further divided into subdimensions. The environmental practices are subdivided into emission reduction (ER), and pollution & waste (PW). The social practices are subdivided into social & political contribution (SPC), and human resources management (HRM). Finally, the governance practices are subdivided into anti-competitive behavior (ACB) and corporate governance (CG). Based on the past studies the measurement items are adapted from [17–19].

On the other hand, the dependent variable is green innovation, which is subdivided into product and process innovation. The items related to the dependent variable are adapted from [20], and [21]. Apart from demographic information, the measurement items were measured on a seven-point Likert scale [22]. [23] stated that a 7-Likert scale is more likely than other Likert scales to reflect a respondent's real subjective usability questionnaire questions.

Table 2 Demographic profile

Variable	Items	Frequency	Percentage (%)
Gender	Male	70	64.22
	Female	39	35.78
Age	20–30	25	22.94
	31–40	33	30.28
	41–50	42	38.53
	≥51	9	8.26
Education	Intermediate	21	19.27
	Graduation	43	39.45
	Master/Doctorate	33	30.28
	Other	12	11.01

5 Analysis and Results

5.1 *Demographic Analysis*

Table 2 contains a comprehensive demographic breakdown of the respondents. According to the gender breakdown, there were more males (64.22%) than females (35.78%). In terms of age distribution, 22.94% of respondents were under the age of 30, 30.28% were between the ages of 31 and 40, 38.53% were between the ages of 41 and 50, and 8.26% were over the age of 51. Finally, 19.27% of respondents have an intermediate qualification, 39.45% have a graduate degree, 30.28% have a master's degree, and the remaining 11.01% have some other qualification.

5.2 *Artificial Neural Networks (ANNs) Analysis*

The artificial neural network technique is considered the most intelligent of the available analytical techniques. The ANN demonstrated a large but complex network that contains multiple neurons distributed into three layers; input, hidden, and output layers. Models like multivariate regression analysis (MRA) and structural equation modeling (SEM) cannot represent the complexity of human decision-making since these analytical methods only find the linear relation. Additionally, MRA and SEM are compensating models assuming that a decline in one variable may be compensated by an addition of another variable [24].

In the research, the independent/exogenous variables are not compensable. This means that a drop in one ESG practice cannot be substituted by an increase in another, because all constructs are distinct in terms of conceptualization and definitions, therefore these constructs are not identical. In addition, ANNs are inappropriate for testing and assessing causal relationships between exogenous and endogenous variables due to their "black-box" nature [25]. The application of ANN in this research is utilized to evaluate each predictor variable's relative importance. This technique outperforms linear models in terms of multicollinearity, homoscedasticity, and non-normality of distribution [24]. ANN models have surpassed traditional statistical techniques like MRA and SEM due to their high degree of prediction accuracy.

Prior research has attempted to provide a more detailed description of ANNs. [26] stated that ANN is a massively parallel distributed processor composed of simple units with a neural propensity for accumulating & storing experimental knowledge and make available for use. In a later study (2004), [26] stated that ANNs analysis is like the human brain to performs a particular function or task. This technique is employed in a variety of research fields like supply chain quality management, blockchain in SME operations, m-commerce, social media addiction, and e-learning. However, its applicability to corporate governance to innovation performance is limited. Therefore, by using ANN analysis to the predictive power of exogenous constructs to explain the endogenous construct, this work intends to make a significant methodological contribution.

An ANN model's architecture is made up of three layers; input, hidden, and output. The root means square errors (RMSE) and normalized significance of the input neurons were determined using the feed-forward-back-propagation technique and multilayer perceptrons. To address model fit, like [27], the researcher assigned 70% of the data for training and 30% for testing. This study used a ten-fold cross-validation process to avoid the possibility of over-fitting and obtained the RMSE values. The average values of sustainable product and process variables of training and testing are shown in Table 3.

Tables 4 and 5 show the sensitivity analysis of each predictor variable according to its relative importance. Based on the findings, the SPC (100%) is the most important factor to achieve sustainable product innovation followed by PW (91%) and

Table 3 RMSE values for the ANNs of product and process innovation

Sustainable product innovation					Sustainable process innovation				
Training		Testing		Total samples	Training		Testing		Total samples
N	RMSE	N	RMSE		N	RMSE	N	RMSE	
74	0.576	35	0.817	119	80	0.558	29	0.515	119
81	0.613	28	0.682	119	80	0.572	29	0.492	119
80	0.642	29	0.447	119	70	0.580	39	0.431	119
71	0.605	38	0.604	119	69	0.563	40	0.555	119
85	0.597	24	0.491	119	77	0.513	32	0.520	119
80	0.636	29	0.594	119	69	0.459	40	0.602	119
68	0.627	41	0.657	119	78	0.529	31	0.488	119

(continued)

Table 3 (continued)

Sustainable product innovation					Sustainable process innovation				
Training		Testing		Total samples	Training		Testing		Total samples
N	RMSE	N	RMSE		N	RMSE	N	RMSE	
70	0.630	39	0.572	119	78	0.523	31	0.446	119
81	0.594	28	0.479	119	63	0.528	46	0.593	119
79	0.639	30	0.571	119	81	0.540	28	0.666	119
Mean	0.616		0.591		Mean	0.537		0.531	
S. D	0.022		0.109		S. D	0.035		0.073	

ER (49%). From Table 4, the PW (100%) is the most important factor to achieve sustainable process innovation followed by ACB (45%), and ER (35%).

Table 4 Sensitivity analysis of product innovation

Neural Network (NN)	ER	PW	SPC	HRM	ACB	CG
1st	0.188	0.344	1.000	0.240	0.183	0.231
2nd	0.202	1.000	0.812	0.037	0.284	0.321
3rd	0.660	1.000	0.795	0.859	0.305	0.310
4th	0.232	0.905	1.000	0.103	0.395	0.346
5th	0.140	1.000	0.559	0.244	0.295	0.306
6th	0.627	0.732	1.000	0.230	0.611	0.268
7th	0.753	0.316	1.000	0.356	0.463	0.268
8th	0.959	0.827	1.000	0.217	0.236	0.103
9th	0.258	1.000	0.730	0.489	0.387	0.288
10th	0.323	0.928	1.000	0.149	0.172	0.195
Mean importance	0.434	0.805	0.890	0.292	0.333	0.264
Normalized importance	49%	91%	100%	33%	37%	30%

Table 5 Sensitivity analysis of process innovation

Neural Network (NN)	ER	PW	SPC	HRM	ACB	CG
1st	0.411	1.000	0.041	0.188	0.714	0.270
2nd	0.115	1.000	0.872	0.478	0.790	0.166
3rd	0.213	1.000	0.127	0.100	0.400	0.153
4th	0.799	1.000	0.485	0.057	0.769	0.433

(continued)

Table 5 (continued)

Neural Network (NN)	ER	PW	SPC	HRM	ACB	CG
5th	0.107	1.000	0.247	0.251	0.246	0.070
6th	0.295	1.000	0.191	0.184	0.131	0.085
7th	0.367	1.000	0.277	0.229	0.229	0.170
8th	0.711	1.000	0.539	0.665	0.560	0.325
9th	0.234	1.000	0.015	0.217	0.037	0.172
10th	0.200	1.000	0.213	0.405	0.626	0.165
Mean importance	0.345	1.000	0.301	0.277	0.450	0.201
Normalized importance	35%	100%	30%	28%	45%	20%

6 Conclusion and Discussion

This study focuses on predicting the influence of Environment, Social, and Governance (ESG) practices on industry 4:0. This study hypothesized that small and medium enterprises (SMEs) could enhance green innovation if they focus on ESG practices. The questionnaire-based survey was completed within two months by the small and medium enterprises. By using the artificial neural network (ANN) approach, this study highlights the major factors of ESG practices that influence to enhance green innovation in SMEs. In this regard, the findings showed the social and political contribution (100%) as the most impactful factor to enhance sustainable product innovation followed by pollution and waste (91%) and emission reduction (49%). Furthermore, to enhance sustainable process innovation, the results showed that pollution and waste (100%) is the most influential factor followed by waste anti-competitive behavior (45%) and emission reduction (35%). Particularly, we find the influence of ESG practices to enhance the green innovation in the small and medium enterprises.

6.1 Practical Implication

From the theoretical perspective, this study enriches the limited literature on ESG practices and green innovation, particularly in SMEs. This research discovered the undiscovered area and tried to minimize the gap by predicting the effect of ESG practices on green innovation.

Moreover, the main result of this study is that ESG practices enhance green innovation which has practical implications. This study should be provided confidence to managers that the ESG practices not only enhance financial performance but also enhance green innovation which would have positive efficiency to employ the sustainable process and designing sustainable products. In this regard, the managers should be encouraged to employ ESG practices that may enhance green innovation in business operations. In this respect, making strategic decisions, managers should consider the influence of ESG practices on green innovation rather than focus on short-term profit activities. Conclusively our results are specified that ESG practices in the organization would drive green innovation.

Furthermore, the results have implications for investors that consider sustainable green practices in the organization while making investment decisions. The investors seek an organizations which is employing sustainable process innovation and design sustainable products. In this way, the investors also exert pressure on the organizations for the employment of green innovation. Finally, these findings can be used by regulators and legislators to outline future legislation and mandate the adoption of ESG practices not just in SMEs but also in other enterprises both in developed and emerging economies to boost green innovation for sustainable development.

Acknowledgements The authors would like to acknowledge Universiti Teknologi PETRONAS (UTP) Short-term Internal Research Funding (STIRF) for providing financial support for this research under cost center: 015LA0-028.

References

1. Wang H et al (2021) Green innovation practices and its impacts on environmental and organizational performance, vol 11, pp 1–15. https://doi.org/10.3389/fpsyg.2020.553625
2. Shad MK, Lai FW, Fatt CL, Klemeš JJ, Bokhari A (2019) Integrating sustainability reporting into enterprise risk management and its relationship with business performance: a conceptual framework. J Clean Prod 208:415–425. https://doi.org/10.1016/j.jclepro.2018.10.120
3. Zhang F, Qin X, Liu L (2020) The interaction effect between ESG and green innovation and its impact on firm value from the perspective of information disclosure. Sustain 12(6). https://doi.org/10.3390/su12051866
4. Weng HHR, Chen JS, Chen PC (2015) Effects of green innovation on environmental and corporate performance: a stakeholder perspective. Sustain 7(5):4997–5026. https://doi.org/10.3390/su7054997
5. Vasiliauskas AV, Yıldız B (2021) Green innovation in environmental complexity: the implication of open innovation
6. Song W, Wang GZ, Ma X (2020) Environmental innovation practices and green product innovation performance: a perspective from organizational climate. Sustain Dev 28(1):224–234. https://doi.org/10.1002/sd.1990
7. Xu J, Liu F, Shang Y (2021) R&D investment, ESG performance and green innovation performance: evidence from China. Kybernetes 50(3):737–756. https://doi.org/10.1108/K-12-2019-0793
8. Suganthi L (2019) Examining the relationship between corporate social responsibility, performance, employees' pro-environmental behavior at work with green practices as mediator. J Clean Prod 232:739–750. https://doi.org/10.1016/j.jclepro.2019.05.295
9. Shad MK, Lai FW, Shamim A, McShane M (2020) The efficacy of sustainability reporting towards cost of debt and equity reduction. Environ Sci Pollut Res 27(18):22511–22522. https://doi.org/10.1007/s11356-020-08398-9
10. Li D, Tang F, Jiang J (2019) Technology Analysis & Strategic Management Does environmental management system foster corporate green innovation ? The moderating effect of environmental regulation. Technol Anal Strateg Manag 0(0):1–15. https://doi.org/10.1080/09537325.2019.1602259

11. Pan X, Sinha P, Chen X (2020) Corporate social responsibility and eco-innovation: the triple bottom line perspective, pp 1–15. https://doi.org/10.1002/csr.2043
12. Shahzad M, Qu Y, Javed SA, Zafar AU, Rehman SU (2020) Relation of environment sustainability to CSR and green innovation: a case of Pakistani manufacturing industry. J Clean Prod 253:119938. https://doi.org/10.1016/j.jclepro.2019.119938
13. Abbas J (2020) Impact of total quality management on corporate green performance through the mediating role of corporate social responsibility. J Clean Prod 242:118458. https://doi.org/10.1016/j.jclepro.2019.118458
14. Guerrero-Villegas J, Sierra-García L, Palacios-Florencio B (2018) The role of sustainable development and innovation on firm performance. Corp Soc Responsib Environ Manag 25(6):1350–1362. https://doi.org/10.1002/csr.1644
15. Chu Z, Wang L, Lai F (2018) Customer pressure and green innovations at third party logistics providers in China. The moderation effect of organizational culture. https://doi.org/10.1108/IJLM-11-2017-0294
16. Memon MA, Ting H, Cheah J-H, Thurasamy R, Chuah F, Cham TH (2020) Sample size for survey research: review and recommendations. J Appl Struct Equ Model 4(2):i–xx. https://doi.org/10.47263/jasem.4(2)01
17. Park SR, Jang JY (2021) The impact of ESG management on investment decision: institutional investors ' perceptions of country-specific ESG criteria, vol 2020, no Unpri
18. Sultana S, Zainal D (2017) The Influence of Environmental, Social and Governance (ESG) on investment decisions: the Bangladesh perspective. Soc Sci Humanit
19. Harvard Law School Forum on Corporate Governance (2020) The stakeholder model and ESG. https://corpgov.law.harvard.edu/
20. Yusr MM et al (2020) Green innovation performance! How to be achieved? A study applied on Malaysian manufacturing sector. Sustain Futur 2:100040. https://doi.org/10.1016/j.sftr.2020.100040
21. Chiou TY, Chan HK, Lettice F, Chung SH (2011) The influence of greening the suppliers and green innovation on environmental performance and competitive advantage in Taiwan. Transp Res Part E Logist Transp Rev 47(6):822–836. https://doi.org/10.1016/j.tre.2011.05.016
22. Ali K, Johl SK (2022) Impact of total quality management on SMEs sustainable performance in the context of industry 4.0. In: Lecture notes in networks systems, vol 299, pp 608–620. https://doi.org/10.1007/978-3-030-82616-1_50
23. Finstad K (2009) Response interpolation and scale sensitivity: evidence against 5-point scales usability metric for user experience view project, November 2009, pp 1–8. https://www.researchgate.net/publication/265929744
24. Wong LW, Leong LY, Hew JJ, Tan GWH, Ooi KB (2020) Time to seize the digital evolution: adoption of blockchain in operations and supply chain management among Malaysian SMEs. Int J Inf Manage 52:101997. https://doi.org/10.1016/j.ijinfomgt.2019.08.005
25. Chan FTS, Chong AYL (2012) A SEM-neural network approach for understanding determinants of interorganizational system standard adoption and performances. Decis Support Syst 54(1):621–630. https://doi.org/10.1016/j.dss.2012.08.009
26. Haykin S, Kacprzyk J (1994) Neural networks: a comprehensive foundation, possibility theory. An approach to the computerized processing of uncertainty (Plenum the Management of Uncertainty, edited by L. A)
27. Sharma SK, Gaur A, Saddikuti V, Rastogi A (2017) Structural equation model (SEM)-neural network (NN) model for predicting quality determinants of e-learning management systems. Behav Inf Technol 36(10):1053–1066. https://doi.org/10.1080/0144929X.2017.1340973

Conceptualizing a Model for the Effect of Entrepreneurial Digital Competencies and Innovation Capability on the Tourism Entrepreneurship Performance in UAE

Mohamed Battour, Mohamed Salaheldeen, Khalid Mady, and Avraam Papastathopoulos

Abstract Entrepreneurship is a prominent topic these days, as technological advancements and developments in infrastructure generate a lot of opportunities for entrepreneurs. Entrepreneurs must work now to prepare the travel industry for a future driven by technology and innovation, as well as to establish digitally, scalable focused business models. In today's organizations such as small and medium tourism enterprises (SMTEs), There is a rising awareness of the gap between the workforce's present and required digital capabilities. SMTEs are assumed as the economic drivers of tourism destinations. One of the most implications of increased tourism in the UAE is the government support to digital entrepreneurs. This paper aims to present a theoretical account of the connection by addressing the structural relations between entrepreneurial digital competencies, innovation capability, and tourism entrepreneurship performance. The findings of this study could help policymakers, tourism operators, entrepreneurs to maximize tourism entrepreneurial performance.

Keywords Entrepreneurial digital competencies · Innovation capability · Tourism entrepreneurship performance

M. Battour · A. Papastathopoulos
College of Business Administration, University of Sharjah, Sharjah, UAE
e-mail: mbattour@sharjah.ac.ae

A. Papastathopoulos
e-mail: apapastathopoulos@sharjah.ac.ae

M. Salaheldeen (✉)
Faculty of Economics and Muamalat, Universiti Sains Islam Malaysia (USIM), Nilai, Malaysia
e-mail: m_salah6000@yahoo.com

Faculty of Commerce, Menoufia University, Shebin El-Kom, Egypt

K. Mady
Faculty of Business and Economics and Social Development, Universiti Malaysia Terengganu, Kuala Terengganu, Malaysia
e-mail: Khaled.Mady@com.kfs.edu.eg

M. Al-Emran et al. (eds.), *International Conference on Information Systems and Intelligent Applications*, Lecture Notes in Networks and Systems 550,
https://doi.org/10.1007/978-3-031-16865-9_43

1 Introduction

The fast growth of technology and digitalization characterize today's social systems. the question of how to foster entrepreneurship performance has become an important topic in public policy debates in most industrial countries [1]. Alford and Jones [2] claim that there is a lack of adoption and ineffective utilization of digital technologies in smaller tourism businesses. Recently, there has been an increasing interest to study the relationship between digital competencies and performance [3–5]. However, research in how to use digital entrepreneurship competencies for fostering entrepreneurial performance in general, and particular in tourism and hospitality industry, is scarce. Ngoasong [6] claimed that there is a need to investigate the direct linkages between specific dimensions of digital entrepreneurship competencies and performance. In line with that, Hallak, Assaker [7] also recommended Future entrepreneurial tourism research would benefit from a more objective and comprehensive measurement of firm performance.

Despite, the prominence of innovation in entrepreneurship literature, there has been little emphasis on service and product innovation in tourism [8–10]. Fu, Okumus, Wu, and Köseoglu [11] claimed that entrepreneurship literature in hospitality and tourism is scarce, particularly in theoretical development. Moreover, there is a lack of consideration devoted to innovation in tourism within academic and political contexts [12]. That is, the sources of knowledge and the processes for innovation in the service industry like tourism companies are either informal or more complex than in industrial companies. Conclusively, technology, innovation, and tourism efficiency would remain important in the tourism industry [13].

According to the tech entrepreneurship ecosystem report in UAE, there is a necessity to maintain a centralized yet loose context of the ecosystem, as well as to give opportunities to entrepreneurs. Establishing the regulatory foundations for innovation would guarantee that plans are realized. SMTEs are considered the economic drivers of tourism destinations [14]. Therefore, a question needs to be answered; what are the digital competencies that enable small and medium tourism enterprises (SMTEs) in UAE to maximize tourism entrepreneurial performance. Thus, this study expands on the existing body of knowledge on tourism entrepreneurship by examining the structural relationships among entrepreneurial digital competencies, innovation capability, and tourism entrepreneurship performance. The findings of this study could help policymakers, tourism operators, entrepreneurs to maximize tourism entrepreneurial performance.

To fill the gap, the objectives are as follows; 1) To explore the entrepreneurial digital competencies that might maximize tourism entrepreneurial performance in SMTEs in UAE, 2) To investigate the relationship between entrepreneurial digital competencies and innovation capability in the entrepreneurial tourism industry in UAE. 3) To investigate the relationship between innovation capability and tourism entrepreneurship performance in UAE. 4) To test the mediating effect of innovation capability between entrepreneurial digital competencies and tourism entrepreneurship performance in UAE. 5) To investigate the relationship between entrepreneurial digital competencies and tourism entrepreneurship performance in UAE.

2 Literature Review

2.1 Entrepreneurial Digital Competencies

Digital entrepreneurship is a process of chasing new venture prospects given by new multimedia or online technologies [15]. A digital entrepreneur uses information and communication technologies (ICTs) to design and provide core business operations and services such as marketing, production, and distribution [16]. ICT is utilized generally to include computers, landline telephones, television and radio, and emerging digital technologies (e.g. online platforms, smartphones, and artificial intelligence) [17]. This dependency on ICTs is important due to the distinctive opportunities and challenges that emerging digital organizations confront in terms of entry mode, production techniques, payment/revenue capture, and stakeholder relationship management [18].

Digitalization is a socio-technical process that involves the application of digitizing methods to broad social and institutional contexts to make digital technology infrastructure-ready [19]. Digital competence is described as the comfortable and analytical use of information society technology for business, pleasure, and communication [20]. Most work opportunities expect at least minimal digital skills [21, 22]. The literature on entrepreneurship defines these competencies in a variety of ways, including skills and knowledge [17]. Digital entrepreneurship tendencies differ in terms of the underlying innovation system situations and conditions.

2.2 Entrepreneurial Digital Competencies and Tourism Entrepreneurship Performance

Entrepreneurship scholars have increasingly recognized the relationship between entrepreneurial competencies and venture performance [23]. Entrepreneurs should build entrepreneurial competencies to develop successful enterprises [24]. In today's businesses and political views, there is an increasing consciousness of the gap stuck between current and required digital competencies of the workforce to control the opportunities and challenges of the digitalized work in the future [25]. Success is determined not only by the business itself but by the technological and architectural decisions made by a platform firm [26]. When a platform lacks a significant high reputation and solid positioning, the success of the constructed business is also limited [21, 27, 28].

Mariani [29] confirmed that within the tourism and hospitality literature, a digital entrepreneurship discipline may evolve. Small and medium tourism enterprises (SMTEs) play a vital role in supporting tourism services, guaranteeing tourist satisfaction, in addition to promoting a favorable image of the destination [30]. SMTEs performance is essential to the success of the tourism industry and the viability for tourism destinations. And this performance is influenced by digital competencies [31].

Hypothesis 1. Entrepreneurial digital competencies positively influence tourism entrepreneurship performance.

2.3 *Entrepreneurial Digital Competencies and Innovation Capabilities*

Entrepreneurial Digital competencies are the synthesis of both entrepreneurial competencies and ICT competencies that influence digital startups' decisions as well as post-entry strategic choices [17, 32]. Using the key competence framework (KCF), the key entrepreneurship Competence refers to entrepreneurs' ability to initiate and turn the ideas into successful ventures, these competencies include creativity, risk-taking, and managerial abilities [33]. The emerging technology paradigm has placed collaborative and collective intelligence at the core of effective and sustainable business initiatives [34, 35]. Given that using digital technologies has become necessary not only in work life, but also in our everyday life, entrepreneurial venture requires diverse groups of individuals who are with heterogeneous backgrounds in knowledge, abilities, and skills [36]. Hence, digital literacy has been one of the key entrepreneurial competencies [37].

Concerning how entrepreneurs can exploit their entrepreneurial digital competencies to innovate, entrepreneurial digital competencies, especially digital literacy and knowledge and skills developed by Online platform, can empower to strengthen innovation capabilities, either product/process innovation capability (the ability to introduce or develop new and existing product) or marketing innovation capability (the ability to commercialize new products/services) [38]. To enrich the existing knowledge about corporate entrepreneurship strategy and product innovation, It appears critical to promote knowledge sharing inside companies through the use of digital platforms [39]. Digital entrepreneurship could be considered as a driving force in innovation development [40]. In the same way, Digital entrepreneurship turns into a challenge (e.g., opportunities and vulnerabilities) for the sustainability and the resilience of the innovation system [41]. As a result, it is critical to comprehend how digital entrepreneurship, as a driver of digital transformation, may impact the innovation systems [42].

Hypothesis 2. Entrepreneurial digital competencies positively influence innovation Capability in the entrepreneurial tourism industry.

2.4 Innovation Capability and Tourism Entrepreneurship Performance

According to Schumpeter [43], innovative enterprises that develop new goods or technologies may achieve the highest levels of financial performance and serve as a driver of company and economic growth. Innovativeness means to "… engage in and support new ideas, novelty, experimentation, and creative processes that may result in new products, services or technological processes" [44], and It is measured as the proportion of innovations introduced by a company in a certain period of time [45]. Typical 4.0 industry technologies like Internet of Things (IoT), Big Data, Artificial Intelligence (AI), Big Data, Virtual Reality (VR), or Augmented Reality (AR) or Virtual Reality (VR), can assist in unlocking innovative potential opportunities in the tourism industry [46].

Innovation is critical for sustaining competitiveness and providing the greatest tourist experience in tourism. It is an essential factor in raising the value of tourism services as well as fostering business performance [47]. The tourism sector will be required to conduct new adjustments, even if the outcomes are questionable, to stimulate demand and ensure a safe environment for tourists as a result of the COVID-19 pandemic [12]. Interestingly, tourism service companies are progressively adopting technology advancements into their service design and development [48]. Nevertheless, the general absence of significant innovation inside the tourism industry is an opportunity that the most innovative tourism entrepreneurs may exploit [49].

Entrepreneurs have essential product innovation capabilities that include: developing new goods with distinct technical features and quality standards than existing products, upgrading products and services by boosting ease of use and improving clients satisfaction, developing new goods with unique components, and minimizing manufacturing costs related to substances and parts of existing products [10, 50]. It is recognized that the owners/managers of Chinese manufacturing SMEs use many forms of innovation capabilities. These capabilities include process, products, organizational, as well as marketing innovation capabilities.

Marketing and product innovation capabilities have a positive and significant impact on the financial performance of SMEs. whereas process and organizational innovation capabilities contribute to enhanced operational performance in SMEs [51]. Innovation capability is often seen as a critical source of long-term competitive advantage. Financial performance and non-financial performance are both positively and strongly correlated with innovation capability [52]. Moreover, The relationship between innovation capability and performance demonstrates that enhancing innovation capability is a critical prerequisite for improving performance (see Fig. 1) [38, 53].

Hypothesis 3. Innovation capability positively influences tourism entrepreneurship performance.

Hypothesis 4. Innovation capability mediates the relationship between entrepreneurial digital competencies and tourism entrepreneurship performance.

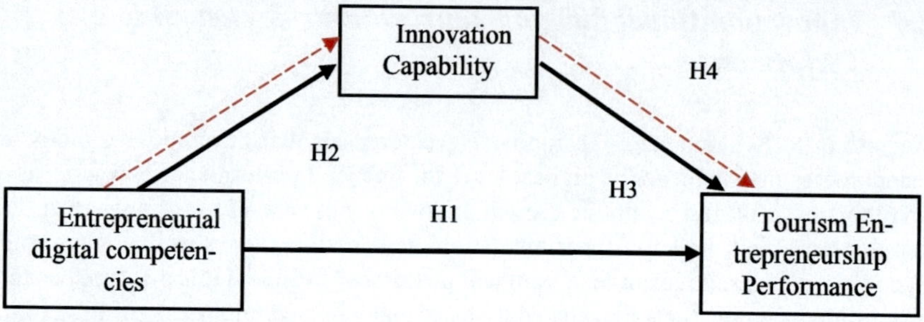

Fig. 1 Theoretical framework

3 Methodology

Conceptual papers primarily propose novel relationships between constructs; the aim is therefore to establish logical and comprehensive arguments regarding these relationships instead of testing them statistically [54]. As a result, the concern of how to develop logical arguments is critical. We not only argued that ideas are related, but we also presented a theoretical explanation for that relationship. That explanation is crucial for theory construction since it reveals the logic of relationships between concepts [55]. According to Jaakkola's [56], the common elements of the research design of conceptual papers are; theory adaptation, typology, theory synthesis, and modelling. A model research design is adopted in this conceptual paper.

Conceptual papers might aim to enhance comprehension of an idea or phenomena in large leaps instead of little stages [57]. To be considered seriously, any such leap should be founded on careful consideration and rational explanation of an adequate research design. That is, one of the potential goals and applications for model papers is developing theoretical propositions that present new relationships between constructs [56]. This research paper is developed to build a theoretical framework that indicates the relations between constructs. Therefore, critical assessment in tourism and entrepreneurship literature is used to investigate entrepreneurial digital competencies. This study suggests a theoretical framework linked between entrepreneurial digital competencies, innovation capability and tourism entrepreneurship performance.

4 Discussion and Conclusion

UAE topped the region in government service automation. It has launched the eGovernment in 2011, and the smart Government in 2013. These days, the UAE witness the digital government as a part of the 4th industrial revolution, which depends on digitization and information technology. According to Global Entrepreneurship Monitor (GEM), United Arab Emirates ranked first globally in entrepreneurship and was deemed "The Most Supportive Environment for Entrepreneurship" [58]. Understanding the contexts and causes that enable digital entrepreneurship is of interest to the scientific discipline. Digitalization influences business practice and government policies aimed at promoting this phenomenon given in the economy. However, society's push for new innovative business models contrasts with a scarcity of research on the prospects, obstacles, and critical factors for digital entrepreneurship [59].

The tourism and hospitality industry has lately gained popularity of using advanced services which are characterized by technological innovation [46, 60]. Simultaneously, travelers want more personalized services and a robust digital experience. UAE has a tourism industry that includes both huge international companies and substantial local operators. Nevertheless, the government is also now supporting smaller businesses to exploit the industry's expansion. The Abu Dhabi Tourism & Culture Authority (TCA Abu Dhabi) is supporting local small and medium tourism enterprises (STEMs) to contribute their knowledge and skills to the tourism sector. The competitiveness of the UAE business environment may be a challenge for entrepreneurs, however, the emirate's commitment to fostering enterprises, backed up by its sophisticated digital transformation drive, may help entrepreneurs achieve success.

Entrepreneurship is considered to be one of the major important outcomes of increased tourism in the UAE. Therefore, this study has opened new gates for academicians to explore and examine factors that are influencing in promoting tourism in the country. Important implications for the entrepreneurial digital competencies are highlighted. This paper gathered the state-of-the-art literature on Entrepreneurial digital competencies. Also, an up-to-date compilation of the theoretical relationships between Entrepreneurial digital competencies, Innovation capability, and tourism entrepreneurship performance is provided.

Digital transformation has an impact on socio-economic systems, resulting in essential transformations to business operations, specifically those connected to resource needs, networking procedures, and communication systems in entrepreneurial activities. For future research, it is expected that data will be collected from entrepreneurs operating in tourism in UAE and test the full model empirically. Also, this model could be tested in other contexts or other countries. further studies should be conducted to investigate the quick evolution in the field of digital entrepreneurship, which adds to theoretical development and also practical usefulness in terms of entrepreneur counseling.

References

1. Bakir C, Gunduz KA (2020) The importance of policy entrepreneurs in developing countries: a systematic review and future research agenda. Public Adm Dev 40(1):11–34
2. Alford P, Jones R (2020) The lone digital tourism entrepreneur: knowledge acquisition and collaborative transfer. Tour Manage 81:104139

3. Malecki EJ (2018) Entrepreneurship and entrepreneurial ecosystems. Geogr Compass 12(3):e12359
4. Kuratko DF, Hoskinson S (2018) The challenges of corporate entrepreneurship in the disruptive age. Emerald Publishing Limited
5. Youssef AB, Boubaker S, Omri A (2018) Entrepreneurship and sustainability: the need for innovative and institutional solutions. Technol Forecast Soc Chang 129:232–241
6. Ngoasong MZ (2018) Digital entrepreneurship in a resource-scarce context. J Small Bus Enterp Dev
7. Hallak R, Assaker G, Lee C (2015) Tourism entrepreneurship performance: the effects of place identity, self-efficacy, and gender. J Travel Res 54(1):36–51
8. Battour M, Salaheldeen M, Mady K (2022) Halal tourism: exploring innovative marketing opportunities for entrepreneurs. J Islamic Mark 13(4):887–897. https://doi.org/10.1108/JIMA-06-2020-0191
9. Battour M, Mady K, Salaheldeen M, Elsotouhy M, Elbendary I, Boğan E (2022) AI-enabled technologies to assist Muslim tourists in Halal-friendly tourism. J Islamic Mark (ahead-of-print). https://doi.org/10.1108/JIMA-01-2022-0001
10. Salaheldeen M, Battour M, Nazri MA, Bustamam USA (2021) Prospects for achieving the sustainable development goals 2030 through a proposed halal entrepreneurship success index (HESI). SHS Web Conf 124:08001. https://doi.org/10.1051/shsconf/202112408001
11. Fu H, Okumus F, Wu K, Köseoglu MA (2019) The entrepreneurship research in hospitality and tourism. Int J Hosp Manag 78:1–12
12. Montañés-Del-Río MÁ, Medina-Garrido JA (2020) Determinants of the propensity for innovation among entrepreneurs in the tourism industry. Sustainability 12(12):5003
13. Mastercard-Crescent Rating (2020) Halal Travel Frontier 2020
14. Getz D, Carlsen J, Morrison A (2004) The family business in tourism and hospitality: CABI
15. Abubakre M, Faik I, Mkansi M (2021) Digital entrepreneurship and indigenous value systems: an Ubuntu perspective. Inf Syst J 31(6):838–862
16. Boellstorff T (2019) The opportunity to contribute: disability and the digital entrepreneur. Inf Commun Soc 22(4):474–490
17. Ngoasong MZ (2018) Digital entrepreneurship in a resource-scarce context: a focus on entrepreneurial digital competencies. J Small Bus Enterp Dev 25(3):483–500
18. Baradaran MS, Yadollahi Farsi J, Hejazi SR, Akbari M (2019) A competency-based typology of technology entrepreneurs: a systematic review of the empirical studies. Iran J Manage Stud 12(2):191–211
19. Caputo A, Pizzi S, Pellegrini MM, Dabić M (2021) Digitalization and business models: where are we going? A science map of the field. J Bus Res 123:489–501
20. Spante M, Hashemi SS, Lundin M, Algers A (2018) Digital competence and digital literacy in higher education research: systematic review of concept use. Cogent Educ 5(1):1519143
21. Salaheldeen M (2015) Management control systems as a package: an application to science & technology parks: UPTEC case study. In: 8th conference on performance measurement and management control, Nice, France
22. Salaheldeen M, Battour M, Nazri MA (eds) (2019) Halal entrepreneurship and its role in sustainable development goals 2030 (SDGs). In: International conference on Dakwah and Islamic management (IC-DAIM 2019), Malaysia
23. Man TW, Lau T, Chan K (2002) The competitiveness of small and medium enterprises: a conceptualization with focus on entrepreneurial competencies. J Bus Ventur 17(2):123–142
24. Gümüsay AA, Bohné TM (2018) Individual and organizational inhibitors to the development of entrepreneurial competencies in universities. Res Policy 47(2):363–378
25. Oberländer M, Beinicke A, Bipp T (2020) Digital competencies: a review of the literature and applications in the workplace. Comput Educ 146:103752
26. Salaheldeen M, Battour M, Nazri MA, Ahmad Bustamam US, Hashim AJCM (2022) The perception of success in the halal market: developing a halal entrepreneurship success scale. J Islamic Market (ahead-of-print). https://doi.org/10.1108/JIMA-10-2021-0341

27. Noureldeen A, Salaheldeen M, Battour M (2022) Critical success factors for ERP implementation: a study on mobile telecommunication companies in Egypt. In: Al-Emran M, Al-Sharafi MA, Al-Kabi MN, Shaalan K (eds) Proceedings of international conference on emerging technologies and intelligent systems ICETIS 2021. Lecture notes in networks and systems, vol 299. Springer, Cham, pp 691–701
28. Salaheldeen M (2017) Artificial intelligence in business research: trends and future. In: Emerging issues and challenges in management conference; Faculty of Commerce, Menoufia University, Egypt
29. Mariani M (2019) Big data and analytics in tourism and hospitality: a perspective article. Tour Rev
30. Battour M, Mady K, Elsotouhy M, Salaheldeen M, Elbendary I, Marie M et al (2022) Artificial intelligence applications in halal tourism to assist Muslim tourist journey. In: Lecture Notes in networks and systems. Proceedings of international conference on emerging technologies and intelligent systems, ICETIS 2021, vol 322. Springer, Cham, pp 861–72
31. Ngoasong MZ (2017) Digital entrepreneurship in a resource-scarce context: a focus on entrepreneurial digital competencies. J Small Bus Enterp Dev
32. Al-Sharafi MA, Arshah RA, Abu-Shanab EA, Alajmi Q (eds) The effect of sustained use of cloud-based business services on organizations' performance: evidence from SMEs in Malaysia. In: 2019 5th international conference on information management (ICIM), pp 24–27
33. Gianesini G, Cubico S, Favretto G, Leitão J (2018) Entrepreneurial competences: comparing and contrasting models and taxonomies. Entrepreneurship and the industry life cycle. Springer, pp 13–32
34. Elia G, Margherita A, Passiante G (2020) Digital entrepreneurship ecosystem: how digital technologies and collective intelligence are reshaping the entrepreneurial process. Technol Forecast Soc Chang 150:119791
35. Hajar MA, Alkahtani AA, Ibrahim DN, Darun MR, Al-Sharafi MA, Tiong SK (2021) The approach of value innovation towards superior performance, competitive advantage, and sustainable growth: a systematic literature review. Sustainability 13(18):10131
36. Yu X, Wang X (2021) The effects of entrepreneurial bricolage and alternative resources on new venture capabilities: evidence from China. J Bus Res 137:527–537
37. Dudin MN, Shakhov OF, Ivashchenko NP, Shakhova MS (2021) Development of entrepreneurial competencies in the economy (evidence from digital entrepreneurship). Revista Inclusiones 2021:54–69
38. Rajapathirana RJ, Hui Y (2018) Relationship between innovation capability, innovation type, and firm performance. J Innov Knowl 3(1):44–55
39. Ben Arfi W, Hikkerova L (2021) Corporate entrepreneurship, product innovation, and knowledge conversion: the role of digital platforms. Small Bus Econ 56(3):1191–204
40. Elmashtawy A, Salaheldeen M (2022) Big data and business analytics: evidence from Egypt In: Proceedings of International Conference on Information Systems and Intelligent Applications. ICISIA 2022 . Lecture Notes in Networks and Systems, (Lecture Notes in Networks and Systems. Cham: Springer 2022, ch. Proceedings of International Conference on Information Systems and Intelligent Applications. ICISIA 2022
41. Steiner G (2018) From probabilistic functionalism to a mental simulation of innovation: by collaboration from vulnerabilities to resilient societal systems. Environ Syst Decis 38(1):92–98
42. Satalkina L, Steiner G (2020) Digital entrepreneurship: a theory-based systematization of core performance indicators. Sustainability 12(10):4018
43. Schumpeter JA (1942) Capitalism, socialism, and democracy
44. Lumpkin GT, Dess GG (1996) Clarifying the entrepreneurial orientation construct and linking it to performance. Acad Manag Rev 21(1):135–172
45. Covin JG, Slevin DP (1989) Strategic management of small firms in hostile and benign environments. Strateg Manag J 10(1):75–87
46. Battour M, Salaheldeen M, Mady K (2021) Exploring innovative marketing opportunities for halal entrepreneurs in hospitality and tourism industry. SHS Web Conf 124:10001. https://doi.org/10.1051/shsconf/202112410001

47. Gomezelj DO (2016) A systematic review of research on innovation in hospitality and tourism. Int J Contemp Hosp Manag 28(3):516–558
48. Correia PÁP, Medina IG, Romo ZFG (2020) Entrepreneurship and innovation in tourism E-businesses: their relationships with their audiences. In: Multilevel approach to competitiveness in the global tourism industry. IGI Global, pp 159–76
49. Ness H, Fuglsang L, Eide D (2018) Networks, dynamics, and innovation in the Tourism industry. Taylor & Francis
50. Mafimisebi OP, Obembe D, Aluko O (2020) Organization and product design pairings: a review of product innovation capabilities, conceptualization, and future directions. Strateg Chang 29(1):13–24
51. Sawaean F, Ali K (2020) The impact of entrepreneurial leadership and learning orientation on organizational performance of SMEs: the mediating role of innovation capacity. Manage Sci Lett 10(2):369–380
52. Al-kalouti J, Kumar V, Kumar N, Garza-Reyes JA, Upadhyay A, Zwiegelaar JB (2020) Investigating innovation capability and organizational performance in service firms. Strateg Change 29(1):103–113
53. Kim K (2021) The interplay between the social and economic human resource management systems on innovation capability and performance. Int J Innov Manag 25(07):2150074
54. Gilson LL, Goldberg CB (eds) (2015) Comment: so, what is a conceptual paper? SAGE Publications Sage, Los Angeles, pp 127–30
55. Ridder H-G (2017) The theory contribution of case study research designs. Bus Res 10(2):281–305
56. Jaakkola E (2020) Designing conceptual articles: four approaches. AMS Rev 10(1):18–26
57. Xin S, Tribe J, Chambers D (2013) Conceptual research in tourism. Ann Tour Res 41:66–88
58. Hill S, Ionescu-Somers A, Coduras A, Guerrero M, Roomi MA, Bosma N et al (eds) (2022) Global entrepreneurship monitor 2021/2022 global report: opportunity amid disruption. Expo 2020 Dubai
59. Salaheldeen M (2022) Opportunities for Halal Entrepreneurs in the Islamic digital economy: future and trends from a cultural entrepreneurship perspective. In: Ratten, V. (eds) Cultural Entrepreneurship. Springer, Singapore. https://doi.org/10.1007/978-981-19-2771-3_9
60. Battour M, Salaheldeen M, Mady K, Elsotouhy M (2021) Halal tourism: what is next for sustainability? J Islamic Tour 1(Inaugural Issue):80–91. https://jistour.org/en-us/makele/halal-tourism--what-is-next-for-sustainability/37/pdf.

Building Information Modelling: Challenges, Benefits, and Prospects for Adoption in Developing Countries

A. H. Al-Sarafi, A. H. Alias, F. M. Jakarni, H. Z. M. Shafri, and Yaser Gamil

Abstract In the fast-expanding construction industry worldwide, building information modelling (BIM) is a robust process. However, to date, developing countries are not very well adopting the techniques proven to help significantly produce effective management of construction projects. This study reviews numerous current studies conducted on the challenges and benefits of adopting BIM. It aims to identify the challenges and benefits of BIM. Additional focus was given to developing countries since fewer documented articles were found in the literature. However, many challenges are identified which hinder BIM adoption to full potential, particularly in developing countries. The most common findings proposed five critical benefits of BIM adoption, namely: i) improved data management (rich) information; ii) improved visualization of project execution; iii) clash detection; iv) reducing waste in the material; v) reducing the financial risk associated with the project in order by obtaining earlier reliable cost estimates. Likewise, the most common findings defined five major BIM adoption obstacles are: i) resilience to change industry culture; ii) high Investment cost; iii) lack of client demand; iv) absence of stakeholder collaboration; v) lack of awareness. It was found that there is a considerable benefit gained by those construction organizations already practicing the information modelling. Most of the organizations that adopted BIM are situated in European countries, followed

A. H. Al-Sarafi (✉) · A. H. Alias · F. M. Jakarni · H. Z. M. Shafri
Department of Civil Engineering, Faculty of Engineering, Universiti Putra Malaysia, UPM, 43400 Serdang, Selangor, Malaysia
e-mail: alsarafiali@gmail.com; gs51839@student.upm.edu.my

A. H. Alias
e-mail: aidihizami@upm.edu.my

F. M. Jakarni
e-mail: fauzan.mj@upm.edu.my

H. Z. M. Shafri
e-mail: helmi@upm.edu.my

Y. Gamil
Building Materials, Department of Civil, Environmental and Natural Resources Engineering, Luleå University of Technology, 97187 Luleå, Sweden
e-mail: yaser.gamil@ltu.se

M. Al-Emran et al. (eds.), *International Conference on Information Systems and Intelligent Applications*, Lecture Notes in Networks and Systems 550,
https://doi.org/10.1007/978-3-031-16865-9_44

by the united states of America. Thus, future work should focus on how to raise the level of awareness and general adaptability, especially in developing nations.

Keywords BIM · Awareness · Adoption · Construction industry · Developing counties

1 Introduction

Building Information Modelling (BIM) is a growing construction industry technology and process that can provide many benefits to construction stakeholders. While some countries such as Estonia still exist, the Information and Communications Technology (ICT) sector and BIM adoption lag in their construction industries [1]. Via building lifecycle phases, the study further demonstrated BIM benefits and examined the BIM adoption rates in different construction industries. In the construction industry, BIM is a ground-breaking breakthrough for the virtual design and operation of projects across the construction lifecycle. BIM has now contributed to a huge percentage of the global improvement of the construction sector. There are, however, several barriers and challenges which hinder its effectiveness and adaptability. A recent report by Ullah et al. (2019) provided an analysis of the current situation in the construction industry of BIM adoption [2].

Many countries widely adopt BIM in the construction industry to plan, design, build and manage their projects. This advanced system can make projects faster, better, safer, cheaper, and greener [3]. To comprehend the real and full benefit of BIM adoption requires the cooperation of both experts and researchers to set up the implementation strategy. The construction industries, both public and private, increase the BIM adoption in several countries. Thus, there is a demand to ascertain the importance, motivations, challenges, and considerations for government policies toward BIM adoption strategies [4].

The implementation of BIM techniques has been reported to have entirely revolutionized the construction industry in Indonesia, and building practices around the world are increasingly becoming traditional [5]. Research looks at the amount of BIM awareness, expertise, perceived benefits, and implementation obstacles faced by Indonesian construction firms. BIM adoption in architecture, engineering, and construction (AEC) organizations has been hampered by the expensive cost of software and hardware. Because BIM has enhanced output and productivity in many nations, it is projected to help the Indonesian construction industry [5].

This study includes information on BIM adoption in the construction industry and will serve as a foundation for additional research. However, it posed several constraints on the successful implementation of BIM strategies previously documented in some current works of literature. The goal of this study is to determine the challenges and benefits of BIM adoption in developing countries. This is a list of factors that have been incorporated into a framework. In order to achieve the study's goal, this study will attempt to answer the following research question: RQ:

What are the challenges and benefits of BIM adoption in developing countries for the construction industry?

The following is an overview of the paper's structure: Sect. 2 BIM Technology and Evolution, whereas Sect. 3 discusses the method used in this study. The findings are presented in Sect. 4 and the Discussion are offered in Sect. 5, while the conclusion and future work are presented in Sect. 6.

2 BIM Technology and Evolution

2.1 Global Awareness of BIM in Construction Industries

BIM in the construction industry is being globally recognized as an essential process to implement construction projects; educators started to teach the importance of BIM and its applications as an integrated process to plan, initiate and execute the construction projects [6]. Besides, construction industry stakeholders appreciate the importance and the success attained from adopting BIM in construction. They can easily manage, oversee and track the project activities on a real-time basis. That has contributed to saving the cost and time of the project. A study by Enegbuma et al. [7] showed that the adoption of BIM in Malaysia is rising due to recent efforts to raise awareness among construction professionals about the need for the strategic implementation of information technology (IT). The study recommends developing grey areas, such as standard contract forms, strengthening communication between construction experts in the construction industry. In evaluating the views of other key stakeholders in the construction industry, the model can be used for future studies: architects, quantity surveyors, and contractors [8].

BIM has been implemented gradually in different countries; in research by Rosli et al., [9], with the beginning of BIM in Malaysia, the disturbing precedent envisaged by construction professionals has garnered more attention on the topic of technology adoption. A model for BIM adoption in Malaysian building projects was established as part of the research. Overall, the model confirmed the conceptual structure's effect on the adoption rate of BIM understanding among Malaysian construction industry experts. The findings also pointed out areas where stakeholders in the building sector should focus their efforts to improve the understanding of BIM technology [10].

Similarly, Rosli et al. [9] investigated the link between numerous constructs that influence BIM adoption. The Structural Equation Modeling (SEM) model fit indices and the association strength within the components were used to investigate this link. It highlighted the seeming insufficiency of the literature on BIM operations in the building industry in Malaysia. It also emphasized the seeming insufficiency of the literature on BIM operations in the Malaysian building industry. A report by Alhumayn et al. [11] The conference highlighted the potential for BIM to have an important impact on the Saudi construction sector. Lack of awareness of the BIM adoption process, managerial support and lack of realistic standards and guidelines were cited as barriers to BIM implementation.

A study by Chileshe [12] about the southern Australian building industry found that the construction industry is hesitant to accept technologies and advancements. The Australian government's Department of Planning, Transport, and Infrastructure of the South Australian government conducted the study. Lack of understanding of BIM, education, training costs, start-up costs, and changing the way firms do business have all been mentioned as significant hurdles.

Likewise, research by Babatunde et al. [13] listed the factors that influence effective BIM adoption in Nigerian businesses and have been studied empirically. Professional organizations, governmental agencies, and non-governmental organizations should all work to increase BIM awareness, according to the report. BIM represents development projects by handling data and predicting it all at once. The factor analysis grouped the identified drivers into three groups: cost and time savings, increased communication, and BIM and government financing expertise. The use of BIM in buildings and its total lack of instruction are the main obstacles to achieving safer working conditions. Based on these results, BIM developers will be able to integrate correct BIM products that are appropriate [14].

Strong government regulations that encourage BIM use should also be in place in developing countries. This medium will give policymakers and construction stakeholders the information they need to make policy decisions that will help AEC businesses and the construction industry as a whole fully implement BIM. The analysis of BIM adoption factors by various construction experts would provide a fuller and more accurate understanding of BIM adoption drivers in Yemen [14]. The limited studies on the level of BIM expertise, usage, and especially in construction organizations, have only investigated BIM to date. This study shows that many respondents have no idea what BIM is or how it works; an application was proper only in a particular situation. BIM shows that the global economy views as critical for development and competitiveness in the built environment [12].

Furthermore, the construction industry makes a major contribution to Saudi Arabia's Gross Domestic Product. Total construction operations in 2014 were $24 billion, second only to the oil industry, according to Deloitte. Saudi Arabia's 2030 vision plan, which was released in 2016, encourages all sectors to be more innovative, efficient, and environmentally accountable. BIM technology has the potential to change Saudi Arabia's building industry [15]. Furthermore, worker training is an important factor in the successful implementation of BIM. Ahmed [16] reported that enough evidence has been presented to suggest that BIM can help Bangladesh improve building efficiency. Owners, consultants, and contractors, among other construction stakeholders, should play a significant role in transforming the paradigm from a conventional to a more innovative approach.

2.2 *The Benefits of BIM in the Construction Industry*

As mentioned earlier, BIM offers many noticeable benefits to the construction industry. A research was undertaken in Malaysia to explore BIM adoption. The findings of a BIM model analysis are presented in this research. People, operation, technology, strategic IT planning, and collaborative planning are the key dimensions for improving BIM adoption [10]. The conceptual model can be used as a central store of information for project managers in the development of teams and the maintenance of cooperation. The model is conceptual, and it must be validated using actual data. Project managers should use care when applying the findings to their project. Because the model is focused, the results may not apply to the whole supply chain of BIM-enabled projects [17].

Another development was made in Singapore by Attarzadeh et al. [18], AEC Singapore BIM helps AEC companies turn their construction value chain into a more technologically advanced process. The purpose of this research was to find out what factors influence BIM acceptance and implementation in the Singapore AEC industry. Government agencies should provide frequent, thorough functional guides, models, and BIM libraries for many industries as supporters of developing technologies.

Furthermore, the biggest challenge to BIM in the construction business is a widespread lack of training and application [19]. In fact, there are plans to hold training sessions on the concept of BIM and the advantages of using it. BIM will be demanded as a contract requirement by clients and other construction companies. By reference to the Chinese AEC market, Hosseini et al. [20] investigated the impact of government subsidies on BIM technology spread in China. BIM has the potential to significantly enhance the global AEC. AEC enterprises that were earlier averse to BIM adoption may become positive as a result of the government's subsidies schemes. From a new perspective, the research leads to a new understanding of BIM adoption behaviors across AEC firms, since it can both shorten the joining time and improve the efficacy of the BIM used.

In the United Kingdom, a study by Ahmed Louay Ahmed et al. [21] identified BIM innovation characteristics. The study used a systematic analysis methodology to identify discrepancies and suggest possible research topics. The findings can be utilized to investigate the impact of various drivers, variables, and determinants of BIM's organizational acceptability in markets with varying macro diffusion patterns. They included perceived innovative characteristics, internal environment fac-tors (innovator or organizational readiness) and external environment components (isomorphic pressures). Furthermore Rakshit [22], investigated the maximum potential of BIM in India, and it was discovered that this potential has yet to be realized on projects by architectural firms. Adoption of this new technology has only reached the third stage of literature. The adoption of BIM in emerging markets is examined in this article. BIM's full potential has been exploited in India's construction sector, although many people are still unaware of it. The findings of the study are assessed and compared to those of other emerging and established markets. Based on the survey results, recommendations for boosting BIM use are made [23].

In different study, Doan et al. [24] looked at the viewpoints of New Zealand construction experts on the Green Star scenario and how it relates to BIM adoption. Experts performed twenty-one interviews with 25 participants for either BIM or Green Star projects. Nonetheless, the benefits of Green Star to the environment are well-known among inhabitants, developers, owners, and interviewees as well as improving social consciousness, was minimal. BIM allows designers to cooperate on a single model rather than having each team member rebuild and provide information, according to the study Davila Delgado et al. [25]. 'Clashes detection,' which resulted in fewer expensive adjustments during construction and consequent delays, was the third highest-rated advantage (frequency = 15) of BIM. According to a study conducted by BIM in Australia, BIM also allows any clashes between different professions or disciplines to be identified. Conflicts or discrepancies between works from different disciplines can be resolved in the virtual environment, resulting in a significant shift in consultants' time commitment from the construction phase to the design phase, with implications for project team management and fee structures.

Alhumayn et al. [11], BIM is a technology-driven concept that allows project stakeholders to share reliable and timely information. The visual integration of building systems is one of the advantages of BIM implementation. According to the results, BIM adoption in Saudi Arabia has been slow but steadily in recent years. In contrast to a traditional CAD workflow, this takes into account the new ideas implemented by BIM. Likewise, Hussain & Choudhry [26] in their report showed that 65 percent of construction projects in Pakistan were not considered for BIM implementation, awareness, benefits, and challenges of adopting BIM in the construction industry. It also introduces a roadmap for further research in BIM, particularly in undeveloped countries.

2.3 *The Challenges of BIM Adoption in the Construction Industry*

BIM is being adopted at a sluggish rate in the construction industry around the world. To describe and assess BIM adoption processes, we propose a new, enlarged model. From substantial survey data, we examine the historical and collaborative features of BIM adoption, as well as current findings [27]. A study outlines a process for implementing BIM in small and medium-sized construction firms (SMOs) [28]. The suggested BIM adoption model examines the advantages, costs, and challenges that SMOs experience while implementing BIM. Non-construction industries such as civil works and services are not included in the research and will need to be assessed separately in the future. It claims that employee participation in the implementation phase should be prioritized [29].

Ma et al. [30] investigated how BIM usage in Chinese AEC firms might be improved. BIM would improve the effectiveness and efficiency of the AEC industry. However, before BIM can be implemented, a few things must be changed. Management of companies and software access are basic considerations. According to the study, a thorough examination would alleviate weaknesses by implementing BIM more deeply and increasing BIM understanding in China.

Hosseini et al. [20] introduced some results of a study effort in Australia where they employed a questionnaire survey to target SMEs in the construction sector were released. The research provides the most up-to-date information on BIM in Australia's small and medium-sized enterprises. It offers and expands upon a framework based on the innovation diffusion concept (IDT). Based on 135 surveys answered by SMEs, the present status of BIM adoption and impediments to BIM adoption for SMEs were examined using partial least square structural equation modeling (PLS-SEM) and the suggested structure. Furthermore, Davila Delgado et al. [25] investigated the inefficiencies and poor productivity in the construction projects which are plaguing the building industry. Robotics technology also has the opportunity to provide the UK building industry with various benefits, but adoption is very poor. This study assists stakeholders in considering the key factors unique to the industry that restricts robotics adoption. There is a widespread expectation that the study says robots and other innovations will replace jobs but still have comparative benefits. The primary tasks that stakeholders want to automate are concrete construction, survey and tracking, drilling, excavation, and demolition.

According to the findings, approximately 42% of Australian SMEs are now adopting BIM at Levels 1 and 2, Only 5% of users have attempted Level 3. Lack of knowledge inside small and medium-sized firms and across the construction supply chain is not a significant obstacle for Australian SMEs. As observed by SMEs' major actors, the biggest hurdles stem from the risks associated with BIM's uncertain return on investment (ROI). The data show that the proposed structure can be used to explain BIM adoption in Australian SMEs. Furthermore, examining BIM adoption in construction enterprises through the lens of innovation adoption is recommended as the most successful way [31]. Hair Jr et al. [32] discussed some numerous restrictions, the analysis' findings can be adopted, taking into account the contributions. Consumers and large corporations that work with SMEs should be the focus of any potential investigations. Furthermore, delivering corrective answers to the significant hurdles identified in the current study will bring tremendous value to the body of knowledge as another potential subject for future research studies.

A conceptual model for the BIM innovation adoption mechanism in the United Kingdom was proposed by Ahmed Louay Ahmed et al. (2017) where they conducted empirical investigations of the BIM Innovation adoption process using the proposed conceptual model's gathered drivers and factors. The study compiles a comprehensive list of factors that influence organizational adoption of the British BIM scheme. The fundamental lenses of the associated theories and models are integrated into the theoretical fundamentals of the conceptual model proposed [21].

Moreover, Munir et al. [33] reached to a conclusion that a constructed asset's life cycle cost is three times more than a construction cost. More process-based problems are challenged than individuals or technology. In the AEC sector, the findings help progress towards improved BIM adoption. Special consideration is needed for the technical difficulties related to systems integration and technological limitations; As concludes that in the realization of BIM, there is value, although the AEC industry must address the challenges found. In another development, Shirowzhan et al., [34] identified BIM problems related to compatibility with the environment in Australia. The study argues that issues of interoperability prevail as the main functional obstacle to the implementation of BIM. Construction companies should also recognize the compatibility principle to determine their requirements, expertise, and infrastructure. According to the report, clear data and model discussions across stakeholders with varied needs and using alternative formats help broaden BIM applications and speed up the pace of adoption. A slew of challenges and roadblocks stand in the way of efficient BIM implementation in the construction industry. For instance, consider survey data analysis.

3 Research Method

In this article, comprehensive literature research is used to extract the essential factors that influence BIM adoption and the benefits and challenges of BIM implementation. An extensive analysis of journal publications, research papers, technical and review papers, books, and articles led to the finalization of the final set. The documents are searched in ScienceDirect, Scopus and Google Scholar. Then assessed, and any unconnected items are eliminated. The remaining studies are then thoroughly scrutinized to extract any BIM-related characteristics. This method resulted in 62 papers passing all of the preceding stages. Several potential factors impacting BIM adoption were found in the 62-research published between 2010 and 2021 (2014 to 2021). These are drivers, benefits, challenges, restrictions, essential success elements, initiatives, and other BIM-related concerns.

Similarity analysis was performed to finalize the extraction of benefits and challenges where some factors were repeated in different articles with similar meanings and phrasings. This helps to avoid repetition while maintaining the comprehensiveness of the review. After that, frequency analysis was introduced to understand the most repeated challenges and benefits of BIM adoption. Higher frequency denotes more attention from previous literature research. The final factors are listed in Tables 1 and 2. The factors affecting BIM adoption in the construction field are ranked based on the frequency of literature [35]. Figure 1 shows the steps of the methodology.

Table 1 The benefits of BIM adoption in construction industry

Code	BIM adoption benefits	Reference	Frequency
BE1	Improved data management (rich) information	[5, 10, 12, 13, 16–19, 22, 24, 27, 29, 33, 36–43]	21
BE2	Improve visualization of project execution	[1, 5, 12–15, 17, 19, 34, 37, 39, 40, 42, 44–49]	19
BE3	Clash detection	[9, 10, 17, 18, 27, 29, 38, 39, 41, 43, 45, 48–51]	15
BE4	Reduce waste in material	[5, 10, 15, 16, 40, 52–59]	13
BE5	Reducing the financial risk associated with the project in order by obtaining earlier reliable cost estimates	[5, 10, 15, 16, 18, 24, 29, 34, 37, 42, 45, 60]	12
BE6	Improve facility management concept	[1, 5, 14, 16, 19, 24, 34, 38, 40, 41, 48, 61]	12
BE7	BIM helps to expedite the decision-making process	[18, 27, 29, 34, 45, 51, 53, 55, 60–62]	11
BE8	Enhanced communication	[10, 17, 19–21, 24, 34, 36, 43, 60]	10
BE9	Time saving (Reduce waste in time)	[15, 16, 18, 24, 29, 33, 34, 42, 45, 59]	10
BE10	Improving the quality control	[1, 12, 25, 46, 54, 57, 60, 61]	8
BE11	Improve progress monitoring of construction projects	[14, 25, 33, 40, 48, 49, 53]	7
BE12	Allows intervention and early errors detection construction	[13, 18, 38, 43, 44]	5
BE13	During the construction process, improving coordination with the owner and design firms	[27, 41, 42, 44, 46, 49]	5

(continued)

Table 1 (continued)

Code	BIM adoption benefits	Reference	Frequency
BE14	Reducing the rework process	[29, 36, 42, 53]	4
BE15	Better performance in production	[47, 60]	2
BE16	Improved audit processes and approval	[63]	1

Table 2 Challenges to adopting BIM in construction industry

Code	Challenges to adopting BIM	Reference	Frequency
CH1	Resilience to change industry's cultural	[1, 7, 9, 10, 13, 25, 27, 33, 34, 36, 38–41, 43–45, 56, 60, 62]	20
CH2	High investment cost	[5, 36, 38, 41, 46, 47, 49, 51, 54–56, 58, 60, 61]	14
CH3	Lack of client demand	[1, 14, 19, 20, 24, 25, 29, 39, 40, 46, 49, 51]	12
CH4	Absence of stakeholder collaboration	[1, 5, 17, 20, 24, 29, 36, 39, 44, 53]	10
CH5	Lack of awareness	[1, 5, 14, 16, 22, 36, 40, 41, 53, 64]	10
CH6	Lack of vision of benefits	[9, 10, 25, 34, 36–38, 40, 46]	9
CH7	BIM guidelines that aren't appropriate	[16, 17, 23, 65–68]	7
CH8	Lack of government policy	[14, 22, 25, 27, 36, 51]	6
CH9	Resistance at operational level	[7, 19, 29, 36, 40, 62]	6

(continued)

Table 2 (continued)

Code	Challenges to adopting BIM	Reference	Frequency
CH10	Protocols and standards are lacking	[69–72]	5
CH11	Lack of infrastructure	[16, 19, 25, 34, 47]	5
CH12	Lack of BIM expertise	[19, 22, 36]	3
CH13	Team members' reluctance to share information	[36, 47, 60]	3
CH14	Return on investment (ROI) issue	[18, 69, 73]	3
CH15	Lack of adequate quality control management	[52]	1
CH16	Limited project funding to support BIM	[74]	1

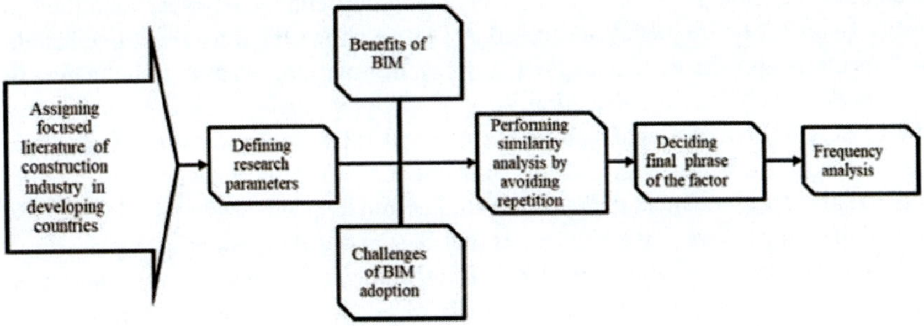

Fig. 1 Research methodology

4 Findings

The current study looked at the literature on the factors that influence the adoption of BIM in the construction industry. This study evaluated the frequency of each used factor in the gathered studies to discover the benefits and challenges variables that influence BIM adoption. Table 1 lists the benefits of BIM adoption in the construction industry, as well as the frequency of the variables collected from the research. Table 1 also illustrate the frequency of benefits of BIM adoption in the construction industry. The top five benefits factors affecting BIM adoption in the construction industry are (1) Improved data management (rich) information. (F = 21), (2) Improve visualization of project execution (F = 19), (3) Clash detection (F = 15), (4) Reduce waste in material (F = 13), (5) Reducing the financial risk associated with the project in order by obtaining earlier reliable cost estimates. (F = 12).

Table 2 summarises the challenges to BIM adoption in the construction industry, as well as the frequency of the characteristics highlighted in the study. Table 2 also illustrate the frequency of challenges of BIM adoption in the construction industry. The top five challenges affecting BIM adoption in the construction industry as seen in Table 2 are (1) Resilience to change in the construction industry's cultural (F = 20), (2) High Investment Cost (F = 14), (3) Lack of Client Demand (F = 12), (4) Absence of stakeholder collaboration (F = 10), (5) Lack of awareness (F = 10).

5 Discussion

This review article focused on the identification of benefits offered by the adoption of BIM in the construction industry and also the challenges that come along to fully accept the technology. A total of 16 benefits and 16 challenges were identified in Table 1 and 2. The frequency addressed in Table 1 shows the most repetitive benefit factors like, improved data management (rich) information, improve visualization of project execution and clash detection. Also, the frequency addressed in Table 2 shows the most repetitive challenges in the literature like; resilience to change industries, cultural high investment cost and lack of client demand. This result helps to conduct more statistical assessment in the future studies and helps to develop model to study the relationship between the challenges and benefits of BIM adoption especially in undeveloped countries which they lack similar studies. The future study is proposed to compare and introduce a road map of BIM adoption taking into consideration lessons learnt from the existing development of BIM implementation in developed countries.

6 Conclusion

Developing countries are found not very well adopting the BIM techniques, which greatly help operate the construction industry's affairs effectively. This paper identified and summarized the benefits and challenges of BIM adoption. It showed that the key findings proposed five significant benefits of BIM adoption, namely: i) improved data management (rich) information; ii) improve visualization of project execution; iii) clash detection; iv) reduce waste in the material; v) reducing the financial risk associated with the project in order by obtaining earlier reliable cost estimates. Also, the key findings proposed five major BIM adoption obstacles are: i) resilience to change industry culture; ii) high Investment cost; iii) lack of client demand; iv) absence of stakeholder collaboration; v) lack of awareness. There are vast benefits gain by the construction companies to know and practicing this information before starting their BIM adoption. Significant parts of the companies that adopted the BIM are situated in European countries, followed by the united states of America. Consequently, future works should focus on how to raise the level of awareness and general adaptability, especially in developing nations. More studies are recommended to statistically assess the benefits and challenges of BIM adoption and that will help to draw attention of project stakeholders and policymakers to address these challenges and appreciate the benefits whilst developing a roadmap for future adoption policies.

References

1. Ullah K, Lill I, Witt E (2019) An overview of BIM adoption in the construction industry: benefits and barriers. In: 10th Nordic conference on construction economics and organization, vol 2, pp 297–303
2. Charef R, Emmitt S, Alaka H, Fouchal F (2019) Building information modelling adoption in the European Union: an overview. Elsevier
3. Hardin B, McCool D (2015) BIM and construction management: proven tools, methods, and workflows. Wiley
4. Sacks R, Eastman C, Lee G, Teicholz P (2018) Facilitators of BIM adoption and implementation. In: BIM handb., pp 323–363
5. Fitriani H, Budiarto A, Ajayi S, Idris Y (2019) Implementing BIM in architecture, engineering and construction companies: perceived benefits and barriers among local contractors in Palembang, Indonesia. Int J Constr Supply Chain Manag 9(1):20–34
6. Latiffi AA, Mohd S, Kasim N, Fathi MS (2013) Building information modeling (BIM) application in Malaysian construction industry. Int J Constr Eng Manag 2(4A):1–6
7. Enegbuma WI, Aliagha GU, Ali KN, Badiru YY (2016) Confirmatory strategic information technology implementation for building information modelling adoption model. J Constr Dev Ctries 21(2):113–129
8. Haron AT (2013) Organizational readiness to implement building information modelling: a framework for design consultants in Malysia. University of Salford
9. Rosli N et al (2015) Effects of perceptions on BIM adoption in Malaysian construction industry. J Teknol 1:1–6
10. Enegbuma WI, Aliagha UG, Ali KN (2014) Preliminary building information modelling adoption model in Malaysia a strategic information technology perspective. Constr Innov 14(4):408–432

11. Alhumayn S, Chinyio E, Ndekugri I (2017) The barriers and strategies of implementing BIM in Saudi Arabia. WIT Trans Built Environ 169:55–67
12. Chileshe N (2012) Awareness, usage and benefits of building information modelling (BIM) adoption-the case of South Australian construction organizations. In: Procs 28th annual ARCOM conference, Edinburgh, UK, 3–5 September 2012. Association of Researchers in Construction Management, pp 2–12, 3–12 May 2012
13. Babatunde SO, Ekundayo D, Adekunle AO, Bello W (2020) Comparative analysis of drivers to BIM adoption among AEC firms in developing countries. J Eng Des Technol 23
14. Gamil Y, Rahman IAR (2019) Awareness and challenges of building information modelling (BIM) implementation in the Yemen construction industry. J Eng Des Technol 17(5):1077–1084
15. Banawi A (2018) Barriers to implement building information modeling (BIM) in public projects in Saudi Arabia. Springer, Cham
16. Ahmed S (2018) Barriers to implementation of building information modeling (BIM) to the construction industry: a review. J Civ Eng Constr 7(2):107
17. Oraee M, Hosseini MR, Edwards DJ, Li H, Papadonikolaki E, Cao D (2019) Collaboration barriers in BIM-based construction networks: a conceptual model. Int J Proj Manag 37(6):839–854
18. Attarzadeh M, Nath T, Tiong RLK (2015) Identifying key factors for building information modelling adoption in Singapore. In: Proceedings of the institution of civil engineers-management procurement and law, vol 168, no 5, pp 220–231
19. Enshassi A, Ayyash A, Choudhry RM (2016) BIM for construction safety improvement in Gaza strip: awareness, applications and barriers. Int J Constr Manag 3599
20. Hosseini MR et al (2016) BIM adoption within Australian Small and Medium-sized Enterprises (SMEs): an innovation diffusion model. Constr Econ Build 16(3):71–86
21. Ahmed AL et al (2017) A conceptual model for investigating BIM adoption by organizations. In: Proceedings of the joint conference on computing in construction (JC3), vol 1, pp 447–455
22. Rakshit (2018) Factors influencing BIM adoption in emerging markets – the case of India. Taylor Fr.
23. Succar B (2010) Building information modelling maturity matrix, pp 65–103
24. Doan DT, Ghaffarianhoseini AAA, Naismith N, Ghaffarianhoseini AAA, Zhang T, Tookey J (2019) Examining Green Star certification uptake and its relationship with Building Information Modelling (BIM) adoption in New Zealand. J Environ Manage 250
25. Davila Delgado JM et al (2019) Robotics and automated systems in construction: understanding industry-specific challenges for adoption. J Build Eng 26
26. Hussain K, Choudhry R (2013) Building Information Modeling (BIM) uses and applications in Pakistan construction industry. In: 13th conference on construction applications of virtual reality, London, UK
27. Herr CM, Fischer T (2019) BIM adoption across the Chinese AEC industries: an extended BIM adoption model. J Comput Des Eng 6(2):173–178
28. Babatunde SO et al (2020) An investigation into BIM-based detailed cost estimating and drivers to the adoption of BIM in quantity surveying practices. J Financ Manag Prop Constr 25(1):61–81
29. Hong Y, Hammad AWAA, Sepasgozar S, Akbarnezhad A (2019) BIM adoption model for small and medium construction organizations in Australia. Eng Constr Archit Manag 26(2):154–183
30. Ma G, Jia J, Ding J, Shang S, Jiang S (2019) Interpretive structural model based factor analysis of BIM adoption in Chinese construction organizations. Sustainability 11(7):1982
31. Murphy ME (2014) Implementing innovation: a stakeholder competency-based approach for BIM. Constr Innov 14(4):433
32. Hair Jr JF, Black WC, Babin BJ, Anderson RE (2014) Multivariate data analysis [Internet]. Sevent. Pearson New International Edition
33. Munir M, Kiviniemi A, Jones S, Finnegan S (2020) BIM business value for asset owners: key issues and challenges. Int J Build Pathol Adapt
34. Shirowzhan S, Sepasgozar SME, Edwards DJ, Li H, Wang C (2020) BIM compatibility and its differentiation with interoperability challenges as an innovation factor. Autom Constr 112:103086

35. Rahman IA, Gamil Y (2019) Assessment of cause and effect factors of poor communication in construction industry. IOP Conf Ser Mater Sci Eng 601(1):12014
36. Oesterreich TD, Teuteberg F (2019) Behind the scenes: understanding the socio-technical barriers to BIM adoption through the theoretical lens of information systems research. Technol Forecast Soc Change 146:413–431
37. Zhang L, Chu Z, Song H (2020) Understanding the relation between BIM application behavior and sustainable construction: a case study in China. Sustainability 12(1):306
38. Rogers J, Chong HY, Preece C (2015) Adoption of Building Information Modelling technology (BIM): perspectives from Malaysian engineering consulting services firms. Eng Constr Archit Manag 22(4):424–445
39. Cao Y, Zhang LH, McCabe B, Shahi A (2019) The benefits of and barriers to BIM adoption in Canada. In: ISARC. Proceedings of the international symposium on automation and robotics in construction, vol 36, pp 152–158
40. Aibinu A, Venkatesh S (2014) Status of BIM adoption and the BIM experience of cost consultants in Australia. J Prof Issues Eng Educ Pract 140:1–10
41. Gu N, London K (2010) Understanding and facilitating BIM adoption in the AEC industry. Autom Constr 19(8):988–999
42. Kuang S, Hore A, McAuley B, West R (2019) Development of a framework to support the effective adoption of BIM in the public sector: lessons for Ireland. In: Conference papers
43. Enegbuma WI, Aliagha UG, Ali KN (2014) Measurement of theoretical relationships in building information modelling adoption in Malaysia. In: Proceedings of the 31st international symposium on automation and robotics in construction and mining (ISARC)
44. A Researcher (2020) BIM: a technology acceptance model in Peru. J Inf Technol Constr 25:99–108
45. Yuan H, Yang Y (2020) BIM adoption under government subsidy: technology diffusion perspective. J Constr Eng Manag 146(1):1–15
46. Gurevich U, Sacks R (2020) Longitudinal study of BIM adoption by public construction clients. J Manag Eng 36(4)
47. Ahmed SHAA, Suliman SMAA (2020) A structure equation model of indicators driving BIM adoption in the Bahraini construction industry. Constr Innov 20(1):61–78
48. Oyewole EO, Dada JO (2019) Training gaps in the adoption of building information modelling by Nigerian construction professionals. Built Environ Proj Asset Manag 9(3):399–411
49. Almuntaser T, Sanni-Anibire MO, Hassanain MA (2017) Adoption and implementation of BIM – case study of a Saudi Arabian AEC firm. Int J Manag Proj Bus 11(2018):608–624
50. Succar B, Kassem M (2015) Macro-BIM adoption: conceptual structures. Autom Constr 57(57):64–79
51. Ayinla KO, Adamu Z (2018) Bridging the digital divide gap in BIM technology adoption. Eng Constr Archit Manag 25(10):1398–1416
52. Kassem MA, Khoiry MA, Hamzah N (2019) Risk factors in oil and gas construction projects in developing countries: a case study. Int J Energy Sect Manag 13(4):846–861
53. Al-Ashmori YY et al (2020) BIM benefits and its influence on the BIM implementation in Malaysia. Ain Shams Eng J 11(xxxx):1013–1019
54. Gamil Y et al (2017) Qualitative approach on investigating failure factors of Yemeni mega construction projects. MATEC Web Conf 103
55. Kassem MA, Khoiry A, Hamzah N (2019) Evaluation of risk factors affecting on oil and gas construction projects in Yemen. Int J Eng Technol 8(1):6–14
56. Gamil Y, Abdul Rahman I (2018) Assessment of critical factors contributing to construction failure in Yemen. Int J Constr Manag 0(0):8
57. Dahmas S, Li Z, Liu S (2019) Solving the difficulties and challenges facing construction based on concurrent engineering in Yemen. Sustain 11(11):3146
58. Gamil Y, Rahman IA, Nagapan S, Nasaruddin NAN (2020) Exploring the failure factors of Yemen construction industry using PLS-SEM approach. Asian J Civ Eng 21(6):967–975
59. Alaghbari W, Al-Sakkaf AAA, Sultan B (2019) Factors affecting construction labour productivity in Yemen. Int J Constr Manag 19(1):79–91

60. Ahmed AL, Kawalek JP, Kassem M (2017) A comprehensive identification and categorization of drivers, factors, and determinants for BIM adoption: a systematic literature review. In: Computing in civil engineering 2017, vol 0. American Society of Civil Engineers (ASCE), pp 220–227
61. Elghdban MG, Azmy NB, Bin Zulkiple A, Al-Sharafi MA (2020) factors affecting the adoption of advanced IT with specific emphasis on building information modeling based on TOE framework: a systematic review. Int J Adv Sci Technol 29(4):3314–3333
62. Hochscheid E, Halin G (2020) Generic and SME-specific factors that influence the BIM adoption process: an overview that highlights gaps in the literature. Front Eng Manag 7(1):119–130
63. Hochscheid E, Halin G (2019) Micro BIM adoption in design firms: guidelines for doing a BIM implementation plan. Proc Creat Constr Conf 119
64. Elghdban MGM et al (2021) A systematic review of the technological factors affecting the adoption of advanced IT with specific emphasis on building information modeling. Int J Adv Sci Technol 29(4):3314–3333
65. Durdyev S, Mbachu J, Thurnell D, Zhao L, Reza Hosseini M (2021) BIM adoption in the Cambodian construction industry: key drivers and barriers. ISPRS Int J Geo-Inf 10(4)
66. Elhendawi A, Omar H, Elbeltagi E, Smith A (2019) Practical approach for paving the way to motivate BIM non-users to adopt BIM. Int J BIM Eng Sci 1–22
67. Kassem M, Succar B, Dawood N (2015) Building information modeling: analyzing noteworthy publications of eight countries using a knowledge content taxonomy. Build Inf Model Appl Pract 61:329–371
68. Vukovic V, Hafeez MA, Chahrour R, Kassem M, Dawood N (2015) BIM adoption in Qatar: capturing high level requirements for lifecycle information flow. Convr 2
69. Hamma-Adama M (2020) Framework for macro building information modelling (BIM) adoption in Nigeria
70. Van Tam N, Diep TN, Quoc Toan N, Le Dinh Quy N (2021) Factors affecting adoption of building information modeling in construction projects: a case of Vietnam. Cogent Bus Manag 8(1):1918848
71. Hamma-Adama M, Kouider T, Salman H (2020) Analysis of barriers and drivers for BIM adoption. Int J BIM Eng Sci 3(1):18–41
72. Hamma-Adama M, Kouider T (2019) What are the barriers and drivers toward BIM adoption in Nigeria? In: Skibniewski MJ, Hajdu M (eds) CCC 2019 proceedings of the creative construction conference (2019) 073, pp 529–538
73. Walasek D, Barszcz A (2017) Analysis of the adoption rate of building information modeling [BIM] and its return on investment [ROI]. Procedia Eng 172:1227–1234
74. Hamma-Adama M, Kouider T, Salman H (2020) State of building information modelling (BIM) adoption in Nigeria. Constr Ind Fourth Ind Revolut

Determinants of the Sustainability of Tech Startup: Comparison Between Malaysia and China

Chin Wai Yin, Ezatul Emilia Muhammad Arif, Tung Soon Theam, Seah Choon Sen, Theresa Chung Yin Ying, and Cham Tat Huei

Abstract Tech startups are critical in sustaining innovation and growth in a country. They need to be nurtured so that they can grow to become viable and sustainable business entity. A suitable ecosystem therefore is vital to the development of startups. The aim of this study is to investigate the relationship between the sustainability of startups and the supporting factors, which are incubators, accelerators, co-working spaces, mentors and events. Comparison between tech startups in Malaysia and China is examined to support factors and the sustainability of startups. The result of this study showed that accelerators, events and mentors have the strongest influence to the sustainability of tech-startups as compared to co-working space.

Keywords Startups · Ecosystems · Incubators · Accelerators · Co-working spaces · Mentors · Events

C. W. Yin (✉) · E. E. M. Arif · T. S. Theam · S. C. Sen · T. C. Y. Ying
Faculty of Accountancy and Management, Universiti Tunku Abdul Rahman, Kajang, Selangor, Malaysia
e-mail: chinwy@utar.edu.my

E. E. M. Arif
e-mail: ezatul@utar.edu.my

T. S. Theam
e-mail: tungst@utar.edu.my

S. C. Sen
e-mail: seahcs@utar.edu.my

T. C. Y. Ying
e-mail: tying1124@1utar.my

C. T. Huei
UCSI Graduate Business School, UCSI University, Kuala Lumpur, Malaysia

M. Al-Emran et al. (eds.), *International Conference on Information Systems and Intelligent Applications*, Lecture Notes in Networks and Systems 550,
https://doi.org/10.1007/978-3-031-16865-9_45

1 Introduction

This research is undertaken to study on the determinants (namely incubators, accelerators, coworking spaces, mentors, and event) of the sustainability of tech start-ups in China and Malaysia. The aim is to study the relationship between these five determinants and the sustainability of tech start-up. The research questions will address whether there is any significant relationship between these determinants and sustainability of start-up. A start-up plays an important role in economic and social growth by providing job opportunities, serving as an incubator for eco-innovation, and creating new markets. But difficulties in sustaining and surviving in the business world faced by startups have led to their high failure rate at 50 to 95% [1]. The difficulties have become more intensive especially in an emerging country like Malaysia. The startup for information industry has registered the highest failure rate at 63% as at 2022 [2]. Startups have contributed around 39% of total revenue in the global industry [1]. However, there are numerous startups failed to recognise the challenges of the operation and the entrepreneurs are incapable to achieve success [3]. Startups are fragile, the failure rate is as high as 90%. Past studied showed that, 20% of startups failed within one year, 30% within two years, 50% within five years and 70% within ten years of operation [2].

Malaysia being the 39th largest economy in the world has fundamentals for startups such as strategic locations, availability of natural resources, and accessible modernised ports at sea routes. Over the last few years, Malaysian government has granted several incentives to help startups, such as Cradle fund, Malaysia Digital Economy Corporation (MDEC), Malaysian Technology Development Corporation (MTDC), and Malaysia Venture capital management Berhad (MAVCAP). Start-up incubators such as MaGIC, and MyCreative Ventures have been established to assist start-ups. These government incentives tend to reduce the high failure rates of startups [4]. In the past several researchers have studied on few determinants individually without adopting a holistic approach.

China is one of the world-leading countries supported by high-speed innovation and a strong economy based. The country has a booming startups ecosystem that is supported by the capital raised, and the figure of the growth on startups, and unicorns [5]. According to statistics, there is around 25% of unicorns are from China, this has shown that the power of startup in China [6, 7]. China has been recognised as a tech giant as the country has shown its success in several aspects like Baidu, Wechat, TikTok, etc. By looking at the list of unicorn companies, there are few companies are from China, such as Bytedance and Sense Time (Artificial Intelligence), Shein (Ecommerce), Cgtz (Fintech), etc. [8]. China has a strong head start on the development of tech startup. Over the last decade, many unicorns have been established in China. Malaysia is still at the infant stage of tech startup in comparison to China.

There are insufficient studies undertaken to develop an appropriate framework on the determinants of the sustainability of tech startup in Malaysia in comparison with China. Hence, the researchers have undertaken this research based on 5 selected determinants with the aim of developing a framework for tech startup in Malaysia. In this study, the phrase "sustainability" is used to determine the success of a tech startup that operating for more than three years [9]. In the era of industry revolution, the involvement of technology could be a direction in determinant of the sustainability of startup [10]. The research findings of this study will help to promote more tech startups in Malaysia in the future using China's success as a guide.

2 Literature Review

A strong ecosystem can help tech startups until they become sustainable enough to survive in the business world. Various authors highlight how a good ecosystem can boost the competitiveness of a region [11].

Past studies highlighted the importance of availability of financial support from government, market support to commercialise their products, technology-related support to the success of startups [1, 12, 13]. Some studies specifically focus on incubators [14, 15], accelerators [9, 16–18], co-working spaces [19–21], mentors [4] and events on their own or some combination [16, 22]. Most studies on startups concentrate on initiatives of specific country such as startup ecosystem in Australia [10, 16], and Czech Republic [23], challenges faced by start-up in South Africa, accelerator programs in US [17, 18], business startup programmes in Scotland [4] and also government support for start-ups in Malaysia [1] to name a few. However, few studies compared the significance of incubators, accelerators, co-working spaces, mentors and events to the sustainability of tech startup in different countries. Therefore, this study compares the factors that contributes to the sustainability of tech startups between Malaysia and China.

2.1 *Incubators and Sustainability of Tech Startup*

Incubators enables the nurturing of startups at the early stage and until the commercialisation of research or product [1, 17, 23]. In the early stages, incubator provide startups with infrastructure services, entrepreneurial skills knowhow and opportunity to crystallise business ideas or prototype through networking and finding potential partners [22]. However, a study by [24], highlighted the challenges Malaysian incubators faced in providing the necessary support to firms being incubated thereby hindering their potential. Thus, the following hypothesis was formed:

H1: There is a significant relationship between incubators and the sustainability of tech startups.

2.2 *Accelerators and Sustainability of Tech Startup*

Accelerator differs from incubators in that it provides a short-term program that helps boost development process of startups usually between three to six months [23]. Several studies proposed that the accelerators programs are giving a positive influence on startups [9, 18]. More than 70% of startups are still operating after participating in the accelerator program [22]. The reasons being that accelerator provides several functions like funding or mentorship which are critical for entrepreneurs. Hence, accelerator programs may contribute to rising the sustainability of startups. The second hypothesis of this study will be:

H2: There is a significant relationship between accelerators and the sustainability of startups.

2.3 *Co-Working Spaces and Sustainability of Tech Startup*

Coworking spaces are a place for individuals or a group of entrepreneurs to work alone or together [19] as well as for culture exchange and opportunities to collaborate [22]. Infrastructure like co-working spaces is provided for startups by incubators and accelerators programs. Organisers can rent an office that provides tables and chairs, or offer some organizational support like networking, printing services, and conference meeting support. A co-working space includes offices space for rent, organisational support like networking, printing services and conference support. Thus, giving participants access to information, knowledge, important resources, social capital, and opportunities for serendipity [21]. Many of the companies found that operating from co-working spaces makes employees feel comfortable as they are not working at a place with a traditional office setup. Co-working spaces also provide opportunities for entrepreneurs to connect with other entrepreneurs who are working at large films. According to [25], saving money can help the startup to keep sustainable of business running till financial break-even point. Therefore, the third hypothesis for this study is:

H3: There is a significant relationship between co-working spaces and the sustainability of tech startups.

2.4 *Mentors and Sustainability of Tech Startup*

The mentor focuses on connecting growing startups with entrepreneurs having expert knowledge and vast experience [26]. Mentor providers in Malaysia have included mentor plus (MDEC), and MaGIC Mentorship. China accelerator has been recognized as one of the most active mentorship networks in China by offering service to several fields of startups [27]. The objective of a mentor is to provide guidance and coaching for both startup founders and team members to learn the skill and knowledge on business and product development [23]. The help provided include information such as legal aspects and user experience tips and, in some cases, they may also become the partners of the companies [25]. Hence, the hypotheses for this study is:

H4: There is a significant relationship between mentors and the sustainability of startups.

2.5 *Events and Sustainability of Tech Startup*

Events served as an activity that happens at a particular time and place to giving chance for collaboration and knowledge sharing among the participants [22]. An event could offer a chance for building a network between entrepreneurs and founders of successful startups, investors, and companies. During events, startups can pitch their ideas to investors [25], media, and new experts [22]. Founding team members can seek advice when attending the events to solve difficulties they faced. Besides, the latest information will be informed disclosed to participants during the pitching event. Thus, the fourth hypotheses are:

H5: There is a significant relationship between events and the sustainability of startups.

Depicts the research model of this study:

Fig. 1 Research model

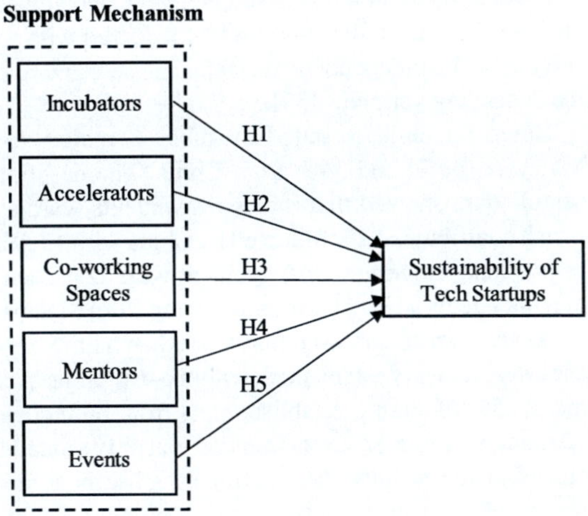

3 Research Method

The target respondence of this research were business startup owners in the tech-related field both in Malaysia and China. In order to produce the most accurate and related responses in this study, targeted job positions were chosen through purposive sampling as suggested by the prior literature [28–31]. The choices of respondents' job position were mainly founder, co-founder or top management of the company. However, other positions were also taken into considerations in this study.

A survey was created using Google form were distributed and collected from the tech startup communities of Facebook from China and Malaysia. The questionnaire consists of two sections, where Section A comprised of questions to screen out non-targeted respondents. Startup owners that are not from the tech-field were not included in the survey. There were demographic questions included in this section that had helped identify respondent's job position, country of origin, startup duration and tech field. In this section, the results of variance, standard deviation, median, mean, and mode will be determined. Section B on the other hand included the five determinants to assess the sustainability of tech startup. All measurement scale used in this study were adapted from past studies of [22].

In addition to the above, a quota sampling technique was used to ensure that each sample has an equal representation in a research study [32]. Therefore, in conducting a study to compare subject matters between two countries like the present study, this sampling technique is deemed fit for the purpose. A total of 200 survey data were received from the targeted audience and subjected to data cleaning. The data analysis for the present study was conducted with the use of IBM SPSS Statistical software. The independent variables are selected to determine the relationship with the dependent variables [33].

Based on findings stated in Table 1, majority of the respondents were from Malaysia (66%) and 34% from China. The descriptive analysis conducted on the sample data showed that major respondents were startup founders from Malaysia which contributed 22% followed by China with 15.5%. The second largest group was respondents among top management with 22% each from both countries. Majority of respondents among co-founders were from Malaysia with 20.5% and China 5.5%.

Analysed result showed that more than half (52.5%) of the startups in both countries were recently established; within 0–6 months. Accumulatively, it was discovered that 20.5% of startup establishments from both countries, have operated within 7–12 months. However, it was founded that only a small fraction among the respondents (12.5%) had sustained their startup establishment for more than 1 year. In this study, it showed that tech-startups are mushrooming in both countries, especially from the field of Information Communication, Technology (ITC) which accounted for 19%, Media Tech field 18.5% and Data field with 17% of the total data collected. Nevertheless, out of 200 respondents and over 10 fields, only 10% had sustained in the market for more than one year.

According to [34], Cronbach's Alpha is a statistic commonly quoted to demonstrate that tests and scales that have been constructed or adopted for research projects

Table 1 Job position and duration of startup establishment

Job position	Malaysia		China	
	Frequency	%	Frequency	%
Founder	44	22	31	15.5
Co-founder	22	11	22	11
Top management	41	20.5	11	5.5
Others	25	12.5	4	2
Total	132	66	68	34
Duration	Malaysia		China	
	Frequency	%	Frequency	%
0–6 months	78	39	27	13.5
7–12 months	23	11.5	18	9
Months	17	8.5	12	6
>19 months	14	7	11	5.5
Total	132	66	68	34

are fit for purpose. Cronbach's Alpha of the variables in this study had achieved more than 0.6, which was proven to support the reliability of this study. Meanwhile, Co-working spaces (0.764), mentors (0.763), events (0.755) and accelerators (0.733) had all achieved more than 0.7 of value; which fell under the good range of reliability towards the sustainability of tech startups. Although the last variable (incubator) was not listed, it is still giving a fair range of reliability towards the sustainability of tech-startups.

The inferential analysis is to test the hypothesis developed in this study by using Pearson Correlation Coefficient analysis and Multiple Regression analysis.

Table 2 illustrates an inferential statistic derived from SPSS showing that all the independent variables are significant to the dependent variables as the Pearson correlation is between 0.534 and 0.676, at the significant level of <0.001. Thus, the variables are positively correlated. Accelerators has the strongest prediction power of 0.676 R-value towards the dependent variable (Sustainability of tech-startups). This proved that there is a strong positive relationship and a high correlation between accelerators and sustainability. This is followed by events with 0.655, mentors with 0.602 and incubators with 0.558 and lastly co-working spaces with 0.534 R-value. Based on Cronbach 1951, it showed that incubators and co-working spaces have the weakest relationship towards the sustainability of tech-startups.

The outcome of R Square, which is 59.5% of the variation in the dependent variable (sustainability of tech-startup) is influenced by the independent variables (accelerators, incubators, co-working spaces, mentors, and events). The R-value is 0.771; R Square is 0.595 and this showed that the independent variables used in this study have the influence power towards the dependent variable.

ANOVA test conducted also proved that all the five independent variables used in this study are significant to explain the dependent variable. The result showed F

Table 2 Pearson correlation analysis

		Incubators_Mean	Accelerators_Mean	CoworkingSpaces_Mean	Mentors_Mean	Events_Mean	DV_Mean
Incubators_Mean	Pearson correlation	1	0.619**	0.438**	0.524**	0.497**	0.558**
	Sig. (2-tailed)		0.000	0.000	0.000	0.000	0.000
	N	200	200	200	200	200	200
Accelerators_Mean	Pearson correlation	0.619**	1	0.426**	0.621**	0.647**	0.676**
	Sig. (2-tailed)	0.000		0.000	0.000	0.000	0.000
	N	200	200	200	200	200	200
CoworkingSpaces_Mean	Pearson correlation	0.438**	0.426**	1	0.502**	0.519**	0.534**
	Sig. (2-tailed)	0.000	0.000		0.000	0.000	0.000
	N	200	200	200	200	200	200
Mentors_Mean	Pearson correlation	0.524**	0.621**	0.501**	1	0.530**	0.602**
	Sig. (2-tailed)	0.000	0.000	0.000		0.000	0.000
	N	200	200	200	200	200	200
Events_Mean	Pearson correlation	0.497**	0.647**	0.519**	0.530**	1	0.655**
	Sig. (2-tailed)	0.000	0.000	0.000	0.000		0.000
	N	200	200	200	200	200	200
DV_Mean	Pearson correlation	0.558**	0.676**	0.534**	0.602**	0.655**	1
	Sig. (2-tailed)	0.000	0.000	0.000	0.000	0.000	
	N	200	200	200	200	200	200

value of 57.009 at 0.000 significant level, thus verifying that the suitability of the model is achieved. The standardised coefficient aims to examine the most important independent variable.

The coefficient test executed in this study aimed to examine the most important independent variable and how one unit change in an independent variable can affect the dependent variable. The unstandardised coefficients result obtained from this study interpreted 0.123 changes of incubators when there is a unit change in the dependent variable. Followed by 0.280 of accelerators, 0.143 of co-working spaces, 0.148 of mentors and lastly 0.256 of events.

Whereas: Y = Sustainability of tech-startup.
= Constant term, Value of Y when X become zero
= Dimension of the sustainability of tech-startup
= Accelerators
= Incubators
= Co-working Spaces
= Mentors
= Event

According to the above illustration, the equation of multiple regression of this study is as the following:

$$Y = a + b1X1 + b2X2 + b3X3 + b4X4 + b5X5$$

$$\text{Sustainabilityoftech-startup} = (0.212) + (0.123)\,(\text{Incubators}) + (0.280)\,(\text{Accelerators}) + (0.143)\,(\text{Co-workingSpaces}) + (0.148)\,(\text{Mentors}) + (0.256)\,(\text{Events}).$$

Based on the standardised coefficients beta, accelerators seemed to have scored the highest value of 0.279 among the other variables; which means, accelerators is the most significant factor that influences the sustainability of tech-startups. This is followed by 0.257 of events; 0.159 of mentors; 0.155 of co-working spaces and lastly 0.107 of incubators. Incubators having the lowest score has the least influence toward the dependent variable as shown in Table 3.

In significance to the study of two countries comparison, A t-test was conducted on the dependent variable to interpret the differences of sustainability of tech-startups between Malaysia and China. Table 4 illustrated the result of data comparison where t = 1.909, which means there is no significant difference of sustainability of tech-startups between Malaysia and China.

Table 3 Summary of hypotheses results

H1	There is a significant relationship between incubators and the sustainability of tech-start-up	p-value: 0.081	Rejected
H2	There is a significant relationship between accelerators and the sustainability of tech-start-up	p-value: 0.000	Accepted
H3	There is a significant relationship between co-working spaces and the sustainability of tech start-up	p-value: 0.007	Accepted
H4	There is a significant relationship between mentors and the sustainability of a start-up	p-value: 0.012	Accepted
H5	There is a significant relationship between events and the sustainability of a start-up	p-value: 0.000	Accepted

Table 4. t-test on sustainability of tech-startups (dependent variable)

Mean_Sustainability of tech-startup	Nationality	Frequency	t	Sig. (2-tailed)
	Malaysia	132	1.909	0.058
	China	68		

4 Discussion

Past study stated that the failure rate of startups has reached 50 to 95% especially in emerging countries [1]. Another study by [35] supported the statement through a study that proved a significant number of startups failed during their first year of operation and most crashed within five years. Hence, this study was conducted to assess the factors that contributed towards the sustainability of tech-startup and what could be learnt from China as one of the pioneers.

The background of this research study had adopted and adapted a start-up ecosystem as the framework. A suitable ecosystem is required to be built to support startups' early stages due to its' fragility. According to the case study in Oulu, several elements like incubators, accelerators, co-working spaces, mentors, and events have been recognized as supporting factors in the startup ecosystem [22]. Past study also mentioned that the startup ecosystem is a regional phenomenon supported by multiple sub-elements. Hence, those supporting factors are playing an important role in building up the ecosystem which is helpful for startups' early-stage development.

The findings of this study have revealed the strongest and most relevant factors that contributes to the sustainability of tech-startup in both countries. These factors include accelerators, events and mentors. It is believed that accelerators are ranked as the most relevant factor because it is a limited-duration program aimed at helping entrepreneurs to define ideas and build their first prototype [36]. Most startup shaped their prototype during the program. On the other hand, event plays a significant role where startup founders use it as a platform to expand their business network and exchanging business idea [37]. Along their journey to sustain the business, some were fortunate to meet a good mentor via the accelerators or events [38]. Advise and guidance given by mentors will help tech-startups in setting an accountable goal, developing their contacts, and identifying the key resources for the startup [39].

Co-working spaces seemed to be the weakest factor but have always been recognised to be one of the supporting factors that contributed to the early-stage development of startup. The co-working spaces act as a catalyst to connect the startup founders by providing collaboration and networking facilities to increase their survivability through open innovation [40].

5 Conclusion

The study on sustainability of tech-startups found many factors that contributes to its ability to thrive. Comparison between the two countries have showed similarities on the role of supporting factors to tech-startups. Although it is also believed that the random sampling technique used in this study was inappropriate to derive a fair number of samples from both countries. It is suggested that in future, researchers should consider cluster sampling technique in order to achieve an equal number of responses from both countries in comparison. More variables should be considered to identify any significant difference between the sustainability of tech-startups in Malaysia and China so that a greater contribution can be benefited by the affected industry.

References

1. Kee DMH, Yusoff YM, Khin S (2019) The role of support on start-up success: a PLS-SEM approach. Asian Acad Manage J 24
2. Failory Homepage. https://failory.com/blog/startup-failure-rate. Accessed 28 Mar 2022
3. Yusuf JE (2014) Impact of start-up support through guided preparation. J Entrep Public Policy 3(1):72–95. https://doi.org/10.1108/JEPP01-2012-0004
4. Deakins D, Sullivan R, Whittam G (2000) Developing business start-up support programmes: evidence from Scotland. Local Econ 15(2):159–167. https://doi.org/10.1080/02690940050122703
5. The Org: The top 10 Chinese Startups you should know about in 2020. https://theorg.com/insights/the-top-10-chinese-startups-you-should-know-about-in-2020. Accessed 30 Mar 2022
6. Embroker: 106 Must-Know Startup Statistics for 2021 (n.d.). https://www.embroker.com/blog/startup-statistics/. Accessed 8 June 2021
7. Cham TH, Low SC, Lim CS, Aye AK, Ling RLB (2019) The preliminary study on consumer attitude towards fintech products and services in Malaysia. Int J Eng Technol 7(2.29):166–169
8. Cheong YS, Seah CS, Loh YX, Loh LH (2021) Artificial Intelligence (AI) in the food and beverage industry: improves the customer experience. In: 2021 2nd international conference on artificial intelligence and data sciences (AiDAS). IEEE, pp 1–6
9. Winston Smith S, Hannigan TJ, Gasiorowski L (2013) Accelerators and crowd-funding: complementarity, competition, or convergence in the earliest stages of financing new ventures. In: University of Colorado-Kauffman Foundation crowd-funding conference, Boulder, CO
10. Har LL, Rashid UK, Te Chuan L, Sen SC, Xia LY (2022) Revolution of retail industry: from perspective of retail 1.0 to 4.0. Procedia Comput Sci 200:1615–1625

11. Tripathi N, Seppänen P, Boominathan G, Oivo M, Liukkunen K (2019) Insights into startup ecosystems through exploration of multi-vocal literature. Inf Softw Technol 105:56–77
12. Cham TH, Easvaralingam Y (2012) Service quality, image and loyalty towards Malaysian hotels. Int J Serv Econ Manage 4(4):267–281
13. Fam KS, Cheng BL, Cham TH, Tan CYM, Ting H (2021) The role of cultural differences in customer retention: evidence from the high-contact service industry. J Hosp Tour Res 10963480211014944. https://doi.org/10.1177/10963480211014944
14. Peters L, Rice M, Sundararajan M (2004) The role of incubators in the entrepreneurial process. J Technol Transf 29(1):83–91
15. Jamil F, Ismail K, Mahmood N (2015) A review of commercialization tools: university incubators and technology parks. Int J Econ Fin Issues 5(S):223–228
16. Bliemel MJ, Flores RG, de Klerk S, Miles MP, Costa B, Monteiro P (2016) The role and performance of accelerators in the Australian startup ecosystem. Department of Industry, Innovation & Science
17. Cohen S, Hochberg YV (2014) Accelerating startups: the seed accelerator phenomenon
18. Hallen BL, Cohen S, Bingham C (2019) Do accelerators work? If so, how? SSRN https://ssrn.com/abstract=2719810 or http://dx.doi.org/10.2139/ssrn.2719810
19. Howell T, Bingham C (2019) Coworking spaces: working alone, together. Kenan Institute working paper, Chapel Hill, NC
20. Kojo I, Nenonen S (2016) Typologies for co-working spaces in Finland–what and how. Facilities 34(5/6):302–313
21. Leclercq-Vandelannoitte A, Isaac H (2016) The new office: how coworking changes the work concept. J Bus Strateg 37(6):3–9
22. Tripathi N, Oivo M (2020) The roles of incubators, accelerators, co-working spaces, mentors, and events in the startup development process. In: Fundamentals of software startups. Springer, Cham, pp 147–159
23. Krajcik V, Formanek I (2015) Regional startup ecosystem. Eur Bus Manage 1(2):14–18
24. Lose T, Rens V, Yakobi K, Kwahene F (2020) Views from within the incubation ecosystem: discovering the current challenges of technology business incubators. J Crit Rev 7(19):5437–5444. https://doi.org/10.31838/jcr.07.19.632
25. Melegati J, Kon F (2020) Early-stage software startups: main challenges and possible answers. Fund Softw Startups 129–143
26. MDEC Homepage. https://mdec.my/gain/mentor-plus/. Accessed 28 Mar 2022
27. Chinaaccelerator homepage. https://chinaaccelerator.com/mentors. Accessed 28 Mar 2022
28. Cheng BL, Cham TH, Micheal D, Lee TH (2019) Service innovation: building a sustainable competitive advantage in higher education. Int J Serv Econ Manage 10(4):289–309
29. Cham TH, Cheng BL, Low MP, Cheok JBC (2020) Brand Image as the competitive edge for Hospitals in Medical Tourism. Eur Bus Rev 31(1):31–59
30. Cham TH, Cheng BL, Ng CKY (2020) Cruising down millennials' fashion runway: a cross-functional study beyond Pacific borders. Young Consumers 22(1):28–67
31. Cham TH, Lim YM, Sigala M (2022) Marketing and social influences, hospital branding, and medical tourists' behavioural intention: before-and after-service consumption perspective. Int J Tour Res 24(1):140–157
32. Cham TH, Lim YM, Aik NC, Tay AGM (2016) Antecedents of hospital brand image and the relationships with medical tourists' behavioral intention. Int J Pharm Healthc Market 10(4):412–431
33. Samsuddin S, Shah ZA, Saedudin RR, Kasim S, Seah CS (2019) Analysis of attribute selection and classification algorithm applied to hepatitis patients. Int J Adv Sci Eng Inf Technol 9(3):967–971
34. Keith ST (2018) The use of Cronbach's Alpha when developing and reporting research instruments in science education. Res Sci Educ 48:1273–1296
35. Ondas A (2021) A study on high-tech startup failure, Masters thesis of Master's Degree in Entrepreneurship and Business Competence, School of Business, JAMK University. https://doi.org/10.13140/RG.2.2.25524.37765

36. Mansoori Y, Karlsson T, Lundqvist M (2019) The influence of the lean startup methodology on entrepreneur-coach relationships in the context of a startup accelerator. Technovation 84:37–47
37. Füller J (2021) The difference in entrepreneurs and non-entrepreneurs perceptions of achievement and enjoyment during a startup event-on the example of Skinnovation (Doctoral dissertation, University of Innsbruck)
38. Aguiar RBD, Silva DS, Caten CST, Silva LCP (2019) Lean Mentorship: fitting external support to entrepreneur needs over the startup development. Production 29
39. What is the role of a mentor? | DO-IT (2021). https://www.washington.edu/doit/what-role-mentor. Accessed 1 May 2022
40. Lestari ED (2020) Is co-working increase survivability? Study on how collaborating and networking facilitates open innovation process for startups. IJNMT (Int J New Media Technol) 7(2):68–75

Mobile-Based Green Office Management System Dashboard (GOMASH) for Sustainable Organization

Naveenam A/P Mayyalgan, Mazlina Abdul Majid, Muhammad Zulfahmi Toh, Noor Akma Abu Bakar, Ali Shehadeh, and Mwaffaq Otoom

Abstract Green is now known for being environmentally friendly and energy-efficient, while sustainability aids in conserving and preserving natural resources and also the environment. Although the higher education institutes (HEI) in Malaysia play a critical part in providing more sustainable place, by introducing green campus initiatives to provide a greener environment, however, the university staffs' high involvement in maintaining sustainability and green practices in the office side is still questionable. Therefore, this project is developed mainly to propose a green office management system dashboard (GOMASH) as an initiative to practice green inside faculty as well as to analyze the paper usage and reduce the total amount of paper wastage inside the office. Furthermore, GOMASH also provides paper limit suggestion to reduce the paper wastage in the office with a goal of sustainability. Agile methodology is adopted as the software development life cycle (SDLC) of GOMASH where the process iteratively caters the requirements from the users which makes way for improvement of the system by using a deductive approach with a general idea that GOMASH helps develop green awareness among users to reduce wastage of paper. This paper provides a conceptual and empirical study on the amount of paper wastage in faculty that is observed in the beginning of the development and

N. A. Mayyalgan · M. A. Majid · M. Z. Toh
Faculty of Computing, Universiti Malaysia Pahang, 26600 Pekan, Pahang, Malaysia
e-mail: mazlina@ump.edu.my

M. Z. Toh
e-mail: zulfahmi@ump.edu.my

N. A. A. Bakar (✉)
Department of Computing and Information System, TAR University College Pahang, Kuantan, Malaysia
e-mail: noorakma@tarc.edu.my

A. Shehadeh · M. Otoom
Faculty for Engineering Technology, Yarmouk University, Irbid, Jordan
e-mail: ali.shehadeh@yu.edu.jo

M. Otoom
e-mail: mof.otoom@yu.edu.jo

M. Al-Emran et al. (eds.), *International Conference on Information Systems and Intelligent Applications*, Lecture Notes in Networks and Systems 550,
https://doi.org/10.1007/978-3-031-16865-9_46

the data collected from the application is then analyzed and visualized in a dashboard. All modules have been tested using software test description (STD) to make sure GOMASH is validated and verified and found appropriate to be implemented in order to improve green office management practice for sustainable organization.

Keywords Green technology · Sustainability · Green office · Green IT/ IS

1 Introduction

Over the last decades, the impact of information technology (IT) on society, environment and economy are increasing. The IT practitioners [1] which includes the higher education institutes who are taking a vast number of initiatives to provide green environment and sustainability development inside the faculty. Green generally is a word that is used to name a color which is a combination of yellow and blue. However, in this era of globalization, the color Green got linked with environmental issues and therefore progressed to have a deeper meaning [1]. Green is now known for being environmentally friendly and energy efficient. Meanwhile, sustainability refers to planning as well as implementing IT infrastructure that aid in attaining the institutes' short-term purposes at the same time conserving and preserving natural resources and also the environment [2]. The extreme growth of the natural resources' consumption, CO2 emissions as well as the emergent awareness of the environmental issues had influenced the IT practitioners to progressively recognize the importance of these sustainable practices. The higher education institutes (HEI) play an important role in transiting to a more sustainable society which has been highlighted and recognized for almost three decades [3]. To make this a success, green campus initiatives have been introduced to gain a significant momentum of greening the environment. Green campus is understood as the model through which higher education institutes develop organizations able to make the interaction between environment and university dwellers live together in a sustainable way [4]. Green campus is a place where there is a combination of environmentally friendly initiatives and education to promote sustainable and eco-friendly practices in the campus. Lecturers, staffs as well as other dwellers of the university can play a crucial part in this initiative by practicing green while they are in the campus. This is because inside campus, the usage of papers and resource consumption eventually depends on these people.

Green practices had been implemented and conducted by the Universiti Malaysia Pahang (UMP) for quite a long time. They are focused more to operate in a sustainable manner with being prudent on using the energy and other natural resources. However, questions have been raised about the staffs' and lecturers' involvement in maintaining sustainability in the green campus, especially in the office side. The number of printed forms utilized and unused in the faculty was counted. When comparing the number of unused printed forms to the number of used forms, it was discovered that the number of unused printed forms was larger. It is because for each semester, the staffs are printing forms without having the proper calculation or study on the average number of forms actually being used. This directly leads to the main issue which is paper wastage. The staffs are lack in awareness of the green and what they need to do to maintain the green inside the faculty UMP organization has been practicing green initiatives in the campus. The organization had adopted and adapted green initiatives from the 17 Sustainable Development Goals, 5 Strategic Trust of National Green Policy as well as the 7 indicators of UI Green metric World University Ranking in many of the activities and decision-making processes. However, the main problem the organization currently faces is the involvement of the staffs and lecturers in these green initiatives and sustainability development.

The staffs and lecturers are somewhat directly involved in maintaining the green inside the faculty. Although some green activities or initiatives are practiced inside the university especially in minimizing the energy or electricity consumptions, conducting the 6R campaign in the aspect of waste management, and providing awareness about green, however there are still exist problems to maintain these green activities, mainly because the staffs or the other dwellers could not really see the cost that has been wasting as a result of not maintaining these activities properly, especially inside the office. These make them to less contribute to providing a greener environment in UMP. This is mainly because of the lack of awareness about the green and sustainability development that could be done inside the office. Not only that, many lecturers and staffs still prefer to use the manual way of using papers for some applications such as internship forms, PA forms, leave application form and many more. There is no proper tool to analyze the amount of money that can be saved by digitalizing these forms for them to refer which will aid them in practicing green more inside the office. Due to this problem, this project is proposed with the main aim of providing a tool or dashboard which can be used by the faculty in UMP to practice green inside the office by analyzing the amount of cost that can be saved by reducing the paper usage in the office as this tool provides the awareness regarding green office and sustainable development which eventually help in providing a greener campus.

2 Related Works

Green Building Studio [5] is a cloud-based system that permits the users to run building performance simulations to optimize energy efficiency and work towards carbon neutrality at the beginning of the design process. It helps to design high-performance buildings fast and at cost of conventional method. It is designed to greatly simplify the task of conducting whole building performance analysis in today's Building Information Modelling (BIM) authoring tools. It uses DOE-2, a proven and validated simulation engine, to provide energy use, water use, and carbon emission results. It is used by the building organization in the earlier design process. This system has a special feature of including the climate data automatically, which is used to analysis the energy usage of the designed building. Not only that, it also analyses the energy used in both conceptual and information model. Besides that, it is also automatically including the building information such as the building's types and sizes of the walls, windows, floors, roofs and other elements from Energy Analytical Model, so it eases the users as they do not need to transfer the building information manually. The data of this system is stored in Autodesk Climate Server. This system is focused on analysis of energy, water and carbon emission green metrics. It is deployed in web server. However, this system has limited availability; requires an internet connection because all data is integrated with the cloud. Minimal analysis type: Not many analysis types are available with this system, so fewer analysis can be conducted.

The next system is GreenDash system which is used to calculate the sustainability for green software design. It is used by the software developers and auditors to give proper suggestion based on the green percentage after it has been analyzed. It provides few special features which includes the analysis of the sustainability of the green for each component namely database, hardware, people and network. Not only that, it also allows the users to view the suggestions from the auditors to help them further improve the green software design. The data of this system is stored in MySQL and this system focused to analyze mainly the performance energy which is the power utilization of the software design. This system is also deployed in the web server. However, it supports on web platform only: No mobile application or platform available. Delayed analysis or calculation: No live calculation of the green percentage using Artificial Intelligence. SGP Impact Tracker [7] is a dashboard system that is used by the printing community that allows the printers to manage their certification thoroughly and make sure a continued progress as they seek to further reduce waste as well as improve the efficiency. This system provides few certified features such as a criterion and to track sustainability success rates and operational improvements against previous performance and other printers available in the community. Besides that, it allows access to all sustainable data in one handy location, which aids on an easy recertification. Not only that, it provides real time graphics to illustrate the status of the certification and progress of the organization.

This system uses a cloud server to store the information of the system such as the feedback and status of the certification. It focuses on analyzing mainly the energy, water, waste, carbon footprints of the printers and it is deployed in a web server. However, this system is high cost; subscription is needed and is required to pay. Complex functionalities: The users must possess strong knowledge about the functionalities of the system to use it efficiently. Table 1 shows the summary of specification existing systems such as Green Building Studio, Green Dash System and SGP Impact Tracker.

3 Methodology

Agile model has been used as Software Development Life Cycle (SDLC) for the proposed system named, Green Office Management Dashboard System (GOMASH). This model provides a combination of iterative and incremental process which breaks down the development of GOMASH into small incremental builds. It consists of five

Table 1 System features comparison

Specification	Green building studio	Green dash system	SGP impact tracker
User	Building organization	Software developers and auditors	Printing community
Special features	Automatic climate data	Database, hardware, people and network analysis	Criterion and track sustainability success rates and operational improvements against previous performance and other printers available
	Analyse of energy used in both conceptual and detailed information model	Direct suggestions from auditors	Access all sustainability data in one handy location
	Automatic input from Energy Analytical Model (EAM)		Real time graphics to illustrate the status of the certification and progress of the organization
Database	Autodesk Climate Server	MySQL	Cloud Server
Focused green metric	Energy, water, carbon emission	Performance-Energy (Power utilization)	Energy, water, waste, carbon footprints
Deployment	Web application	Web application	Web application

phases including requirement analysis, design, development, testing and deployment. This method is adopted in this study as this process works well for small projects and it allows to assess the users' satisfaction at each phase, easing the project development and ensuring great user experience, especially when the given duration of this project development is short. Figure 1 shows the flow of the GOMASH System. First, the system will receive the form type details from admin and these details will be sent to the staff by displaying a list of form type along with their details. Next, the staff will add the used paper details for the current semester. Then, the system will process the details entered by the staff and calculate the total cost of the paper used for the current semester. The processed details are then analyzed and displayed in the dashboard where the staff can see the total cost of the paper used for the current semester in a visualized graph. They also can use the graph to compare the cost of the paper used from previous semesters. If the current semester cost is higher compared to the previous semester, the system will calculate and display a paper limit suggestion for the staff to use for the upcoming semester. These details are displayed at the analysis dashboard for the staff.

In the requirement analysis phase, requirement scope and functions of the system are elicited and identified. The first goal of this project which is usage of papers

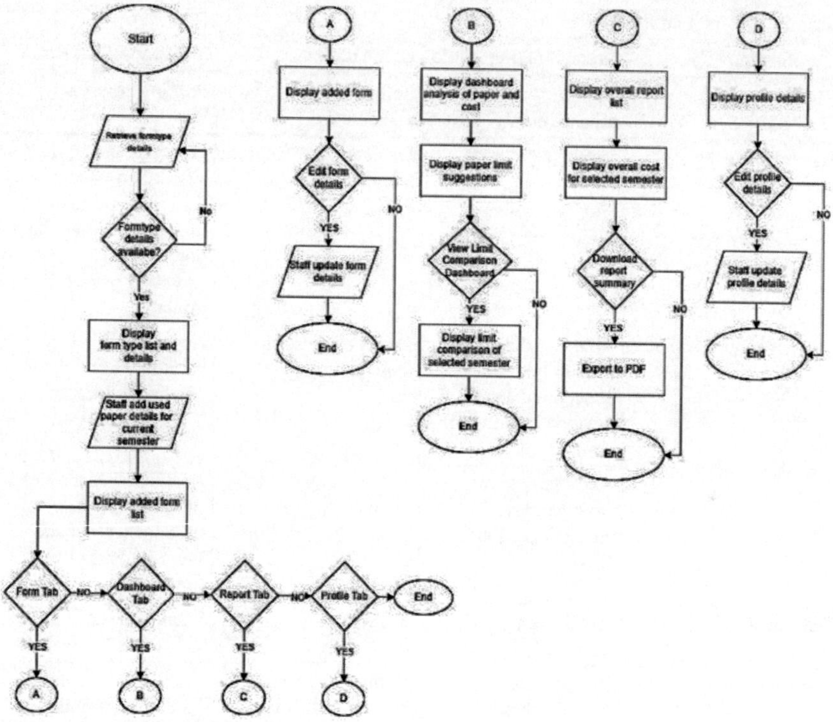

Fig. 1 GOMASH system flow

in the faculty and ways to digitalize them is done during this phase. Besides, few previous works and existing systems are studied and compared for further reference to develop the GOMASH system. Besides that, Software requirement specification is prepared to describe how GOMASH is developed consisting of context diagram, data flow diagram, use case diagram and its description. During the designing phase, the prototype interfaces of GOMASH are designed. Furthermore, software design documentation (SDD), is also prepared to define the system architecture, detailed design, and data dictionary. For GOMASH, Model-View-Controller (MVC) architecture is used which consists of three components—model, view, controller.

GOMASH is designed using the Android Studio development platform. The second goal of this work is to develop a green office dashboard that contains relevant information that provide awareness of green by analyzing the usage of papers. The AnyChart Android Chart API is used as the data visualization library. It is useful for creating the interaction/ interacting charts for the dashboard analysis purposes. The next phase is the testing phase. In this phase, each of the function of the system is tested followed by the testing of the overall flow of the system. If any error/ bug is found, it will be fixed before going on the next phase which is deployment. User Acceptance Test (UAT) is performed to test this system with the end user which is the UMP Faculty of Computing (FK) staffs. Deployment is the final for GOMASH and if it is fully functional and satisfied by the end user, it will be deployed to real environment. Otherwise, if the system is not ready to be deployed the Agile cycle will be repeated until satisfaction is achieved.

GOMASH System is made up of two separate mobile application systems. One of the mobile application systems is used by the admin to verify the staff account. This application is used by the admin to manage the form type that is available in the UMP faculty. The admin may add, edit and delete any form type from the application which will be sent to the staff to add the used form information according to the form type. Figure 2 depicts the interfaces of admin add form type and admin form type list respectively.

The other GOMASH part will be used by the staff. They use this system to view the dashboard analysis of the total cost of the paper used in the office. The staff can add the details of the used papers in the office as well as edit these details. The staff then can view the analysis of the cost of the papers in a visualised graph and use it to compare the cost of the papers used from previous semesters. Not only that, the staff also can view the paper limit suggestion calculated by the system to reduce the paper usage in the upcoming semesters. Lastly, the system will also generate a summary report of the analysis to display it to the staff. Figure 2 shows the dashboard analysis of the paper used and total cost. Agile methodology provides an effective and flexible way to continuously improve the requirements and the functionality of GOMASH to make sure the system provides a useful tool to analyse the paper usage and increase the green awareness among users in the faculty.

Fig. 2 GOMASH with Staff Analysis Dashboard for admin

Table 2 System specifications and requirements

Component	Description
Android Studio	The official integrated development environment for Android's operating system that is used to run emulations of mobile application systems on the computer
Android Emulator	A software that is used to simulates android device and run android applications on computer
AnyChart Android Charts API	Data visualization library for creating interacting charts in Android
MPAndroid Chart API	Data visualization library for creating interacting charts in Android
Smartphone	To run the developed mobile application system

3.1 System Specification

Table 2 describes the software and hardware requirements to develop the GOMASH system. The software that are required includes Android Studio, Android Emulator and AnyChart (Android Charts API). Meanwhile, the hardware that is required is a smartphone to run the developed application.

3.2 GOMASH Modules

Figure 3 depicts the interfaces of GOMASH system. This system is developed using Java in Android Studio 4.2.1 for mobile application development. Structured code modules in Android Studio allows the project to split into functionality units where we can develop, test and debug separately. Android Studio is used to code and design the interfaces and the functionalities in the application. Furthermore, Firebase database is used to record all the data for the output and input of this application. Realtime database and storage database is selected to insert, update, retrieve and delete data for both staff and admin modules as data can be shared in real time across all devices and remain accessible even if the device goes offline.

3.2.1 Form Module

The first module of GOMASH is manage form for Staff. User can select the semester and form name using the spinner, and insert the sheets amount to create a new form and the added form will be displayed in a list as shown.

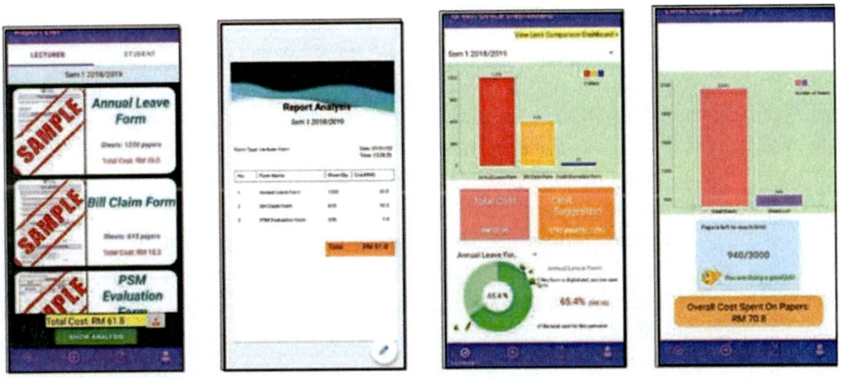

Fig. 3 GOMASH report and dashboard modules

3.2.2 Report Module

Information regarding the added forms including the form name, form picture, number of sheets used and the total cost of respective forms are displayed. Moreover, user can view the total cost of all the forms used for the selected semester according to the user type. A PDF button is located at bottom which downloads the report summary in a PDF format as shown.

3.2.3 Dashboard Module

User can view visualized graphs to provide dashboard analysis of the papers used in the office according to semester. Next, a total cost section which provides the total cost for printing forms for the selected semester is shown. Not only that, a limit suggestion is also calculated as an initiative to reduce the number of papers printed for upcoming semesters. The user also can view a circular chart at bottom which contain information of cost that can be saved if the selected form is digitalized for the selected semester. The user can navigate to overall dashboard interface by clicking onto the 'View Overall Dashboard'.

4 Result and Discussion

This tool provides awareness regarding green office and sustainable development which eventually help in providing a greener campus. The primary function of the GOMASH is to provide an analytic dashboard by calculating the cost of the papers used by using the number of papers that have been printed by the faculty for the respective semester. Functional testing is used as the testing strategy for this project as it aids in testing of functionality of every function of the proposed system such as manage account, manage form, manage report and manage dashboard. It mainly involves black box testing, and it is carried out by providing appropriate inputs and verifying the output against the requirements. The unit testing is done on modules of the system which includes the registration module, login module, account module, form module, report module and dashboard module. The summary of the unit testing are including the registration function is tested fully to provide a pass results as the requirement indicated, the login function provides a result of pass as the required requirements after the testing, the account function which is tested provides a pass result as requirement indicated, the form list function is tested provides a result of pass as the requirement indicated, the report list result which is tested provides a result of pass as the requirement needed and lastly the dashboard function which is tested completely provides a pass result as the requirement needed. All unit components which are tested completely for functionalities are integrated appropriately as a complete system to ease the users in using the GOMASH system without any errors or bugs. Conclusively, all the modules have been tested using software test description (STD) and tester has summarize that: display all the modules successfully with errorless. Thus, GOMASH system has been validated and verified and passed all the STD tests.

5 Conclusion

The main aim of this work is to propose a green office management system dashboard (GOMASH) as an initiative to practice green inside faculty using a mobile-based application. Few objectives have been achieved throughout this research work in order to accomplish the goal of this work. First objective is to identify the usage of papers in the faculty and analyze ways to digitalize them that will aid in promoting green inside office and is achieved; the usage of papers and the problems arising from using too many papers in the faculty was identified and considered. The second objective is to develop GOMASH system which contains relevant information and provide awareness of green by analyzing the usage of papers in the office such as UMP faculty; it was successfully developed with all the requirement specified and can be used by two types of users. The third objective is to validate the functionality of the GOMASH using functional testing. Thus, GOMASH has been validated and verified successfully with errorless and suggested to be applied to any office to increase awareness and implementation of green technology. However, this project is possesses some limitation which are: the system is developed for minimum SDK API 19 which is Android 4.4(KitKat) and only runs on phone which supports Android 4.4 and above and also the operating system such as iOS users could not use this application. The Mobile-based Green Office Management System Dashboard can be improved and enhanced in the future for betterment of the system. Future studies with more UI Green Metric categories need to be conducted to establish a greener and sustainable environment inside the faculty. Wider analysis variables and approach could be used in order to provide a better and meaningful output result from the dashboard. Some other improvements that can be considered are; To increase the scope of the users by including other faculties (as the case study) and wide scope of users for practicing green inside the campus. To made available this system for iOS users so that they can use it in the campus in the future as well and to increase the number of dashboard analysis in other viewpoint to give more fruitful analysis and information.

Acknowledgements This research is supported by the University Malaysia Pahang Research Grant (RDU190167), Malaysia National Research Grant (FRGS/1/2018/ICT04/UMP/02/4) and supported by Tunku Abdul Rahman University College (Malaysia) and Yarmouk University (Jordan).

References

1. Pereira Ribeiro J, Santa S, Andrade Guerra JB (2019) Green campuses and sustainable development
2. Pa NC, Karim F, Hassan S (2017) Dashboard system for measuring green software design. In: Proceeding—2017 3rd international conference on science in information technology, theory application of IT for education, industry, and society in big data era, ICSITech 2017, vol. 2018-January, pp. 325–329. https://doi.org/10.1109/ICSITech.2017.8257133
3. SGP Sustainable Green Printing Partnership (2018) SGP impact tracker: early impressions of the SGP impact tracker, pp. 1–2
4. Bakar NAA, Fauzan AIA, Majid MA, Allegra M (2019) The simulation models for human pedestrian movement of a departure process in an airport institute for educational technology. IOP conference series
5. Ismail KA, Majid MA, Zain JM, Bakar NAA (2016) Big data prediction framework for weather temperature based on map reduce algorithm. IEEE conference on open systems (ICOS), pp. 13–17
6. Masitry AK, Majid MA, Toh MZ (2013) An investigation on learning performance among disabled people using educational multimedia software: a case study for deaf people. Int J Biosci Biotechnol 5(6):9–20
7. Alsariera YA, Majid MA, Zamli KZ (2015) SPLBA: an interaction strategy for testing software product lines using the Bat-inspired algorithm. International conference on software engineering and computer systems (ICSECS), Pages 148–153

The Determinants of the Self-disclosure on Social Network Sites

Research-in-Progress

Lina Salih, Ahlam Al-Balushi, Amal Al-Busaidi, Shaikha Al-Rahbi, and Ali Tarhini

Abstract With the emergence of new technologies and trends, the habits of people and how they are dealing with these technologies has changed. Social Network Sites (SNS) have become a big part of most people's daily life. It is important to understand the factors that contribute to self-disclosure on SNS. A dramatic shift has been noticed concerning user privacy, where people's tendency to disclose information has been increasing. The disclosure of personal information can be very dangerous. This research-in progress aims to propose a conceptual framework that considers the inhibitors and enablers of self-disclosure on SNS. The conceptual framework will be tested via a large-scale survey of teens, students, employees and others. This research-in-progress will help the user to question their online behavior critically in order to protect themselves from oversharing personal information.

Keywords Self-disclosure · Social network sites · Fear of missing out · Privacy paradox · Trust · Self-esteem · Entertainment

1 Introduction

Social media platforms have become an essential part of daily life. Specifically, most human interactions have been shifted to virtual platforms like Facebook, Instagram, Twitter and LinkedIn [1, 2]. These Social Network Sites (SNS) have not only affected the type and amount of social interaction but have also expanded the necessity for self-disclosure [3, 4]. Most social media research fosters the "ideology of openness", which recommends that operative communication build upon a maximum level of transparency [5]. Accordingly, on a regular basis, people are disclosing information to their friends and beyond through posts on SNS [6]. Nabity-Grover, et al. [7] define self-disclosures as revealing any personal information to another person. This continued behavior has enabled and contributed to the tremendous growth of SNS

L. Salih · A. Al-Balushi · A. Al-Busaidi · S. Al-Rahbi · A. Tarhini (✉)
Department of Information Systems, Sultan Qaboos University, Muscat, Sultanate of Oman
e-mail: ali.tarhini@hotmail.co.uk

M. Al-Emran et al. (eds.), *International Conference on Information Systems and Intelligent Applications*, Lecture Notes in Networks and Systems 550,
https://doi.org/10.1007/978-3-031-16865-9_47

[8]. The concept of self-disclosure is not limited to the amount of information that has been revealed, but also the ease with which real users can be identified [9].

However, even when the visibility of information on SNS can be restricted to a specific targeted group, the options controlling that information do not always exist [10, 25]

Furthermore, the volume of information being revealed is the fuel that feeds SNS machines and users are invited to disclose their information voluntarily [9]. On the other side, SNS providers monetize this user information by selling it to third parties who have the desire to reach demographic users [9]. Consequently, the practice of self-disclosure threatens privacy aspects and creates risks for the individuals [10]. Negative experiences like harassment and threats, resulting in disappointment, distrust and fear have already been reported by SNS users [10]. Despite the associated risks with self-disclosure, empirical research has demonstrated repeatedly that the behavior of people contradicts their privacy assessments and preferences [10]. Literature reveals that many users are afraid of committing online privacy violations, but only a few of them implement the required steps to safeguard sensitive information [11]. Users are not reflecting the result of their online behavior, whether negative or positive because they are valuing the benefits of sharing rather than thinking about the possible risks [11]. As the number of online users increases, the privacy concern rises [9]. Due to the parallel advances in mobile technology, users are tending to be in a state of being always online, which has resulted in habituating the community to information exchanging and normalizing the heightened degree of information disclosures, even in voluntary settings [8].

The aim of the research is to propose a conceptual framework that considers the inhibitors and enablers of self-disclosure on SNS. Despite the associated risks and threats, individuals are still practicing self-disclosure. There is a lack of literature related to the influencing factors of self-disclosure on SNS [12]. The drivers of the acceleration towards a digital and social economy need to realize these factors in order to understand the reason why people are still revealing their information regardless of the risks.

The study suggests focusing on different types of online users such as teens, students, employees, and others. Followed by the introduction, Sect. 2 proposes a conceptual framework that addressed the influencing factors on the self-disclosures on SNS as well as the suggested hypotheses. To address these issues and understand the factors that lead to this kind of behavior, we propose a model which includes seven independent variables which can lead to self-disclosure. We expect the model, if supported, to identify the most relevant factors which lead to self-disclosure on SNS. Furthermore, this model can be used to explain this behavior and spread awareness on a more detailed foundation. Before concluding the paper, the third section discusses the methodology of this research, and the fourth section deliberates the potential contribution of this study.

2 Conceptual Framework

The study is based on a conceptual model that incorporates the determinants that may impact the disclosing of information on SNS among teens, students, employees, and others as shown in Fig. 1. It is worth mentioning that this study did not consider any moderating effect of age, gender and education level.

These factors are privacy concern, trust, Fear of missing out (FOMO), relation development, benefit, self-esteem and entertainment as detailed below:

2.1 Privacy Concerns

Privacy can be defined as fulfilling generic human needs in terms of self-evaluation, protected communication and autonomy [13]. It is related to capturing the degree to which the user is caring about his personal information flow, including the exchange of that information [14]. Personal perception of privacy is linked with personal experiences, values, culture, and beliefs [15]. In online communities, users who realize higher risks to privacy are disposed to revealing their personal information and tend to protect themselves. In contrast, whenever users perceive lower privacy threats, they tend to reveal more personal information [3]. As discussed previously, researchers

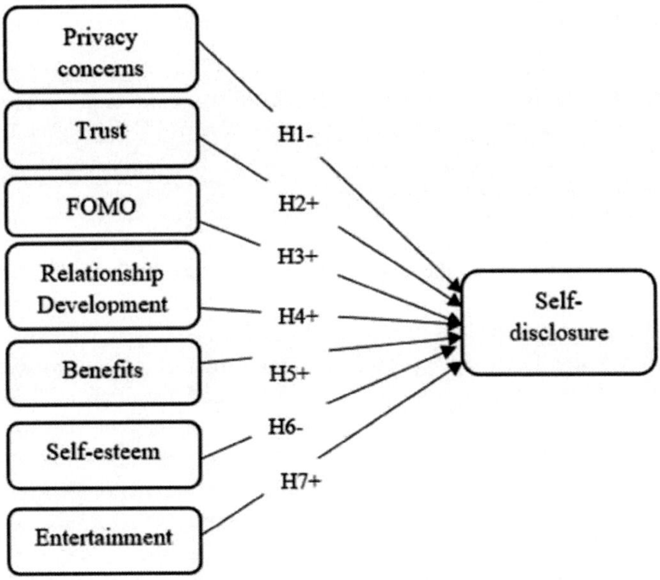

Fig. 1 Proposed conceptual model: self-disclosure on SNS

agreed that privacy concerns are indeed significantly related to self-disclosure, thus the following hypothesis is presented:

***H1:** Privacy concerns will negatively influence self-disclosure on social network sites.*

2.2 Trust

According to Xie and Kang [16], one of the most significant factors that influence self-disclosure is trust. Online trust can be defined as the level to which a user feels that those within their online environment are trustworthy and reliable with information that makes them vulnerable [17]. Scholars found that trust is a prerequisite for self-disclosure because it may minimize the perceived threats involved in disclosing sensitive and private information [17]. Trust is considered to be one of the elements that improve and build on the bonds between individuals [12, 18]. From another perspective, Krasnova et al. [19] argue that trust is related to mitigating user concerns with regard to the platform used, which will lead to more self-disclosure. The relationship between self-disclosure and trust is an element of privacy calculus [9]. To elaborate, whenever users have the faith that the SNS platform used by them is sufficiently safeguarded, they have a stronger intention to disclose their information on that platform [9]. Hence, whenever the degree of trust exceeds a certain threshold, and it's less risky, users tend to perceive that the environment is safe and feel comfortable revealing information [20]. In keeping with prior literature, it is hypothesized that:

***H2:** Trust will positively influence self-disclosure on social network sites.*

2.3 Fear of Missing Out (FOMO)

Fear of missing out can be defined as the continuous desire to be connected to what others are doing. The online activities provided by SNS such as sharing photos, videos and content with friends promote the development of an addiction habit of always being online to remain updated and avoid missing out on trending issues and news [21]. FOMO is driven by the individuals' need for satisfaction. Whenever individuals experience chronic deficits in obtaining satisfaction, they try to self-regulate by looking for information about what's going on in their environment to avoid missing out [21]. Since individuals with a high degree of FOMO look for more bonds, they are expected to use SNS to enhance their sense of self-presentation by disclosing their information [24]. According to Sultan [21, 23], the behavior of disclosing strengthens the relationship between a person and other individuals, where it has been identified as a critical factor in personal relationship development due to its positive results on social connections, intimacy, and closeness. As a basic psychological need of humans, there is a need to be connected, to feel close, and to belong, so that FOMO is considered to be a convenient outlet in SNS that fulfills individuals' desire to stay continually in touch with others [24]. Talwar, et al. [25] clarify that there is a positive relationship between FOMO and self-disclosure. In keeping with prior literature, it is hypothesized that:

H3: *FOMO will positively influence self-disclosure on social network sites.*

2.4 Relational Development

Disclosure can be influenced by relational development as a way to increase closeness and intimacy with others [6]. It is associated with continuous self-disclosure in SNS and anticipates more honest, positive, intentional disclosure [6]. As it was discussed in social compensation theory, people with psychological distress are more prominent in disclosing to achieve their relational goal [6]. They are aware of their defects in social relationships and have a stronger need for affiliation and connection than their counterparts [6]. For instance, lonely people were more willing to reveal their information on SNS than social people. Comparing people with different self-esteem in disclosing information, people who have low esteem disclose more due to the desire for social networking and social compensation [6]. There is ample proof of the existence of a link between relational outcomes and the intimacy of disclosure. In a study that was conducted by Utz [26], he mentions that people tend to like the people who disclose more, and people disclose more to the people they like. In another study that was conducted by Huang [27], the researcher mentions that the social penetration theory highlights that disclosing information is a key concept that contributes to relationship development. Hence such an action not only enables the development of a close relationship, but also helps to maintain it [27]. In keeping with prior literature, it is hypothesized that:

H4: *The relational development factor will positively influence self-disclosure on social network sites.*

2.5 Benefit

Benefit can be defined as the outcome individuals derive from having the ability to easily and efficiently stay in touch with others on SNS [19]. On SNS people tend to disclose their information seeking the benefits that may occur [9]. Accordingly, industry executives, policymakers and scholars are all invested in finding out why individuals go online and reveal huge amounts of their information without obvious benefits such as financial gain [9]. One branch of reasoning for such behavior is the feeling that they are gaining through their revealing information [9]. Nevertheless, benefits in this regard are but researchers have a common view that the practice of disclosing information is done because of the perceived benefit [9]. For instance, self-expression and social validation can be perceived benefits which encourage users to reveal their information on SNS [8]. Furthermore, benefits include the desire to build a relationship, the convenience of maintaining a relationship, and enjoyment [19].

Maintaining relationships means keeping in contact with people via technological features such as sharing moments, photos, exchanging posted comments with others, following friends' events and news. Enjoyment is done by practicing pleasant activities such as playing games, watching videos, and reading motivating articles where the behavior of disclosing is required to gain these entertainments [19, 22]. Whenever there are greater benefits and fewer risks, the behavior of disclosing increases [7]. Krasnova, et al. [19] argue that a typical result of convenience contributes to motivating users to self-disclose. Users are ready to jeopardize their privacy to gain more convenience through decreased frictional costs or personalization. Rosen and Sherman [28] argue that the construct or form of perceived enjoyment of SNS has a higher influence than the form of usefulness. Sledgianowski and Kulviwat [29] share the same idea, claiming that SNS usage is positively affected by individuals' sense of playfulness and enjoyment in using SNS, and their degree of self-disclosure reflects this. In keeping with prior literature, it is hypothesized that:

H5*: The benefit factor will positively influence self-disclosure on social network sites.*

2.6 Self-esteem

Self-esteem is the grade to which persons have feelings about themselves (positive or negative) and their own value. It is a comparatively stable personality trait and varies from person to person [30]. According to Jozani, et al. [31], users participate in SNSs to "build social capital, improve their self-worth and self-esteem, and satisfy their enjoyment needs" [32]. This factor has been examined in previous studies and has been a significant factor in the usage of SNS and therefore self-disclosure in SNS [12, 30, 33]. Scholars argue that self-esteem is related to the individuals' response to potential threats. Individuals with high self-esteem are usually more self-protective about themselves and are more concerned about their privacy than users with low self-esteem [30, 34, 39]. Therefore, SNSs users with low self-esteem are more likely to disclose information about themselves than users with high self-esteem. In keeping with prior literature, it is hypothesized that:

H6*: High self-esteem is negatively associated with self-disclosure.*

2.7 Entertainment

According to Verhagen et al. [36], entertainment is defined as "the degree to which the use of an information system is a fun and pleasant experience and lifts the user's spirits". When using SNS users can be entertained while getting away from the pressure of daily life [37, 38]. Getting in touch with friends or gaining recognition while sharing the latest events of personal life can be examples of entertainment on SNS. Studies have noted that entertainment is a significant factor regarding the involvement of the users [36, 39, 40]. Entertainment is considered to disclose personal information as a result of the interaction and entertainment with other users [39]. Hence, entertainment is a significant factor for self-disclosure on SNS [39]. In keeping with prior literature, it is hypothesized that:

***H7**: Entertainment positively influences self-disclosure.*

3 Research Methodology

An online survey questionnaire has been designed to evaluate the research model's predictive power. The questionnaire will be distributed by e-mail with a link to the online questionnaire. Our sample will include all individuals who are currently present or have been present on SNS. The questionnaire consists of two main sections. The first section collects data about the demographic characteristics of the individuals (age, gender, and current employment status). This allows us later to conduct an inter-group analysis. The second section measures the items of the seven constructs from the proposed model. The constructs used in the proposed research model were adapted from previous studies where they have been proved to be valid. Specifically, *Privacy Concern* was adopted from the work of Trepte, et al. [14], Wang, et al. [33] and Xie and Kang [16]. *Self-esteem* was adopted from the work of Wang, et al. [33] and Tran, et al. [2]. *Trust* was adopted from Thompson and Brindley [8] and Liu, et al. [20]. *Entertainment* was adopted from the work of Lin and Chu [37] and Mouakket [39]. *FOMO* was adopted from the work of Law [20] and Talwar, et al. [25]. *Benefit* was adopted from the work of Thompson and Brindley [8] and Contena, et al. [41] and *Relationship Development* was adopted from the work of Luo and Hancock [6] and Liu, et al. [20]. Privacy concerns, self-esteem, trust, FOMO, benefit and relationship development will be measured using five items, whereas entertainment will be measured using four items. These items were anchored on a five-point Likert scale, ranging from 1 = "strongly disagree" to 5 = "strongly agree". Prior to the full-scale data collection a pilot-test has been conducted and the survey has been verified.

4 Potential Contributions of the Study

This research is expected to contribute mainly to the body of knowledge on the factors that influence self-disclosure in social network sites. Additionally, other contributions may include the following:

(1) The development of a model that determines the number of factors that have a direct influence on self-disclosure in social network sites.

(2) The testing of the degree of existing awareness about privacy-related issues associated with self-disclosure.
(3) Further research into helping to understand human behavior and the tendency of disclosing on social network sites.
(4) Providing a summary list of influencing determinants that may be addressed by future researchers.

5 Conclusion and Future Research

The advanced social technologies such as SNS have radically affected the type and amount of the user's experience of social interaction and disclosing behavior [3]. The concept of self-disclosure is not limited to the amount of information that has been revealed, but also to the ease with which real users can be identified. The disclosure of personal information is associated with various risks such as social-engineering, harassment, and threats. The literature reveals that many users are afraid about their online privacy violations, although few of them implement the required steps to safeguard sensitive information.

To address these issues and understand the factors that lead to this kind of behavior, this paper aims to propose a conceptual framework which includes seven independent variables which can lead to self-disclosure. We expect the model, if supported, to identify the most relevant factors which lead to self-disclosure on SNS. Furthermore, this model can be used to explain this behavior and spread awareness on a more detailed foundation.

This paper proposed a conceptual framework. Hence, in the future, studies can empirically validate this proposed model.

References

1. Alalwan AA, Rana NP, Algharabat R, Tarhini A (2016) A systematic review of extant literature in social media in the marketing perspective. In: Yogesh K et al (eds) Social Media: The Good, the Bad, and the Ugly. I3E 2016. LNCS, vol 9844, pp 79–89. Springer, Cham. https://doi.org/10.1007/978-3-319-45234-0_8
2. Tran TTH, Robinson K, Paparoidamis NG (2022) Sharing with perfect strangers: the effects of self-disclosure on consumers' trust, risk perception, and behavioral intention in the sharing economy. J Bus Res 144:1–16
3. Towner E, Grint J, Levy T, Blakemore S-J, Tomova L (2022) Revealing the self in a digital world: a systematic review of adolescent online and offline self-disclosure. Curr Opin Psychol 45(6):101309
4. Walsh RM, Forest AL, Orehek E (2020) Self-disclosure on social media: the role of perceived network responsiveness. Comput Hum Behav 104:106162
5. Richey M, Gonibeed A, Ravishankar M (2018) The perils and promises of self-disclosure on social media. Inf Syst Front 20(3):425–437
6. Luo M, Hancock JT (2020) Self-disclosure and social media: motivations, mechanisms and psychological well-being. Curr Opin Psychol 31:110–115

7. Nabity-Grover T, Cheung CM, Thatcher JB (2020) Inside out and outside in: how the COVID-19 pandemic affects self-disclosure on social media. Int J Inf Manag 55:102188
8. Thompson N, Brindley J (2020) Who are you talking about? Contrasting determinants of online disclosure about self or others. Inf Technol People 34(3):999–1017
9. Taddei S, Contena B (2013) Privacy, trust and control: which relationships with online self-disclosure? Comput Hum Behav 29(3):821–826
10. Kroll T, Stieglitz S (2021) Digital nudging and privacy: improving decisions about self-disclosure in social networks. Behav Inf Technol 40(1):1–19
11. Hallam C, Zanella G (2017) Online self-disclosure: the privacy paradox explained as a temporally discounted balance between concerns and rewards. Comput Hum Behav 68:217–227
12. Lin C-Y, Chou E-Y, Huang H-C (2020) They support, so we talk: the effects of other users on self-disclosure on social networking sites. Inf Technol People 34(3):1039–1064
13. Krämer NC, Schäwel J (2020) Mastering the challenge of balancing self-disclosure and privacy in social media. Curr Opin Psychol 31:67–71
14. Trepte S, Scharkow M, Dienlin T (2020) The privacy calculus contextualized: the influence of affordances. Comput Hum Behav 104:106115
15. Koohikamali M, Peak DA, Prybutok VR (2017) Beyond self-disclosure: disclosure of information about others in social network sites. Comput Hum Behav 69:29–42
16. Xie W, Kang C (2015) See you, see me: teenagers' self-disclosure and regret of posting on social network site. Comput Hum Behav 52:398–407
17. Posey C, Lowry PB, Roberts TL, Ellis TS (2010) Proposing the online community self-disclosure model: the case of working professionals in France and the UK who use online communities. Eur J Inf Syst 19(2):181–195
18. Albanna H, Alalwan AA, Al-Emran M (2022) An integrated model for using social media applications in non-profit organizations. Int J Inf Manag 63:102452
19. Krasnova H, Spiekermann S, Koroleva K, Hildebrand T (2010) Online social networks: why we disclose. J Inf Technol 25(2):109–125
20. Liu Z, Min Q, Zhai Q, Smyth R (2016) Self-disclosure in Chinese micro-blogging: a social exchange theory perspective. Inf Manag 53(1):53–63
21. Sultan AJ (2021) Fear of missing out and self-disclosure on social media: the paradox of tie strength and social media addiction among young users. Young Consum 22(4):555–577
22. Law M (2020) Continuance intention to use Facebook: understanding the roles of attitude and habit. Young Consum 21(3):319–333
23. Sultan AJ (2021) User engagement and self-disclosure on snapchat and Instagram: the mediating effects of social media addiction and fear of missing out. J Econ Administrative Sci ahead-of-print
24. Roberts JA, David ME (2020) The social media party: fear of missing out (FoMO), social media intensity, connection, and well-being. Int J Hum Comput Interact 36(4):386–392
25. Talwar S, Dhir A, Kaur P, Zafar N, Alrasheedy M (2019) Why do people share fake news? Associations between the dark side of social media use and fake news sharing behavior. J Retail Consum Serv 51:72–82
26. Utz S (2015) The function of self-disclosure on social network sites: not only intimate, but also positive and entertaining self-disclosures increase the feeling of connection. Comput Hum Behav 45:1–10
27. Huang H-Y (2016) Examining the beneficial effects of individual's self-disclosure on the social network site. Comput Hum Behav 57:122–132
28. Rosen P, Sherman P (2006) Hedonic information systems: acceptance of social networking websites
29. Sledgianowski D, Kulviwat S (2008) Social network sites: antecedents of user adoption and usage
30. Apaolaza V, Hartmann P, D'Souza C, Gilsanz A (2019) Mindfulness, compulsive mobile social media use, and derived stress: the mediating roles of self-esteem and social anxiety. Cyberpsychol Behav Soc Netw 22(6):388–396

31. Jozani M, Ayaburi E, Ko M, Choo K-KR (2020) Privacy concerns and benefits of engagement with social media-enabled apps: a privacy calculus perspective. Comput Hum Behav 107:106260
32. Heravi A, Mubarak S, Choo K-KR (2018) Information privacy in online social networks: uses and gratification perspective. Comput Hum Behav 84:441–459
33. Wang L, Yan J, Lin J, Cui W (2017) Let the users tell the truth: self-disclosure intention and self-disclosure honesty in mobile social networking. Int J Inf Manag 37(1):1428–1440
34. Bearden WO, Hardesty DM, Rose RL (2001) Consumer self-confidence: refinements in conceptualization and measurement. J Consum Res 28(1):121–134
35. Al-Qaysi N, Mohamad-Nordin N, Al-Emran M (2020) Employing the technology acceptance model in social media: a systematic review. Educ Inf Technol 25(6):4961–5002
36. Verhagen T, Feldberg F, van den Hooff B, Meents S, Merikivi J (2012) Understanding users' motivations to engage in virtual worlds: a multipurpose model and empirical testing. Comput Hum Behav 28(2):484–495
37. Lin Y-H, Chu MG (2021) Online communication self-disclosure and intimacy development on Facebook: the perspective of uses and gratifications theory. Online Inf Rev 45(6):1167–1187
38. Al-Qaysi N, Mohamad-Nordin N, Al-Emran M (2020) What leads to social learning? Students' attitudes towards using social media applications in Omani higher education. Educ Inf Technol 25(3):2157–2174
39. Mouakket S (2018) Information self-disclosure on mobile instant messaging applications: uses and gratifications perspective. J Enterp Inf Manag 32(1):98–117
40. Ifinedo P (2016) Applying uses and gratifications theory and social influence processes to understand students' pervasive adoption of social networking sites: perspectives from the Americas. Int J Inf Manag 36(2):192–206
41. Contena B, Loscalzo Y, Taddei S (2015) Surfing on social network sites: a comprehensive instrument to evaluate online self-disclosure and related attitudes. Comput Hum Behav 49:30–37

Determinants of Consumers' Acceptance of Voice Assistance Technology: Integrating the Service Robot Acceptance Model and Unified Theory of Acceptance and Use of Technology

Research-in-Progress

Lhia Al-Makhmari, Abrar Al-Bulushi, Samiha Al-Habsi, Ohood Al-Azri, and Ali Tarhini

Abstract Users are increasingly using AI-based applications in their daily activities, with voice assistants playing a more advanced and pervasive role. However, little is known about the specific drivers of user acceptance of voice assistance technology. Therefore, a conceptual model was developed that integrates factors from the Service Robot Acceptance Model (sRAM), and Unified Theory of Acceptance and Use of Technology (UTAUT). The model will be tested with a large-scale survey of respondents of different ages, genders, and educational levels. This ongoing research focuses on users' acceptance of the use of digital voice assistants in their daily lives and suggests a model that considers the factors that enable their acceptance.

Keywords Consumer behavior · Voice assistance · Siri · Augmented reality · Virtual reality · Technology adoption · UTAUT

1 Introduction

In recent decades, customers have been increasing using technology to help them plan their daily tasks [1]. Nowadays, the service sector has undergone substantial changes, such as the increased deployment of artificial intelligence (AI) service robots, for example applications and automated technology, virtual assistants or chatbots [2]. Voice assistants (VA) are a sort of artificial intelligence that may be activated through voice commands [10]. The ability of algorithms to simulate intelligent human behavior is referred to as AI [5]. Moreover, [31] AI illustrates a certain

L. Al-Makhmari · A. Al-Bulushi · S. Al-Habsi · O. Al-Azri · A. Tarhini (✉)
Department of Information Systems, Sultan Qaboos University, P.O Box: 20, P.C: 123 Muscat, Sultanate of Oman
e-mail: ali.tarhini@hotmail.co.uk

M. Al-Emran et al. (eds.), *International Conference on Information Systems and Intelligent Applications*, Lecture Notes in Networks and Systems 550,
https://doi.org/10.1007/978-3-031-16865-9_48

level of intelligence defined by digital interfaces. However, AI refers to the "cognitive" jobs such as problem solving and learning which we associate with the human brain [3]. Apple's Siri, Amazon's Alexa, Google Assistant, and Microsoft Cortana are examples of virtual assistants available as smartphone apps, while Amazon's Echo, Google's Home, and Apple's Home are examples of smart speaker services.

In detail, VA is changing the culture of users' interaction and being a part of users' social life in many ways. Users use VA to travel, listen to music, send text messages, operate smart home devices, make phone calls, place orders such as booking a Uber ride or ordering as pizza, and many other things. According to the NPR and Edison study, 21% of Americans (53 million individuals) own a smart speaker, up from 14% in 2018. In addition, the Google Assistant mobile app has been downloaded 500 million times, according to Hoffman, vice president of Google Assistant [4].

Nowadays, developers are developing many algorithms that will provide VA with personality and social features [31]. In a recent study [5], it is reported that that people have a social response to robots that behave like human beings. AI-powered virtual VA can learn about users' needs and favorite topics effortlessly and without having to type, read, or carry a device or gadget [5]. Furthermore, VA helps individuals as customers to contact to many businesses, search for some information and place orders. Furthermore, VA helps individuals as customers contact many businesses, search for information and place orders. Recently, in order to win customers many companies are working on their own voice chat services, while others are making agreements with specialized companies to provide them with these services [13]. Due to the exponential rise in voice-based technology [6, 41], its human-like features, and its first appearance in mobile devices [7], many users are interacting with VA in meaningful, convenient, and purposeful ways and techniques [8].

With Apple's Siri, AI voice assistant software can understand voice messages, glean keywords that users employ, and, therefore, perform a series of built-in instructions and responses [9]. In fact, Siri is no longer in its infancy, VA is now online and able to access personal information that is stored on the phone, which helps to enable Siri to predict and react to user inquiries and requests and provide users with accurate information [10].

Based on the growth of voice assistance technologies, according to [10] adoption of Siri in developed countries is very high. In the United States, the number of voice assistant users grew from 47.3 million in 2020 to 90 million in 2021. However, the context of the research carried out in developing countries research has shown that it is still in its infancy. The purpose of this research is to evaluate the factors that influence the use of digital voice assistance technologies and how these technologies could be made to look more attractive for a wider range of people to utilize. In view of this, the objective of this research is to address the following question: What are the factors that contribute towards encouraging people to use digital voice assistance technology? In view of this, the objective of this research is to address the following question: What are the factors that contribute towards encouraging people to use digital voice assistance technology?

To answer this research question, the Service Robot Acceptance model (sRAM), and Unified Theory of Acceptance and Use of Technology UTAUT were used as the conceptual model in order to understand the different factors that can encourage people to take advantage of the technology.

The research will discuss the factors that affect people the most in accepting voice assistance technology, as it is very useful and is an interactive feature available in everyone's mobile. Further, the model will help us to discover the limitations of the voice assistance technology which cause people not to use it effectively and efficiently. The study will focus mainly on the functional, social, and rational dimensions towards the acceptance of digital voice assistance technology.

2 Conceptual Framework

This study proposes a conceptual model that considers the factors that may affect the acceptance of the voice assistance technology Siri. The model has three dimensions (functional, social, and relational). Figure 1 illustrates the proposed conceptual framework, and the subsections that follow explain each of these relationships in detail.

2.1 *Functional Elements*

The functional dimensions of the model for technology will include Performance Expectancy, Efforts Expectancy, Subjective Social Norms, and Perceived Enjoyment.

Performance Expectancy and Effort's Expectancy. "Performance expectancy is defined as the degree to which an individual believes that using the system will help

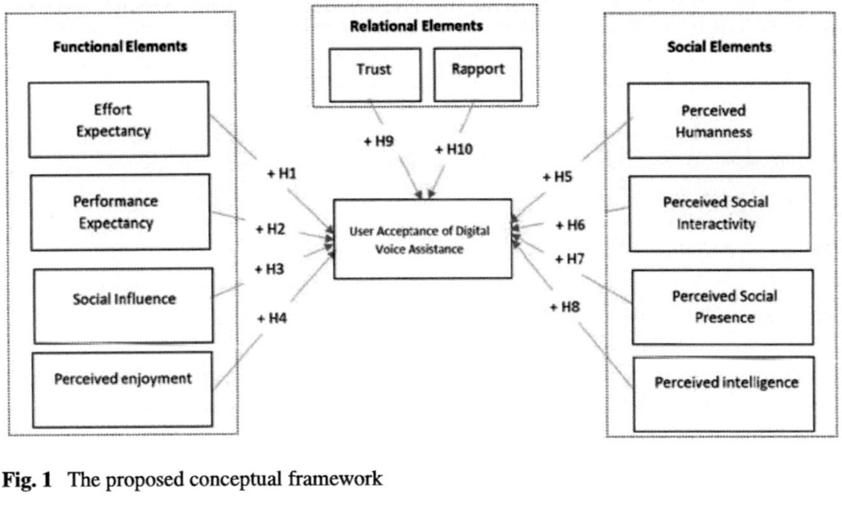

Fig. 1 The proposed conceptual framework

them to attain gains in job performance" and "Effort expectancy is defined as the degree of ease associated with the use of the system" these being the main elements of the Unified Theory of Acceptance and Use of Technology (UTAUT) developed by [11]. UTAUT was developed to measure the factors that can be determined to assess a technology's acceptance and use in many applications [12]. The UTAUT Performance expectancy and Effort expectancy was used to examine the acceptance of mobile banking & internet banking [13, 14]. Other studies were conducted to test the acceptance of mobile health services [15]. In addition, the factors were used to examine the student's acceptance of the virtual learning environment [17]. The UTAUT model factors were evaluated to study the acceptance of voice commerce using smart speakers for making online purchases [18]. In this research, the UTAUT model performance and effort expectancy will be used to investigate the acceptance of digital voice assistance. Therefore, Performance expectancy & Effort's expectancy could help positively in the adoption of digital assistance agents.

Hypothesis 1: Performance Expectancy has a positive influence on customer acceptance of digital voice assistant agents 'Siri.

Hypothesis 2: Effort Expectancy has a positive influence on customer acceptance of digital voice assistant agents 'Siri.

Subjective Social Norms. Subjective social norms are known as "individual perception that most people who are important to them think they should or should not perform the behavior in question" [19]. People usually become influenced by the opinions and behaviors of the people within their social network such as family and friends [20]. Therefore, social norms are associated with the influence of personal behaviors towards the adoption of the new technologies based on the social group surrounding an individual [21, 26]. In previous studies subjective norms had a positive impact on influencing individual behaviors in using e-commerce [22]. Further, they were used as a factor in measuring a user's adoption and acceptance of e government systems [23]. Social influence was used as a variable that has a direct influence on perceived ease of use and perceived usefulness that will indirectly affect the user's intention towards the online learning environment [21]. In other studies, it was also used to measure individual perception concerning the intention of using mobile banking [20].

Hypothesis 3: Subjective Social Norms have a positive influence on a customer's acceptance of digital voice assistant agents 'Siri.

Perceived Enjoyment. Perceived enjoyment is defined as "the extent to which the activity of using the computer is perceived to be enjoyable in its own right, apart from any performance consequences that may be anticipated" [24]. In a study conducted by [25] to investigate the trends that affect secondhand online shopping, the study found that perceived enjoyment has a positive impact on users' satisfaction. In addition, another study was conducted to investigate the factors that contribute to the user's intention of joining a customer service chatbot. The study found that perceived enjoyment is one of the drivers that has a positive impact on the acceptance of the service [26]. In addition, a study was conducted to examine the factors that affect users' acceptance of augmented reality (AR) smart glasses and the study found that perceived enjoyment directly influenced attitude in accepting the technology [27]. Another study conducted by [28] aimed to understand the virtual personal assistance (VPA), the study results showed that perceived enjoyment has a significant impact on the usage intention of artificial intelligence technology such as the VPA. Therefore, perceived enjoyment can help positively in the adoption of digital assistance agents.

Hypothesis 4: Perceived Enjoyment has a positive influence on customers and their acceptance of digital voice assistant agents 'Siri.

2.2 Social Elements

Venkatesh et al. [11] defines social elements by the scope where an individual perceives that an older employee or (anyone who can impact manner) believes they should use the information system. According to the sRAM model, there are four social dimensions: Social Presence, Social Interaction, Perceived Humanity and Perceived Intelligence.

Perceived Humanity. On the perceived humanity dimension, [29] note that this dimension is important because the massive development of artificial intelligence will make robots indistinguishable from humans. Moreover, this will enable users to build strong relationships between users and anthropomorphic robots. On the other hand, some researchers are of the opinion that this dimension has more disadvantages than advantages. According to [30] they believe in the "Elissa effect". They state that there would be a frightening and unhelpful interaction between robots and humans because the robot is not able to be 100% human. However, we expect that:

Hypothesis 5: Perceived humanity has a positive influence on a customer and their acceptance of digital voice assistant agents 'Siri.

Social Interaction. The perceived social interaction dimension is that users perceive that the robot understands their emotions and gives them appropriate responses based on societal norms [31]. Thus, the social attractiveness of the robot will increase because the user feels that they are able to interact with the robot as if they were a real person. This leads to increased user interest and interaction with VA as Siri. Thus, we expect:

Hypothesis 6: Social Interaction has a positive influence on a customer and their acceptance of digital voice assistant agents 'Siri.

Social Presence. In the dimension of social presence, McLean and Osei-Frimpong [8] described it as the ability of an AI robot to communicate with individuals and make them feel that they are interacting with a real social entity. Thus, social presence may lead individuals to interact with Siri in the same way that they would talk with a real human being [32]. Therefore, our hypothesis:

Hypothesis 7: Social Presence has a positive influence on a customer and their acceptance of digital voice assistant agents 'Siri.,

Perceived Intelligence. Perceived intelligence is defined by [33] as "individuals' perception that the personal intelligent agent's behavior is efficient and autonomous with the ability to process and produce natural language and deliver effective output". In this research we refer to the intelligence of digital assistance agents which encourage users to utilize them. This factor was used to measure the voice digital assistance in completing work productivity [34]. In addition, in the context of online shopping, the system intelligence leads to an increase in customer purchases [35]. Previously, system intelligence was known for solving complex mathematical problems. This, however, has changed, with artificial intelligence now being defined as the ability to generate and formulate human-like behaviors. In the case of robot behavior, the robot will interact with a human by interpreting a set of patterns, which can be boring if it was developed using a limited vocabulary [36]. Therefore, the perceived intelligence is linked with anthropomorphism. Mostly, when people find intelligent devices, they try to assign human-like characteristics to them. Further, previous research recommended that the intelligence of a system will result in delivering a useful and effective service [37]. Therefore, perceived intelligence can help positively in the adoption of digital assistance agents.

Hypothesis 8: Perceived Intelligence has a positive influence on customer and acceptance of digital voice assistant agents 'Siri.

2.3 Relational Elements

Trust. The trust dimension is an important dimension in sRAM. According to [31] trust makes the user feel confident and makes sure that the AV or robots work reliably when interacting with them. In the case of Siri, it is important to have the user's trust because Siri often handles private user data. As the user gains trust, they become less suspicious of Siri and more willing to adopt it. Therefore, we assume:

Hypothesis 9: Trust has a positive influence on a customer and their acceptance of digital voice assistant agents 'Siri.

Rapport. Rapport is defined as "personal connection between the two interactants" [38]. It is characterized by the customer perception of having an enjoyable interaction with a robot. For example: the user's feeling of receiving care and friendliness, also the ability of robots to simulate curiosity and establish connection. It is essential to build rapport and social closeness for services in the field of elderly care, education, and financial services [31].

According to [39] verbal acknowledgment and hand motion helps to enhance rapport between robot and human. Further, in previous studies, task execution using robots was improved by developing the participant's rapport, engagement, and collaboration. Other studies found that personalized and interactive digital agents were facilitated for game playing, re-habitation services, conversation, and exercises in elderly care centers [40]. Therefore, the adoption and acceptance of voice assistance robots such as voice assistance technologies will depend on rapport to accomplish customer needs [31].

Hypothesis 10: Rapport has a positive influence on a customer and their acceptance of digital voice assistant agents 'Siri.

3 Research Methods

This section will address the sample and data collection procedure. Based on research context and nature, a quantitative approach will be adopted to answer the questions of the research considering a non-probability convenience sampling approach which will be targeting digital voice assistance users. The aim of the research is to test three main elements: Functional, Rational, Social in the context of the acceptance of digital voice assistance technology. The first section contains the functional elements which include Performance Expectancy and Effort Expectancy (Venkatesh et al., 2003), Subjective Social Norms (Fishbein and Ajzen, 1977) and Perceived Enjoyment (Davis et al., 1992). The second section includes the rational factors which are the Perceived humanness, perceived social interactivity, Perceived social presence (Wirtz et al., 2018), and Perceived intelligence (Davis et al., 1992). The Rational factors are Trust and Rapport (Wirtz et al., 2018). In addition, the factors, with their relationships that have been used in the model, were validated, and developed using many models and theories. Therefore, using a cross-sectional survey will be the most applicable method for data collection. The survey will be designed using google forms to make it easy and quick to collect data from a large number of participants.

A Seven-point Likert scale is adopted to assess the response on the acceptance of voice digital assistance. Furthermore, the data confidentiality and privacy will be addressed using the survey cover page which will be sure to receive a high response rate. The survey was pre-tested by four candidates and some minor changes were made in reference to their comments in order to ensure the validity and reliability of the questionnaire items. The data analysis for the survey responses will be done using an SPSS tool for data frequencies and reliability analysis, as well as to perform the regression testing for the dependent and independent constructs.

4 Future Research

The next stage is to conduct a pilot-test of the survey. Ethical approval was obtained by Sultan Qaboos University for completing the initial online questionnaire. After we check the efficiency of the questionnaire, a large-scale data collection exercise will be conducted. To this end, the survey will be shared with various members of the community using social email and social media platforms.

5 Concluding Remarks

Voice Assistances are examples of emerging technologies that are rapidly progressing and becoming increasingly more human-like. A further investigation is required to explore the factors that influence consumer adoption and acceptance in encounters with virtual assistants. Furthermore, the research has developed a model based on integrating factors from the Service Robot Acceptance Model (sRAM), and Unified Theory of Acceptance and Use of Technology (UTAUT) by external factors and did not consider the role of moderate variables such as gender, educational level, and age as being determinants of consumer acceptance to voice assistance technology. It is believed that defining the role of these intermediaries adds significant value to potential future work.

In addition, this research will use a quantitative approach by aggregating data through questionnaire surveys, which will not provide an in-depth exploration of the determinants of consumer acceptance of voice assistance technology. Further experiments are encouraged to use an in-depth mixed method to gain a better and deeper exploration of the topic. This study recognizes that this is an area that needs further research and attempts to contribute to knowledge in this field.

References

1. Kunz WH, Heinonen K, Lemmink JG (2019) Future service technologies: is service research on track with business reality? J Serv Market 33(4):479–487
2. Gummerus J et al (2019) Technology in use—characterizing customer self-service devices (SSDS). J Serv Market
3. Syam N, Sharma A (2018) Waiting for a sales renaissance in the fourth industrial revolution: machine learning and artificial intelligence in sales research and practice. Ind Mark Manag 69:135–146
4. Poushneh A (2021) Humanizing voice assistant: the impact of voice assistant personality on consumers' attitudes and behaviors. J Retail Consum Serv 58:102283
5. Horstmann AC et al (2018) Do a robot's social skills and its objection discourage interactants from switching the robot off? PLoS ONE 13(7):e0201581
6. Tuzovic S, Paluch S (2018) Conversational commerce—a new era for service business development? In: Bruhn M, Hadwich K (eds) Service Business Development, pp 81–100. Springer Gabler, Wiesbaden. https://doi.org/10.1007/978-3-658-22426-4_4
7. Guzman AL (2019) Voices in and of the machine: source orientation toward mobile virtual assistants. Comput Hum Behav 90:343–350
8. McLean G, Osei-Frimpong K (2019) Hey Alexa… examine the variables influencing the use of artificial intelligent in-home voice assistants. Comput Hum Behav 99:28–37
9. Hoy MB (2018) Alexa, Siri, Cortana, and more: an introduction to voice assistants. Med Ref Serv Q 37(1):81–88
10. Lima L et al (2019) Empirical analysis of bias in voice-based personal assistants. In: Companion Proceedings of the 2019 World Wide Web Conference
11. Venkatesh V et al (2003) User acceptance of information technology: toward a unified view. MISQ 27(3):425–478
12. Venkatesh V, Thong JY, Xu X (2016) Unified theory of acceptance and use of technology: a synthesis and the road ahead. J Assoc Inf Syst 17(5):328–376

13. Abbas SK et al (2018) Integration of TTF, UTAUT, and ITM for mobile banking adoption. Int J Adv Eng Manag Sci (IJAEMS) 4(5):375–379
14. Tarhini A et al (2016) Extending the UTAUT model to understand the customers' acceptance and use of internet banking in Lebanon: a structural equation modeling approach. Inf Technol People 29(4):30–849
15. Alam MZ, Hu W, Barua Z (2018) Using the UTAUT model to determine factors affecting acceptance and use of mobile health (mHealth) services in Bangladesh. J Stud Soc Sci 17(2):137–172
16. Phaosathianphan N, Leelasantitham A (2019) Understanding the adoption factors influence on the use of intelligent travel assistant (ITA) for eco-tourists: an extension of the UTAUT. Int J Innov Technol Manag 16(08):1950060
17. Gunasinghe A et al (2020) The viability of UTAUT-3 in understanding the lecturer's acceptance and use of virtual learning environments. Int J Technol Enhanc Learn 12(4):458–481
18. Zaharia S, Würfel M (2021) Voice commerce–studying the acceptance of smart speakers. In: Ahram T, Taiar R., Langlois K, Choplin A (eds) Human Interaction, Emerging Technologies and Future Applications III. IHIET 2020. Advances in Intelligent Systems and Computing, vol 1253. Springer, Cham. https://doi.org/10.1007/978-3-030-55307-4_68
19. Fishbein M, Ajzen I (1977) Belief, attitude, intention, and behavior: an introduction to theory and research. Philos Rhetor 10(2):177–189
20. Altin Gumussoy C, Kaya A, Ozlu E (2018) Determinants of mobile banking use: an extended TAM with perceived risk, mobility access, compatibility, perceived self-efficacy and subjective norms. In: Calisir F, Camgoz Akdag H (eds) Industrial Engineering in the Industry 4.0 Era, pp 225–238. LNMIE. Springer, Cham. https://doi.org/10.1007/978-3-319-71225-3_20
21. Rejón-Guardia F, Polo-Peña AI, Maraver-Tarifa G (2020) The acceptance of a personal learning environment based on Google apps: The role of subjective norms and social image. J Comput High Educ 32(2):203–233
22. Ramadania S, Braridwan Z (2019) The influence of perceived usefulness, ease of use, attitude, self-efficacy, and subjective norms toward intention to use online shopping. Int Bus Account Res J 3(1):1–14
23. Chen L, Aklikokou AK (2020) Determinants of E-government adoption: testing the mediating effects of perceived usefulness and perceived ease of use. Int J Public Adm 43(10):850–865
24. Davis FD, Bagozzi RP, Warshaw PR (1992) Extrinsic and intrinsic motivation to use computers in the workplace 1. J Appl Soc Psychol 22(14):1111–1132
25. Ashfaq M et al (2019) Customers' expectation, satisfaction, and repurchase intention of used products online: empirical evidence from China. SAGE Open 9(2):2158244019846212
26. Ashfaq M et al (2020) I, Chatbot: Modeling the determinants of users' satisfaction and continuance intention of AI-powered service agents. Telemat Inform 54:101473
27. Holdack E, Lurie-Stoyanov K, Fromme HF (2020) The role of perceived enjoyment and perceived informativeness in assessing the acceptance of AR wearables. J Retail Consum Serv 65(3):1–11
28. Yang H, Lee H (2019) Understanding user behavior of virtual personal assistant devices. Inf Syst e-Bus Manag 17(1): 65–87
29. Van Pinxteren MM et al (2019) Trust in humanoid robots: implications for services marketing. J Serv Market 33(4):507–518
30. Kim SY, Schmitt BH, Thalmann NM (2019) Eliza in the uncanny valley: anthropomorphizing consumer robots increases their perceived warmth but decreases liking. Mark Lett 30(1):1–12
31. Wirtz J et al (2018) Brave new world: service robots in the frontline. J Serv Manag 29(5):907–931
32. Chattaraman V et al (2019) Should AI-Based, conversational digital assistants employ social-or task-oriented interaction style? A task-competency and reciprocity perspective for older adults. Comput Hum Behav 90:315–330
33. Moussawi S, Koufaris M (2019) Perceived intelligence and perceived anthropomorphism of personal intelligent agents: scale development and validation. In: Proceedings of the 52nd Hawaii International Conference on System Sciences

34. Marikyan D et al (2022) Alexa, let's talk about my productivity: the impact of digital assistants on work productivity. J Bus Res 142:572–584
35. Balakrishnan J, Dwivedi YK (2021) Conversational commerce: entering the next stage of AI-powered digital assistants. Ann Oper Res 2021(3):1–35
36. Bartneck C et al (2009) Measurement instruments for the anthropomorphism, animacy, likeability, perceived intelligence, and perceived safety of robots. Int J Soc Robot 1(1):71–81
37. Moussawi S, Koufaris M, Benbunan-Fich R (2021) How perceptions of intelligence and anthropomorphism affect adoption of personal intelligent agents. Electron Mark 31(2):343–364
38. Gremler DD, Gwinner KP (2000) Customer-employee rapport in service relationships. J Serv Res 3(1):82–104
39. Wilson JR, Lee NY, Saechao A, Hershenson S, Scheutz M, Tickle-Degnen L (2017) Hand gestures and verbal acknowledgments improve human-robot rapport. In: Kheddar A et al (eds) Social Robotics. ICSR 2017. LNCS, vol 10652. Springer, Cham. https://doi.org/10.1007/978-3-319-70022-9_33
40. CreativeDigital. This human-like robot is lending a helping hand in aged care homes (2017). [cited 10 Mar 2022]. www.createdigital.org.au/human-like-robot-aged-care-homes/
41. Fernandes T, Oliveira E (2021) Understanding consumers' acceptance of automated technologies in service encounters: drivers of digital voice assistants adoption. J Bus Res 122:180–191

Factors Affecting Students Behaviroal Intention Towards Using E-learning During COVID-19: A Proposed Conceptual Framework

Research-in-Progress

Muaath AlZakwani, Ghalib AlGhafri, Faisal AlMaqbali, Sadaf Sadaq, and Ali Tarhini

Abstract Nowadays, e-learning has become significant in distance teaching. Nevertheless, factors influencing students' behavioral intentions toward different e-learning platforms are still not well understood. This research-in progress aims to propose a conceptual framework that consider the key influencing factors that may enable or hinder the adoption of e-learning tools during the COVID-19 pandemic. Data will be collected from Sultan Qaboos University students by using an online survey to test the conceptual framework. The data gathered will be analyzed using structural equation modeling approach. This research-in-progress will help the decision makers formulate strategies to improve the adoption of online learning systems during the COVID-19 pandemic.

Keywords Behavioral intention · Technology adoption · e-learning platforms · COVID-19 · UTAUT2

1 Introduction

Learning online is described as a teaching and learning strategy that is entirely dependent on the use of the Internet to improve learning, interaction, and collaboration [1, 2]. In the twenty-first century, technology is becoming increasingly crucial in our daily lives, requiring professionals, educators, and learners to reevaluate their core belief in the use of technology to redesign or re-engineer the education and training system [3].

The COVID-19 epidemic poses a significant challenge in terms of controlling the educational process, whether theoretical or practical [4]. As seen throughout the world, it is compelling educational organizations such as universities and colleges

M. AlZakwani · G. AlGhafri · F. AlMaqbali · S. Sadaq · A. Tarhini (✉)
Department of Information Systems, Sultan Qaboos University, Seeb, Sultanate of Oman
e-mail: ali.tarhini@hotmail.co.uk

M. Al-Emran et al. (eds.), *International Conference on Information Systems and Intelligent Applications*, Lecture Notes in Networks and Systems 550,
https://doi.org/10.1007/978-3-031-16865-9_49

to move swiftly to remote and online e-learning platforms. It has forced universities worldwide to embrace various online learning systems. Nevertheless, we are currently in a state of emergency, and we must use a variety of readily available learning methods, such as e-learning systems and mobile learning applications, to respond. Both online and remote learning are familiar to students and learners. On the other hand, COVID-19 has rekindled an interest in exploring online teaching and learning opportunities [5].

According to [6], school and university suspensions have many detrimental consequences for students, including disrupted study and depriving learners and youth of the possibilities for maturing and development. However, e-learning platforms can handle this problem by enabling admission to these technologies using speedy and stable internet connections. In actuality, e-learning technologies are forming a vital part of this pandemic. E-learning platforms can oblige learning providers to organize, plan, execute, and measure various learning and teaching activities. Indeed, E-learning platforms aim to aid lecturers, academics, and schools in making learning more convenient during university and school closing hours. Moreover, most of these systems are free, which will aid in continual learning during the pandemic. During the COVID-19 pandemic, several scholars have discussed e-learning acceptability in more formal education [5, 7–9].

However, according to our knowledge, there is still a scarcity of research on students' intentions to utilize e-learning platforms during COVID-19, including e-learning background and system quality as exterior variables, especially in developing countries. For example, e-learning was not widely used in universities in the Sultanate of Oman before the epidemic. Therefore, decision-makers in the Sultanate may not find detailed research on this aspect when a decision is needed which motivate us to provide something helpful. Accordingly, this study aims to propose a model that help the researchers to examine the factors that may enable or hinder the adoption of online learning during the COVID-19 Pandemic.

In addition, following the introduction, Section two includes a literature review and conceptual framework and the suggested hypotheses. The third part discusses the methodology that was used in the study. In the fourth section, the outcomes are provided. Finally, Section five propose some implications of the study and conclude the paper.

2 Literature Review and Conceptual Framework

2.1 Literature Review

To prevent the spread of the COVID-19 pandemic, most colleges worldwide have moved away from face-to-face instruction and towards online learning via virtual platforms. However, the virtual platforms employed differ between institutions and even from country to country [10, 11]. All students and teachers have been considered self-quarantined in their houses during this pandemic. Moreover, online learning has become a popular method of acquiring education, supported by the Internet and intelligent terminal devices such as smartphones and tablet PCs. Online learning alters traditional learning practices by allowing people to learn whenever and wherever they choose [12]. E-learning is a teaching and learning method that is dependent on the Internet. It can increase the efficiency of communication, and interaction between students and teachers [13]. E-learning platforms provide significant advantages for both teachers and students, including cost savings, learning process enhancement, accommodating learning techniques, and dynamic course material [14].

In the same context, the adoption of e-learning is dependent on a student's judgment to utilize technology, which is referred to in the literature as "behavioural intention" [15]. In regard to answering the research question, this examination uses the UTAUT2 (unified theory of acceptance and use of technology) model created by [16], but including one more factor, which is awareness of COVID-19 [17], in order to analyze behavioural intentions towards the adoption of e-learning platforms. Prior studies have discovered that many cultural, organizational, individual, or technological factors could affect students' intention to embrace a specific technology [14, 18].

Nonetheless, studies examining university students' behavioural intentions toward e-learning platforms were rare. Instead, they used their high school e-learning experience as an external reference. Furthermore, a substantial number of prior research studies on system quality influence students' behavioural intentions in embracing e-learning [19]. [20] identified three primary pillars: the simplicity of delivering course material content, the student, and the system's vital component. In other words, efficient communication channels and new ways are required for e-learning systems to assess the efficacy of academic activities, thereby assuring students' desire to embrace the e-learning system and motivate their behavioural intention to embrace it.

According to [21], students find it challenging to access an online environment due to a lack of internet access and tools promoting online learning. Performance expectations, effort expectations, and social influences are directly related to behavioural intents. In contrast, the final enabling conditions are related to actual usage. Furthermore, gender, age, experience, and voluntariness influence behavioural intentions [22]. According to the UTAUT2 model, social influences, performance expectancy, habit, effort expectancy, trust, price value, hedonic motivation, and facilitating conditions are directly related to behavioural intentions, whereas the final enabling circumstances are actual usage. Gender, age, experience, and voluntariness all influence behavioural intentions [22, 23].

2.2 Conceptual Framework

This study extends the UTAUT2 model by including awareness of COVID-19 with other factors affecting Sultan Qaboos University students' intent to utilize E-learning

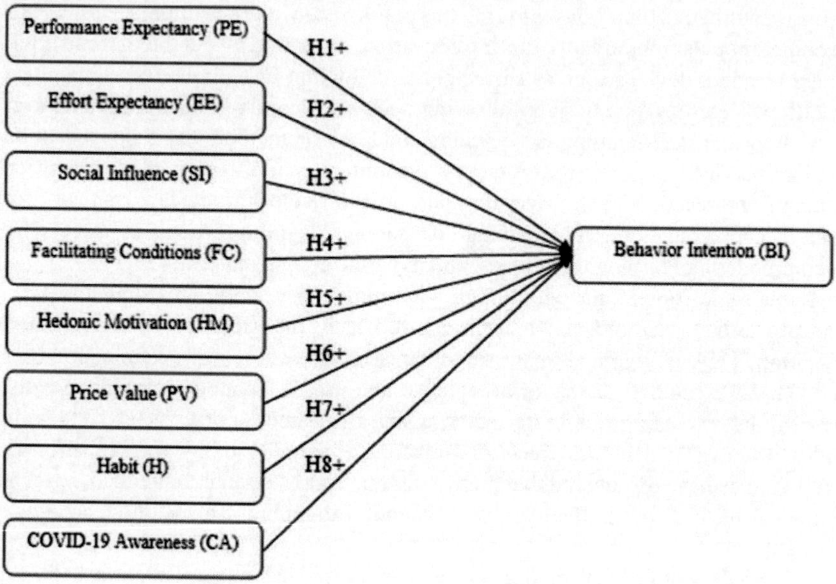

Fig. 1 Conceptual model: behavioural intention on using e-learning during COVID-19

platforms during COVID-19. The reason for choosing the UTAUT2 model is that it is the most comprehensive of all the models mentioned in the previous paragraph. Moreover, it has been tested for validity and reliability in several prior research papers [24–26]. In the same context, we mention COVID-19 awareness, which was used in [17] study as one of the factors affecting behavioural intention to answer the question of this study. Furthermore, this study will not include any moderating effects as the target has almost the same age and experience. Figure 1 illustrates the proposed research model.

The following sections explain each factor separately with the hypothesis to show the relationship between the factors.

2.2.1 Performance Expectancy (PE)

This is defined as the level of confidence there is that utilizing the technology will enhance career performance [28]. The student expects to benefit at the academic level when using e-learning technology. This benefit is distinct from the formal study that the students of lectures are accustomed to, by which we mean here the attendance the study seats on the university campus. Based on the research papers talk about e-learning [23] and mobile learning [29], there is a direct relationship between students' performance expectations and their behavioural intentions. The following hypothesis illustrates this:

***H1**. PE influences students' BI positively to use e-learning platforms.*

2.2.2 Effort Expectancy (EE)

EE is described as the rank of comfort associated with the utilization of technology. [16] This study suggests that learners can embrace an e-learning system if it is straightforward. This factor describes the importance of aligning students' expectations with the technology offered, which is e-learning in this case. When a student finds that using the e-learning system is easy and that they do not need help, that should motivate students' behavioural intention to use the platform. That is what was stated in the research papers that talked about e-learning platforms [23][23]. It results the following hypothesis, which is equivalent to what [27] proposes:

***H2**. EE influences students' BI positively to use e-learning platforms.*

2.2.3 Social Influence (SI)

The SI factor reflects how others influence the individual to use a specific technology, regardless of their desire. It is expressed as an individual's perception of how significant it is that others think they should adopt the new e-learning system [28]. The following hypothesis summarizes the positive association between social influence and behavioural intention. The same hypothesis is mentioned in the research papers that talk about e-learning [23] and mobile learning [29].

***H3**. SI positively and significantly influences students' BI to use e-learning systems.*

2.2.4 Facilitating Conditions (FC)

This is described as an individual's conviction in the presence of organizational and technological needs to allow the use of the platform [28]. An environmental element impacts users' perception of how challenging or accessible a task is to perform—knowing that the effects vary according to the students' outer environment. Often the student needs a solid base provided by the educational institution to help them use the technology. Reviewing [30] literature, regarding the acceptability of mobile restaurant apps, there appears to be a linkage between facilitating conditions and student behavioural intention. Therefore, this research proposes the theory below.

***H4**. Students' BI to operate e-learning platforms is influenced positively by FC.*

2.2.5 Hedonistic Motivation (HM)

This can be represented as the enjoyment an individual derives from utilizing a technology, which directly impacts their desire to operate that technology in the future [16]. When students find pleasure in employing the e-learning system, they are motivated and eager to use this technology continuously. After reviewing the research paper about adopting social networks, the subsequent hypothesis explains the association between behavioural intention and hedonistic motivation [31].

H5. *HM has a positive influence on students' BI to employ e-learning platforms.*

2.2.6 Price Value (PV)

The price value is considered positive; when the advantages of using a technology are more significant than the financial cost. [16]. This occurs when the benefit from using a technology outweighs the cost to a person to provide it. For example, e-learning technology saves students several costs. The students may find that e-learning benefits them more than the cost spent on the technology. The research [32] found that the elements influencing customers to use food delivery apps are that the price value and BI have a positive relationship. On the other hand, [15] mentions that the price value could represent the intangible expenses for workplace learning, such as time and effort. Thus, this research assumes the following theory.

H6*: PV positively impacts students' BI to use e-learning platforms.*

2.2.7 Habit (H)

Limayem [33] have illustrated habit as the degree to which individuals perform behaviours automatically due to learning. Therefore, the practices that a person does continuously and automatically can be a reason for the continuity of work on the technology. Moreover, the postgraduate students' habit of using e-learning technology may affect their Behavioural Intention. The following hypothesis, which comes from the research paper [31], is that the habit of utilizing technology to share user-generated scope positively influences behavioural intention.

H7*: H positively influences students' BI to utilize e-learning platforms.*

2.2.8 COVID-19 Awareness (CA)

During the pandemic, awareness included various aspects such as necessary precautions and practical alternatives. The Ministry of Health issued several alerts in the beginning, such as maintaining a safe distance, staying at home, and sanitary isolation. In addition, the universities spread awareness of the mechanism of using e-learning systems and verified that students were using the systems with complete comfort and without problems. Therefore, this factor affects the severity of behavioural intention. [34] explain this point by saying, "Without regular awareness, the adoption of e-learning systems cannot be carried out smoothly.". Thus, this research concludes with the following theory:

H8: Awareness of COVID-19 positively influences students' BI to utilize e-learning platforms.

3 Research Methodology

This study will employ a quantitative approach and will gather the data using an online survey to validate the conceptual framework. In light of the study's nature, a probability random sampling strategy will be employed to choose students from Sultan Qaboos University. The data and findings will be evaluated by using structural equation modeling. The research questionnaire will consist of two parts. In the first part questions are related to the respondent's demographic information. On the other hand, in the second part questions are related to each model factor.

It should be noted that all the factors employed in this study were adopted from previous literature. For the measurement of the items, a five-point Likert scale will be used. The Likert scale reaches from "strongly disagree" to "strongly agree". Prior to the full-scale data collection, a pilot-test was conducted and the survey was verified.

4 Potential Theoretical and Practical Contributions of the Study

From theoretical perspective, this research is planned to primarily contribute to the body of information on the factors influencing students' intentions to use the E-learning platform by university students. From practical perspective, this research in progress will help decision-makers at higher educational institutions improve the e-learning systems until it reaches the required level that can better serve students during the pandemic. In addition, the research will help to discover the weaknesses in the current system to focus on them and fix them to the extent required. The Ministry of Higher Education can also benefit from this experience to develop solutions for the learning system and make the appropriate decisions to improve the level of e-learning in the Sultanate.

5 Future Research

The researchers worked to organize the content of the study which contained valuable information to promote the research and graduate it correctly. The researchers see the importance of applying this study to more than one university institution in order to compare the data between the various university institutions in the Sultanate. Since this research was limited to Sultan Qaboos University students only, the researchers recommend the importance of carrying out this study to more than one university in the Sultanate, thus enhancing the results of this study and diversifying it.

6 Conclusion

Throughout the epidemic, this study strived to construct a comprehensive model to determine the factors influencing students' behavioural intention to utilize the E-learning platform during the COVID-19 pandemic. The research model was built using components from the UTAUT2 model with another factor related to COVID-19 awareness to accomplish this purpose. Moreover, the study provides valuable information and perspective to support educational institutions in their decision-making process regarding online learning during the COVID-19 pandemic. In short, we propose that e-learning platforms and other educational tools be assessed holistically through student perspectives before being extensively employed during any epidemic. Furthermore, testing and examination should continue even after COVID-19 has waned.

References

1. Krishnan KST, Hussin H (2017) E-learning readiness on Bumiputera SME's intention for adoption of online entrepreneurship training in Malaysia. Management 7(1):35–39. https://doi.org/10.5923/j.mm.20170701.04
2. Salloum SA, Alhamad AQM, Al-Emran M, Monem AA, Shaalan K (2019) Exploring students' acceptance of e-learning through the development of a comprehensive technology acceptance model. IEEE Access 7:128445–128462
3. Kumar Basak S, Wotto M, Belanger P (2018) E-learning, M-learning and D-learning: conceptual definition and comparative analysis. E-Learn Digit Media 15(4):191–216
4. Ratten V, Jones P (2020) Covid-19 and entrepreneurship education: Implications for advancing research and practice. Int J Manag Educ 19(1):1–10
5. Almaiah MA, Al-Khasawneh A, Althunibat A (2020) Exploring the critical challenges and factors influencing the E-learning system usage during COVID-19 pandemic. Educ Inf Technol 25:5261–5280
6. UNESCO (2022) Education: from disruption to recovery. https://en.unesco.org/covid19/educationresponse. Accessed 26 Feb 2022
7. Siron Y, Wibowo A, Narmaditya BS (2020) Factors affecting the adoption of e-learning in Indonesia: lesson from Covid-19. JOTSE J Technol Sci Educ 10(2):282–295
8. Vladova G, Ullrich A, Bender B, Gronau N (2021) Students' acceptance of technology-mediated teaching–how it was influenced during the COVID-19 pandemic in 2020: a study from Germany. Front Psychol 12(1):1–15
9. Sukendro S et al (2020) Using an extended technology acceptance model to understand students' use of e-learning during Covid-19: Indonesian sport science education context. Heliyon 6(11):e05410
10. Altameemi AF, Al-Slehat ZAF (2021) Exploring the students' behavior intentions to adopt e-learning technology: a survey study based on COVID-19 crisis. Int J Bus Manag 16(6):31–41

11. Alzahrani L, Seth KP (2021) Factors influencing students' satisfaction with continuous use of learning management systems during the COVID-19 pandemic: an empirical study. Educ Inf Technol 26(6):6787–6805
12. Alghamdi AM, Alsuhaymi DS, Alghamdi FA, Farhan AM, Shehata SM, Sakoury MM (2022) University students' behavioural intention and gender differences toward the acceptance of shifting regular field training courses to e-training courses. Educ Inf Technol 27(1):451–468
13. Liu N, Pu Q (2020) Factors influencing learners' continuance intention toward one-to-one online learning. Interact Learn Environ:1-22 (ahead-of-print)
14. Taghizadeh SK et al (2021) Factors influencing students' continuance usage intention with online learning during the pandemic: a cross-country analysis. Behav Inf Technol 41(9):1998–2017
15. Mehta A, Morris NP, Swinnerton B, Homer M (2019) The influence of values on E-learning adoption. Comput Educ 141:103617
16. Venkatesh V, Thong JYL, Xu X (2012) Consumer acceptance and use of information technology: extending the unified theory of acceptance and use of technology. MIS Q 36(1):157–178
17. Alea LA, Fabrea MF, Roldan RDA, Farooqi AZ (2020) Teachers' COVID-19 awareness, distance learning education experiences and perceptions towards institutional readiness and challenges. Int J Learn Teach Educ Res 19(6):127–144
18. Unal E, Uzun AM (2021) Understanding university students' behavioural intention to use Edmodo through the lens of an extended technology acceptance model. Br J Educ Technol 52:619–637
19. Mailizar M, Burg D, Maulina S (2021) Examining university students' behavioural intention to use e-learning during the COVID-19 pandemic: an extended TAM model. Educ Inf Technol 26(6):7057–7077
20. Costa GJM, Silva NSA (2010) Knowledge versus content in e-learning: a philosophical discussion. Inf Syst Front 12(4):399–413
21. Maphosa V (2021) Factors influencing student's perceptions towards e-learning adoption during COVID-19 pandemic: a developing country context. Eur J Interact Multimed Educ 2(2):e02109
22. Abbad MM (2021) Using the UTAUT model to understand students' usage of e-learning systems in developing countries. Educ Inf Technol 26(6):7205–7224
23. El-Masri M, Tarhini A (2017) Factors affecting the adoption of e-learning systems in Qatar and USA: extending the unified theory of acceptance and use of technology 2 (UTAUT2). Educ Tech Res Dev 65(3):1–21. https://doi.org/10.1007/s11423-017-9526-1
24. Altameemi AF, Al-Slehat ZAF (2021) Exploring the students' behavior intentions to adopt e-learning technology: a survey study Based on COVID-19 Crisis. Int J Bus Manag 16(6):31–41
25. Marlina E, Tjahjadi B, Ningsih S (2021) Factors affecting student performance in e-learning: a case study of higher educational institutions in Indonesia. J Asian Financ Econ Bus 8(4):993–1001
26. Samsudeen SN, Mohamed R (2019) University students' intention to use e-learning systems: a study of higher educational institutions in Sri Lanka. Interact Technol Smart Educ 16(3):219–238
27. Al-Emran M, Al-Maroof R, Al-Sharafi MA, Arpaci I (2020) What impacts learning with wearables? An integrated theoretical model. Interact Learn Environ:1–21 (ahead-of-print)
28. Venkatesh V, Morris MG, Davis GB, Davis FD (2003) User acceptance of information technology: toward a unified view. MIS Q 27(3):425–478
29. Arain AA, Hussain Z, Rizvi WH, Vighio MS (2019) Extending UTAUT2 toward acceptance of mobile learning in the context of higher education. Univers Access Inf Soc Int J 18(3):659–673. https://doi.org/10.1007/s10209-019-00685-8
30. Palau-Saumell R, Forgas-Coll S, Sánchez-García J, Robres E (2019) User acceptance of mobile apps for restaurants: an expanded and extended UTAUT-2. Sustainability 11(4):1210
31. Herrero Á, San Martín HM, García de los Salmones MDM (2017) Explaining the adoption of social networks sites for sharing user-generated content: a revision of the UTAUT2. Comput Hum Behav 71:209–217217. https://doi.org/10.1016/j.chb.2017.02.007

32. Ramos K (2021) Factors influencing customers' continuance usage intention of food delivery apps during COVID-19 quarantine in Mexico. Br Food J 124(3):833–852
33. Limayem M, Hirt SG, Cheung CMK (2007) How habit limits the predictive power of intention: the case of information systems continuance. MIS Q 31(4):705–737
34. Almaiah MA, Al-Khasawneh A, Althunibat A (2020) Exploring the critical challenges and factors influencing the e-learning system usage during COVID-19 pandemic. Educ Inf Technol 25(6):5261–5280

An Approach to Enhance Quality of Services Aware Resource Allocation in Cloud Computing

Yasir Abdelgadir Mohamed and Amna Omer Mohamed

Abstract A new technology called cloud computing has revolutionized the way services are delivered to businesses and consumers. As an online service, it offers a variety of options to registered users. Quality of service (QoS) requirements must be reached in order for the customer to be completely satisfied. As a result of its impact on other issues faced by cloud users and providers alike, QoS-aware resource allocation is the most essential issue in resource allocation. There is no effective solution that meets both the needs of the service provider and the consumer, yet it is still regarded a difficulty by many. This research aims to reduce the amount of time needed to assign cloud resources, improving overall performance. The social spider algorithm (SSA) is presented to map resources with the suitable job in order to fulfill the specified objectives and handle the complexity of the resource allocation issue. In order to simulate spider foraging behavior, SSA created an algorithm. It focuses on the spider, its prey, and the strength of its vibrations. This is how a victim gets out of the spider web: by attempting to release itself from the web, which creates vibrations in the web. At that point, every spider in that web was able to pick up on the vibration. The more fit the sufferer is, the greater the strength of the vibrations. Vibration intensity created on the web determines the victim's potential. In the cloud, the job is the spider, and the resource is the prey. In terms of resource fitness, task fitness is seen as the ability to make effective use of available resources. Using DEV-C++ to construct the suggested technique, tests have shown that it saves execution time by up to 10% while simultaneously improving service quality. In terms of execution time, the SSA algorithm with first fit exceeds the SSA algorithm with best fit, while the best fit excels in terms of utilization. Furthermore, when the SSA algorithm is compared to the SSCWA method, the SSA algorithm performs better in terms of execution time, usage, and throughput. The SSA results in improved resource allocation, which results in higher QoS parameters and performance. Additional QoS

Y. A. Mohamed (✉)
College of Business, Alburaimi University, MIS, Al Buraimi, Oman
e-mail: Yasir.a@uob.edu.om

A. O. Mohamed
University of Gezira/Computer Engineering, Medani, Sudan

M. Al-Emran et al. (eds.), *International Conference on Information Systems and Intelligent Applications*, Lecture Notes in Networks and Systems 550,
https://doi.org/10.1007/978-3-031-16865-9_50

considerations, such as resource dependability, are a conceivable possibility in the future. Additional research may be done to speed up the execution time even more.

Keywords Cloud computing · Resource allocation · Social spider algorithm

1 Introduction

Without human control, data storage and processing power are made accessible on demand. The term "cloud computing" refers to the use of internet data centers. Core server functions are frequently distributed in today's massive clouds. [1] Storage, hardware, and software may be used by both public and private companies. A cloud computing ecosystem is dominated by cloud providers and clients. To optimize revenues via high resource utilization, cloud providers have a big number of pay-per-use computing resources in the cloud. Cloud customers with erratic loads and leased resources may operate their applications at the lowest possible cost. The emphasis of cloud computing is on IaaS resource management (IaaS). Resource allocation, modification, and scheduling are examples of these.

To put it another way, resource allocation is the process through which Internet-based cloud applications are systematically given access to available resources. The end user of a cloud service has access to the complete computer stack, from hardware to software. Pay-as-you-go services are common in cloud computing. The end user of a cloud service may adjust the amount of resources accessible to their apps. Users may be liable for paying extra fees for this benefit, which is one of the key benefits of cloud computing [2].

Users' resource availability and distribution preferences must be included into cloud computing business models. Client satisfaction should improve with IaaS cloud computing [3]. Concerns of cloud computing resource allocation include QoS-aware, dynamic, and price-sensitive allocation. Cloud computing relies on service-aware resource allocation. This requires allocating resources based on the quality-of-service requirements of cloud users (QoS). The most critical is resource allocation based on quality of service (QoS). This has an impact on both cloud consumers and providers. Most (if not all) solutions take service providers or customers into account, but not both. It is tough to find a solution that meets both needs. The goal of this research is to improve cloud computing resource allocation based on quality of service (QoS), specify a framework to achieve high availability and throughput for both cloud providers and customers, and design and build a system that efficiently distributes randomly received requests to dynamically updated resources.

2 Background

QoS-aware resource allocation is critical to cloud computing. When it comes to cloud computing, this implies assigning resources to satisfy the requirements and expectations of cloud users in terms of service levels and availability (SLA). In order to prevent increasing failure rates, lack of resources, poor resource use, and SLA violence, it is necessary to allocate resources carefully [4]. Resource suppliers may give on-demand services to clients in a transparent manner by using cloud computing. Providers must be able to dynamically optimize available resources in order to assure quality of service without limiting the number of accepted requests. Virtualized resources might be difficult to manage when it comes to deploying and monitoring programs [5].

The diverse client demands make it challenging to provide a cloud-based QoS guarantee. The same service may be provided by many companies using different technologies. Vibrations may be used by spiders to detect prey or other spiders on their webs. Spiders forage by sharing information and using a variety of methods. Vibrations in the web may be used by each spider to assess its fitness. The more the spider is used, the higher the person's vibration. As a result, vibration intensity may be used to assess fitness. The victim's potential is determined by the strength of the vibration. The intensity of web vibration improves the victim's fitness.

A spider or victim on the spider web will record parameters such as location, vibration intensity, and current fitness following assessment [6]. Because social spiders use vibration intensity to communicate, they may detect alien species/victims ensnared in the web. A victim ensnared in the web struggles to get free, causing vibrations in the web.

This vibration will be felt by every spider on the web. The intensity of vibration increases with the victim's fitness. The spiders attack the victim based on their need to feed. The resource management module will adjust the location of the spiders whose use is higher. So they get closer to the victim and seize it [7].

Cloud computing relies on resource management and placement. Resources are diversified and changing, making resource allocation difficult [8]. Many academics have worked on QoS-aware resource allocation. In a dynamic environment, it is necessary to provide assured services to the customer.

With grid computing, RAAJS was built. In this case, the grid matched the resources, as cloud and grid environments share resources. Kumar [9] recommends using weight metrics for jobs and resource decisions. WM reorganized jobs, enhancing the algorithm's performance. As the algorithm's resource allocation percentage grows, job completion time and the number of attempts to access particular services decrease.

Also, [10] focuses on the long/short term virtual machine renting issue. A statistical learning approach for resource need was presented, as was a dynamic virtual machine renting mechanism. These methods reduced operating costs while maintaining QoS specifications. To replicate the real-world load, a Markov Modulated

Poisson Process (MMPP) was used to create end-user arrivals. Extensive numerical studies validated the suggested algorithms' efficacy.

Horri [11] proposed QoS-aware virtual machine consolidation based on resource utilization and virtual machine history. Using virtual machine resource history reduces energy utilization and SLAV. Since VMs don't peak concurrently, energy consumption and the number of times a host hits 100% utilization reduce (SLAV). The goal is to train a SLA-aware algorithm, minimize SLA violations, and save operating costs. These algorithms balance efficiency and performance.

In addition, Gawali [12] offers a heuristic strategy for task scheduling and resource allocation that combines MAHP, BATS + BAR optimization, LEPT, and divide-and-conquer techniques. Prior to allocating each task to cloud resources, each job was (MAHP) assessed. The allocation of resources using (BATS + BAR) optimization, that considers the bandwidth and load of the cloud resources as constraints. The recommended method employs (LEPT) preemption to minimize wasting resources. The divide-and-conquer technique improves the proposed system in comparison to the existing (BATS) and improved multi - objective evolutionary algorithm (IDEA) frameworks. Cooperative game theory was used to address network resource distribution. Fair Allocation guarantees bandwidth and divides it based on network weights using online and offline algorithms. Flexible dependability and load balancing are offered by the recommended technique [13]. Also, resource management module (ReMM) was suggested for optimal resource usage, QoS and workload balancing, computing resources are given to cloud users with varying workloads and are specified during performance analysis. This method, the guidance of setups of varying user demand may be calculated. The simulation results suggest that the proposed module may meet changing resource demands while maintaining QoS [14].

The researcher in [15] created EQUAL, an energy- and QoS-conscious virtual machine resource allocation technique using AntLion optimization. Both energy and QoS are considered (VMs). EQUAL may operate in power-aware, performance-aware, or balanced modes. CloudSim was used to develop and test virtual machines (VMs) and resource-intensive activities.

The findings of the experiments have shown that the strategy may cut energy consumption by up to 15%, as depicted in Fig. 2. Additionally, the quality of service is improved as a consequence of a decrease in the number of activities whose due dates were not met [16].

Model predictive controller (MPC) was used by the authors of [17] to provide a dynamic QoS-aware resource allocation in a FaaS platform. Resource allocation choices are based on forecasts of future occurrences as well as user requests for Quality of Service (QoS) enforcements. When making decisions, the controller takes into account: (1) an estimate of the rate of events associated with each function as a service (FaaS) function that will occur in the near future, (2) the amount of QoS violation incidents that have occurred in the past epochs as feedback, and (3) the reconfiguration cost. As compared to the best-effort method, the controller achieves an average increase of 21% resource utilization and a three-fold decrease in QoS-violation occurrences, while retaining the mean latency of actions 19.9% lower than the best-effort strategy.

P. Abroll [18] introduced a QoS-aware re-source placement algorithm based on the social spider mating strategy, which uses QoS metric optimization to automatically manage and allocate workloads for re-source computation. The social spider cloud web algorithm (SSCWA) was created in order to map out cloud workloads and the resources that are conveniently accessible to them. It is also recommended that a QoS-aware cloud orchestrated framework for efficient resource placement (COFER) be developed, which provides an orchestrated framework for assessing and deploying required workloads on available resources depending on different QoS characteristics. The framework's use of resource management and placement approaches resulted in consistent resource consumption. According to a study, the suggested framework beats competitors in terms of cost, execution time, and throughput, as well as cloud resource availability and dependability and optimum utilization.

According to [19], resources must be allocated such that each application receives the resources it needs while without going over the cloud environment's maximum capacity. the starvation of apps may be dealt with by effective resource allocation, which allows service providers to distribute resources for each module individually.

The social spider algorithm simulates spider foraging behavior [20, 21]. It emphasizes the spider's fitness and vibration intensity. Spiders utilize the spider web to communicate, causing vibrations [22, 23]. The spider with the most vibrations is the best fit on the spider web. The fittest spider in the spider web has more needs than the less fit spiders, thus it gets precedence in attacking the victim.

When a person gets entangled in a spider web, they naturally want to break free. The vibration was felt by all spiders on the web. The intensity of vibration increases with the victim's fitness. The victim's potential is decided by the web's vibration intensity. The victim's fitness increases as their web vibration intensity increases.

The spiders' senses intensify with distance from the victim. Because vibration strength decreases with distance, spiders near the victim will detect higher vibrations on the spider web than spiders farther away [24]. The spiders assault the victim depending on their hunger, with the hungriest spiders attacking first. So, they get closer to the victim and catch it. A spider or victim on the spider web will record parameters such as location, vibration intensity, and current fitness following assessment. In the cloud, the job is the spider and the resource are the victim. The job with the highest QoS requirements gets precedence. Resource fitness refers to resource capacity, whereas task fitness refers to task use.

A hybrid machine learning strategy for organizing workloads and distributing cloud resources described in [25]. The researchers improved feline population development, simplified the basic deep brain structure, and employed a lightweight confirmation plan to increase memory, CPU, resources, and data transmission. Based on asset usage, RATS-HM effectively detects high-utility assets. A maximum use of the CPU, memory, and data transfer was attempted. The proposed framework improves memory and CPU transfer. More about security in distributed environment on [26, 27].

3 Methodology

When the clients send requests as shown in the flow chart in Fig. 3, it is received by the cloud providers and then stored in the task table. The next step is comparing the requested resources for the task with the available resources. If the available resources can execute the task, the SSA algorithm will be executed, if not the task will be stored in waiting queue and the cloud providers would be waiting for new vibration generated by resources.

The SSA algorithm calculates the fitness and the vibration intensity of both the tasks and the resources. Then the algorithm chooses the appropriate pair of resource-task by using the best fit-first fit algorithm. In best fit, allocate the strongest resource vibration for execution of the task. In first fit, allocate the nearby resource for execution of the task. Then the algorithm compares the vibration intensity of the resource with the vibration intensity of task. If the resource satisfies the task requirement, the task would be executed, else the task searches for another resource.

In order to replicate social spider foraging, a scheduling algorithm based on the comparison of task utilization and resource capacity has been developed. The initialization step of the algorithm is when the job and resource lists are set up, as well as the algorithm's properties. Determinc the Quality of Service (QoS) parameter, as well as populate the web page (the dimension of the web).

The second stage is evaluating the fitness of both the task and resources, and evaluates the vibration intensity.

The capacity of the task/vibration intensity of the task It(Tn) is calculated as:

$$\mathrm{F(Tn)} = \text{task length/capacity of Vm} \tag{1}$$

$$\mathrm{It(Tn)} = (\mathrm{U\ max} - \mathrm{F(Tn)})\hat{\mathrm{p}}\ \text{ max utilization} \tag{2}$$

$$\mathrm{It(Tn)} = \mathrm{Log}((1/\,\mathrm{F(Tn)} - \mathrm{U\ min}) * \mathrm{p})\ \text{ min utilization} \tag{3}$$

where F(Tn) = fitness function of task, It(Tn) = vibration intensity of the task, Umax = maximum task utilization constant, Umin = minimum task utilization constant, P = population of spider web.

The capacity of the resource/vibration intensity of the resource Iv(Vm) is calculated as:

$$\mathrm{F(Vm)} = \sum \text{Capacity of Vm} \tag{4}$$

$$\mathrm{Iv(Vm)} = (1/\mathrm{Umax} - \mathrm{F(Vm)}) * \mathrm{p}\ \text{ max capacity} \tag{5}$$

$$\mathrm{Iv(Vm)} = \log((1/\mathrm{F(Vm)} - \mathrm{Umin}) * \mathrm{p})\ \text{ min capacity} \tag{6}$$

where F(Vn) = fitness function of resource, Iv(Vm) = vibration intensity of the resource, P = population of spider web.

The third stage is the vibration generation. To generate the resource vibration, the software agent has been used. The software agent is a light weight piece of software that functions as agent and it can interact with its environment. The task of this agent

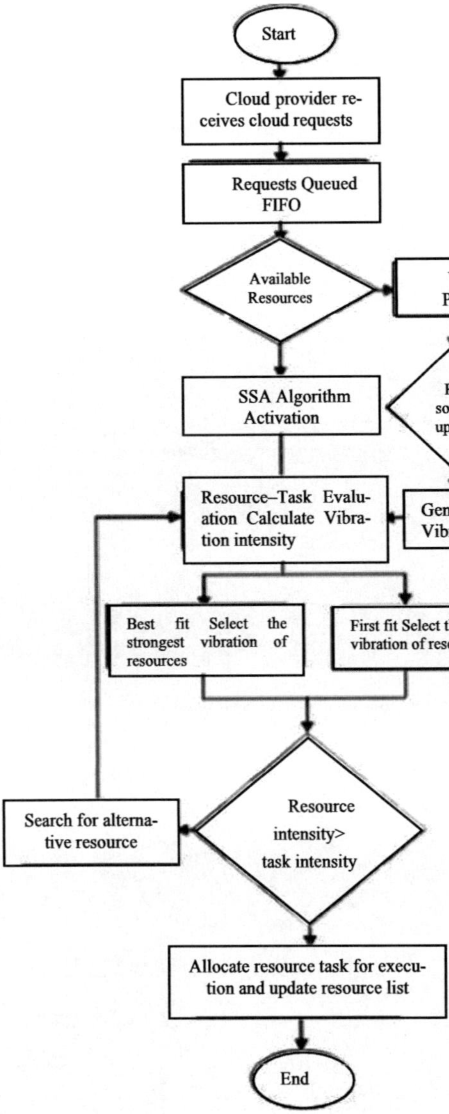

Fig. 1 SSA flowchart

is to send a message to the admin when the resource is idle. The fourth stage is allocating the resources for the task (Fig. 1).

There are two ways to choose which resource allocated to which task by using the concept of the first fit, best fit algorithm. In the first fit, task compares its vibration intensity with the neighboring resource vibration intensity, if that resource satisfies the utilization of the tasks then the resource allocated to the task, otherwise the task will search for another resource. In best fit, the task looks for the strongest vibration of resource, not for the nearby resource, and allocates it for the execution.

The final stage is calculating Attenuation and QoS metrics. The attenuation in the intensity of vibration is the loss in the vibration from the source, because its fade away, due to the distance factor. Attenuated intensity of vibration duo to distance is as follows:

$$\text{Att Iv(Vm)} = \text{Iv(Vm)} \cdot e - D/R \tag{7}$$

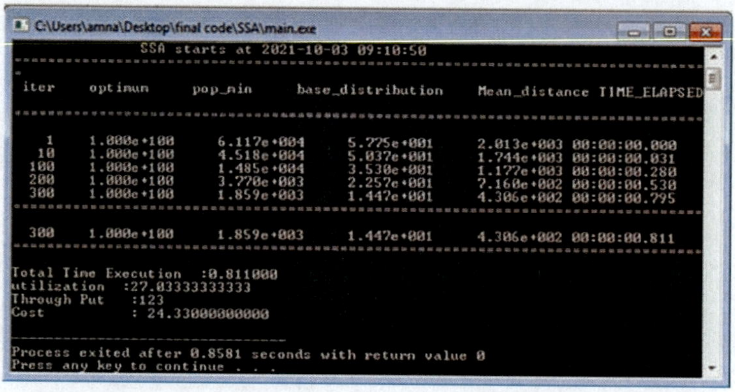

Fig. 2 Execution time, utilization and throughput of 100 tasks

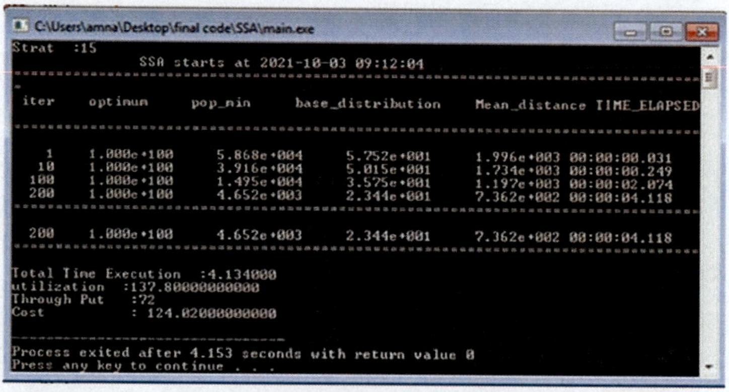

Fig. 3 Execution time, utilization and throughput of 500 tasks

where Att Iv(Vm) attenuated intensity of vibration duo to distance, Iv(Vm) is vibration intensity of the resource, D is the distance between task and resource, R is the attenuation rate. This procedure is then continuing until the entire task set is appropriately allocated with the requested resources. The algorithm steps can be summarized as follows:

Initialize resource list
Initialize task list
Create the population pop
Enter the QoS parameter
While (task list! = 0) do
For each task T in pop do
Evaluate the fitness of task
F(Tn) = task length/Σ capacity of Vm
Calculate the vibration intensity of the task
*It(Tn) = Log((1/ F(Tn) – U min)*p)*
End for
For each resource V in pop do
Evaluate the fitness of resource Vm
F(Vm) = Σ Capacity of Vm
Generate vibration as per capacity of the resource
Calculate the vibration intensity of the resource
*Iv(Vm) = log((1/F(Vm) – U min)*p)*
End for
(Best fit algorithm)
Select the strongest vibration of resource
If1 the intensity vibration of resource is greater than the intensity vibration of the task
Then allocate the resource for task
End If1
(First fit algorithm)
Compare the intensity vibration of resource with the intensity vibration of the task
If2 the intensity vibration of task is smaller than the intensity vibration of the resource
Then allocate the resource for task
Else find other resource with greater vibration intensity
End If2
Calculate vibration attenuation over distance
Att Iv(Vm) = Iv(Vm) · e − D/R
Update resource list
End while
Calculate execution time T = burst time / n (n: number of task.
Calculate utilization = total execution time / total resource time in work
Calculate throughput = total task set / total execution time
Output the best solution

4 Implementation

The cloud environment is modeled in C++ using DEVC++. These experiments used 30 resources with varied job counts. The code has five classes: Position, Problem, SSA, Spider, and Vibration, each with several functions. Two situations are implemented. Part 2 compares the (SSA) and (SSCWA) algorithms. Steps can be summarized as follows:

- Specify the position of spiders (tasks) and the victims (resources) in the web.
- Then evaluate the capacity of both spiders and victims.
- Find the appropriate resource and allocate it for the task execution. Using of SSA algorithm reduce the execution time of the tasks and enhance the overall performance.

The Evaluation Metrics considered to benchmark the performance are as follows:

- Execution Time (Measured (msec)): It is the interval time in which the n numbers of tasks get executed.
- Execution time = Burst Time / n
- Throughput (Measured (Process per Time)): It can be calculated as the total number of Tasks to be executed to the total execution time.
- Throughput = Total Task Set/Execution Time
- Utilization (%): it can be defined as ratio of the total execution time to the total time the resource was in working condition.

Utilization = Execution Time/Total Resource Time

In order to measure how long it takes for resources to be allocated, you may use the execution time. The ratio of execution time to total resource use is referred to as the utilization rate. How many tasks can be completed in a given length of time? This is called throughput. Many tests have been carried out to see how the number of jobs affects execution time, utilization, and throughput. The performance evaluation is done as per two scenarios:

4.1 QoS Parameter Analysis Using First Fit Algorithm

In this experiments, 100, 300 and 500 tasks and 30 resources were taken. When the task number increasing the execution, time is also increasing as shown in the figures: Figs. 2, 3, and 4. As consequence the resource utilization increase.

In Fig. 2 the task number is 100, the execution time was 0.811 ms, throughput was 123 and utilization was 27.

In Fig. 3 the task number is 300, the execution time was 4.13 ms, throughput was 72 and utilization was 137. In Fig. 4 the task number is 500, the execution time was 11.12 ms, throughput was 44 and utilization was 37. As check from these

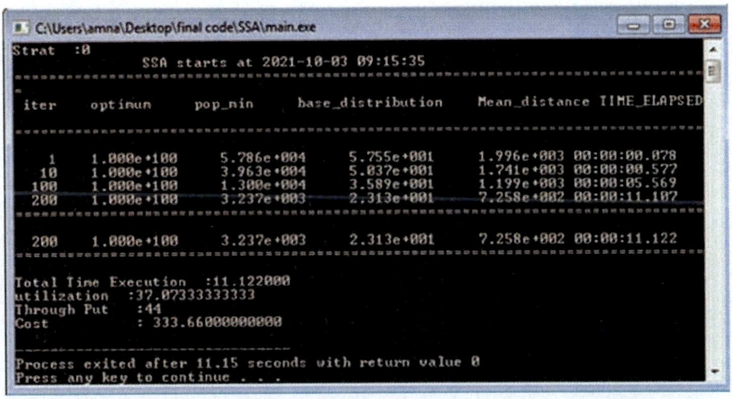

Fig. 4 Execution time, utilization and throughput of 300 tasks

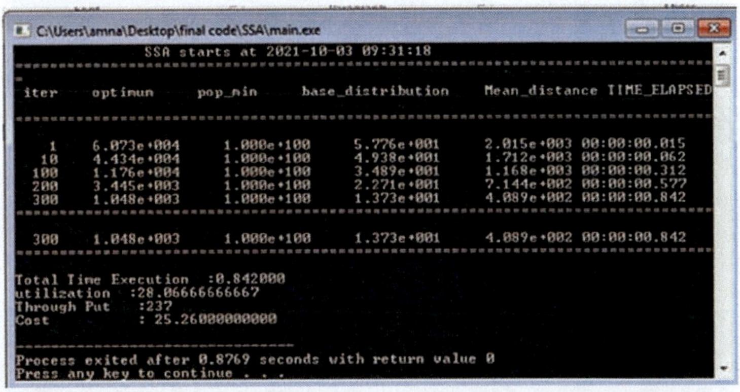

Fig. 5 Execution time. Utilization and throughput of 100 tasks

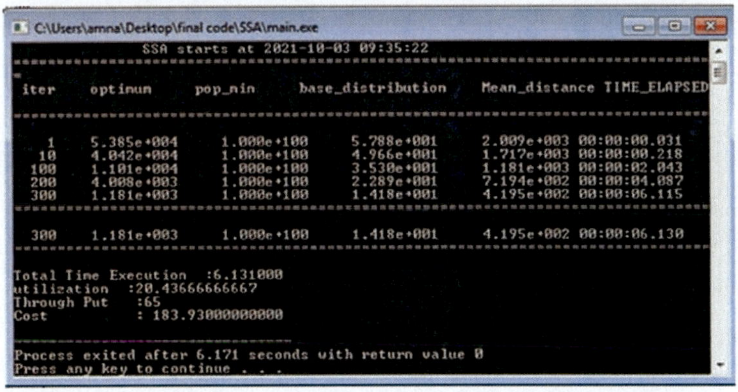

Fig. 6 Execution time, utilization and throughput of 300 tasks

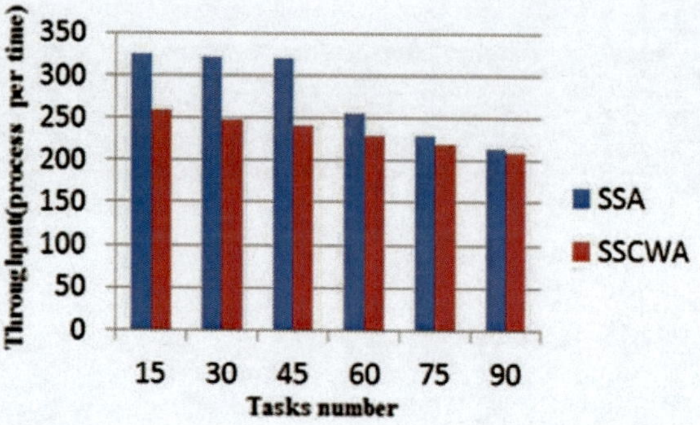

Fig. 7 Throughput analysis

experiments, when the task number increase the execution time and the utilization increase, also the throughput decrease.

4.2 QoS Parameter Analysis Using Best Fit Algorithm

The same number of resources (30) is used with varying numbers of tasks. In Fig. 5 the task number is 100, the execution time was 0.842 ms, throughput was 237 and utilization was 28.

In Fig. 6 the task number is 300, the execution time was 6.131 ms, throughput was 65 and utilization was 20.

The first fit algorithm outperforms the best fit algorithm in terms of execution time for the same number of jobs, as indicated in the preceding figures. There is a difference in performance when it comes to actual use, though.

The execution time for Best Fit and First Fit is shown in Table 1.

Table 1 Execution time for best fit and first fit

Task number	1st fit execution time (msec)	Best fit execution time (msec)
100	0.811	0.842
300	4.13	6.13
500	11.12	16.63

4.3 *Comparison of the SSA and the Social Spider Cloud Web Algorithm (SSA) (SSCWA)*

The proposed framework is compared against social spider cloud web algorithm (SSCWA). SSCWA is an algorithm mimics the concept of spider mating behavior. Three metrics, namely throughput, measured (process per time), execution time, measured (msec), utilization (%) are selected for the evaluation.

1) **Throughput**
   Throughput is the ratio of the total number of tasks to the total execution time required to execute. Figure 7 shows the throughput analysis for the two algorithms SSA and SSCWA, the value of throughput is 55% for SSA and 45% for SSCWA, which indicate that the SSA has a 10% higher performance. The tasks number ranging from 15 to 90 tasks.
2) **Execution Time**
   Figure 7 specifies the execution time analysis for SSA and SSCWA. When the number of the task increase the execution time also increase. Also, it shows that the SSA outperforms when compared with SSCWA. At 15, 30, 45 tasks the execution time at SSA is lesser 50% than SSCWA, after 60 tasks the execution time increase abruptly in SSA algorithm but still the SSA performs better.

   The analysis prove that the increasing of task number reduces the execution time in comparison to SSCWA algorithm, hence improving the overall performance.
3) **Utilization**
   While the tasks number increase, the utilization also increases as shown in Fig. 8. At first, SSCWA use exceeds SSA utilization; however, as the number of tasks approaches 60, the SSA consumption jumps unexpectedly and significantly.

The evaluation of SSA algorithm with first fit proves that it is better than SSA algorithm with best fit in term of execution time, and the best fit outperforms in term of utilization. Additionally, the comparison of SSA with SSCWA algorithm it is observed that the SSA algorithm performs better in execution time, utilization, and throughput. The SSA leads to better resource allocation, hence satisfying the QoS parameters and better performance.

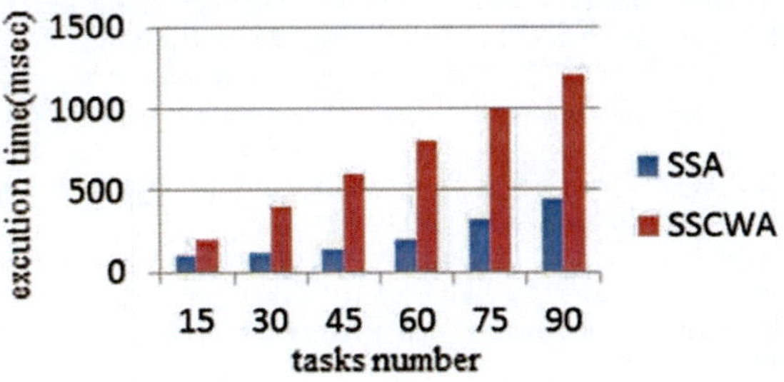

Fig. 8 Utilization analysis

5 Conclusion

When it comes to satisfying the needs of users and guaranteeing optimal cloud performance, the problem of resource allocation is very essential. In this study, a QoS-enabled resource allocation algorithm (SSA) is presented to map the resources with the right task. This algorithm, which is patterned after the social spider foraging strategy for resource allocation in the cloud, is intended to map the resources with the appropriate task. The primary objective of this study is to find ways to speed up the completion of cloud-based tasks by using less time. The findings of the simulation indicate that the SSA algorithm surpasses its competitors by reducing the amount of time required for execution in a cloud environment. The algorithm is compared to the social spider cloud web algorithm (SSCWA), and the results show that the proposed framework reduces the execution time by up to 30%, and performs better in terms of throughput by up to 10%, and consumption of cloud resources and utilizes these resources optimally. By implementing the suggested method, it is possible to speed up the process of allocating resources in the cloud while also supporting a wide range of resource types that may be requested. It is possible that in further work, the idea of resource reliability, along with other QoS factors, will be added. Additional analysis may be carried out to cut down on the amount of time needed for the execution.

References

1. Mohamed YA (2013) A novel mechanism for securing cloud computing. In: ACIT 2013 Proceedings (The International Arab Journal of Information Technology). Sudan University. Khartoum
2. Abdulhamid SM, Latiff MSA, Bashir MB: Scheduling techniques in on-demand grid as a service cloud: a review. J Theor
3. Yasir A, Mohamed M, Aziz A (2017) A novel approach for data integrity protection in cloud. Int J Comput Sci Inf Technol (ijcsit) 5(07–12):1–5
4. Ayadi I, Simoni N, Diaz G (2013) QoS-aware component for Cloud computing. In: ICAS 2013, The Ninth International Conference on Autonomic and Autonomous Systems, pp 14–20
5. Batista B et al (2015) Performance evaluation of resource management in cloud computing environments. PLoS ONE 10(11):e0141914
6. Mutasim Elsadig Adam and Yasir Abdalgadir Ahmed Hamid (2022) A two-stage assessment approach for QoS in internet of things based on fuzzy logic. Int J Adv Comput Sci Appl (IJACSA) 13(4). https://doi.org/10.14569/IJACSA.2022.0130480
7. Abrol P, Gupta S, Singh S (2020) A QoS aware resource placement approach inspired on the behavior of the social spider mating strategy in the cloud environment. Wirel Pers Commun 113(4):2017–2065
8. Abrol P, Gupta S (2018) Social spider foraging-based optimal resource management approach for future cloud. J Supercomput 76(3):1880–1902
9. Kumar S, Stecher G, Tamura K (2016) MEGA7: molecular evolutionary genetics analysis version 7.0 for bigger datasets. Mol Biol Evol 33(7):1870–1874
10. Li et al (2014) QoS-aware dynamic virtual resource management in the cloud. Appl Mech Mater 556–562:5809–5812
11. Horri A, Mozafari MS, Dastghaibyfard G (2014) Novel resource allocation algorithms to performance and energy efficiency in cloud computing. J Supercomput 69(3):1445–1461

12. GawaliSubhash MB, Shinde K (2018) Task scheduling and resource allocation in cloud computing using a heuristic approach. J Cloud Comput 7(1):1–16
13. Guo J, Liu F, Lui J, Jin H (2016) Fair network bandwidth allocation in IaaS datacenters via a cooperative game approach. IEEE/ACM Trans Netw 24(2):873–886
14. Li J, Li D, Ye Y, Lu X (2015) Efficient multi-tenant virtual machine allocation in cloud data centers. Tsinghua Sci Technol 20(1):81–89. https://doi.org/10.1109/TST.2015.7040517
15. Kılıç H, Yüzgeç U (2019) Tournament selection based antlion optimization algorithm for solving quadratic assignment problem. Eng Sci Technol Int J 22(2):673–769
16. Mencagli G (2015) Adaptive model predictive control of autonomic distributed parallel computations with variable horizons and switching costs. Concurr Comput Pract Exp 28. https://doi.org/10.1002/cpe.3495
17. Abrol P, Gupta S, Singh S (2020) A QoS aware resource placement approach inspired on the behavior of the social spider mating strategy in the cloud environment. Wirel Pers Commun 113:2027–2065. https://doi.org/10.1007/s11277-020-07306-1
18. Madni SHH, Latiff MShA, Coulibaly Y, Abdulhamid ShM (2017) Recent advancements in resource allocation techniques for cloud computing environment: a systematic review. Clust Comput 20(3):2489–2533. https://doi.org/10.1007/s10586-016-0684-4
19. Sathya GSM, Swarnamugi M, Dhavachelvan P (2017) Evaluation of QoS based web- service selection techniques for service composition. J Int J Softw Eng 110(9):73–90
20. Abu-safe AN, Elrofai SE (2020) An efficient QoS-aware services selection in IoT using a reputation improved- social spider optimization algorithm. Res Sq.https://doi.org/10.21203/rs.3.rs-38596/v1
21. Kaewunruen S, Ngamkhanong C, Xu S (2020) Large amplitude vibrations of imperfect spider web structures. Sci Rep 10:19161
22. Mortimer B, Soler A, Siviour CR, Vollrath F (2018) Remote monitoring of vibrational information in spider webs. Naturwissenschaften 105(5–6):37. https://doi.org/10.1007/s00114-018-1561-1
23. Zak M, Ware J (2020) Cloud based distributed denial of service alleviation system. Ann Emerg Technol Comput 4:44–53. https://doi.org/10.33166/AETiC.2020.01.005
24. Gonzalez NM et al (2017) Cloud resource management: towards efficient execution of large-scale scientific applications and workflows on complex infrastructures. J Cloud Comput Adv Syst Appl 6:13. https://doi.org/10.1186/s13677-017-0081
25. Bal PK, Mohapatra SK, Das TK, Srinivasan K, Hu Y-C (2022) A joint resource allocation, security with efficient task scheduling in cloud computing using hybrid machine learning techniques. Sensors 22(3):1242. https://doi.org/10.3390/s22031242
26. Mohamed YA, Abdullah AB (2010) Implementation of IDS with response for securing MANETs. In: 2010 International Symposium on Information Technology, pp 660–665. https://doi.org/10.1109/ITSIM.2010.5561608
27. Mohamed YA, Abdullah AB (2009) Immune-inspired framework for securing hybrid MANET. In: 2009 IEEE Symposium on Industrial Electronics & Applications, pp 301–306. https://doi.org/10.1109/ISIEA.2009.5356451

Sentiment Analysis to Extract Public Feelings on Covid-19 Vaccination

Yahya Almurtadha, Mukhtar Ghaleb, and Ahmed Mohammed Shamsan Saleh

Abstract Covid-19 (Corona virus) hits the world with wildness, affecting various sectors of life. The whole world has united to confront the virus, and different vaccines were developed to vaccinate the largest possible percentage as an effort to reach community immunity to limit its spread. Governments seek to measure public opinion about vaccination campaigns to improve the quality of services provided. One of the most effective ways to do this is to use artificial intelligence to sense and analyze what the public is posting on social media such as Twitter to ensure that their opinion is known without bias. The study used Twitter API to retrieve Arabic tweets then measured public acceptance of vaccination against Covid-19 disease by using sentiment analysis combined with deep learning as a technique that ensures access to people's opinions quickly and at a very low cost. The results of this study showed that most people are having a positive opinion on the vaccination with different percentages vary from a vaccine type to another.

Keywords Opinion mining · Sentiment analysis · Covid-19 Vaccination · Twitter text analysis

1 Introduction

The world witnessed one of the most powerful health pandemics that led to general closures, curfews, and social distancing, which was reflected in many life magazines that affect people's lives such as work, economy, health and education. COVID-19

Y. Almurtadha (✉) · A. M. S. Saleh
Faculty of Computing and IT, University of Tabuk, Tabuk, Saudi Arabia
e-mail: y.murtadha@ut.edu.sa

A. M. S. Saleh
e-mail: ah_saleh@ut.edu.sa

M. Ghaleb
College of Sciences and Arts, University of Bisha, Al-Namas, Saudi Arabia
e-mail: mghaleb@ub.edu.sa

M. Al-Emran et al. (eds.), *International Conference on Information Systems and Intelligent Applications*, Lecture Notes in Networks and Systems 550,
https://doi.org/10.1007/978-3-031-16865-9_51

was the cause of this pandemic. COVID-19 is a very serious respiratory disease that was first discovered in December 2019 [1]. The governments of the world tried to unite to confront the outbreak of the epidemic by imposing a set of health precautionary measures and measures represented in "isolation"/social distancing/travel bans/complete closure of all state institutions: schools, universities, companies, factories, places of entertainment and tourism. It became clear from the great and continuous impact of the pandemic until now that the world is going through great and influential changes that have great repercussions due to its spread among all countries. Such lockdown and isolations are negatively affecting the economics, social and phycology system of the globe. As a result, this brought the global system into a state of recession. In short, the world after Covid 19 will not be the same as before.

As a natural result of the survival instinct, all governments of the world have sought to fund medical research to find a vaccine to vaccinate against Covid 19, in addition to psychological and social research to avoid the negative effects of general closure and isolation on people. Medical research focused on finding vaccines as quickly as possible to boost immunity and produce antibodies to protect against the virus. Vaccination is considered recently a modern and strongly affecting preventative health measures [2]. Strengthening immunity is an effective and powerful way to protect against diseases and prevent their spread, especially those that do not become extinct. Immunity enhancement aims to strengthen the human immune system, which encourages the body to resist disease and infection. The idea behind immunization is focused on teaching the human immune system to form antibodies that fight vigorously any viral attack, especially those that come in waves of spread, as is the case with Covid 19. Given the infection of a large number of the world's population with the virus and the rapid and frightening waves of spread, the increase in the number and frequency of infections, and the large deaths resulting from that, it is imperative for scientists to search for a vaccine to prevent the virus and limit its spread by teaching the immune system to develop an immune response that protects the body, which reduces infection. Finding an effective vaccine and proceeding with the vaccination process will result in reducing infection and preventing spread, which will facilitate the decision to lift the ban globally and gradually return to normal life.

The remarkable development in information technology has played a major and essential role as an effective weapon to help doctors and scientists in their research by processing huge data and simulating the various components to find an effective vaccine. Therefore, artificial intelligence and data science experts struggled alongside research and medical bodies, resulting in appreciation and admiration from various segments and showing them as sciences that effectively contributed to accelerating the creation of a vaccine that without them would have taken longer to find it. Many artificial intelligence techniques appeared that health and research authorities relied on and explained how these technologies helped accelerate the development of a vaccine, for example:

- Track the spread of the Coronavirus with machine learning
- Using artificial intelligence to diagnose people infected with the Coronavirus

- Relying on robots for sterilization and patient handling
- Artificial intelligence is helping to accelerate the creation of a vaccine

At the same time as the world tried to confront the epidemic in a healthy way, another big dilemma emerged, which is the horror caused by spreading rumors in social media, which increased the burden on governments in summarizing the possible means to reassure people and guide them to take appropriate precautionary measures… This burden also appeared later, after finding the vaccine in an attempt to reach the public and convince them of the necessity of vaccination to reduce the spread of infection and combat the epidemic. Given the spread and abundance of social media, it formed two-dimensional channels to send awareness messages and study the public's response and interactions with the vaccine through the use of artificial intelligence techniques to analyze what they write and express in their accounts.

2 Related Works

Opinion mining is "the process of extracting human thoughts and perceptions from unstructured texts, which with regard to the emergence of online social media and mass volume of users' comments, has become to a useful, attractive and also challenging issue" [3]. Opinion Mining (OM) or Sentiment Analysis (SA) can be defined as the task of detecting, extracting, and classifying opinions [4] on unstructured, large and rich natural language texts. Many research have been conducted on using sentiment analysis in various areas such as movie review [5], product review [6], recommender systems [7], Exploring students' feedback in online assessment system [8], hotel review [9] and many more areas. Authors in [10] analyzed the algorithms of sentiment analysis and opinion mining for social multimedia.

In the educational system, student opinion is crucial for assessing the quality of instruction. As a result, the authors in [11] used a lexicon-based approach to demonstrate the learners" positive and negative behavior. To assess the polarity of words as a lexical source, the authors created a set of English sentiment words. The sentiment terms dictionary includes words associated with academia field to achieve a better result. Almurtadha uses SA to mine trending hash tags on Twitter in [12]. The authors of [13] suggested using SA in youth tweets as a genuinely effective means of assessing the educational problems they face to make improvements. In [13], the author proposes a novel approach to using SA to discover public reaction and views on Twitter as a new poll tool for reviewing academic educational seeking academic accreditation.

The authors of [14] applied Random Forest Algorithm and SA on a social media network. Their job aids both the supplier and the consumer in tracking product sentiment. In this study, sentiment analysis is used to identify consumer feelings from their feedback. In [15], the author improved a method for sentiment analysis. For Arabic SA, he proposed a corpus-based approach to label the tweets as negative or positive in Twitter. In [16], the authors proposed a sentiment analysis system based on deep learning. The suggested mechanism for categorizing positive and negative customer feedback. Their approach is based on supervised learning, which necessitates the collection of training data.

Several studies investigated using opinion mining to support health sector. [17] investigated using opinion mining on issues related to health. [18] studied using opinion mining in online microblogging for supporting public health initiatives. Patient feelings and expressions to investigate their drugs gratified using supervised learning has been analyzed in [19]. The authors in [20] examine how the community accepts distance learning during Covid-19 pandemic as a precaution. A necessity for understanding the threat occurred by anti-vaccination efforts on social media is vital for helping the global COVID-19 vaccination programs [21]. The movement to delay vaccination has been growing, which has backed to eruptions of vaccine-halt diseases [22]. Vaccine hesitancy on social media was a major issue affecting public health triggering the alarm to raise the attention as explained by [23]. The results of an opinion mining investigation on vaccination conducted on Twitter from September 2016 to August 2017 in Italy was presented in [21]. Mistrust and social media echo chambers forecast COVID-19 vaccine delay has been explored in [24]. This study investigates using of opinion mining to extract public emotions during Covid-19 vaccination campaign in Arabic language.

3 Methodology

This study aims at measuring the acceptance of the public to Covid-19 vaccination in Arabic language. To accomplish that we will use the opinion mining technique whereby we don't need to follow the traditional survey methods such as questionnaires. Giving the advantage that people like posting freely in microblogging platforms such as Twitter, this study used the sentiment analysis to analysis these tweets which reflect the public opinions without any bias in a timely manner and low cost. This section explains the steps followed to accomplish the study objective.

3.1 Dataset

The purpose was to obtain public tweets in the Arabic language regarding the Covid-19 vaccination. We used Twitter streaming API during the first week of May 2021 to retrieve what people write pertaining Covid-19 vaccination in Twitter in real time. We set the Twitter API to retrieve the tweets in Arabic language and to retrieve 10,000 tweets at a time. As they are posting freely in the microblogging environment Twitter, therefore those tweets comprise their real feelings and reflect their expression on the vaccination. The retrieved tweets were compiled in records where each record consisted of {source (device), text, geo-location latitude, geo-location longitude, retweet count}. We concentrate on the tweets' texts for opinion mining. All the tweets were pre-processed to prepare them for the sentiment analysis to extract public acceptance of the Covid-19 vaccine.

3.2 Sentiment Analysis Model

Twitter provides a research service for the researchers. Upon subscription, an API streaming token is allocated to the researcher to extract the tweets pertaining particular subject so that the researcher may apply different data science techniques to mine, investigates and extract the useful knowledge from those tweets. Figure 1 elaborates the general methodology of the research. Retrieve the tweets by feeding the search twitter operator with the keywords needed to be found in the tweets. The settings also include choosing the geographical area and the language of the tweets to be retrieved. We set the language to be Arabic. As the retrieved tweets are extracted with many details, set role operator is used to choose the field to be assigned as a label for the classifications and the fields to be considered as predictions. Apply preprocessing to the tweets to exclude some contents from the tweets' tests such as http and URLs.

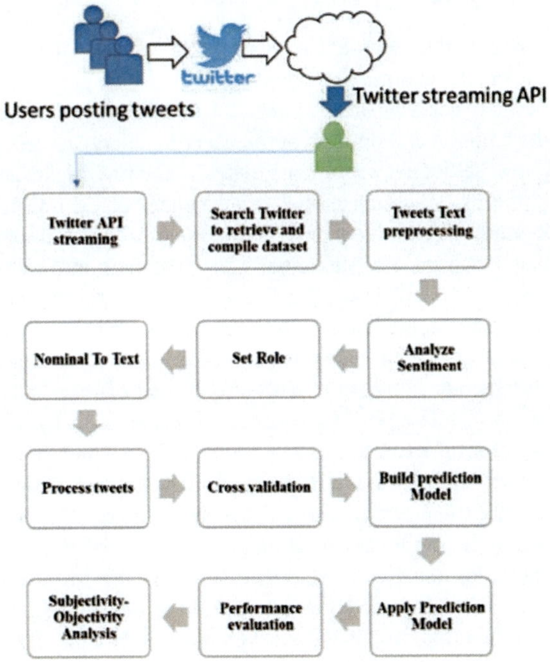

Fig. 1 Twitter sentiment analysis methodology

Fig. 2 Tweets' texts pre-processing steps

Prepare the analyzing environment by providing the needed authorization token provided by Twitter API streaming and AYLN text processing supported by Rapid-miner tool to apply text processing techniques to the needed source of data. This tool is integrated with RapidMiner platform for data science. Nominal to text step is required to change the type of selected nominal attributes to text to ease the subsequent step of tweets' texts processing. Apply tweets test processing as illustrated by Fig. 2 which include several steps:

- Tokenization: break the texts into small items for easier processing.
- Remove the stop-words from the tweets. As this study retrieved the Arabic tweets, therefore we removed the Arabic stop-words from the texts. These stop-words removed since they have no influence on the text meanings such as prepositions, pronouns, … etc.
- Stemming: Return the remaining tokens to their original root so that the word like "القاح، تلقيح، يلقح"considered as one word-instead of three words- with the same meaning.

Apply the sentiment analysis model to divide the retrieved tweets into two groups: subjective, neutral, and objective tweets. The purpose is to identify whether the tone of the tweet is positive or negative and whether the tweet's text is subjective (reflect the emotion and opinion of the user) or objective (reflects some facts). Normally we prefer subjective sentiments to dig in and investigate associated opinions. Following is the classification model development based on H2O deep learning [25]. Deep learning-based algorithms "show great promise in extracting features and learning patterns from complex data" [26]. The authors in [27] provides a comprehensive review on deep learning and recent usage in sentiment analysis. The aim of this step is to build a predictive model that learns how to classify the retrieved tweets (the tweets with their status as subjective, neutral or objective). This model should be able then to assign any new tweet to one of these classes based on the constructed model. while a positive tweet reflects a positive feeling, a negative tweet opens the door for the existence of a negative feeling which in return should be considered for further study by the health sector policy makers. Cross validation is used to estimate the statistical performance of a learning model. The goal is to estimate the extent to which the model can work later with different data. In this study, we applied cross validation for dividing the data set into 70% for training and 30% for testing with cross validation of 10 folds. Each time one fold was employed as a validation set and the remaining k-1 folds were engaged as the training set. The average of these validation is calculated to give a single value indicated the validation accuracy of the classification model. Finally calculate the accuracy, precision and recall for the classification tasks to evaluate the performance.

$$Precision = TP/(TP + FP) \tag{1}$$

$$Recall = TP/(TP + FN) \tag{2}$$

$$Accuracy = TP + TN/(TP + TN + FP + FN) \tag{3}$$

where: TP = True positive, FP = False positive, TN = True negative and FN = False negative.

4 Results and Discussion

The study aims to dig into Twitter tweets and extract what the tweeters write and apply sentiment analysis to evaluate their feelings about vaccination against Corona disease. The use of sentiment analysis as a smart treatment of what the tweeters write is a multi-advantageous method. Previously, people's opinions would be extracted by distributing questionnaires, which take a lot of time to reach them, not to mention the need to use different analyzes to ensure the validity and reliability of the answers. Another way through interviews, and here it will be reached to a small number of people because it is difficult to do interviews with a large number to inquiry their opinions, which requires relying on a clear criterion in choosing the sample. With the development of artificial intelligence and sentiment analysis research, it became possible to quickly reach the opinions of the public, not to mention ensuring that they actually express what they want without guidance in an unbiased manner. By applying sentiment analysis on the dataset collected during the first week of May 2021, Tables 1, 2, 3, 4, 5 and 6 elaborates the accuracy, precision, and recall of the retrieved tweets for each topic of search round using different key terms to investigate public opinion on Corona vaccination and different known types of vaccines. As mentioned in the methodology, each tweet is classified as subjective, neutral, or objective. Sentiment analysis gives attention to subjective tweets as they may be divided into positive or negative. Those tweets classifieds as negative tweets are crucial to look for criticism or negative feelings. It can be seen from the tables that the number of subjective tweets is so small compared to objective tweets giving indications that most people at that time -the time of announcing of discovering vaccination for the Covid-19- are having a positive opinion on the vaccination with different percentages differ from a vaccine type to another. This can be explained by the desire of the those people to have a vaccination to avoid the severe symptoms of infection.

Table 1 Corona vaccination

لقاح كورونا	True objective	True subjective	Class precision (%)
pred. objective	75	13	85.23
pred. subjective	4	8	66.67
class recall	94.94%	38.10%	
Accuracy: 83.00% ± 8.23% (micro average: 83.00%)			

Table 2 Corona vaccination (Synonym)

تطعيم كورونا	True objective	True subjective	Class precision (%)
pred. objective	81	9	90.00
pred. subjective	2	8	80.00
class recall	97.59%	47.06%	
Accuracy: 89.00% ± 9.94% (micro average: 89.00%)			

Table 3 Pfizer vaccination

لقاح فايزر	True subjective	True objective	Class precision (%)
pred. subjective	0	0	0.00
pred. objective	3	97	97.00
class recall	0.00%	100.00%	
Accuracy: 97.00% ± 4.83% (micro average: 97.00%)			

Table 4 Astra vaccination

لقاح أكسفورد	True objective	True subjective	Class precision (%)
pred. objective	53	15	77.94
pred. subjective	8	24	75.00
class recall	86.89%	61.54%	
Accuracy: 77.00% ± 9.49% (micro average: 77.00%)			

Table 5 Sputnik vaccination

لقاح سبوتنك	True objective	True subjective	Class precision (%)
pred. objective	32	4	88.89
pred. subjective	1	12	92.31
class recall	96.97%	75.00%	
Accuracy: 89.00% ± 16.63% (micro average: 89.80%)			

Table 6 Chinese vaccination

اللقاح الصيني	True objective	True subjective	True neutral	Class precision (%)
pred. objective	436	64	1	87.03
pred. subjective	27	24	0	47.06
pred. neutral	2	0	1225	99.84
class recall	93.76%	27.27%	99.92%	
Accuracy: 94.72% ± 0.89% (micro average: 94.72%)				

5 Conclusions

The world has started vaccination campaigns recently, using vaccines, most known are the German-American Pfizer, the Chinese Sinopharm, the British Oxford AstraZeneca, and the Russian Sputnik. Studies have proven that all societies must vaccinate 65–70% of their population to reach herd immunity to combat and eliminate the spread of the virus. This research aims at measuring the public response to the vaccination of Covid-19 by applying sentiment analysis to their posts in Twitter the most famous microblogging environment. The results proved that a positive acceptance for the different types of vaccination. Future works will highlight on studying in detail the public response to the vaccination including elaborating the tweets with criticism or looking for improvements.

References

1. W. H. Organization: Novel Coronavirus (COVID-19) Situation, WHO, 11 June 2020
2. Wilde BB, Park DJ (2019) Immunizations primary care–clinics in office practice. https://doi.org/10.1016/j.pop.2018.10.007
3. Hemmatian F, Sohrabi MK (2019) A survey on classification techniques for opinion mining and sentiment analysis. Artif Intell Rev. https://doi.org/10.1007/s10462-017-9599-6
4. Saberi B, Saad S (2017) Sentiment analysis or opinion mining: a review. Int J Adv Sci Eng Inf Technol. https://doi.org/10.18517/ijaseit.7.5.2137
5. Balahadia FF, Fernando MCG, Juanatas IC (2016) Teacher's performance evaluation tool using opinion mining with sentiment analysis. https://doi.org/10.1109/TENCONSpring.2016.7519384
6. Bhat S, Garg S, Poornalatha G (2018) Assigning sentiment score for twitter tweets. https://doi.org/10.1109/ICACCI.2018.8554762
7. Da'u A, Salim N, Rabiu I, Osman A (2020) Weighted aspect-based opinion mining using deep learning for recommender system. Expert Syst Appl. https://doi.org/10.1016/j.eswa.2019.112871
8. Wook M, Vasanthan S, Ramli S, Razali NAM, Hasbullah NA, Zainudin NM (2020) Exploring students' feedback in online assessment system using opinion mining technique. Int J Inf Educ Technol. https://doi.org/10.18178/ijiet.2020.10.9.1440
9. Hu YH, Chen YL, Chou HL (2017) Opinion mining from online hotel reviews–a text summarization approach. Inf Process Manag. https://doi.org/10.1016/j.ipm.2016.12.002

10. Li Z, Fan Y, Jiang B, Lei T, Liu W (2019) A survey on sentiment analysis and opinion mining for social multimedia. Multimed Tools Appl.https://doi.org/10.1007/s11042-018-6445-z
11. Tripathi P, Vishwakarma SK, Lala A (2016) Sentiment analysis of English tweets using rapid miner. https://doi.org/10.1109/CICN.2015.137
12. AlMurtadha Y (2018) Mining trending hash tags for Arabic sentiment analysis. Int J Adv Comput Sci Appl. https://doi.org/10.14569/IJACSA.2018.090227
13. Alqarni HA, AlMurtadha Y, Elfaki AO (2018) A twitter sentiment analysis model for measuring security and educational challenges: a case study in Saudi Arabia. J Comput Sci. https://doi.org/10.3844/jcssp.2018.360.367
14. AlMurtadha Y (2018) Public response sentimental analysis model to review educational program seeking academic accreditation. https://doi.org/10.1145/3232174.3232184
15. Karthika P, Murugeswari R, Manoranjithem R (2019) Sentiment analysis of social media network using random forest algorithm. https://doi.org/10.1109/INCOS45849.2019.8951367
16. Alsalman H (2020) An improved approach for sentiment analysis of Arabic tweets in Twitter social media. https://doi.org/10.1109/ICCAIS48893.2020.9096850
17. Seetharamulu B, Reddy BN.K, Naidu KB (2020) Deep learning for sentiment analysis based on customer reviews. https://doi.org/10.1109/ICCCNT49239.2020.9225665
18. Kim JC, Chung K (2020) Discovery of knowledge of associative relations using opinion mining based on a health platform. Pers Ubiquitous Comput. https://doi.org/10.1007/s00779-019-01231-2
19. Zhan Q et al (2019) Opinion mining in online social media for public health campaigns. J Med Imaging Health Inform. https://doi.org/10.1166/jmihi.2019.2742
20. Gopalakrishnan V, Ramaswamy C (2017) Patient opinion mining to analyze drugs satisfaction using supervised learning. J Appl Res Technol. https://doi.org/10.1016/j.jart.2017.02.005
21. Almurtadha Y, Ghaleb M (2021) Sentiment analysis to measure public response to online education during coronavirus pandemic. https://doi.org/10.1109/NCCC49330.2021.9428838
22. Tavoschi L et al (2020) Twitter as a sentinel tool to monitor public opinion on vaccination: an opinion mining analysis from September 2016 to August 2017 in Italy. Hum Vaccines Immunother. https://doi.org/10.1080/21645515.2020.1714311
23. Al-Regaiey KA et al (2021) Influence of social media on parents' attitudes towards vaccine administration. Hum Vaccines Immunother. https://doi.org/10.1080/21645515.2021.1872340
24. Piedrahita-Valdés H et al (2021) Vaccine hesitancy on social media: sentiment analysis from June 2011 to April 2019. Vaccines. https://doi.org/10.3390/vaccines9010028
25. Jennings W et al (2021) Lack of trust and social media echo chambers predict COVID-19 vaccine hesitancy. medRxiv
26. Jamshidi M et al (2020) Artificial intelligence and COVID-19: deep learning approaches for diagnosis and treatment. IEEE Access 8:109581–109595. https://doi.org/10.1109/ACCESS.2020.3001973
27. Cao C et al (2018) Deep learning and its applications in biomedicine. Genomics, Proteomics and Bioinformatics. https://doi.org/10.1016/j.gpb.2017.07.003
28. Zhang L, Wang S, Liu B (2018) Deep learning for sentiment analysis: a survey. Wiley Interdiscip Rev Data Min Knowl Discov. https://doi.org/10.1002/widm.1253

QR Codes Cryptography: A Lightweight Paradigm

Heider A. M. Wahsheh and Mohammed S. Al-Zahrani

Abstract A QR Code is a two-dimensional barcode scanned by a digital device or smartphone that holds data as a sequence of pixels in a square-shaped pattern. QR codes are widely employed in commercial tracking systems, encoding URLs, contact information, map coordinates, and physical and digital documents. Nowadays, several smartphones have built-in QR readers; they are often employed in marketing and advertising campaigns. More recently, QR codes have recreated a critical role in tracing COVID-19 pandemic exposure and slowing the spread of the virus. Web attackers can encode malicious URLs of custom malware or phishing site into a QR code, which could violate or disclose personal or financial information on a smartphone's data when scanned. This study investigates several symmetrical lightweight cryptography (LWC) algorithms to enhance QR code protection. Modern well-defined LWC features (performance and security) are compared and evaluated. The results adopt reliable and safe mechanisms for QR codes' security issues.

Keywords QR codes · Authentication · LWC cryptography · ANOVA · Satisfaction level

1 Introduction

A QR Code is a two-dimensional barcode scanned by a digital device or smartphone that holds data as a sequence of pixels in a square-shaped pattern. They are considered free, simple, and practical tools available to all and capable of storing up to 2,953 bytes and retrieving the stored data quickly [1]. QR codes allow users to navigate among

H. A. M. Wahsheh (✉)
Department of Information Systems, College of Computer Science and Information Technology, King Faisal University, P.O. Box 400, Al-Ahsa 31982, Saudi Arabia
e-mail: hwahsheh@kfu.edu.sa

M. S. Al-Zahrani
Department of Computer Networks and Communications, College of Computer Science and Information Technology, King Faisal University, P.O. Box 400, Al-Ahsa 31982, Saudi Arabia
e-mail: malzahrani@kfu.edu.sa

M. Al-Emran et al. (eds.), *International Conference on Information Systems and Intelligent Applications*, Lecture Notes in Networks and Systems 550,
https://doi.org/10.1007/978-3-031-16865-9_52

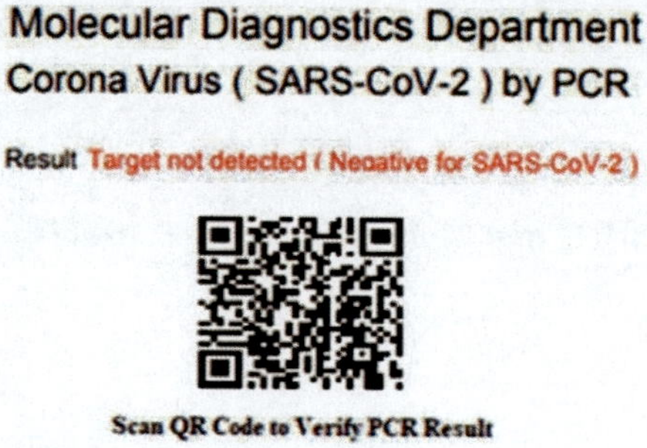

Fig. 1 An example of QR code usage in PCR test

different resources in three main modes; online, offline, or combination [2]. Users can access the online website, send an email or read SMS, save contact numbers, find map coordinates, listen to audio, or watch video [3]. QR codes support robust four levels (percentages) of error correction capabilities for restoring destructive data [4]. Unfortunately, there is no standard for covering all the QR Code scans for all products worldwide [5].

QR codes could be attached to any screen, poster, or product surface. QR codes can encode the URL of the advertiser to facilitate interaction with users or retrieve additional product information [6]. In addition, QR codes can link physical objects to electronic resources [2], which can be effectively used in coastal zone management, education, transportation, ticketing services, and tourism promotion [1, 7–9]. Furthermore, E-health services can employ QR codes to store patients' information, drugs, and medical reports [10, 11]. The QR code has been widely used during and besides the COVID-19 pandemic in the patient identity system, electronic permits, electronic prescribing, and verifying PCR tests and vaccine certificates [5]. Figure 1 shows an example of a QR code used to verify the PCR test result.

QR codes allow high-speed component scanning in factories [12]. They have become popular in storing one-time passwords, Wi-Fi login information, bank account information, and credit card numbers [13]. QR codes indicate multiple issues arising about the security, privacy, and ethical problems related to or influenced by QR codes that should be appropriately countered [14–16]. The QR code may be used as a medium to hold phishing links such as QRishing [17], aiming to steal users' sensitive information. Furthermore, QR codes may be utilized to propagate spam URLs [18–22], leading users to malicious pages [23] and fake SSL certificates [24] or retrieving irrelevant and phony content. Because of the limited size of the QR code, we need to employ lightweight security mechanisms to protect QR code content with usability considerations [15]. In this context, lightweight cryptography (LWC) is a concept that protects the information in an enhanced security mode employing low assets and providing higher throughput, conservativeness, and low power utilization [25–27]. The lightweight cryptographic mechanisms are classified into symmetric and asymmetric algorithms. This study investigated symmetric lightweight algorithms and analyzed and compared the security and performance considerations.

Symmetric lightweight algorithms are, for the most part, employed as a part of QR code innovation for more standard security with the least memory and power capabilities for smartphone reader applications [15, 25]. The paper is organized as follows: Sect. 2 presents the experimental structure for the QR code scanning experience. Section 3 illustrates the lightweight cryptographic mechanisms' security and performance evaluation discussion. Lastly, Sect. 4 concludes the paper and suggests future work.

2 Experimental Structure for QR Code Scanning Experience

To assess the satisfaction of the QR code reader process, we have conducted comprehensive investigations that examine the users' satisfaction. Here we have developed Barcode Satisfaction Tester (BarSTest) [2], an Android application that utilizes the ZXing library [28] to scan QR codes. BarSTest poses various user questions and gathers feedback and statistics to assess the barcode scanning experience. Figure 2 illustrates the experimental structure for the QR code scanning experience. When reading a barcode, there are three possible scanning results:

- Correct: the barcode is correctly interpreted as a QR code.
- Failure: the barcode is incorrectly interpreted as a QR code; the scanner reads another barcode format from a QR code image.
- Cancel: the user aborts the scan.

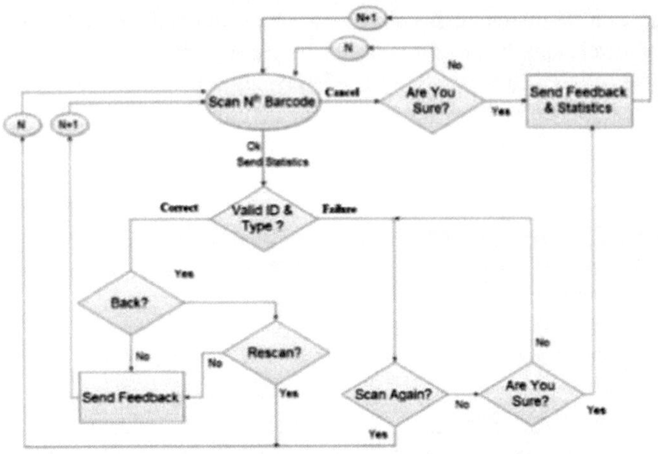

Fig. 2 Experimental structure for QR code scanning experience

After each Correct (successful read), the user is invited to give their satisfaction level. The positive response is expressed on a scale of three levels and might be: High, Middle, or low. The BarSTest will collect the feedback and user opinion and use the sentiment analysis technique proposed in [29, 30] to understand the reasons for the problematic QR code scanning experience.

We have printed various QR code images to conduct the tests of five data sizes: 100–400, 500–800, 900–1200, 1300–1600, and 1700–2000 bytes. We used image sizes: 200×200 pixels that, visualized on a 96 Dots per inch (DPI) screen, fit to 5.29×5.29 cm. One hundred forty-nine students (from Italy and Jordan) used their smartphones to scan 1000 different barcodes containing random data. The camera resolution for most of the devices was around 8–16 MP, which provides variety to our sample. Devices were at least CPU Octa-core (2×2.0 GHz Cortex-A75 $6 \times$ 1.8 GHz Cortex-A55) and 4 GB RAM.

2.1 *Scanning Time (ST) and ANOVA Analysis*

The Histogram of the Scanning Time indicates skewed to the right side distributions for all the three satisfaction outputs (High, Middle, and Low). This denotes that the median and InterQuartile Range (IQR) are more appropriate illustrative measures than the standard mean and standard deviation. Since the data distribution is skewed, the mean is usually not in the middle. The median is a better estimate of the center for this distribution [31]. We have estimated the median Scanning Time, the first and third quartiles Q1, Q3, the IQR (Q3 − Q1), the minimum non-outlier, and the maximum non-outlier and corresponded these metrics for the users' three satisfaction levels: High, Middle, and Low as shown in Table 1.

The question we invited at this point is, are these marked differences between the satisfaction levels significant? Or is it that the essence of these practical disparities is only because of chance variation? The explanation for these queries comes from running a test for comparing population means that is the Analysis of Variance (ANOVA) test [31]. A one-way analysis of variance (ANOVA) is a statistical method employed to test the differences between population means. Mathematically, the analysis of variance F statistic for ANOVA has the form of $F = MSB/MSE$ with a corresponding p-value that reveals the likelihood of occurrence [2, 31]. P-values are utilized to decide whether a null hypothesis will be accepted or rejected. The

Table 1 Descriptive overview of the ST for users satisfaction levels (seconds)

Satisfaction levels	Minimum non-outlier	Q1	Median	Q3	IQR	Maximum non-outlier
High	0.85	2.7	4.1	5.9	3.1	10
Middle	1.4	4.7	7.3	14.4	9.7	26.7
Low	0.78	5.6	14.4	27.4	21.8	56

most investigations refer to statistically significant as p-value < 0.05 [32]. For our concern, we performed a one-way ANOVA (a one-way ANOVA is a design in which there is only one factor; in our topic, the users' satisfaction level) to compare the distributions of the reaction variable Scanning Time for the three-factor levels we have: High, Middle, and Low. The goal is to decide whether the population disparities over the three levels are statistically significant or not [31].

We transformed the data (the logarithmic transformation, Log base 10) to hold the normality condition reasonably; our data indicated skewed to the right distributions with few outliers [33]. The logarithmic transformation was involved in each sample of the data. Then, the distributions were assessed and revealed less skewed and more normal distributions. At this phase, we used the ANOVA on the Log-transformed data hypothesizing these null and alternative hypotheses:

1. The null hypothesis: the population means of all levels under consideration are equal. Mathematically, it can be described as:
   H_0: Log(μHigh) = Log(μMiddle) = Log(μLow)
2. The alternative hypothesis: At least one of the population means different.
   H_a: not all Log(μHigh), Log(μMiddle), Log(μLow) are equal

ANOVA was performed to test the hypothesis of whether the means of the Log-transformed (ST) of the three users' satisfaction levels would greatly differ or not. Note that F parameters are (K-1, N-K), where K is the number of groups and N is the total number of reads. The analysis outcome demonstrated a significant difference F(2, 352) = 52.23, p-value = 0.000. Even though we have done the ANOVA test on the Log-transformed variable, the results are back-transformed (raised 10 to the power of each number) and declared in the actual units for more useful understanding, as guided in [31, 33]. Table 2 shows the outcomes of the one-way ANOVA test we executed, where N is the number of reads and C% is the Confidence Interval. CI is possible that the interval will catch the true population value in repeated instances. That is, the confidence level is the sensation rate of the technique. As we evaluate the value of population parameters, the statistical hypothesis (measured by the Confidence Interval) delivers a way of pulling population findings from the sample data. C% value is usually user-defined; a 90% or higher is preferred. The most standard confidence level used is 95% [31–33]. A 95% CI was selected here for our ANOVA test. So, we can be confident 95% of the time that the population means of the High users' satisfaction level lies within this interval (3.7, 4.6). High user satisfaction levels included mainly data sizes groups 100–400 and 500–800 bytes.

The p-value corresponding to the F statistics indicated a significant difference of 0.000 when corresponded to α = 0.05 (α is the error level we selected along

Table 2 Back transformed mean and 95% CI of the ST for users satisfaction levels

Satisfaction levels	N	Mean	95% CI
High	209	4.1	(3.7, 4.6)
Middle	99	8	(6.9, 9.4)
Low	47	12.4	(9.9, 15.4)

with the 95% CI). This directs us to reject the null hypothesis and figure that there is strong evidence that the three population means of the user's satisfaction levels are significantly different. We can detect that the Confidence Interval of the mean per of the three levels of users' satisfaction does not coincide with the Confidence Interval of the different groups. This means that the user's satisfaction levels are distinguishable and differentiated. Due to the importance of QR code security, this study takes an inclusive outlook on symmetric key lightweight cryptography (LWC) algorithms, also known as secret-key cryptography, which uses a single shared secret key to encrypt content between groups. The symmetric encryption methods can be categorized into Block Ciphers and Stream Ciphers. In Block Ciphers, a plaintext is processed in blocks (groups) of bits at a time, then a sequence of functions are executed on this block to generate a block of ciphertext bits, in which the number of bits in a block is appointed. In-Stream Ciphers, the plaintext is processed one bit at a time, then a sequence of functions is performed on it to generate one bit of ciphertext [27]. This study illustrates software performance metrics depending on specified key aspects of LWC and provides a classification of LWC depending on their internal structure.

Security is estimated via the number of key bits, so the delivered security will be higher by increasing the size of the key. Performance (speed) is evaluated based on the total clock cycles to achieve an operation balanced with throughput. Among these factors, a trade-off causes optimizing all of them jointly in one design challenge. For example, security is balanced with performance [27, 34]. Latency and throughput will be used to estimate software requirements as follows [35]:

1. Latency: the minimum processing time to produce the cipher from the original text for one block independently of others.
2. Throughput: the average total plaintext in k bytes divided by the average encryption time. It is processed per CPU clock cycle at a 4 MHz frequency.

3 Lightweight Cryptographic Mechanisms Evaluation

Depending on the modern smartphones used in the experiment, which we referred to in the previous section, the performance characteristics (latency and throughput) are achieved in an optimal situation [36, 37].

Based on recent studies [34, 35, 38], we found that the most famous attacks on encryption algorithms are divided into the following:

- A linear cryptanalysis is a general form of cryptanalysis established on discovering affine approximations to the action of a cipher. Attacks have been produced for block ciphers and stream ciphers.
- Integral cryptanalysis is particularly suitable to block ciphers with substitution-permutation grids. Reported with two other names, Square and saturation attacks. It employs selected plaintexts of which position is held constant, and another piece ranges through all possibilities.

- Algebraic cryptanalysis is established on equation-solving algorithms and is sufficient for lightweight implementation due to its straightforward format (less number of rounds with less algebraic complexity).

Recent studies [27, 38–40] confirm that the following commonly LWC algorithms did not suffer from the mentioned attacks:

1. Advanced Encryption Standard (AES): AES is a block cipher officially adopted by the National Institute of Standards and Technology (NIST) in 2001. AES used three key lengths: 128, 192, and 256 bits, while the block size is 128. AES has presented as a highly secure algorithm. AES is used for confidentiality, with three modes; Cipher Block Chaining (CBC), Output Feedback (OFB), and Cipher Feedback (CFB). AES with Galois/Counter Mode (GCM) mode guarantees authentication and data integrity [2].
2. PRESENT: is a lightweight block cipher, invented by the Orange Labs, Ruhr University Bochum, and the Technical University of Denmark in 2007 and approved by the ISO/IEC 29,192 standard. PRESENT uses a block size of 64 bits, and the key length can be 80 bits or 128 bits. The non-linear layer is founded on a single 4-bit S-box developed with hardware optimizations in mind. PRESENT is suitable where low-power consumption and high chip efficiency are desired and used for confidentiality [39].
3. Camellia: is a symmetric key block cipher in which the block size is 128 bits, and it has three key lengths of 128 (required 18 rounds), 192, and 256 (required 24 rounds) bits. Mitsubishi Electric and NTT of Japan together produced it and used it for confidentiality. The cipher has security levels and processing capabilities similar to the Advanced Encryption Standard [34].
4. SPECK: is one of the common lightweight block ciphers released by the National Security Agency (NSA) in 2013, used to provide confidentiality. SPECK is used to optimize software performance, while its sister algorithm, SIMON, has been optimized for hardware executions. SPECK adopts several blocks and key size alternatives. The most efficient software performance needs 599 cycles with 186 bytes of ROM for a 64-bit block with a 128-bit key [38, 39].

We developed a lightweight paradigm QR code security tool (LWC-QR) that employs lightweight symmetric mechanisms (see Fig. 3). LWC-QR guarantees confidentiality, authentication, and data integrity and adds Access Control List (ACL) in a particular frame, including username, password, encryption mode, and the data content (see Fig. 4). ACL will authorize numerous safe layers of data with dynamic QR code content. ACL will utilize the available QR code space to achieve a High user satisfaction level and support QR code sustainability.

Fig. 3 Main interface of LWC-QR tool

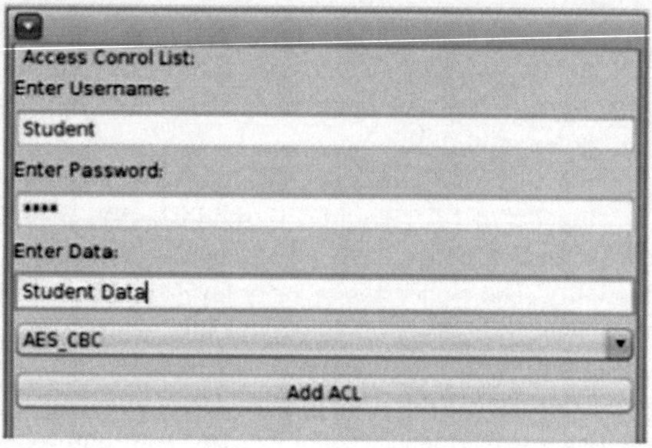

Fig. 4 ACL for LWC-QR code contents

4 Conclusion and Future Works

QR codes' interoperability has delivered advantages for multiple enterprises where the need for them dramatically increased. Like other technologies, QR Codes content is dangerous if misrepresented to contain malicious and irrelevant content. This work assessed the barcode scanning experience by analyzing the users' feedback for the scanning time and discussed security features for some symmetrical lightweight cryptography (LWC) algorithms. The results recommended AES, PRESENT, Camellia, and SPECK mechanisms to generate safe QR codes according to the needed security and feasibility trade-off. We can extend results for more block ciphers in future work and compare their performance with other enciphering methods such as stream ciphers.

Acknowledgements The authors acknowledge King Faisal University for the financial support.

References

1. Akta C (2017) The Evolution and Emergence of QR Codes, 1st edn. Cambridge Scholars Publishing, United Kingdom
2. Wahsheh HA (2019) Secure and usable QR codes. PhD thesis, Universita Ca Foscari Venezia
3. Al-Zahrani MS, Wahsheh HA, Alsaade FW (2021) Secure real-time artificial intelligence system against malicious QR code links. Secur Commun Netw 2021:1–11
4. ISO/IEC Standard. ISO/IEC 18004:2015, Information technology–Automatic identification and data capture techniques–QR code 2005 Bar code Symbology Specification (2015)
5. Wahsheh HA, Al-Zahrani MS (2021) Secure real-time computational intelligence system against malicious QR code links. Int J Comput Commun Control 16(3):1–9
6. Demir S, Kaynak R, Demir KA (2015) Usage level and future intent of use of quick response (QR) codes for mobile marketing among college students in Turkey. Procedia Soc Behav Sci 181:405–413
7. Al-Zahrani MS, Wahsheh HAM (2022) Secure real-time artificial intelligence system against malicious QR code links an environmental approach. Fresenius Environ Bull 2:1618–1623
8. Palazón J, Giráldez A (2018) QR codes for instrumental performance in the music classroom. Int J Music Educ 36:447–459
9. Pérez-Sanagustín M, Parra D, Verdugo R, García-Galleguillos G, Nussbaum M (2016) Using QR codes to increase user engagement in museum-like spaces. Comput Hum Behav 60:73–85
10. Mira JJ et al (2015) Use of QR and EAN-13 codes by older patients taking multiple medications for a safer use of medication. Int J Med Inform 84:406–412
11. Uzun V, Bilgin S (2016) Evaluation and implementation of QR code identity tag system for healthcare in Turkey. Springerplus 5:1–24
12. Ventura C, Aroca R, Antonialli A, Abrão A, Rubio JC, Câmara M (2016) Towards part lifetime traceability using machined quick response codes. Procedia Technol 26:89–96
13. Zhu X, Hou Z, Hu D, Zhang J (2016) Secure and efficient mobile payment using QR code in an environment with dishonest authority. In: Wang G, Ray I, Alcaraz Calero J, Thampi S (eds) Security, Privacy, and Anonymity in Computation, Communication, and Storage. SpaCCS 2016. LNCS, vol 10066, pp 452–465. Springer, Cham. https://doi.org/10.1007/978-3-319-49148-6_37
14. Focardi R, Luccio FL, Wahsheh HAM (2018) Security threats and solutions for two-dimensional barcodes: a comparative study. In: Daimi K (eds) Computer and Network Security Essentials, pp 207–219. Springer, Cham. https://doi.org/10.1007/978-3-319-58424-9_12
15. Focardi R, Luccio F, Wahsheh H (2019) Usable Security for QR code. J Inf Secur Appl 48:1–9
16. Focardi R, Luccio F, Wahsheh HAM (2018) Usable cryptographic QR codes. In: Proceedings of the 19th International Conference on Industrial Technology, pp 1664–1669. IEEE
17. Vidas T, Owusu E, Wang S, Zeng C, Cranor LF, Christin N (2013) QRishing: the susceptibility of smartphone users to QR code phishing attacks. In: Adams AA, Brenner M, Smith M (eds) Financial Cryptography and Data Security. FC 2013. LNCS, vol 7862, pp 52–69. Springer, Berlin, Heidelberg. https://doi.org/10.1007/978-3-642-41320-9_4
18. Elnouby, Mohamed Abdelbasset.: GitHub–OWASP/QRLJacking. https://github.com/OWASP/QRLJacking. Accessed 4 Jan 2022
19. Al-Kabi MN, Alsmadi IM, Wahsheh HA (2015) Evaluation of spam impact on Arabic websites popularity. J King Saud Univ Comput Inf Sci 27:222–229
20. Al-Kabi MN, Wahsheh HA, Alsmadi IM (2014) OLAWSDS: an online Arabic web spam detection system. Int J Adv Comput Sci Appl 5:105–110
21. Al-Kabi M, Wahsheh H, Alsmadi I, Al-Shawakfa E, Wahbeh A, Al-Hmoud A (2012) Content-based analysis to detect Arabic web spam. J Inf Sci 38:284–296
22. Wahsheh HA, Al-Kabi MN, Alsmadi IM (2013) A link and content hybrid approach for Arabic web spam detection. Int J Intell Syst Appl (IJISA) 5:30–43

23. Alsmadi M, Alsmadi I, Wahsheh HAM (2022) URL links malicious classification towards autonomous threat detection systems. In: Al-Emran M, Al-Sharafi MA, Al-Kabi MN, Shaalan K (eds) Proceedings of International Conference on Emerging Technologies and Intelligent Systems. ICETIS 2021. LNNS, vol 322, pp 497–506. Springer, Cham. https://doi.org/10.1007/978-3-030-85990-9_40
24. Ukrop M, Kraus L, Matyas V, Wahsheh HAM (2019) Will you trust this tls certificate? Perceptions of people working in it. In: Proceedings of the 35th Annual Computer Security Applications Conference, pp 718–731
25. Wahsheh HAM, Al-Zahrani, MS (2022) Lightweight cryptographic and artificial intelligence models for anti-smishing. In: Al-Emran M, Al-Sharafi MA, Al-Kabi MN, Shaalan K (eds) Proceedings of International Conference on Emerging Technologies and Intelligent Systems. ICETIS 2021. LNNS, vol 322, pp 483–496. Springer, Cham. https://doi.org/10.1007/978-3-030-85990-9_39
26. Li L, Fan M, Wang G (2018) LWSQR: lightweight secure QR code. In: Li F, Takagi T, Xu C, Zhang X (eds) Frontiers in Cyber Security. FCS 2018. Communications in Computer and Information Science, vol 879, pp 241–255. Springer, Singapore. https://doi.org/10.1007/978-981-13-3095-7_19
27. Thakor VA, Razzaque MA, Khandaker MR (2021) Lightweight cryptography algorithms for resource-constrained IoT devices: a review, comparison and research opportunities. IEEE Access 9:28177–28193
28. GitHub: ZXing Project Home. https://github.com/zxing/zxing/. Accessed 1 Apr 2022
29. Al-Kabi M, Al-Qudah NM, Alsmadi I, Dabour M, Wahsheh H (2013) Arabic/English sentiment analysis: an empirical study. In: The Fourth International Conference on Information and Communication Systems (ICICS 2013), pp 23–25
30. Al-Kabi MN, Wahsheh HA, Alsmadi IM (2016) Polarity classification of Arabic sentiments. Int J Inf Technol Web Eng (IJITWE) 11:32–49
31. McDonald JH (2014) Handbook of Biological Statistics. Sparky House Publishing, Baltimore, MD
32. Limited S (2018) P-value . https://www.statsdirect.com/help/basics/p_values.htm. Accessed 1 Apr 2022
33. Moore DS, Kirkland S (2007) The Basic Practice of Statistics. WH Freeman, New York
34. Panahi P, Bayılmış C, Çavuşoğlu U, Kaçar S (2021) Performance evaluation of lightweight encryption algorithms for IoT-based applications. Arab J Sci Eng 46:4015–4037
35. Shah A, Engineer M (2019) A survey of lightweight cryptographic algorithms for IoT-based applications. In: Tiwari S, Trivedi M, Mishra K, Misra A, Kumar K (eds) Smart Innovations in Communication and Computational Sciences. Advances in Intelligent Systems and Computing, vol 851, pp 283–293. Springer, Singapore. https://doi.org/10.1007/978-981-13-2414-7_27
36. Wahsheh HA, Luccio FL (2019) Evaluating security, privacy and usability features of QR code readers. In: ICISSP, pp 266–273
37. Wahsheh HA, Luccio FL (2020) Security and privacy of QR code applications: a comprehensive study, general guidelines and solutions. Information 11:217
38. Dhanda SS, Singh B, Jindal P (2020) Lightweight cryptography: a solution to secure IoT. Wirel Pers Commun 112:1947–1980
39. Thakor VA, Razzaque M, Khandaker MR (2020) Lightweight cryptography for IoT: a state-of-the-art. arXiv preprint arXiv:2006.13813
40. Dastidar A, Mishra S (2022) Encryption and decryption algorithms for IoT device communication. Electron Devices Circuit Des Chall Appl Internet Things 3:97–112

Comparative Analysis of USB and Network Based Password Cracking Tools

Mouza Alhammadi, Maryam Alhammadi, Saeed Aleisaei, Khamis Aljneibi, and Deepa Pavithran

Abstract Passwords play a major role in securing various applications. It is used for authentication in Operating system login, web services, ATM machines and many more. Even though we educate the end users to choose a complex password, there are several attacks that can break through such password. With passwords also being a very attractive asset for the attackers to have their hands on, the methods of attacks are many. In this paper, we provide a comparative analysis of USB based and Network based attacks on passwords. We provide detailed description and comparison of various network based and USB based password cracking tools. For the comparison, we benchmarked the passwords into Easy password, Hard password, and Medium Password. We then measured the time required to crack these passwords using various tools.

Keywords Window OS · USB-based · Network-based

1 Introduction

The area of password attacks on the Windows operating system is being updated by the minute since the windows OS is by far the most popular. Passwords are stored in an encrypted manner in all current safe computer systems. When a user signs in,

M. Alhammadi (✉) · M. Alhammadi · S. Aleisaei · K. Aljneibi · D. Pavithran
Abu Dhabi Polytechnic, Abu Dhabi, UAE
e-mail: A00049936@adpoly.ac.ae

M. Alhammadi
e-mail: A00044139@adpoly.ac.ae

S. Aleisaei
e-mail: A00050014@adpoly.ac.ae

K. Aljneibi
e-mail: A00044533@adpoly.ac.ae

D. Pavithran
e-mail: deepa.pavithran@adpoly.ac.ae

M. Al-Emran et al. (eds.), *International Conference on Information Systems and Intelligent Applications*, Lecture Notes in Networks and Systems 550,
https://doi.org/10.1007/978-3-031-16865-9_53

the password supplied is first encrypted, then compared to the encrypted password connected with the user's login name that is saved. It's as easy as that: a match succeeds, and a mismatch fails.

Setting a user password is scarcely safer than having no password at all in some instances, which may surprise people. The actual value of a password for an account is to prevent unauthorized network access. The user password does not prevent unauthorized access to any of the device's data (unless the password is used in conjunction with some additional protection such as encryption) This is true not only for Microsoft Windows, but also for Mac OS X, Linux, and most other current operating systems [1].

2 Theoretical Background

2.1 Password Stored as OWF

The passwords stored in windows in two different ways by default, first way is LAN Manager one-way function, and the other is NT OWF. This One-way function is a one-way mathematical translation of data is referred to as a one- way transformation. The data being changed can only be encrypted in one direction, and it cannot be reversed, and the cryptographic hash is the most common type of the one- way function. A hash is a tiny collection of data that is computationally linked to a larger set of data that is used to create the hash. When the bigger set of data is altered, the hash is altered as well [1].

Password Stored in Active Directory
Active Directory is the default authentication solution. It was easier to use common techniques of achieving a task than to create a custom solution. In a diversified and spread context like the university, Active Directory also provides many other important capabilities such as group rules and delegation of power. The challenge would be either having Active Directory route authentication requests to a Unix system or automating the creation of Active Directory accounts that matched the Unix accounts.

Users' access to network resources is regulated by a login procedure in which they must enter their credentials to obtain access to services and applications. Kerberos protocol is used to authenticate users in Active Directory. The Kerberos protocol is a versatile authentication security technology. Instead of transferring user credentials over the network, a session key is produced and utilized for a limited period. [1]

Passwords Stored in the Local SAM
The Security Accounts Manager (SAM)registry file stores hashed local passwords. These hashes may be retrieved, and the accounts' passwords changed to blank in the VM snapshot using information from the SYSTEM registry hive. When a client

Table 1 Comparison on windows and Linux passwords

Password located	Windows [1]	Linux [2]
Where Password Stored	SAM passwords are stored local. On domain members and workstations, password hashes are saved in a local Security Account Manager (SAM) Database in the registry	In Linux, passwords were originally saved in /etc./passwd (which is public), but were later relocated to /etc./shadow (and backed up in /etc./shadow-)
How Password stored in file	Windows password hashes are stored in the SAM file (C:WindowsSystem32Config)	The /etc./shadow file maintains the actual password for the user's account in encrypted format (more like a hash of the password)
Limit for Password	Passwords are stored in Windows as 256-character UNICODE strings. However, the logon dialog is limited to 127	There is no maximum password length limit

account's password is less than 15 characters long, Windows creates a LAN Manager hash (LM hash) and a Windows New Technology LAN Manager hash (NTLM hash). The fact that the Windows operating systems have left a duplicate of the SAM file in some other folder in the windows folder, which has no user accounts or passwords save for the administrator account with a blank password, which may be exploited to get into the system, is unknown to the programmers. The backup copy of the SAM file is used to replace the active SAM file [1].

Cashed Credential's

When a domain user logs in to a domain member, Windows additionally saves a password verifier on that domain member. If the computer cannot access the domain controller, this verifier can be utilized to authenticate a domain user. A cached credential is another name for the password verifier. It's calculated by concatenating the username with the NT hash, then hashing the result with the MD4 hash function [1]. Table1 provides a comparison on windows and Linux passwords.

3 Related Work

Authors in [3], shows the GPU-based brute force cracking tools, it's weakness and algorithm. As a result, the cracking takes less than 1 s for the 6-digit password. So the users must have a longer and complex password to be in the safe side.

Authors in [4], claims that five instruments were evaluated in two groups based on predetermined characteristics. The speed comparison was offered for both tools running on the same computer and each tool running on two distinct machines, and the speed test analysis will aid in identifying the best tools in both categories. The verdict was Cain and Abel is the winner in the offline category whereas in the online category HTC-Hydra is the winner.

4 Password Cracking Tools

4.1 John the Ripper

The most well-known password cracking program is John the Ripper (John). It is a free program that supports both brute force and dictionary assaults. It is a lengthy password breaking tool. This program employs a dictionary of terms in a range of languages as well as the most common password character sequences. John can easily break into networks and user accounts protected by weak passwords thanks to its comprehensive dictionary. John can also be used to crack passwords using basic numerical or special character permutations in the password phrase [5, 6].

A crypto hash of each user's password is maintained with the user's information in many cryptographic and data protection systems to establish the user's identity when they log in. Because the unencrypted password is never saved in the system, this is also a way to protect the password.

But even so, attackers were able to pre-generate hashes for a significant number of passwords using this easy technique to password protection [7].

Ophcrack

Ophcrack is one such program that accelerates password cracking by using a pre-generated database of hashes and passwords. Instead of keeping a comprehensive list of all hashes of all passwords, Ophcrack employs the notion of "rainbow tables," which require only a small fraction of all passwords and hashes to be maintained in a pre-generated database, minimizing the amount of storage required. A "reduction" function is used in the rainbow table technique to generate a new password from a given hash.

Ophcrack begins generating another chain of passwords and hashes beginning with the provided hash when given a hash saved on a hard disk. If a created password matches one of the rainbow table's chain terminating passwords. To determine the password corresponding to the hash on the disk, Ophcrack regenerates the appropriate chain using the password that initiates the chain. When a "salt" is used to randomize the has generation, this innovative implementation of Martin Hellman's "time-memory trade-off" fails, just like previous hash lookup attacks do [7, 8].

Elcomsoft

Elcomsoft was one of the first businesses to create a password cracking tool that could not only be distributed across numerous computers, but also used the graphical processor unit (GPU) of the machine to hash password guesses. Moreover, unlike the previously stated password cracking products, Elcomsoft's flagship password cracking program EPDR is designed to crack both file encryption and Windows log-in passwords [9].

Cain & Able
Cain & Able is a well-known password cracking application with a sizable following. In terms of cracking, it can perform brute force assaults, cryptanalysis attacks, and revealing cached passwords. Its appeal stems from the fact that it runs on Windows systems, is free, has a simple graphical interface, and, most importantly, integrates a variety of additional tools directly into it. A network sniffer is a common function that automatically captures passwords and password hashes it encounters. If it captures a password hash, it can run a password cracking attack on it automatically. Cain & Able also contains the ability to launch an ARP poisoning assault, which makes this function considerably more powerful [10].

Cracking SAM Passwords
First collect the hashes stored within the operating system in order to break passwords. The Windows SAM file stores these hashes. This file is available at C:WindowsSystem32config on the machine, but it is not accessible when the operating system is running. These settings are also kept in the registry at HKEY LOCAL MACHINESAM, but this section of the registry is likewise unavailable while the operating system is booting.

Then try to break them using various methods to obtain the password for a Windows account [16].

5 Different Types of USB Attacks on Windows Password

5.1 *Password Protection Bypass Patch*

If you have password-protected files on your USB, they will be vulnerable to the Password Protection Bypass Patch malware. This USB infection, as its name implies, breaks the security of your encrypted files wide open. Password Protection Bypass Patch accomplishes this by modifying the firmware of your USB drive [17].

Rubber Ducky
Rubber Ducky is a ransomware threat that was first discovered in 2010. Its main goal is to encrypt your files by impersonating a keyboard and entering pre-programmed keystrokes. It works with any operating system that recognizes a USB flash drive as the primary input devicc (keyboard) [17].

USB-Driveby
The USBdriveby is a sophisticated USB development board that can be connected to a USB flash drive. In 60 s, this gizmo can hack any computer. The USBdriveby

disguises itself as a mouse or keyboard when plugged onto your PC or laptop. It then disables your computer's firewall using pre-programmed keystrokes. When your firewall is off, USBdriveby begins to attack your computer's DNS settings [18].

Evilduino
Evilduino takes an Arduino microcontroller, reprograms it, and uses it to infect your computer with malicious keyboard and mouse strokes. is a hack tool that makes use of Arduino microcontrollers to perform cursor movements on the host device in line with a preloaded script. The tool may be made for a low cost out of outdated electronic components and can build and run complex scripts in seconds [17].

USB Hardware Trojan
Kernel-space and user-space channels are used by this Trojan and are not protected by endpoint security safeguards. USB ports are commonly seen in modern computer systems. A USB hardware Trojan horse device can leverage such unintentional channels to establish two-way interactions with a targeted network endpoint, compromising the integrity and confidentiality of the data stored on the endpoint [19].

USB Thief
USB Thief is a type of malware that runs invisibly on USB drives and uses portable programs like Firefox or TrueCrypt to do so. It features a powerful self-protection system that prevents it from being replicated. This malware's goal is to collect information from computers that aren't connected to the internet [20].

6 Design and Implementation

6.1 USB Based Tools

Windows lockpicker script can steal hashes from a locked fully patched Windows 10 system with a working firewall, The script can then try to brute force the hashes by using john the ripper. It grabs many requests from different protocols including the NTLM authentication. The steps in getting the password are (Fig. 1):

What the victim machine sees: As soon as the hash is grabbed, keystrokes are sent to wake up the machine. The password that was cracked is typed in the login screen. A notepad file is opened, and the password is written in the notepad.

Fig. 1 Victim machine, attacker device and connection in the USB attacks

The script doesn't just type a list of passwords on the login screen since that can be blocked by Windows after a few failed attempts and seeing your computer trying different password is suspicious, instead this all happens in the background.

What the attack does: The first thing is that the raspberry pi is going to introduce itself as a network device to the target and then send DHCP configurations. A WPAD (Web Proxy Auto-Discovery) entry is sent to the target and then the raspberry is going to redirect all the traffic into itself using several methods for them to be sent to Responder.py. Responder tries to grab hashes that were requested for authentication, at that time the raspberry is going to blink 3 times showing that the hash is grabbed, now there's 2 options either to un-plug it and crack offline or keep it connected to be sent to John The Ripper. If the hash was successfully cracked, then it will be typed in and written in a notepad.

6.2 *Network Based Tools*

John the ripper [12] is on the first password cracking tools to ever exist, it was first introduced back in 1996 to test password strength and brute force hashed passwords. John the ripper supports some common encryption methodologies for UNIX and windows systems, it detects the encryption of the hash and compares it against a plain text of common passwords.

Ophcrack [13] is a free password recovery or cracking tool, it uses rainbow tables to achieve its goal. Ophcrack uses LM hashes and compares then against rainbow tables this can be done by directly dumping the SAM file of windows or many other ways, The rainbow tables are already provided by the tools, but additional tables can be found online some free and some are paid.

Cain and Abel [15] is a little bit different than other password cracking tools, It uses a ton of methods to try to get the password. These methods are sniffing the network, dictionary attacks versus hashed passwords, Brute forcing, cached password, analyzing different routing protocols and many more. Cain and Abel is no longer supported by its developers but still has some support from other security enthusiast's (Fig. 2).

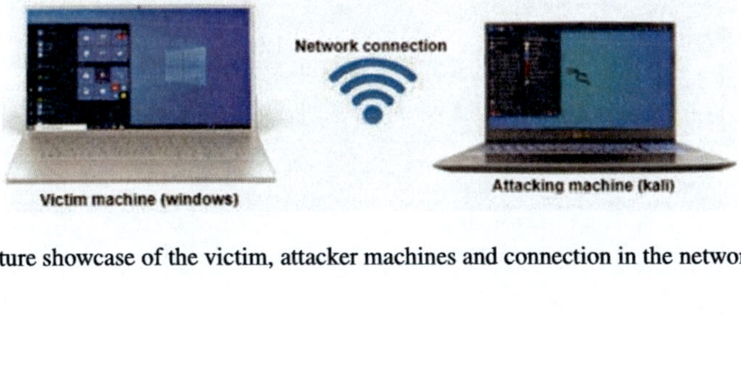

Fig. 2 Picture showcase of the victim, attacker machines and connection in the network attacks

7 Result and Discussion

We Benchmarked the passwords as Easy, Medium, and Hard passwords and measured the time required to crack these passwords using USB based and network-based methods. Easy passwords are those that use either (A to Z) or (0 to 9). Medium passwords are combination of both characters and numbers whereas Hard passwords are combination of characters, numbers, and special characters with length up to 12. Table 2 provide comparison of the USB and network-based tools. Password used for the analysis and its hash values are given in Table 3. The same dictionary file is used for all the tools.

In Table 3 we can see Easy passwords taking an average of 27 s with Cain and Able taking the longest. Medium passwords with an average of 76 s with John taking much less than the others. Hard passwords were only cracked by John and it took less than the other tools in medium passwords, John clearly came in top in our testing.

Table 2 Comparison of various password cracking tools

Types of tools	*OS*	Advantages	Disadvantages	Static or Live Analysis	References
John the Ripper	UNIX and Windows	Parallelization and incremental segmentation are two characteristics of JtR that can be useful	It is a little bit difficult to use [11]	Live & static	[5, 12]
Ophcrack	Windows, Linux and Mac OS	It's compatible with Linux and Mac. It may be used to crack simple passwords quickly or complex passwords over the course of several hours	Passwords with more than 14 characters are not recoverable Windows 10, 8.1, and 8 are not compatible	Live & static	[7, 8, 13]
Elcomsoft	Window, Mac OS	The most versatile password cracking tool for Windows. it also supports multiple operating systems	The free edition has restricted capability, which is the biggest downside	Live	[14]

(continued)

Table 2 (continued)

Types of tools	*OS*	Advantages	Disadvantages	Static or Live Analysis	References
Cain & Abel	Windows OS	This is a completely free utility To breach the computer's password, various approaches are used	It must obtain the necessary "Rainbow Tables" from the internet	Live	[9, 15]

Table 3 Table comparison of USB and network tools

Brute force data (Tools)							
Password	Character used	length	Attempts/hashes	OpHcrack	Cain and Able	John	Windows Lockpicker
Easy	(A-Z) or (0–9)	1–6	1	14 s	56 s	11 s	2 min 40 s
Medium	(0–9) + (A-Z)	1–6	1	1 min, 36 s	1 min, 47 s	27 s	2 min 55 s
Hard	(0–9) + (A-Z) + Special Character	1–12	1	Fail	Fail	48 s	Fail

When it comes to USB attacks, the average time is much higher than a network attack, but that's because it considers two factors: the time to grab the hash and the time to crack it. The cracking of the hash depends on the list being used. If it's not in the list, then the attacker must try to crack it offline since the hash has been grabbed but not successful, therefore easy and medium passwords were successful but not the hard ones. Another thing to notice is the hashing function, MD5 is usually considered less secure than SHA. MD5 has collision problems which means that different password might generate the same MD5 and that lowers the security. MD5 needs 2^(power 64) bit operations to break but that also means it's faster.

8 Conclusion

Before adapting any password or authentication mechanisms, users and application developers should be aware of various password attacks and apply appropriate solution to it. In this paper, we provide a survey of various password attacks. We listed several Network based attacks, USB-based attacks and Tools used for such attacks. A performance comparison of USB-based and network-based password cracking tools is provided.

References

1. Ethical hacking: breaking windows passwords–InfoSec Resources. https://resources.infosecinstitute.com/topic/ethical-hacking-breaking-windows-passwords/. Accessed 11 Mar 2022
2. What is the max length of password on Unix/Linux system?–Super User. https://superuser.com/questions/148971/what-is-the-max-length-of-password-on-unix-linux-system. Accessed 28 Feb 2022
3. Vu AD, Han J IL, Nguyen HA, Kim YM, Im EJ (2011) A homogeneous parallel brute force cracking algorithm on the GPU. In: 2011 International Conference on ICT Convergence (ICTC 2011), pp 561–564. https://doi.org/10.1109/ICTC.2011.6082661
4. Islam S (2021) Security auditing tools: a comparative study. Int J Comput Sci Res 5:407–425. https://doi.org/10.25147/ijcsr.2017.001.1.49
5. Sykes ER, Lin M, Skoczen W (2010) MPI enhancements in John the Ripper. J Phys Conf Ser 256. https://doi.org/10.1088/1742-6596/256/1/012024
6. Kakarla T, Mairaj A, Javaid AY (2018) A real-world password cracking demonstration using open source tools for instructional use. In: 2018 IEEE International Conference on Electro/Information Technology (EIT) 2018-May, pp 387–391. https://doi.org/10.1109/EIT.2018.8500257
7. Dandass YS (2008) Using FPGAs to parallelize dictionary attacks for password cracking. In: Proceedings of the 41st Annual Hawaii International Conference on System Sciences (HICSS 2008), pp 1–8. https://doi.org/10.1109/HICSS.2008.484
8. Kumar J, Farik M, Kumar J, Farik M (2017) Cracking advanced encryption standard-a review. Int J Sci Technol Res 06:101–105
9. Weir CM (2010) Florida State university libraries password cracking attacks
10. Chester JA (2015) Analysis IdeaExchange@UAkron Analysis of Password Cracking Methods & Applications Recommended Citation Analysis of Password Cracking Methods & Applications
11. What is one of the disadvantages of using John the Ripper?—Colors-NewYork.com. https://colors-newyork.com/what-is-one-of-the-disadvantages-of-using-john-the-ripper/. Accessed 25 Sep 2021
12. John the Ripper password cracker. https://www.openwall.com/john/. Accessed 14 Mar 2022
13. Ophcrack. https://ophcrack.sourceforge.io/. Accessed 14 Mar 2022
14. Rafique M, Khan MNA (2013) Exploring static and live digital forensics: methods, practices and tools. Int J Sci Eng Res 4:1048–1056
15. A Full Review of Cain and Abel Password Recovery. https://www.tunesbro.com/blog/cain-and-abel-review/. Accessed 26 Sep 2021
16. Sanders C (2010) How I Cracked your Windows Password (Part 1). WindowSecurity.com, pp 1–7
17. Here's a List of 29 Different Types of USB Attacks. https://www.bleepingcomputer.com/news/security/heres-a-list-of-29-different-types-of-usb-attacks/. Accessed 29 Sep 2021

18. Cannoles B, Ghafarian A (2017) Hacking experiment using USB rubber ducky scripting. In: IMCIC 2017–8th International Multi-Conference Complexity, Informatics Cybern Proc 2017-March, pp 73–78
19. Clark J, Leblanc S, Knight S (2011) Compromise through USB-based hardware Trojan horse device. Futur Gener Comput Syst 27:555–563. https://doi.org/10.1016/J.FUTURE.2010.04.008
20. 9 Different Types of Malware That Can Attack Your Unprotected USB Drive. https://promotionaldrives.com/blog/types-of-malware/. Accessed 29 Sep 2021

Low-Cost Home Intrusion Detection System: Attacks and Mitigations

Meera Alblooshi, Iman Alhammadi, Naema Alsuwaidi, Sara Sedrani, Alia Alaryani, and Deepa Pavithran

Abstract In the last decades, people headed to build smart homes to increase the security surrounding the homes and to prevent unauthorized users from entering secure places without permission based on using biometrics and many sensors. The Ultrasonic Sensor is one of the sensors that has widely been used to detect and measure the distance precisely of any intruder or object by using ultrasonic sound waves, and it's considered as a low-cost system to build home IDS. By exploiting the home IDS vulnerabilities, the attackers can overcontrol the home and easily enter for different goals such as stealing, damaging their property, violating their privacy using only a single attack. Also, by exploiting the home IDS vulnerabilities the attackers can spread different attacks across the home IDS. Enhancing the security using smart technologies will not be enough, homeowners must use detection and prevention techniques and methods to reduce the effect of the possible attacks and prevent them from occurring. This paper addresses the detection of intruders in homes, the possible attacks on home IDS, and how to mitigate them as well.

Keywords Home IDS · Ultrasonic range sensor · Attacks · Mitigation

M. Alblooshi (✉) · I. Alhammadi · N. Alsuwaidi · S. Sedrani · A. Alaryani · D. Pavithran
Abu Dhabi Polytechnic, Abu Dhabi, UAE
e-mail: A00053204@adpoly.ac.ae

I. Alhammadi
e-mail: A00044141@adpoly.ac.ae

N. Alsuwaidi
e-mail: A00051122@adpoly.ac.ae

S. Sedrani
e-mail: A00043437@adpoly.ac.ae

A. Alaryani
e-mail: A00051576@adpoly.ac.ae

D. Pavithran
e-mail: deepa.pavithran@adpoly.ac.ae

M. Al-Emran et al. (eds.), *International Conference on Information Systems and Intelligent Applications*, Lecture Notes in Networks and Systems 550,
https://doi.org/10.1007/978-3-031-16865-9_54

1 Introduction

Home safety is mandatory to any single person especially if the house is left unattended, Layering the house security to feel secure and safe is important. The home IDS would act as an extra layer to protect the house and it's going to mainly focus on detecting if a stranger has entered the premises. The system is a low-cost IDS system build using Arduino, 1Sheeld, and an Ultrasonic Sensor. Often houses are at risk of having malicious people intrude into their houses to steal or damage their property. This issue can be addressed by implementing a camera that will sense the intruder. Once the intruder is sensed by the sensor an alert will go off and a picture of the intruder will be taken. This picture will be sent to the owner of the house via 1Sheeld application and email as well. A simple implementation of the process is applied, and the use of all the hardware and software devices used during the implementation is provided. By using the Arduino platform, we would attach an ultrasonic sensor that could detect if an obstacle passed through the chosen house within 10 cm or less. After an object or a stranger intrudes on the house, an alert will begin using the 1Sheeld application and a picture of the intruder will be taken at the spot instantly and it would be saved on the phone, and it will be sent via email as well. This would not only send an alert of a stranger breach, but it will send the picture of the intruder as well which will benefit to catching the intruder much faster. In this project, we implemented a low-cost home intrusion detection system, and then we identify all possible attacks on it. We classified the attacks into four classifications: Physical attacks, Network attacks, Malware attacks, and Ultrasonic attacks. After classifying the attacks, we provided their mitigation techniques.

2 Background

An ultrasonic sensor is a device that has the ability to scale the distance to the object, it uses an ultrasonic sound wave to transfer and pick the ultrasonic pulses which carry the object's data and information, an example of an ultrasonic sensor is that it can expose any vehicles that are beside each other in the parking area, streets and it alerts the person who is controlling the vehicle about their surroundings [1, 2]. Arduino is an open-source electronics platform based on easy-to-use hardware and software. The Arduino board can read the sensor's input light, button finger, or Twitter message, convert it to output, activate the motor, turn on the LED, and post something online. Home alerts are one of the most inefficient security measures that are created to guard smart devices, which make the internet of things weak with a bind to break a very secure infrastructure although, it's going to be a very engaging target that allows the attackers to hack easily. There are many possible intrusions that may detect our homes and destroy them. For example, network-based attacks such as denial of service, a man in the middle, spoofing, reconnaissance, and replay [3]. In addition, an Ultrasonic sensor allows us to equipped robots with a means of perceiving surroundings objects alternative to technical vision with sound waves of a frequency that is high no human can hear it. Ultrasonic works with four components VCC- connect with supply 5 V, ground the ultrasonic sensor to make sure that the sound is working without it there will be no sound, echo, and trigger where echo emits the sound wave, a trigger is receiving the sound waves so, they are the transmitter and the receiver [4]. It emits short and high-frequency bases into a regular interface where there is an object in front of the ultrasonic sensor, the sound base will be reflected back from the echo to the trigger and the ultrasonic sensor will compute the distance according to the time spent to the waves to reach back.

3 Related Work

Many studies on the use of ultrasonic sensors with Arduino have been undertaken, and this system represents a technology innovation. When an ultrasonic sensor identifies a barrier in front of the robot, it will immediately seek a path that is not blocked. Research chooses to enhance technology through maintaining and staying current with new advancements [5]. In [6] the authors devised a better technique to improve driver safety by identifying blind spots with an ultrasonic sensor and automatically directing the automobile in the right direction. In [7] the researchers proposed a solution that works by detecting the overall level of water in percentages using an ultrasonic sensor. The focus of their study is to look into water level management utilizing an ultrasonic sensor that detects the amount of water in a tank and provides the proportion of water present.

According to the authors [8], ultrasonic sensors are utilized to produce an accurate map of a vehicle's exterior.

Some of the cybersecurity attacks in vehicular sensors such as spoofing attack, acoustic cancellation attack, jamming attack, sensor interference attack, cloaking attack, physical tampering attack, and blind-spot exploitation attack has been listed in [9] in [10] have developed ultrasonic sensor defensive strategies that can withstand spoofing and jamming attacks. Replay attack, tampering attack, DoS, Injection attack a concrete approach for attack surface assessment has been listed in [11]. The authors [12] of this study proposed an eight-category classification of threat vectors including, possible threat countermeasures.

Ultrasonic sensors, according to the majority of current research, are subject to attacks. Attacks against automotive ultrasonic sensors were the subject of a previous study. In this study, we use the Arduino system as a framework to investigate flaws in ultrasonic sensors and identify attacks on home intrusion detection systems that use ultrasonic sensors.

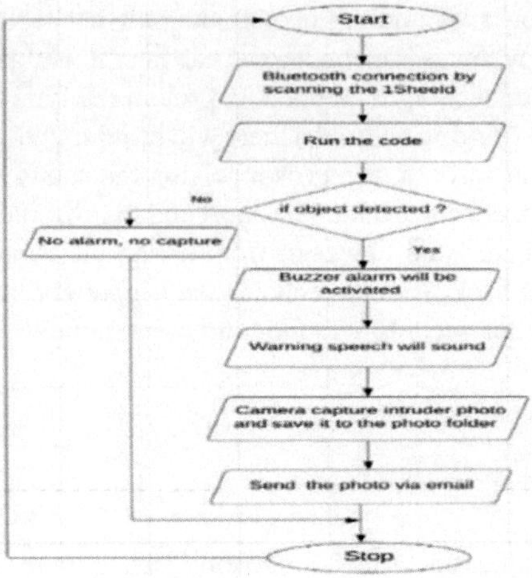

Fig. 1 Home IDS flowchart

4 Project Design

The workflow design of the home IDS is shown in Fig. 1. Starts with scanning the Sheeld device with the Sheeld application in the phone, then it is connected to the computer where the code will be written and run within the Android device after switching Sheeld mode to load mode. Once the intruder breaks into the house, the alert buzzer will be activated, the camera will take the photo of the intruder and save the photo and send it by email. On the other hand, If the intrusion is not detected, there will be no buzzer alarm and no warning message also camera will not capture the intruder. Attackers can use several attack techniques to evade the home IDS. We classified these attacks into four different categories, physical attacks, network attacks, malware attacks, and ultrasonic attacks. Mitigation techniques can be used to prevent the Physical, Network, Ultrasonic, and Malware attacks from happing or to reduce the effect of the possible attacks and prevent them from occurring.

5 Implementation Details

For the Home IDS system implementation, we used Arduino and an Ultrasonic sensor. The Fig. 3 shows the experimental setup of the project, the hardware components that have been used are 1Sheeld, Arduino, Male/Female Jumper Wires, HC-SR04 Ultrasonic Sensor, USB-A To B Cable and the software devices used in the Intrusion detection system are Sheeld application and Arduino IDE software.

5.1 Hardware Design Implementation

The initial connection of wires start with connecting the Sheeld hardware on top of the Arduino, the male\femlae jumper wires connect the ultrasonic and the Sheeld hardware together, to insert the code a USB-A wire is connected to the computer where the code will be inserted after switching the Sheeld mode into upload mode, once the code is uploaded the implementation mode will start [13].

5.2 Software Design Implementation

Using the 1Sheeld application will help us to over-control our sensors. Also, it will allow us to use different security systems techniques along with detecting intrusions only by using one application [14]. A mobile application is used to control the ultrasonic sensor. First of all, once we run the application using an Android device or IOS, we have to scan the 1Sheeld board so we can use all the possible capabilities and techniques of the virtual shields provided by the application. We used a camera, buzzer, terminal, text to speech, and email shields in our scenario. All of these functions will be combined together and ready to test. Once the intruder breaks into the house the buzzer alarm will be activated, the camera will capture the intruder photo and save the photo, and sent it via email, and at the same time, there will be a warning speech as a sound [15] (Fig. 2).

The home intrusion detection system demonstrates an app on 1Sheeld's camera function that will hunt any malicious or threat from any human sneaking into the house and capture its photo and alert us as well. Figure 3 shows the software we used to write our code. Arduino IDE is a software tool that can be connected to the Arduino hardware to write the code and import it on the Arduino hardware. We downloaded Arduino IDE on our PC then, we connected it with the Arduino hardware which is connected to the 1Sheeld hardware and the ultrasonic sensor as well. The IDE application is free software for Arduino, which can work with multi-operating systems such as Windows, Linux, and Mac as well and it supports C++ language [16].

Fig. 2 Experimental setup

Fig. 3 Arduino IDE software

6 Classification of Attacks

6.1 Network Attacks

Jamming Attack. Is a wireless network assault in which an attacker purposefully sends out interfering signals to interfere with current wireless communication [16] and makes noise to reduce the sensor signals [17]. As an ultrasonic sensor has been used to detect intruders or objects entering the home, the attacker will manipulate and control the frequency signal that has been sent from the sender and create a strong noise to prevent the sensor from detecting intruders [18].

Acoustic Cancellation Attack. This attack is also known as active noise cancellation, and it's accomplished by sending a 180-degree phased wave on the receiver end. The echo's amplitude will be reduced, and the sensor will not receive an echo signal. In other words, this attack work by using a signal to cancel the sensor's real signal which will result in hiding the object as if the object was transparent. intruders can enter the home without being detected by the sensor because the sensor won't be able to detect any objects [19].

Bluebugging Attack. This attack is one of the sophisticated attacks on Bluetooth devices, the attacker can manipulate the phone and execute whatever he wants, this attack enables the attacker to have full control over the phone [20]. The attacker can achieve this attack by using the Bluebugger tool [21], this tool enables the attacker to perform penetration testing on the phone that is used to capture the intruder picture. Once the attacker gains full control, he can disable the connection from the phone to the Sheeld hardware which will prevent the Ultrasonic sensor from capturing pictures.

Bluesmacking Attack. In this attack, a numerous number of the packet is sent to the Bluetooth device which will result in Denial of service. This attack is also known as the ping of death attack on Bluetooth devices [22]. The attacker will implement this attack to prevent the ultrasonic sensor from working. The attacker can use a Kali Linux machine in order to implement this attack by using a tool called hci tool [21]. This tool allows the attacker to get the targeted Bluetooth device and then using the L2ping command to perform the ping of death attack.

Blueprinting Attack. This attack is performed to gather information about Bluetooth devices, which will allow the attacker to gain data and information about the device itself in order to exploit it, the attacker will know what the device model, firmware and manufacturer is as well [23]. Gaining information about the device would help the attacker in the script an attack against the device to make it stop working and to make the Ultrasonic sensor unresponsive. Bluediving is one of the tools the attacker can use to perform this attack [24].

Dolphin Attack. Dolphin Attack was among the first to demonstrate inaudible attacks towards voice-enabled devices by injecting ultrasound signals over the air. Commercial speech recognition systems like Siri, Google Now, and Alexa detected inaudible voice instructions. Experiments using smartphones from different vendors have been used to validate the attacks [25, 26].

6.2 Malware Attacks

Mobile Ransomware Attack. This attack is a type of malware that targets mobile phones and tablets. A cybercriminal can use a mobile virus to steal sensitive data from a smartphone or lock it, then demand payment to unlock it or restore the data to the owner. In a home intrusion detection system, the object will be detected, captured, and send the photo via email. however, email is inexpensive and easy to use, so it makes a convenient way for attackers to spread ransomware. Users are accustomed to receiving documents over email and have no qualms about opening a file attached to an email. The malicious macro executes, downloading ransomware to the local device before delivering its payload [26].

Memory or Non-malware Attacks. This a non-malware or fileless cyberattack is one in which the harmful code has no physical presence in the file system [27]. File-free malware can be downloaded from an infected email or presented as malicious code from an infected application. In the place of this attack, the email files and accompanying photos are destroyed, so they are stolen and encrypted [28].

Code Injection Attack. Code injection is the flaw that happens in the system when installing vulnerable and unveiled data to the system which allows the attacker to gain the access to the mechanisms of the client code by implementing injected code inputs with no sense [29]. This attack allows the attacker to make changes on the buzzer of the client code and destroy the functionality of the sensor and make it vulnerable for everyone to see the data and steal the information of the client's house [30].

6.3 Physical Attacks

Search-based Physical Attacks. Initially, the attacker scans the network for sensors, utilizing appropriate resources to detect signals sent by the sensors. Following detection, the attacker manually destroys the identified sensors. Physical force, radiation, as well as other hardware/circuit tampering tactics, are commonly also used destruct small size sensors [31].

Physical Tampering Attack. Tampering is a technique used by an attacker to obstruct or detect unauthorized entry to a certain device or spoof the security system physically. So, this attack may affect the code of the system if it gains access attacker will have supplementary knowledge by interacting with vulnerable devices and intended to destroy the security of it also, can modify the memory [32].

6.4 Ultrasonic Attacks

Spoofing Attack. The attacker can modify the measured distance of intruders or objects either by making the object very close or very far from the sensor. For example, if we set the object detected distance to 10 cm, the attacker can change the distance and set it to 1 cm which means the intruder must be too close to the sensor to be detected.

Signal Injection Attacks. Signal injection attacks target, the usually unprotected, analog sensing interface of the sensors and induce arbitrary signals in them [29] (Table 1).

Table 1 Possible attacks on ultrasonic sensor & mitigation techniques

Attack name	Attack type	Attacking mode	Mitigation	References
Jamming attack	Network attack	Sending signals to interfere with the sensor signal, make a noise to make the sensor unstable	Use timestamp and anti-jamming techniques	[19]
Spoofing attack	Ultrasonic attack	Modifying the measured distance of objects	Sensor fusion	[19]
Acoustic cancellation attack	Network attack	Cancelling the sensor real signal which will result in hiding the objects		[19]

(continued)

Table 1 (continued)

Attack name	Attack type	Attacking mode	Mitigation	References
Bluebugging attack	Network attack	Attacking the Bluetooth operated devices, it allows full control of a certain device	Add non-discoverable feature while Bluetooth is on	[20]
Bluesmack attack	Network attack	DOS attack on Bluetooth devices	Add a pairing pin to avoid getting paired by anyone via bluetooth	[21]
Blueprinting attack	Network attack	Gaining information about the Bluetooth device by using bluediving tool	Bluetooth Firewall	[23]
Mobile ransomware attack	Malware attack	steal sensitive data from a smartphone or lock it	Backup all files and Stay informed about the latest threats	[26]
memory or non-malware attacks	Malware attack	harmful code has no physical presence in the file system	Install security patches	[27]
Code injection attack	Malware attack	Allows to injects the code of the system because of the flaw	Control the activity by validate user inputs through the creation of an allowed list	[30]
Dolphin attack	Network attack	An attacker can give an arbitrary voice command to a digital assistant without the recipient's knowledge	By using Inaudible voice command cancellation	[26]
Signal injection attacks	Ultrasonic attacks	analog sensing interface of the sensors and induce arbitrary signals in them	By use the physical closeness of the intended signal to the sensor and the ability to elicit response to distinguish between real and fake signals	[29]
Search-Based physical attack	Physical attack	The attacker's goal is to find then physically destroy networking sensors in order to damage system performance	Locking network equipment in rooms or secure areas	[33]

(continued)

Table 1 (continued)

Attack name	Attack type	Attacking mode	Mitigation	References
Physical tampering attack	Physical attack	Method used to detect unlicensed to a specified device to change in the security features	Use autonomous network transaction, use distributed mobile agent for the exposure	[33]

7 Conclusion

Improving physical security is mandatory for everyone. So, we created a camera that can detect if someone has broken into the house. We utilized the 1Sheeld program to operate our sensors using the Arduino platform. Our proposal added an extra layer of security to keep criminals out of our homes. We demonstrated mainly the function on 1Sheeld's application that uses our code to sense if a person passed through the ultrasonic sensor and once the intruder passes the ultrasonic sensor a snapshot will be taken of the intruder, and an alarm will be activated. In our scenario, we employed a camera, alarm, terminal, text-to-speech, and email shielding. Once the low-cost home intrusion detection was applied, we searched for possible attacks on the intrusion detection system and we classified the attacks into 4 classifications. Physical attacks, Network attacks, malware attacks and ultrasonic attacks and we found the mitigation techniques for these attacks.

References

1. Dimitrov A, Minchev D (2016) Ultrasonic sensor explorer. In: International Symposium on Electrical Apparatus and Technologies (SIELA)
2. Carullo A, Parvis M (2001) An ultrasonic sensor for distance measurement in automotive applications
3. Anthi E, Williams L, Slowinska M, Theodorakopoulos G, Burnap P (2019) A supervised intrusion detection system for smart home IoT devices. IEEE Internet Things J 6:9042–9053. https://doi.org/10.1109/JIOT.2019.2926365
4. Zhmud VA, Kondratiev NO, Kuznetsov KA, Trubin VG, Dimitrov LV (2018) Application of ultrasonic sensor for measuring distances in robotics. J Phys Conf Ser. Institute of Physics Publishing
5. Irawan Y, Muhardi, Ordila R, Diandra R (2021) Automatic floor cleaning robot using arduino and ultrasonic sensor. J Robot Control (JRC) 2:240–243. https://doi.org/10.18196/jrc.2485
6. Ajay TS, Ezhil R (2016) Detecting blind spot by using ultrasonic sensor. Int J Sci Technol Res 5:5
7. Varun KS, Kumar KA, Chowdary VR, Raju CSK (2018) Water level management using ultrasonic sensor (Automation). Int J Comput Sci Eng 6:799–804. https://doi.org/10.26438/ijcse/v6i6.799804
8. Madhavan VR, van der Sande TPJ: Ultrasonic sensor-based mapping for an autonomous vehicle
9. El-Rewini Z, Sadatsharan K, Sugunaraj N, Selvaraj DF, Plathottam SJ, Ranganathan P (2020) Cybersecurity attacks in vehicular sensors. IEEE Sens J 20:13752–13767. https://doi.org/10.1109/JSEN.2020.3004275

10. Xu W, Yan C, Jia W, Ji X, Liu J (2018) Analyzing and enhancing the security of ultrasonic sensors for autonomous vehicles. IEEE Internet Things J 5:5015–5029. https://doi.org/10.1109/JIOT.2018.2867917
11. Zelle D, Plappert C, Rieke R, Scheuermann D, Krauß C (2022) ThreatSurf: A method for automated threat surface assessment in automotive cybersecurity engineering. Microprocess Microsyst 90. https://doi.org/10.1016/j.micpro.2022.104461
12. Rugo A, Ardagna CA, el Ioini N (2023) A security review in the UAVNet era: threats, countermeasures, and gap analysis. ACM Comput Surv 55:1–35. https://doi.org/10.1145/3485272
13. Security System Using Arduino Bluetooth Camera–Arduino Project Hub. https://create.arduino.cc/projecthub/amrmostaafaa/security-system-using-arduino-bluetooth-camera-616c4d. Accessed 5 Mar 2022
14. Mahammad FS, Sudireddy SS (2021) An efficient home automation and security system using Arduino and 1-Sheeldin 2
15. Software I Arduino. https://www.arduino.cc/en/software. Accessed 5 Mar 2022
16. Bhaskar M, Manjunatha P (2021) Signal jamming autonomous rover view project smart sericulture system using image processing view project signal jamming autonomous rover
17. Baanav B, Ravi Y, Kabir R, Mishra N, Boddupalli S, Ray S: AUTOHAL: an exploration platform for ranging sensor attacks on automotive systems
18. He Q, Meng X, Qu R (2020) Towards a severity assessment method for potential cyber attacks to connected and autonomous vehicles. J Adv Transp 2020. https://doi.org/10.1155/2020/6873273
19. Gluck T, Kravchik M, Chocron S, Elovici Y, Shabtai A (2020) Spoofing attack on ultrasonic distance sensors using a continuous signal. Sensors (Switzerland) 20:1–19. https://doi.org/10.3390/s20216157
20. Pandey T, Khare P: Bluetooth hacking and its prevention
21. Nasim R (2012) Security threats analysis in bluetooth-enabled mobile devices. Int J Netw Secur Its Appl 4:41–56. https://doi.org/10.5121/ijnsa.2012.4303
22. Browning D, Kessler GC: Bluetooth hacking: a case study bluetooth hacking: a case study bluetooth hacking: a case study
23. Ibn Minar NBN (2012) Bluetooth security threats and solutions: a survey. Int J Distrib Parallel Syst 3:127–148. https://doi.org/10.5121/ijdps.2012.3110
24. OConnor MT, Reeves D (2008) Bluetooth network-based misuse detection. In: Proceedings - Annual Computer Security Applications Conference, ACSAC. pp 377–391
25. Yan Q, Liu K, Zhou Q, Guo H, Zhang N (2020) SurfingAttack: interactive hidden attack on voice assistants using ultrasonic guided waves. Internet society
26. Park Y, Choi H, Cho S, Kim YG (2019) Security analysis of smart speaker: security attacks and mitigation. Comput Mater Contin 61:1075–1090. https://doi.org/10.32604/cmc.2019.08520
27. Jin R, Wang B (2013) Malware detection for mobile devices using software-defined networking. In: 2013 Second GENI Research and Educational Experiment Workshop. 10.1109/.15
28. Canfora G, Mercaldo F, Medvet E, Visaggio CA (2015) Detecting Android malware using sequences of system calls. In: 3rd International Workshop on Software Development Lifecycle for Mobile, DeMobile 2015 - Proceedings. Association for Computing Machinery, Inc., pp 13–20
29. Riley R, Jiang X, Xu D: CERIAS Technical report 2007-01 an architectural approach to preventing code injection attacks an architectural approach to preventing code injection attacks
30. Snow KZ, Krishnan S, Monrose F: Google NP SHELLOS: enabling fast detection and forensic analysis of code injection attacks
31. Wang X, Chellappan S, Gu W, Yu W, Xuan D (2005) Search-based physical attacks in sensor networks. In: Proceedings–International Conference on Computer Communications and Networks, ICCCN 2005, pp 489–496. https://doi.org/10.1109/ICCCN.2005.1523922

32. Aida K, Yamada K, Hotchi R, Kubo R (2021) Dynamic network path provisioning and selection for the detection and mitigation of data tampering attacks in networked control systems. IEEE Access 9:147430–147441. https://doi.org/10.1109/ACCESS.2021.3124024
33. Wang X, Chellappan S, Gu W, Yu W, Xuan D: Search-based physical attacks in sensor networks

Relationship Between Consumer's Social Networking Behavior and Cybercrime Victimization Among the University Students

Yousuf Saif Al-Hasani, Jasni Mohamad Zain, Mohammed Adam Kunna Azrag, and Khalid Hassan Mohamed Edris

Abstract Social media is broadly used for various reasons and one being for academic purposes in universities. Although it has its own repercussion such as cyber-crime victimization. This study intends to investigate the relationship between social networking and cybercrime victimization among university students. It identifies illegal behaviours carried through computers or other internet-connected devices. These illegal behaviours range from institutional hacking to individual victimization. In this paper, questionnaires were given to Sultan Qaboos University students of various ages to find out why and how personal attributes, social networks, user technical efficacy, and online bullying risk add to these illegal activities. The study found that consumers with high use are more prone to be targets and victims of cyber-crime. Regulating multifunction Social Networking services SNS (like Facebook) use and victimization have a mathe matically essential relationship. Also, online victimization and business social networking sites have statistically significant positive links (such as LinkedIn). Finally, cyberbullying requires social media awareness because social media sites and services save personal data.

Keywords Social media · Cybercrime victimization · Information

1 Introduction

The internet has evolved into a basic existence requirement, posing possible risks to those who use it. Although the incredible progress and convenience of dealing,

---

Y. S. Al-Hasani
Faculty of Computer Science and Mathematics, Universiti Teknologi Mara (UITM), 40450 Shah Alam, Selangor Darul Ehsan, Malaysia

J. M. Zain · M. A. K. Azrag (✉)
Institute of Big Data Analytics and Artificial Intelligence (IBDAAI), Universiti Teknologi Mara, 40450 Shah Alam, Selangor Darul Ehsan, Malaysia
e-mail: adamkunna@uitm.edu.my

K. H. M. Edris
Faculty of Computer, University of Medical Science and Technology, Khartoum, Sudan

M. Al-Emran et al. (eds.), *International Conference on Information Systems and Intelligent Applications*, Lecture Notes in Networks and Systems 550,
https://doi.org/10.1007/978-3-031-16865-9_55

access, and consumption to others has provided huge benefits, it has also worked in tandem to drive down prices to engage in potentially unlawful online activities [1]. The study's main objective is to shine light on victimization experienced by young users individually on the internet at a national level. Previous research into people's perceptions of cybercrime has identified popular online crimes like harassment, rates, and fraud [1, 2]. Overall, defamation identity theft of online bullying is flawed; however, the ratio of victimization of younger users is far greater than mature internet users.

1.1 Background

The In modern society, social media is quickly taking regular person-to-person communication [5]. The growing use of social networking has created distinct implications for the criminal justice system, like the corruption of proof by consumers and ensuring the proper to a fair trial [6]. Although various leading and contemporary criminological theories can explain crime. Thus, the regular use of social networking platforms has resulted in new kinds of criminal activity and victimization. Consequently, to deal with the new cybercrime domain, traditional theories may require broadening or perhaps re-envisioning.

Victimization has been connected to users' acts or inactions on social networking platforms. Inactions are frequently linked to security or privacy settings and the over-sharing of information, providing a great opportunity for determined criminals. For example, Facebook®, which is essentially a commonly used social networking platform with an international average of 1.01 billion active users a day [8], offers users two privacy/security options. Users may also make their profiles private or public; a public profile implies that anybody may see anything the user has put on their profile. Users who prefer a more private profile can choose which information they want to share with those they consider as "friends" on the network site [8].

Furthermore, the security feature is known as "user control" affords users the capability to accept or perhaps decline a friend request(s) to be related to another user's profile page [9–16]. In other words, a person's online habits and lifestyle might increase the chances of being a victim. This research's core hypothesis is that Facebook® utilization (measured as an online activity or several hours spent on the web) will impact online victimization. Concerns about cybercrime are not merely issues of an individual but concerns of both the government and business community. Businesses get pressured to spend a considerable amount of money improving their Information and Communication Technology (ICT) security and making it safe from online crimes. As, in Canada, the "whole government approach to cybersecurity" was seen as an approach by Public Safety Emergency Preparedness Canada in its 2009–10 report, the Ministry of Public Safety (2009,10). A step taken by Ontario Provincial Police 4 Canada to deal with telemarketing frauds was Phone Busters Smyth (2010). Phone Buster is a central Canadian agency where information about identity theft and telecommunication fraud gets compiled [11–13].

Since creating applications like MySpace, Friendster, Facebook, Blogspot, YouTube, and Flickr in 2008, social media has begun to take a foothold in the hearts of network users. Another will be the Science of Information and Technology (IPTEK) with the interconnected network system (the internet). Users learn new things due to the user-friendliness of the platform. The improvement in the number of media users per year is 7.6%. Mobile users are growing with the inclusion of new and modern more designs. Through the Ministry of Information and Communications Technology, the government has given legal products on March twenty-five, 2008, Law No.11 of the Yr. 2008 on Information and Electronic Transactions. It also gave Law No.19 of the Yr. 2016 Amendment of Law Number Eleven (from now on abbreviated as UU ITE) meant users arranged to remain wise to use social platforms. Due to a lack of understanding related to legislation, teenagers become victims or criminals of social media [17–22].

2 Methodology

Present research follows the procedure of interpretivism paradigm to explore illegal activities range from hacking attacks at the institutional level to experiencing victimization individually. SPSS was used as a method of analyzing data.

2.1 Research Model

The research model involves the reasons for victims of cybercrime, Heading, and their effects. People become victims of Cybercrime since somehow, they are enabling themselves to be targeted. The reason is that they spend so much unnecessary time using social media, and some using insecure WIFI networks [3, 4, 6, 13]. Also, most people use "weak" passwords to secure their accounts, making it easy for attackers to access their information. Figure 1 displays the reasons why people face victimization nowadays.

This study considers five hypotheses:

- High social media usage has a positive impact on the risk of online victimization for users
- High perception of control over data maintained through social media has a negative effect on the risk of user victimization
- Improved ICT skill possession has a negative impact on the risk of user victimization
- Lower perception risk has a positive effect on the risk of user victimization
- Higher propensity risk imposes a positive impact on the risk for user victimization.

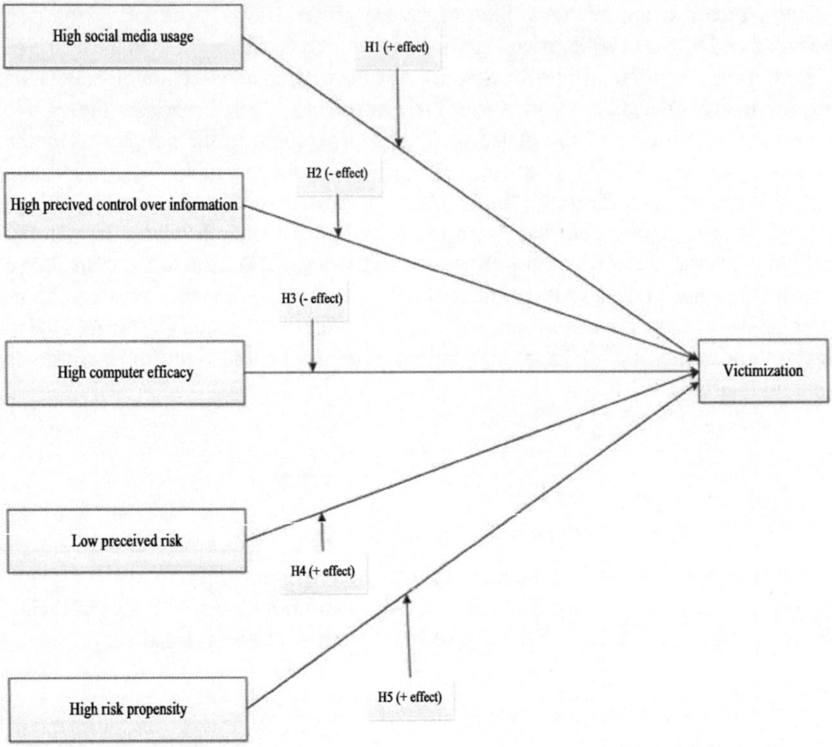

Fig. 1 Research models for this study

A total of 400 respondents between the ages of 20 and 30 years old are considered, highlighting the highest percentage of victimizations are stated in Table 1.

For citations of references, we prefer the use of square brackets and consecutive numbers. Citations using labels or the author/year convention are also acceptable. The following bibliography provides a sample reference list with entries for journal articles [1], an LNCS chapter [2], a book [3], proceedings without editors [4], as well as a URL [5].

3 Result

The study began with a sample survey of 400 respondents, which was taken into consideration. The descriptive statistical analysis of the data is the subject of the next section. All descriptive approaches that establish any differences or similarities between the results have one goal in mind: to provide quantitative data (products). The total male is 197, equal to 49.3%, while the female's total is 203, equal to 50.8%.

Table 1 Measure items used as an independent variable measure

Construct	Question adopted for this study	Measurement scale in the original study
Control over per sonal information	I may control over private information collected by social services I may control what SNS releases personal information I opt to have control over how SNS uses private data I may control the personal information provided to SNS	Strongly disagree to strongly agree (5 points)
Technical efficacy	I am confident working on a PC I pretty understand terms related to computer hardware I pretty understand terms related to computer software I feel it vital to analyze computer problems	Strongly disagree to stronly agree (5 points)
Risk perception	It has been purported to be risky to give data to SNS It has been noted that there is a high loss potential with giving data to SNS There might be huge uncertainty related to giving data to SNS Giving SNS data would result in many unexpected problems	Strongly disagree to strongly agree (5 points)
Risk propensity	There is a substantially high risk of me doing online buying, and I will take it I will willingly accept some probabiity of losing money if online buying will probably involve not much amount of risk There might be a high probability of loss by providing data to SNS I am familiar with SNS than others I am not sure about I am cautious when testing the latest SNS	Strongly disagree to strongly agree (5 points)

In research, most of the respondents are female. In respect to Table 2 which shows the age of respondents (20 to 30 yrs.), (31 to 40 yrs.) and (41 to 50 yrs.), (51 to 60 yrs.) and 61 or above of respondents are 24.3, 23.3, 17.5, 16.0 and 19.0 % respectively. In this table, the age between 20–30 years highlights the maximum percentage setting of being victimized. The table presents the stages of professions of the respondents. Information from the table illustrates that 24.3% of respondents are students; 18.3% of respondents are doing jobs in govt. The sector, 24.55 doing jobs in semi-govt. In the sector, 14.0% of respondents are doing private jobs, and others are retired. The

Table 2 Presents the stages of professions of the respondents

Factors	Sample Respondents	% Sample respondents'
Gender		
Male	197	49.2
Female	203	50.8
Profession		
Student	97	24.3
Govt. Job	73	18.3
Semi Govt. Job	98	24.5
Private Job	56	14.0
Age of respondents		
From 20–30 Years	97	24.3
From 31–40 years	93	23.3
From 41–50 years	70	17.5
From 51–60 years	64	16.0
Form 61 or above	76	19.0
Total	400	100.0

data also highlights that 24.3% and 24.5% of respondents get victimized by social media's excessive usage and vice versa.

The first hypothesis is that high social media usage has gotten accepted since it assumed that it would positively impact victimization. This hypothesis proves that if social media usage increases, the chances of being the victim of any circumstances would also increase since people are increasing with the number of social media users worldwide, which directly increases their privacy gets invaded. Thus, it can be seen in Table 3.

In Table 3, the correlation coefficient was 0.655, with a Sig. value of 0.000. It indicates that victimization rises due to social media usage, even to the point where a substantial value of 0.000 is desirable. It is a positive and strong correlation if the value is more than the 0.5 mid-value of correlation. In other words, social media use has raised the risk of victimization. As a result, increased social media usage raises the risk of victimization. Therefore, the relationship is positive, indicating that these variables continue to increase in tandem. There will be an increase in personal information leaking in public as social media usage among professionals rises. The primary hypothesis that underpins our research is accepted as expected. However, there is a comparison on our findings to those of other previous researchers. Looked at the same outcomes that our initial hypothesis anticipated. The authors [17] concluded that as more individuals worldwide join the social sphere, including various social media users, the danger of being a victim grows. Governments all across the world are attempting to control the digital world; yet, efforts to shield this generation from different cyber-crimes have proven ineffectual. Regulatory methods such as court action and laws frequently fail because they cannot keep up with

Table 3 The correlation between victimization and high social media usage

		Providing information to SNS, it got found high ambiguity connected to it	Profession
By giving information to SNS, a high amount of uncertainty get expected	Colleration of Pearson Sig., 2-tailed N	1 400	0.655** 0.0000 400
Profession	Colleration of Pearson Sig., 2-tailed N	0.655* 0.0000 400	1 400
Age of respondents	Colleration of Pearson Sig., 2-tailed N	Age of respondents 1 400	
I believe that I have the authority to see if someone sees my personal information collected from SNS	Colleration of Pearson Sig., 2-tailed N	−0.387** 0.000 400	The information collected −0.387** 0.000 400

the fast-changing cyber world environment; hence, combating crime is a difficult undertaking. As a result, all countries must take tangible steps to combat the increase of cybercrime. The following hypothesis will be tested, which suggests that if social media has control over the information, it will negatively influence the chance of user victimization. This link was found between victimization and personal information control. In Tables 3 and 4, the result of −0.387 shows a negative correlation. The statistical p-value is 0.000. As a result, the researcher believes that it may avoid being a victim of cybercrime if there is maintain control over the social networking system. On the one hand, the results indicated that carelessness was the cause of victimization since research has shown that teenagers' negligent attitude in using social networking sites is directly responsible for being targets of cybercrime and being blackmailed. Moreover, they do not create complex passwords with several characters to make it harder for others to guess or crack them. Setting a strong password with different personalities is critical for safeguarding oneself from cyber-attacks such as harassment and blackmail, and many youngsters' accounts lack these passwords.

A sample of correlation coefficients ranges from –ve 1 to +ve 1. Moreover, correlation might be as low as 0.2 (or −0.2) for +Ve (−Ve) groups. The selected correlation coefficient says "r" is equal to (−0.387), which indicates a negative correlation. The statistical p-value is 0.000. The study concluded that if the social networking system is controlled, individuals will be protected from being victims of cybercrime. However, variables have a negative influence because if there is minimal control over those with access to personal information acquired from social media sites, a

Table 4 The correlation between Victimization and control over personal information

		Age of respondents	The information collected from SNS get controlled if someone sees it
	Correlation of Pearson Sig., 2-tailed N	1	−0.387** 0.000 400
I believe that I have the authority to see if someone sees my personal information collected from SNS	Pearson Sig. (2-tailed) N	−0.387** 0.000 400	1 400

certain age group of respondents becomes victimized. As a result, there should be contemporary practices for individual data control.

The researcher has concluded that some ICT skills training should save people from such savage being victims of victimization. Thus, the second and third Hypotheses get interlinked to each other. The third has extracted from the second because of acceptance of the second hypothesis directing us toward the provision of ICT Skills of the people to save them from cyber-attacks. The correlation between Victimization and Possession of better ICT skills was −0.273 in Table 5. The hypothesis was accepted but this time it supports the research in a double way. The 2nd hypothesis again got a vital flag by this hypothesis acceptance.

As a result, the researcher can describe the relationship's direction based on Table 5, emphasizing that ICT Skills will protect them from victimization. [18] considered that one of our schools' vital aspects should be taught a little bit of ICT skills. As the researcher notes in society, there are several institutions where ICT skills are taught as a course or a vocation to educate individuals. The analysis of the correlation got used to examine the connection between professional victimization and ICT skills. Researcher can state that the direction of the relationship is negative. However, as the researcher observes in society, there are many institutions where ICT skills learn as a

Table 5 Correlation between victimization and possession of better ICT skills

		Having a complete grasp of words or terms in computer hardware makes me feel confident enough	Profession
Having a complete grasp of computer hardware, I feel entirely confident over its terms or correlation words	Correlation of Pearson Sig., 2-tailed N		−0.273** 0.000 400.000
Profession		−0.273** 0.000 400.000	1 400.000

course or professionalism to aware people through proper education. The P-value in Table 6, the correlation coefficient is 0.000, and which value is less than 0.05; thus, the researcher can finalize a statistically significant correlation.

The relationship between professional victimization and ICT skills was investigated using correlation analysis. As a result, the researcher might claim that the relationship's direction is negative. However, as the researcher notes in society, there are several institutions where ICT skills are taught as a course or a career to educate students and professionals. Therefore, if every age group is aware of computer concerns and encounters less victimization concerns, the researcher can draw this conclusion. As shown in Table 7, the correlation coefficient P is 0.027 less than 0.05, indicating that the connection is statistically significant.

The third hypothesis, however, suggested acquiring ICT skills as a solution. So, it has been redirected toward some answers from the midpoint of the investigation pattern. Hypothesis 4: A low likelihood impression has a substantial immediate impact on the likelihood of being a victim. The "r" value of 0.202 with Sig. = 0.000 for this applied correlation between Victimization and Risk Probability Perception supports our outcome assumption in Table 8 once again. The findings enhance the danger probability perception of SNS among respondents of a particular age group and assert that the association is positive in nature, implying that these variables tend to rise in tandem. This correlation coefficient has a P-value of 0.000. The outcome enhances the age group's sense of SNS danger. The association is positive, indicating that the variables are more likely to overlap. This correlation coefficient has a P-value of 0.000.

Table 6 The correlation between victimization and possession of better ICT skills

		Profession	I feel confident in troubleshooting computer problems
Profession	Correlation of Pearson Sig., 2-tailed N	1	−0.268** 0.000 400.000
I feel confident in troubleshooting computer problems	The Correlation of Pearson Sig., 2-tailed N	−0.268** 0.000 400	1 400

Table 7 The correlation between victimization and possession of better ICT skills

		I feel confident in troubleshooting computer problems	Age of respondents
I feel confident in troubleshooting computer problems	Correlation of Pearson Sig., 2-tailed N	1 400	−0.110* 0.027 400
Age of respondents	Correlation of Pearson Sig., 2-tailed N	−0.110* 0.027 400	1 400

Table 8 The correlation between victimization and perception of risk

		A high level of risk is associated with giving information to SNS	Age of respondents
A high level of risk is associated with giving information to SNS	Correlation of Pearson Sig., 2-tailed N	1 400	0.202** 0.000 400
Age of respondents	Correlation of Pearson Sig., 2-tailed N	0.202** 0.000 400	1 400

Researchers stated that [19] Social Media Users are usually undisciplined. They do not perceive seriously that to which extent they can get harmed via cyber-attacks. Therefore, it is essential to make potential victims aware of the hazards of the internet. For instance, law enforcement agencies are usually unable to look over every crime occurring in a state; thus, it is crucial to educate.

Table 8 shows that respondents' perceptions of SNS risk are increasing among their professions. The relationship direction might begin at this positive point, which indicates that both variables are increasing simultaneously. Because the correlation coefficient has a P-value of 0.000 and is less than 0.05, it is statistically significant.

In Table 9, the last hypothesis is that high probability risk has a direct effect on the risk of user victimization. Victimization and risk propensity has a 0.147 correlation, indicating a positive relationship. The statistical p-value is 0.003, which is less than 0.005, indicating that Professional respondents use social media at every stage of life to obtain knowledge, develop relationships, and develop strong bonds with people, as we conclude that there is a need to increase social media usage. All of our hypotheses support our study. The assumptions we made were not only accepted but also rejected by researchers who were opposed to their idea.

Table 9 The correlation between victimization and risk propensity

		A significant amount of loss gets expected in giving the information to SNS	Profession
That may be a high probability for loss, which ascertained with giving (information) to SNS	Correlation of Pearson Sig., 2-tailed N	1 400	0.147** 0.003 400
Profession	Correlation of Pearson Sig., 2-tailed N	0.147** 0.003 400	1 400

3.1 Discussion

Overall, the discussion aims to give a general view of cybercrime, particularly in terms of defining cybercrimes perpetrated against youth and then determining what measures a student in a cyber-environment should take. One of them, for example, was using social media sites less or more cautiously. In addition, adopting security backdrops on social networking sites and the system also protects users against phishing attacks. Another fact is that, as indicated in the third and second hypothesis, not revealing their credentials with anyone may be harmful. It does not imply that the next individual is also not trustworthy; it means that the sites and platforms were not. People should have technical abilities to protect themselves from these attacks. The number of familiar friends, recognition of the individuals who issue friend requests and consumer control are all socially relevant. As a result, the topic of cybersecurity is attracting an increasing number of researchers. Research organizations should encourage and try to bring together the community to report cybercrimes. These players are trusted and employ different techniques to prevent these suspicious acts. They can also inform other organizations investigating cybercrimes to flag and block the sites containing questionable data [20–23].

4 Conclusion

Based on this research, there is a concern, and a particular age group is a target if information obtained from Social Networking Sites has no true privacy and there is no control over this information disclosed. Regardless of whether they do it voluntarily or just for the sake of discussion. After all, they do so, and as a result, their information is disclosed, potentially putting them in danger. In the fourth component when risk perception rises, victimization rises with it. As a result, numerous crimes may occur in various forms that are less well investigated and regulated, encouraging the perpetrators. The last hypothesis, like the others, is found to be supportive of our assumptions. Since all the assumptions have been accepted, it can be inferred that there is a significant need to strengthen security measures for online consumers that visit various websites.

Acknowledgements The authors thank the Universiti Teknologi Mara and the Director of Institute for Big Data Analytics and Artificial Intelligence (IBDAAI) for their great support of this research.

References

1. Benson V, Saridakis G, Tennakoon H, Ezingeard JN (2015) The role of security notices and online consumer behaviour: an empirical study of social networking users. Int J Hum Comput Stud 80:36–44

2. Saridakis G, Benson V, Ezingeard JN, Tennakoon H (2016) Individual information security, user behaviour and cyber victimisation: an empirical study of social networking users. Technol Forecast Soc Chang 102:320–330
3. Mohammed AM, Benson V, Saridakis G (2020) Understanding the relationship between cybercrime and human behavior through criminological theories and social networking sites. In encyclopedia of criminal activities and the deep web (pp. 979–989). IGI Global
4. Kirwan GH, Fullwood C, Rooney B (2018) Risk factors for social networking site scam victimization among Malaysian students. Cyberpsychol Behav Soc Netw 21(2):123–128
5. Evers CW, Albury K, Byron P, Crawford K (2013) Young people, social media, social network sites and sexual health communication in Australia:" This is funny, you should watch it". Int J Commun 7:18
6. Subramanian KR (2017) Influence of social media in interpersonal communication. Int J Sci Prog Res 38(2):70–75
7. Rege-Patwardhan A (2009) Cybercrimes against critical infrastructures: a study of online criminal organization and techniques. Crim Justice Stud 22(3):261–271
8. Laleh N, Carminati B, Ferrari E (2016) Risk assessment in social networks based on user anomalous behaviors. IEEE Trans Depend Secur Comput 15(2):295–308
9. Wu JCW (2019) Resolving the privacy paradox: bridging the behavioral intention gap with risk communication theory
10. Azad S et al (2017) VAP code: a secure graphical password for smart devices. Comput Electr Eng 59:99–109
11. Smith RG (2008) Coordinating individual and organisational responses to fraud. Crime Law Soc Chang 49(5):379–396
12. Goodstein JD, Wicks AC (2007) Corporate and stakeholder responsibility: making business ethics a two-way conversation. Bus Ethics Q 17(3):375–398
13. Yunus MM, Zakaria S, Suliman A (2019) The potential use of social media on Malaysian primary students to improve writing. Int J Educ Pract 7(4):450–458
14. Fisher M, Boland R Jr, Lyytinen K (2016) Social networking as the production and consumption of a self. Inf Organ 26(4):131–145
15. Garcia N (2018) The use of criminal profiling in cybercrime investigations (Doctoral Dissertation, Master's Thesis). Available from ProQuest dissertations & theses global database. (Accession Order No. AAT 10839020)
16. Dhahir DF (2018) Internet adoption of Indonesian remote society: integrated broadband village versus commercial mobile broadband. J Penelitian Komunikasi 21(2):145–158
17. Denecke K, Bamidis P, Bond C, Gabarron E, Househ M, Lau AYS, Hansen M (2015) Ethical issues of social media usage in healthcare. Yearb Med Inf 10(1):137
18. Kokkinos CM, Antoniadou N, Asdre A, Voulgaridou K (2016) Parenting and Internet behavior predictors of cyber-bullying and cyber-victimization among preadolescents. Deviant Behav 37(4):439–455
19. Kokkinos CM, Kipritsi E (2012) The relationship between bullying, victimization, trait emotional intelligence, self-efficacy and empathy among preadolescents. Soc Psychol Educ 15(1):41–58
20. Cheng C, Chan L, Chau CL (2020) Individual differences in susceptibility to cybercrime victimization and its psychological aftermath. Comput Hum Behav 108:106311
21. Deka GC, Zain JM, Mahanti P (2012) ICT's Role in e-Governance in India and Malaysia: a review. ArXiv preprint arXiv:1206.0681
22. Zain JM, Herawan T (2014) Data mining for education decision support: a review. Int J Emerg Technol Learn 9(6):4–19
23. Azrag MAK, Kadir TAK (2019) Empirical study of segment particle swarm optimization and particle swarm optimization algorithms. Int J Adv Comput Sci Appl 10(8):480–485

Modeling for Performance Evaluation of Quantum Network

Shahad A. Hussein and Alharith A. Abdullah

Abstract Quantum networks are emerging sciences and are anticipated to be the core networking technologies in the future. Due to the difficulty of implementing quantum networks in a real way, because quantum devices are not widely available, they only exist within their laboratories. In addition, they are costly and also need special environments that are not easy to obtain in other than laboratories. In this paper, the authors build a simulator using the language of Python programming to simulate quantum networks in terms of quantum devices, such as repeaters, final nodes and channels, where the behavior of these elements within the network is simulated for the purpose of sending quantum information represented by quantum bits, and therefore the work will be within the principle of the graph and finally facilitate experiments on networks Quantum devices without the need for real physical devices. The most remarkable result that emerged from the simulated data generated and detected is that the modeling process provides guidance for quantum networks design, characterization of their protocols, and their behavior. As a result of this study, one could simulate a quantum network repeater and end node as well as a quantum link (entanglement link) and implement some of the quantum protocols like Quantum Key Distribution (QKD), Teleportation and quantum protocol. In the end, it is concluded the possibility of simulating the behavior of the quantum network, its devices, and protocols, as well as implementing it and developing the quantum applications, an integrated study about the quantum internet and its routing in it. In addition, we were able to develop a quantum repeater protocol in order to enable end-to-end entanglement.

Keywords Quantum network · Quantum internet · Entanglement probability · Quantum simulation · Qubits

S. A. Hussein · A. A. Abdullah (✉)
College of Information Technology, University of Babylon, Babil, Iraq
e-mail: alharith@itnet.uobabylon.edu.iq

M. Al-Emran et al. (eds.), *International Conference on Information Systems and Intelligent Applications*, Lecture Notes in Networks and Systems 550,
https://doi.org/10.1007/978-3-031-16865-9_56

1 Introduction

Quantum networks are the networks through which devices are interconnected, where these devices communicate through quantum communications by exploiting quantum mechanics, where the success of quantum information technologies lies at the heart of quantum networks [1]. Also, it provides great capabilities unparalleled in classical networks, and therefore, these capabilities are used in the formation of penetrations on classical networks [2].

The nature of the work and behavior of quantum networks in transferring and sharing information is not an easy process that can be imagined. It depends on the principle of quantum entanglement between quantum bits that exist within quantum memories [3]. At the same time, the delivery of this information within the best path to the recipient is based on the principle of entanglement swapping according to the criteria adopted by the routing protocol used in a quantum router (currently adopted quantum repeater), which details will be clarified in the following sections.

On the other hand, imagining quantum networks does not stop there. It also extends to the architecture of the quantum Internet, as it specializes in what is contained in quantum networks [3]. Accordingly, the development of quantum networks makes the implementation of the quantum internet in the near future. Due to this, there is a need to test everything related to quantum networks, such as testing protocols and the nature of the work of devices and evaluating their performance before implementing them in a real way to avoid problems that may appear at that time, and this can be done at the laboratory. Still, because quantum devices are not widely available, need special environments, and are very expensive, it is challenging to apply quantum networks to real devices for the purpose of testing. Therefore, the trend is toward building a simulator for quantum networks [4].

As for quantum networks, such simulators (classical network simulators) are not compatible with them. Through this, it can be said that there are no basic simulators for working with quantum networks, but with that, some sources have shown the existence of some simulators for quantum networks built using programming languages like QuNetSim [4]. NetSquid is often closer to the SQUANCH than the SimulaQron in terms of behavior, as it stimulates the physical property of quantum devices like noise in the link, but it is not available to everyone yet [5]. In comparison, SQUANCH is similar to the SimulaQron in terms of performing tasks, except that it adds the possibility of simulation within the physical layer, which allows simulating the processing of quantum information. In addition, it allows the user to add the error model within the physical layer [6]. Besides, SimulaQron simulates software that runs within the application layer on quantum devices, and thus, it simulates quantum internet programs [7]. Some of which are open source while the others are not. Therefore, this research aims to build a simulator for quantum networks, where it is easy for developers to develop quantum applications and protocols and their ease of implementation on network devices, in addition to developing a quantum repeater protocol and allocating it to each of the nodes specialized in implementing this protocol based on the cumulative pre-entanglement possibilities. The use of this entanglement is over long distances to carry out the transmission of quantum information in the process of teleportation.

The present research is divided as follows: Sect. 2 of the research reviews some Quantum Network considerations, while the proposed Quantum Network Platform Design is revealed in Sect. 3, and Sect. 4 shows the result and discussion of the research, finally, the conclusions are revealed Sect. 5.

2 Consideration of Quantum Network

There are many considerations that are specific to quantum networks, such as no-cloning, quantum measurement, quantum entanglement, and others. This section will summarize the most important principles that the present research deals with.

2.1 Qubits and Entanglement

A single quantum bit carries two possibilities for the information it transmits [1]. Therefore, Entanglement indicates that two quantum particles (qubits) are in a common state. Also, these two particles, regardless of the distance between them, one of them is completely affected by the other immediately [8]. Through this, one of the qubits of the entanglement pair can be directly fixed once the other qubit of this quantum state is measured according to quantum mechanics. It represents the core of quantum internet; besides, quantum entanglement is a unique concept, and it is impossible to find anything like it in the classical physics upon which the current classical networks depend [2].

2.2 Quantum Entanglement Swapping

Entanglement Swapping is a recent phenomenon considered a fundamental key to the realization of quantum networks, especially the transmission and routing of quantum data [3, 3]. Through which Einstein–Podolsky–Rosen (EPR) pairs can be shared over long distances [10], so quantum networks based on repeaters can overcome the problem of loss during the quantum information transmission process [11] due to the absence of the possibility of signal amplification within these repeaters due to quantum mechanics [3, 11]. Figure 1.(a) represents a short link (short distance entanglement) between the qubits of quantum memories of the quantum repeater and the end nodes. While (b) represents forming the long-distance entanglement after swapping on repeater [8, 8], as shown in Fig. 1.

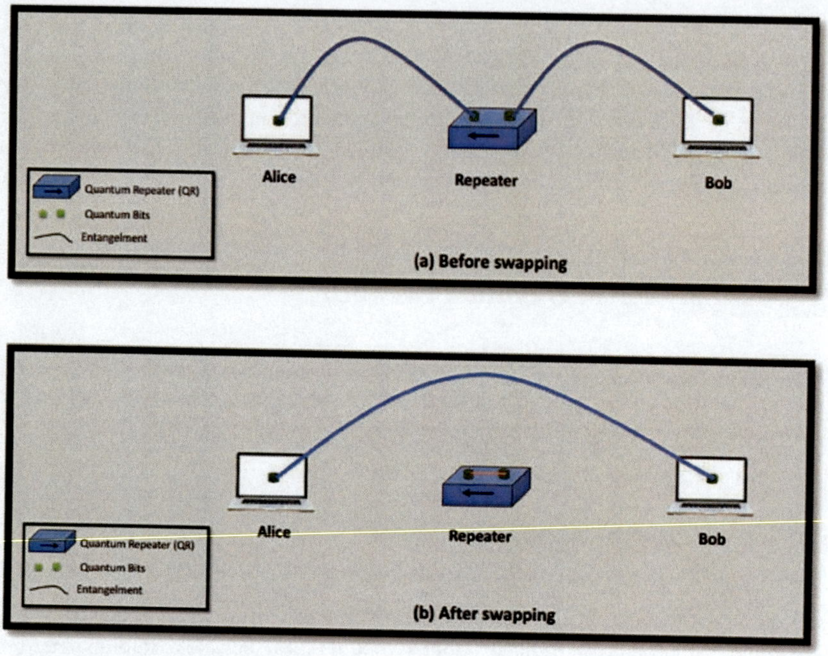

Fig. 1 Build a long-distance entanglement by entanglement swapping Alice represents the sender, and Bob represents the receiver

2.3 *Quantum Repeater*

The heart of the quantum internet is the repeater node. This node makes it possible to establish long-term communication between the sender and the receiver [3].

While quantum repeaters are subject to quantum laws like the theory of non-cloning [12], their work is limited to quantum information and through which control messages can be exchanged between nodes by linking repeaters with other repeaters and quantum processors through the classical internet [3, 3].

In the end, since all types of networks are not free from loss resulting in the channel that connects the network as a result of the surrounding physical conditions and others, the presence of repeaters is necessary, as it is installed at distances commensurate with the amount of loss existing in the channel to improve the performance of the network, whether quantum was it classic [4]. The next section will explain more details about how quantum repeaters work.

2.4 Network Stack

Computer networks are complex, whether they are classical or quantum networks. The OSI model is followed to break the complexity found in classical networks. Then, the TCP/IP model is used, where the network becomes operating in several layers, each layer in the model plays a role and serves. The next layer, and therefore each layer, receives information messages (packets) from the previous layer. The necessary information is added to the message packet and passed to a higher layer than it at one end of the network, while the other end reverses the operations carried out by the first end. Thus, it obtains the basic information sent from the sender to the recipient [1]. In addition to that, each layer of the model has its own set of protocols, and the data has a specific name in each layer. Also, every physical device works according to a certain layer in the network. For example, the data in the network layer is called packets, referred to as a segment at the transport layer, and while the physical switch device operates at the data link layer, the physical router device operates at the network layer, and so on.

Quantum networks are not conceptually different from classical networks, as they also have their difficulty. Thus, quantum network concepts, such as entanglement, session communication techniques and error scaling, lead us to a layered architecture of their but remain lab-scale. Due to the radical quantum field limitations, it does not constitute a complete network architecture because it is still under experiment [5].

While classical network software and hardware work in the TCP/IP model, quantum network software and hardware operate within a model almost similar to what is used in current networks, which is called the layer model (as shown in Fig. 2.). It consists of four layers arranged from top to bottom: application layer, transport layer, network layer, and network access layer. Each has its tasks, and perhaps some of them are similar to what is found in the classic networking model [13], as follows:

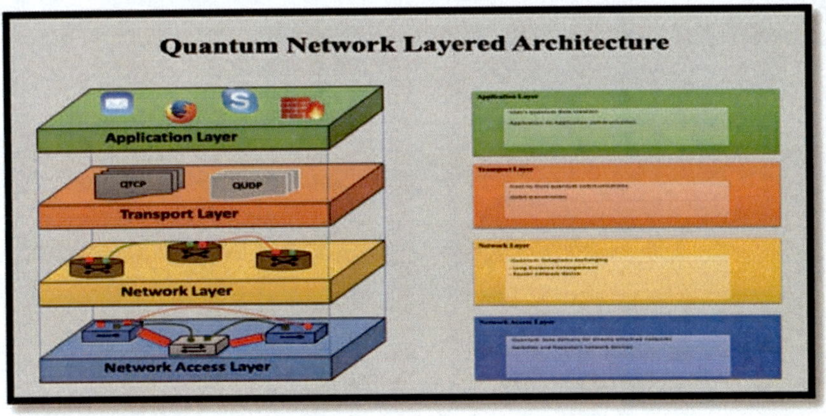

Fig. 2 Layered model of quantum network architecture

3 Quantum Network Platform Design

This section deals with the proposal of the current study to implement the GUI Model through which the work of the quantum network is simulated in terms of graphic design, interfaces and devices control, and the proposed quantum repeater protocol work mechanism.

3.1 GUI Modeling

The proposed simulation model consists of sections, where the following Fig. 3 represents the way to organize our simulator into four sections: A: the network preview frame, B: the tools frame, C: the protocols frame, and D: the processing follow-up frame.

Where this interface was designed on the PyCharm platform from JetBrains software company by using the Tkinter library of graphic interface design in the programming language Python. Therefore, this library provides many tools that help build interactive interfaces, such as the button, the dropdown list, and others. Still, on the other hand, the Networkx library was used mainly to provide the requirements for building the network based on this on the principle of graph theory. This library is considered one of the best and most widely used libraries in existence libraries. Moreover, this library can provide many functions that facilitate building networks and dealing with them, such as adding nodes, linking between nodes, and analyzing the network.

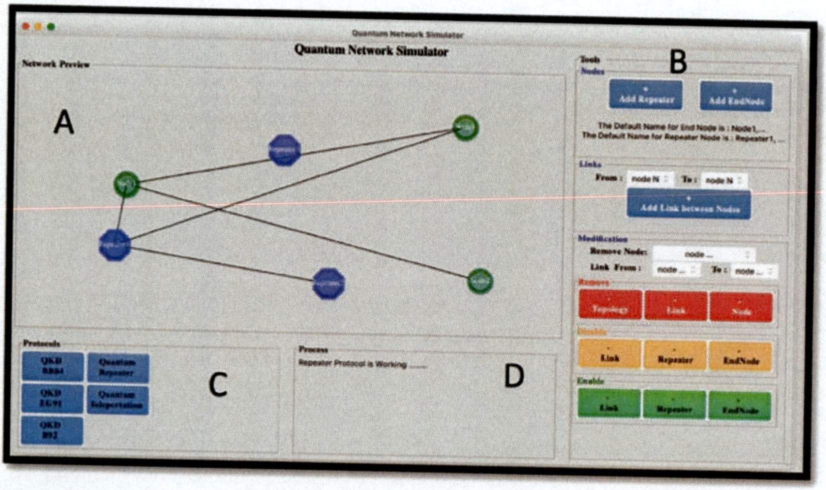

Fig. 3 GUI of quantum network simulation

3.1.1 The Network Preview Frame

This window includes the presentation of network design, where the contents of the network are explained in an easy-to-understand manner, and these components can be distinguished. In addition to that, the network devices that are added from the tool window can be identified, as well as distinguishing whether the devices or channels are activated or not. This framework includes a graphic illustration of what is being done within this network.

3.1.2 The Tools Frame

This frame includes all the tools necessary to deal with quantum networks and is divided into a group of subframes. Each of which provides for specific tools, where the first subframe offers the ability to add quantum devices (End nodes and quantum repeaters Nodes) and the links between nodes. While the deletion sub-frame includes the possibility to delete any previously added component. Besides, the tools for enabling and disabling devices and links give the capability to delete the entire network topology.

3.1.3 The Protocols Frame

. This frame contains the most important protocols in quantum networks, such as the quantum key distribution protocols, the quantum teleportation protocol, and the proposed quantum repeater protocol, which will be discussed in the next section. Later, the authors will add more quantum protocols that help in performing the work of the network, such as routing tangle.

3.1.4 The Processing Follow-Up Frame

The processing framework enables the follow-up of some of the processing that takes place during the network design process. For example, displaying the number of devices in the network or deleting a specific device or link and the details that take place during the implementation of a protocol.

3.2 *Protocols Modeling*

The quantum repeater is considered the heart of quantum networks, as it is responsible for connecting devices with each other. In addition, the quantum repeater overcomes the challenge of long distances connections in quantum networks [11]. The developed quantum repeater protocol that will increase the end-to-end entanglement probability

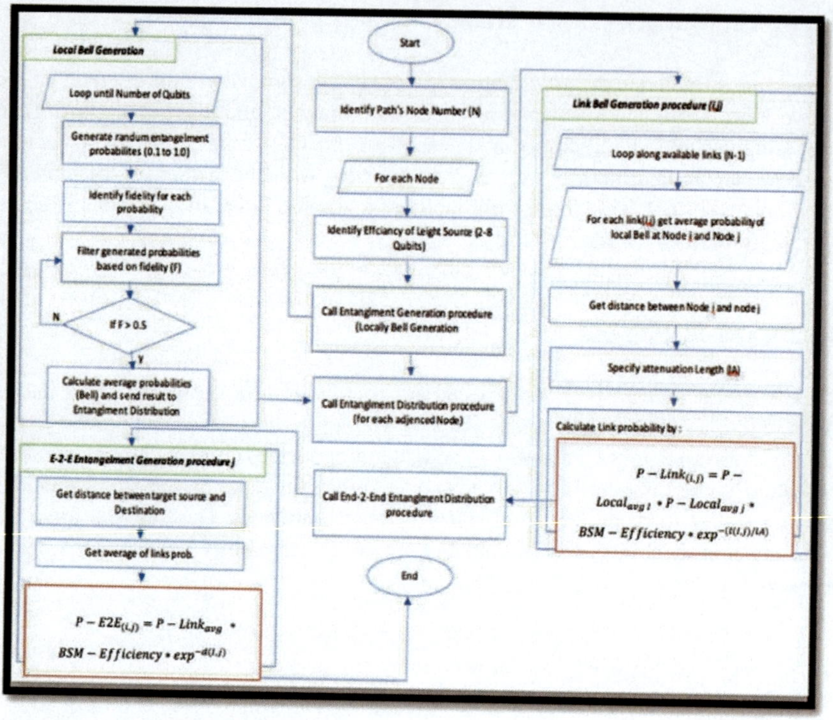

Fig. 4 Quantum repeater protocol workflow

between connected ends, where the probability of end-to-end entanglement success depends on the pre-entanglements cumulatively. Figure 4. represents the workflow diagram of the proposed protocol.

Since the principle of operation of the quantum repeater depends on the property of entanglement, it is assumed that the light source generates entangled qubits on each node with a probability (P-Local), as shown in Fig. 5. The number of qubits generated within the nodes ranges from 2 to 8 [15]. However, the maximum number of entangled qubits generated is 8 qubits [14], where we chose the maximally entangled qubits based on high fidelity for each probability [4].

Then, according to Eq. (1), the authors calculate the probability of entanglement distribution (P-Link) between adjacent nodes (see Fig. 5b). Where, the probability of entanglement between nodes (Pi, j) depends on the two local probabilities that were previously generated for both neighboring nodes, in addition to the length of the optical link connecting the two nodes (l (i, j)). However, the probability decreases exponentially with the increase in the length of the optical link [5, 16], and the probability is also affected by the efficiency of quantum devices, such as Bell-state Measurement (BSM) and others.

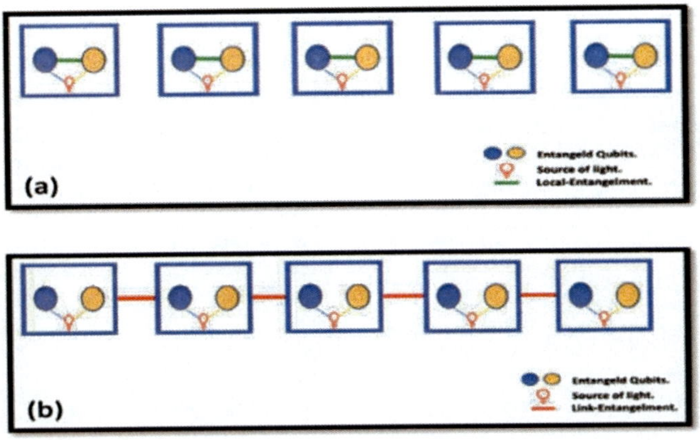

Fig. 5 Entanglement generation **a** within quantum repeater **b** between quantum repeater

$$P - Link_{(i,j)} = P - Local_{avg\ I} * P - Local_{avg\ j} * BSM - Efficiency * exp^{-(l(I,j)/lA)} \quad (1)$$

where the BSM-Efficiency will concede constant, and lA represents fiber optical length attenuation. After completing the calculation of the p-link between adjected nodes along the path between source and destination, one can calculate the end-to-end entanglement probability (P-E2E) according to Eq. (2), which is also affected by distance (di, j) between target source and destination, BSM-Efficiency that performs entanglement swapping [17], and probability of Link entanglement (P-link), as

$$P - E2E_{(i,j)} = P - Link_{avg} * BSM - Efficiency * exp^{-d(I,j)} \quad (2)$$

Thus, we can choose a path with a high end-to-end probability of exchanging quantum information using teleportation.

4 Results and Discussion

The proposed quantum network simulator is generally conducted in several stages. First, a quantum network is modeled using a set of basic network components: quantum repeater devices and end-user devices and the physical links are represented by optical links. Then, the quantum protocols in the network are programmed and assigned to the devices, such as the quantum repeater protocol, which is assigned to work on the quantum repeater device. Finally, this network is implemented and turned on. Table 1 is a comparison between previous simulations and this work in simulating a quantum network.

Table 1 General comparison of quantum network simulators

SimulaQron (2018) [7]	SQUANCH (2018) [6]	NetSquid (2021) [5]	Our Qu-Net-simulator (2022)
• Possibility of simulation within the Application layer • The simulation is open source • Programmed in python • Don't have GUI • Available for everyone	• Possibility of simulation within the Application and physical layer • The simulation is open source • Programmed in python • Don't have GUI • Available for everyone	• Possibility of simulation within the Application and physical layer • The simulation is open source • Programmed in python • Don't have GUI • Doesn't available for everyone yet	• Possibility of simulation within the Application, network and physical layer • The simulation is open source • Programmed in python • Has GUI • Doesn't available for everyone yet

The simulator works on simulating a large part of the physical layer. In addition to the network and application layers of the quantum layered model, quantum bits have been generated within the quantum repeater device with random possibilities to take all the possibilities. Besides, the quantum repeater can work with in terms of the presence of entanglement between the qubits inside it according to the source light used (laser). Then, the selection of maximally entangled qubits within each repeater creates a distribution of entanglement among the memories of quantum qubits. However, the authors have depended on the highest fidelity between the generated entangled qubits. On the other hand, after applying Eqs. 1 and 2, the researchers have obtained a long-distance entanglement between distant quantum devices, which is then used to transfer quantum information between sending and receiving devices.

5 Conclusion

In conclusion, the current work presents the most important basics that lie in quantum networks, especially in the quantum internet. In addition to building a quantum network simulator in terms of devices and protocols that work within different layers of the layered model of the quantum internet, it provides the possibility of developing applications for quantum networks. The proposed quantum network simulator is under development, and new features, protocols and tools will be added over time. Besides, as long as the quantum internet is still at the lab level, i.e., under test, the simulator will remain in a state of constant updating, as it has been built and implemented using software libraries belonging to Python, and it is possible to update and add to them as needed in future.

References

1. Pirker A, Dür W (2019) A quantum network stack and protocols for reliable entanglement-based networks. J Phys 21(3):033003. https://doi.org/10.1088/1367-2630/ab05f7
2. Cacciapuoti AS, Caleffi M, Tafuri F, Cataliotti FS, Gherardini S, Bianchi G (2019) Quantum internet: networking challenges in distributed quantum computing. IEEE Network 34(1):137–143
3. Basso Basset F et al (2019) Entanglement swapping with photons generated on demand by a quantum dot. Phys Rev Lett 123(16). https://doi.org/10.1103/PhysRevLett.123.160501
4. DiAdamo S, Nötzel J, Zanger B, Beşe MM (2021) Qunetsim: A software framework for quantum networks. IEEE Transactions on Quantum Engineering 2:1–12
5. Coopmans T et al (2021) Netsquid, a network simulator for quantum information using discrete events. Commun Phys 4(1):1–15. https://doi.org/10.1038/s42005-021-00647-8
6. Bartlett B (2018) A distributed simulation framework for quantum networks and channels.arXiv preprint arXiv:1808.07047
7. Dahlberg A, Wehner S (2018) SimulaQron—a simulator for developing quantum internet software. Quantum Sci Technol 4(1):015001. https://doi.org/10.1088/2058-9565/aad56e
8. Caleffi M, Cacciapuoti AS, Bianchi G (2018) Quantum internet: from communication to distributed computing! Proceedings of the 5th ACM international conference on nanoscale computing and communication, pp. 1–4
9. Behera BK, Seth S, Das A, Panigrahi PK (2019) Demonstration of entanglement purification and swapping protocol to design quantum repeater in IBM quantum computer. Quantum Inf Process 18(4):1–13
10. Ma L, Slattery O, Tang X (2020) Optical quantum memory and its applications in quantum communication systems. J Res Natl Inst Stan 125:125002
11. Shirichian M, Tofighi S (2018) Protocol for routing entanglement in the quantum ring netword. 2018 9th International symposium on telecommunications (IST), pp. 658–663. IEEE
12. Kozlowski W, Dahlberg A, Wehner S (2020) Designing a quantum network protocol.Proceedings of the 16th international conference on emerging networking experiments and technologies, pp. 1–16
13. Yu N, Lai CY, Zhou L (2021) Protocols for packet quantum network intercommunication. IEEE Transactions on Quantum Engineering 2:1–9
14. Shi S, Qian C (2020) Concurrent entanglement routing for quantum networks: Model and designs. Proceedings of the annual conference of the ACM special interest group on data communication on the applications, technologies, architectures, and protocols for computer communication, pp. 62–75
15. Dahlberg A, Skrzypczyk M, Coopmans T, Wubben L, Rozpędek F, Pompili M, Wehner S (2019) A link layer protocol for quantum networks. Proceedings of the ACM special interest group on data communication, pp. 159–173
16. Shi S, Qian C (2019) Modeling and designing routing protocols in quantum networks. arXiv preprint arXiv:1909.09329
17. Pant M et al (2019) Routing entanglement in the quantum internet. npj Quantum Information 5(1):1–9. https://doi.org/10.1038/s41534-019-0139-x

SQL Injection Detection Using Machine Learning with Different TF-IDF Feature Extraction Approaches

Mohammed A. Oudah, Mohd Fadzli Marhusin, and Anvar Narzullaev

Abstract The risk of attacks on web systems increased with the reliance of web systems in a wide range of businesses, and attackers invent new techniques to crack these systems. According to OWASP SQL injection stays one of the top 10 web applications security risks. This research use machine learning to detect SQL injection attacks, we used four machine learning models to detect SQL injection attacks. An insight into the data showing that data preparation and feature extraction have influenced the detection accuracy. The used training dataset is a combination of live requests extracted from user requests log file and a training dataset contains records of benign and malicious SQL queries. Then we compared the use of these models in term of detection quality and speed of training, results showed that Support Vector Model achieved highest detection accuracy with .997 accuracy followed by Extreme Gradient Boosting with .995 accuracy. In other hand Naïve Bayes using N-gram level feature extraction model was the fastest model it required 6 ms to train the classifier.

Keywords SQL injection · Cyber security · Machine learning · Web applications security

1 Introduction

Web security importance increased with reliance on web applications in different businesses especially commercial systems, banking systems, educational systems,

M. A. Oudah (✉) · M. F. Marhusin · A. Narzullaev
Faculty of Science and Technology, Universiti Sains Islam Malaysia, 71800 Nilai, Malaysia
e-mail: mohammedoudah@raudah.usim.edu.my

M. F. Marhusin
e-mail: fadzli@usim.edu.my

A. Narzullaev
e-mail: anvar@usim.edu.my

M. Al-Emran et al. (eds.), *International Conference on Information Systems and Intelligent Applications*, Lecture Notes in Networks and Systems 550,
https://doi.org/10.1007/978-3-031-16865-9_57

and other different systems. Different types of threats and attacks penetrates security standards of web applications and leads to malfunction of these applications. SQL injection is one of the most common cybersecurity concerns that infect web applications. According to Open Web Application Security Project OWASP, SQL Injection Attack (SQLIA) is the first of top 10 web application security risks for 2021 [1]. SQL Injection is inserting malicious SQL commands utilizing programming vulnerabilities in web systems to gain access to the database, then the attacker can retrieve, update, or delete data [2]. SQL injection attacks still the most popular data attacks and represent about two-third of all web attacks. In 2019, SQL injection attacks represented 77% of all application attacks of 3.1 billion alerts on Akamai's security platform [3].

Researchers proposed different techniques to detect and prevent SQL injection attacks on web applications. In this paper, we investigate the use of four different machine learning models in SQL injection detection enhanced with multi tokenization levels and compare their results in terms of accuracy and time, to determine which classifier model is more suitable for SQL injection detection.

2 SQL Injection

SQL Injection Attack (SQLIA) is a serious attack that targets web applications, which use database to store and retrieve information. SQLIA exceed the core principles of information security availability, confidentially and integrity, it can bypass systems authentications, or change data in database, even erase data by inserting malicious SQL code [4]. SQL injection vulnerability results when the developer gives an attacker unintentionally the ability to influence passes to back-end database. It can be any code that accepts input from user to form dynamic SQL statements [5]. SQL injection found since SQL databases were connected to web applications, so different web applications exposed to SQL injection attacks such as governmental services, healthcare systems, educational institutions etc. Beside OWASP classification of SQL injection, a provider of enterprise security solutions company called Positive Technologies published a report in 2017 about most common types of web application attacks in different industries [6]. The report stated that SQL injection is the top of 5 attacks that aimed health care web applications with 46% percentage, and 48.1% of top 5 attacks aimed energy manufacturing companies.

2.1 *SQL Injection Classifications*

SQLIA is classified to two main types, classical and advanced SQLIA, each type has various subtypes.

Classical SQL injection attacks

- Tautology attacks: In this technique, the attacker inject a code request to make the query result always true to get data. For Example, attacker add "or 1 = 1" after where clause for the input field.
- Union query: an attack done by injecting UNION to a select query and get data from other tables, ex. " Select * from Table 1 UNION Select * from Table 2".
- Blind SQL injection: In this type, attacker depends on asking true and false SQL queries to gain information about database.
- Piggy-backed query: In this, the attacker injects malicious code with traditional queries and performs data manipulation operation like INSERT, UPDATE and DELETE clause for manipulating a record.
- Illegal / logical incorrect queries: the attacker in this type, exploit information showed in error messages generated by database server.
- Alternate encoding: here, the attackers use alternative encoding for SQL injection attack strings, such as hexadecimal, ASCII and Unicode.
- Stored Procedure: In this type, attacker focuses on the stored procedures, which are executable functions present in the database system by calling these procedures in sent SQL queries.

Advanced SQL Injection Attacks
Advanced SQLIA are modern SQL injection attacks that overcome many detection and prevention techniques [7], such as:

- **Fast Flux**—this SQLIA is used to improve data extraction and phishing attacks. In the case of traditional phishing, host can be detected easily by tracking down the public Domain Name Server or the IP address, therefore, Fast Flux is used to hide phishing and malware distribution sites behind a changing network of compromised host [7].
- **Compounded SQL Injection Attack** (SQLIA + web application attacks) [8]—in this type of attack the SQLIA is integrated with other web attacks such as DDoS or XSS to drop down the server.

SQL Injection Detectors
Cyber security specialists described two main types of SQL injection detectors, the first type is internal detector which embedded in the web system and built for specific web system. In this type, different detectors should be built for diverse web systems,

Table 1 Confusion Matrix

	Actual		
		Positive (Malicious)	Negative (Benign)
Predicted	**Positive (Malicious)**	TP	FP
	Negative (Benign)	FN	TN

Table 2 Results using Word Level Vectors

	Accuracy	Precision	Recall	F1_score
Naive Bayes	0.984	0.981	0.982	0.981
Linear classifier	0.973	0.980	0.960	0.969
SVM model	0.987	0.990	0.980	0.985
Extreme gradient boosting	0.993	0.995	0.990	0.992

for example if we have three web systems, we must build three different detection tools for each one up to programming language used and functionality.

The second type is external detector or black box, in this type the detector is built separately from the web system regardless of web architecture and programming language so it can be used for diverse web systems independently, it is also called dynamic analysis because it works in runtime [9, 10].

3 Machine Learning

Machine Learning (ML) algorithms used for different purposes, now ML widely used in SQL injection attacks detection and prevention. Many research papers utilized ML to gather and process data to make decisions and predictions based on built rules without human intervention. ML is a technique that enables machines to make predictions and decisions about new data using an existing dataset [11]. Therefore, the advantage of using machine learning is the ability to improve detection accuracy and predicting future attacks with training models. ML generally classified as Supervised Learning algorithms and Unsupervised Learning algorithms. Other researchers [11] added two classifications are semi-supervised learning and reinforcement learning. Supervised learning is the simplest form, it uses a labelled (classified) training dataset to learn the relationship between the data and the label then use this data to judge and classify new data, which called test dataset, and this dataset determines the accuracy of supervised learning. In the Unsupervised learning, the dataset is unlabelled used to find something common hidden beyond this data and build rules base on these findings.

The Semi-Supervised learning use combination of classified (labelled) and unclassified (unlabelled) dataset, its goal is to train classification model with unlabelled data first, then train with labelled data. In Reinforcement learning the learning model, the algorithm utilizes the information gathered from interaction with the environment and acts based on this interaction and produces intelligent agents.

3.1 Dataset Preparation

An important factor to get high quality results when using ML classifiers is the quality and size of training dataset, so dataset preparing and text pre-processing is an important step to improve feature extraction results in next step.

In our paper we applied four steps to clean data as the following:

- Drop Null value records.
- Convert all text to lower case.
- Remove special characters such as ?,!, ‘,” … etc.
- Stemming, which is the process of removing prefixes and suffixes, to reduce the number of extracted features [12].

3.2 Feature Extraction

In text mining, we should do Natural Language Processing NLP to better understand and process text. There are different techniques for NLP, one of these techniques called Bag of Words (BoW) [13] which extract features from text content by converting text to numerical features. To do this we used three different levels of Term Frequency – Inverse Document Frequency (TF-IDF) feature extraction algorithm, TF-IDF score represents the importance of a term in the document by showing relevance degree of words in a document. TF-IDF Vectors can be generated at different levels of input tokens:

- **Word level**—This level gives a matrix representing occurrence of each word in the document.
- **Characters level**—It results a matrix of characters occurrence in the document.
- **N-gram level**—In this model, the tokens are grouped according to n consecutive words. It is used in NLP to predict the sequence of the next item of word in a text. For example, in the sentence “The weather is fine” when 2-g used it will be split as the following “The weather”, “weather is”, “is fine”. Therefore, the size of n is important because each n size will create different text sequence and text predicting results will differentiate [14].

4 Related Work

For years SQL injection attacks remain one of the most cyber security interesting issues, because new SQLIA techniques are invented and continually improved by attackers to compromise security of web applications. Different research used machine learning classifiers for SQL injection detection and prevention, in the following we will present some related research that use different ml classifiers with various data preparation techniques.

In [15], Pham et al. they used ML algorithms to detect SQL injection attacks on client side. The proposed model includes five steps, first is the data preparation, with removing noise and unnecessary data, the second is splitting data into training and testing data sets, then text parsing process using tokenization to split queries into token words because data is non-structured texts. After that, a Natural Language Processing technique is used to look up the term and find out how frequently the term searching and figuring out a frequency that measure the occurrence of words. They tested their model using five ML algorithms and evaluated results using Precession, Recall, F1-score, Weighted Average and Accuracy. The experiment results showed that three of five tested algorithms, which are Logistic Regression, Random Forest and Extreme Gradient Boosting, achieved 100% accuracy. Their experiment should be tested out with larger datasets to improve results reliability, also it did not mention how to deal with new types of attacks and what are injection types that can be detected.

Mishra [16] used Naïve Bayes ML model for SQL injection detection. The researcher mentioned that using Naïve Bayes ML for SQL injection is simple to implement, requires less computational resources, and can be trained even on small datasets. In the other hand, the Naïve Bayes sometimes fails when a new SQL injection type is used for the first time, so he used Ensample Learning to predict a value which help in reducing Bias Error and Variance Error. A dataset includes 6000 SQL injections is used for training the model, it consists of Plain Text Dataset, and SQL injection Dataset. Another SQL injection is created for testing using an open-source tool called Libinjection. For the text-based dataset, Mishra did a tokenization that includes removing certain characters and sequence of characters. First step in this approach is feature extraction to find the G-test scores for all token values, which is the sum of actual score of data occurrence, and number of tokens in a particular row. The second step is to find the Entropy that measures the randomness of the data. If the data is very similar to each other, entropy of such dataset will be low. In this research they used different feature extraction method than we used in our research and, and their experiments also showed that using Gradient Boosting classifier achieve better accuracy than Naïve Bayes classifier, but it needs more computational resources.

Azman et al. [17] used ML utilizing the user access log files for SQL injection detection. Their proposed system architecture consists of three phases, the first is extracting attribute values from log files by searching unusual keywords, then extracted log file is separated into training and testing datasets. After that, the classifier is trained and creates a Knowledge Base KB of benign and malicious queries, then the classifier uses the KB to detect injection. They used Boyer's Moore string matching algorithm to compare malicious features in log strings to detect injections. In experiments, they used Damn Vulnerability Web Application (DVWA) to collect training dataset and make their experiments. They divided testing sets to five small sets. The accuracy test shows that 93% accuracy for the first set and 100% the other four sets. The tested data sets considered small sets and better to test the model against larger datasets, also it should consider real time web requests in addition to reading user access log files.

Uwagbole et al. [18] proposed predictive analytics on big data to build a dataset with patterns and historical data items to train ML classifier, using supervised Support

Vector Machine model. They generated the dataset by extracting known attack patterns, including SQL tokens. For testing, they built a.net web system with a web proxy API that intercepts and checks user requests. Then the requests are moved to the trained SVM to classify requests, classification works by labelling SQL statement elements which subdivided to clauses (Update, Set, Where … etc.), predicates (e.g. UserNname = 'moh') and expressions (as in 'moh'). They have specified that the injection point is after the WHERE clause in SQL query and it is the location for SQLIA predictive analysis. The success of their proposed approach depends on the correctness of their suggestion of the injection point, so it can be effective with some types of SQL injection attacks but not others.

Cheon et al. [19] proposed to use ML pattern classifiers to detect injection attacks and protect web applications. Their proposed system consists of three modules are Monitor, Converter and Classifier. First, they convert the http/https request to numeric attributes then classify it based on Bayesian classifier. The Monitor captures the requests, decodes characters, and turns each letter into upper case. Then Converter converts web request parameters into numeric attributes and use length of parameters and number of keywords parameters as pattern attributes. After that Classifier determine each pattern's class with Bayesian classifier which calculate the probability of each class of an objects. If SQLIA detected, it will warn the user and write this pattern to injection log file then block and terminate the connection. For experiments training and testing, they developed a program to generate controllable legitimate patterns and injection patterns. Researchers used SQLmap to test the system in real environment and evaluate detection of various types of injection patterns. In some applications, some special symbols (single quotes) can be legal in name strings but are illegal in password strings, so system cannot forbid these symbols in all inputs, and it will be more complicated.

Abdulmalik [20] in his paper focuses on extraction semantic features that can indicates SQL injection Attack using ML techniques. It consists of three phases, dataset phase, static phase, static and dynamic analysis, and model construction phase. Researcher tested his approach using Random Forest (RF), Artificial Neural Network (ANN), Support Vector Machine (SVM), and Logistic Regression (LR). The researcher evaluated the results based on two factors; Detection Overhead which is the time required to detect SQL injection, and the Error Rate which depends on the of rate False Positives and False Negatives, but he didn't show the evaluation results in this paper.

As shown in related researches, while using ML in SQL injection detection the importance of preparation and feature extraction appears in the quality of detection results, so in our research, we will apply different data preparation and feature extraction levels to show their effect and performance in SQL injection detection.

5 Methodology

Our proposed SQL injection detection system consists of six main steps as shown in Fig. 1, first we extract and decode SQL queries from user access log file, then we make the data preparation process to get clean data. After that we apply three different levels of TF-IDF feature extraction technique are character level, word level and n-gram level to find the best suitable for SQL injection detection. Then the extracted features added to ML classifier dataset, this dataset is split into training and testing dataset 80% and 20% of the dataset. After training and testing step the classifier will be able to compare and classify new queries and specify if it is either normal or malicious request.

To build the classifier we did 7 steps: first step is dataset import, second is data cleaning, then split data into training and testing datasets, 80% for training and 20% for testing, we set the split random state to 10 to get same results every test. After that is the feature extraction step, the fifth step is the model building, in this step we built four different ML classifiers to compare their quality results with different text feature extraction levels in the evaluation step.

5.1 *Dataset*

For experiment, we used a dataset found on Kaggle.com [21] that contains 37,093 records of web requests collected from different websites, these records are labeled to benign and malicious queries.

5.2 *ML Classifier*

The ML classifier is a trained classifier to detect SQL injection requests using the prepared dataset, it classifies the incoming web request to benign or malicious and redirect request regarding results as shown in Fig. 1. Different ML algorithms can be used to detect SQL injection, the role of these algorithms to classify SQL queries in web requests into benign or malicious query. Main points in training the ML classifiers that it depends on classifier threshold value which set and tested during classifier training. Furthermore, the type and size of data used in training and testing affect the classifier detection quality. For choosing better ML classifier we tested our detection system with four classifiers are Naïve Bayes, Linear Classifier, Support Vector Machine, and Extreme Gradient Boosting Model and compare each result.

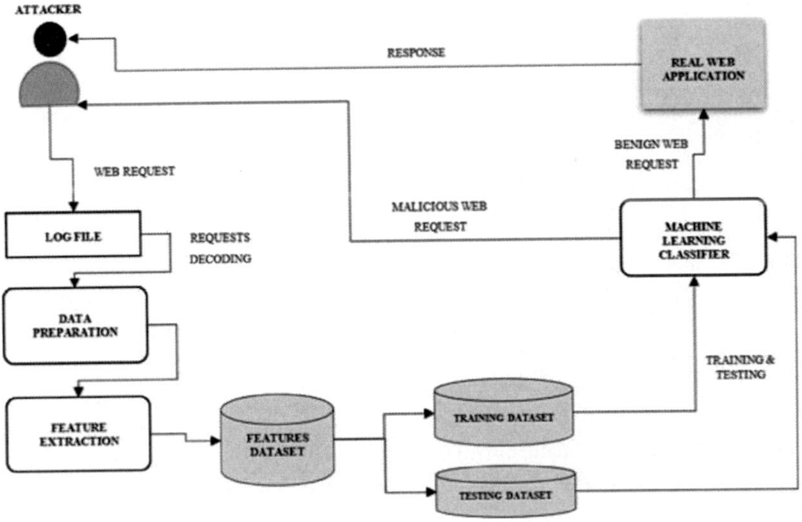

Fig. 1 Proposed system architecture

5.3 *Evaluation*

To evaluate ML model classifying performance, there are different measurement metrics can be used are: Accuracy, Precision, Recall and F1-score. Accuracy is ratio of sum of true positive values and true negatives to total sum of true negatives, false positives, true negatives, and false negatives as shown in Eq. 1.

$$Accuracy = \frac{TP + TN}{TP + FP + TN + FN} \quad (1)$$

Precision is ratio of true positives to sum of true positives and false positives, and *Recall* is the ratio of true positives to sum of true positives and false negatives, in our case Recall is how many of the malicious queries we were able to predict correctly with our model. For our evaluation process we will present each of these metric values then compare Accuracy and precision because it is enough to show model performance [2]. Different machine learning algorithms can be used to detect SQL injection, the role of these algorithms to classify SQL queries in web requests into benign or malicious query.

6 Experiments and Results

We implemented this classifier using Python 3 in Jupyter platform, because of its predefined Pandas and Scikit-learn ML libraries. The used laptop device processor is Core i3 2.13 Ghz and 8 GB RAM. Classification results are categorised to four types are TP, FP, TN, and FN. TP are the injection queries that were really predicted as injection, while TN are benign queries that the classifier really classifies as benign queries. As shown in Table 1, two types of errors may occur while classifying the queries, the FN values are the injection queries that the classifier predicts as benign queries, and the FP are the values which are predicted as benign but are actually being SQL injection.

To compare the quality of different model results we used *accuracy, precision,* and *recall.* report metrics. In addition, we implemented different feature extraction levels are *word level, character level,* and *N-gram level* in data preparation step. Within experiments we found that setting *n-gram* size range from 2 to 4 gets better results. Finally, we compared detection results when using each feature extraction model to determine which is suitable for SQLIA detection in our system.

The following tables shows measurements results of the four classification models with different extraction feature levels. Table 2 shows that Extreme Gradient Boosting (EGB) model achieve the highest results regarding to Recall and Precision using Word Level feature extraction method.

From Table 3, we see that Extreme Gradient Boosting model gets the highest Accuracy with 0.978 and Recall is 0.966 metrics using N-gram level feature extraction.

Table 4 shows that Support Vector Machine SVM achieve highest Recall and Precision results with character level vector features followed by EGB in the second level with a little difference.

From the above results we compared Recall of each model to decide which extraction feature extraction level is better to use. As shown in Table 5 that classification Recall results using Character level is better than using Word level and N-gram level in each classifier model.

Table 3 Results using N-Gram level vectors

	Accuracy	Precision	Recall	F1_score
Naive Bayes	0.969	0.978	0.952	0.964
Linear classifier	0.960	0.972	0.939	0.953
SVM model	0.963	0.974	0.943	0.956
Extreme gradient boosting	0.978	0.984	0.966	0.974

Table 4 Results using character level vectors

	Accuracy	Precision	Recall	F1_score
Naive Bayes	0.993	0.994	0.991	0.993
Linear classifier	0.992	0.994	0.988	0.991
SVM model	0.997	0.997	0.995	0.996
Extreme gradient boosting	0.995	0.996	0.992	0.994

Table 5 Recall comparison using different extraction feature levels for different classifiers

	Word level	N-Gram level	Character level
Naive Bayes	0.982	0.952	0.991
Linear classifier	0.960	0.939	0.988
SVM model	0.980	0.943	0.995
Extreme gradient boosting	0.990	0.966	0.992

Table 6 Accuracy comparison using different extraction feature levels for different classifiers

	Word level	N-Gram level	Character level
Naive Bayes	0.984	0.969	0.993
Linear classifier	0.973	0.960	0.992
SVM model	0.987	0.963	0.997
Extreme gradient boosting	0.993	0.978	0.995

Table 7 Different classifiers training performance time

	Word level	N-Gram level	Character level
Naive Bayes	0:00:00.07	0:00:00.06	0:00:00.08
Linear classifier	0:00:01.67	0:00:01.61	0:00:01.58
SVM model	0:02:57.98	0:02:28.33	0:06:07.56
Extreme gradient boosting	0:00:07.66	0:00:06.35	0:00:38.21

In Table 6 we compare the accuracy of the four ML classifiers when using the different levels of feature extraction and it shows that using the character level achieved the best accuracy results in all the models compared to N-gram and word level.

Table 7 shows the required time to train and test each model with the used dataset. It shows that Naïve Bayes required 6 ms to train the classifier using N-gram level and 7 ms with word level followed by 8 ms in character level. In the second order is the Linear classifier with 1 s and 58 ms using the character level, 1 s 61 ms with N-gram level, and 1 s 67 ms using the word level. in the third speed level is the EGB model, and the last level is the SVM model which requires 6 min 7 s and 56 ms in character level, 2 min 28 s and 33 ms in N-gram level and 2 min 57 s and 98 ms in Word level.

7 Discussion and Challenges

The results showed that the accuracy is best while using TF-IDF character level feature extraction using the SVM model with 0.997 accuracy followed by Extreme

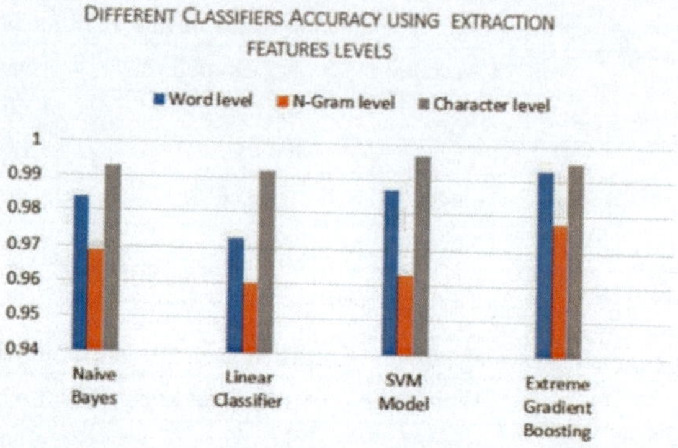

Fig. 2. Tested ML classifiers accuracy using different feature extraction levels

Gradient Boosting with 0.995, in the third level is the Naive Bayes model with 0.993 accuracy and in last level comes the Linear classifier model with 0.992 accuracy. This means that using character level is better than N-gram level and World level feature extraction.

As shown in Table 7, the Naïve Bayes is the fastest model, it required 6 ms to train classifier with N-gram level and 7 ms with word level followed by 8 ms in character level. Experiment results in Table 6 showed that SVM model achieved best accuracy results using character level TF-IDF feature extraction compared to other models but regrading to required performance time it consumes the longest time to train the model. The results also showed that Extreme Gradient Boosting model gets the highest classification quality with N-gram and word TF-IDF feature extraction levels compared to other tested models, Fig. 2 shows a comparison chart of accuracy results for the tested models with different feature extraction levels.

One of the challenges we faced during the research, is how to use NLP in text preprocessing for SQL injection detection and identifying what are the useless text in the data that should be removed, because in SQLIA removing such as stop words and clauses like 'where' clause might result inaccurate results, also removing some characters like '?' may also change results.

8 Conclusion

In our research, we presented the use of different NLP techniques to extract features of text to prepare data for SQL injection detection. In general sense, results based on our experiment showed that using Character level is better than Word level and N-gram level in terms of accuracy and performance time. In addition, we found that Extreme Gradient Boosting ML classifier achieved the highest accuracy results followed by Naïve Bayes in the second level and Linear Classifier in the third level. Therefore, using EGB classifier is preferred for SQL injection detection using TF-IDF Character level feature extraction.

There are some future works for this research, we look forward to test our approach against modern SQL injection attacks and compare other ML classifiers such as Neural Networks. In addition, we will use more advanced data preparing techniques and feature engineering techniques, because better training data yields a better detection result.

References

1. OWASP, "OWASP Top 10 web application security," OWASP foundation (2021). https://owasp.org/www-project-top-ten/. Accessed 15 Feb 2021
2. Jemal I, Cheikhrouhou O, Hamam H, Mahfoudhi A (2020) SQL injection attack detection and prevention techniques using machine learning. Int J Appl Eng Res 15(6):569–580 (**ISSN 0973-4562**)
3. Fakhreddine A (2019) State of the internet. Akamai Technologies, Inc, Cambridge
4. Binu S, Ashish K (2018) Proposed method for SQL injection detection and its prevention. Int J Eng Technol 7:213–216
5. Clarke J (2012) SQL Injection Attacks and Defense, vol 2. Elsevier, Waltham
6. Positive technologies, "Web Application Attack Statistics: Q2 2017," Positive Technologies, 14 Sep 2017. https://www.ptsecurity.com/ww-en/analytics/web-application-attack-statistics-q2-2017/. Accessed 19 23 2020
7. Puneet SJ (2016) Analysis of SQL injection detection techniques. ArXiv preprint arXiv:1605.02796
8. Alwan ZS, Younis MF (2017) Detection and prevention of SQL injection attack : a survey. Int J Comput Sci Mob Comput 6(8):5–17
9. Ramasamy P, Abburu DS (2012) SQL injection attack detection and prevention. Int J Eng Sci Technol (IJEST) 4:1396–1401
10. Shegokar AM, Manjaramkar AK (2014) A survey on SQL injection attack, detection and prevention techniques. Int J Comput Sci Inf Technol (IJCSIT) 5(2):2553–2555
11. Mohammed MMZE, Khan MB, Mohammed Bashier EB (2017) Machine learning: algorithms and applications. Taylor & Francis Group, LLC, NewYork
12. Kadhim AI, Cheah Y-N, Hieder IA, Ali RA (2017) Improving TF-IDF with singular value decomposition (SVD) for feature extraction on twitter. 3rd International engineering conference on developments in civil & computer engineering
13. Kumawat D (2019) 7 Natural Language Processing Techniques for Extracting Information, AnalyticsSteps, 18 November 2019. https://www.analyticssteps.com/blogs/7-natural-language-processing-techniques-extracting-information. Accessed 21 Sep 2021
14. Marhusin F, Lokan CJ (2018) A preemptive behaviour-based malware detection through analysis of API calls sequence inspired by human immune system. Int J Eng Technol 7(4):113–119
15. Pham BA, Subburaj VH (2020) An experimental setup for detecting SQLi attacks using machine learning algorithms. J Colloquium Info Syst Secur Educ 8(1):1–5
16. Mishra S (2019) SQL injection detection using machine learning, master's projects. SJSU ScholarWorks
17. Azman MA, Marhusin MF, Sulaiman R (2021) Machine learning-based technique to detect SQL injection attack. J Comput Sci 17:296–303
18. Uwagbole S, Buchanan WJ, Fan L (2017) Applied machine learning predictive analytics to SQL injection attack detection and prevention. 3rd IEEE/IFIP workshop on security for emerging distributed network technologies (DISSECT), Lisbon, Portugal

19. Cheon EH, Huang Z, Sik Lee Y (2013) Preventing SQL injection attack based on machine learning. Int J Adv Comput Technol (IJACT) 5(9):967–974
20. Abdulmalik Y (2021) An improved SQLInjection attack detection model using machine learning techniques. Int J Innov Comput 11(1):53–57
21. Shah SSH (2020) Kaggle.com 03 Mar 2020. https://www.kaggle.com/syedsaqlainhussain/sql-injection-dataset?select=SQLiV3.csv. Accessed 10 May 2021

Analysis of Data Mining Algorithms for Predicting Rainfall, Crop and Pesticide Types on Agricultural Datasets

Mustafa Omer Mustafa, Nahla Mohammed Elzein, and Zeinab M. SedAhmed

Abstract Data mining is a classification technique that can be used to handle large volumes of data. Hence, data mining has evolved as an excellent solution for large agricultural datasets. This is partly because it can predict categorical class labels, classify data based on training set and class labels, and it can also evaluate new data. In agricultural production, farmers and agribusiness representatives need to make daily decisions. However, accurate yield estimate of the various crops related to the planning is a critical issue for agricultural plannings. Data mining technique is therefore required for achieving realistic and effective outcomes. The aim of this study is to classify different data features and implement various algorithms as it relates to agricultural big data. Additionally, a given dataset is preprocessed to ensure that relevant data is present in all datasets. Algorithms such as Rule JRIP, Tree LMT, and Naive Bayes are implemented. Then, the Mean Absolute Error (MAE) and Relative Absolute Error (RAE) were compared, and the performance error of the resulting classification algorithm is performed on each dataset. The overall results indicates that JRIP has the highest efficiency with a value of 96%. This is followed by Naive Bayes which has 84% efficiency, whereas tree LMT has 78% efficiency. The result of this study can help to advance current research, as well as benefit future research in the agricultural sector.

Keywords Data mining · Classification · Rule JRIP · Tree LMT · Naive Bayes

1 Introduction

Data mining is a method used to extract, analyze, and generate viable information from a large amount of data. The data processed through data mining might have been generated from many different data sources. Therefore, there is sometimes need for preprocessing which has to do with identifying inaccurate or missing data [1]. The algorithms used for classification in data mining are called classifiers [2],

M. O. Mustafa (✉) · N. M. Elzein · Z. M. SedAhmed
Faculty of Computer Science, The Future University, Khartoum, Sudan
e-mail: alfaki9397@gmail.com

M. Al-Emran et al. (eds.), *International Conference on Information Systems and Intelligent Applications*, Lecture Notes in Networks and Systems 550,
https://doi.org/10.1007/978-3-031-16865-9_58

and data mining techniques are basically employed to improve performance. Data mining can help to eliminate any noise or unwanted data that might interfere with the testing process by converting or removing the data into a format that can be easily understood [3]. In particular, the predictive data mining method can be used to evaluate the accuracy of classification rules. It also helps to accurately predict the behavior of intra-group units [2]. Here in Fig. 1 is shown Data mining models.

Classification is one of the data mining techniques mainly used to evaluate a specific dataset and assign each occurrence to a specific class with the lowest classification error. It is used to extract models from datasets that correctly identify important data classes. Classification is a two-stage process. In the first stage, the model is constructed by applying a classification algorithm to the training dataset. Then in the second stage, the extracted model is evaluated against a preset test dataset. This is to determine the training performance, and accuracy of the model as shown in Fig. 2. Hence, classification has to do with assigning class labels to datasets with unknown class labels[4].

Classification algorithms are used to improve forecast accuracy. This is particularly useful for identifying risks, trends, and performance [5]. Over the years, the

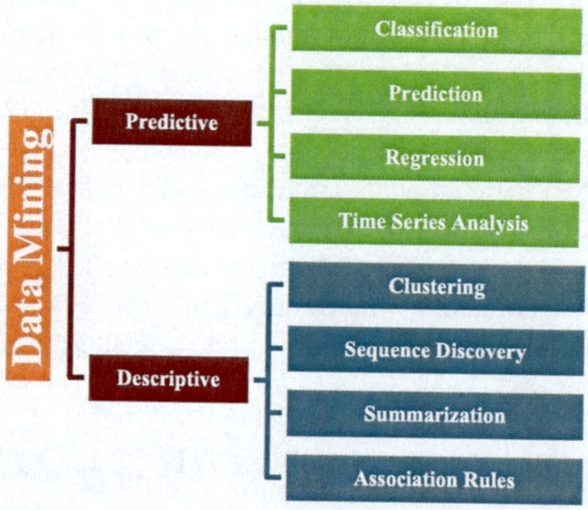

Fig. 1 Data mining models

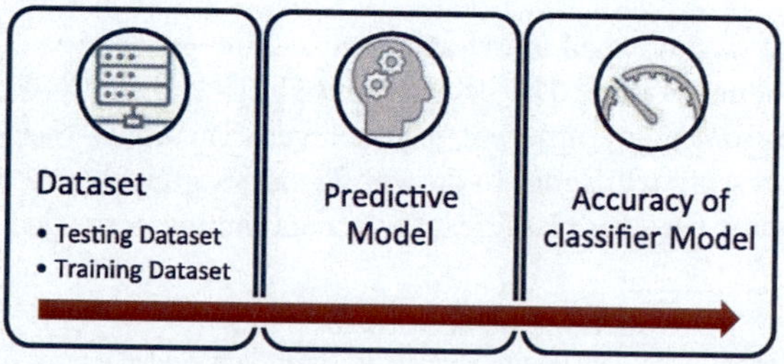

Fig. 2 Classification process

volume of recorded data has grown, and people are often bogged down in retrieving performance metrics and trends. This makes the implementation of data mining highly expedient as in predicting outcomes and evaluating the result. This in turn, facilitates crop yield, and helps to optimize factors related to it, using algorithms that can further be visualized using graphs. The use of graphs would make results easy to understand [6].

The aim of this paper is to conduct an experimental analysis of various data mining algorithms, to select the most efficient classifier for each dataset, and to propose the highest efficient classifier on all datasets.

2 Literature Review

Data mining in crop prediction is becoming a trend among scientific agricultural research. The use of datasets to predict yield helps to optimize crop production. On the other hand, classification is a set of supervised data mining learning techniques used by researchers in agricultural data mining. This approach helps to provide critical knowledge about different studies, its context, and the contribution of various researchers to the agricultural sector.

Different data mining algorithms were presented by [7], such as JRip, J48, and Naive Bayes for forecasting soil types. These classifier algorithms were used to extract knowledge from soil data, and two soil types were considered: red soil and black soil. On their dataset, the JRip classification method outperforms the others with 98% accuracy. Hence, it was concluded that JRip is a good tool for predicting soil types. In another study, [8] investigated the potential of six key classifiers for predicting classification accuracy. They used JRip, SVM, ANN, NB, J48, and KNN. Their result showed that the classifier accuracy ranged from 97% for JRip to 95% for J48 and 91% for ANN, with all having superior classification accuracy above 90%.

In another study, Rules-based Decision Table, PART, JRip, Tree-based J48, Random Forest, RandomTree, LMT, REP Tree, Bayesian-based Naive Bayes, Lazy-based IBK, and KStar were investigated [9]. Based on their findings, it was discovered that the KStar model, which has a 93 % accuracy rate, is the most suitable for prediction. Therefore, based on these studies, it can be inferred that the use of a single dataset to compare classifiers help to evaluate how the classifiers perform, and to recommend the optimal classifier. This can then be used for future agricultural systems, which might use multiple datasets. A comparison has been made in previous studies in terms of the parameters achieved, the techniques used, and the final results as shown in Table 1.

Table 1 Comparative between techniques and parameters and final outcome for previous study

Ref	Used Techniques	Achieved Parameters	Final outcome
Rajeswari and Arunesh [7]	JRip, J48, and Naive Bayes are used to forecast soil type	Two types of soil are examined when using classifier algorithms to extract knowledge from soil data: red soil and black soil	On this dataset, the JRip classification algorithm outperforms the others with 98% accuracy. JRip is a good tool for predicting soil types
Sai and Sathiaseelan [8]	JRip, SVM, ANN, NB, J48, and KNN are six main classifiers used to predict classification accuracy	Soil records are compiled from soil surveys undertaken in several districts of Tamil Nadu, India, in agricultural regions	Classifiers with greater classification accuracy above 90% are JRip (97%), J48 (95%), and ANN (91%)
Vegad, Parmar [9]	Rules-based Decision Table, PART, and JRip, Tree-based J48, Random Forest, Random Tree, LMT, and REPTree, Bayesian-based Naive Bayes, and Lazy-based IBK and KStar are all classification models	Scholars pursuing post-graduate studies in agricultural extension at SAUs are listed in this database	Classifiers with the highest prediction accuracy include JRip (84%), LMT (84%), J48 (90%), and KStar (93%)
Baskar, Arockiam [10]	Different classification methods, such as Nave Bayes, J48 (C4.5), and JRip, were evaluated	Soil datasets are created from soil surveys conducted in agricultural areas throughout Tamil Nadu, India, and are based on properties such as EC and pH	J48 is a fairly simple classifier for creating a decision tree, yet it produced the best results in the study, scoring 93%. JRip, on the other hand, produced findings that were 90% accurate
Gholap, Ingole [11]	A comparison of several classification algorithms, such as Nave Bayes, J48 (C4.5), and JRip	Surveys are conducted on a regular basis in the Pune District. Field sampling is used to collect primary data for the soil survey	With 91% accuracy, the J48 classifier is the best. JRip came close with 90% accuracy

3 Methodology

This research will employ a predictive data mining model, which forecasts data values based on previous results from various datasets. The primary goal of the predictive data mining model is to foresee the future based on prior data. Data Mining predictive model includes classification, prediction, time series analysis and regression.

In this research, data is collected from the FAOSTAT database for crop production [12], pesticide types [13] and rainfall [14]. This is an international organization concerned with food security. Hence, across the world, the set of data found at FAOSTAT could differ depending on different factors. These factors include region, country, years and so on. The data collected contains various columns in CSV format ranging from crop types, crop production, precipitation amount and pesticide types. The following section provides a quick overview of the Naive Bayes, Tree LMT, and JRip algorithms:

JRip (RIPPER): One of the most fundamental and widely used algorithms is RIPPER (Repeated Incremental Pruning to Produce Error Reduction). In this algorithm, classes are evaluated in increasing sizes, and an initial set of criteria for the class is constructed using incrementally reduced error. JRip carries on by taking all the examples of a certain decision in the training data as a class and determining a set of recommendations that apply to all the individuals in that class. Following that, it moves on to the next class and repeats the process until all classes have been covered. It is divided into four stages: developing a rule, pruning, optimization, and selection [7].

Tree LMT: Logistic model tree (LMT) is a supervised training algorithm that combines logistic regression and diction tree learning. It produces higher results compared to other algorithms such as C4.5. LMT produces a single tree with binary splits on numeric attributes, multiway breaks on nominal attributes, and logistic regression models at the leaves. In addition, missing values, binary and multi-class variables, numeric and nominal properties are all supported. It generates small, precise trees using the CART pruning method, but it does not require the use of any tuning options [15].

Naive Bayes: Based on the Bayes theorem, the Naive Bayes algorithm has interdependence characteristics that can differentiate between the missing or present feature in a particular class and other classes. It generally reliable results by dealing with complex real-world data. The Naive Bayesian classifier is quick and incremental, it can handle both discrete and continuous attributes, it works well in real-world problems, and can justify its conclusions in terms of informational benefits [16].

3.1 Project Resource Requirement

The study used a WEKA tool environment (version 3.8.6) of knowledge famous for data mining with its user-friendly GUI including massive learning libraries and publications across the internet. The PC requirements include an Intel Core i7 4600 M CPU, 6 GB DDR3 RAM, and an HGST 500 GB 7200RPM 32 MB Cache SATA 6 GB/s hard drive.

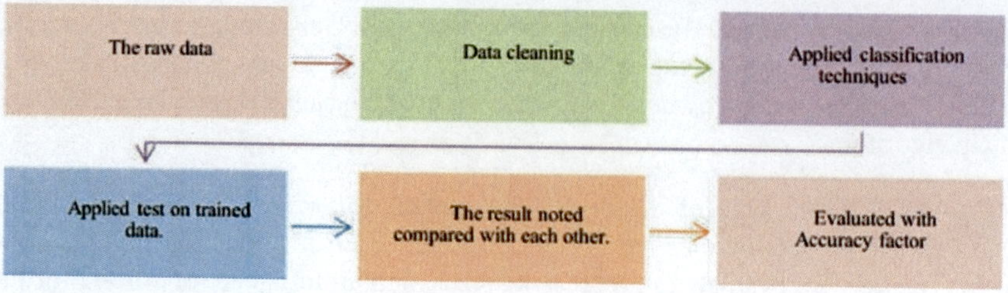

Fig. 3 The process of model prediction system

3.2 Methods

Figure 3 shows the entire process of our model prediction system. The raw data used were cleaned, the attributes were selected and then sorted. The classification techniques such as JRip, LMT, and Naive Bayes were then implemented over the trained data. The result of each algorithm was noted from WEKA and compared with each other. Root Mean Squared Error (RMSE), Mean Absolute Error (MAE) and Relative Absolute Error (RAE) values were taken into consideration for each case. Thereafter performance was measured using Accuracy factors used for "year" which contains the year in which each event happened. The other factors used includes "item" which describes the types of pesticides and crops, "range" which indicates the range of rainfall, and "value" which represents the number of pesticides and crop types per Tons [17].

4 Experiment and Analysis

To start the experiment, preprocessing was applied to all the three datasets. Initially, the "rainfall" dataset contained five columns. Then, an additional column named "range" was added which represented the class variable. Likewise, the "crop production" and "pesticide types" datasets were added. Columns such as "flag" and "flag description" were removed to avoid noise, which may affect the efficiency of the classifiers. In contrast, column "item" was shifted ahead of the "value" column to represent the class variable.

4.1 Data Preprocessing and Attribute Selection

The collected datasets are as follows: the "rainfall" dataset contained 288 instances and six attributes, the "crop production" dataset has 1223 instances and twelve attributes, while the "pesticide types" has 247 instances and twelve attributes. The datasets were modified from "CSV" format to "ARFF" format, and the steps used to explore the dataset includes: 1) Load data, 2) Save data into ARFF format, 3) Preprocess data, and 4) Implement classifiers.

The datasets were first loaded into the WEKA tool in "CSV" format, then saved as "ARFF" format to make it compatible with the official WEKA tool format. Preprocessing was then implemented, which includes removing and shifting columns up or down, as well as the implementation of classifiers to obtain specific measurements about the datasets. Following data import, preprocessing methods were used to provide output in terms of discretization, re-sampling, normalization, attribute selection, and so on.

5 Experiment Result

The testing options on the classifiers were put on cross-validation with ten folds. Three classifiers were implemented which are JRip, Naive Bayes and tree LMT. The results of the classifiers are shown in tables and figures. Figure 4. shows the rainfall dataset, which shows that LMT produces a high (97.9%) accuracy, but with a larger MAE and RAE value than JRip. In contrast, JRip although coming in second at 97.2%, has the presents the greatest error reduction on a pruning set. The Naive Bayes gave 95.1% to rank last. As presented in Fig. 5. for crop production dataset, JRip gave 97.8% thereby ranking first, much ahead of both Naive Bayes with 73.9% and LMT with 70.5%. Similarly, as illustrated in Fig. 6 for the pesticide type dataset, it is evident that JRip gave 94.3%, Naive Bayes produces 82.9%, and LMT has 68% accuracy.

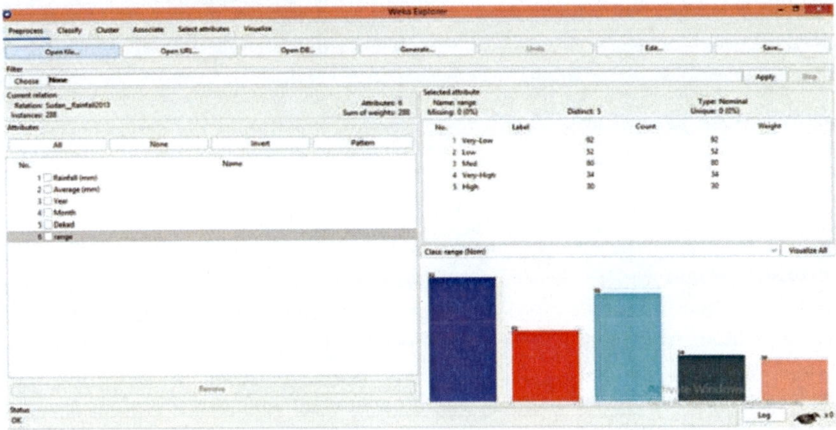

Fig. 4 WEKA explorer window rainfall

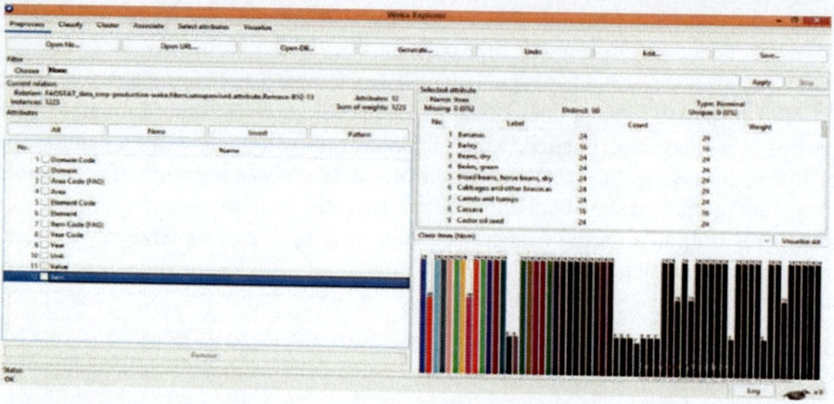

Fig. 5 WEKA explorer window crop production

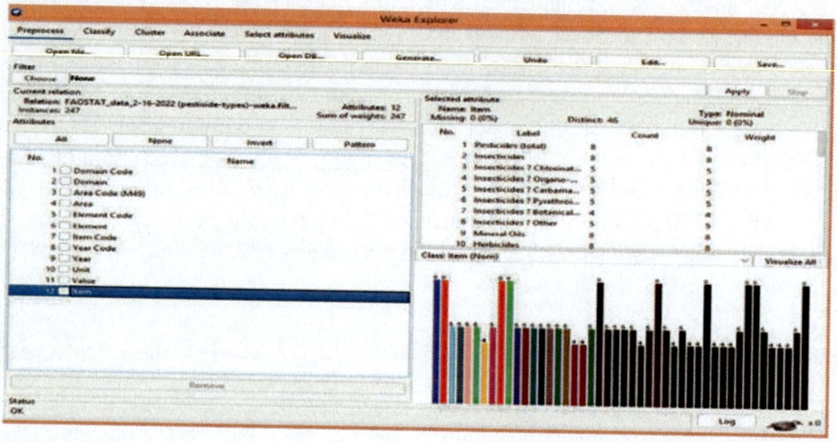

Fig. 6 WEKA explorer window pesticide types

5.1 Performance Error of Classification Algorithm

Results of the investigated classifiers are shown in Tables 2 and 3 for "rainfall", "crop production", and "pesticide types" datasets. The overall analysis of the classifiers is also presented. As seen in Table 2, tree JRip has a better accuracy than LMT and Naive Bayes classification algorithms. The WEKA tool's results show that the tree LMT classifier has greater accuracy and a lower error rate on the rainfall dataset. These results are entirely dependent on the dataset.

Table 2 Efficiency analysis of each data classification model

Efficiency Analysis	Different Model Algorithms								
	Rules Based			Tree Based			Bayesian Based		
	JRip			LMT			Naïve Bayes		
Dataset	RF	CP	PT	RF	CP	PT	RF	CP	PT
No of selected Attributes	288	1223	247	288	1223	247	288	1223	247
Kappa Statistic	0.963	0.978	0.941	0.972	0.700	0.672	0.936	0.735	0.825
Correctly Classified Instances	97.222	97.874	94.332	97.916	70.564	68.016	95.138	73.998	82.996
Incorrectly Classified Instances	2.777	2.125	5.668	2.083	29.435	31.983	4.861	26.001	17.004
TP Rate	0.972	0.979	0.943	0.979	0.706	0.680	0.951	0.740	0.830
FP Rate	0.010	0.000	0.002	0.004	0.006	0.007	0.009	0.005	0.004
Precision	0.973	0.980	0.945	0.979	0.699	0.692	0.952	0.737	0.841
Recall	0.972	0.979	0.943	0.979	0.706	0.680	0.951	0.740	0.830

Rainfall (RF), Crop Production (CP), Pesticide Type (PT), True positive (TP), False positive (FP)

Table 3 Classification algorithms performance error for datasets

Performance Error	Classifier	JRIP	Naive Bayes	LMT
Rainfall Dataset	MAE	0.0124	0.0282	0.0234
	RRSE(%)	26.771	31.383	23.113
	RAE(%)	4.065	9.227	7.642
	Accuracy(%)	97.222	95.138	97.916
Crop Production Dataset	MAE	0.0008	0.009	0.0141
	RRSE(%)	19.578	58.224	62.361
	RAE(%)	2.344	27.438	43.096
	Accuracy(%)	97.874	73.998	70.564
Pesticide Type Dataset	MAE	0.0024	0.0073	0.0161
	RRSE(%)	31.287	51.680	65.609
	RAE(%)	5.648	17.097	39.017
	Accuracy(%)	94.332	82.996	68.016
Overall	Accuracy(%)	96.476	84.044	78.832

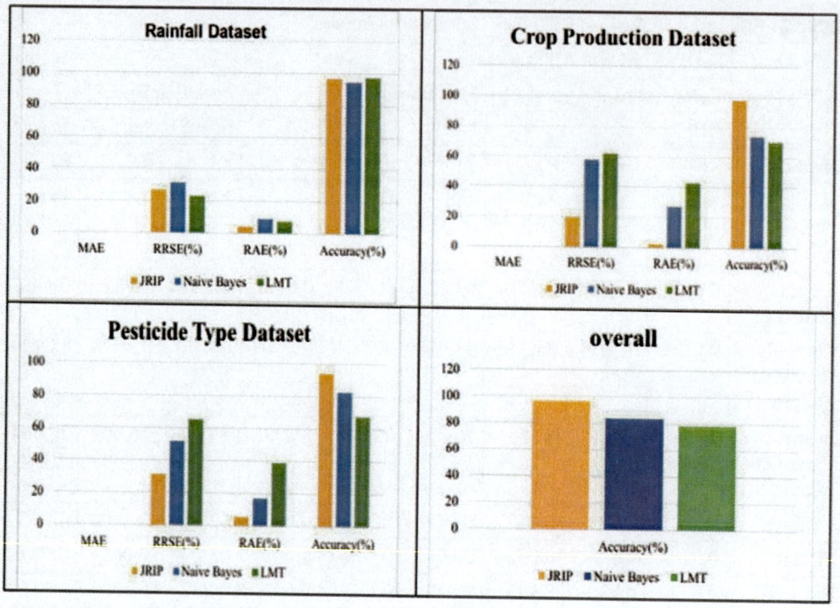

Fig. 7 Accuracy graphs of different algorithms for all datasets

6 Discussion

The classifiers presented in this study have been used in previous studies for agricultural applications, but with different datasets. In [7, 8], the JRip classifier achieved high accuracy results. Similarly, the JRip achieved excellent results with 90% accuracy but fell short of J48 with 93% in the study conducted by [10]. In like manner, J48 produced nearly the best results, with 90% in [9] and 91% in [11] whereas JRip scored 84% and 90% in these studies, respectively. As presented in Table 3, JRip achieved the highest overall dataset accuracy of 96%, as illustrated in Fig. 7. Therefore, based on this study and the results reported in previous studies, it can be inferred that for agricultural dataset prediction, the JRip classifier is highly efficient and thereby highly recommended.

7 Conclusion

Different datasets related to agriculture, were analyzed in this research and a predictive data mining model was implemented. The classifiers used for analysis includes Tree LMT, Rule JRip and Naive Bayes. Overall, JRIP came in top with a 96% efficiency, Naive Bayes came in second with an 84% efficiency, and tree LMT came in third with a 78% efficiency. Agriculture has always been affected by low to high crop yield. Hence, a good understanding of which algorithm is optimal for prediction is highly valuable for the agriculture community. Therefore, this study confirms the possibility of analyzing agricultural data using data mining, and JRip is considered as a more suitable classifier, based on the observation from this study.

8 Future Research

Based on the observation and result from this study, it is recommended that subsequent studies may include the implementation of classifiers in wider datasets related to the agricultural sector in FAOSTAT or other relevant data sources. Furthermore, the model results given in this experiment can be implemented based on real-time data from agricultural software systems for predictions and recommendations to make agricultural data-driven decisions.

Acknowledgements This research work has been funded under the Faculty of Computer Science, The Future University, Khartoum, Sudan.

References

1. Aher SB, Lobo L (2011) Data mining in educational system using WEKA. International conference on emerging technology trends (ICETT)
2. Gorade SM, Deo A, Purohit P (2017) A study of some data mining classification techniques. Int Res J Eng Technol 4(4):3112–3115
3. Alasadi SA, Bhaya WS (2017) Review of data preprocessing techniques in data mining. J Eng Appl Sci 12(16):4102–4107
4. Nikam SS (2015) A comparative study of classification techniques in data mining algorithms. Oriental J Comput Sci Technol 8(1):13–19
5. Arcinas MM et al (2021) Role of data mining in education for improving students performance for social change. Turkish J Physiotherapy Rehabil 32(3):204–226
6. Raorane A, Kulkarni R (2013) Role of data mining in Agriculture. Int J Comput Sci Info Technol 4(2):270–272
7. Rajeswari V, Arunesh K (2016) Analysing soil data using data mining classification techniques. Indian J Sci Technol 9(19):1–4
8. Sai R, Sathiaseelan J (2018) Comparison of classifiers to predict classification accuracy for soil fertility. Int J Adv Stud Sci Res 3(9):75–79
9. Vegad N, Parmar R, Chauhan N (2020) E-extension employability of scholars pursuing post graduation in agricultural extension in SAUs: using data mining techniques. Guj J Ext Edu 31(1):11–17
10. Baskar S, Arockiam L, Charles S (2013) Applying data mining techniques on soil fertility prediction. Int J Comput Appl Technol Res 2(6):660–662
11. Gholap J et al (2012) Soil data analysis using classification techniques and soil attribute prediction. ArXiv preprint arXiv:1206.1557
12. FAOSTAT, datasets for crop production (2022). https://www.fao.org/faostat/en/#data/QCL
13. FAOSTAT, dataset for pesticide types (2022). https://www.fao.org/faostat/en/#data/RP

14. FAOSTAT, dataset for rainfall (2022). https://dataviz.vam.wfp.org/seasonal_explorer/rainfall_vegetation/visualizations
15. Fayaz SA, Zaman M, Butt MA (2021) An application of logistic model tree (LMT) algorithm to ameliorate Prediction accuracy of meteorological data. Int J Adv Technol Eng Explor 8(84):1424
16. Bhargavi P, Jyothi S (2009) Applying naive Bayes data mining technique for classification of agricultural land soils. Int J Comput Sci Netw Secur 9(8):117–122
17. FAOSTAT, Food and agriculture organization of the united nation (2022)

Survey on Enabling Network Slicing Based on SDN/NFV

Suadad S. Mahdi and Alharith A. Abdullah

Abstract Network slicing has surfaced as one of the most promising technologies for enabling a variety of services in order to satisfy the demands of fifth-generation networks in new-generation networks. Although there are many surveys on network slicing, there is no comprehensive study of all aspects including slicing implementation scenarios and network slicing security. In this paper, we give an overview of network slicing based on software-defined networks and network function virtualization, and also a comprehensive analysis of the fundamental concepts of slicing architecture. On the other hand, we highlight the different scenarios for implementing the network slicing concept, after which we discuss the security concept of network slicing, threats and weaknesses in the network slice architecture, and possible solutions for this. Finally, we review the latest research findings in the field of achieving basic security objectives for network slicing. This paper paves the way for academics to work on attaining network slicing security in a variety of slicing scenarios.

Keywords Network slicing · Software-defined networking · Network virtualization · Network function virtualization · Slicing scenarios · Slice security

1 Introduction

Over the past decades and day by day, the need for various services via the Internet is increasing. Where the Internet has become the main interface to provide this huge amount of services through countless devices and communications [1].

On the other hand, these various services require a balance between their requirements, as some services require high reliability, while there are services that require high productivity or low latency, depending on the type of application [2]. However, there is no network architecture capable of satisfying this many service simultaneously and the requirements for services in the future cannot be predicted. Therefore, the need led to the introduction of a new concept, which is network slicing (NS) [3].

S. S. Mahdi · A. A. Abdullah (✉)
College of Information Technology, University of Babylon, Babil, Iraq
e-mail: alharith@itnet.uobabylon.edu.iq

M. Al-Emran et al. (eds.), *International Conference on Information Systems and Intelligent Applications*, Lecture Notes in Networks and Systems 550,
https://doi.org/10.1007/978-3-031-16865-9_59

A network slice is an isolated network that provides one or more network services to fulfill the needs of users [4]. Each slice is provided with network resources according to the needs of the application, and usually the resources are allocated by reorganizing the distribution of network resources independently according to the requirements of the application scenario.

The advantages of two core technologies, virtualization and software-defined networks, inspired the creation of NS (SDN). Whereas virtualization abstracts the entire network's resources to meet the flexibility of offering diverse resources for heterogeneous services, software-defined networks separate the control plane and the data plane to provide for greater flexibility and efficiency in managing virtual networks [2, 5].

1.1 Existing Network Slicing Surveys

After the concept of network slicing was introduced, it was utilized to address a variety of challenges in 5G networks in order to meet the various service requirements for heterogeneous networks [6–10]. On the other hand, there are those who are interested in implementing network slicing in different environments [11–16].

The researchers presented in [17] a brief article on SDN and NFV architectures to achieve network slicing, in addition, they reviewed open research issues in order to stimulate this direction.

As for the researchers in [18], they presented a survey of the technology in the field of network slicing in the fifth-generation networks. On the other hand, they discussed the proposals, evaluated the proposed work, and then identified the open research challenges.

In [2] the researchers present a comprehensive study on network slicing, its history, and basic concepts, in addition to the various uses and purposes of network slicing and presenting the challenges before slicing. It also presented the general challenges of implementing the concept of network slicing and suggested some solutions that limit the impact of these issues. While Kaloxylos, A. from the University of Peloponnese [19] presented a survey on how to manage network slicing in most fields of networks, as well as identify issues that need to be addressed in the future. In the paper [20], a study was reviewed on the importance of new generation networks and network slicing, followed by the areas and projects of using SDN and NFV in network slicing in 5G networks, in addition to discussing methods of managing network slicing to suit multiple domains.

The paper [1] discussed enabling different Internet of Things (IoT) applications through network slicing, followed by presenting the basic requirements for implementing network slicing, and finally, the researchers presented the challenges associated with network slicing and possible solutions. While the researchers in the paper [11] focused in particular on analyzing network slicing on the Internet of Things, where they discussed the technical challenges that can be solved through the implementation of network slicing, and finally reviewed the integration of network slicing

Table 1 Summary of existing surveys on network slicing based on SDN/NFV

Covered range	[17]-2017	[18]-2017	[2]-2018	[19]-2018	[20]-2019	[1]-2020	[11]-2021	Our paper-[2022]
Network slicing concepts	✔	✔	✔	✔	✔	✔	✔	✔
Background on key concepts for network slicing	✔	✖	✔	✖	✔	✖	✖	✔
Network slicing architecture	°	✔	✖	°	✔	°	✔	✔
Network slicing implementation scenarios	✖	✖	✖	✖	✖	✖	✖	✔
Virtualization hypervisors	✖	✖	✔	✖	✔	✖	✖	✔
Network slicing security	✖	✖	°	✖	°	°	°	✔

* *

✔: indicates that the attributes are presented in the paper.
✖: indicates that the attributes are not presented in the paper.
°: indicates that the attributes are not detailed in the paper.

with modern technologies, blockchain, artificial intelligence, and machine learning in IoT networks. Table 1 shows a summary of survey papers related to the use of SDN and NFV technology in network slicing.

1.2 *Paper Contribution*

In this paper, we present a comprehensive study of the basic techniques that enable the concept of network slicing in a way that makes it easier for the reader to understand the general concept of network slicing and its architecture as well as the life cycle of slice. We also explain possible scenarios for implementing network slicing and provide the reader with the latest implementations of each scenario. We summarize the contributions of this paper in the following points:

- We describe the basic concepts that enable network slicing.
- We offer a survey for network hypervisors.
- We explain in detail the concept and architecture of network slicing.
- We review different scenarios for implementing network slicing.
- We show the importance of each scenario and its fields of application and explore the field of research in each scenario.
- We highlight the security aspects of network slicing and the research field in this direction.

1.3 Paper Organization

This study divided into six sections. The basic techniques associated with network slicing are discussed in Sect. 2. Section 3 delves into the notion of network slicing and its architecture, while Sect. 4 covers network slicing implementation scenarios. The security aspects of network slicing are discussed in Sect. 5. Finally, in Sect. 6, we wrap up the paper with a conclusion.

2 Background on Basic Techniques to Enable Network Slicing

In this section we highlight the techniques necessary to realize the concept of network slicing.

2.1 Software-Defined Network

Software Defined Networking (SDN) is a new and rapidly evolving networking technology that decouples the control plane from the data plane, allowing for more network control flexibility based on specific policies and security enforcements [21].

Previously, network devices had a control and data plane implemented in static hardware appliances in traditional networks. SDN, on the other hand, isolates the control logic from the network devices and places it in a separate entity known as the SDN controller or Network Operating System (NOS) [22] as shown in Fig. 1.

As a result, the SDN architecture is made up of three layers and three application programming interfaces (APIs) [23]. Figure 2 depicts the SDN architecture.

The forwarding devices (network devices) in the infrastructure layer (Data Plane) are responsible for forwarding traffic flows according to control plane choices, while the controller software in the control plane is responsible for creating and controlling traffic flow forwarding methods [21].

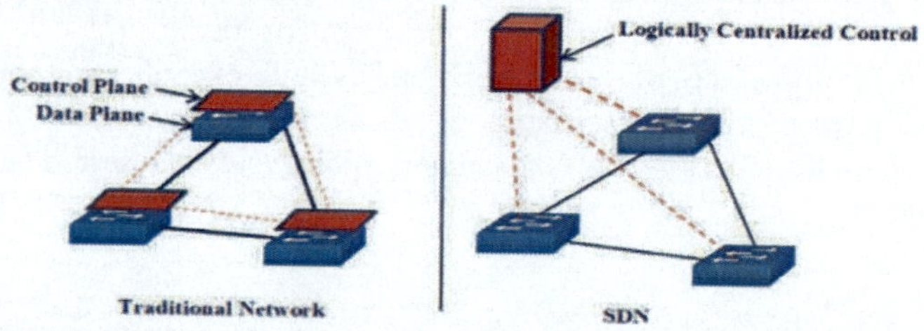

Fig. 1 Traditional network vs SDN

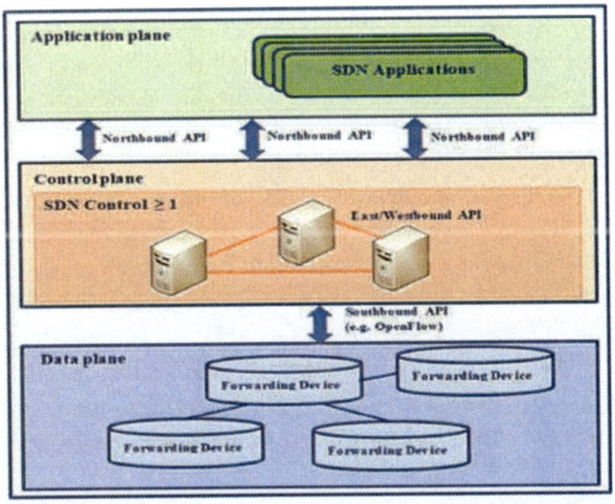

Fig. 2 SDN architecture

The control plane is the network's brain, and it usually consists of one or more controllers. It is responsible for managing and controlling the network because it has a whole view of the network, also this layer is permitted to make any required changes on the forwarding elements like update, install and delete the forwarding rules responsible for making decisions and rules on which the network is based in guidance and policies. Also, it collects live information about the status of the network, traffic and performance in order to change paths in the case of link saturation or projection and make the network as effective as possible and so on [21]. At this time there are many controllers that have been written in different programming languages. The most popular controllers used in many surveys have been shown [24, 25].

The last layer is an application plane that includes all the network applications such as a QoS, Routing, and Security applications. The APIs can connect these three layers and there are two different types of APIs used. The first one is the open southbound API which is the communication interface that is responsible for managing communication between the control and infrastructure layers in order to allow the controller to configure, manage, and send the flow forwarding decisions to the forwarding devices. OpenFlow [26] is a new SDN protocol (southbound API) for establishing data plane communication between the controller and network forwarding devices like Open vSwitch. On the other hand, the SDN applications that are found in application layer can communicate with the controller via Northbound API, also the controller provides relevant information about network elements to the SDN applications. While East-Westbound API is used to communicate between the controllers in distributed SDN controller environment [23].

There are many articles that have covered scanning on the general principles of SDN [27–30]. On the other hand, there were many articles concerned with a comprehensive survey of the security challenges of SDN networks [31–33], while these challenges were addressed in many articles [31, 34]. While there are researchers focused on implementing SDN architecture with IOT, sensor networks, and the cloud [35–37]

OpenFlow Protocol

One of the first SDN standards, OpenFlow [26], defines the communication between the controller and network forwarding devices in the data plane, such as switches and routers. OpenFlow [38] was recently created to handle both hardware and software switches. The initial component of an OpenFlow-enabled switch is a FlowTable, which is setup and programmed using the OpenFlow protocol. The Secure Channel is the second component, and it's in charge of establishing a secure link between the switches and the remote controller so that orders may be exchanged between the forwarding devices and the controller to add or remove flow entries from the FlowTable using the OpenFlow protocol.

A set of messages is sent from the controller to the switch and vice versa in this protocol. The messages allow to control in the switching of user traffic by allowing the controller to program the switch in a way that makes it define, modify, and delete flows.

In general, three categories of messages can be classified: symmetric, async and controller-switch [22]. Symmetric messages are sent from the switch to the controller or from the controller to the switch while the async messages are sent from the switch to the controller. In the controller-switch category, messages are sent from the controller to switch and some messages may be replied by the switch.

The OpenFlow protocol has progressed from version 1.0, which had only 12 match fields and one flow table, to version 1.5, which has several tables, over 41 matching fields, and a number of new functions [26].

Distributed SDN Controller

The idea of forming multiple controllers is an ideal solution to solve the problem of a single point of failure when the controller fails, whether the failure is in communication or with the controller itself [39]. Also, to solve the scalability problem to handle multiple forwarding path requests from switches. There are various strategies for SDN controllers as shown in Fig. 3.

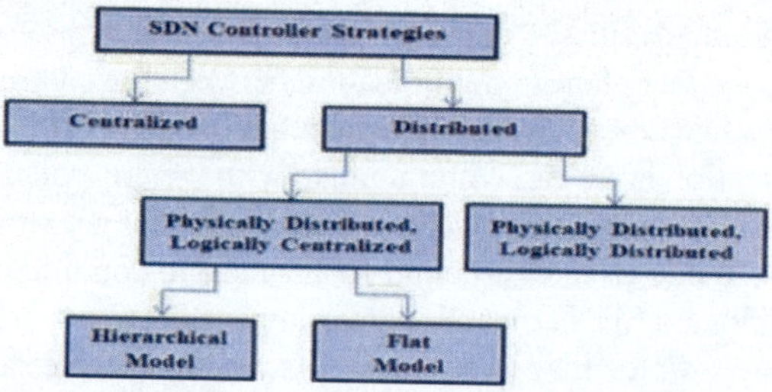

Fig. 3 SDN controller strategies

In order to overcome the problems of centralization of control, the researchers proposed to apply the control to clusters. Each cluster represents a domain and is a network block with its own controller [39]. The fundamental goal of this architecture, as illustrated in Fig. 4, is to reduce communication requests from switches on the SDN controller by shrinking the network size for one domain, which is referred to as physically and logically distributed SDN controllers.

In a logically centrally distributed controller architecture, multiple controllers' function as a central but physically separate controller, acting as if they are connected to each other (communication between the controllers are through the East-Westbound APIs) and all have the same control over the data but far from each other [40].

As shown in Fig. 5, SDN distributed control architectures are often split into flat SDN controller architecture and hierarchical SDN controller architecture based on the physical arrangement of SDN controllers.

In a hierarchical architecture one or more (not all) controllers have the state of the global network while in a flat architecture all the SDN controllers have the overview of the entire network state.

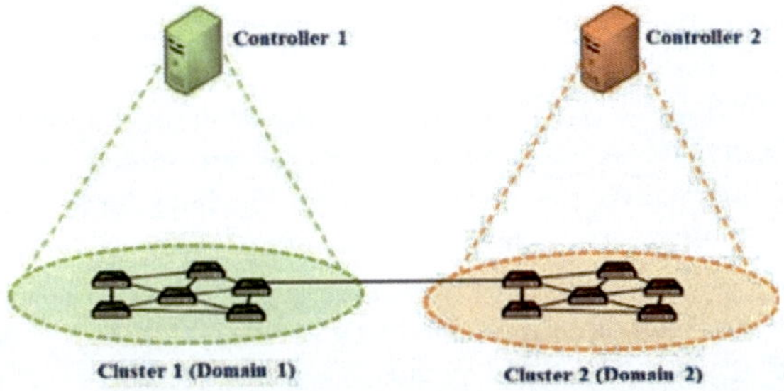

Fig. 4 Physically and logically distributed SDN controllers

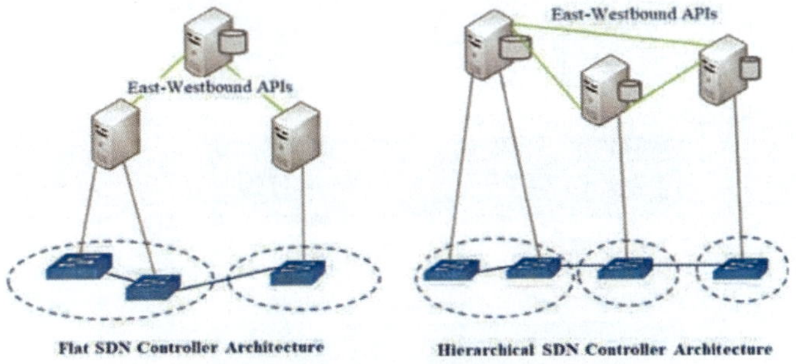

Fig. 5 Flat and hierarchical SDN controller architecture

2.2 Virtualization and Network Virtualization

Virtualization is one of the technologies that have made a big difference in information technology in the last decade [41]. Essentially, virtualization provides a layer of abstraction for physical resources, including storage devices, computing, networking, etc., and thus the application layer runs on top of it.

In general, virtualization is defined as the process of separating software from the basic hardware to create virtual instances of physical resources, and this concept is prevalent in the fields of virtualization in computing for servers, networking devices, and services [41].

Virtualization has become an important research project in communication networks as a result of its success in the field of computing. One of the most important aspects of implementing network virtualization is dividing the network infrastructure into multiple virtual networks referred to as slices [2]. Where the basic work of virtualization is focused on summarizing the underlying physical network and then creating separate virtual networks (slices) through specific abstraction and isolation functional blocks that are will discuss in detail in next section.

The Fig. 6 shows the architecture for virtualization, which consists of the infrastructure layer, the virtualization layer, and finally the virtual infrastructure that creates and opens virtual networks (VNs) over it.

In general, the architecture of network virtualization (NV) consists of a network infrastructure layer that consists of nodes and links where the node represents network devices such as routers, switches, and servers while the links are wire lines or wireless connections [42]. While the virtualization layer is through which the process of abstraction of the physical network infrastructure, and this layer is considered as the backbone of a virtual infrastructure consisting of virtual resources. Whereas virtual networks (VNs) are created and run on virtual network resources, a VN is defined as

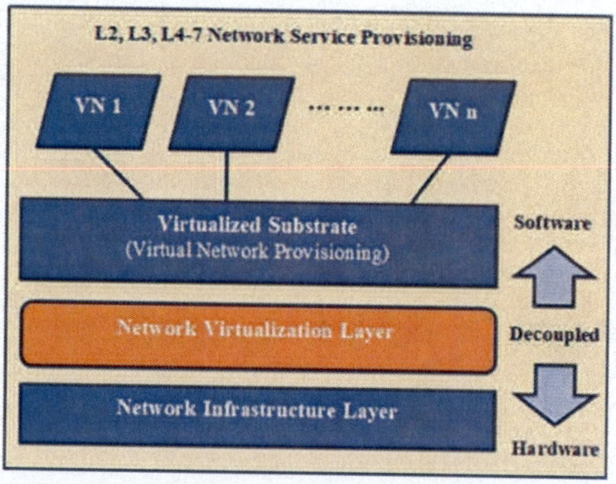

Fig. 6 Network virtualization architecture

a set of virtual nodes that are interconnected through virtual links to form a virtual topology.

Network Function Virtualization

Network function virtualization (NFV) technology has received great attention due to significant reductions in operating expenses (OPEX) and capital expenditures (CAPEX) as well as facilitating rapid deployment of new services [43].

Network Function Virtualization is defined as the separation of network functions (NFs) from the physical devices on which they operate. In other words, implementing many network functions (in multiple virtual machines VMs) such as firewall, load balancer, and intrusion detection in one server instead of implementing each function on a separate hardware device, thus reducing the number of devices required to perform the tasks as shown in Fig. 7.

Figure 8 shows the three major components of the NFV architecture: (1) Network Function Virtualization Infrastructure (NFVI), (2) Virtualized Network Functions (VNFs), and (3) Management and Network Orchestration (MANO) [44].

NFVI includes physical resources, computing devices, storage, processing, etc., while virtual resources are abstractions of the resources themselves. This abstraction is done through a virtual layer (hypervisor).

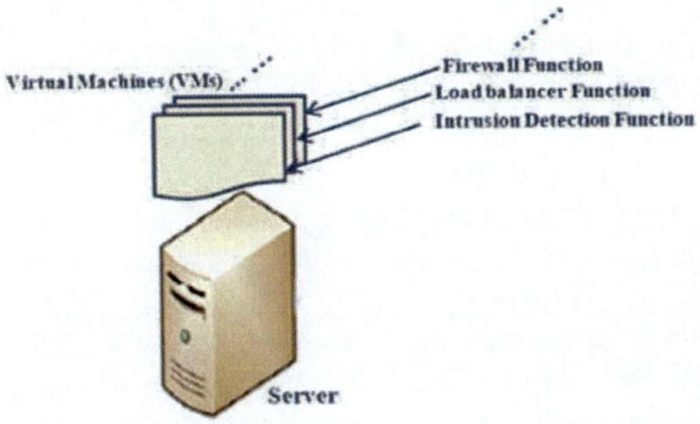

Fig. 7 Implement multiple network functions in one server in NFV

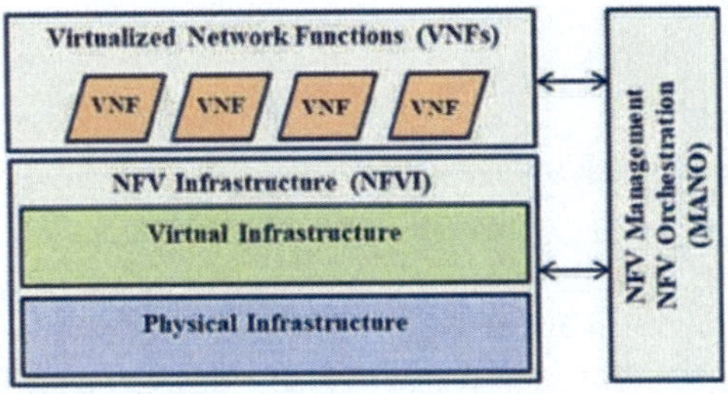

Fig. 8 NFV architecture

The VNFs represent software applications for network functions such as firewall, key management, load balance, and others. Whereas Management and Network Orchestration (MANO) is the main controller responsible for managing and regulating the VNFs and NFVI.

Network Hypervisors

In this section, we review the most important network elements responsible for the abstraction of physical infrastructure, such as communication links, network elements, and services. The hypervisor in a physical SDN provides APIs that substantially simplify the effort of designing complicated network services by abstracting the physical layer into logically isolated virtual network slices.

In the context of the concept of network monitoring software (hypervisor) [45], there are many software that work on network slicing such as OpenVirteX [46], FlowVisor [47], CoVisor [48] OpenSlice [49], MobileVisor [50], RadioVisor [51] and HyperFlex [52].

One of the oldest hypervisors for slicing a fixed and wired SDN network is FlowVisor. It provides the abstraction of physical switch ports and because it acts as a transparent proxy, it cannot abstract the intermediate keys. Unlike OpenVirtex which is an alternative to FlowVisor, it can provide topology abstractions Complete, which makes tenants the ability to freely implement the control function across the virtual network topology to meet special needs, including network topology and network IP addresses. However, two virtual switches for the same tenant cannot be represented by a single physical key.

According to [53] a performance comparison was made between FlowVisor and OpenVirtex, and it was found that FlowVisor is the best in terms of delay, Jitter, Throughput, and computing resources. On the other hand, OpenVirtex has network failover capability, full network virtualization, flexible network topology, and easy configuration.

In contrast with most hypervisors, CoVisor [48] enables multiple controllers to optimize SDN network performance and collaborate in managing the same data plane traffic using topology abstractions. While there is optical network monitoring software such as Open Slice and Optical FlowVisor and others for mobile networks such as MobileVisor. For more details on SDN hypervisors, we recommend the paper [45].

2.3 *SDN-Based Network Virtualization*

Because it works to integrate the physical resources of the network between the various services provided, network virtualization is one of the most important topics in the field of networks. However, the solidity of the network architecture has led to significant challenges in controlling each virtual network [54]. This difficulty was overcome with the introduction of the SDN architecture, which separated the control and forwarding operations of network devices, allowing each virtual network to be controlled independently.

The network hypervisor plays the primary role in the virtualization architecture of the network based on SDN, as it leads to the formation of virtual networks consisting of switches and random links from the basic physical network, as shown in Fig. 9.

The infrastructure layer of the virtual SDN architecture, which includes network devices such as OpenFlow switches and end devices, is responsible for data transmission and forwarding. The control layer, which is the brain of the SDN network and consists of one or more controllers, manages the infrastructure layer.

The hypervisor, on the other hand, sits between the infrastructure and control layers and is responsible for virtualizing networks as well as allocating resources to each virtual network (usually each virtual network is called a network slice).

Finally, there's the application layer, which consists of a collection of network management apps and delivers services. On top of a single SDN console, each application runs in its own network slice, which only shows the network view that corresponds to a single virtual network. Northbound APIs are used to communicate between the application layer and the control layer, while southbound APIs are used to interact between the control layer and the infrastructure layer.

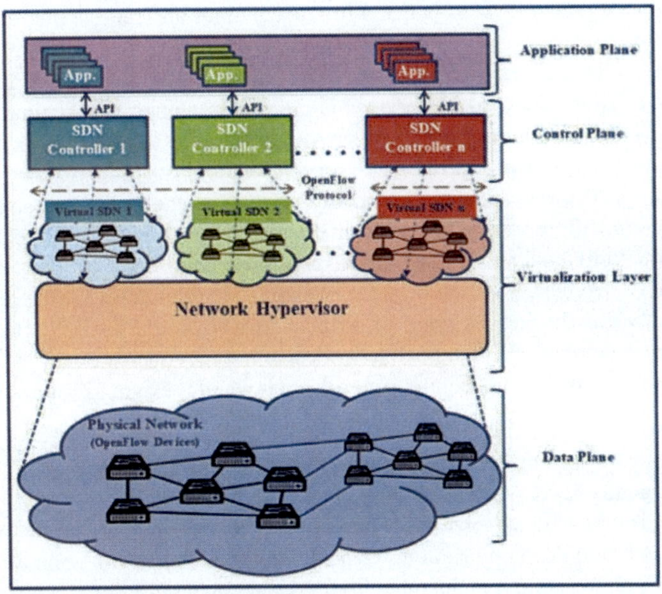

Fig. 9 SDN-based network virtualization

3 Network Slicing Concept and Architecture

The network slicing came up to address the problem of growing network services [55]. Previously, the prevailing concept in networks was "one size fits all", but this concept does not apply to the fifth-generation networks and beyond, the reason is due to different network requirements of heterogeneous applications.

Therefore, in this section, the concept of network slicing and its architecture will be explained in detail.

3.1 Network Slicing Concept and History

The notion of network slicing was first introduced in the past with the concept of network overlay, which groups diverse network resources to form virtual networks from the same core resources [56]. Virtual Local Area Networks (VLANs) [57] arose from this notion, however it lacks the benefit of programmability.

Today, with network virtualization technology and software-defined networks, and the ability to abstract resources, the concept of network slicing is ready to create programmable network slices isolated from each other and release them to the real world.

Slicing a physical network into several logical networks (slices) and distributing resources to each slice and its services is the notion of network slicing [2]. As a result, the network operator is able to provide optimized solutions for a variety of market scenarios requiring services with varying levels of functionality, performance, and isolation. As the slices are conceptually segregated, the resources can be shared between them, different network slices can be created from the same physical network to fulfill the individual networking demands of different users.

Figure 10 depicts the notion of network slicing, which allows for the creation of logical networks for various types of services as well as the flexibility to scale up and down on demand. Each logical network will have its own strength.

There are many benefits behind network slicing and they are as follows:

- Each slice is allotted a configurable number of resources and may be reserved to handle different traffic classes with varied security concerns, allowing infrastructure-level service differentiation [55], as shown in Fig. 11.
- Slicing is managed by software components that allow for the formation, reconfiguration, and decommissioning of network slices in real time and on demand in order to respond to changing traffic demand and/or meet Service Level Agreements (SLAs).
- Underutilized network slices can be leased to virtual network operators, maximizing resource usage and generating new revenue streams for infrastructure providers.

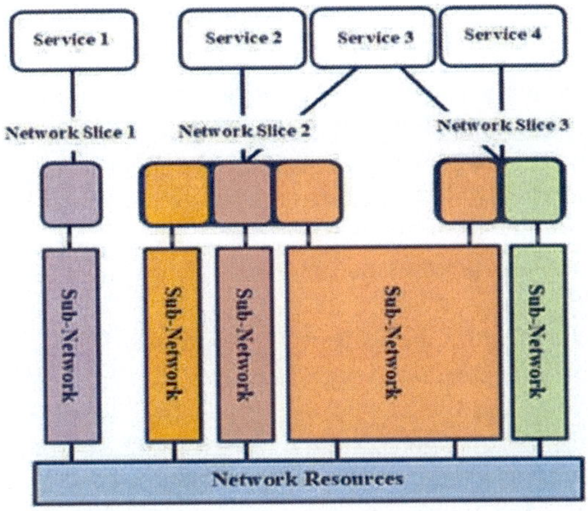

Fig. 10 Network slicing concept

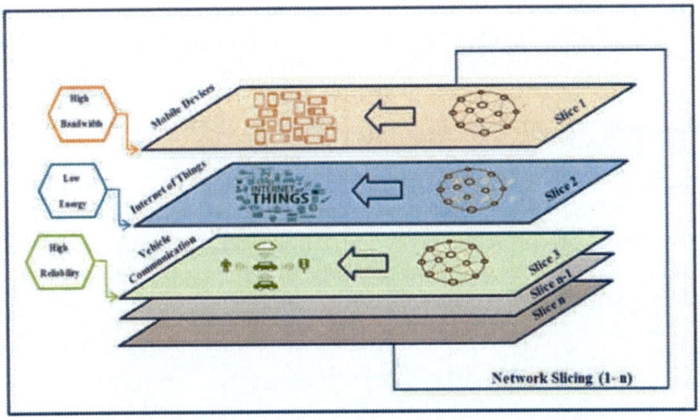

Fig. 11 Example of network slicing

3.2 *Network Slicing Requirements*

Network slicing is based on seven basic requirements that represent the concept of network slices [58], which are as follows:

- Automation: The network is divided dynamically based on the request to create a slice from the tenant, considering the start and end time of the slice, the duration, as well as its life cycle.
- Isolation: The network is divided in such a way that each slice is isolated from the other to ensure the performance and security of each tenant, whereby a slice is prevented from excessive use and error of resources, thus avoiding damage

to the performance and stability of other slices. The concept of isolation can be implemented in different ways (1) by using different physical resources, (2) or by implementing virtualization technology on network resources, and (3) by sharing a resource with specific policies that define access rights for each tenant.

- Customization: This feature ensures that the resources needed by the tenant are provided in order to fulfill the requirements of the services.
- Programmability: It is the basic key for network slicing to enable the control of the resources of each slice programmatically through open application programmable interfaces (APIs).
- End-to-End: This feature makes provision of service all the way from service providers to end user/customer(s) in an easy way.
- Elasticity: This feature is related to another feature, which is the allocation of resources to each slice, where flexibility allows achieving the required service level agreement without causing a significant impact on the services of this slice or other slices.
- Simplification: Simplification refers to lowering the complexity of operating network slices by simplifying the architecture. Flexibility needs, on the other hand, add complexity to network slice design and operation, therefore the problem is to strike the correct balance between flexibility and cost-cutting simplification.

3.3 *Network Slicing Architecture*

In general, the network slicing architecture can be summarized into four layers, which are the virtualized infrastructure layer, the network slice instance layer, the service instance layer, and network management and orchestration layer as shown in Fig. 12 with the following main components [59]:

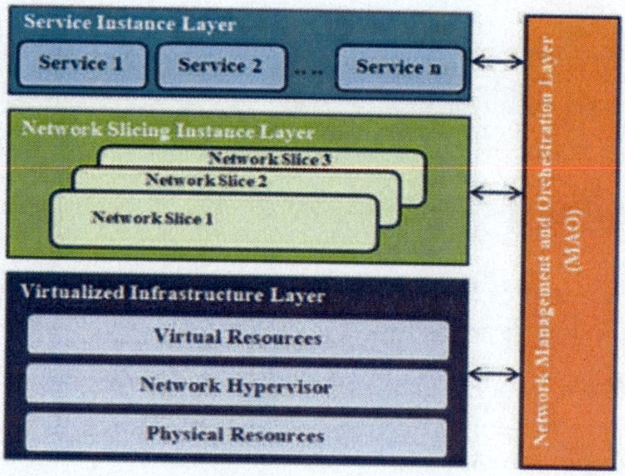

Fig. 12 Network slicing architecture

i. Virtualized Infrastructure Layer: this layer provides virtual instances of network resources to map to one or more slices. Virtualization plays the main role in this layer, as a network hypervisor, as it is managed through the virtual infrastructure manager.
ii. Network Slice Instance Layer: this layer represents the logical slices that run on top of a virtualized infrastructure layer. Infrastructure resources are organized on slices in proportion to the requirements of each network slice and are managed through the management and orchestration layer.
iii. Service Instance Layer: represents the different services that run on all other layers and that these services are provided by the network operator or by a third party.
iv. Layer of Network Management and Orchestration: It is the most important component of network administration, and it is made up of the following sub-modules:

- Virtualized Infrastructure Manager (VIM) To support the virtualization of infrastructure resources, each VIM comprises one or more network hypervisor software.
- NFV Orchestrator and Manager (Network Function Virtualization Orchestrator (NFVO), and Network Function Virtualization Manager (NFVM)).
- For managing network slices, the Software-Defined Networking Orchestrator (SDNO) can contain one or more SDN controllers.

3.4 Network Slicing Life Cycle

As indicated in Fig. 13, each network slice has a life cycle that may be separated into four stages: preparation, commissioning, operation, and decommissioning [60].

– Preparation stage: There is no network slice at this stage, but through this stage, the requirements of the slice are evaluated and any other requirements are prepared, and then the necessary network environment is created.

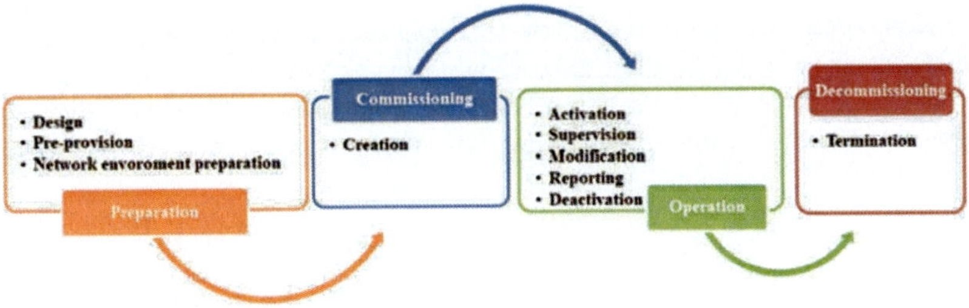

Fig. 13 Network slice life cycle

- Commissioning stage: At the end of this phase, the network chip will be ready to run, during which the required network resources are allocated.
- Operation stage: This stage includes many sub-tasks related to the network slice, which are:
  - Activation: The network slice is activated through some operations such as transferring traffic to the slice.
  - Supervision: Supervision of the network slice will be on an ongoing basis.
  - Monitoring: The performance indicators of the network slice will be constantly monitored.
  - Modification: will be reconfiguration, and changes in the topology of the network slice to suit the necessary requirements.
  - Deactivation: Here the network slice is taken out of active service.
- Decommissioning stage: After this phase, the network slice no longer exists, because during this phase, the resources and settings assigned to the network slice will be released.

3.5 *Network Slicing Types*

Network slicing is classified according to the use case into two categories, namely, vertical network slicing and horizontal network slicing [12].

In vertical network slicing, all nodes within a particular network slice perform similar functions, whereby infrastructure resources are shared between various services and applications to improve quality of service (QoS). In the other word, vertical network slicing separates traffic depending on each service or application.

Whereas in horizontal network slicing the infrastructure resources are divided into horizontal layers where the device can operate on more than one slice. On the other hand, horizontal network slicing separates computing resources thus providing capacity scaling. Traffic typically travels end-to-end with horizontal network slice locally between the access network and the end device.

Another classification of network slicing is static network slicing and dynamic network slicing.

The slice sets are pre-created in the static network slicing and only the devices need to specify which slice it will have a connection to. Whereas in dynamic network slicing, operators define slice design and dynamically allocate and optimize resources to suit service requirements or slice conditions [61].

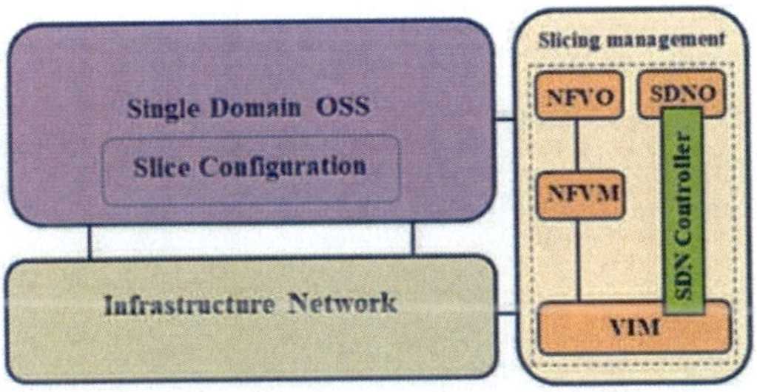

Fig. 14 Architecture of network slicing with single controller/orchestrator

4 Network Slicing Implementation Scenario

In this section, the scenarios for implementing network slicing will be explained. Each scenario will address the positive aspects and disadvantages of this scenario, which will be explained in detail.

4.1 Single Owner, Single Controller

A single controller is utilized to manage network slices in this scenario, with each controller focusing on arranging resources from a specific sort of network infrastructure resource domain. The northbound API interface is used to implement management and coordination tasks on top of the SDN control.

The SDN controller serves as the network slicing orchestrator and manager in this situation (VIM and SDNO), Fig. 14 shows the architecture of the network slicing with a single controller/orchestrator. This scenario is for a limited range of the network because there is one controller that controls all the different network slices and this leads to problems in terms of performance in addition to representing a single point of failure and affecting the reliability and availability of network tasks.

4.2 Single Owner, Multiple Controller

This scenario supports an SDN proxy through which the network infrastructure is divided into many virtual networks, and usually, the infrastructure owner is the one who controls the SDN proxy.

Multiple virtual tenants can use this scenario to deploy their own SDN controllers on the infrastructure to manage network slices and preserve isolation between them. The Fig. 15 shows architecture of SDN proxy in the slicing environment.

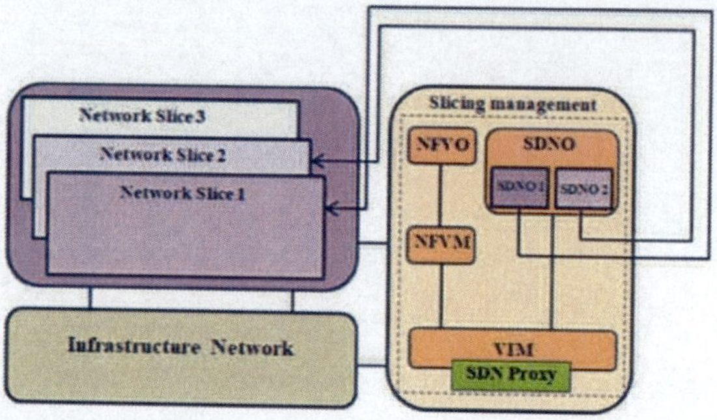

Fig. 15 Architecture of network slicing with multiple controller/orchestrator

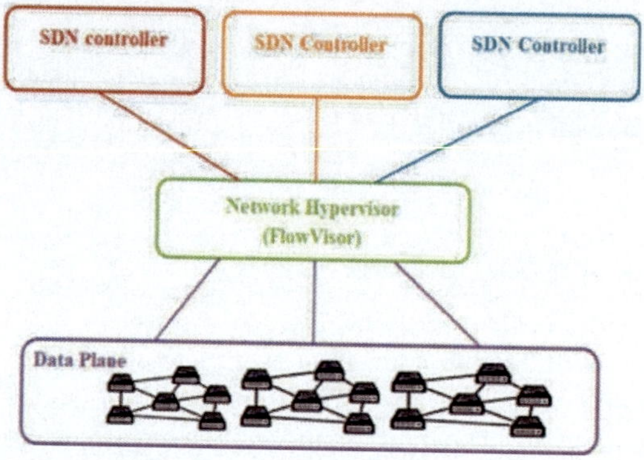

Fig. 16 Location SDN proxy in network architecture

One of the most important hypervisor software used to implement this scenario is FlowVisor, where FlowVisor acts as SDN proxy that intercepts messages between the data layer and the control layer [47], as shown in Fig. 16.

The FlowVisor represent an infrastructure resource virtualization layer and allow multiple controllers to be controlled so that each controller has a view of only the part it is responsible for. In other words, FlowVisor cuts the infrastructure into logical network slices isolated from one another, and communication between the infrastructure and the FlowVisor is done through the OpenFlow protocol, as well as between the FlowVisor and the controllers.

4.3 Multiple Owner, Multiple Controller

This scenario gives the tenants complete freedom to select the resources they need from the infrastructure layer through the virtualization layer as it gives flexibility to the tenant.

The OpenVirteX network virtualization software is one of the most essential tools for achieving this situation, although the infrastructure owner retains control over their SDN virtual resources [46].

5 Security of Network Slicing

This section will deal with network slicing security. This section begins by defining basic principles of slice security, discussing potential threats to network slicing architecture, and finally reviewing studies to achieve security for network slices.

5.1 Security Principles of Slice

The isolation characteristic of network slices is its most important feature, and it enhances the network slicing architecture in terms of slice security and privacy [62]. It is important that a network slice does not affect other slices so that if a particular slice is attacked, other slices are not affected by it, and information about the status of the slice is not shared with the other slices.

However, there are basic objectives of security, which are confidentiality, integrity, availability authentication and, authorization, which can be defined in the concept of network slices as follows:

- Confidentiality: The confidentiality of the network slice is achieved when packets are available only within the same slice or slices that are allowed to communicate with it, in addition to that, the information loaded within the packets is not available to anyone except authorized persons or end-users.
- Integrity: meaning that only network slice owners have the ability to change applications, specify flows, slice configurations, and so on.
- Availability: This means that the infrastructure is available to the network segments as agreed upon or as specified by MANO and this requires that NSMs and NFs remain accessible at all times.
- Authentication: Verifies the authenticity of people and devices connected to each network slice, as well as the validation and authentication of communication with the NFV Management and Orchestrator (MANO), in order to manage network elements only from authorized individuals.
- Authorization: This refers to allowing users to access certain slices, with access to each slice controlled by the slice owners' administration. The infrastructure providers, on the other hand, have complete control over network slicing administration and accounting.

5.2 *Network Slicing Security Threats*

In this section, we will describe five major threats to the network slicing environment, as summarized in Fig. 17.

Threat Vector 1: Attack on Slice-Service connection, which can cause an attack on one of the services as DoS attacks and damage the service as well as monitoring traffic. The destruction of the service may lead to the destruction of the network slice through direct communication between the services and the network slices, and it may also lead to the destruction of other services that operate on the same slice.

Possible solutions: Using a two-way authentication mechanism (mutual authentication) to secure communication between services and network slices through the use of secure protocols to achieve this connection. As well as the use of traffic analysis and behavioral analysis techniques within or between different slices and components, in order to investigate unauthorized communication and detect anomalies [63].

Threat Vector 2: Attack on Slices, by exploiting a less secure slice by the attacker in order to attack an important and more secure slice. This threat is when there is communication between network slices and this threat vector leads to unauthorized access and unauthorized disclosure of confidential information transmitted within the network slice.

Possible solutions: isolation between different network slices in addition to taking security measures for distributing confidential communication parameters (such as encryption and authentication key) within the network slice and not sharing them with other slices through the creation and use of unique and special security parameters for each network slice.

Threat Vector 3: Attack on Slice-Resource layer connection, an attacker can attack the communication channel and modify on the network slice requirements from the resources, thus losing integrity. On the other hand, the communication channel can

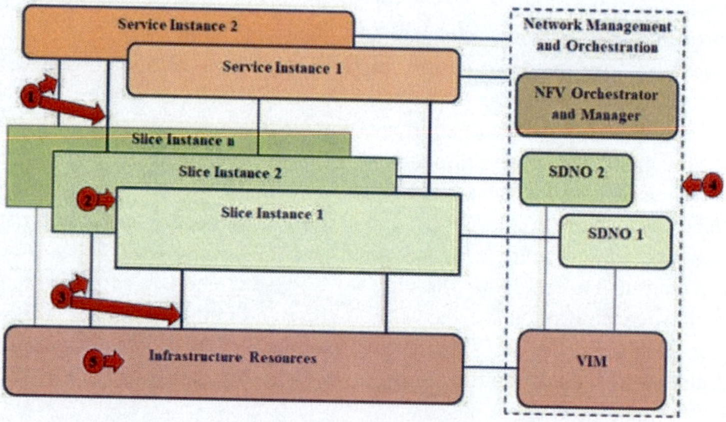

Fig. 17 Network slicing main threat vectors

Table 2 Impact threats vector on security goals

Threats	Confidentiality	Integrity	Availability	Authentication	Authorization
Vector 1	✔		✔	✔	
Vector 2	✔				
Vector 3		✔	✔	✔	
Vector 4		✔	✔	✔	✔
Vector 5			✔	✔	

be exploited to launch Denial of Service (DoS) attacks on infrastructure resources and resource drain.

Possible solutions: Use of security mechanisms and protocols to achieve minimum security requirements of confidentiality, integrity, data authentication, and peer-to-peer authentication.

Threat Vector 4: Attack on Slice Management, by which a tenant (the administrator who manages their slice) tries to gain access to network functions that are out of agreement. It also includes the point of attack when a tenant of a particular slice attempts to gain unauthorized access to other slices that usually belong to different tenants.

Possible solutions: Provide good isolation between different slice managers through robust procedures for authentication and access control.

Threat Vector 5: Attack on Infrastructure Resource, the attacker focuses on depleting physical resources through DoS and DDoS attacks, thus destroying network slices and related functions.

Possible solutions: Mutual authentication, enforcement of strict credential access policies, physical security, and safety checks are among the most important measures that reduce the effects of physical attacks. In Table 2 summarizes the effect of threat vectors on main security goals.

5.3 *Security in Network Slicing*

Recently, the security of network slicing has received wide attention, but despite that, not many works have been published on it. Most of the work presented in the past takes to focuses on the aspects of authentication, encryption, and key management, as well as monitoring the general behavior of the network slices.

Ni, Jianbing, Xiaodong Lin, and Xuemin Sherman Shen [63]: offered a solution for achieving effective and secure service-oriented authentication for 5G IoT applications, including network slicing and fog computing, to assure anonymity, user credibility, and service data confidentiality. Users are authenticated by utilizing access credentials produced by the IoT server, which allow them to access the IoT service. Otherwise, the attacker would be unable to do so without a legitimate access credential.

Liu, Jingwei, et al., [64]: developed a hybrid strategy to protect communications between 5G network slices in distinct public cryptosystems, and two heterogeneous cipher schemes to achieve reciprocal communications between the public key infrastructure (PKI) and Certificate Less Public Key Cryptography (CLC) environments.

The researchers Porambage, Pawani et al., [65]: propose a key-distribution scheme suitable for the network slicing architecture when the slices are accessed by third-party applications. The proposed scheme consists of two technologies, the first is Shamir's secret sharing to distribute and rebuild private key shares, and the second technique is ElGamal cryptosystem to encrypt and decrypt the separator keys.

Bonfim, Michel, et al., [66]: proposes a scenario for real-time attack detection in network slices 5G based on FrameRTP4. FrameRTP4 is a P4-based framework that provides an attack detection method based on an efficient and scalable ACL to detect known attacks and control channel monitoring to reduce channel overhead.

Thantharate, Anurag, et al., [67]: explored the concerns of a distributed denial of service attack on a network slicing and presented a model based on deep learning to create a robust network slicing framework to proactively combat DDoS attacks and eliminate overburdened connections before they impact and invade 5G networks.

Wang, Weili, et al., [68]: propose a new algorithm based on one-class support vector machine (OCSVM) to detect anomalies in real time, while they used another algorithm to detect link anomalies based on canonical correlation analysis. The two algorithms are proposed to protect the core network and thus protect multiple network slices from anomalies within a short time. Table 3 presents a summary of the related work.

Table 3 Summary of the related work

Ref.	System feature				
	Authentication service	Traffic monitoring	Key management	Encryption	Implementation environment
[63]	✔	✖	✔	✖	The results of implementing the proposal did not appear in the Network Slice environment
[64]	✔	✖	✖	✖	Execution of the simulation on client and server using two separate raspberry pi platform

(continued)

Table 3 (continued)

Ref.	System feature				
	Authentication service	Traffic monitoring	Key management	Encryption	Implementation environment
[65]	✔	✖	✔	✔	–
[66]	✖	✔	✖	✖	Use the Python programming language to implement the proposal to monitor traffic, but not mention its implementation in a network slicing environment
[67]	✖	✔	✖	✖	–
[68]	✖	✔	✖	✖	Synthetic and real-world network datasets are implemented to evaluate anomaly detection algorithms in node and links

6 Conclusion

In this paper, we analyze the concept of network slicing in terms of basic slicing enabling techniques and the tools used for it. The paper summarized scenarios for implementing the slicing concept in line with the tenant's requirements. It also focused on the security aspect of slats, where the principles and objectives of security were discussed on the slicing architecture and threats, and later it reviewed the works that dealt with this aspect. Network slicing security is expected to occupy a large area in the future, just as the concept of network slicing has become widespread in new generation networks.

References

1. Khan LU, Yaqoob I, Tran NH, Han Z, Hong CS (2020) Network slicing: recent advances, taxonomy, requirements, and open research challenges. IEEE Access 8:36009–36028

2. Afolabi I, Taleb T, Samdanis K, Ksentini A, Flinck H (2018) Network slicing and softwarization: a survey on principles, enabling technologies, and solutions. IEEE Commun Surv Tutor 20(3):2429–2453
3. Bozakov Z, Papadimitriou P (2014) Towards a scalable software-defined network virtualization platform. In: 2014 IEEE network operations and management symposium (NOMS). IEEE, pp 1–8
4. Chen Q, Wang X, Lv Y (2018) An overview of 5G network slicing architecture. In: AIP conference proceedings, vol 1967, no 1. AIP Publishing LLC, p 020004
5. Sivarajan KN (2020) Network slicing and SDN: new opportunities for telecom operators. CSI Trans ICT 8(1):15–20
6. Yousaf FZ, Gramaglia M, Friderikos V, Gajic B, Von Hugo D, Sayadi B, Sciancalepore V, Crippa MR (2017) Network slicing with flexible mobility and QoS/QoE support for 5G Networks. In: 2017 IEEE international conference on communications workshops (ICC Workshops). IEEE, pp 1195–1201
7. Richart M, Baliosian J, Serrat J, Gorricho JL (2016) Resource slicing in virtual wireless networks: a survey. IEEE Trans Netw Serv Manage 13(3):462–476
8. Zhang H, Liu N, Chu X, Long K, Aghvami AH, Leung VC (2017) Network slicing based 5G and future mobile networks: mobility, resource management, and challenges. IEEE Commun Mag 55(8):138–145
9. Caballero P, Banchs A, De Veciana G, Costa-Pérez X, Azcorra A (2018) Network slicing for guaranteed rate services: admission control and resource allocation games. IEEE Trans Wireless Commun 17(10):6419–6432
10. Muñoz R, Vilalta R, Casellas R, Martinez R, Szyrkowiec T, Autenrieth A, López V, López, D (2015) Integrated SDN/NFV management and orchestration architecture for dynamic deployment of virtual SDN control instances for virtual tenant networks. J Opt Commun Netw 7(11):B62–B70
11. Wijethilaka S, Liyanage M (2021) Survey on network slicing for Internet of Things realization in 5G networks. IEEE Commun Surv Tutor 23(2):957–994
12. Li Q, Wu G, Papathanassiou A, Mukherjee U (2016) An end-to-end network slicing framework for 5G wireless communication systems. arXiv preprint arXiv:1608.00572
13. Shen X, Gao J, Wu W, Lyu K, Li M, Zhuang W, Li X, Rao J (2020) AI-assisted network-slicing based next-generation wireless networks. IEEE Open J Veh Technol 1:45–66
14. Popovski P, Trillingsgaard KF, Simeone O, Durisi G (2018) 5G wireless network slicing for eMBB, URLLC, and mMTC: a communication-theoretic view. IEEE Access 6:55765–55779
15. Kalør AE, Guillaume R, Nielsen JJ, Mueller A, Popovski P (2017) Network slicing for ultra-reliable low latency communication in industry 4.0 scenarios. arXiv preprint arXiv:1708.09132
16. Delgado C, Canales M, Ortín J, Gállego JR, Redondi A, Bousnina S, Cesana M (2017) Joint application admission control and network slicing in virtual sensor networks. IEEE Internet Things J 5(1):28–43
17. Ordonez-Lucena J, Ameigeiras P, Lopez D, Ramos-Munoz JJ, Lorca J, Folgueira J (2017) Network slicing for 5G with SDN/NFV: concepts, architectures, and challenges. IEEE Commun Mag 55(5):80–87
18. Foukas X, Patounas G, Elmokashfi A, Marina MK (2017) Network slicing in 5G: survey and challenges. IEEE Commun Mag 55(5):94–100
19. Kaloxylos A (2018) A survey and an analysis of network slicing in 5G networks. IEEE Commun Stand Mag 2(1):60–65
20. Barakabitze AA, Ahmad A, Mijumbi R, Hines A (2020) 5G network slicing using SDN and NFV: a survey of taxonomy, architectures and future challenges. Comput Netw 167:106984
21. Schaller S, Hood D (2017) Software defined networking architecture standardization. Comput Stand Interf 54:197–202
22. Goransson P, Black C, Culver T (2016) Software defined networks: a comprehensive approach. Morgan Kaufmann
23. Latif Z, Sharif K, Li F, Karim MM, Biswas S, Wang Y (2020) A comprehensive survey of interface protocols for software defined networks. J Netw Comput Appl 156:102563

24. Paliwal M, Shrimankar D, Tembhurne O (2018) Controllers in SDN: a review report. IEEE Access 6:36256–36270
25. Ahmad S, Mir AH (2021) Scalability, consistency, reliability and security in SDN controllers: a survey of diverse SDN controllers. J Netw Syst Manage 29(1):1–59
26. Braun W, Menth M (2014) Software-defined networking using OpenFlow: protocols, applications and architectural design choices. Future Internet 6(2):302–336
27. Lara A, Kolasani A, Ramamurthy B (2013) Network innovation using openflow: a survey. IEEE Commun Surv Tutor 16(1):493–512
28. Benzekki K, El Fergougui A, Elbelrhiti Elalaoui A (2016) Software-defined networking (SDN): a survey. Secur Commun Netw 9(18):5803–5833
29. Rowshanrad S, Namvarasl S, Abdi V, Hajizadeh M, Keshtgary M (2014) A survey on SDN, the future of networking. J Adv Comput Sci Technol 3(2):232–248
30. Chica JCC, Imbachi JC, Vega JFB (2020) Security in SDN: a comprehensive survey. J Netw Comput Appl 159:102595
31. Maleh Y, Qasmaoui Y, El Gholami K, Sadqi Y, Mounir S (2022) A comprehensive survey on SDN security: threats, mitigations, and future directions. J Reliab Intell Environ 1–39
32. Ahmad I, Namal S, Ylianttila M, Gurtov A (2015) Security in software defined networks: a survey. IEEE Commun Surv Tutor 17(4):2317–2346
33. Dacier MC, König H, Cwalinski R, Kargl F, Dietrich S (2017) Security challenges and opportunities of software-defined networking. IEEE Secur Priv 15(2):96–100
34. Li W, Meng W, Kwok LF (2016) A survey on OpenFlow-based Software Defined Networks: security challenges and countermeasures. J Netw Comput Appl 68:126–139
35. Manguri KH, Omer SM (2022) SDN for IoT environment: a survey and research challenges. In: ITM web of conferences, vol 42. EDP Sciences, p 01005
36. Ghonge MM (2022) Software-defined network-based vehicular ad hoc networks: a comprehensive review. Softw Defined Netw Ad Hoc Netw 33–53
37. Mohamed A, Hamdan M, Khan S, Abdelaziz A, Babiker SF, Imran M, Marsono MN (2021) Software-defined networks for resource allocation in cloud computing: a survey. Comput Netw 195:108151
38. Vestin J (2018) SDN-enabled resiliency in computer networks. Doctoral dissertation, Karlstads universitet
39. Oktian YE, Lee S, Lee H, Lam J (2017) Distributed SDN controller system: a survey on design choice. Comput Netw 121:100–111
40. Bannour F, Souihi S, Mellouk A (2017) Distributed SDN control: survey, taxonomy, and challenges. IEEE Commun Surv Tutor 20(1):333–354
41. Chaudhari S, Mani RS, Raundale P (2016) SDN network virtualization survey. In: 2016 International conference on wireless communications, signal processing and networking (WiSPNET). IEEE, pp 650–655
42. Schaffrath G, Werle C, Papadimitriou P, Feldmann A, Bless R, Greenhalgh A, Wundsam A, Kind M, Maennel O, Mathy L (2009) Network virtualization architecture: proposal and initial prototype. In: Proceedings of the 1st ACM workshop on virtualized infrastructure systems and architectures, pp 63–72
43. Veeraraghavan M, Sato T, Buchanan M, Rahimi R, Okamoto S, Yamanaka N (2017) Network function virtualization: a survey. IEICE Trans Commun 2016NNI0001
44. Mijumbi R, Serrat J, Gorricho JL, Bouten N, De Turck F, Boutaba R (2015) Network function virtualization: state-of-the-art and research challenges. IEEE Commun Surv Tutor 18(1):236–262
45. Blenk A, Basta A, Reisslein M, Kellerer W (2015) Survey on network virtualization hypervisors for software defined networking. IEEE Commun Surv Tutor 18(1):655–685
46. Jin B, Guo B, Huang H, Li S, Shang Y, Huang S (2017) An implementation of optical network virtualization based on OpenVirteX. In: 2017 16th international conference on optical communications and networks (ICOCN). IEEE, pp 1–3
47. Sherwood R, Gibb G, Yap KK, Appenzeller G, Casado M, McKeown N, Parulkar G (2009) Flowvisor: a network virtualization layer. OpenFlow Switch Consortium, Tech. Rep 1:132

48. Jin X, Gossels J, Rexford J, Walker D (2015) {CoVisor}: a compositional hypervisor for {software-defined} networks. In: 12th USENIX symposium on networked systems design and implementation (NSDI 2015), pp 87–101
49. Liu L, Muñoz R, Casellas R, Tsuritani T, Martínez R, Morita I (2013) OpenSlice: an OpenFlow-based control plane for spectrum sliced elastic optical path networks. Opt Express 21(4):4194–4204
50. Van Giang N, Kim YH (2014) Slicing the next mobile packet core network. In: 2014 11th international symposium on wireless communications systems (ISWCS). IEEE, pp 901–904
51. Gudipati A, Li LE, Katti S (2014) RadioVisor: a slicing plane for radio access networks. In: Proceedings of the third workshop on Hot topics in software defined networking, pp 237–238
52. Blenk A, Basta A, Kellerer W (2015) HyperFlex: an SDN virtualization architecture with flexible hypervisor function allocation. In: 2015 IFIP/IEEE international symposium on integrated network management (IM). IEEE, pp 397–405
53. Nurkahfi GN, Mitayani A, Mardiana VA, Dinata MMM (2019) Comparing flowvisor and open virtex as SDN-based site-to-site VPN services solution. In: 2019 international conference on radar, antenna, microwave, electronics, and telecommunications (ICRAMET). IEEE, pp 142–147
54. Chowdhury NMK, Boutaba R (2010) A survey of network virtualization. Comput Netw 54(5):862–876
55. Napolitano A, Giorgetti A, Kondepu K, Valcarenghi L, Castoldi P (2018) Network slicing: an overview. In: 2018 IEEE 4th international forum on research and technology for society and industry (RTSI). IEEE, pp 1–4
56. Al-Asfoor M, Abed MH (2022) The effect of the topology adaptation on search performance in overlay network. In: Expert clouds and applications. Springer, Singapore, pp 65–73
57. Vakharkar S, Sakhare N (2022) Critical analysis of virtual LAN and its advantages for the campus networks. In: Mobile computing and sustainable informatics. Springer, Singapore, pp 733–748
58. Devlic A, Hamidian A, Liang D, Eriksson M, Consoli A, Lundstedt J (2017) NESMO: network slicing management and orchestration framework. In: 2017 IEEE international conference on communications workshops (ICC workshops). IEEE, pp 1202–1208
59. Habibi MA, Han B, Schotten HD (2017) Network slicing in 5G mobile communication architecture, profit modeling, and challenges. arXiv preprint arXiv:1707.00852
60. Abbas K, Khan TA, Afaq M, Song WC (2021) Network slice lifecycle management for 5g mobile networks: an intent-based networking approach. IEEE Access 9:80128–80146
61. S Staff (2017) What is dynamic network slicing? What is dynamic network slicing? https://www.sdxcentral.com/5g/definitions/dynamic-network-slicing/
62. Cunha VA, da Silva E, de Carvalho MB, Corujo D, Barraca JP, Gomes, D., Granville LZ, Aguiar RL (2019) Network slicing security: challenges and directions. Internet Technol Lett 2(5):e125
63. Ni J, Lin X, Shen XS (2018) Efficient and secure service-oriented authentication supporting network slicing for 5G-enabled IoT. IEEE J Sel Areas Commun 36(3):644–657
64. Liu J, Zhang L, Sun R, Du X, Guizani M (2018) Mutual heterogeneous signcryption schemes for 5G network slicings. IEEE Access 6:7854–7863
65. Porambage P, Miche Y, Kalliola A, Liyanage M, Ylianttila M (2019) Secure keying scheme for network slicing in 5G architecture. In: 2019 IEEE conference on standards for communications and networking (CSCN). IEEE, pp 1–6
66. Bonfim M, Santos M, Dias K, Fernandes S (2020) A real-time attack defense framework for 5G network slicing. Softw Pract Exp 50(7):1228–125
67. Thantharate A, Paropkari R, Walunj V, Beard C, Kankariya P (2020) Secure5g: a deep learning framework towards a secure network slicing in 5g and beyond. In: 2020 10th annual computing and communication workshop and conference (CCWC). IEEE, pp 0852–0857
68. Wang W, Liang C, Chen Q, Tang L, Yanikomeroglu H (2022) Distributed online anomaly detection for virtualized network slicing environment. arXiv preprint arXiv:2201.01900

Development and Initial Testing of Google Meet Use Scale (GMU-S) in Educational Activities During and Beyond the COVID-19 Pandemic

Mostafa Al-Emran, Ibrahim Arpaci, and Mohammed A. Al-Sharafi

Abstract Google Meet has been identified as one of the effective virtual meeting platforms that has the potential to deliver the learning materials to students during the COVID-19 pandemic. However, a scale evaluating its use for instructional activities has yet to be developed. Therefore, we developed the Google Meet use scale (GMU-S) and evaluated its characteristics among two samples with a total of 560 participants. The results indicated that the GMU-S has initial evidence of internal consistency reliability, construct, convergent, and discriminant validity. This study provides evidence that the developed scale is sound to evaluate the use of Google Meet in educational activities during and beyond the COVID-19 pandemic and other emergencies that might affect the education sector. Theoretically, this research is believed to be one of the pioneered studies that reported the development and initial testing of a new scale (GMU-S). Practically, the developed scale can be generalized to evaluate the use of other virtual meeting platforms (e.g., Skype, Zoom, Microsoft Teams, etc.).

Keywords Google meet · Scale development · GMU-S · Education · COVID-19

M. Al-Emran (✉)
Faculty of Engineering and IT, The British University in Dubai, Dubai, UAE
e-mail: mustafa.n.alemran@gmail.com

Department of Computer Techniques Engineering, Dijlah University College, Baghdad, Iraq

I. Arpaci
Department of Software Engineering, Faculty of Engineering and Natural Sciences, Bandirma Onyedi Eylul University, 10200 Balıkesir, Turkey

M. A. Al-Sharafi
Department of Information Systems, Azman Hashim International Business School, Universiti Teknologi Malaysia, 81310 Skudai, Johor, Malaysia

Department of Business Analytics, Sunway University, 47500 Bandar Sunway, Selangor, Malaysia

M. Al-Emran et al. (eds.), *International Conference on Information Systems and Intelligent Applications*, Lecture Notes in Networks and Systems 550,
https://doi.org/10.1007/978-3-031-16865-9_60

1 Introduction

A new coronavirus (COVID-19) has appeared in late December 2019 in Wuhan, China [1]. On March 11, 2020, the World Health Organization (WHO) has declared the outbreak of COVID-19 as a global pandemic [2]. In responding to this global crisis, a number of preventive measures were taken all over the world to flatten the curve of COVID-19 outbreak [3]. One of such precautionary measures was the lockdown of educational institutes [4–6]. Students who are majored in engineering, medicine, IT, and other majors that are practical in nature, were left idle. This has motivated educational institutes to undertake urgent decisions to continue providing education to students [7]. To compromise between social isolation and the delivery of education to students, the transition to online learning was the ideal solution in such pandemics [8].

Educational institutes have paid a lot of efforts to prepare the necessary platforms and resources to support the continuity of education. The penetration of educational technologies into the academic environment has provided effective solutions for educators to facilitate the delivery of education during the quarantine period. To enable online learning, educational institutes have used different tools, such as learning management systems (e.g., Moodle, Blackboard), virtual meeting platforms (e.g., Skype, Zoom, Microsoft Teams, Google Meet), social media (e.g., WhatsApp groups, Facebook groups, YouTube channels), and other educational platforms [9].

Among the virtual meeting platforms, Google Meet has been adopted by several educational institutes to deliver live streaming lectures to students. Google Meet facilitates the delivery of online learning, with up to 250 participants accessing the lecture simultaneously [9]. Besides, the lectures can be recorded and stored on Google Drive to be accessed later for those who didn't attend the live classes. The adoption of Google Meet in delivering the learning materials to students has just emerged with the appearance of the COVID-19 pandemic; thus, it is regarded as new technology. Adopting new technologies in education requires the understanding of the determinants affecting their sustainability [10–13]. The main criteria for adopting any technology are the ease of use and usefulness perceived by end-users [14–16]. Besides, individuals would feel positive toward using technology if it provides them with sufficient quality characteristics, including service, content, and information [17].

To evaluate the use of Google Meet for educational activities, there is a need to understand the main determinants affecting its use, including ease of use, usefulness, and quality features. When the students perceive that Google Meet is user-friendly, easy to use, and useful, they would exhibit high adoption levels. Likewise, when the quality of learning material is delivered to the students through Google Meet at the same capability as face-to-face classrooms, their attitudes toward using Google Meet would increase. While the importance of ease of use, usefulness, and quality features has already been examined in the previous literature [18–20], it is believed that Google Meet has distinct characteristics, and the determinants affecting its use would also be different. Due to its recency, an instrument has yet to be developed that measure the aforementioned features as qualities of evaluating the use of Google Meet for instructional activities.

Given the importance of the above-mentioned features and the lack of a measurement scale for evaluating the use of Google Meet for educational activities during the COVID-19 pandemic, this research aims to develop such a measure called Google Meet use scale (GMU-S). To measure the impact of ease of use, usefulness, and quality features, the GMU-S is developed on the basis of the Technology Acceptance Model (TAM) [14] and DeLone and McLean information systems success model [21]. The developed GMU-S is believed to provide a valuable contribution to the educational technology domain and serve as an instrument for evaluating the use of other virtual meeting platforms.

2 Method

2.1 *Study 1*

2.1.1 Sample

The sample of the first study, used for the exploratory factor analysis (EFA), consists of 250 participants (159 males (63.6%) and 91 females (36.4%)) from Malaysia. The participants' characteristics for this sample are demonstrated in Table 1.

Table 1 Participants' characteristics in study 1

Characteristics	Items	Frequency	Percentage (%)
Gender	Male	159	63.6
	Female	91	36.4
Education	Undergraduate	141	56.4
	Graduate	25	10.0
	MSc/PhD	84	33.6
Age	16–24	138	55.2
	25–35	55	22.0
	36–45	45	18.0
	46 and above	12	4.8
Google Meet use	Yes	32	12.8
	No	218	87.2

2.2 Procedure

The data were collected through a questionnaire survey through Google Forms, and the informed consents were obtained electronically from the participants before filling out the survey. The participants were informed about the aim of the study and asked to indicate their level of agreement on the statements using a five-point Likert scale ranging from "1 = strongly disagree" to "5 = strongly agree". The exploratory and confirmatory factor analyses were employed in the items development and testing phase. A total of 32 initial scale items were developed by the scholars and assessed independently by a jury of three experts. The experts' assessment was based on a scale ranges between 1 and 10, where 1 indicates that the item can't measure the use of Google Meet for learning activities. The redundant items and those rated lower than 0.80 by the experts were eliminated from the scale. This has resulted in a total of 14 items, which were then administered online to collect data from participants. The first dataset was subjected to an EFA. Finally, 310 participants rated the 14 items (participants of the second study were not involved in the first study), and the second dataset was subjected to a confirmatory factor analysis (CFA). Besides, the differences between the low and high groups were compared by 27% low and high groups item analysis as evidence of discriminant validity.

3 Results

3.1 Face Validity

A total of 32 items were developed based on the DeLone and McLean IS success model and TAM. The items were reviewed by the scholars and evaluated independently by three experts (i.e., 2 IS experts and 1 psychometrician) using a 10-point Likert scale. Those rated with an average of 0.80 and above were regarded to have sufficient face validity. Accordingly, a total of 14 items were retained for the EFA.

3.2 Exploratory Factor Analysis

The EFA was carried out with varimax rotation and principal components extraction method to figure out the factor structure. The EFA results indicated that the 14-items were loaded on a single factor and loaded more than the threshold value of 0.40. The one-factor solution accounted for 66.209% of the total variation. The Kaiser–Meyer–Olkin Measure of Sampling Adequacy was 0.961, and the Bartlett's test of sphericity was significant ($\chi^2_{(df=91)} = 3026.249$, $p < 0.001$), which shows that the GMU-S is a good candidate for factor analysis [22]. The communalities were ranged

Table 2 Reliability and validity results

Items	Communalities	Loadings	Corrected item-total correlation	Cronbach's alpha if item deleted
Item1	0.625	0.791	0.755	0.958
Item2	0.682	0.826	0.794	0.957
Item3	0.691	0.831	0.802	0.957
Item4	0.685	0.828	0.796	0.957
Item5	0.738	0.859	0.831	0.956
Item6	0.690	0.831	0.799	0.957
Item7	0.468	0.684	0.642	0.960
Item8	0.660	0.813	0.779	0.957
Item9	0.493	0.702	0.658	0.960
Item10	0.719	0.848	0.819	0.956
Item11	0.714	0.845	0.814	0.956
Item12	0.666	0.816	0.782	0.957
Item13	0.745	0.863	0.835	0.956
Item14	0.692	0.832	0.800	0.957

between 0.468 and 0.745, through which all values were greater than the suggested value of 0.40. Table 2 indicates the reliability and validity results.

3.3 *Normality and Internal Consistency*

The normality testing showed that the skewness and kurtosis statistics are ranged within the suggested values of ± 3 [23], and thus, the data were normally distributed. The Cronbach's alpha of the total scale was 0.96. Table 3 presents the descriptive statistics of the 14 items.

Table 3 Descriptive statistics

Items	Min.	Max.	Mean	Std. Dev.	Skewness (SE = 0.154)	Kurtosis (SE = 0.307)
Item1	1.00	5.00	3.8640	1.07803	−0.715	−0.084
Item2	1.00	5.00	3.9200	1.03435	−0.936	0.528

(continued)

Table 3 (continued)

Items	Min.	Max.	Mean	Std. Dev.	Skewness (SE = 0.154)	Kurtosis (SE = 0.307)
Item3	1.00	5.00	3.8440	1.05457	−0.720	−0.028
Item4	1.00	5.00	3.7440	1.11510	−0.619	−0.359
Item5	1.00	5.00	3.5880	1.13817	−0.524	−0.323
Item6	1.00	5.00	3.5720	1.12868	−0.501	−0.357
Item7	1.00	5.00	3.7880	1.04086	−0.600	−0.113
Item8	1.00	5.00	3.7120	1.04375	−0.426	−0.397
Item9	1.00	5.00	4.0840	1.03209	−1.164	0.930
Item10	1.00	5.00	3.8160	1.01700	−0.686	0.049
Item11	1.00	5.00	3.8680	0.96243	−0.659	0.139
Item12	1.00	5.00	3.8280	1.05207	−0.694	−0.053
Item13	1.00	5.00	3.6920	1.02437	−0.415	−0.308
Item14	1.00	5.00	3.8640	0.95559	−0.671	0.200

3.4 *Study 2*

3.4.1 Sample

The sample of the second study consists of 310 participants (189 males (61%) and 121 females (39%)) from Malaysia. The descriptive statistics for the participants in the second study are indicated in Table 4.

Table 4 Participants' descriptive statistics in study 2

Characteristics	Items	Frequency	Percentage (%)
Gender	Male	189	61.0
	Female	121	39.0
Education	Undergraduate	164	52.9
	Graduate	45	14.5
	MSc/PhD	101	32.6
Age	16–24	177	57.1
	25–35	72	23.2
	36–45	48	15.5
	46 and above	13	4.2
Google Meet use	Yes	39	12.6
	No	271	87.4

Table 5 Normality, reliability, and discriminant validity

Items	Skewness (SE = 0.138)	Kurtosis (SE = 0.276)	Corrected item-total correlation	Cronbach's alpha if item deleted	Item discrimination indices (*t*)
Item1	-1.156	0.590	0.605	0.933	12.809*
Item2	-0.359	-0.435	0.723	0.930	12.756*
Item3	-1.347	1.431	0.574	0.934	16.264*
Item4	-0.245	-0.755	0.732	0.930	16.224*
Item5	-1.030	0.392	0.678	0.931	9.630*
Item6	-0.069	-0.787	0.752	0.929	9.610*
Item7	-1.141	0.603	0.484	0.936	16.866*
Item8	-0.138	-0.799	0.739	0.929	16.841*
Item9	-1.867	2.744	0.549	0.935	14.585*
Item10	-0.247	-0.396	0.737	0.929	14.538*
Item11	-0.173	-0.311	0.723	0.930	18.488*
Item12	-0.565	-0.086	0.757	0.929	18.465*
Item13	-0.461	-0.152	0.747	0.929	9.998*
Item14	-0.372	-0.213	0.807	0.927	9.970*

Note: * $p < 0.001$

3.5 Normality, Reliability, and Discriminant Validity

The normality testing indicated that the skewness and kurtosis statistics were ranged within the suggested values of ±3, and thus, the data were normally distributed. The Cronbach's alpha for the overall scale was 0.935. The discriminant validity was investigated by 27% low and high groups item analysis. The independent sample *t*-test results indicated that the items can significantly discriminate the subjects ($t(167) = 28.740, p < 0.001$). Thus, the discriminant validity of the scale is ascertained. Table 5 shows the skewness, kurtosis, reliability coefficients, and item discrimination indices.

3.6 Construct Validity

The confirmatory factor analysis (CFA) was employed through SPSS AMOS (v.23) to validate how well the one-factor structure fits the data. Several criteria, including goodness of fit index (GFI), adjusted goodness of fit (AGFI), comparative fit index (CFI), normed fit index (NFI), incremental fit index (IFI), Tucker-Lewis fit index (TLI), and root mean squared error of approximation (RMSEA) were used to assess the fit of the model to data [24]. The results provided adequate model fit, including $\chi^2_{(df=26)} = 62.733$, $\chi^2/df = 2.413$, $p < 0.001$, GFI = 0.972, AGFI = 887, NFI =

Table 6 Model fit indices

	Model	Threshold Value(s)
χ^2	62.733	
p value	< 0.001	$0.05 \leq p \leq 1.00$
χ^2/df	2.413	< 3
GFI	0.972	≥ 0.90
AGFI	0.887	≥ 0.80
NFI	0.981	≥ 0.90
TLI	0.959	≥ 0.90
CFI	0.988	≥ 0.90
IFI	0.989	≥ 0.90
SRMR	0.0313	≤ 0.10
RMR	0.027	< 0.05
RMSEA	0.068	< 0.08

0.981, IFI = 0.989, TLI = 0.959, CFI = 0.988, and RMSEA = 0.068 [90% confidence interval = 0.046 and 0.089]. Table 6 shows the fit indices of the model.

4 Discussion

The existing literature on Google Meet is limited by the scarcity of a usage scale. Therefore, this study was conducted due to the lack of a scale for using Google Meet for instructional activities. Such a scale would provide a thorough understanding of using Google Meet for educational activities, specifically for delivering live-streaming lectures during the COVID-19 pandemic and other similar crises. Based on that, we developed the Google Meet use scale (GMU-S) and carried out an assessment of its initial characteristics across two studies. The developed GMU-S in this research is based on 14-items, which were measured using a five-point Likert scale to evaluate the use of Google Meet among students with different educational levels and age groups. The 14-items of the developed GMU-S are listed in the Appendix.

The EFA and CFA results provided evidence that the developed GMU-S could be employed to evaluate the use of Google Meet for educational activities. The scale structure constructed in study one (n = 250) was also confirmed in study two (n = 310). The results also indicated that the developed GMU-S has adequate internal consistency and convergent and discriminant validity.

The developed GMU-S is believed to satisfy the main determinants for system use (i.e., usefulness and ease of use) and quality features in one single scale, which in turn, contributes to a more fine-grained understanding of using Google Meet for educational purposes. For instance, it was suggested that ease of use and usefulness have a direct influence on using various educational technologies [25]–[27]. Likewise, if students acknowledge how Google Meet facilitates the delivery of lectures and improves their learning performance as with traditional classrooms, they might be more likely to keep using the platform during the COVID-19 pandemic and other future emergencies that might affect the education sector. In addition, the previous literature showed that quality features (e.g., content quality, system quality, and information quality) have a significant influence on using educational technologies [28]–[31]. Similarly, when the quality of the lectures and learning activities delivered through Google Meet is the same as with the physical classes, students would keep using the platform during the COVID-19 pandemic. To evaluate the usefulness, ease of use, and Google Meet quality features in any educational context, scholars could use the GMU-S that was developed on the basis of 14-items embracing the aforementioned features in a single scale.

5 Conclusion

Numerous IT innovations were employed in the education field [32, 33]. Google Meet appeared as an emerging technology to deliver online learning to students during the COVID-19 pandemic. This study demonstrated the development and initial testing of a new scale called the Google Meet use scale (GMU-S). The results indicated that the developed GMU-S has sufficient reliability and validity, and thus, it could be a valid tool for evaluating the use of Google Meet for learning purposes among the general population. The developed GMU-S could provide a plentiful explanation of using Google Meet for educational activities based on the main characteristics of any educational technology (i.e., usefulness, ease of use, and quality features).

From the theoretical perspective, this research is believed to be one of the pioneered studies that reported the development and initial testing of a new scale (GMU-S). From the practical perspective, educational institutions that are currently running the Google Meet for delivering the learning materials to students can use the developed scale to evaluate the effectiveness of the platform and improve their practices accordingly. Further, since the developed scale is based on the three main features that any educational technology might afford, it could be further employed to evaluate the use of other virtual meeting platforms (e.g., Skype, Zoom, Microsoft Teams, etc.).

Nonetheless, this research has some limitations that need to be reported and considered in future attempts. First, while the samples underlying this research embraced students with different age groups and various educational levels, it was limited to one geographical area (i.e., Malaysia). Therefore, further research trials are encouraged to replicate this study with more diverse cultures and backgrounds. Second, although the samples are considered satisfactory, specifically for the development and initial testing of new scales, it is suggested to test the scale using larger samples to further generalize the effectiveness of the tool. Third, the developed

GMU-S was developed based on 14-items written in English language. It is, therefore, suggested to develop the GMU-S versions in other languages and evaluate its cross-cultural equivalences.

Appendix: GMU-S items

Item1. Learning through Google Meet is easy for me.

Item2. I would find Google Meet to be flexible to interact with.

Item3. It would be easy for me to become skillful at using Google Meet for learning activities.

Item4. Using Google Meet in my university/college would enable me to accomplish learning activities more quickly.

Item5. Using Google Meet would improve my learning performance.

Item6. Using Google Meet in learning activities would increase my productivity.

Item7. Google Meet is suitable for my particular needs.

Item8. Google Meet is secured and protects information privacy.

Item9. Information delivered through Google Meet is rich and useful.

Item10. I could use the Google Meet services at anytime, anywhere I want.

Item11. Google Meet is a well-structured platform for learning purposes.

Item12. Google Meet provides high-speed information access.

Item13. The information delivered through the Google Meet meets my educational needs.

Item14. The information delivered through the Google Meet is reliable.

References

1. Zhu N et al (2020) A novel coronavirus from patients with pneumonia in China, 2019. N Engl J Med. https://doi.org/10.1056/NEJMoa2001017
2. Cucinotta D, Vanelli M (2020) WHO declares COVID-19 a pandemic. Acta Biomed 91(1):157–160. https://doi.org/10.23750/abm.v91i1.9397
3. Arpaci I, Huang S, Al-Emran M, Al-Kabi MN, Peng M (2021) Predicting the COVID-19 infection with fourteen clinical features using machine learning classification algorithms. Multimed Tools Appl 80:11943–11957. https://doi.org/10.1007/s11042-020-10340-7
4. Verma A, Verma S, Garg P, Godara R (2020) Online teaching during COVID-19: perception of medical undergraduate students. Indian J Surg 82(3):299–300. https://doi.org/10.1007/s12262-020-02487-2
5. Mishra L, Gupta T, Shree A (2020) Online teaching-learning in higher education during lockdown period of COVID-19 pandemic. Int J Educ Res Open. https://doi.org/10.1016/j.ijedro.2020.100012
6. AL-Nuaimi NM, Al Sawafi OS, Malik SI, Al-Emran M, Selim YF (2022) Evaluating the actual use of learning management systems during the covid-19 pandemic: an integrated theoretical model. Interact Learn Environ, 1–26. https://doi.org/10.1080/10494820.2022.2055577

7. Moszkowicz D, Duboc H, Dubertret C, Roux D, Bretagnol F (2020) Daily medical education for confined students during coronavirus disease 2019 pandemic: a simple videoconference solution. Clin Anat, 1–2. https://doi.org/10.1002/ca.23601
8. Al-Emran M (2020) Mobile learning during the era of COVID-19. Rev. Virtual Univ Católica del Norte 61:1–2
9. Machado RA, Bonan PRF, Da Cruz Perez DE, Martelli Júnior H (2020) COVID-19 pandemic and the impact on dental education: discussing current and future perspectives. Braz Oral Res 34:1–6. https://doi.org/10.1590/1807-3107BOR-2020.VOL34.0083
10. Granić A, Marangunić N (2019) Technology acceptance model in educational context: A systematic literature review. Br J Edu Technol. https://doi.org/10.1111/bjet.12864
11. Al-Emran M, Mezhuyev V (2019) Examining the effect of knowledge management factors on mobile learning adoption through the use of importance-performance map analysis (IPMA). In: International Conference on Advanced Intelligent Systems and Informatics, pp. 449–458. https://doi.org/10.1007/978-3-030-31129-2_41
12. Al Shamsi JH, Al-Emran M, Shaalan K (2022) Understanding key drivers affecting students' use of artificial intelligence-based voice assistants. Educ Inf Technol, 1–21. https://doi.org/10.1007/S10639-022-10947-3
13. Al-Sharafi MA, Al-Emran M, Iranmanesh M, Al-Qaysi N, Iahad NA, Arpaci I (2022) Understanding the impact of knowledge management factors on the sustainable use of AI-based chatbots for educational purposes using a hybrid SEM-ANN approach. Interact Learn Environ, 1–20. https://doi.org/10.1080/10494820.2022.2075014
14. Davis F (1989) Perceived usefulness, perceived ease of use, and user acceptance of information technology. MIS Q 13(3):319–340. https://doi.org/10.2307/249008
15. Al-Emran M, Al-Maroof R, Al-Sharafi MA, Arpaci I (2020) What impacts learning with wearables? An integrated theoretical model. Interact Learn Environ, 1–21. https://doi.org/10.1080/10494820.2020.1753216
16. Al-Saedi K, Al-Emran M, Abusham E, El-Rahman SA (2019) Mobile payment adoption: a systematic review of the UTAUT model. https://doi.org/10.1109/ICFIR.2019.8894794
17. Govender I, Rootman-le Grange I (2015) Evaluating the early adoption of moodle at a higher education institution. In: European conference on e-learning, p. 230
18. Teo T, Zhou M, Fan ACW, Huang F (2019) Factors that influence university students' intention to use Moodle: a study in Macau. Educ Technol Res Dev. https://doi.org/10.1007/s11423-019-09650-x
19. Almaiah MA, Jalil MA, Man M (2016) Extending the TAM to examine the effects of quality features on mobile learning acceptance. J Comput Educ 3(4):453–485. https://doi.org/10.1007/s40692-016-0074-1
20. Al-Sharafi MA, Al-Emran M, Arpaci I, Marques G, Namoun A, Iahad NA (2022) Examining the impact of psychological, social, and quality factors on the continuous intention to use virtual meeting platforms during and beyond COVID-19 pandemic: a hybrid SEM-ANN approach. Int J Human-Comput Interact. https://doi.org/10.1080/10447318.2022.2084036
21. DeLone WH, McLean ER (1992) Information systems success: the quest for the dependent variable. Inf Syst Res 3(1):60–95
22. Tabachnick BG, Fidell LS (2001) Using multivariate statistics. HarperCollins, New York
23. Brown TA (2006) Confirmatory factor analysis for applied research. New York
24. Tabachnick BG, Fidell LS, Ullman JB (2007) Using multivariate statistics. Pearson, London
25. Sheppard M, Vibert C (2019) Re-examining the relationship between ease of use and usefulness for the net generation. Educ Inf Technol 24:3205–3218. https://doi.org/10.1007/s10639-019-09916-0
26. Saroia AI, Gao S (2019) Investigating university students' intention to use mobile learning management systems in Sweden. Innov Educ Teach Int 56(5):569–580. https://doi.org/10.1080/14703297.2018.1557068
27. Al-Emran M, Mezhuyev V, Kamaludin A (2019) An innovative approach of applying knowledge management in m-learning application development: a pilot study. Int J Inf Commun Technol Educ 15(4):94–112. https://doi.org/10.4018/IJICTE.2019100107

28. Calisir F, Gumussoy CA, Bayraktaroglu AE, Karaali D (2014) Predicting the intention to use a web-based learning system: perceived content quality, anxiety, perceived system quality, image, and the technology acceptance model. Hum Factors Ergon Manuf Serv Ind 24(5):515–531
29. Alsabawy AY, Cater-Steel A, Soar J (2016) Determinants of perceived usefulness of e-learning systems. Comput Human Behav 64:843–858
30. Damnjanovic V, Jednak S, Mijatovic I (2015) Factors affecting the effectiveness and use of moodle: students' perception. Interact Learn Environ 23(4):496–514
31. Wongvilaisakul W, Lekcharoen S (2015) The acceptance of e-Learning using SEM approach: a case of IT Literacy development for PIM students. In: Electrical engineering/electronics, computer, telecommunications and information technology (ecti-con), 2015 12th international conference on, pp. 1–6
32. Al-Emran M, Alkhoudary YA, Mezhuyev V, Al-Emran M (2019) Students and educators attitudes towards the use of M-learning: gender and smartphone ownership differences. Int J Interact Mob Technol 13(1):127–135. https://doi.org/10.3991/ijim.v13i01.9374
33. Saa AA, Al-Emran M, Shaalan K (2019) Mining student information system records to predict students' academic performance. In: International conference on advanced machine learning technologies and applications, pp. 229–239